SECOND EDITION HUMAN BIOLOGY, ANATOMY & PHYSIOLOGY

FOR THE HEALTH SCIENCES

WENDI A. ROSCOE

Fanshawe College

NELSON

NELSON

Human Biology, Anatomy, and
Physiology for the Life Sciences,
Second Edition
by Wendi A. Roscoe

VP, Product Solutions, K–20:
Claudine O'Donnell

Director, Hardside Publishing
Paul Fam

Marketing Manager:
Tia Nguyen

Content Manager:
My Editor Inc.

Photo and Permissions Researcher:
Jessie Coffey

Senior Production Project Manager:
Imoinda Romain

Production Service:
MPS Limited

Copy Editor:
Carolyn Jongeward

Proofreader:
MPS Limited

Indexer:
Julie Grady

Design Director:
Ken Phipps

Higher Education Design PM:
Pamela Johnston

Interior Design:
Ken Cadinouche

Cover Design:
Courtney Hellam

Cover Image:
Claus Andersen Photography

Art Coordinators:
Pam Johnston, Suzanne Peden

Illustrators:
Allan Moon, Articulate Graphics,
Annie Tseng, Ann Sanderson,
Chelsea Canlas, Crowle Art Group,
Jerry Gu, MPS Limited, Natividad
Chen

Compositor:
MPS Limited

Library and Archives Canada Cataloguing in Publication

Roscoe, Wendi A. (Wendi Annette), author
      Human biology, anatomy, and physiology for the health sciences / Wendi A. Roscoe, Fanshawe College. — Second edition.

Includes index.
Issued in print and electronic formats.
ISBN 978-0-17-673915-7 (softcover).
—ISBN 978-0-17-685129-3 (PDF)

      1. Human biology—Textbooks.
2. Human anatomy—Textbooks.
3. Human physiology—Textbooks.
4. Textbooks. I. Title.

QP34.5.R68 2019   612
C2018-905260-0
C2018-905261-9

ISBN-13: 978-0-17-673915-7
ISBN-10: 0-17-673915-7

To my sons, Adam and Josh.

Brief Contents

Table of Contents

CHAPTER 7 Evolution 120

E. R. Degginger/Science Source

CHAPTER 8 DNA Structure and Function 137

Andrea Danti/Shutterstock.com

CHAPTER 9 Gene Expression and Regulation 149

Stockbyte/Comstock Images/Getty Images

CHAPTER 10 Biotechnology 164

© iStockphoto.com/DNY59

CHAPTER 15 The Nervous System 291

nobeastsofierce/Shutterstock.com

CHAPTER 16 The Sensory System 337

HQuality/Shutterstock.com

CHAPTER 17 The Endocrine System 362

Preface

NOTE TO THE STUDENT

This text includes the foundational knowledge of human biology, anatomy, and physiology that is relevant for students entering healthcare programs. I wrote this text so students could learn about and understand how their own body functions: from the macromolecules that we eat to the structures and function of our cells and to how groups of cells work together to form our organ systems. As well, I've included many examples of what can happen when our molecules, cells, or systems do not function properly. It is well known that people learn best when the information matters to them, when it is interesting, personal, and they can form connections to information they already know. When students understand their own health, they will live a healthier lifestyle and also become better healthcare professionals. My goal was to write this text for students to learn and understand that what we eat, think, and do affects us at every level, from the receptors on our cell membranes to the development of disease. I tried to continually interrelate the topics from each chapter so that previous information is reinforced, and so that by the end of the text students can see how everything fits together into a big picture.

NOTE TO THE INSTRUCTOR

In healthcare programs, it is most common for students to learn anatomy and physiology. In general and introductory programs, it is common for students to learn biology, which usually includes topics such as biodiversity, cells, genetics, plant structure and function, evolution of organisms, and organ systems. In this textbook, I have chosen biology topics that specifically relate to human health, evolution and inheritance, metabolism, genetics, biotechnology in medicine, infections and the importance of normal flora, as well as the anatomy and physiology of the body systems. In each chapter, I have included "Did You Know?" boxes that provide interesting facts that can increase students' interest and understanding of the content. Throughout the text I have included the most recent research related to nutrition that affects cell chemistry and physiology so students can see more clearly how environmental factors play a role in the regulation of our body systems and the imbalances that lead to disease. For example, because many students have difficulty with the cellular metabolism unit, I have included a discussion about how the vitamins we eat in foods act as the coenzymes that allow our cells to produce energy. So the content is not just a list of chemical reactions that students must memorize; they can start to connect what they do know about food with their own energy level and body weight, making the topic much more interesting and relevant. I have written this text in a way that's clear, concise, engaging, and easy to understand. I have included review questions at the end of each section of each chapter so that students can frequently test their knowledge and level of understanding. At the ends of appropriate chapters, I have also included a review of medical terminology and a number of diagrams for students to label so this text has the feel of a workbook. It is my hope that the combination of topics covered in this text as well as the student-focused approach will give students a good starting point for understanding human health as they prepare for their healthcare program.

WHAT'S NEW IN THIS EDITION?

The second edition of this textbook is written in the same easy-to-understand manner as the first edition. I have edited some sections to be even more concise in order to provide a manageable amount of reading material while maintaining the required detail for understanding. I have changed the sentence structure in some areas to point-form lists or to charts that make it easier for students to discern the important points.

I have added two major topics that are important and relevant for students entering healthcare programs: nutrition and the sensory system. Nutrition information has been added to Chapter 2: Biological Molecules and Nutrition. Included is information about micronutrients, vitamins, and minerals as well as current information regarding fibre, the glycemic index, and how starchy carbohydrates affect insulin production, which impacts weight gain. These topics are significant in healthcare, since obesity is now considered one of the leading causes of death and affects 40% of the population: approximately 14 million Canadians. In addition, a new Chapter 15: The Nervous System has been added, covering touch, temperature and pain, vision, hearing, taste, and smell. This chapter describes how various sensory information contribute to our awareness of our environment so that our brains can interpret and respond to changes and keep our body in homeostasis.

This edition includes information about aging at the ends of the organ system chapters. The Effects of Aging boxes are concise, thought-provoking blocks of information that help students understand how body systems can change and ways that they themselves can prevent or slow down some undesirable processes. I wrote this information so that it connects and reinforces concepts students have learned about normal physiology in each chapter.

Lastly, this edition includes an application question in each chapter. Recent changes in teaching methodology focus on active learning, where students can integrate knowledge in a meaningful way so that they are more prepared for healthcare programs that require students to think critically. For example, in Chapter 14: The Muscular System, Dave is doing a squat workout and students must think about which muscles are being used, whether is this an aerobic or anaerobic workout, what types of fibres are primarily being used, and how those cells would make energy based on the type of exercise. The students can assimilate a lot of information from this chapter into a situation they are familiar with, which will improve overall comprehension of the material. Professors can use the application questions to stimulate class discussions that further increase interest and understanding.

INSTRUCTOR RESOURCES

The **Nelson Education Teaching Advantage (NETA)** program delivers research-based instructor resources that promote student engagement and higher-order thinking to enable the success of Canadian students and educators. Be sure to visit Nelson Education's Inspired **Instruction website** at **http://www.nelson.com/inspired** to find out more about NETA.

The following instructor resources have been created for *Human Biology, Anatomy and Physiology for the Health Sciences*, Second Edition. Access these ultimate tools for customizing lectures and presentations at www.nelson.com/instructor and/or within the Möbius Content Repository.

NETA Test Bank

This resource was written by Trish Morrow, St. Lawrence College. It includes over 1,000 multiple-choice and true/false questions written according to NETA guidelines for effective construction and development of higher-order questions. The test bank was copyedited by a NETA-trained editor.

The NETA Test Bank is available in a new, cloud-based platform. **Testing Powered by Cognero®** is a secure online testing system that allows you to author, edit, and manage test bank content from any place you have Internet access. No special installations or downloads are needed, and the desktop-inspired interface, with its drop-down menus and familiar, intuitive tools, allows you to create and manage tests with ease. You can create multiple test versions in an instant and import or export content into other systems. Tests can be delivered from your learning management system, your classroom, or wherever you want. Testing Powered by Cognero for *Human Biology, Anatomy, and Physiology for the Health Sciences,* Second Edition can also be accessed through **http://www.nelson.com/login** and **http://login.cengage.com**.

NETA PowerPoint

Microsoft® PowerPoint® lecture slides for every chapter have been created by Wendi A. Roscoe, Fanshawe College. There is an average of 50 slides per chapter, many featuring key figures, tables, and photographs from *Human Biology, Anatomy, and Physiology for the Health Sciences*, Second Edition. NETA principles of clear design and engaging content have been incorporated throughout, making it simple for instructors to customize the deck for their courses.

Image Library

This resource consists of digital copies of figures, short tables, and photographs used in the book. Instructors may use these jpegs to customize the NETA PowerPoint or create their own PowerPoint presentations.

Video PowerPoint Slides

Through a select number of video animations now embedded within PowerPoint slides, instructors can bring their classrooms to life with visualizations.

Polling Questions

Another valuable resource for instructors are polling questions created specifically for *Human Biology, Anatomy, and Physiology for the Health Sciences*, Second Edition, by Wendi A. Roscoe, Fanshawe College. Using your favourite polling solution, access a bank of questions created in PowerPoint for use with the second edition.

eLabs

New to the second edition are 12 laboratory activities to accompany specific chapters of the book. Each lab has an option for a wet-lab environment with equipment such as microscopes and one for a classroom environment that does not have such equipment. The labs are designed in modules so that you can choose to use as many or as few of the modules as suit the needs of your course and the materials you have available. These labs have gradable exercises that can be downloaded and modified to suit your specific laboratory environment as well as online questions that can be autograded. Instructor resources are provided as Microsoft Word files so that you can modify them for your own purposes.

Möbius

Möbius allows you to integrate powerful, dynamic learning and assessment tools throughout your online course materials, so your students receive constant feedback that keeps them engaged and on track.

- Integrate meaningful, automatically graded assessment questions into lessons and narrated lectures, in addition to formal assignments, so students can test their understanding as they go.
- Provide interactive applications for exploring concepts in ways not available in a traditional classroom.
- Leverage more questions to provide practice for students as they master concepts, as well as individual summative assessments.
- Incorporate engaging, enlightening visualizations of concepts, questions, and solutions, through videos and animations that students can modify and explore.
- Bring your online vision to life, including online courses, open-access courses, formative testing, placement and remediation programs, independent learning, outreach programs, and flipped or blended classrooms.
- Provide exactly the content you want, from individual lessons and textbook supplements, to full courses, remedial materials, enrichment content, and more.
- Choose the learning experience by allowing students open access to your course material or guiding them along a specific learning path.
- Stay in control of your content, creating and customizing materials as you wish to suit your needs.

STUDENT ANCILLARIES

Möbius

Bring online learning to life for your students with a platform that integrates the text, animations, interactives and additional questions with Möbius. It is a total learning environment for your standard and advanced stream of students. Bring human biology to life TODAY!

Möbius is an HTML5-native online courseware environment that takes a "learn-by-doing" philosophy to STEM education, utilizing the highly interactive DigitalEd visualization engine that drives online applications for immediate learning-outcome development and assessment. The power of assessments is immediate confirmed understanding of difficult STEM-based topics in real time. This type of power is necessary to ensure the high level of learning outcomes that is possible within the environment. Instructors can easily create and share their own assessments and modify any lesson, assessment, or interactive activity, and share with their students or the wider Möbius user community. In addition, unlike traditional learning technologies, textbook exposition, interactives, and assessment are all "in line" so that students are presented with a unified learning environment, keeping them firmly focused on the topic at hand.

ACKNOWLEDGMENTS

I would like to thank everyone who has helped in the development of this textbook. Thank you to Katherine Goodes, developmental editor; Imoinda Romain, senior production project manager; Pam Johnston, design project manager; Sue Peden, art coordinator; Carolyn Jongeward, copyeditor; Sharib Asrar, project manager; and Jessie Coffey, permissions researcher. Thanks to Karen Nancekivell, senior sales and editorial representative, and Paul Fam, senior publisher, who encouraged me to originally write this text and supported me during every step of the way. Finally, I would like to thank the reviewers:

Josephine Anthony, Simon Fraser University
Michele Baron, Georgian College
Nick Cacciato, St Clair College
Lori Carre, Seneca College
Leslie A. Dafoe, Sault College
Barbara Dobson, Lambton College
Kari Draker-Fortis, Fleming College
Tara Hayes, Sheridan College
Bernard Ho, Humber College
Cathy Hughes, Seneca College
Maria Ieluzzi, Mohawk College
Carol Gibbons Kroeker, University of Calgary
Louis Mattar, Vancouver Island University
Patty McCabe, Lambton College
Sharon McCleave, Seneca College
Iain McKinnell, Carleton University
Kelly Miles, University of New Brunswick
Trisha Morrow, St. Lawrence College
Maria Romain, Mohawk College
Clint Sheehan, Northern College
Barbara Thistle, George Brown College
Stephen Turnbull, University of New Brunswick

Their comments were honest, valuable, and helpful. They challenged me to do my best work and encouraged me to find the best information possible. Their willingness to share their ideas and resources continue to make this book a first-class resource.

Wendi Roscoe, 2019

About the Author

Wendi Roscoe (H.B.Sc., M.Sc., B.Ed., Ph.D.) is a faculty member in the Department of Health Sciences at Fanshawe College, in London, Ontario. Wendi has an H.B.Sc. (genetics), a M.Sc. (physiology), a B.Ed., and a Ph.D. (physiology and pharmacology) from the University of Western Ontario, London, Ontario. Her Ph.D. research was focused on neuroinflammation in a mouse model of multiple sclerosis. She has published in several peer-reviewed journals, such as *Journal of Neuroimmunology* and *Journal of Neuroscience Research*. Wendi has been teaching physiology, biology, anatomy, and nutrition courses at the university and college level since 2004 and is currently teaching biology and physiology at Fanshawe College. She is also a Registered Holistic Nutritionist, graduating from the Canadian School of Natural Nutrition, in 2017. Recently, she spent 9 months at Griffith University on Australia's Gold Coast researching active learning and flipped classroom teaching methods in higher education healthcare programs. Wendi has an excellent understanding of student challenges and needs within this course.

Lilia Korolova

1 Introduction to Biology, Anatomy, and Physiology

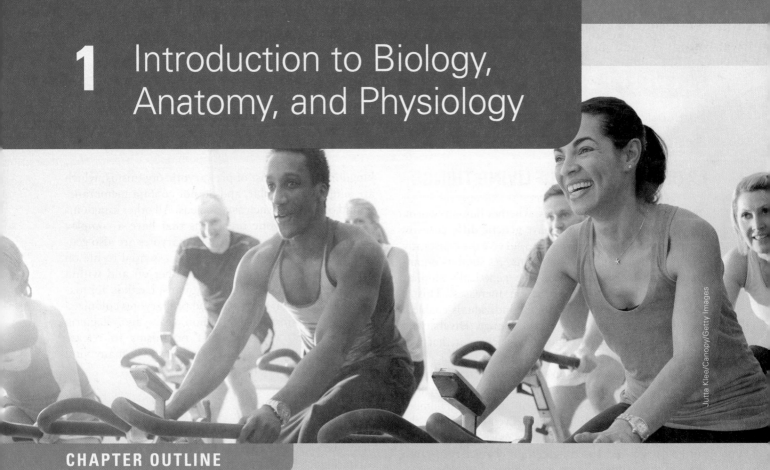

CHAPTER OUTLINE

1.1 OVERVIEW

Biology, anatomy, and physiology are areas of science that are often studied together because of the overlap of the content in each subject. **Biology** is the study of all living things, including their structure, function, growth, and interrelationships. The focus of this textbook is on humans and microorganisms. **Anatomy** is the study of physical structures within an organism, and **physiology** is the study of how those anatomical parts function individually as well as in conjunction with other structures of the body. For example, the respiratory system consists of the anatomical structures of the bronchi, bronchioles, and alveoli, which contain specific types of epithelial and connective tissues. Because those cells and tissues are structured in a particular way, they allow oxygen to diffuse into the bloodstream. The respiratory system functions together with the cardiovascular system to bring oxygen to the cells. Throughout this text, we look at different aspects of molecular and cellular biology, anatomy, and physiology to learn how the human body works—and learn how to maintain optimal health. Although this introductory textbook is intended for students entering a health care program, most of the information will provide a useful guide for anyone who wants to understand their own health.

Education is the most powerful weapon which you can use to change the world.

Nelson Mandela

CONCEPT REVIEW 1.1

1. Define biology, anatomy, and physiology.

1.2 ORGANIZATION OF LIVING THINGS

Organisms within each species, whether human, worm, potato, or bacterium, must have **genetic differences** in order for the species as a whole to survive the constantly changing environmental conditions. As soon as organisms within a species become too genetically similar, the probability of their extinction increases. This is because genetic differences among individuals enable a species to survive environmental change. **Diversity** of living organisms is a very important concept that will be discussed throughout this textbook.

Approximately 8.7 million species are currently known to be living on earth, which is estimated to be only 8 to 15% of the total. Some experts estimate the total number of species is close to 100 million. Even considering the vast amount of genetic variation that exists, all organisms can be categorized based on certain characteristics into one of the following six **kingdoms**: Archaebacteria, Eubacteria, Protista, Fungi, Plants, or Animals.

The Kingdoms

Each kingdom includes organisms with specific traits (Figure 1.1). **Archaebacteria** and **Eubacteria** are the two kingdoms that consist of **prokaryotic** organisms, which are simple, unicellular, and do not contain membrane-bound organelles, such as a nucleus. All other kingdoms consist of **eukaryotic** organisms that have a complex cellular structure (Chapter 3). Prokaryotes are also generally termed *bacteria*. Bacteria are essential to life on earth; they live cooperatively together on and within every other species, including humans. Earth is approximately 4.5 billion years old, and prokaryotes colonized Earth approximately 3.5 billion years ago. Bacteria (cyanobacteria) were the first organisms to release oxygen into the atmosphere, and they play many vital roles in maintaining life on earth.

Archaebacteria

Archaebacteria are prokaryotic, unicellular organisms that were most likely the first living cells on earth (Figure 1.2). They still exist today, some surviving in extreme environments, such as hot springs, volcanoes, the highly salty Dead Sea, soil, and the human digestive tract. Archaea have an independent evolutionary history and, compared to prokaryotic organisms, have very different DNA sequences, lipid structures, cell wall structures, and metabolism.

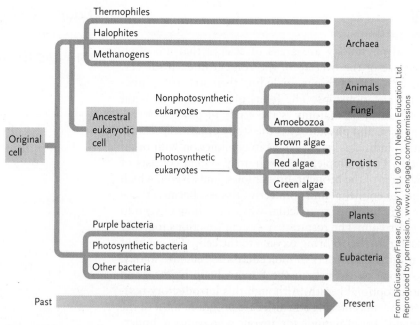

From DiGiuseppe/Fraser. *Biology* 11 U. © 2011 Nelson Education Ltd. Reproduced by permission. www.cengage.com/permissions

Figure 1.1 The Six Kingdoms

All living organisms can be categorized into one of six kingdoms.

Figure 1.2 Archaebacteria

Archaebacteria are prokaryotes that live in many extreme environments. This example is a species of methanogen that produces methane gas in anoxic conditions.

Figure 1.3 Eubacteria

Eubacteria are prokaryotes that live in many environments, including inside our digestive tract, and some species can cause disease. This example is *E. coli*, which lives in the large intestines of animals.

Eubacteria

Eubacteria are also prokaryotic, single-celled organisms, although sometimes they cluster together, such as the common *Staphylococcus aureus* that inhabits human skin and nasal passages (Figure 1.3). Eubacteria survive in many different environments, including the human body; they constitute much of our body's normal flora that is extremely beneficial and important for our health. A small percentage of bacteria can cause infections in humans, including tuberculosis, salmonella, *E. coli*, and chlamydia (Chapter 24).

Protists

Organisms in the Protista kingdom are the most numerous and most diverse of any kingdom and can be found in any environment containing water. **Protists** are mostly unicellular and are eukaryotic (Figure 1.4). Some species of protists are multicellular and very large, such as the giant kelp, which grows to 65 feet (20 metres). Eukaryotic cells are complex and contain many membrane-bound organelles—such as mitochondria, a nucleus, and endoplasmic reticulum (Chapter 3)—that are not found in prokaryotic cells. Species in the Protista kingdom can have features of the other three eukaryotic kingdoms, and can be divided into fungus-like, animal-like, and plant-like. It is thought that the multicellular

DID YOU KNOW?

The human body contains 10 times more bacterial cells than human cells.

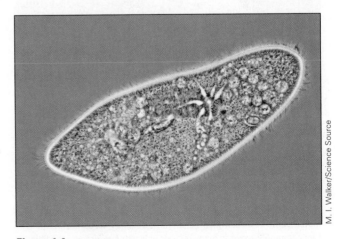

Figure 1.4 Protists

Protists are the eukaryotic, single-celled organisms that gave rise to the eukaryotic, multicellular organisms. This example is a paramecium, which is commonly found in fresh water.

kingdoms arose through the evolution of protists. As do bacteria, protists survive almost everywhere on earth: in the oceans, fresh water, and on land. Some examples of protists include algae, slime moulds, and plankton.

Fungi

Organisms in the **Fungi** kingdom are eukaryotic and multicellular, with the exception of yeast, which is unicellular (Figure 1.5). They are not photosynthetic and most often absorb their nutrients from other organisms, such as trees or dead organisms. Some examples of fungi include mould, mildew, mushrooms, and yeast. Fungi play an important role in **decomposition** (as do some archaea and eubacteria) and nutrient recycling. Yeast is beneficial to humans in the production of bread, alcohol, and soy sauce (*Saccharomyces cerevisiae*), and other fungal species are used to produce antibiotics, such as penicillin. Some fungal species can cause infection in humans, for example, athlete's foot or yeast infections (*Candida albicans*).

DID YOU KNOW?

Crimini mushrooms are small, nutritious, portabella mushrooms that contain vitamins C, B6, B12, and D.

Plants

Species in the Plants kingdom are eukaryotic and multicellular (Figure 1.6). This kingdom includes a wide

Figure 1.6 Plants

Plants are eukaryotic, multicellular organisms that use photosynthesis to acquire energy. This example is romaine lettuce, which is commonly eaten in salads.

Courtesy of Garry Roscoe

range of organisms, many of which are important food sources for humans. **Plants** have **chloroplasts** in their cells, which are membrane-bound organelles that contain the pigment **chlorophyll** that is used in **photosynthesis**. During photosynthesis, plants combine carbon dioxide from the air with hydrogen from water, and energy from sunlight to produce glucose. Many plants provide a healthy source of food, including broccoli, spinach, carrots, apples, and blueberries. Some plants or parts of plants can be poisonous to humans; for example, the leaves of cherry trees and peach trees contain toxic molecules that can be converted into cyanide poison.

DID YOU KNOW?

Many bean species contain toxic chemicals and can cause severe and possibly fatal effects in humans. Raw beans, such as lima beans and kidney beans, must be soaked in water for at least five hours and then boiled for a minimum of 10 minutes to ensure the toxin is degraded. Canned beans have already undergone this process and can be eaten without cooking.

Animals

The **Animals** kingdom includes organisms that are eukaryotic, multicellular, and are capable of **locomotion**; that is, they can move from one location to another (Figure 1.7). Examples of organisms in this kingdom are humans, dogs, fish, sponges, roundworms, and birds. Many animal species are food sources for humans—and this varies among cultures around the world—such as cows, chickens, turkey,

Courtesy of Garry Roscoe

Figure 1.5 Fungi

Fungi are eukaryotic, multicellular organisms that include yeasts, moulds, and mushrooms. This example is a shaggy mane mushroom that commonly grows in many regions of North America and is edible; however, it can easily be confused with the magpie fungus, which is poisonous.

fish, pigs, lamb, shark, snake, buffalo, moose, duck, and deer.

By looking at the organization of living things in terms of six kingdoms, we can gain perspective on where humans fit within this classification. The rest of this chapter and the textbook as a whole focus on human cell biology, anatomy, and physiology. Chapter 23 describes some organisms from other kingdoms that can potentially cause human disease.

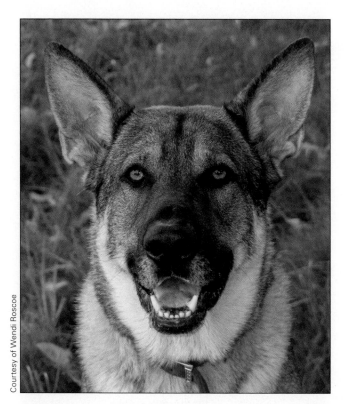

Courtesy of Wendi Roscoe

Figure 1.7 Animals

Animals are eukaryotic, multicellular organisms that have locomotion. Dogs and humans are both examples of animals.

DID YOU KNOW?

Dogs have been called "man's best friend" and have been domesticated for thousands of years. The companionship and loyalty between dogs and humans may be due in part to some specific genetic similarities. For example, humans and dogs live cooperatively in groups, and both are exceptionally good at reading body language and bonding; this is due to their high levels of secretion of oxytocin (Chapter 17). Humans have used selective breeding extensively with dogs (Chapter 7) to perpetuate certain desired traits in dogs.

CONCEPT REVIEW 1.2

1. Make a chart to organize the characteristics of organisms in each kingdom.

2. Give examples of species from each kingdom.

1.3 ORGANIZATION OF THE HUMAN BODY

The human body can be organized into different levels from the microscopic to the entire organism: chemicals, molecules, cells, tissues, organs, organ systems, and organism (Figure 1.8). **Chemicals** are the individual elements, such as nitrogen, hydrogen, carbon, and oxygen, that combine to form small molecules, such as glucose, water, amino acids, or nucleotides, that also combine to form **macromolecules**, such as complex carbohydrates, DNA, proteins, and fats (Chapter 2). The small molecules combine in various ways to produce the large macromolecules that form part of every **cell** of every living organism (Chapter 3).

Cells combine to form the various **tissues** in the body: epithelial, connective, muscle, and nerve (Chapter 11). Different combinations of those tissues form **organs**. For example, the innermost layer of the stomach is composed of epithelial tissue, which is surrounded by strong collagen-containing connective tissue, which binds to three smooth muscle layers that contract to digest food. Neurons that send signals to the smooth muscle cells cause them to contract when food is in the stomach. All organs are composed of different amounts of all four tissue types. Certain organs function with other organs and tissues to form **organ systems**, such as the digestive system, respiratory system, or cardiovascular system (Chapters 12–22). All organ systems combined make up the entire human **organism**.

Figure 1.8 Organization of the Body

The body is organized from the smallest atoms and molecules to form tissues, organs, and organ systems.

CONCEPT REVIEW 1.3

1. Describe how the human body is organized into levels.

1.4 ORGAN SYSTEMS

There are 11 organ systems in the body, and they all interact with each other. Note that you will need to be able to relate information from one chapter to the next, so by the end of the book you will have a working knowledge of how the various parts of the body work *together*. For example, when we breathe air, the oxygen diffuses across the alveoli (respiratory system) into the bloodstream (cardiovascular system), which carries that oxygen to all the cells in the body so that each cell can produce energy (metabolism) in the form of ATP. The cardiovascular system also carries nutrients (molecules) absorbed by the digestive system, which is regulated by the nervous system and the endocrine system. Some organs are considered part of more than one organ system: for example, the pancreas is part of the digestive system because it produces digestive enzymes that are secreted into the small intestine, and it is also part of the endocrine system because it produces insulin, which is a hormone that regulates blood sugar levels. This section gives a brief overview of the structures and functions of each organ system, which will be discussed in detail in later chapters.

The Integumentary System

The integumentary system (Chapter 12) is made up of the skin, hair, nails, sebaceous glands, and sweat glands (Figure 1.9). The major functions of this system are to protect the body from infection, physical damage, and dehydration; help regulate body temperature through production of sweat and dilation or constriction of blood vessels in the dermis layer of the skin; excrete some waste products in sweat; produce vitamin D from cholesterol during exposure to sunlight; and detect sensations, such as touch, pain, pressure, vibration, and temperature.

The Skeletal System

The skeletal system (Chapter 13) consists of bones, cartilage, and the joints (Figure 1.10). The function of this system is to provide support for the body and to protect internal organs: for example, the skull protects the brain. It works in combination with the **muscular system** to allow body movement. The skeletal system also stores minerals, such as calcium, phosphorus, and magnesium, and plays an important role in blood mineral homeostasis. It stores fats in the yellow bone

Figure 1.9 The Integumentary System

marrow and produces red and white blood cells in the red bone marrow.

The Muscular System

The muscular system (Chapter 14) consists of skeletal muscles, smooth muscle (in organs), and cardiac muscle

Figure 1.10 The Skeletal System

Figure 1.11 The Muscular System

(the heart) (Figure 1.11). The function of the skeletal muscles is to enable the body to move and maintain posture. Smooth muscles contract and move substances through organs, blood vessels, or bronchial tubes. Cardiac muscle contractions move blood from the heart to the blood vessels. Muscles also function in heat production because heat is produced as a by-product of cellular respiration, and, compared to all other cell types in the body, muscle cells produce the most ATP.

The Nervous System

The nervous system (Chapter 15) consists of the brain, spinal cord, nerves, and receptors. Receptors include tactile receptors in the skin, baroreceptors, osmoreceptors, chemoreceptors found throughout the body,

and also those of the special senses (Chapter 16)—visual, auditory, equilibrium, gustatory (taste), and olfactory (smell) (Figure 1.12). The function of the nervous system is to *detect* changes within the body and the environment, *interpret* those stimuli, and then *respond* to those changes. The nervous system plays an important role in regulating certain biological factors, such as body temperature, blood pressure, and hormone levels; this regulation is called **homeostasis** (Section 1.6).

The Endocrine System

The endocrine system (Chapter 17) consists of the glands that produce hormones that travel through the bloodstream to highly specific target cells that have

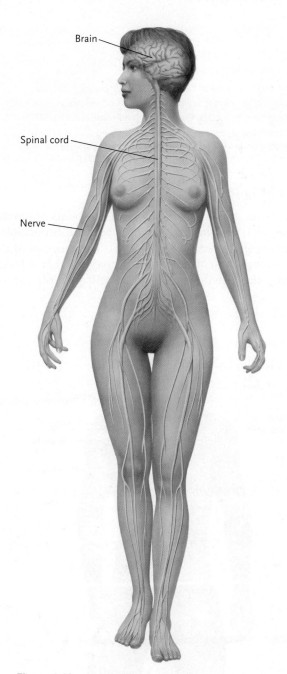

Figure 1.12 The Nervous System

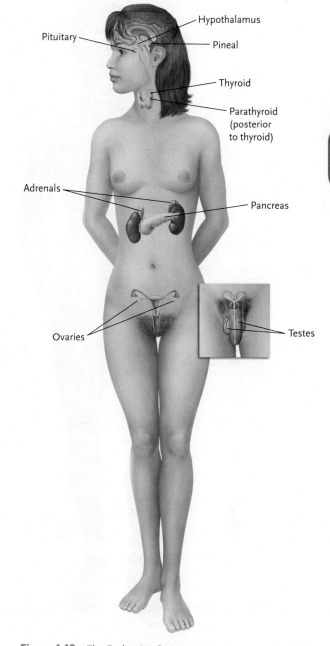

Figure 1.13 The Endocrine System

very specific functions (Figure 1.13). Like the nervous system, the endocrine system plays an important role in homeostasis. The glands of the endocrine system include the hypothalamus, pituitary gland, pineal gland, thyroid gland, parathyroid gland, adrenal glands, pancreas, ovaries, and testes. There are also non-endocrine cells that produce and secrete hormones, such as in the digestive system and kidneys. The numerous functions of the endocrine system depend on which gland is secreting which hormone. Very briefly, some functions include the regulation of the autonomic nervous system (hypothalamus), regulation of growth and reproduction (hypothalamus and

pituitary gland), regulation of metabolism (thyroid gland), regulation of blood calcium levels (parathyroid gland), regulation of stress hormones (adrenal gland), regulation of blood sugar (pancreas), and production of sperm or eggs (testes or ovaries).

The Cardiovascular System

The cardiovascular system (Chapter 18) consists of the heart, blood vessels, and blood (Figure 1.14). The function of the cardiovascular system is to transport oxygen, nutrients, hormones, and waste products in the bloodstream to the various tissues of the body, or

Heart

Blood vessels

Blood (within vessels)

Figure 1.14 The Cardiovascular System

The Respiratory System

The respiratory system (Chapter 19) consists of mouth, nose, pharynx, larynx, trachea, bronchi, bronchioles, and alveoli (air sacs) (Figure 1.15). The function of the respiratory system is to take up oxygen that diffuses across the alveoli into blood vessels; give off carbon

Nasal cavity

Oral cavity

Larynx

Pleural membranes

Pharynx

Trachea

Lung

Bronchi

Bronchioles

Alveoli

Figure 1.15 The Respiratory System

to the kidneys for excretion. The blood is composed of red blood cells that carry oxygen; white blood cells that perform immune functions (Chapter 23); platelets, which function in blood clotting; and plasma, the water portion of the blood that contains many dissolved ions, nutrients, and molecules that act as chemical buffers: for example, bicarbonate regulates blood pH. The cardiovascular system—in conjunction with the endocrine system and the nervous system—plays an important role in the regulation of body temperature and blood pressure by dilating or constricting certain blood vessels.

dioxide; regulate blood pH by expelling carbon dioxide; and produce vocal sounds through the larynx, commonly called the *voice box*.

The Digestive System

The digestive system (Chapter 20) consists of many organs that are directly involved with transporting and digesting food, including the mouth, esophagus, stomach, small intestine, large intestine, rectum, and anus (Figure 1.16). The digestive system also includes accessory organs that assist food digestion, such as the salivary glands, liver, pancreas, and gallbladder. The functions of the digestive system are to break down food macromolecules (Chapter 2), such as

proteins, complex carbohydrates, and fats into their smaller components—amino acids, simple carbohydrates, and fatty acids—that can then be absorbed into the bloodstream or the lymphatic vessels for circulation to all cells in the body; it also functions in the elimination of undigested waste. Accessory glands such as the liver, gall bladder, and pancreas have roles in assisting digestion—for example, by producing bile or digestive enzymes—or can have related roles, such as regulating blood sugar.

The Urinary System

The urinary system (Chapter 21) consists of the kidneys, ureters, bladder, and urethra (Figure 1.17). The

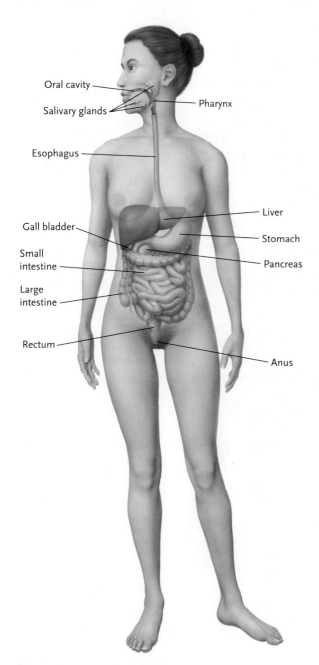

Figure 1.16 The Digestive System

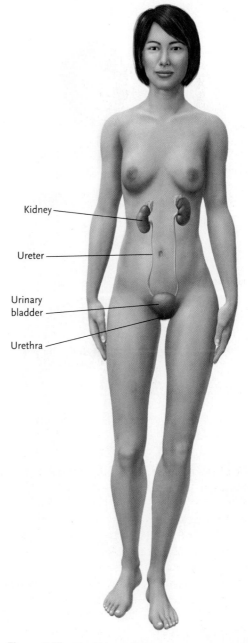

Figure 1.17 The Urinary System

functions of the urinary system are to filter the blood that circulates through the kidneys, reabsorb essential nutrients and water, and excrete waste such as excess ions, water, and urea (produced from the breakdown of proteins). In conjunction with hormonal signals from the endocrine system, the kidneys regulate blood pressure by excreting or reabsorbing water and ions. The kidneys also help regulate blood pH by the excretion of excess hydrogen ions. The urinary system is involved in regulating the production of red blood cells—by producing the hormone erythropoietin—and the production of vitamin D3 (the active form of

vitamin D), which regulates blood calcium levels (Chapters 17 and 21).

The Reproductive System

The female reproductive system (Chapter 22) consists of mammary glands, ovaries, fallopian tubes, uterus, and vagina. The male reproductive system includes the testes, vas deferens, epididymis, prostate gland, and penis (see Figure 1.18). The functions of the reproductive system are to produce gametes (oocytes or sperm) in the gonads (ovaries or testes) and to produce milk in

Mammary glands

Fallopian tube

Ovary

Uterus

Vagina

Vas (ductus) deferens

Seminal vesicle

Prostate

Testis

Epididymis

Penis

Figure 1.18 The Reproductive System

the mammary glands. The reproductive system is part of the endocrine system because the ovaries produce the hormones estrogen and progesterone, and the testes produce testosterone. Both males and females produce each of these hormones, but males make more testosterone, and females make more estrogen and progesterone.

DID YOU KNOW?

Prostate cancer is the most common type of cancer in men in Canada; it affects one in seven men and kills over 300 000 men in Canada and the U.S. every year. Men at highest risk are over the age of 65, overweight, have a family history of prostate cancer, and/or are of African descent. Research shows that diet is a major factor in this and many other cancers (Chapters 2 and 5).

Figure 1.19 The Immune and Lymphatic System

The Immune System and the Lymphatic System

The immune system (Chapter 23) consists of the white blood cells, bone marrow, lymph nodes, spleen, thymus, and lymphatic system, which is composed of lymph nodes, lymph vessels, lymphoid tissue (such as tonsils), and lymph fluid (Figure 1.19). The main functions of the lymphatic system are to take up excess fluids that have leaked from the cardiovascular system (plasma) into the interstitial or intercellular space (interstitial fluid) and to carry that fluid through the lymphatic vessels (lymph) to be deposited back into the circulatory system. The lymphatic system is also involved in absorbing fats from the digestive tract and then circulating them to the bloodstream. The immune system also functions to destroy infectious microorganisms such as bacteria, viruses, fungi, or parasites (Chapter 24).

CONCEPT REVIEW 1.4

1. Compile a list of the structures that make up each organ system.

2. Are any organs associated with more than one organ system? If so, which ones?

3. What are the main functions of each organ system?

4. Match the organ system to the following functions: (A) temperature regulation, (B) absorption of nutrients, (C) regulation of blood sugar, (D) elimination of waste, (E) body movement, (F) storage of minerals and fat, (G) transportation of oxygen, (H) production of vitamin D, (I) release of carbon dioxide. Note that some functions may involve more than one organ system.

    a. Respiratory system _____
    b. Cardiovascular system _____
    c. Endocrine system _____
    d. Muscular system _____
    e. Urinary system _____
    f. Digestive system _____
    g. Skeletal system _____

1.5 ANATOMICAL TERMS

The parts of the human body are named in combination with directional terms (Figure 1.20). This makes it possible for specific regions of the body to be consistently identified among health care professionals and in-patient records. For example, if someone has a burn on their right forearm near the wrist, the location of the burn would be identified as the right, anterior, antebrachial region, proximal to the carpal region. An understanding of medical anatomical terminology is very important in the health profession. The terms applied to any region of the body refer to the body in what's known as the **anatomical position**: standing up, facing forward, toes forward, and palms facing forward.

For the purpose of anatomical terminology, the body is divided into the following general regions: head; neck; trunk, including thoracic, abdominal, and pelvic regions; upper limbs; and lower limbs. In each region of the body, the underlying bones, muscles, nerves, or blood vessels often have similar names. For example, the thigh is the femoral region, the thigh bone is the femur, and the thigh contains a femoral artery and a femoral nerve. The temporal region of the head contains a temporal bone, and the temporal lobe of the brain. An understanding of these anatomical terms will help in your study of the organ systems.

Directional Terms

Directional terms are used to describe the regions of the body in relation to each other, and these terms refer to the body in the anatomical position (Figure 1.21). For example, the skin is superficial to the muscles, and the

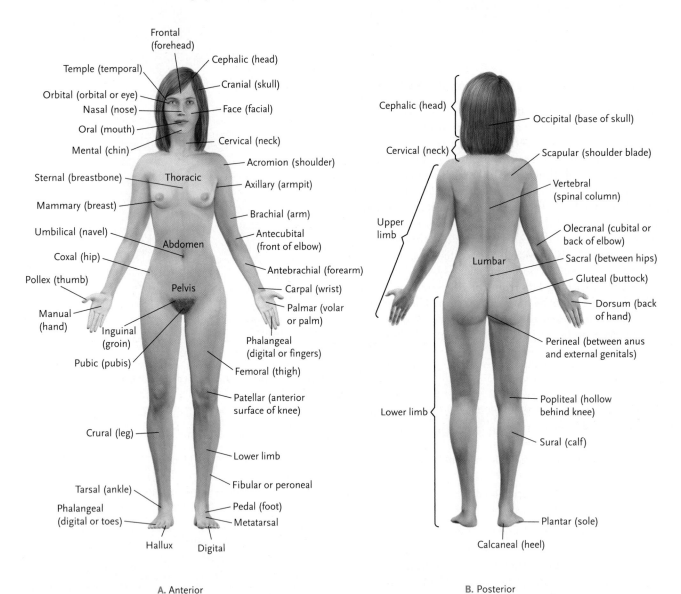

A. Anterior

B. Posterior

Figure 1.20 Anatomical Terms

Anatomical terms are applied to various regions of the body as shown in the anatomical position.

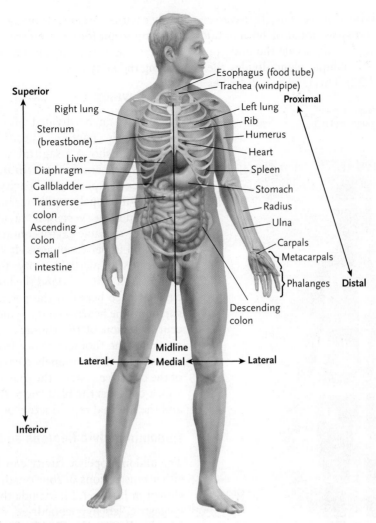

Figure 1.21 Directional Terms

The location of each body part can be described relative to other body parts by using directional terms.

patellar region is distal to the inguinal region. See Table 1.1 for a description of each directional term and to develop an understanding of where the major body organs are located in relation with each other. For example, the esophagus is posterior to the trachea: you could also say the trachea is anterior to the esophagus. Most of the liver is located in the right abdominal region, the heart is intermediate to the lungs, and the pancreas is inferior (below) and posterior (behind) to the stomach. This information is medically significant, especially if you or your patient has pain in a certain location; for example, pain in the upper right or lower right abdominal region could indicate either appendicitis or gall stones.

Body Planes

The body, or parts of the body, can also be divided into sections: sagittal, frontal, transverse, or oblique. A **sagittal plane** divides the body into left and right sections; equal sections are **midsagittal**, and unequal sections are **parasagittal**. A **frontal plane** divides the body into anterior and posterior sections. A **transverse plane** divides

TABLE 1.1

Directional Terms

Directional Term	Definition
Superior	Above, over, toward the head
Inferior	Below, under, further from the head
Anterior	In front, toward the front of the body
Posterior	Behind, toward the back of the body
Proximal	Closer to the attachment point of a limb to the trunk
Distal	Further from the attachment point of a limb to the trunk
Medial	Toward the midline
Lateral	Further from the midline
Intermediate	In between two structures
Superficial	Toward the surface of the body
Deep	Toward the interior of the body

the body into superior and inferior sections. Transverse sections are also called a **cross-sectional** or **horizontal plane**, and these do not necessarily divide the body or organ into equal halves. An **oblique plane** divides the body on an angle (Figure 1.22). This textbook describes body cavities, organs, or tissues in a variety of sections. It is important for you to learn what type of section you are looking at so you can understand the relationships among the body parts.

Body Cavities

The body can be divided into specific **body cavities**, which each contain specific organs or tissues. Figure 1.23 shows a sagittal and a frontal view of the body cavities: **cranial**; **vertebral**; **thoracic**, including the **pleural cavity**, **pericardial cavity**, and the **mediastinum**; and **abdominopelvic**. The cranial cavity contains the brain, and the vertebral cavity contains the spinal cord. The thoracic cavity contains the lungs, heart, trachea, and upper esophagus. It also contains two other cavities—the pleural cavity (space between pleural membranes that surround the lungs) and the pericardial cavity (space between the pericardial membranes that surround the heart)—and the mediastinum (anatomical central region of the thoracic cavity). The diaphragm separates the thoracic cavity from the abdominopelvic cavity. The abdominopelvic cavity contains the organs of the digestive system, the spleen, and the pelvic region, which contains the bladder, parts of the large intestine, and the internal reproductive organs.

Abdominopelvic Regions and Quadrants

The abdominopelvic cavity can be further divided into either nine regions or four quadrants. The nine regions, shown in Figure 1.24, include the right hypochondriac, epigastric, left hypochondriac, right lumbar, umbilical, left lumbar, right inguinal, hypogastric, and left inguinal. The upper horizontal division line is at the distal end of the rib cage; the lower horizontal line is just superior to the pelvis (iliac crest). The two vertical division points are approximately one inch medial to the nipples. Figure 1.24 shows some of the organ structures in each region. A patient with epigastric pain could have a problem with several different organs, including the liver, stomach, distal esophagus, or transverse colon.

The abdominopelvic region is also commonly divided into quadrants. Figure 1.25 shows the quadrants; the right and left sides are divided into equal halves, and the horizontal plane is divided at the umbilicus.

Figure 1.22 Planes of the Body and Planes of an Organ
The body (A) and the organs (B) can be described in terms of frontal, sagittal, transverse, or oblique planes.

DID YOU KNOW?

Sometimes pain in a certain organ can be felt in very different areas of the body; this is called *referred pain*. Pain associated with a heart attack can be felt in the neck, upper arm, or back, even though the heart is located in the intermediate thoracic region. A "brain freeze" from eating or drinking cold foods or liquids quickly can be felt in the frontal sinuses because the trigeminal nerve carries temperature and pain sensory information from the hard palate, which is the back of the roof of the mouth.

Figure 1.23 Body Cavities and Thoracic Cavity

The body cavities (A) include the cranial, vertebral, thoracic, and abdominopelvic cavities. The thoracic cavity (B) includes the pleural and pericardial cavities.

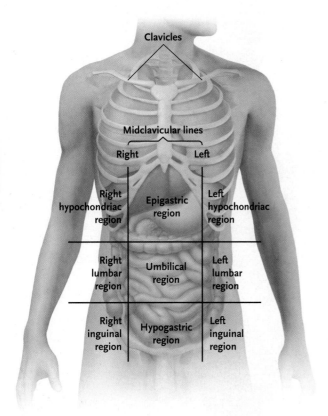

Figure 1.24 Abdominopelvic Cavity Regions
The abdominopelvic cavity can be divided into nine regions.

Figure 1.25 Abdominopelvic Cavity Quadrants
The abdominopelvic cavity can also be divided into four quadrants.

CONCEPT REVIEW 1.5

1. Why is the anatomical position important?
2. Make a chart comparing the common name and anatomical name for each body region.
3. Describe the different planes that are used to divide the body or a body part.
4. What organs are found in each of the following cavities: cranial, vertebral, thoracic, and abdominopelvic?

5. Where would you find pleural membranes, pericardial membranes, and the peritoneum?
6. What is the mediastinum?
7. List the major organs that are located within each of the nine abdominopelvic regions, or the four quadrants.

1.6 PROPERTIES OF LIVING THINGS

Living organisms have certain characteristics that differentiate them from non-living things. These characteristics are introduced below and covered in detail in this textbook as follows: all living things are made up of the same macromolecules (Chapter 2); they have a cellular structure (Chapter 3); they grow and metabolize nutrients (Chapter 4); they have mechanisms of homeostasis (below); they reproduce at a cellular and organism level (Chapters 5 and 22); they have DNA that they pass on through inheritance (Chapter 6); and all species change over time through a process called **evolution** (Chapter 7).

Application

As you read through the rest of this chapter, consider this scenario. Freda has just consumed a large piece of chocolate cake. The sugar is absorbed from the digestive tract into her bloodstream. The rise of blood sugar causes her pancreas to release insulin, which causes the body cells, such as the liver, to take up the blood sugar and convert it into fat to be stored. Freda's blood sugar rises after eating but then decreases back to normal levels. Freda is healthy, and homeostasis is maintaining her blood sugar levels.

a. What is the stimulus?
b. What type of receptor is involved?

c. What is the integrating centre?

d. What are the effectors?

e. Is this positive or negative feedback?

f. What would happen to Freda if her body was not able to regulate blood sugar?

Macromolecules

Every living thing is made up of the basic elements, carbon, oxygen, hydrogen, and nitrogen. All living things combine these basic chemical elements into small molecules, such as sugars, amino acids, fatty acids, and nucleotides. The small molecules combine into various macromolecules: polysaccharides (complex carbohydrates), proteins, fats, and nucleic acids.

Cellular Structure

All living organisms consist of one or more cells, which are composed of phospholipids, proteins, carbohydrate groups, and cholesterol. Single-cell organisms can be prokaryotes or eukaryotes (Section 1.2), and complex multicellular organisms are eukaryotic. **Viruses** have a protein capsule instead of a cell membrane and, therefore, are not technically considered living things, but viruses are significant to all living things and will be discussed in detail in Chapter 24.

Growth and Metabolism

All living organisms grow and undergo metabolism, producing energy in the form of ATP. Depending on the particular species and conditions, ATP can be produced either with oxygen—aerobically—or without oxygen—anaerobically. Humans produce ATP both with and without oxygen, and we can produce ATP from glucose, amino acids, fatty acids, and nucleotides.

Reproduction

All living organisms reproduce at a cellular level through mitosis and at an organism level through meiosis, by producing sperm or eggs, spores, or seeds.

Hereditary Material

Every living organism passes on some of their genetic material to their offspring. Humans have 46 DNA molecules (chromosomes): 23 from one parent and 23 from the other parent. Offspring are similar and yet different from their parents, and certain traits are passed on through generations.

Evolution

All populations of organisms continuously change genetically from one generation to the next. Humans have 46 chromosomes, but the particular genes on those chromosomes vary: such as the genes that encode skin pigmentation, or body height. Although humans today differ from humans 100 000 years ago, the distinctly human DNA remains intact.

Homeostasis

Homeostasis is the regulation of multiple factors within the **internal environment** of the body. The internal environment includes whatever is in close contact with the body's cells, such as the blood and the interstitial fluid that surrounds all cells and contains certain amounts of ions and nutrients. The **external environment** includes everything exterior to the body that the body cannot regulate, such as temperature, food acidity, humidity, air pollutants, and microorganisms that may become inhaled or swallowed. Any part of the body that is in contact with the outside world is considered part of the external environment, including the digestive tract, external ear canals, nose and sinuses, inside the trachea, bronchi, and air sacs. No matter what kind of changes occur in that external environment—such as changes in oxygen levels, nutrients ingested, and inhaled or ingested toxins—the internal environment is highly regulated so that the body's cells function optimally.

Homeostasis is a continual process that's necessary for the normal functioning of every living thing. Some factors are strictly regulated within a narrow range for the human body, including oxygen, carbon dioxide, pH, H^+, blood glucose, core body temperature, blood pressure, water, and electrolytes (sodium, potassium, chloride, calcium, etc.). When these factors are in balance, cells function properly, and we feel healthy.

If homeostasis is not maintained for a short time, you will feel certain symptoms. For example, if your blood sugar decreases and your liver cells are not efficient at breaking down fat to produce glucose (gluconeogenesis) and if you don't consume food right away to increase your blood sugar, you might feel tired, weak, or perhaps dizzy.

If homeostasis cannot be maintained at all, you will have a disease, and without medical intervention death could occur. For example, if your pancreas was not able to make insulin, then cells would not be able to take up blood sugar and therefore make ATP, and blood sugar levels would be too high; this describes diabetes, which is fatal without injections of insulin.

Every organ system is involved in homeostasis in some way, and this will be discussed in each of the textbook's chapters on organ systems.

Components of Feedback Systems

Homeostasis regulates most of the factors discussed above through a process of **negative feedback** (Figure 1.26), which brings about a reverse in the level of a controlled factor. As well, a process of **positive feedback** sometimes occurs, in which a response caused by an initial stimulation continues to increase.

All of the body's feedback systems consist of the following components: **receptors** that detect the level of some factor (such as temperature); an **integrating centre** that

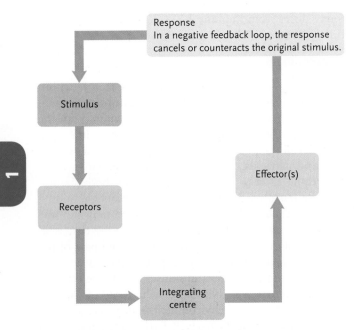

Figure 1.26 Negative Feedback Mechanism

A negative feedback mechanism consists of receptors that detect a change and signal the integrating centre, which then sends a signal to effectors to correct the change.

compares information from the receptors to a standard set point (such as 37°C for normal body temperature); and **effectors**, which respond to signals from the integrating centre: for example, sweat glands. The receptors constantly send information about the body's internal environment to the integrating centre, which constantly signals the effectors to increase or decrease a specific response.

Receptors The receptors are either extensions of a sensory neuron that sends information to the integrating centre about a specific stimulus, or individual receptor cells that communicate directly with a sensory neuron. The following list gives the names and functions of the common types of sensory receptors: **chemoreceptors** detect chemical concentrations such as neurotransmitters, drugs, or hormones; **osmoreceptors** detect changes in osmolarity, that is, water and ion concentrations; **tactile receptors** detect touch, pressure, and vibration; **baroreceptors** detect blood pressure; **photoreceptors** detect light and are found only in the retina; **mechanoreceptors** detect stretching (e.g., spindle fibres in muscles); **proprioceptors** detect body position; **nociceptors** detect pain; and **thermoreceptors** detect temperature.

Integrating Centre The integrating centre interprets the signals coming from the receptors and determines when any deviation occurs from the standard set point. Certain parts of the brain act as an integrating centre, and certain glands also act as an integrating centre: for example, the **hypothalamus** assesses temperature, thirst, hunger, and sleep; the **medulla oblongata** regulates breathing, blood pressure, and heart rate; and **pancreas** determines blood glucose levels.

Effectors The effectors include any part of the body that responds to stimulation from the integrating centre and brings an altered factor back to normal. They include **skeletal muscles** (e.g., shivering when cold); **smooth muscles** (e.g., vasodilation, bronchodilation, increased gastric motility); **cardiac muscle** (e.g., change in heart rate or its force of contraction); and the **endocrine glands**.

Negative Feedback Mechanism

Negative feedback means there is a reverse in the level of a controlled factor. For example, exercise causes the body temperature to increase, which is then regulated by the production of sweat that cools the body. The higher the increase in body temperature, the more sweat is produced in an effort to maintain the normal body temperature. When normal body temperature cannot be maintained in extreme conditions, illness and possibly death will occur.

The following example describes what's involved in a negative feedback mechanism to regulate body temperature:

stimulus (cold external environment)—receptors (thermoreceptors)—integrating centre (hypothalamus determines and signals that body temperature has dropped below the 37°C set point)—effectors (skeletal muscles shiver and produce heat)—new stimulus (increased body temperature)—receptors (thermoreceptors)—hypothalamus (stops signalling that body temperature is below normal)—effectors (skeletal muscles stop shivering)

Positive Feedback Mechanism

A positive feedback mechanism continues to increase a response caused by an initial stimulation (Figure 1.27).

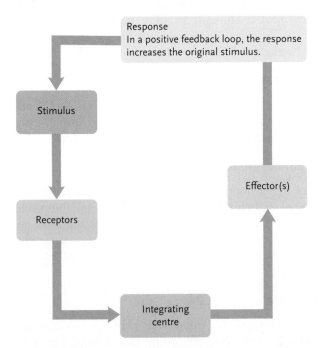

Figure 1.27 Positive Feedback Mechanism

Childbirth is an example of a positive feedback mechanism, where the response is continually increased by the stimulus until the baby is born.

There are very few examples of positive feedback systems in the human body. One example is childbirth, where the stimulus is pressure of the head of the fetus, the receptors are mechanoreceptors in the uterus, and the integrating centre is hypothalamus, which produces oxytocin that signals the effectors—the uterine muscles—to contract; this response leads to further stimulation of the mechanoreceptors, which signals the hypothalamus to produce more oxytocin.

CONCEPT REVIEW 1.6

1. Make a list of the different features that are found in all living organisms.
2. What is homeostasis? Describe the difference between negative feedback and positive feedback mechanisms, and give examples.
3. Explain the function of a receptor, an integrating centre, and an effector.
4. What are the different types of receptors, and what kind of stimulus do they detect?
5. Name some common effectors in the body.

CHAPTER REVIEW

Medical Terminology

TERM	DEFINITION	EXAMPLE
ab-	away from	abduction
ad-	toward, near	adduction
-algia	pain	neuralgia
ambi-	positioned on both sides	ambidextrous
amphi-	around, on both sides	amphipathic
ana-	upward, backward	anaplasia
anomal/o	irregular	anomaly
ante-	before, forward	anteflexion
anter/o	front	anterior
anthrop/o	man, human being	anthropology
apo-	away, separation	apoptosis
bio-, bi/o	life, living	biology
blast/o, -blast	early embryonic stage, immature	blastocysts
calor/i	heat	calorie
carcin/o	cancer	carcinoma
cata-	down, downward	cataract
caud/o	tail	caudal
-cele	pouching	varicocele
chem(o)	chemistry, drug	chemotherapy
chron/o	time, timing	chronological
circum-	around	circumference
cis	on this side	cis fatty acid
co-	together	coenzyme
contra-	against, opposite	contralateral
corpor/o	body	corporeal

TERM	DEFINITION	EXAMPLE
cyt/o, -cyte	cell	erythrocyte
de-	away from or cessation	dehydration
dextr/o	right, on the right side	dextrose
dia-	through, apart	diaphysis
dif-	separation	differentiate
dis-	apart, to separate	dislocate
dors/o	back	dorsal
-dynia	pain	gastrodynia
dys-	difficult, abnormal	dyspnea
ec-, ecto-	outside, out	ectopic
-ectomy	surgical removal	appendectomy
en-, endo-	inside, within	endocrine
epi-	above, over, upon	epigenetics
equi-	equality, equal	equilibrium
eso-	within	esophagus
eti/o	cause	etiology
exo-	outside of, outward	exocytosis
extra-	outside	extracellular
fore-	before, in front of	forehead
-genesis	production, formation	gluconeogenesis
-genic	producing	cardiogenic
hemi-	one half	cerebral hemisphere
hetero-	the other of two, different	heterozygote

(Continued)

TERM	DEFINITION	EXAMPLE
hist/o	tissue	histology
homo-	similar	homozygous
hyper-	above, excessive, beyond	hypertonic
hypo-	under, deficient, below	hypothalamus
-ic	pertaining to	hepatic
-icle	small	follicle
infra-	below, beneath	infraspinatus
inter-	between	interstitial
intra-	within	intracellular
ipsi-	same	ipsilateral
juxta-	near	juxtaglomerular
kary/o	nucleus	eukaryote
kel/o	fibrous, growth	keloid
later/o	side	lateral
levo-	left	levodopa
-logist	specialist	neurologist
-logy	study of	cardiology
macr/o	large	macromolecule
media/o	middle	medial
meso-	middle	mesoderm
meta-	after, behind	metacarpal
micro-	small	microorganism
mon/o	single	mononucleosis
morph/o	shape, form	polymorphic
nucle/o	nucleus	nucleoplasm
nutri/o	to nourish	nutrition
para-	alongside, near, beyond, abnormal	parathyroid
per-	through, throughout	periodic table
peri-	around, surrounding	peritubular
physi/o	nature	physiologist
-plasm	formation, growth	neoplasm
-poiesis	formation	hematopoiesis
post-	after, behind	post-translation
poster/o	behind, toward the back	posterior
pre-	before, in front of	premature
pro-	before	prophylactic
proxim/o	near	proximal
retro-	behind, backward	retrolingual
-sect	to cut	section
somat/o	body	somatoscopy
-some	body	chromosome
-stomy	surgical opening	colostomy
sub-	under, beneath	subcutaneous

TERM	DEFINITION	EXAMPLE
super-	above, beyond	superior vena cava
supra-	above, beyond	supraorbital
-therapy	treatment	therapeutics
therm/o	heat	thermometer
-tomy	surgical incision	appendectomy
trans-	across	transmembrane
-type, typ/o	class, representative form	phenotype
ventr/o	belly, front of the body	ventrolateral
-verse	turn	transverse
viscer/o	internal organs	visceromegaly

Labelling Activity

Label the following diagrams.

1.

2. a.

b.

3. a.

b.

Answer to Application Question

a. Stimulus. Rise in blood sugar is the stimulus.

b. Type of receptor. Chemoreceptors detect chemicals, such as sugar; the chemoreceptors are in the pancreas.

c. Integrating centre. Normally the integrating centre is the brain, but in this example, the pancreas detects the blood sugar and also determines that the blood sugar is increased.

d. Effector. The pancreas is the effector because it is releasing insulin.

e. This is negative feedback because the effect (release of insulin) causes the blood sugar levels to decrease.

f. If Freda could not regulate her blood sugar, she would have diabetes.

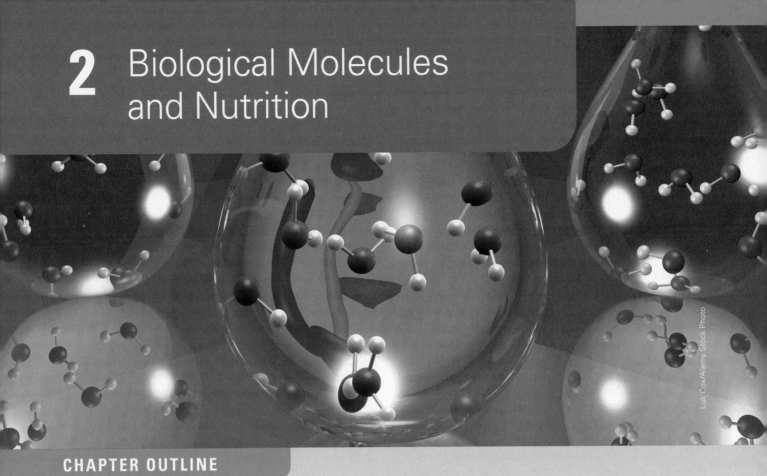
Luk Cox/Alamy Stock Photo

2 Biological Molecules and Nutrition

CHAPTER OUTLINE

2.1 WATER

Universal Solvent

All living things from the smallest single-celled bacterium to multicellular plants and animals are composed of anywhere from 45 to 95% water. In humans, infants are approximately 75% water, and this amount decreases with age: mostly in the first 10 years of life, but continuing into old age. Healthy adult males are on average 60% water, and females are 55% water. Healthy females compared to healthy males of appropriate body weight have slightly less water because they tend to carry more body fat. Anyone who is obese can have as little as 45% body water content. One litre of water weighs 1 kg. An average 70 kg male (154 lb) has approximately 42 L (60%) of body water. Body water is distributed among three main compartments: one intracellular and two extracellular.

1. **Intracellular**—40% of body weight (approximately 28 L)
2. **Extracellular**—20% of body weight (approximately 14 L)
   a. **Interstitial fluid**—13% (approximately 9 L)
   b. **Blood plasma**—7% (approximately 5 L)

It is the mark of an educated mind to be able to entertain a thought without accepting it.

Aristotle

Many important nutrients, ions, and molecules are dissolved in the water inside and outside of cells. Water is the fluid of the body, the **solvent**, and it contains dissolved molecules, the **solutes**. Due to its unique properties, water is the universal solvent in all living things.

Polarity

Oxygen has eight positively charged protons compared to hydrogen's single proton. Oxygen and hydrogen form a **covalent bond** in which negatively charged electrons are shared between the two atoms. With the formation of the covalent bond, the negatively charged electrons are pulled slightly closer to the protons in the oxygen atom, giving the oxygen atoms a slightly negative charge and the hydrogen atoms a slightly positive charge (Figure 2.1). This charge difference causes water molecules to be **polar**. Because of this polarity, water molecules form **hydrogen bonds**, where the slight positive and slight negative charges attract (Figure 2.2). Hydrogen bonds cause water molecules to form droplets and stay close together rather than spread apart; this is called **surface tension**.

Hydrophilic and Hydrophobic Molecules

Hydrophilic means "water loving." Any molecule or atom that has a slight charge or a full charge, such as a sodium ion or a glucose molecule, which is polar, can dissolve in water (Figure 2.3a). Substances such as

Figure 2.2 Structure of Water

(A) Polar water molecule. (B) Hydrogen bonds hold water molecules close together because of the slight charge differences in the oxygen and hydrogen of water molecules. (C) Water droplet.

From DiGiuseppe/Sanader. *Chemistry 12U.* © 2012 Nelson Education Ltd. Reproduced by permission. www.cengage.com/permissions

Courtesy of Garry Roscoe

Figure 2.3 Charged, Polar, and Amphipathic Molecules

(A) Charged molecules dissolve easily in water because their charges are attracted to slight charges in the water molecules. (B) Molecules that contain both polar and nonpolar regions are amphipathic.

(A) From Haberer/Salciccioli/Sanader. *Chemistry 11U, 1E.* © 2011 Nelson Education Ltd. Reproduced by permission. www.cengage.com/permissions; (B) From DiGiuseppe/Fraser. *Biology 12U, 1E.* © 2012 Nelson Education Ltd. Reproduced by permission. www.cengage.com/permissions

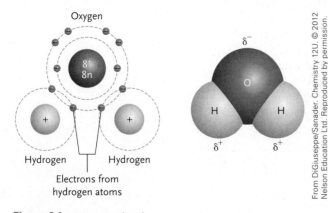

Figure 2.1 Water Molecule

Oxygen and hydrogen atoms share electrons in a water molecule and form covalent bonds.

From DiGiuseppe/Sanader. *Chemistry 12U.* © 2012 Nelson Education Ltd. Reproduced by permission. www.cengage.com/permissions

oils that do not have a charge are **nonpolar**, and they cannot dissolve easily in water; these molecules are **hydrophobic**, which means "water fearing." Both hydrophilic and hydrophobic molecules are essential to the specific functioning of cells, the formation of cell membranes, and the ways that nutrients and wastes move across those membranes. Sometimes molecules can have regions that are polar and regions that are nonpolar—such as soap, or the phospholipids in cell membranes. These molecules are called **amphipathic** because they have *both* hydrophilic and hydrophobic components (Figure 2.3b).

CONCEPT REVIEW 2.1

1. Approximately how much of the human body is composed of water?

2. What is the difference between a solute and a solvent?

3. Explain the difference between hydrophobic and hydrophilic, polar and nonpolar.

4. What is surface tension?

5. Describe the properties of an amphipathic molecule.

6. Why is water so biologically important?

2.2 MACROMOLECULES

All living things are based on four primary atoms—carbon, oxygen, hydrogen, and nitrogen—that combine to form molecules. **Macromolecules** are very large molecules made up of combinations of many smaller molecules; for example, protein macromolecules are composed of many amino acids. Molecules combine to form macromolecules by a process called **dehydration synthesis**. The four macromolecules that make up all living cells are proteins, carbohydrates, nucleic acids, and lipids. These are all formed from the combination of the small individual molecules of amino acids, monosaccharides, nucleotides, and fatty acids. There are many different kinds of proteins, carbohydrates, lipids, and nucleic acids in different organisms. The proteins in a fish or an almond are different from the proteins formed in our cells after digestion and absorption, but the individual amino acids that a protein is made up of are the same in every living thing. The differences are in how those small molecules combine together. We will look at each macromolecule in more detail in the following sections.

When we eat food, we eat the polymers that plants or animals produced. These are called **organic molecules** because they come from living organisms. Our digestive system breaks these down into **monomers** (building blocks) that are absorbed into the bloodstream and circulated to the cells. Our cells use these molecules either to produce energy in the form of ATP (adenosine triphosphate) (Chapter 4) or to build human proteins, carbohydrates, nucleic acids, and lipids, which all have a multitude of cellular functions in various body systems.

Molecules combine by the removal of a water molecule, hence the term, *dehydration synthesis*. For example, during protein synthesis in our cells, the protein is the macromolecule, and it is made up of **amino acids**. A hydroxyl group is removed from one amino acid, and hydrogen is removed from the other amino acid, and then a covalent bond forms (Figure 2.4).

When **polymers** are broken down in our cells or in our digestive system, the reaction known as hydrolysis occurs, which is the reverse of dehydration synthesis. A water molecule is added with the aid of a hydrolytic enzyme, and the covalent bond is broken. Hydrogen and a hydroxyl group are added to form the individual monomers (Figure 2.5).

From Jenkins et al. Nelson Chemistry 12, 1E. © 2003 Nelson Education Ltd. Reproduced by permission. www.cengage.com/permissions

Figure 2.4 Dehydration Synthesis

Macromolecules are formed by a process called dehydration synthesis, where a water molecule is produced and a covalent bond is formed.

$$H_2N-\overset{\overset{H}{|}}{\underset{H}{C}}-\overset{\overset{O}{||}}{C}-\overset{\overset{H}{|}}{N}-\overset{\overset{H}{|}}{\underset{CH_3}{C}}-\overset{\overset{O}{||}}{C}-OH + H_2O \longrightarrow H_2N-\overset{\overset{H}{|}}{\underset{H}{C}}-\overset{\overset{O}{||}}{C}-\boxed{OH + H}\;N-\overset{\overset{H}{|}}{\underset{CH_3}{C}}-\overset{\overset{O}{||}}{C}-OH$$

Dipeptide   Glycine   Alanine

Figure 2.5 Hydrolysis

Macromolecules are broken down by a process called *hydrolysis*, where a water molecule is added and a covalent bond is broken.

Application

As you read through the rest of the sections in this chapter, consider the following meals and snacks and think about which foods would contain the highest amount of protein, nucleic acids, carbohydrates (simple and complex), and fat (saturated or unsaturated). Which foods would have the most fibre? Which foods would have the most micronutrients? Which foods could contain trans fat?

- One hard boiled egg, piece of whole grain toast with butter
- Tim Horton's blueberry muffin
- Salad with mixed greens, hemp seeds, green pepper, radish, cucumber, homemade olive oil dressing
- Bag of potato chips and a coffee with cream and sugar
- Chicken, white rice with legumes, asparagus
- Crackers with almond butter

CONCEPT REVIEW 2.2

1. Explain the difference between dehydration synthesis and hydrolysis.

2.3 PROTEINS

Amino Acids

The human body produces over 100 000 different proteins. Amino acids are the monomers that are the building blocks of the thousands of different types of proteins. Amino acids that are incorporated into proteins are called **proteinogenic**. There are 20 different amino acids. These amino acids differ only by their functional group, and they can be arranged in multiple ways to produce proteins (Figure 2.6). Every cell in the body can build the correct proteins in the correct amount throughout our lives, and different cells can produce different proteins. For example, after we eat and blood sugar levels increase, the pancreas beta islet cells are the only cells that synthesize the protein hormone insulin (Chapters 17 and 20). Whereas every cell in the body produces enzyme proteins involved in the production of ATP, every protein is formed by very specific amino acid sequences. A typical protein contains anywhere from 100 to 1000 amino acids. The *order* of amino acids in a protein is critical for the proper folding of the protein, which is based on interactions between the functional groups. The protein can function only if it contains the correct amino acids and is folded into the correct

Amino acids   Polypeptide   Protein

Figure 2.6 Protein Synthesis

Proteins are macromolecules made up of amino acids, and the order of amino acids in a protein determines the structure and function of that protein.

conformational structure. For example, a single amino acid change in a chloride-channel protein causes improper folding, with the result that the membrane protein cannot function, and this causes the disease cystic fibrosis. However, not all changes in the amino acid sequence will cause disease (Chapters 6 and 8).

An individual amino acid has a **carboxyl group**, an **amino group**, and a functional group. The functional group is what differs among all 20 amino acids and gives each amino acid its distinctive characteristics (Figure 2.7). The covalent bond that links the amino

Figure 2.7 Amino Acids

The 20 amino acids differ only in their functional groups.

Nonpolar amino acids

Alanine — Ala — A
Valine* — Val — V
Leucine* — Leu — L
Isoleucine* — Ile — I
Glycine — Gly — G
Cysteine — Cys — C
Phenylalanine* — Phe — F
Tryptophan* — Trp — W
Methionine* — Met — M
Proline — Pro — P

Uncharged polar amino acids

Serine — Ser — S
Threonine* — Thr — T
Tyrosine — Tyr — Y
Asparagine — Asn — N
Glutamine — Gln — Q

Negatively charged (acidic) polar amino acids

Aspartic acid — Asp — D
Glutamic acid — Glu — E

Positively charged (basic) polar amino acids

Lysine* — Lys — K
Arginine — Arg — R
Histidine — His — H

acids is called a **peptide bond**. When many amino acids are covalently bonded by peptide bonds, a **polypeptide**—a protein—is formed.

Essential Amino Acids

The human body can synthesize 11 of the 20 amino acids from other molecules. The remaining nine, which are called **essential amino acids**, cannot be synthesized by the human body and need to be acquired from our diet: phenylalanine, valine, threonine, tryptophan, isoleucine, methionine, leucine, histidine, and lysine. The most common food sources that contain *all* amino acids are meat, fish, eggs, chicken, soy products, quinoa, hemp seeds, and dairy products. Other foods contain some, but not all, of the essential amino acids. See Table 2.1.

Protein Structure

The sequence of amino acids in a protein is called the *primary structure* (Figure 2.8). For example, the formation of the protein insulin requires the correct amino acids in the correct order, and the protein has to fold

TABLE 2.1

Major Categories of Proteins in the Human Body

Type of Protein	Function	Examples
Enzyme	Increases the rate of chemical reactions in all cells of the body	Lipase breaks down lipids (Chapter 20). Polymerase combines nucleotides during DNA replication (Chapter 8).
Protection	Involved in protecting the body from pathogens or cellular damage	IgG is an antibody involved in acute infections (Chapter 23). Fibrinogen is a protein involved in blood clotting (Chapter 18).
Membrane proteins	Allows various molecules to cross cell membranes, act as signalling molecules, or adhere to surrounding tissues	Na^+/K^+ pump transports sodium out of the cell and potassium into the cell (Chapter 3). Calcium channels transport calcium into neurons to release neurotransmitters (Chapter 15).
Cell recognition	Acts as markers that allow immune cells to recognize self-cells and pathogens	Red blood cells contain markers known as blood type, A, B, AB, and O (Chapter 6). Major histocompatibility proteins are involved in presenting foreign antigens to immune cells (Chapter 23).
Contractile	Involved in muscle contraction and movement	Actin and myosin are the contractile proteins in muscle tissue (Chapter 14).
Transporter	Involved in moving substances throughout the body	Lipoproteins transport fats through the bloodstream (Chapters 18 and 20). Hemoglobin transports oxygen through the bloodstream (Chapters 6, 18, and 19).
Structural	Maintains the structure of the cytoskeleton or extracellular matrix	Collagen is an extracellular matrix protein that holds cells together (Chapters 3 and 11). Actin is a protein that forms intracellular microfilaments that allow cells to move (Chapter 3).
Regulatory	Involved in signalling molecules that regulate cell functions	Insulin is a hormone that regulates blood sugar (Chapter 17). Growth hormone is involved in maintaining bone and muscle density (Chapter 17). Troponin and tropomyosin are regulatory proteins involved in muscle contraction (Chapter 14).

Figure 2.8 Protein Structure

(A) Primary structure is the amino acid sequence. (B) Secondary structure occurs when the primary sequence begins to fold into beta sheets or alpha helices. (C) Tertiary structure is the completely folded protein. (D) Quaternary structure is formed when more than one protein combine. (E) Interactions between functional groups cause protein folding.

From DiGiuseppe/Fraser. *Biology 12U*, 1E. © 2012 Nelson Education Ltd. Reproduced by permission. www.cengage.com/permissions

into the correct shape so that it will bind specifically to insulin receptors. The body's cells know the correct order of amino acids because of the DNA sequences called genes, which is discussed in Section 2.4.

As amino acids become arranged in the correct order on the ribosomes in each cell, they take on a *secondary structure*; this is based on the interactions of the functional groups of the individual amino acids (Figure 2.8e). Hydrogen bonds form between polar amino acids, just like hydrogen bonds form between water molecules. Hydrophobic amino acids are attracted to other hydrophobic amino acids, and the secondary structure develops. Most of the secondary structures formed are called **alpha helices** or **beta sheets**.

The protein takes on a three-dimensional shape based on the primary amino acid sequence and the secondary structures that are formed. Eventually, the protein forms a three-dimensional *tertiary structure*, which constitutes the functional structure of the protein. To function properly, every distinctive protein has to have not only the correct amino acid sequence but also the correct tertiary structure, which is determined by the interactions between functional groups. Nonpolar amino acids are attracted to one another, polar or charged amino acids are attracted to one another, and sulfur groups will form disulfide bonds. Sometimes multiple tertiary structures combine to form a final functional protein product, called a *quaternary structure*. An example of a quaternary protein in our body is hemoglobin, where four protein subunits combine to form the final functional macromolecule.

Denaturation

When you cook an egg, the increased temperature causes the egg proteins to unfold; this explains why the egg changes from a semi-liquid to a semi-solid state. The increased temperature adds enough energy to break the weak hydrogen bonds that form the secondary and tertiary protein structures, but the covalent peptide bonds remain intact. A change in the pH can also affect the bonds that hold proteins in their conformational shape: for example, the acidic environment of the stomach. This process of breaking the interactions between functional groups in a protein is called **denaturation**. Denaturation is an irreversible reaction—for example, cooling a cooked egg does not change it back to its original state—and denatured proteins no longer function (Figure 2.9) because their folded structure is critical for protein function. However, denaturation does not change the structure of individual amino acids. When we eat proteins, digestive enzymes break the proteins into the amino acids that we absorb, so cooking food or digesting it in our acidic stomach does not affect the structure of the amino acids.

Figure 2.9 Denaturation

Proteins unfold and become denatured when exposed to heat during cooking or to hydrochloric acid in the stomach.

DID YOU KNOW?

Having an extremely high fever is fatal because cellular proteins can start to denature. Without the thousands of proteins doing their normal cellular functions, death could occur quite rapidly.

Enzymes

Enzymes are very important globular proteins that facilitate thousands of chemical reactions in our body's cells. In humans, more than 75 000 different enzymes are produced in various types of cells and perform a myriad of essential functions. Enzymes have grooves that are specific tertiary structures that form **active sites** whereby an enzyme can **catalyze**, or speed, the rate of chemical reactions. Enzymes increase the contact between specific regions of molecules so that a particular chemical reaction can occur: for example, bringing amino acids in close proximity so that dehydration synthesis can occur (Figure 2.10).

Figure 2.10 Enzyme Function

Amino Acids That Have Other Roles

Amino acids are not always incorporated into proteins, in which case they are referred to as *non-proteinogenic amino acids*. Non-proteinogenic amino acids have a variety of important functions, such as the production of neurotransmitters. See Table 2.2.

TABLE 2.2

Examples of Non-proteinogenic Amino Acids	
Amino Acid	**Function**
Tryptophan	Tryptophan is the precursor for serotonin, an important mood molecule.
Tyrosine	Tyrosine is the precursor for the stress hormones epinephrine and norepinephrine, the thyroid hormones T3 and T4, and dopamine, which is a neurotransmitter involved in reward-motivated behaviour.
Glycine	Glycine can function as an inhibitory neurotransmitter, making us feel calm and decreasing anxiety. Glycine can be found in high amounts in gelatin and bone broth.
Glutamic acid	The ion form, glutamate, is the most abundant excitatory neurotransmitter; it is important for thinking, learning, and memory. Excess glutamate can cause feelings of anxiety. Glutamate is in the food additive MSG.
Leucine, isoleucine, and valine	These are the branched-chain amino acids (BCAAs). They are sometimes consumed as an exercise supplement because they play a role in glucose and fat metabolism and ATP production in the muscles and they promote protein synthesis and muscle repair after high-intensity exercise.

CONCEPT REVIEW 2.3

1. How many amino acids are there? How many of them are essential?

2. Name three components of an amino acid. Which of these components are the same in every amino acid?

3. What types of bonds are formed in proteins? Which bonds hold amino acids together, and which bonds are important for the formation of secondary and tertiary structures?

4. Describe the different structures of proteins.

5. What are alpha helices and beta sheets?

6. Give three examples of quaternary proteins in the human body.

7. What is denaturation?

8. Why are enzymes important?

9. What are the branched-chain amino acids, and why do some people take them as a supplement?

10. Which amino acid is required for the body to make serotonin?

2.4 NUCLEIC ACIDS

Nucleic acids provide cellular "instructions" that determine the correct amino acid sequence for every protein produced by every cell in the body. This is true for every living organism.

DNA

Each cell in the human body contains a nucleus with 46 **chromosomes**. We have two copies of 23 chromosomes:

one set received from each parent. Each chromosome consists of one nucleic acid macromolecule called DNA: deoxyribonucleic acid. Each DNA macromolecule contains hundreds to thousands of **genes**, which give the blueprint (instructions) for the production of a specific **protein**. Large chromosomes can have over 2000 genes, such as chromosome 1. Smaller chromosomes, such as chromosome 21, have less than 400 genes (Figure 2.11). The X chromosome has over 2000 genes, and the male Y chromosome has only 78 genes. In total, the human genome is made up of 25 000 genes.

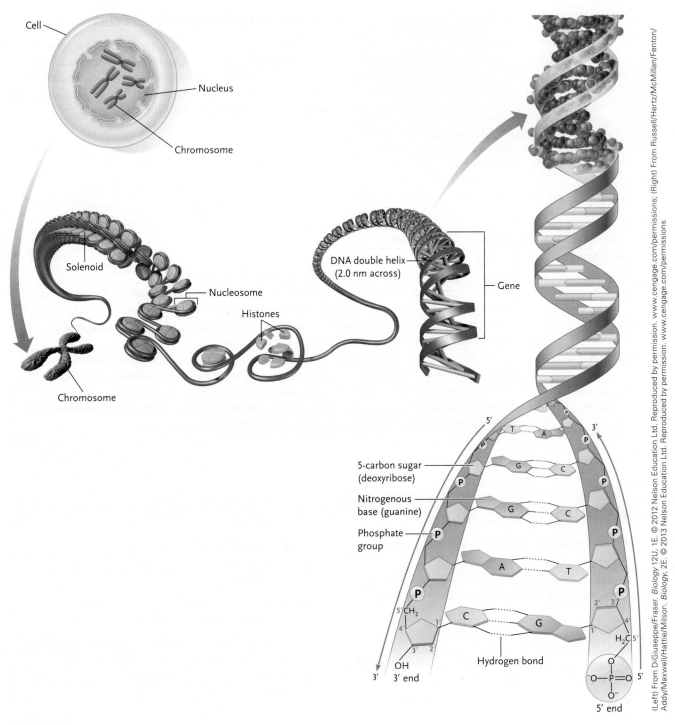

Figure 2.11 DNA

Each chromosome in every cell nucleus is a single macromolecule of DNA, which contains hereditary material: 500 to 1000 genes per chromosome.

DID YOU KNOW?

A single DNA molecule, if stretched to its full length, is approximately two inches or 50 mm long? All 46 chromosomes must be extremely condensed to fit into each cell nucleus, which is only about 10 micrometres in diameter!

DNA Structure

As discussed in Section 2.3, the *order* of amino acids in the protein determines the structure and function of each of the thousands of proteins in the human body. Every cell produces the correct proteins with the correct order of amino acids because of the sequence of **nucleotides** in each gene. When red blood cells produce the hemoglobin protein, the order of amino acids is determined by the nucleotides in the hemoglobin gene, which all humans have on chromosome 11. Exactly how nucleotide sequences lead to amino acid sequences is discussed in detail in Chapter 8. Each DNA nucleotide is an organic monomer made up of three components: a five-carbon sugar (deoxyribose), a phosphate group, and a nitrogen-containing base (Figure 2.12a). Adenine and guanine are nitrogenous bases with a double-ring structure; they are called **purines**. Cytosine and thymine are single-ring bases called **pyrimidines** (Figure 2.12b).

It is often said that DNA contains the "stored information"—or is the "blueprint"—for protein synthesis. This is because it carries the specific order of the nucleotides (A, T, C, and G) in every gene, and this translates into the amino acid sequence of every protein that's synthesized. Imagine that our nucleotides are like letters of the alphabet; from letters, we make words. The sequence of nucleotide "letters" has to be correct in order to spell amino acid "words" accurately and to make the proper protein "sentences." A misspelled word can dramatically change the meaning of a sentence. Such a misspelling of an amino acid sequence corresponds to a DNA mutation (discussed in Chapters 6 and 8).

DID YOU KNOW?

Some genes are expressed only at certain times in our life, such as the gene for fetal hemoglobin, which is useful for transporting oxygen in a fetus. After birth, the red blood cells make a different hemoglobin protein from a different gene.

DNA is a **double helix** whereby two strands of DNA bind together by the nitrogenous bases. A gene can be as small as 100 nucleotides, or as long as 2000 nucleotides on one strand of DNA. On each strand of DNA the sugar and the phosphates are bound together by very **strong phosphodiester bonds**; these make up the "backbone" of the DNA strand and the nitrogenous bases extend from each sugar molecule. The sugar is bound to the base by a covalent **glycosidic bond**, also very strong. Glycosidic bonds are found between any sugar and some other group that could be another sugar or something else, such as a nitrogenous base. Each base binds with only one specific base on the opposite strand. Adenine always binds with thymine, and cytosine always binds with guanine (Figure 2.13). The nucleotide bases bind to the opposite complementary nucleotide by weak hydrogen bonds. Two hydrogen bonds form between adenine and thymine, and three hydrogen bonds hold guanine and cytosine together. The two strands of DNA are antiparallel because they are oriented in opposite directions. Note that each carbon on the deoxyribose sugar shown in Figure 2.13 is numbered. The first carbon is bound to the base, and the third and fifth carbons are bound to phosphate groups. The DNA strand on the left has a free phosphate group attached to carbon number 5: this is called the 5' (five prime) end. The opposite strand has a free phosphate on carbon number 3: this is called the 3' (three prime) end. How DNA polymerase enzymes replicate DNA molecules is covered in Chapter 8.

RNA

The other type of nucleic acid macromolecule found in cells is called RNA: ribonucleic acid. The function of DNA and RNA is quite different. RNA is the intermediate step between the nucleotide sequence of each gene

Figure 2.12 (A) Nucleotide Structure and (B) Nitrogenous Bases
(A) Each nucleotide contains a five-carbon sugar, a phosphate, and a nitrogenous base. (B) The five nitrogenous bases of the nucleotides are adenine, cytosine, guanine, thymine, and uracil (found only in RNA).

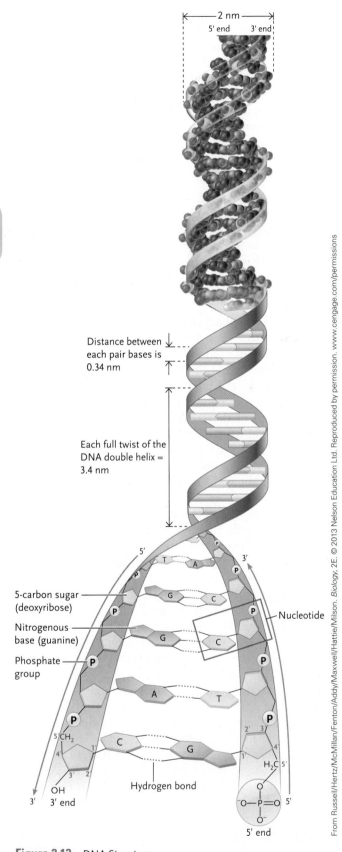

Figure 2.13 DNA Structure

DNA is a double helix; adenine from one strand always binds with thymine on the opposite strand, and guanine always binds with cytosine on the opposite strand.

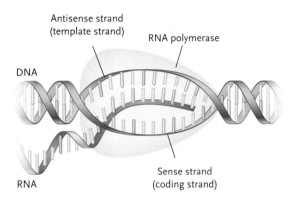

Figure 2.14 RNA

RNA is a single-stranded nucleic acid molecule transcribed from a DNA gene sequence that codes for the synthesis of a protein.

and the eventual completed protein. Suppose that a pancreas cell is stimulated by increased blood sugar levels to produce the protein hormone insulin. Inside each pancreatic beta islet cell nucleus, the insulin gene sequence becomes *transcribed* into an RNA sequence (Figure 2.14). That RNA molecule then leaves the nucleus and moves into the cytoplasm of the cell to a ribosome, where it acts as the template for the assimilation of the amino acid sequence that produces insulin. The insulin gene consists of 153 nucleotides that lead to the production of the small, 51-amino acid, insulin protein. The process of RNA transcription is discussed in detail in Chapter 9 (see Table 2.3).

TABLE 2.3

Comparison of DNA and RNA

	DNA	RNA
Structure	Double-stranded helix	Single-stranded and can have various folding patterns
Nucleotides	A, T, C, and G	A, U, C, and G
Sugar	Deoxyribose	Ribose
Location	Nucleus only	Transcribed in the nucleus and moves into cytoplasm
Function	Stores genetic information	Used as a template for protein synthesis

From Russell/Hertz/McMillan/Fenton/Addy/Maxwell/Hattie/Milson. *Biology*, 2E. © 2013 Nelson Education Ltd. Reproduced by permission. www.cengage.com/permissions

CONCEPT REVIEW 2.4

1. What is a chromosome? How many do humans have?

2. What is the function of DNA? Where is it located?

3. Describe the components of a nucleotide.

4. What three types of chemical bonds would you find in nucleic acids?

5. What is the complementary sequence of the following strand of DNA?

   A T G G C T A G T C

6. What is the function of RNA?

7. What are the differences between DNA and RNA?

2.5 CARBOHYDRATES

Simple and Complex Carbohydrates

Carbohydrates can be generally classified as simple carbohydrates (sugars)—the **monosaccharides** and **disaccharides**—and complex carbohydrates, the **poly-saccharides**. (*Saccharide* is derived from the Greek *saccharin*, which means *sweet*.) All carbohydrates have a ratio of 1:2:1, referring to the ratio of carbon:hydrogen:oxygen. The carbon-hydrogen bonds are what make carbohydrates particularly useful for energy production (Chapter 4). The molecular formula for each monosaccharide is $C_6H_{12}O_6$. However, the structures are slightly different for each of the three monosaccharides: **glucose**, **fructose**, and **galactose** (Figure 2.15). Fructose gives fruit its sweet taste. Honey consists of both glucose and fructose monosac-charides, and it is sweeter than table sugar, which is composed of glucose-fructose disaccharides.

DID YOU KNOW?

Fructose is the sweetest-tasting sugar. As high-fructose corn syrup (HFCS), it is added to many foods, particularly soft drinks, canned fruits, and concentrated juices. Fructose is processed by the liver and converted directly into fat. It does not decrease the hunger response and is a significant contributor to weight gain, cardiovascular disease, and type II diabetes. HFCS is 55% fructose and 45% glucose. Sucrose is 50% fructose and 50% glucose.

Glucose

Fructose

Galactose

Figure 2.15 Monosaccharides

The three monosaccharides—glucose, fructose, and galactose—have the same chemical formula $C_6H_{12}O_6$ and slightly different chemical structures.

Disaccharides

Each of the monosaccharides can be paired by cova-lent glycosidic bonds to form disaccharides. Table sugar, or **sucrose**, is formed from the dehydration syn-thesis reaction between glucose and fructose. **Maltose**, found in grains, is formed from two glucose molecules, and **lactose**, milk sugar, is formed from glucose and galactose. These are the most common disaccharides (Figure 2.16).

Lactose Intolerance

Why do some people become bloated and have stomach pains and diarrhea after drinking milk? Lactose is a disaccharide that requires a specific enzyme, **lactase**, to break it down into glucose and galactose. Some people lose the ability to produce this enzyme as they get older. The lactose stays in the digestive tract instead of being broken down and absorbed. The lactose molecules are then taken up by the normal bacteria that reside in the

Figure 2.16 Disaccharides
Three important disaccharides are (A) sucrose, (B) maltose, and (C) lactose.

digestive tract; these bacteria break down the lactose to make ATP and produce methane gas as a waste product. The lactose remaining in the intestine also increases osmotic concentration in the intestine, which causes water to move into the intestine, and this causes diarrhea.

Polysaccharides

Polysaccharides, the complex carbohydrates, are long chains of glucose. Three major complex carbohydrates are relevant to humans: starch, glycogen, and cellulose (Figure 2.17). Starch, a polysaccharide that we can eat, is the *stored energy* in plant cells. Starchy plant foods include rice, oatmeal, potatoes, yams, and grains such as corn, wheat, barley, rye, millet, and buckwheat. The carbohydrates in our diet are an important source of energy. The digestive system breaks down the long chains of glucose molecules through **hydrolysis**, and individual glucose molecules are absorbed into the bloodstream. The body's cells then take up these glucose molecules and use them to make ATP (Chapter 4). When we eat more carbohydrates than needed for ATP production, the excess is stored as either glycogen or fat. The polysaccharide **glycogen** is the stored energy in animal liver and muscle cells; it can be used later when blood sugar levels start to decrease with exercise or between meals.

Another important plant polysaccharide is **cellulose**. Cellulose is a structural plant carbohydrate and is not used to produce energy. When we eat plants such as fruit, vegetables, whole grains, and legumes, we cannot digest the cellulose. However, cellulose is important in our diet as **fibre**. The consumption of fibre contributes significantly to our health: it regulates bowel movements and prevents constipation; it plays a role in the prevention of bowel cancer, hemorrhoids, and diverticulitis; it lowers blood cholesterol; and it helps regulate blood sugar.

DID YOU KNOW?

Chitin is a structural polysaccharide found in crustaceans and insects; it gives them a tough exoskeleton. Carrageenan is a gelatinous polysaccharide extracted from edible red seaweed that is often added to processed foods as a thickening agent.

Polysaccharides are formed by dehydration synthesis between glucose molecules. However, each polysaccharide has very different properties due to the way the bonds are formed. For example, in cellulose the glucose molecules form bonds between the first carbon of one glucose and the fourth carbon of the next glucose, which leads to a straight chain structure; in glycogen, the glucose molecules form bonds between the first and fourth or the first and sixth carbons of adjacent glucose molecules, which leads to a branched structure (Figure 2.17).

Dietary Carbohydrates

Since starches are made from sugar, why isn't it just as healthy to eat chocolate bars and candy that are made from sugar? One reason why candy, a simple carbohydrate, is less healthy than complex carbohydrates relates to the rate of digestion and absorption from the intestines into the bloodstream. Sugar can be readily

A. Starch

B. Glycogen

C. Cellulose

Figure 2.17 Starch, Glycogen, and Cellulose

Three polysaccharides are important to humans: (A) starch, (B) glycogen, and (C) cellulose. All three are composed of glucose, but each one is a very different molecule because of slight differences in their bond formation.

absorbed into the blood, and this causes a rapid increase in blood sugar level. When this occurs, the pancreas secretes insulin, causing a lot of that sugar to be converted into fat, rather than used for energy production. Starch takes much longer to digest because every bond must be broken by enzymes in the digestive tract before each sugar molecule can be absorbed. A second reason why sugar is not as healthy as starch relates to the composition of the sugar and its usefulness in energy production. Most simple carbohydrates are a combination of glucose and fructose, whereas starch is made up of only glucose. The body's cells prefer to use glucose for ATP production, and the liver converts most fructose into fat. Therefore, eating a lot of simple sugars results in weight gain.

DID YOU KNOW?

People who have high amounts of fructose in their diet can develop fatty liver disease, which has symptoms that are similar to cirrhosis of the liver caused by alcoholism.

Glycemic Index

The glycemic index (GI) is a measure of how rapidly various foods increase blood sugar. Higher GI foods increase blood sugar more than lower GI foods, which causes more insulin to be produced. Insulin stimulates cells to convert sugar into fat. Excess insulin production leads to insulin resistance and eventually to type 2 diabetes. Eating lower GI foods helps you to maintain more stable blood sugar levels, control appetite, control blood LDLs (fats made by the liver from excess sugar and that increase risk of heart disease), and lower your risk of developing type 2 diabetes. Sugars and starchy carbohydrates have a higher glycemic index than proteins or fats. Table 2.4, with data from the Canadian Diabetes Association, provides information on the glycemic index of certain foods.

Fibre

Complex carbohydrates that are not broken down by our digestive enzymes are known as fibre. Dietary fibre is important for intestinal health.

1. Fibre adds bulk to large intestine. Bulk stimulates the large intestine to contract and causes bowel movements, called **defecation**. It is important for undigested materials to be defecated on a daily basis so that unwanted molecules are not absorbed into the bloodstream. Build-up of waste materials in the large intestine and difficulty defecating is called **constipation**. Frequent episodes of constipation increase the risk of inflammatory bowel diseases and colon cancer.

2. Fibre acts as a **prebiotic**. Prebiotics are nutrients that feed our intestinal bacteria. Probiotics are foods, such as yogurt, that contain healthy bacteria. It is important to have healthy bacteria in the intestines because they help in the digestion of food and absorption of nutrients, play a role in regulating immune functions, stimulate cell regeneration in the GI tract, and fight off pathogenic organisms.

Fibre is classified into soluble and insoluble types. **Soluble fibre** is gelatinous and dissolves in water,

TABLE 2.4

GI Food Table

Many starchy foods have a high glycemic index (GI). Choose medium- and low-GI foods as often as possible.

LOW GI (55 or less)*+ Choose most often	MEDIUM GI (56–69) *+ Choose more often	HIGH GI (70 or more)*+ Choose less often
Breads		
100% stone ground whole wheat	Whole wheat	White bread
Heavy mixed-grain	Rye	Kaiser roll
Pumpernickel	Pita	Bagel, white
Cereal		
All Bran™	Grapenuts™	Bran flakes
Bran Buds with Psyllium™	Puffed wheat	Corn flakes
Oat Bran™	Oatmeal	Rice Krispies™
	Quick oats	
Grains		
Barley	Basmati rice	Short-grain rice
Bulgar	Brown rice	
Pasta/noodles	Couscous	
Parboiled or converted rice		
Other		
Sweet potato	Potato, new/white	Potato, baking (Russet)
Yam	Sweet corn	French fries
Legumes	Popcorn	Pretzels
Lentils	Stoned Wheat Thins™	Rice cakes
Chickpeas	Ryvita™ (rye crisps)	Soda crackers
Kidney beans	Black bean soup	
Split peas	Green pea soup	
Soy beans		
Baked beans		

*expressed as a percentage of the value for glucose
+Canadian values where available

Source: Diabetes Canada, 2018. The Glycemic Index, http://www.diabetes.ca/diabetes-and-you/healthy-living-resources/diet-nutrition/the-glycemic-index. Reprinted with permission from Diabetes Canada.

thereby increasing the fluid content of the stool and providing lubrication that makes bowel movements easier. Examples of soluble fibres include psyllium, pectin in fruits, and chia seeds. **Insoluble fibres** do not dissolve in water. They make up plant cell walls, including cellulose and hemicellulose, which is found in whole grains, legumes, and bran.

It is recommended that adults consume at least 25g of fibre per day. Here are some examples of fibre content in foods.

- ½ cup spinach—6g
- ½ cup bran cereal—7g
- ½ cup almonds—8g
- 1 medium apple—4g
- 1 medium pear—5g
- 1 cup baked beans—16g
- 1 slice whole wheat bread—3g
- 2 tbsp chia seeds—11g

- 2 tbsp flaxseeds—4g
- 2 tbsp hemp seeds—2g
- 1 cup lentils—16g
- 1 cup peas—9g
- 1 cup broccoli—5g
- ½ avocado—6g
- 1 cup cooked oatmeal—4g

DID YOU KNOW?

Fibre slows the absorption of sugars and therefore lowers the glycemic index of starchy foods. Whole grain bread has a lower GI than white bread because of the fibre content. If you are craving something sweet, you could lower the GI by adding fibre, such as adding bran, oatmeal, and chia seeds to your chocolate chip cookie recipe.

CONCEPT REVIEW 2.5

1. What are the primary atoms that make up carbohydrates?
2. What is the difference between a monosaccharide, a disaccharide, and a polysaccharide?
3. What are the three primary monosaccharides, and how can they combine to form disaccharides?
4. What three polysaccharides are important for human health? Describe how they differ from each other and how they are important.
5. Give examples of foods that contain maltose, sucrose, lactose, starch, and cellulose.
6. What is the glycemic index?

2.6 LIPIDS

Function and Solubility

Lipids are an important component of our body and include molecules such as triglycerides (fats), cholesterol, and phospholipids. The body uses fats for many vital functions in every cell. Lipids consist of many carbon-hydrogen bonds, which make them very useful molecules for storing energy. See Table 2.5.

Recall from Section 1.1 that water molecules are polar because of a slightly unequal sharing of electrons in the covalent bonds. In fatty acids, the carbon and hydrogen covalent bonds have electrons equally distributed between

TABLE 2.5

Important Functions of Fats

Courtesy of Mark Nielsen, AnatBooks, LTD

Brain. Fats compose 60% of the brain and are essential to its functioning, including learning, memory, and mood.

Cells. Fatty acids and cholesterol are responsible for building cell membranes, and they help cells stay flexible.

(Continued)

TABLE 2.5 (continued)

From Sherwood/Kell/Ward. *Human Physiology*, 2E. (c) 2013 Nelson Education Ltd. Reproduced by permission. www.cengage.com/permissions	**Heart.** Sixty percent of the energy for the heart to contract comes from burning fats.
From DiGiuseppe/Fraser. *Biology 12U*, 1E. (c) 2012 Nelson Education Ltd. Reproduced by permission. www.cengage.com/permissions	**Nerves.** Fats compose the material that insulates and protects the nerves, thereby isolating electrical impulses and speeding their transmission.
	Lungs. Lung **surfactant** enables the lungs to work properly.
	Digestion. Eating fats slows down the digestion process, so the body has more time to absorb nutrients. Fats help provide a constant level of energy and also keep the body satiated for longer. Fat-soluble vitamins (A, D, E, and K) can be absorbed only if fat is present in the digestive tract when the vitamins are ingested.
	Organs. Fats cushion and protect the internal organs.
	Immune system. Certain fats decrease inflammation and increase metabolism.

the atoms that make them nonpolar, or hydrophobic. When fats are placed in water they cluster together as droplets in the water. This property is extremely important in the formation of cell membranes.

Types of Lipids

There are three main categories of lipids: triglycerides, cholesterol, and phospholipids. Triglycerides are composed of one **glycerol** molecule bonded to the carboxylic acid end of three **fatty acids** (Figure 2.18).

Glycerol backbone   Fatty acids

A. Fat molecule (triglyceride)

Figure 2.18 Triglyceride

Triglycerides, the body's main source of stored energy, are found in fat cells (**adipocytes**). Groups of these cells are called **adipose tissue** (Chapter 11)—also called **subcutaneous fat** because it occurs mainly in the subcutaneous layer under the dermis of the skin (Chapter 12). Many people have an excess of this type of fat, and it can only be decreased when the cells use this fat as an energy source. The cells that use the most stored energy to make ATP are the skeletal muscle cells; this explains why exercise helps maintain a healthy body weight.

DID YOU KNOW?

Sixty-four percent of North Americans are overweight, an 18% increase since the 1970s. Obesity is closely correlated with an increase in many diseases, such as cardiovascular disease, diabetes, breathing disorders, depression, and cancer.

Saturated and Unsaturated Fats

Fats can be solid or liquid at room temperature. Generally, fats that are liquid at room temperature are **unsaturated**, and fats that are solid at room temperature are **saturated**. The reason for this is the chemical structure of the fatty acids. If there are only single bonds between the carbon atoms of the fatty acid chain (Figure 2.19), the structure of the fat molecules will be straight and can pack together easily; this makes the

A. Saturated

B. Unsaturated

Courtesy of Garry Roscoe

Courtesy of Garry Roscoe

Figure 2.19 Saturated Fat and Unsaturated Fat

The chemical structures differ for (A) saturated fats, such as butter, and (B) unsaturated fats, such as olive oil.

fatty acid a solid. If there are one or more double bonds between the carbons, the fat molecules will be bent and unable to pack closely together; this makes the fatty acid a liquid.

Foods that contain healthy saturated fats are meats, fish, avocados, coconut oil, and dairy products. Foods that contain healthy unsaturated fats—those in the cis formation (Figure 2.20a)—include plant oils, such as olive oil and flaxseeds, nuts, seeds, and vegetables, and also fish oil. Unsaturated fats that have one double-bonded carbon are called **monounsaturated**. Those that have two or more double bonds in the fatty acid chains are called **polyunsaturated**.

Trans Fats

Margarine is a vegetable fat that has been converted into a solid by a process called **hydrogenation**. During this process hydrogen is added to the fatty acids to essentially convert the double-bonded carbon atoms into single-bonded atoms. This changes the fat into a saturated fat that becomes solid at room temperature. The problem is that not every double bond is fully hydrogenated, so margarine is still partially unsaturated. Furthermore, the way the hydrogen atoms are added is not in a chemical structure that can be easily used by the body. These incompletely hydrogenated fats are called **trans fats** (Figure 2.20b); they are the very worst fats for human health. Some labels claim the margarine is non-hydrogenated, and so it is healthier.

However, the margarine remains semi-solid at room temperature because it contains palm oils, which are saturated fats. This type of margarine is healthier than hydrogenated margarines, but it contains saturated fats, as does butter, except that these are plant-based. Trans fats are found in fast foods and deep-fried foods, because the high cooking temperature converts the unsaturated vegetable oils into trans fats.

Enzymes in our cells that oxidize fatty acids to make energy (Chapter 4), preferentially break down saturated fatty acids and unsaturated fatty acids in the *cis* conformation, but not those in the *trans* conformation. Therefore, trans fats cannot be broken down as easily, and they can remain circulating in the bloodstream much longer than other fats. Trans fats can also bind to cell membrane receptors on the endothelial cells that line blood vessels; this results in these fats being taken into the lining of the blood vessels. Trans fats then stimulate the immune response, causing immune cells to also enter the blood vessel wall in an attempt to rid the body of these toxic molecules. The combination of fat and immune cells in vessel walls is the plaque that causes heart disease (Chapter 18). Higher consumption of trans fats has been linked to increases in risk of Alzheimer's disease, cancers, obesity, liver dysfunction, and mood disorders such as anxiety and depression. Trans fat can be found in packaged and processed foods because they contain refined polyunsaturated oils.

A. Cis fatty acid

Courtesy of Garry Roscoe

B. Trans fatty acid

Ari N/Shutterstock.com

Figure 2.20 (A) Cis Fatty Acid and (B) Trans Fatty Acid

Very healthy unsaturated fats such as olive oil can be transformed into trans fats if they are exposed to high temperatures during cooking. This happens because the double bonds between the carbons are delicate and can be damaged by light and heat. Saturated fats are best for cooking because they cannot be converted into trans fats: for example, coconut oil (which is saturated even though it is called an oil). Some oils, such as peanut oil, have a higher "smoking" point and can be cooked at higher temperatures. If the oil produces smoke when you cook, your food will contain trans fats.

Refined Oils

Refined oils, such as canola, corn, and vegetable oils, go through several processes that can damage the double bonds in unsaturated fats. The Canola Council of Canada describes the general process used in refining oils:

- Cleaning—to remove weeds, stems, and other materials
- Flaking—rolling the seeds to break them open
- Cooking—heating to approximately 90 to 120°C to open the oil cells in the crushed seeds, denature hydrolysis enzymes, and volatize sulphur compounds that can cause odours
- Pressing—to extract the oils
- Solvent extraction—using hexane to extract the remaining 40 to 50% of the oils left in the seeds after heating and pressing
- Desolventizing and toasting—to remove the solvent by heating to 95 to 115°C for up to 90 minutes.

- Refining—putting the oil through a series of processes—to increase shelf life—by using water or organic acids to remove compounds such as phospholipids, free fatty acids, and colour pigments
- Bleaching—filtering the oil through clay to remove any colour compounds
- Deodorizing—by steam distilling the oil to remove any compounds that could cause any unpleasant odour or taste
- Further processing—such as using hydrogenation to solidify canola oil to produce margarine or shortening, or mixing with other oils that are more solid, such as palm kernel oil.

Cold-pressed olive oil is mechanically extracted without the use of solvents, and it is not heated beyond 28°C because heat damages the oil and alters the flavour. "Pure" and "light" olive oils are heat treated, and so they no longer contain beneficial antioxidants. Heat and light can damage the double bonds in vegetable oil, so buy oil that is in dark glass bottles and has a pressed date of less than one year.

Essential Fatty Acids

We need **essential fatty acids** in our diet because the body's cells cannot produce them from other sources. There are three essential fatty acids: alpha-linolenic acid, which is **omega 3**; linoleic acid, which is **omega 6**; and oleic acid, which is **omega 9** (Figure 2.21). These fatty acids are essential for immune-cell regulation and for the production of mood hormones; deficiency in essential fatty acids can cause immune disorders and depression.

A. Alpha-linolenic acid

B. Linoleic acid

C. Oleic acid

Figure 2.21 Alpha-Linolenic Acid, Linoleic Acid, and Oleic Acid

The three essential fatty acids are (A) alpha-linolenic acid (omega 3), (B) linoleic acid (omega 6), and (C) oleic acid (omega 9).

Figure 2.22 Cholesterol

Alpha-linolenic acid is an 18-carbon fatty acid chain with three double bonds. This makes it a polyunsaturated fatty acid: having its first double bond at the third carbon gives the term *omega 3*. Linoleic acid is also an 18-carbon fatty acid chain, but it has two double bonds, is polyunsaturated, and the first double bond is at the sixth carbon, making it omega 6. Dietary sources of essential fatty acids include fish, shellfish, olive oil, peanuts, flaxseed, leafy green vegetables, walnuts, and eggs.

Cholesterol

Most cholesterol in our body is produced by the liver (Figure 2.22). Cholesterol plays very important roles in human physiology: (1) it is a major constituent of cell membranes; (2) it is used by the liver to make bile, which helps in the digestion of fats; (3) it is used to produce low-density lipoproteins (LDLs) that transport fats through the bloodstream; and (4) it is used as the starting material for the production of steroid hormones, such as estrogen, progesterone, testosterone, and aldosterone (Chapter 17).

DID YOU KNOW?

"Bad" blood cholesterol levels (which are LDLs and not just cholesterol) increase more by eating trans fats and sugar than by eating cholesterol. This occurs because eating trans fats and sugar causes the liver to produce more LDLs that contain cholesterol and other fats so that the trans fats—which are not easily taken up by cells—can be transported. This results in the trans fats staying longer in the circulatory system and therefore playing a significant role in the development of cardiovascular disease (Chapter 18).

Phospholipids

The last category of lipids is **phospholipids**, the primary fats found in **cell membranes**. Phospholipids are unique because they have two nonpolar fatty acid chains, called *tails*, and a polar *head* that contains a phosphate group. Because they contain both polar and nonpolar regions, phospholipids are amphipathic. This important feature of these fats explains why they can produce cell membranes. All cell membranes consist of a bilayer of phospholipids: the polar heads are directed toward the water-filled cell **cytoplasm** and the exterior of the cell, the interstitial fluid; and the tails are directed toward the interior of the membrane (Figures 2.23 and 2.24).

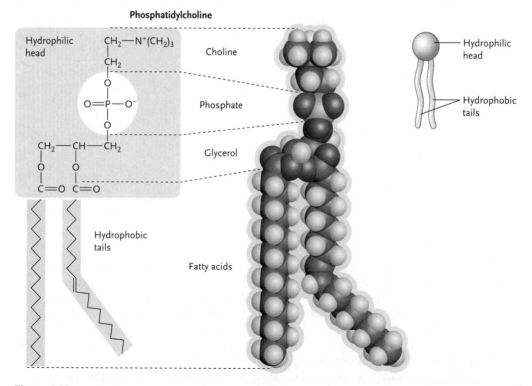

Figure 2.23 Structure of a Phospholipid

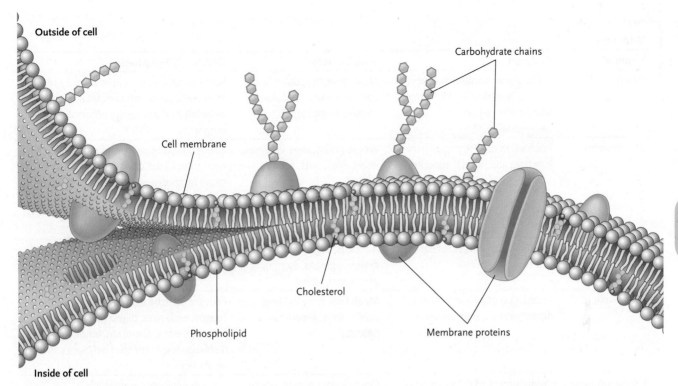

Figure 2.24 Cell Membrane

Since phospholipids are mostly unsaturated, the cells are liquid. In addition to the phospholipids, most animal cell membranes contain cholesterol, which helps the membrane remain flexible. The cell membrane is discussed in more detail in Chapter 3.

CONCEPT REVIEW 2.6

1. Explain how lipids are important for the functioning of three parts of the body (refer to Table 2.3).
2. What is adipose tissue?
3. What is the difference between saturated and unsaturated triglycerides?
4. List food sources of healthy fats, and name the fats you should minimize in your diet. Why?
5. What is hydrogenation?
6. Name four functions of cholesterol.
7. What is the primary type of fat that makes up cell membranes?

2.7 MICRONUTRIENTS

Vitamins

Vitamins are micronutrients obtained from eating plants and animals. Most vitamins cannot be made in our body and must be obtained from our diet, with a few exceptions. We can make vitamin D from cholesterol in our skin with exposure to sunlight, intestinal bacteria can produce some of the B vitamins, and we can produce vitamin A from beta-carotene. Vitamins are used by the body's cells as coenzymes, which enable hundreds of cellular chemical reactions to occur. A lack of vitamins prevents these chemical reactions from occurring and eventually causes deficiency symptoms and disease. Vitamins are classified as either water soluble or fat soluble. The water-soluble vitamins are vitamin C and the B vitamins. These vitamins are found in a variety of plant and animal foods; they are not stored in the body and must be consumed on a daily basis. Water-soluble vitamins can be damaged or lost during cooking or storage. The fat-soluble vitamins are A, D, E, and K. These vitamins are found in the fatty components of plants and animals, and they can be stored in the liver. Table 2.6 contains the most common functions, food sources, and deficiency symptoms for each vitamin. Notice that the deficiency symptoms are logical, given the vitamin's main functions.

TABLE 2.6

Vitamins

Vitamin	Function	Food Sources	Deficiency Symptoms
Vitamin A	Night vision; immune function, especially during viral infections; skin health; cell growth and development	Meat, liver, fish, dairy, beta-carotene (sweet potatoes, carrots, pumpkin)	Night blindness, increased severity of viral infections, dry rough skin, fetal deformities if deficiency occurs during pregnancy
B1—Thiamine	Production of ATP, proper nerve function, and normal appetite	Whole grains, liver, legumes, vegetables, nuts, seeds	Beriberi (severe nerve and muscle damage from total B1 deficiency). Low B1 levels can cause fatigue, weakness, loss of appetite, impaired nerve function, and emotional and cognitive disturbances. This is a common deficiency in alcoholics.
B2—Riboflavin	Production of ATP, skin health, eye health	Almonds, organ meats, whole grains, legumes, eggs, dark green vegetables	Fatigue, slowed growth, dermatitis, cataracts
B3—Niacin	Production of ATP and health of digestive tract, skin, and nerves	Meat, fish, liver, legumes, mushrooms, asparagus, peanuts	Pellagra (diarrhea, dermatitis, dementia), fatigue, weakness, digestive upset, skin lesions, mental confusion, emotional disturbances; a common deficiency in alcoholics
B5—pantothenic acid	Production of ATP; formation of hormones; metabolism of fat, proteins, and sugars	Organ meats, meats, whole grains, eggs, legumes, dairy; also produced by intestinal bacteria	Fatigue, weakness, neurological symptoms. This deficiency is uncommon due to abundance of B5 in many foods.
B6—Pyridoxine	Production of ATP, formation of hemoglobin, skin health	Meat, fish, liver, legumes, nuts, seeds, whole grains	Fatigue, anemia, dermatitis
Biotin	Metabolism of fat, proteins, and sugars	Egg yolk, liver, almonds, legumes, whole grains, vegetables; also produced by intestinal bacteria	Fatigue, changes in mood. This deficiency is uncommon due to the abundance of biotin in many foods.
B12—Cobalamin	DNA synthesis, production of red blood cells, health of nervous system	Meat, fish, organ meats, dairy, eggs. Vegans need a supplement.	Anemia, fatigue, weakness, neurological degeneration (numbness and tingling), mood problems. Pernicious anemia results from lack of B12 absorption, due to lack of intrinsic factor production.
Folic acid	Protein metabolism, red blood cell production, cell growth and differentiation	Meat, fish, lentils, beans, avocado, dark green vegetables, whole grains, citrus fruit	Neural tube defects (such as spina bifida in babies) anemia, fatigue
Vitamin C	Collagen production and connective tissue health, wound healing, immune function, increase in absorption of iron, acts as an antioxidant	Peppers, citrus fruit, colourful vegetables	Scurvy (loss of collagen causes loose teeth, bleeding, poor wound healing); increased infections, easy bruising, poor wound healing; common deficiency in smokers and alcoholics
Vitamin D	Calcium regulation, bone density, immune function	Sunlight exposure, liver, fish oil, fortified foods	Rickets (bone deformities in children); increased risk of infections, autoimmune diseases, and cancer (all immune related)
Vitamin E	Antioxidant, immune function	Meat, organ meats, fish, nuts, seeds, eggs, wheat germ, dark leafy greens, avocado	Nerve problems, impaired immune function, slow wound healing

Minerals

Minerals are micronutrients obtained from eating plants and animals. Minerals are elements that cannot be broken down into a simpler form and must be obtained from our diet. Minerals act as cofactors that help enzymes perform hundreds of functions in our cells. Minerals, found mostly in our skeleton, are responsible for approximately 4 to 5% of our body weight.

Hydrochloric acid production in the stomach is necessary for minerals to dissociate from food proteins or other components so that the food can be absorbed. For example, calcium, magnesium, iron, zinc, and selenium can bind to **phytates** in whole grains, legumes, and nuts, or **oxalates** in leafy greens. People with low stomach acid production or who take antacid medications may have impaired mineral absorption. Healthy intestinal bacteria also help the release of minerals from food. Table 2.7 contains the most common functions, food sources, and deficiency symptoms for some important minerals.

DID YOU KNOW?

People that take diuretic medications or consume diuretics such as alcohol or coffee have increased mineral loss in the urine. Magnesium deficiency is common in alcoholics for this reason. Ironically, magnesium is a mineral that can decrease anxiety and feelings of stress because of how it functions in the nervous system, so a magnesium deficiency can increase feelings of anxiety, which increases the desire for alcoholics to consume alcohol.

TABLE 2.7

Minerals

Mineral	Function	Food Source	Deficiency Symptoms
Calcium	Formation of bones and teeth, nerve transmission, and muscle contraction. Vitamin D is required for calcium absorption in the small intestine.	Dairy, salmon with bones, leafy green vegetables, nuts, seeds	Decreased bone density (osteoporosis), muscle twitching, heart palpitations, mood problems
Magnesium	"Anti-stress" mineral; balances calcium in nervous system and muscles (including smooth muscle); important cofactor for cellular enzymes	Dark green vegetables, seafood, nuts, seeds, legumes	Muscle cramping, heart palpitations, fatigue, mood problems, poor memory, numbness and tingling; common deficiency in alcoholics
Phosphorus	Makes up phospholipids, nucleotides, and ATP	All plant and animal foods	Deficiency is rare due to abundance of phosphorus in many foods. Malnourished children can have growth and bone problems from low phosphorus.
Iron	Forms part of heme in hemoglobin and myoglobin in muscles to carry O_2	Meat, liver, wheat germ. Also, less easily absorbed non-heme iron is found in lentils, nuts, seeds.	Anemia, muscle fatigue, general fatigue, memory and cognitive problems
Zinc	Immune function, metabolism, wound healing	Shellfish (highest in oysters), meat, liver, nuts, seeds	Increased infections, fatigue, poor wound healing, loss of sense of taste, white spots on fingernails
Iodine	Formation of thyroid hormones	Seafood, fish, seaweed, iodized salt	Decreased metabolism, fatigue, weakness, hypothyroidism

CONCEPT REVIEW 2.7

1. What are the water-soluble vitamins?
2. What are the fat-soluble vitamins?
3. Match the deficiency symptoms with the correct nutrient:

   a. night blindness
   b. neural tube defects
   c. scurvy
   d. osteoporosis
   e. fatigue
   f. muscle cramping
   g. hypothyroidism
   h. anemia

4. What foods contain vitamin B12?
5. What foods are high in vitamin C?
6. Which vitamin is made in our skin during sun exposure?
7. What is the micronutrient that stomach acid and intestinal bacteria help the body absorb?

CHAPTER REVIEW

Medical Terminology

TERM	DEFINITION	EXAMPLE
cis	on this side	cis fatty acid
de-	away from or cessation	dehydration
dextr/o	right, on the right side	dextrose
dif-	separation	differentiate
dis-	apart, to separate	dissociate
macr/o	large	macromolecule
micro-	small	micronutrient
nutri/o	to nourish	nutrition
trans-	across	trans fatty acid

Labelling Activity

Identify each of the following molecules.

1.

From DiGiuseppe/Sanader. *Chemistry* 12U. © 2012 Nelson Education Ltd. Reproduced by permission. www.cengage.com/permissions

2.

3.

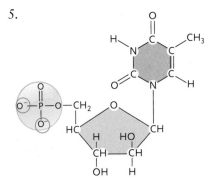

From DiGiuseppe/Fraser. *Biology* 12U, 1E. © 2012 Nelson Education Ltd. Reproduced by permission. www.cengage.com/permissions

4.

From Russell/Hertz/McMillan/Fenton/Addy/Maxwell/Hattie/Milson. *Biology*, 2E. © 2013 Nelson Education Ltd. Reproduced by permission. www.cengage.com/permissions

5.

6.

7.

6 CH₂OH   6 CH₂OH
5   O   5   O
H   H   H   H
H   H   1   H   H   1
4   OH   H   4   OH   H
HO   O   OH
3   2   3   2
H   OH   H   OH

_____

8.

(A triacylglycerol structure with three fatty acid chains)

_____

9.

CH₂OH   H   OH   CH₂OH   H   OH
H   O   OH   H   H   O   OH   H
H   O   H   H   O   H
OH   H   OH   H   O—
H   H   O   H   H   O
H   OH   CH₂OH   H   OH   CH₂OH

_____

10.

C — C — C — C — C — C = C — C — C — H
(fatty acid chain with OH, H on carboxyl end and H's along chain, one double bond)
O
H

_____

11.

CH₃   CH₃
CH
H₂C
CH₂
H₂C
HC — CH₃
H₂C   CH₃
H₂C   C   CH
H₂C   CH₂   CH   CH   CH₂
H₂C   C   CH
HO   CH   C   CH₂
H₂   CH   H

_____

12.

CH₂—N⁺(CH₃)₃
CH₂
O
O=P—O⁻
O
CH₂—CH—CH₂
O   O
C=O   C=O

_____

Answer to Application Question

- Foods that contain protein—egg, hemp seeds, chicken, almond butter
- Foods that contain nucleic acids—any whole food that contains plant or animal cells; includes egg, greens, hemp seeds, green pepper, radish, cucumber, chicken, rice, legumes, asparagus
- Foods that contain simple carbohydrates—blueberry muffin, coffee with sugar
- Foods that contain complex carbohydrates (starch)—whole grain toast, blueberry muffin, rice, crackers

- Foods that contain saturated fat—egg, butter, chicken
- Foods that contain unsaturated fat—blueberry muffin, olive oil, potato chips, crackers, almond butter
- Foods that contain fibre—whole grain toast, mixed greens, hemp seeds, green pepper, radish, cucumber, legumes, asparagus, crackers, almond butter
- Foods that could contain trans fat—blueberry muffin, potato chips, crackers

3 Cell Structure and Function

CHAPTER OUTLINE

3.1   Cell Theory

3.2   Plasma Membrane

3.3   Prokaryotic and Eukaryotic Cells

3.4   Organelles of the Cell

3.5   Intracellular and Extracellular Proteins

3.6   Movement across Membranes

3.1 CELL THEORY

Many different shapes and sizes of cells exist in humans and every living organism. Most human cells are between 5 and 20 μm in diameter (a micrometre (μm) is 1/1000th of a millimetre) (Figure 3.1). The head of a pin is approximately 2 mm wide, which is 2000 μm.

Cell Size

Cells need oxygen and nutrients, and they need to get rid of waste. As a cell gets bigger its volume increases, and oxygen and CO_2 have to diffuse further to enter and leave the cell. Therefore, the *number* of cells in larger organisms increases, not the *size* of their cells. The largest cell in the human body is the egg cell, which is approximately 100 μ. However, some cells have special features that increase surface area: for example, extensions such as microvilli or dendrites. Cells called *neurons* transmit electrical signals within the nervous system and can be as long as one metre in length; however, because they are very narrow, the diffusion distance to the cell surface remains very small (Figure 3.2). Epithelial cells in the small intestine have microvilli, which dramatically increase the surface area of those cells, and so nutrients can be absorbed into the bloodstream efficiently.

Education is not the filling of a pail, but the lighting of a fire.

William Butler Yeats

David M. Phillips/Science Source

Figure 3.1 Human Sperm and Egg Cell

In humans, the female egg cell is the largest cell type, and sperm cells are the smallest.

Courtesy of Wendi Roscoe

A.

B.

From Russell/Hertz/McMillan. *Biology*, 4E. © 2017 Brooks/Cole, a part of Cengage, Inc. Reproduced by permission. www.cengage.com/permissions

Figure 3.2 (A) Microvilli and (B) Neuron

Cells such as microvilli (A) and neurons (B) have specialized structures that give them a greatly increased surface area.

Cell Theory

Cells were seen for the first time under the microscope invented by Robert Hooke in 1665. Through years of research by many scientists, cell theory has come to mean several things:

1. All living organisms are composed of cells (including bacteria, protists, fungi, plants, and animals, but not including viruses, which are not composed of cells and are not considered "living").

2. The cell is the basic unit of all living organisms and is the smallest single unit that can survive independently.

3. All cells come from pre-existing cells.

4. All cells contain genetic material, DNA, which is passed on to new cells that arise from cell division.

5. All cells are either prokaryotic or eukaryotic and are composed of basically the same types of phospholipids, proteins, carbohydrates, and nucleic acid molecules.

CONCEPT REVIEW 3.1

1. Why do some cells have structures such as microvilli or dendrites?

2. List the five main components of the cell theory.

3.2 PLASMA MEMBRANE

All cell membranes, also called **plasma membranes**, are formed from amphipathic phospholipids arranged in a **bilayer** (Chapter 2). The bilayer is formed from two rows of phospholipids arranged so that the hydrophobic tails (water-fearing) are directed inward toward the centre of the cell and the hydrophilic heads (water-loving) are directed outward toward the cytoplasm or extracellular fluid. The membranes that surround organelles in eukaryotic cells also consist of a bilayer formed from phospholipids. There may be some subtle differences between plant, fungus, and animal membranes, but they are all made of essentially the same types of molecules. Recall the structure of a phospholipid; it has a hydrophilic (polar) head, including phosphate, choline, and glycerol, and two hydrophobic (nonpolar), unsaturated fatty acid tails (Figure 3.3).

Hydrophilic head

Hydrophobic tails

Figure 3.3 Structure of a Phospholipid

Phospholipids are made up of a five-carbon sugar, a phosphate group, and two fatty acid tails.

DID YOU KNOW?

Since most of the phospholipid fatty acid tails are unsaturated, cell membranes are fluid, like oil. Human tissues have their particular shape and structure because of structural proteins in the connective tissues.

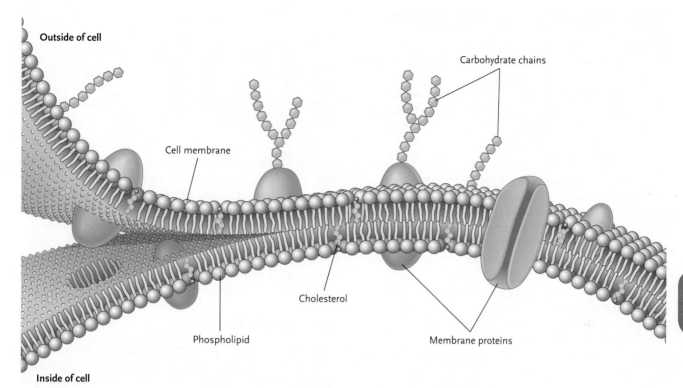

Figure 3.4 The Cell Membrane
The cell membrane is formed by a bilayer of phospholipids and contains cholesterol and many different proteins.

Fluid Mosaic Model

Since cell membranes are fluid and contain many other molecules in addition to the phospholipids, we use the term **fluid mosaic model** to describe a typical cell membrane. Membranes have cholesterol molecules embedded within the membrane between bent, unsaturated fatty acid tails. Cholesterol, which is mostly hydrophobic, increases the strength of the membrane and decreases its fluidity. Cell membranes also contain proteins, glycolipids (lipids with carbohydrate groups), and **glycoproteins** (proteins with carbohydrate groups) (Figure 3.4). The following describes some of the most common membrane molecules:

- **Receptors**—proteins that bind regulatory molecules such as neurotransmitters, hormones, cell-signalling molecules, or nutrients

- **Antigens**—usually glycoproteins that act as cell markers

- **Enzymes**—proteins that act as catalysts for assorted chemical reactions that may occur on cell membranes

- **Ion channels**—proteins that span the membrane and allow the transport of specific ions into or out of the cell

- **Carrier proteins**—proteins that move substances like nutrients across cell membranes by changing their conformational structure

- **Adhesion molecules**—membrane proteins that play an important role in intercellular interactions

- **Gap junction channels**—protein membrane channels that play an important role in intercellular communication

CONCEPT REVIEW 3.2

1. Explain the difference between nonpolar and polar (hydrophobic vs. hydrophilic).

2. Draw a simple phospholipid, and label fatty acids, choline, glycerol, and phosphate.

3. Describe the general structure of a cell membrane.

4. List all of the components of a typical cell membrane and their main function.

3.3 PROKARYOTIC AND EUKARYOTIC CELLS

There are two major types of cells: prokaryotic and eukaryotic. The prokaryotic kingdoms include the archaebacteria and eubacteria, and the eukaryotic kingdoms include the protists, fungi, plants, and animals.

Prokaryotic Cells

Prokaryotic cells—the simplest cell type—lack a nucleus and any other membrane-bound organelles (Figure 3.5). A nucleoid region in the centre of the cell contains the genetic material, which is usually a single circular DNA molecule. Prokaryotic cell membranes consist of a phospholipid bilayer and many different membrane proteins, glycoproteins, and glycolipids (as described in Section 3.2).

Most bacteria have extra protective layers. The cell wall of most eubacteria consists of sugars and amino acids, called **peptidoglycan**, which is important for maintaining cell strength and regulating osmotic pressure inside the cell. Eubacteria can be classified into two main groups based on the structure of the cell wall: (1) **gram-positive** bacteria have many layers of peptidoglycan and stain purple when exposed to the stain crystal violet; (2) **gram-negative** bacteria have a smaller layer of peptidoglycan in addition to an external **lipopolysaccharide** layer and stain pink (Chapter 24). It is important to note that bacterial cell walls are not the same as the cellulose cell wall found in plant cells. Some bacteria may also have an external protein or carbohydrate **capsule** that further protects them from environmental changes such as ion concentrations, temperature, dehydration, and pH. This capsule makes their destruction by our immune cells more difficult and therefore increases the ability of certain bacteria to cause human diseases.

Prokaryotic cells contain cytoplasm with dissolved ions and nutrients. Prokaryotes also have **ribosomes**, which are essential for protein synthesis. Ribosomes consist of RNA and protein and do not have a membrane. Prokaryotes may also have **pili** (singular is pilus) and/or **flagella** (singular is flagellum). A pilus is a hair-like structure that helps bacteria adhere to structures, and, through a process called **conjugation**, the pili function in the transfer of small pieces of DNA between bacteria. A flagellum is an external protein structure used for locomotion.

Eukaryotic Cells

Eukaryotic cells are much larger, approximately 10 times larger, than most prokaryotic cells. They have a more complex cell structure, including a phospholipid bilayer membrane (Section 3.2), cytoplasm, ribosomes, and membrane-bound organelles (Section 3.4). Not all eukaryotic cells are exactly the same; there are some differences between protist, fungal, plant, and animal cells. Protists, fungi, and plants have a cellulose cell wall that provides extra protection against dehydration and other environmental factors. The cells of plants and some protists contain organelles called *chloroplasts*; these contain the pigment chlorophyll, which allows plant cells to undergo photosynthesis.

> **DID YOU KNOW?**
>
> Because some antibacterial drugs such as penicillin block the ability of bacteria to produce a cell wall, the bacteria cannot reproduce. Animal cells do not have a cell wall, so penicillin does not damage our cells.

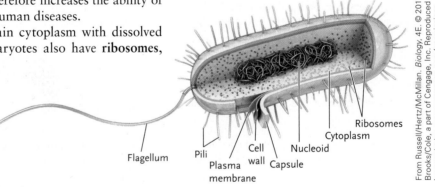

From Russell/Hertz/McMillan. *Biology*, 4E. © 2017 Brooks/Cole, a part of Cengage, Inc. Reproduced by permission. www.cengage.com/permissions

Flagellum Pili Plasma membrane Cell wall Capsule Nucleoid Cytoplasm Ribosomes

Figure 3.5 Prokaryotic Cell Structure
Prokaryotes have the simplest cell structure.

CONCEPT REVIEW 3.3

1. Name the structures in prokaryotic cells.
2. Describe the functions of each structure in prokaryotic cells.
3. Why is the cell wall and capsule important for prokaryotes?
4. What is the difference between gram-positive and gram-negative bacteria?
5. How do eukaryotes differ from prokaryotes?

3.4 ORGANELLES OF THE CELL

This section describes the typical structures and functions of organelles in animal cells (see Figure 3.6).

Nucleus

The **nucleus** is a membrane-bound organelle that houses the genetic material. In humans, there are 46 chromosomes of DNA within every nucleus of every cell in the body: except for mature red blood cells and keratinocytes, which have no nucleus. The nucleus is considered the control centre of the cell because it contains the DNA and is therefore the location of transcription of the genes. During cell division, the DNA becomes condensed into tightly packed chromosomes (Chapter 5). During the regular growth phase when transcription occurs, the DNA exists as loose strands called **chromatin**. The fluid within the nucleus—called **nucleoplasm**—is very similar to the composition of the cytoplasm (Figure 3.7).

The membrane surrounding the nucleus is similar to the plasma membrane: a phospholipid bilayer containing cholesterol and proteins. In addition, the **nuclear**

Figure 3.6 Eukaryotic Cell

Eukaryotic animal cells are complex and contain many membrane-bound organelles.

Figure 3.7 Nucleus

The nucleus consists of a double membrane that contains pores and houses the nucleolus and DNA.

Figure 3.8 Endoplasmic Reticulum

The rough endoplasmic reticulum contains ribosomes and is the site of protein synthesis; the smooth endoplasmic reticulum is the site of carbohydrate and fat synthesis.

membrane, also called the **nuclear envelope**, is double layered and contains pores that allow large molecules such as RNA to move out of the nucleus. After transcription, messenger RNA molecules move into the cytoplasm to connect with ribosomes so that protein synthesis can occur.

The nucleus contains a small non-membrane-bound organelle called the **nucleolus**. The nucleolus contains DNA that consist of genes that code for ribosomal RNA transcripts and proteins that are important in the production of a new ribosome subunit.

Endoplasmic Reticulum

The **endoplasmic reticulum** (ER) is a membrane network found within the cytoplasm of every cell type (Figure 3.8). There are two types of endoplasmic reticulum: smooth and rough. The function of both types varies depending on the cell type. **Smooth ER** has a smooth surface and does not have any ribosomes. The function of smooth ER is to synthesize lipids, such as phospholipids and cholesterol, for the maintenance of the cell membrane. The smooth ER also produces steroid molecules and metabolizes carbohydrates. The smooth ER in liver cells converts excess glucose into glycogen or fat. The liver's smooth ER can also break down

medications. In muscle cells the smooth endoplasmic reticulum (called the *sarcoplasmic reticulum*) stores calcium that is released during muscle contraction. **Rough ER** is mainly involved in protein synthesis on the ribosomes; it is also involved in protein folding and the packaging of newly synthesized proteins that become transported to the Golgi bodies for further processing and modification. Messenger RNA (mRNA) molecules—transcribed in the nucleus—are transported through the nuclear pores to a ribosome that's either free in the cytoplasm or on rough ER; then protein synthesis (translation) can occur. In this sense, the ER is continuous with the nuclear envelope.

Golgi Body

The Golgi body, also called the *Golgi apparatus* or *Golgi complex*, is a separate membrane network found in the cell near the ER. It looks like a stack of semi-circular membranes (Figure 3.9). Proteins that have been newly synthesized on the rough ER are transported to the Golgi for further processing. The proteins are transported within **vesicles** that are membrane-bound sacs that can merge with the membranes of the Golgi. The side of the Golgi closest to the ER is called the **cis Golgi**, and the side farthest from the ER is the **trans Golgi**. When new proteins arrive within the Golgi several things happen, depending on the cell type. After translation occurs, most proteins become modified to include carbohydrate groups or phosphate molecules, so the finished

Figure 3.9 Golgi Complex

In the Golgi complex, proteins are modified and packaged into vesicles.

Figure 3.10 Mitochondrion

Mitochondria are organelles that produce ATP.

functional proteins are often combinations of amino acids plus carbohydrate or lipid groups. **Glycosylation** enzymes in the Golgi cause the addition of **oligosaccharides** (small strands of monosaccharides), and **phosphorylation enzymes** add phosphate groups to some proteins. The Golgi body also functions in transporting lipids and creating **lysosomes**. An example of an important product of the Golgi is **proteoglycan**, which is a glycosylated protein that constitutes a significant portion of the extracellular matrix of most tissues.

Mitochondria

Mitochondria are organelles that produce cell energy, ATP, in the presence of oxygen. Mitochondria consist of two membrane layers; **cristae** is the name of the folded inner membrane (Figure 3.10). It is widely believed that mitochondria originated through a process of **endosymbiosis**—where primitive bacteria engulfed by another cell gave rise to eukaryotic organisms—approximately 1.5 billion years ago. The inner membrane is probably derived from the prokaryotic cell membrane, and the outer mitochondrial membrane likely derived from the plasma membrane of the cell that engulfed the prokaryote. Mitochondria share many similarities with prokaryotic organisms:

1. They are approximately the same size, 1 to 5 μm.
2. They have a small circular DNA molecule that can self-replicate.
3. Mitochondria replicate by binary fission as do bacteria.

4. Mitochondria have ribosomes, which are similar to bacterial ribosomes.

The mitochondrial DNA molecule has 37 genes, and 13 of these are responsible for producing molecules related to **cellular metabolism** and ATP production (Chapter 4). Processes that involve the 24 other mitochondrial genes include regulation of calcium, programmed cell death (apoptosis), cellular aging, the cell cycle, and cell growth.

DID YOU KNOW?

All mitochondria in your body, and therefore the mitochondrial DNA, came from your mother's egg cell during fertilization. The sperm only rarely contributes mitochondria. Mitochondrial DNA can be studied to determine relationships and migration of humans.

Ribosomes

Ribosomes contain subunits of protein and the ribosomal RNA (rRNA) molecules produced in the nucleolus region of the nucleus (Figure 3.11). The small subunit and the large subunit join together in the presence of a messenger RNA (mRNA) transcript. Ribosomes are the site of **translation**, which is protein synthesis (Chapter 9). Ribosomes are found in all

Large subunit

Small subunit

Figure 3.11 Ribosomes
Ribosomes are the site of protein synthesis.

prokaryotic and eukaryotic cells, although the structure in each cell type is slightly different. The ribosomes in mitochondria more closely resemble the ribosomes in prokaryotes.

Lysosomes

Lysosomes can be referred to as "the digestive system of the cell." Lysosomes are small membrane-bound organelles that contain digestive enzymes and play an important role in breaking down macromolecules and old organelles inside the cell. Lysosomes are produced by the Golgi bodies. Membrane proteins that pump hydrogen ions into the lysosome make the pH inside the lysosome more acidic than the cytoplasm. Many molecules that come from this breaking down process are recycled and used by other parts of the cell. Any resulting waste products that are not reused are excreted from the cell either by diffusion or exocytosis; they then enter the bloodstream and are eliminated from the body through the kidneys. In our white blood cells, a large number of lysosomes are involved in breaking down infectious organisms such as bacteria and viruses.

Peroxisomes

A major function of **peroxisomes** is to break down long-chain fatty acids, which are then shuttled to the mitochondria for oxidation and the production of ATP. Peroxisomes in cells of the nervous system, specifically, oligodendrocytes and Schwann cells, break down fatty acid molecules to produce myelin—the protective covering on many neurons (Chapter 15). Peroxisomes also remove hydrogen ions from certain substrates and, in the presence of oxygen, produce **hydrogen peroxide** (H_2O_2). With the help of catalase enzymes, peroxisomes break down toxic substances; for example, liver cells break down alcohol and formaldehyde, and immune cells can be effective in destroying infectious organisms. We can buy hydrogen peroxide to use on cuts and scrapes to help decrease infection by killing bacterial cells.

Centrioles

Centrioles, found in most eukaryotic cells except for plants and fungi, are involved in the *organization* of spindle fibres (microtubules) that transport chromosomes during cell division (Chapter 5). Centrioles produce spindle fibres that help to separate chromosomes during mitosis and meiosis. The position of the centrioles within a cell also determines the position of the nucleus and other organelles in the cell (Figure 3.12).

See Table 3.1 for a summary of the main functions of all the organelles.

Microtubule triplet

Figure 3.12 Centrioles
Centrioles are composed of microtubules and are involved in spindle formation during mitosis.

TABLE 3.1

Summary of the Main Functions of Organelles

Organelle	Main Functions
Nucleus	Stores DNA; location of transcription of mRNA
Nucleolus	Contains genes that code for ribosomal RNA
Rough endoplasmic reticulum (rough ER)	Contains ribosomes that are the location of protein synthesis (translation)
Smooth endoplasmic reticulum (smooth ER)	Location of production of lipids and carbohydrates; calcium storage in muscle cells
Golgi bodies	Modification, packaging, and processing of proteins; production of lysosomes.
Mitochondria	Production of ATP (oxidative phosphorylation)
Ribosomes	Located on RER or free in cytoplasm; location of protein synthesis
Lysosomes	Contain digestive enzymes that break down old organelles and macromolecules
Peroxisomes	Break down fatty acids; produce peroxide to break down toxic substances, such as alcohol, and can be used by immune cells to kill pathogens
Centrioles	Form spindle fibres in animal cells during mitosis and meiosis

CONCEPT REVIEW 3.4

1. Make a chart and summarize the main function of each eukaryotic organelle.

2. What characteristics of mitochondria support the endosymbiosis theory?

3.5 INTRACELLULAR AND EXTRACELLULAR PROTEINS

All cells have an internal structure composed mainly of intracellular proteins and molecules that hold organelles in place, give the cell shape, and help transport other molecules within the cell; these proteins make up the **cytoskeleton**. The three main categories of cytoskeletal proteins are **microtubules**, **intermediate filaments**, and **microfilaments**.

Microtubules

The largest intracellular protein structures, microtubules, are large hollow tubes formed from individual proteins called **tubulin**. Approximately 25 nm thick, microtubules are attached at each end onto the cell membrane. Their role is to help maintain cell structure and to provide a means for transporting molecules throughout the cell. The spindle fibres that help transport chromosomes during cell division are composed of microtubules. Many different motor proteins can bind to microtubules, which help move molecules or vesicles around the cell. Figure 3.13 shows the three main cytoskeletal proteins, and Figure 3.14 shows the motor protein, dynein, transporting a vesicle along a microtubule.

Intermediate Filaments

Intermediate filaments, slightly smaller than microtubules and approximately 10 nm in diameter, are composed of intertwined "ropes" of protein. Different types of intermediate filaments consisting of many different proteins occur in different kinds of cells. The main function of each type is to support cell structure and help organelles remain in place. Some examples of intermediate filaments include **keratin**, found in hair and nails, and **neurofilaments**, found in the axons of neurons.

Microfilaments

The smallest cytoskeletal proteins, the microfilaments, are approximately 7 nm in diameter and composed mostly of **actin** proteins. The primary role of microfilaments is cell movement, such as crawling. Our white blood cells have a large number of microfilaments; they continually move into and out of the bloodstream to locations of inflammation and cellular damage. Movement, also called *migration*, is a property of many different cell types. For example, astrocytes in the central nervous system move toward damaged areas, stem cells migrate to new locations in the body during growth and repair of tissues, and cancer cells can migrate to different locations in the body—a process called **metastasis**.

Figure 3.13 Cytoskeletal Proteins

Cytoskeletal proteins are important for giving cells their shape, supporting the organelles, and helping molecules move around within the cell.

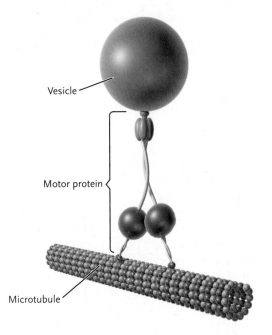

Figure 3.14 Microtubules and Motor Proteins

Microtubules and motor proteins are used for transporting vesicles within the cell.

Extracellular Proteins

Every cell produces **extracellular proteins** that are either secreted out into the extracellular matrix or integrated into the cell membrane; they are involved with intercellular interactions such as adhesion, signalling, or cellular communication (Figure 3.15). **Collagen** surrounds many cell types, particularly cells in connective tissues (Chapter 11), and mostly in cartilage, skin, tendons, and ligaments. Collagen provides strength to many tissues

and resists stretching. **Elastin** is an extracellular matrix protein that allows tissues to be flexible and stretchy and also to re-form to their original shape after being stretched. Tissues with high elastin content include the skin, bladder, lungs, and blood vessels. **Integrins** are

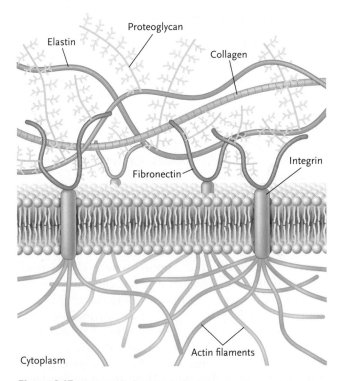

Figure 3.15 Extracellular Matrix Proteins

Extracellular matrix proteins are very important for connecting cells together, and giving tissues strength and flexibility. Different tissues have different amounts and types of extracellular matrix proteins, and this image shows some of the most common ones.

membrane-spanning proteins that are involved in attaching the cell membrane to proteins in the extracellular matrix. They also function in transferring signals from outside to inside the cell, and vice versa, which allows for very rapid responses to environmental changes. When a blood clot forms, integrins on platelets bind to the extracellular fibrin protein during clot formation. **Fibronectin** is a protein that connects integrins to other extracellular matrix proteins. Fibronectin is very important during human growth and development and also in the healing of wounds; it aids cell adhesion, migration, and differentiation. Proteoglycans are glycoproteins found in connective tissues. The carbohydrate groups on these proteins hold water and keep connective tissues hydrated. Some skin care products contain proteoglycans to hold moisture in the skin.

Cellular Junctions

Desmosomes consist of cadherin proteins that make cell membranes adhere together by connecting to the intermediate filaments of the cytoskeleton (Figure 3.16). Desmosomes, found in stratified epithelial tissue and in cardiac muscle, prevent shearing damage to tissues. **Tight junctions,** composed of proteins called **occludens,** also connect epithelial cells. They bring cells very close together so there is minimal extracellular fluid in that

DID YOU KNOW?

Gelatin is used to make jams, Jell-O, marshmallows, gummy candy, and many other foods. Gelatin is hydrolyzed (broken down) collagen protein, usually isolated from pig skin or cow bones. Consuming gelatin is very beneficial for the health of our skin and connective tissues (joints, tendons, ligaments), and it is easy to make on your own by boiling bones (pig, chicken, cow, lamb) with spices to make bone broth. Collagen proteins are high in the amino acid glycine, which also acts as a calming neurotransmitter; drinking homemade bone broth can help decrease stress.

area. In this way they create a barrier that prevents molecules moving between cells—this means intracellular movement can occur only via membrane transport proteins. Tight junctions are important in the formation of the blood-brain barrier and in controlled absorption in the small intestine (Chapters 15 and 20). **Gap junctions** are channels composed of **connexin** proteins that allow the direct transfer of small molecules between cells. Gap junctions, important in cardiac muscle cells, allow the direct and rapid transfer of ions so that heart muscle cells can contract together (Chapter 18).

Cells

Cadherin proteins

Intermediate filaments

Connexin proteins

Occluden proteins

Anchoring junction

Gap junction

Tight junction

Figure 3.16 Cell Junctions

The main types of cell junctions are desmosomes (anchoring junctions), tight junctions, and gap junctions.

From Russell/Hertz/McMillan. *Biology,* 4E. © 2017 Brooks/Cole, a part of Cengage, Inc. Reproduced by permission. www.cengage.com/permissions

CONCEPT REVIEW 3.5

1. Describe the three main cytoskeletal proteins.
2. What are the main functions of integrins, fibronectin, collagen, and elastin?
3. What are the main functions of desmosomes, tight junctions, and gap junctions?
4. Give an example of a tissue that contains each type of cellular junction.

Application

As you read through the rest of this chapter, consider this scenario. John is going to cycle 75 km on his road bike. Suppose it takes him three hours to complete the ride on a warm day. Considering that the average person will sweat approximately 1 L/hr, which type of drink would be the best for rehydration, and why? A hypertonic, isotonic, or hypotonic solution. For the purpose of this question, let's consider a sport drink that contains water, sugar (for energy), and minerals ("electrolytes") that replace the minerals lost in sweat. Body fluids are approximately 300 mOsmol (milliosmoles).

3.6 MOVEMENT ACROSS MEMBRANES

The function of cell membranes is to regulate what goes into and out of the cell. Recall that cell membranes are composed of phospholipids that have a polar head (facing toward the cytoplasm or extracellular space) and nonpolar tails that make up the inner portion of the cell membrane. Cell membranes are **semi-permeable** because only certain molecules can easily cross the hydrophobic tails of the inner membrane; everything else is regulated by proteins in the cell membrane. Molecules that can cross easily include small, nonpolar molecules such as O_2, CO_2, N_2, glycerol, and H_2O. Water can cross the membrane even though it is polar because the water molecule is very small, and osmotic pressure is a strong force that regulates the water balance inside cells. Water can also move across cell membranes through protein pores called *aquaporins* (Chapter 21).

Diffusion

Diffusion refers to the spontaneous movement of any molecule from an area of high concentration to an area of low concentration until equilibrium is reached without any input of energy (Figure 3.17). Once equilibrium is reached, the molecules continue to move, but there is no further *net* movement in any direction (all molecules constantly move due to thermal energy). Diffusion in our body includes the movement of small, nonpolar molecules across a cell membrane and down a concentration gradient (Figure 3.18). For example, oxygen can easily move from a point of higher concentration in the bloodstream toward that of lower concentration inside cells.

Factors That Affect the Rate of Diffusion across a Membrane

- *Concentration gradient.* A large difference in solute concentration on either side of the membrane increases the rate of diffusion.

Figure 3.17 Diffusion
Diffusion is the movement of molecules from an area of higher concentration to an area of lower concentration.

- *Temperature.* An increase in temperature increases the rate of diffusion.
- *Surface area.* An increased surface area on a cell membrane increases the rate of diffusion; for example, microvilli in the small intestine increase the surface area for nutrients to diffuse into the bloodstream.

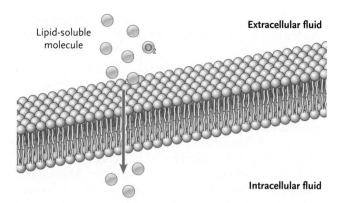

Figure 3.18 Diffusion across a Membrane

Diffusion across a membrane involves the movement of molecules from a region of higher concentration to that of lower concentration without requiring any membrane proteins.

- *Diffusion distance.* A shorter diffusion distance increases the rate of diffusion; tissues may have many or few capillaries for oxygen and nutrients to diffuse from.
- *Size of molecule.* Small molecules can diffuse faster than large molecules through the lipid bilayer.

Osmosis

Osmosis is the diffusion of water across a semi-permeable membrane to an area of lower water concentration (toward a higher solute concentration). The movement of water into or out of a cell depends on the concentration of other molecules (solutes) inside and outside the cell. Charged and polar molecules attract water molecules. The higher the concentration of solutes on one side of a membrane, the higher the osmotic concentration, and this will cause water to move toward the solute. If we compare the solute concentration in the interstitial fluid surrounding a cell (extracellular space) to the solute concentration inside the cell, the **osmotic concentration** is either (1) the same in both, **isotonic;** (2) higher in the interstitial fluid, **hypertonic;** or (3) less in the interstitial fluid, **hypotonic** (see Figure 3.19). The movement of

water into a cell due to the osmotic concentration is called the *osmotic pressure.* If the osmotic pressure is very great the cell may burst. Likewise, an increase in osmotic concentration outside of cells can cause cells to lose so much water that they shrivel up. Because the ion concentration surrounding single-celled organisms, such as protists or bacteria, can be extremely variable, those organisms have cell walls that help control the osmotic pressure inside the cell. In our body, many different organ systems play a role in homeostasis to ensure that the osmotic pressure of our interstitial fluid is highly regulated.

Facilitated Diffusion

Facilitated diffusion is the movement of molecules across a cell membrane and down a concentration gradient without the input of energy, but with the help of a membrane protein (Figure 3.20). The membrane protein can be a channel or a carrier. Channels are specific protein openings in the membrane that allow very specific ions to move across. A sodium channel will not allow calcium to move across the membrane. Many ions are very important for certain cell functions, primarily, Na^+, K^+, Ca^{2+}, and Cl^-. For various reasons, many cell functions require higher concentrations of certain ions either inside or outside the cell, and certain mechanisms ensure those gradients remain intact (Figure 3.21). Cell types that require an influx of sodium ions—such as neurons during an action potential—increase the expression of sodium-channel membrane proteins. Chapters 14 and 15 discuss this concept in more detail.

Carrier Proteins

Similar to ion channels, carrier proteins are membrane proteins that regulate the movement of certain molecules

Hypotonic	Isotonic	Hypertonic
less than 0.9% NaCl or less than 300 mOsm	0.9% NaCl	greater than 0.9% NaCl or greater than 300 mOsm

Figure 3.19 Hypertonic, Hypotonic, and Isotonic Solutions

In a hypertonic solution, water moves out of the cell. In a hypotonic solution, water moves into the cell. In an isotonic solution there is no net movement of water into or out of the cell.

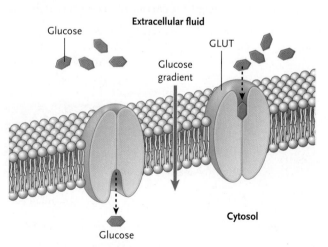

Figure 3.20 Facilitated Diffusion

Facilitated diffusion involves the movement of polar or charged molecules across a cell membrane by means of a membrane transporter protein that is either a channel or a carrier.

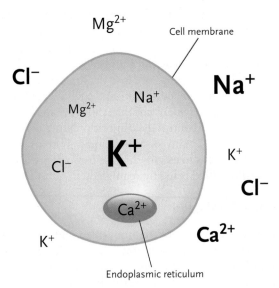

Figure 3.21 Ion Concentrations within and Surrounding a Typical Cell

Every cell must have a specific balance of ions inside and outside the cell. The concentration of sodium and chloride ions is higher outside the cell, that of potassium is higher inside the cell, and that of calcium is higher inside the smooth ER within the cell.

into or out of the cell, down a concentration gradient, without ATP. The difference with carrier proteins is that they are not channels that can allow a continuous stream of molecules to cross the membrane. Instead, they are proteins that bind a specific molecule and then change conformational shape; this causes the specific molecule to move across the membrane. Most membrane proteins that regulate the movement of substances such as nutrients, amino acids, glucose, and fatty acids are membrane carrier proteins. Some ions can be moved across the membrane by carrier proteins, such as co-transporters, discussed below. All the factors that affect the rate of diffusion are true for facilitated diffusion, except for carrier molecules, which can

reach **saturation.** At that point, further increases in the concentration gradient do not further increase the rate of diffusion.

Active Transport

Active transport refers to the movement of molecules across a membrane *against* the concentration gradient, by means of a membrane protein, and *with* the input of energy (ATP). For molecules to move against the concentration gradient, energy is required. Although a membrane protein similar to a carrier protein serves this function, in this case, ATP causes a conformational change in the protein and brings the molecule or ion across the membrane. The most common and important example is the **sodium-potassium pump.** The sodium-potassium pump ensures that a gradient—*not* equilibrium—is maintained across the membrane. Na$^+$ ions need to be in higher concentration outside every cell type for many reasons. For example, neurons and muscles cells use sodium to cause action potentials, and other cells use the sodium gradient to help bring nutrients into the cell: a process called **coupled transport,** or cotransport. All cells in the entire body have sodium-potassium pumps to keep sodium concentrations higher in the interstitial fluid than inside the cell.

Sodium-Potassium Pump

The sodium-potassium pump (Na$^+$/K$^+$ pump) is a membrane protein that uses one molecule of ATP to move exactly three sodium ions out of the cell and two potassium ions into the cell (Figure 3.22). Sodium moves into the cell gradually through a number of mechanisms, and the Na$^+$/K$^+$ pump ensures that the concentration gradients remain intact. Almost one-third of all the ATP used in our body fuels the Na$^+$/K$^+$ pumps! That is equivalent to at least 500 calories per day; this varies based on body weight because more cells mean more

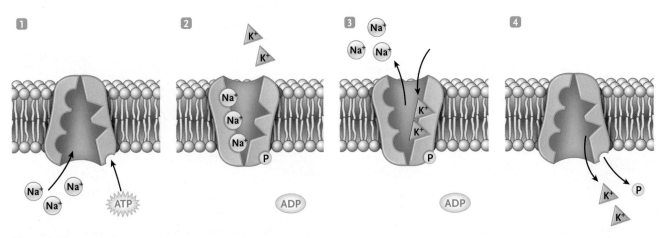

Figure 3.22 Sodium-Potassium Pump

The sodium-potassium pump is an important example of active transport. Every cell in the body must maintain a higher concentration of sodium outside the cell and potassium inside the cell.

Na$^+$/K$^+$ pumps to fuel. Along with maintaining an ion concentration gradient across the cell membrane, an increase in positively charged sodium ions outside the cell causes an **electrical gradient** to occur. Therefore, the inside of every cell is slightly more negative compared to the outside; this is called the **resting membrane potential** and is approximately −70 millivolts in most cell types. *All* cells have a resting membrane potential; only neurons and muscle cells have action potentials or graded potentials (Chapter 15).

Proton Pump

Membrane proteins called **proton pumps** can be found on the membrane of some cells, such as the parietal cells of the stomach, which produce hydrochloric acid (Chapter 20), and within the inner membrane of every mitochondrion in every cell (Figure 3.23). The proton pump is another example of an active transport membrane protein that moves protons (H$^+$) *against* its concentration gradient. In the mitochondria, as protons move into the intermembrane space, the protons' positive charge brings about an electrochemical gradient. Whenever a charge or concentration difference occurs across a membrane, the molecules "want" to reach equilibrium, so the H$^+$ ions tend to move back across the membrane into the matrix of the mitochondria. However, this can only occur by means of a very specific membrane protein called **ATP synthase**. When protons move down their concentration gradient and back into the matrix through ATP synthase enzymatic activity, phosphate groups are added onto the ADP molecules, and new ATP molecules are formed; this process is called **chemiosmosis** (Chapter 4). Note that the purpose of the concentration gradient in this situation is to have the electrochemical gradient perform work, which in this example produces ATP molecules through ATP synthase.

Coupled Transport

Coupled transport is the process of a single carrier protein moving two substances at the same time. **Symport** occurs when both substances move in the same direction; **antiport** is when the substances move in opposite directions. An example is the use of the sodium gradient to move nutrients into the cell (Figure 3.24). Since sodium is always in higher concentration outside the cell than inside, due to the Na$^+$/K$^+$ pump, sodium naturally tends to move back into the cell. Using coupled transport membrane proteins, glucose or amino acids can move into the cell along with sodium. This example again shows that cell membranes are semi-permeable, and allow only certain substances to cross the membrane because they regulate which types and how many membrane proteins will be integrated. Cell membrane proteins are very dynamic and change rapidly, depending on the environmental conditions surrounding the cells. For example, the hormone insulin stimulates the transcription and translation of membrane carrier proteins that allow glucose to be transported across membranes.

Transport of Large Molecules

Some molecules must enter or exit the cell, but they are too big to move through membrane proteins (Figure 3.25). For example, the insulin produced by our pancreatic cells and made up of 51 amino acids is transported from the ER to the Golgi and then from the Golgi to the plasma membrane in membrane-bound vesicles. These insulin-carrying vesicles merge with the plasma membrane because the membrane composition

Figure 3.23 Proton Pump

The proton pump is an example of active transport. Proton pumps are membrane proteins located in the inner mitochondrial membrane and are important for the production of ATP.

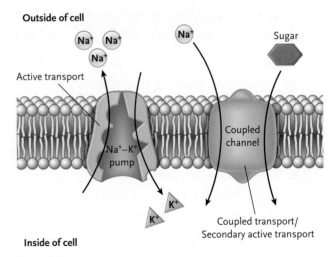

Figure 3.24 Coupled Transport

Coupled transport involves the simultaneous movement of two molecules through the same membrane protein. Two molecules moving in the same direction is called *symport*, and two molecules moving in opposite directions is called *antiport*.

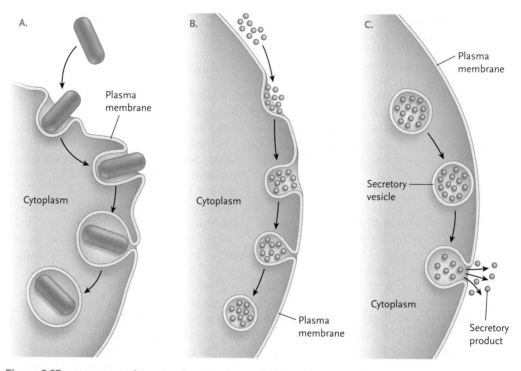

Figure 3.25 Movement of Large Molecules across a Cell Membrane

Phagocytosis (A), pinocytosis (B), and exocytosis (C) are processes that allow the movement of large molecules or large amounts of fluid to enter or leave the cell.

of the vesicle and the cell is the same, and the large insulin proteins are then released from the cell and transported by the bloodstream to various tissues. The release of large molecules from vesicles across the cell membrane is called **exocytosis.** Movement of large molecules into the cell is called **endocytosis,** and this can involve **pinocytosis**—the movement of fluids—or **phagocytosis**—the movement of particles. When a large particle enters a cell, it is engulfed by a portion of the plasma membrane. This process forms a vesicle that

usually merges with lysosomes, which break down the contents of the vesicle. For example, phagocytosis occurs when white blood cells such as **macrophages** engulf bacteria and viruses and any other dead cells in order to break down and destroy such invading micro-organisms (Chapter 23). Pinocytosis involves the non-specific engulfing of large amounts of **extracellular fluid** brought into the cell by the vesicle. Most of the content of these vesicles is already broken down or consists of small nutrients that can be immediately used by the cell.

CONCEPT REVIEW 3.6

1. What does *semi-permeable* mean?
2. Explain how diffusion works. What kinds of molecules can cross the membrane easily?
3. What kinds of molecules require a membrane protein to cross into or out of a cell?
4. What is osmosis?
5. What is solute and solvent?
6. Explain why water would move into or out of a cell.
7. What are hypertonic, hypotonic, and isotonic solutions? What happens to cells if they are placed in each of those solutions?
8. What is facilitated diffusion? Explain the difference between channels and carrier proteins.
9. What does it mean when carrier proteins become saturated?

10. What is active transport? Explain how it differs from facilitated diffusion.
11. Give two examples of active transport in our cells.
12. Explain the steps involved in moving sodium out of cells and potassium into cells. Why is ATP required?
13. What are the two main reasons why cells need Na^+/K^+ pumps? Which cells have Na^+/K^+ pumps?
14. Where are proton pumps located in cells? What is their purpose? What is chemiosmosis?
15. Explain how sugars and amino acids move into the cell using coupled transport.
16. Explain the difference between endocytosis (two types) and exocytosis.
17. What is receptor-mediated endocytosis?

EFFECTS OF AGING

Different cells in the body have different lifespans. Skin cells live for a few weeks, red blood cells live for four months, bone cells can live for ten years, and heart muscle cells and neurons can live for our entire life.

1. One reason cells age is because of the length of the **telomeres** found on the ends of every chromosome in every cell. As cells go through the cell cycle, the telomeres shorten (Chapter 9), and cells that have chromosomes with shorter telomeres will die sooner than cells that have chromosomes with longer telomeres. It is normal for cells to gradually slow their cell cycle and reach cellular **senescence**, a dormant period where they no longer go through mitosis; this is beneficial because it prevents the formation of cancer cells. Senescent cells have decreased cellular functions. Cells can also undergo a process called **apoptosis**, programed cell death, as a result of excessive DNA damage. If cells accumulate DNA damage and do not undergo apoptosis or become senescent, then cancer cells are produced (Chapter 5). DNA damage occurs in cells that are exposed to toxins (see Chapter 5, Section 5.4 for list) and increases the rate of cell aging.

2. *Advanced glycation end products* (AGEs) are glycated proteins or fats, such as those in cell membranes. **Glycated** molecules have sugars attached to them. Glycated membrane lipids and membrane proteins cannot function normally, which leads to decreased function of the entire cell and plays a significant role in cellular aging and many degenerative diseases, such as heart disease, type 2 diabetes, and Alzheimer's disease. Glycation occurs simply from eating too much sugar. Glycated cell membranes (Figure 3.26) cannot properly cross-link collagen in the extracellular matrix and contributes to tissue stiffening and loss of flexibility and elasticity. This is very noticeable in aging skin cells, and also occurs in the epithelial cells of blood vessels, which contributes to "hardening of the arteries." It is possible to measure the amount of glycation in your body using a blood test called *HbA1C*, which measures the amount of glycation on your hemoglobin molecules. An HbA1C value higher than 5.7% indicates prediabetes, a value of 6.5% or higher indicates diabetes.

3. Another major factor involved in cellular aging is the mitochondria—the organelles of the cell that make energy in the form of ATP. A decline in mitochondrial function has been associated with many age-related diseases, and mitochondria regulate a number of aspects of cellular aging. The majority of research studies have shown that cellular aging is increased when there is a decrease in mitochondrial enzymes, a decrease in the ability to make ATP, an increase in reactive oxygen species (free radicals), and a decrease in creatine-phosphate recovery time (Chapter 14). Most research has focused on skeletal muscle cells because they require the most ATP production; however, recall that every cell in the body has mitochondria and requires energy for all cellular functions. To support mitochondria and slow cellular aging, we need to force mitochondria to make energy and to replicate themselves, and this happens in all cells when we increase our metabolic rate, which we can do only by exercising. Mitochondria also require many B vitamins and other

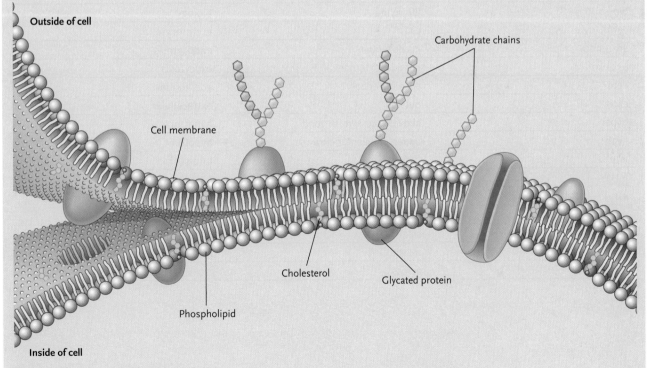

Outside of cell

Carbohydrate chains

Cell membrane

Cholesterol

Glycated protein

Phospholipid

Inside of cell

Figure 3.26 Glycated Membrane Proteins

(Continued)

nutrients for making ATP; therefore, it is important to eat foods that supply the cell with the required building blocks (amino acids, glucose, fatty acids) and cofactors (vitamins and minerals) (Chapter 2).

4. Free radicals are molecules that have an unpaired electron, making them highly reactive. Free radicals include molecules such as reactive oxygen species (ROSs) and hydrogen peroxide, which are produced by many cell types: specifically, by the immune system during chronic inflammation, such as atherosclerosis, autoimmune diseases, Alzheimer's, and cancer. Antioxidants, molecules found in plant foods, act as reducing agents that minimize the damage that can be caused by free radicals.

Aging of individual cells relates to aging of the entire organism. Although every organism has a set lifespan, most humans die long before this point due to accelerated cellular aging processes. The maximum human life expectancy is estimated to be approximately 115 to 120 years old. Many studies show that people continue to lose muscle mass and therefore strength if they don't exercise. A normal decline for a healthy person is approximately 1% per year; however, the average 70-year-old today is declining at a rate of 4% or more.

We can decrease the rate at which our cells age by taking the following action:

1. Decrease exposure to toxins, such as those acquired from smoking, dioxins in herbicides and pesticides, radiation, and common chemicals.
2. Engage in exercise, which increases the function of mitochondria in every cell, not just muscle cells. The benefits of exercise include an increase in the number of mitochondria; improvement in mitochondrial function in neurons of the brain, including significantly increasing cognitive function and mood; and an increase in the production of growth hormone, which helps cells to undergo mitosis and replicate organelles.
3. Consume foods that contain the nutrients required for cellular functions (Chapter 2), including antioxidants, which reduce free radicals and are found in fruits and vegetables.
4. Decrease consumption of sugar to decrease the production of AGEs.

CHAPTER REVIEW

Medical Terminology

Term	Definition	Example
bio-, bi/o	life, living	biology
cyt/o, -cyte	cell	erythrocyte
en-, endo-	inside, within	endocytosis
ex-, exo-	outside, external	exocytosis
equi-	equality, equal	equilibrium
extra-	outside	extracellular
hyper-	above, excessive, beyond	hypertonic
hypo-	under, deficient, below	hypotonic
inter-	between	interstitial
intra-	within	intracellular
kary/o	nucleus	eukaryote
-logy	study of	biology
micro-	small	microscopic
nucle/o	nucleus	nucleoplasm
trans-	across	transmembrane

Labelling Activity

Label the following diagrams.

1.

2.

Answer to Application Question

Hypertonic solution. This solution contains a higher concentration of sugar and electrolytes than body fluids. Water is attracted to any solutes that are charged or polar (as are sugar and electrolytes). A hypertonic solution in the digestive tract causes water to be pulled from the bloodstream into the digestive tract and actually makes you to feel thirstier because water is not moving quickly into the bloodstream. Over time the body can absorb the hypertonic solution, but initially it can reduce hydration. A hypertonic solution is not the most effective means for rehydration. As an extreme example, ocean water has approximately three times the osmolarity of our body fluids. Drinking ocean water causes such a dramatic movement of water out of the bloodstream and into the GI tract that vomiting occurs, making the person more dehydrated than before drinking the ocean water.

Isotonic solution. This solution contains a concentration of sugar and electrolytes that is equivalent to body fluids. An isotonic solution absorbs very easily into the bloodstream. As the electrolytes and sugar are absorbed by facilitated transport mechanisms, water quickly follows. An isotonic solution is the most effective solution for rehydration for middle- and long-distance athletes because it contains sugar and the minerals that need to be replaced during exercise, and it also promotes osmosis of water from the GI tract into the bloodstream.

Hypotonic solution. Drinking plain water would be hypotonic to our body fluids. We drink plain water with no problem all the time. Water is easily absorbed from the intestinal tract into the bloodstream and can quickly bring about rehydration. This is very effective for quickly replacing water and is beneficial for short-duration physical activities, such a weight lifting. Plain water is not be beneficial for a middle- or long-distance athlete who is sweating and thereby losing electrolytes. It is important to note that drinking an excessive amount of plain water can be dangerous because it dilutes the body's electrolytes. We need minerals such as sodium, potassium, calcium, and magnesium for the proper functioning of our nervous system and muscles. As an extreme example, if a person were to receive a hypotonic IV solution, excess water would move into the red blood cells, making them burst; this would be fatal.

4 Cellular Metabolism

CHAPTER OUTLINE

4.1 OVERVIEW OF CELLULAR RESPIRATION

Cellular Energy

Organisms from every kingdom use adenosine triphosphate (ATP) as an energy supply for thousands of chemical reactions that occur in every cell. ATP consists of an adenine (one of the four nitrogenous bases), a ribose sugar, and three phosphates (Figure 4.1). Energy is transferred from the ATP when the last phosphate bond is broken, and the phosphate (inorganic phosphate, Pi) binds to another molecule; this converts the ATP into ADP (adenosine diphosphate): ATP → ADP + Pi.

When the phosphate from ATP binds to a protein, such as the sodium-potassium pump membrane protein (Chapter 3), the protein changes its conformational shape, which then aids a cellular process, such as moving sodium ions out of the cell.

Cellular Production of ATP

Our cells make ATP from the foods we eat. Recall from Chapter 2 that we eat macromolecules: proteins, complex carbohydrates (polysaccharides), and fats (lipids). When we digest these foods, they are broken down into

Education is not the preparation for life; education is life itself.

John Dewey

Figure 4.1 Adenosine Triphosphate (ATP)

Adenosine triphosphate (ATP) is the energy molecule used by cells.

temperature of 1 mL of water 1°C. This is a very small amount, so kilocalorie (Calorie, Cal) is used as the unit of measure. Depending on their age, body size, genetic factors, fitness level, and amount of exercise, adults require anywhere from 1500 to 3000 Cal per day to maintain their body weight (fewer calories are required as we get older).

DID YOU KNOW?

Consuming 3500 extra kilocalories can lead to a weight gain of 0.45 kg (1 lb, same as a block of butter). If you add one can of regular cola per day (130 Cal per can) to your diet without changing your food intake or amount of exercise, you will gain 0.45 kg of body weight after approximately 23 days.

their building blocks—amino acids, monosaccharides (sugar), and glycerol and fatty acids—and then taken up by our cells. **Food energy** (such as glucose) is converted into **chemical energy** (ATP) through the process of cellular respiration. Cells use ATP to perform thousands of functions every second. Excess food energy that is not used for ATP production becomes stored as triglycerides (fat) in the liver or adipose tissue until needed. Therefore, to maintain our body weight, we need to eat the same amount of food energy that our cells use in chemical energy.

DID YOU KNOW?

- Carbohydrates contain 4 Calories (Cal) per gram; 1 tsp of sugar is 4 g and therefore has 16 Cal.
- Protein contains 4 Cal per gram.
- Fats contain 9 Cal per gram; 1 tsp butter is approximately 4 g and therefore has 36 Cal.
- Alcohol contains 7 Cal per gram.
  - 1 beer (12 oz) is approximately 150 Cal and contains alcohol, carbohydrates, and a small amount of protein.
  - 1 glass of wine (6 oz) is approximately 120 Cal and also contains carbohydrates and some protein.
  - 1 shot (1 oz, which is 28 g but contains water and is not 100% alcohol) is approximately 65 Cal.

Food Energy

Nutritionists measure the amount of energy in the food we eat in terms of **Calories** (kilocalories). A

Conversion of Food Energy into ATP Molecules

Eukaryotic cells primarily use glucose to produce ATP; however, our cells can also use fatty acids, amino acids, and nucleotides (Section 4.5). Through a series of **oxidation** and **reduction** reactions the glucose is eventually completely broken down; the hydrogens move from the glucose to a coenzyme, leading to the production of ATP. The carbon-hydrogen bonds in methane are the source of energy in natural gas, just as the carbon-hydrogen bonds in glucose are used in our cells (Figure 4.2). In oxidation-reduction reactions, electrons become transferred from the nutrient

Methane

Glucose

Figure 4.2 Methane and Glucose

Both methane and glucose can be used to make energy because they both have carbon-hydrogen bonds.

being broken down, such as glucose, to the coenzyme NAD^+ or FAD^+; these are then **reduced** to produce NADH or $FADH_2$. Oxidation and reduction reactions always go together, and are referred to as **redox reactions** (Figure 4.3). You can remember oxidation and reduction using the phrase "LEO the lion says GER": Loss of Electrons is Oxidation; Gain of Electrons is Reduction.

The process of oxidizing food molecules (glucose, amino acids, fatty acids, glycerol, and nucleotides) and reducing NAD^+ ions leads to the production of ATP; this process is called **cellular respiration**. The oxidation of glucose in the presence of oxygen is called *oxidative phosphorylation* or **aerobic respiration**. This is why we need to breathe oxygen. When the level of oxygen in a cell is temporarily low, ATP is produced through **anaerobic respiration**: for example, when muscle cells have low oxygen levels during strenuous exercise such as sprinting (Section 4.6).

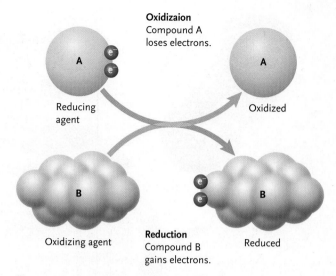

Figure 4.3 Oxidation-Reduction (Redox) Reaction

An oxidation-reduction (redox) reaction involves the transfer of electrons from one molecule to another. Losing electrons is oxidation, and gaining electrons is reduction.

CONCEPT REVIEW 4.1

1. What molecule is used as energy in our cells?
2. How does that molecule give us energy?
3. What nutrients are used in our cells to produce energy?
4. What do *oxidation* and *reduction* mean?
5. Why are redox reactions important for cellular respiration?

6. What is a Calorie?
7. Why are the carbon-hydrogen bonds in food molecules important?
8. What is the difference between aerobic and anaerobic respiration?

4.2 GLYCOLYSIS

Stages of Aerobic Cellular Respiration

1. Glycolysis occurs in the cytoplasm.
2. Pyruvate oxidation occurs in the matrix of the mitochondria.
3. Krebs cycle occurs in the matrix of the mitochondria.
4. Electron transport proteins are found in the inner membrane of the mitochondria.

Aerobic respiration occurs in four main stages, and each stage occurs in different areas of the cell (Figure 4.4). During the four stages, one glucose molecule becomes entirely oxidized (broken down) and yields many ATP molecules and heat. The reaction equation is as follows:

$$C_6H_{12}O_6 + 6O_2 \rightarrow 6CO_2 + 6H_2O + \text{Energy (ATP and heat)}$$

Oxidization of Glucose in the Cytoplasm

Glycolysis is the process of oxidizing glucose to produce two molecules of pyruvate (Figure 4.5). Glycolysis occurs in the cytoplasm of all cells in the body in all living organisms; it is the oldest evolutionary process used to make ATP. Glycolysis—a 10-step process of chemical reactions—involves the oxidation of glucose (six carbon atoms) and the reduction of NAD^+, which results in the production of two pyruvate molecules (three carbon atoms each). Once glucose enters the cell (see coupled transport, Chapter 2), it is phosphorylated with two phosphates from two ATP molecules; therefore, two ATP molecules are required to start glycolysis.

Once glucose is phosphorylated, oxidation-reduction (redox) reactions produce two three-carbon pyruvate molecules. During this process, two NAD^+ molecules are reduced to NADH, and four ATP molecules are formed

Figure 4.4 Overview of Aerobic Cellular Respiration

Cellular respiration involves the following stages: glycolysis, pyruvate oxidation, Krebs cycle, and the electron transport chain.

Figure 4.5 Glycolysis

Glycolysis involves the oxidation of one molecule of glucose to form two molecules of pyruvate, two NADH molecules, and a net amount of two ATP molecules.

(a process called **substrate-level phosphorylation**). Since two ATP molecules are required to begin the process, a *net* amount of two ATP molecules are produced during glycolysis. The NADH molecules are used by the electron transport chain to produce many more ATP molecules.

4.3 PYRUVATE OXIDATION AND KREBS CYCLE

Pyruvate Oxidation

The end result of glycolysis is two pyruvate molecules, each containing three carbons. These pyruvate molecules can diffuse through the mitochondrial membrane into the matrix, where further oxidation can occur. Pyruvate oxidation results in electrons and hydrogen ions moving from pyruvate to the electron acceptor NAD$^+$ to produce more NADH molecules (Figure 4.6). Also, during this process, a carbon atom is removed from each pyruvate, which leaves a two-carbon molecule (an acetyl group) that combines with

coenzyme A. The final product is a molecule called *acetyl-Coenzyme A* (acetyl-CoA). The enzyme that removes the carbon atom from pyruvate is called **pyruvate dehydrogenase**. The carbon atom is removed as a carbon dioxide (CO_2) molecule. The process of removing this carbon as CO_2 is called **decarboxylation**. This explains why we exhale carbon dioxide.

Cellular ATP Capacity

Depending on what cell functions are occurring at a particular time, cells require more or less ATP. During exercise, for example, muscle cells require very high levels of ATP for muscle contraction; however, during periods of rest, those cells do not need to make as much ATP. All cells use ATP for many chemical reactions, including moving substances into or out of the cell, producing proteins, transporting substances within the cell, producing new organelles, breaking down and recycling old organelles, replicating DNA, and facilitating mitosis. Muscle cells require the most ATP for contraction, which is why exercise is the primary way to burn calories. Cells store extra nutrients as fat that can be broken down later when more ATP is needed. When fats are used as energy, they do not go through glycolysis, they go through a process called *beta-oxidation*. Because they are long carbon-hydrogen chains, fats can be converted into two-carbon acetyl groups (Section 4.5). Acetyl-CoA, formed in combination with coenzyme A, enters the Krebs cycle (the next stage of the cellular respiration process).

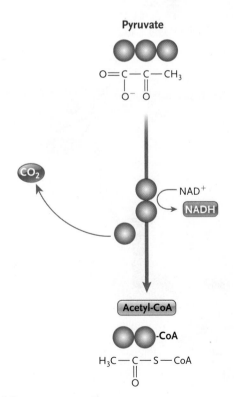

Figure 4.6 Pyruvate Oxidation

Pyruvate oxidation involves the oxidation of pyruvate to form acetyl-CoA.

Application

As you read through the rest of this chapter, consider this scenario. Fred is trying to lose 9 kg (about 20 lb). He currently does not exercise. He is a student that spends many hours a day sitting in class or studying at a computer. He drinks three cans of Coke every day, but otherwise generally eats healthy food. Considering that consuming more calories than are required for energy production results in increased storage of triglycerides, and several hormones are involved in the regulation of metabolism, what advice can you give Fred? If he stopped drinking Coke containing 140 Cal per can, how long would it take Fred to lose 9 kg (20 lb) if he didn't change anything else? What other advice can you give Fred to help him lose weight faster? What vitamins are necessary for the production of ATP?

Krebs Cycle

The Krebs cycle involves a series of nine reactions.

1. Acetyl-CoA enters the cycle and binds to a four-carbon molecule, oxaloacetate, forming a six-carbon molecule, citrate, also called *citric acid*. Because of the formation of this molecule, the Krebs cycle is also called the *citric acid cycle* (Figure 4.7).

2. The two-carbon molecules are removed as carbon dioxide. Recall from the initial equation that six CO_2 molecules are formed for each glucose molecule that is oxidized: two are formed when pyruvate is oxidized, and the other four are formed at this point during the Krebs cycle. During the process of decarboxylation (removal of carbon to produce CO_2), electrons and hydrogen ions are transferred to NAD^+, forming NADH and

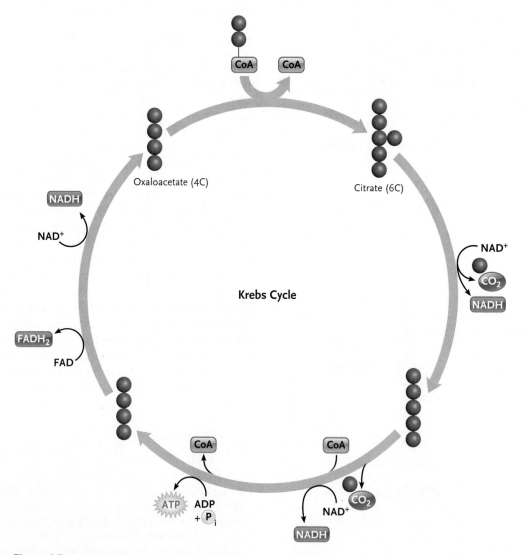

Figure 4.7 Krebs Cycle

The Krebs cycle involves the production of one FADH2 and three NADH molecules for each acetyl-CoA that enters the Krebs cycle. These molecules carry the hydrogens that enter the electron transport chain.

FADH$_2$. Since two carbons have been removed as CO$_2$, in the third step one four-carbon molecule remains: oxaloacetate. This is the initial starting material, which is regenerated and ready for the next cycle.

3. For each acetyl-CoA that enters the Krebs cycle, one molecule of ATP is produced. Since two molecules of acetyl-CoA enter Krebs, two molecules of ATP are produced.

The most important result of the Krebs cycle is the production of the six NADH and two FADH$_2$ molecules. These reduced coenzymes are important because they carry the electrons and hydrogen ions that are required for the electron transport chain to produce a significant amount of ATP.

CONCEPT REVIEW 4.3

1. Where do pyruvate oxidation and the Krebs cycle occur?
2. What is the end product of pyruvate oxidation?
3. What waste product is formed during pyruvate oxidation and the Krebs cycle?
4. What is the important result of the Krebs cycle?
5. How much ATP is formed during the Krebs cycle?
6. Why are NADH and FADH$_2$ molecules so important?

4.4 ELECTRON TRANSPORT CHAIN AND CHEMIOSMOSIS

Importance of NADH and FADH$_2$

The coenzymes NAD$^+$ and FAD—reduced to form NADH and FADH$_2$ during the processes of glycolysis, pyruvate oxidation, and Krebs cycle—provide the necessary electrons and hydrogen ions (protons) for the electron transport chain. Recall from Chapter 3 that proton pumps are membrane proteins that move hydrogen ions across a membrane. The proton pump membrane proteins are located in the cristae—the inner layer of the mitochondrial membrane. These hydrogen ions move across the inner mitochondrial membrane to form a H$^+$ gradient in the **intermembrane space** (Figure 4.8). It is important to have a high concentration of hydrogen ions in the intermembrane space because the concentration difference is used as energy for the production of ATP (secondary active transport). Because molecules always go toward equilibrium (Chapter 3), the high concentration of H$^+$ in the intermembrane space makes the H$^+$ ions "want" to move down the concentration gradient and back into the matrix of the mitochondria. The only way these H$^+$ can move back into the matrix is through a specific membrane protein called **ATP synthase**. Chemiosmosis is the term used to describe the process of using a chemical (H$^+$) gradient for moving across the membrane, with the result of producing many ATP molecules. The enzyme activity of ATP synthase phosphorylates ADP to form ATP.

Electron Movement through Membrane Proteins

When NADH and FADH$_2$ split so that the H$^+$ can be pumped into the intermembrane space, the electrons are transferred through the membrane proteins (hence the term, **electron transport chain**). Recall from Chapter 3 that active transport requires energy to move molecules toward an area of higher concentration. The movement of H$^+$ ions into the intermembrane space is **active transport**, and the energy to move the H$^+$ ions comes from the transfer of electrons through the electron transport proteins (Figure 4.8).

Electrons at the End of the Electron Transport Chain

After being transported through the membrane proteins, the hydrogen ions and the electrons need to combine with something—oxygen. If they didn't, they would build up in the matrix of the mitochondria, the electrons would react with other molecules, and the protons would produce acidic conditions. The final electron acceptor in our cells is oxygen. This is why we need to breathe. The electrons and the protons (H$^+$) that have moved through the electron transport chain and ATP synthase combine with oxygen to form water molecules (see Figure 4.9).

DID YOU KNOW?

People die in fires primarily due to inhaling toxic chemicals produced from burning vinyl, fabrics, and plastics. One such chemical is cyanide—fatal within minutes if inhaled—because it inhibits the enzymes involved in the production of ATP.

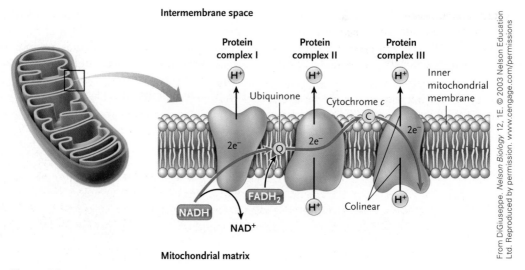

Figure 4.8 The Electron Transport Chain

The electron transport chain contains the proton pumps that create the proton gradient in the intermembrane space. This hydrogen ion gradient provides the energy for the production of ATP through ATP synthase.

Figure 4.9 Overview of Electron Transport Chain

CONCEPT REVIEW 4.4

1. What is the purpose of the NADH and FADH$_2$ molecules that are created during glycolysis, pyruvate oxidation, and the Krebs cycle?

2. Where are the H$^+$ ions transported to? Why?

3. What do the proton pumps use as energy for actively transporting the H$^+$ ions?

4. What happens to the electrons after they move through the electron transport chain?

5. What is the function of ATP synthase?

6. Define chemiosmosis.

4.5 OTHER NUTRIENTS THAT PRODUCE ATP

Our cells primarily use glucose as the starting material for ATP production, but most cell types can also produce ATP from fatty acids, glycerol, amino acids, and nucleotides (Figure 4.10). Glucose is the only nutrient that undergoes glycolysis to produce pyruvate. Other nutrients enter the process at various stages.

When eaten, proteins are digested into their building blocks: amino acids. Amino acids can be used to produce NADH and FADH$_2$ molecules through oxidation, which leads to ATP production. Amino acids contain **nitrogen groups** (amino groups, see Chapter 2), that have to be removed through a process called **deamination**. The nitrogen groups are removed as **ammonia (NH$_3$)**, which is then converted

to **urea** and excreted by the kidneys. The deaminated amino acids can then be converted into pyruvate or acetyl-CoA, which will become oxidized (as described in Section 4.3).

DID YOU KNOW?

Our brain cells Our brain primarily uses glucose to produce to produce energy; for this reason, it is so important to maintain stable blood sugar levels. If blood sugar levels fluctuate significantly, brain functions are affected more significantly than the functions of other cell types and can cause a myriad of symptoms, including fatigue, depressed mood, irritability, decreased ability to concentrate, sugar cravings, nervousness, shakiness, dizziness, or nausea. In extreme conditions, such as type I diabetes, if blood sugar is altered dramatically, coma and death can occur.

Figure 4.10 Overview of the Breakdown of Molecules to Produce ATP

Nucleic acids, found in all foods made of cells—vegetables and meats—are digested into their building blocks, nucleotides, which can also be used to produce ATP. Nucleotides also contain nitrogen, and since nitrogen is not used to make energy, it must be removed. Nitrogen removed from pyrimidine nucleotides is converted into urea, and nitrogen from purines is converted into uric acid; both are excreted in the urine by the kidneys.

Fats stored in adipose tissue, muscles, and the liver are triglycerides and can be used to make large amounts of ATP. Lipase is the cell enzyme that breaks the bonds between the fatty acids and the glycerol; this involves a hydrolysis reaction called **lipolysis**. Fats in adipose tissue are carried through the bloodstream by high density lipoproteins (HDLs) to cells—mostly the muscle cells—that need fat for energy (Chapter 18). A high level of blood HDLs is associated with a decreased risk of heart disease because it is an indication that the body is breaking down stored fat. Fatty acids can also be carried to the muscle cells by the plasma protein albumin. Glycerol can be converted into pyruvate.

Since fatty acids are long chains of carbon-hydrogen bonds, they can be oxidized into two-carbon acetyl-CoA molecules and then enter the Krebs cycle. This process is called **beta-oxidation** (Figure 4.11). Recall that fatty acid chains can be formed from acetyl-CoA if ATP levels are sufficient in the cell and if excess nutrients are present.

Fatty acids

Figure 4.11 Beta-Oxidation

The catabolic process of breaking fatty acid chains into two-carbon compounds that can bind CoA, forming acetyl-CoA that enters the Krebs cycle

DID YOU KNOW?

Healthy body fat percentages for males are quite different from females, as shown in this chart based on data from the American Council on Exercise.

Classification	Men	Women
Essential fat	2–5%	10–13%
Athletic	6–13%	14–20%
Fit	14–17%	21–24%
Average	18–24%	25–31%
Obese	25%	32%

CONCEPT REVIEW 4.5

1. What molecules can be used by cells to produce ATP?
2. Explain how each macromolecule is broken down into building blocks, and describe how they enter the process differently from glucose.
3. What is deamination?
4. What waste products are produced during cellular respiration?
5. What enzyme is required for lipolysis?
6. What is beta-oxidation?

4.6 ANAEROBIC RESPIRATION

ATP Production without Oxygen

In some situations, certain body cells may temporarily lack oxygen: for example, during sprinting when muscle contraction requires bursts of energy. The muscle cells quickly deplete oxygen during strenuous, high-intensity activity, and the cell relies only on glycolysis for ATP production. Recall that a small amount of ATP is produced through substrate-level phosphorylation in the cytoplasm during glycolysis. When oxygen is absent, ATP is produced through anaerobic respiration. The small amount of ATP produced means that fatigue happens very rapidly. Through glycolysis, it is possible to produce enough ATP for 95 to 100% effort for 20 to 30 seconds before fatigue prevents further muscle contraction. Endurance

Lactic acid fermentation in muscle cells

Alcohol fermentation in yeast

Figure 4.12 Anaerobic Respiration

In humans, anaerobic respiration leads to the formation of lactic acid. In yeast, ethanol is produced.

activities such as hiking use aerobic respiration, and approximately 60 to 70% of maximum effort can last for more than an hour (depending on fitness level) before muscle fatigue occurs.

Since oxygen is required as the final electron acceptor in the electron transport chain, when oxygen is absent the electron transport process cannot occur. Therefore, the pyruvate produced during glycolysis is not oxidized to produce acetyl-CoA; instead, the three-carbon pyruvate is converted into a three-carbon **lactate**, also called **lactic acid**. You can feel the burning sensation of lactic acid in your muscles during high-intensity anaerobic exercise. The NADH molecules formed in this process must be converted back into NAD$^+$ because the H$^+$ is not used to build a proton-concentration gradient. The process of moving the H$^+$ from NADH to pyruvate to form lactate is called **fermentation**. Fermentation in humans can be summarized by the following equation:

pyruvate + NADH → lactate + NAD$^+$

Once oxygen is again available, the lactate can be converted back to pyruvate, and aerobic respiration can resume.

Anaerobic Respiration in Yeast

In yeast, the three-carbon pyruvate goes through a decarboxylation step, where a carbon is removed as carbon dioxide, producing **acetaldehyde**. Then the H$^+$ from NADH is transferred to acetaldehyde, forming a two-carbon **ethanol** molecule. Ethanol is the alcohol molecule found in wine and beer (Figure 4.12). The production of CO_2 by yeast makes it a useful organism for baking bread; the CO_2 makes the bread fluffy, and the ethanol evaporates during baking (Figure 4.13).

Figure 4.13 Examples of Fermentation

1. Describe how anaerobic respiration is similar or different in humans and in yeast.
2. When does anaerobic respiration occur in human cells?
3. What is fermentation?
4. What happens to lactate formed in muscle cells?

4.7 REGULATION OF METABOLISM

Since muscle cells require more ATP than any other cells in the body, we can stimulate the breakdown of stored fat with exercise. Slow-twitch muscle fibres preferentially break down fatty acids for energy during aerobic, endurance exercise (Chapter 14). Fast-twitch fibres preferentially use glucose (and break down glycogen) for energy during anaerobic, high-intensity exercise; however, this type of exercise still stimulates fat breakdown because the liver converts fat into glucose to supply these cells (gluconeogenesis). Therefore, any type of exercise will stimulate fat breakdown in some way. The longer the duration or the higher the intensity, the more total calories will be burned.

The following hormones affect fat storage or promote fat breakdown (see also Chapter 17):

Insulin—signals cells to take up blood glucose and convert it into fat or glycogen for storage. Eating a meal causes blood sugar to increase, which stimulates the production of insulin, which causes blood sugar levels to decrease.

Glucagon—signals the liver to break down glycogen and fat to produce glucose. This process, called *gluconeogenesis*, occurs during periods of fasting or exercise and causes blood sugar levels to increase.

Epinephrine, norepinephrine, and cortisol—is produced during a stress response, fasting, and during exercise to stimulate the breakdown of glycogen and fat to increase blood sugar levels. Cortisol also causes the up-regulation of receptors that bind epinephrine and norepinephrine and therefore act synergistically during a stress response.

Growth hormone—is produced during deep sleep, periods of fasting, and exercise to stimulate the breakdown of fat and increases blood sugar.

DHEA (testosterone-like hormones)—are produced by the adrenal cortex during stress and exercise to stimulate the breakdown of fat to increase blood sugar. DHEA also stimulates the production of growth hormone, which is required for protein synthesis during and after exercise to repair muscle cells.

Thyroid hormones (T3 and T4)—bind to receptors in the cell nucleus and regulate the expression of genes that promote cellular respiration and utilization of oxygen. Thyroid hormones are required for the metabolism of carbohydrates, fats, and protein.

1. Which hormone causes cells to convert nutrients into fat?
2. Which hormones cause cells to break down fat?
3. Name the three main stimuli that increase the production of the hormones that break down fat and increase blood sugar.

CHAPTER REVIEW

Labelling Activity

Label the following diagram.

1.

Macromolecules   Nucleic acids   Proteins   Polysaccharides   Lipids

Answer to Application Question

If Fred eliminated Coke (140 Cal per can) entirely from his diet, the reduction in calories consumed would be 420 Cal per day. There are 3500 Cal in 0.45 kg (1 lb) of fat, so removing 420 Cal per day would cause Fred to lose 0.45 kg (1 lb) in 8.3 days. If this was the only change he made, he would lose 9 kg (20 lb) in 166 days (approximately 5 months). When we eat sugar, we produce insulin, which tells our body to store fat. So, if Fred cuts out Coke, this will help decrease his insulin level. Cortisol is a stress hormone that causes the body to store belly fat. Another piece of advice for Fred would be for him to deal with stress: for example, to become organized and not get overwhelmed with school work. Lastly, if Fred also added exercise to his daily routine, he could burn more calories. It is possible to lose 0.45 kg (1 lb) per week by creating a deficit of 3500 Cal per week (500 Cal per day) by consuming 250 Cal less than usual and exercising to burn an extra 250 Cal. Exercise also helps decrease the effects of the cortisol. Vitamins that are important for the production of ATP are the B vitamins.

5 Cellular Reproduction

CHAPTER OUTLINE

*The only person who is edu-
cated is the one who has
learned how to learn and
change.*

Carl Rogers

5.1 PROKARYOTIC REPRODUCTION

Bacteria (prokaryotes) replicate asexually, producing identical copies of themselves. Bacteria have one circular DNA molecule and sometimes a small extra piece of DNA called a *plasmid* (Chapters 10 and 24) (Figure 5.1). The bacterial chromosome may have millions of nucleotides: for example, *E. coli* bacteria have 4.6 million nucleotides on their chromosome. A plasmid is much smaller than a bacterial chromosome and may contain approximately 2000 to 3000 nucleotides. Often the plasmid contains antibiotic resistance genes (Chapter 24).

Prokaryotic cell division involves two steps: **DNA replication** and **binary fission.** This is the simplest type of cell division because prokaryotes have only one main DNA molecule to replicate, and there are no membrane-bound organelles to copy and separate into two new cells. Eukaryotic cells have a much more complex cell division process due to the presence of multiple chromosomes and organelles.

Prokaryotic DNA replication begins at a single point of origin where enzymes begin to separate the strands of DNA—hydrogen bonds are broken between bases (Chapter 2). Once the DNA strands begin to separate, **DNA polymerase** proceeds in both directions in a process that adds new complementary nucleotides to each strand until two new molecules of DNA are formed (see also Chapter 8). After the DNA is copied, the new chromosomes move to each end of the cell, the cytoplasm separates,

Bacterial cell

Figure 5.1 Structures in a Typical Bacterial Cell

the membrane pinches inward through binary fission, and two new identical cells are formed (Figure 5.2).

Genetic variation is extremely important and necessary for the survival of every population of organisms (Chapters 1 and 7), so every living thing must have a method for acquiring new DNA. Humans acquire half of their DNA from each parent during fertilization (meiosis, Section 5.5). Bacteria acquire genetic variation through the following mechanisms of **DNA transfer:**

- Conjugation—occurs when bacteria directly exchange pieces of DNA with another bacterium through a pilus that is usually a plasmid, but may also be a piece of their chromosomal DNA.

- Transduction—occurs when bacteria acquire new DNA fragments from viruses.

- Transformation—occurs when bacteria acquire fragments of DNA or plasmids that happen to be near them.

For example, many different species of bacteria live in our digestive tract, and they can take up DNA from each other: from DNA that might be in our food, from other bacteria that have died and left DNA behind, or from viruses that infect bacteria. So although bacterial reproduction (binary fission) is **asexual**, bacteria have extensive resources for constantly increasing their genetic variation (Figure 5.3). Bacterial genetic variation is discussed further in Chapters 7, 10, and 24.

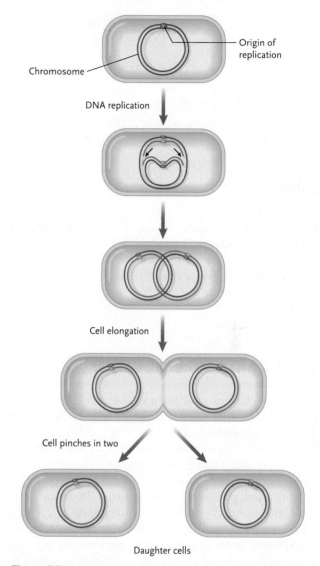

Figure 5.2 Binary Fission

Binary fission, the process of bacterial cell division, involves replication of the DNA and the splitting of the cytoplasm to form two new identical cells.

Figure 5.3 Conjugation, Transduction, and Transformation

Conjugation (A), transduction (B), and transformation (C) are the main mechanisms that allow bacteria to acquire new DNA.

CONCEPT REVIEW 5.1

1. Describe the two main components of bacterial replication.

2. What is a plasmid?

3. Explain the steps in bacterial DNA replication.

4. How do bacteria acquire genetic variation if they replicate asexually?

5. Explain the difference between conjugation, transduction, and transformation.

Application

As you read through the rest of this chapter, consider this scenario. The yellow fever (*Aedes aegypti*) mosquito has a diploid number of 6. Draw one of its cells during metaphase. How many chromosomes are in the cell after S phase? How many chromosomes will be in each new daughter cell? How many chromosomes will be in its gametes after meiosis? If a DNA mutation occurred during replication, at what checkpoint would it be discovered in a healthy organism?

5.2 EUKARYOTIC CELL CYCLE

All cells, whether prokaryotic or eukaryotic, have a cell cycle that involves a growth phase, followed by DNA replication, followed by cell division. Eukaryotic cells contain more DNA than prokaryotic cells, and their DNA must be condensed into transportable chromosomes for the more complex cell division process called **mitosis**. In eukaryotic cells, the cell cycle consists of **interphase** (G_1, S, G_2) and mitosis (M). Cells spend 98% of the time in interphase, both growing and performing normal cell functions or preparing for cell division; only 2% of the time is spent in mitosis (Figure 5.4). The following outlines the steps of the cell cycle in interphase and mitosis.

- **G_1 phase.** This is the main **gap phase**, also called the **growth phase** of the cell. During G1 normal cell functions occur and also preparation for DNA replication. Most cells spend most of their lifespan in this phase of the cell cycle. During this phase, the cell replicates organelles such as mitochondria and centrioles; however, some organelles don't necessarily replicate before each cell division, instead they are simply redistributed into the two new cells.

- **S phase.** During this phase, DNA replication occurs, also called *DNA synthesis* (Chapter 8). In humans, all 46 chromosomes are replicated before the cell divides, so each new cell has a copy of all 46 chromosomes.

- **G_2 phase.** During the second gap phase or growth phase, the cell continues to prepare for mitosis.

In some cells this phase can be very short, and in some cases G_2 does not occur at all. The main event that takes place is the **G_2 checkpoint**, where DNA is checked for mutations after DNA replication. Many different proteins help with this process, specifically the one called **p53** (Section 5.4). Once the DNA has been replicated during the S phase, it quickly starts to condense; this continues through G_2, and the DNA is fully condensed by the time mitosis begins.

- **M phase.** During mitosis protein synthesis stops, and no more cell growth occurs; all of the cell's energy is directed toward the separation of the chromosomes and organelles into two new daughter cells. At the end of mitosis, the process of splitting the cytoplasm occurs, called **cytokinesis**. Mitosis results in two new daughter cells that are exactly identical to the original cell.

Chromosomes

To understand mitosis, it is important to take a closer look at the DNA. Humans have 46 chromosomes; we acquire 23 chromosomes from each parent. Those chromosomes are the largest macromolecules in the body, measuring up to two inches or 50 mm in length if they were to be removed from a cell and stretched to full length. After DNA replicates during S phase, each human cell nucleus (~10 μm) has a total of 92 chromosomes. For this to be possible, the DNA must be highly condensed (Chapter 2, Figure 2.14). All DNA must not only be condensed but also be organized so that each new daughter cell receives an exact duplicate

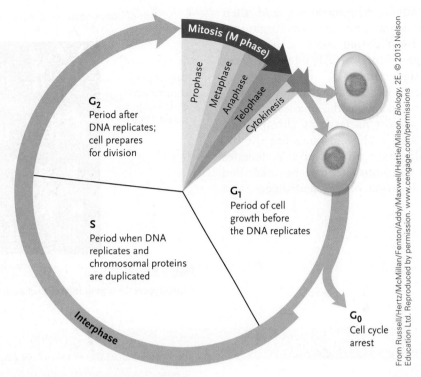

From Russell/Hertz/McMillan/Fenton/Addy/Maxwell/Hattie/Milson. *Biology*, 2E. © 2013 Nelson Education Ltd. Reproduced by permission. www.cengage.com/permissions

Figure 5.4 The Cell Cycle

The steps of the cell cycle include G1, S, G2, mitosis, and cytokinesis.

of the original 46 chromosomes. Each pair of 23 chromosomes are called **homologous chromosomes**, or **homologues**. After DNA replication, each chromosome has an identical copy, which is called a **sister chromatid** (Figure 5.5). It is very important

to recognize the difference between homologous chromosomes and sister chromatids in order to understand how chromosomes are separated during mitosis and how this process differs from meiosis (Section 5.5).

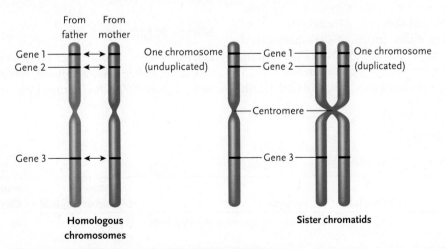

Figure 5.5 Homologous Chromosomes and Sister Chromatids

Homologous chromosomes contain the same genes, but they may have different variations of those genes, which are called *alleles*. One homologous chromosome is obtained from each parent. Sister chromatids are exactly the same and are a result of DNA replication.

When all the human chromosomes and their homologous pairs from one cell are organized according to size from chromosome 1 to 23, the result is a **karyotype** (Figure 5.6). The longest chromosome is designated number 1, and the rest are approximately ordered in decreasing size to chromosome 22. The Y chromosome is the smallest, and the X chromosome is almost as large as chromosome number 1. The X and Y chromosomes are the sex chromosomes. Males have an X and a Y, females have 2 Xs. Every chromosome can be identified based on its size, location of the centromere, and banding pattern.

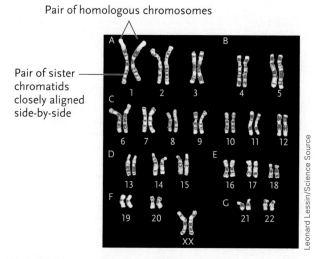

Pair of homologous chromosomes

Pair of sister chromatids closely aligned side-by-side

Leonard Lessin/Science Source

Figure 5.6 Karyotype

A karyotype is the organization of all of a cell's chromosomes.

CONCEPT REVIEW 5.2

1. What is the cell cycle?
2. In what phase do most cells spend the most time?
3. Explain generally what is occurring in G_1, S, G_2, and mitosis.
4. What is cytokinesis?
5. What is the main purpose of mitosis?

6. How many chromosomes do humans have in each cell nucleus, and how many do we receive from each parent?
7. What is the difference between homologous chromosomes and sister chromatids?
8. What is a karyotype?

5.3 MITOSIS

Somatic Cells and Germ Cells

Mitosis occurs in eukaryotic **somatic cells**, which are all the body's cells that are not sperm or eggs: for example, liver cells, bone cells, and skin cells. *Meiosis* refers to the process of generating either sperm or eggs (also called **gametes**) from **germ cells** (Section 5.5). Note that somatic cells have all 46 chromosomes—both copies of the 23 homologous pairs—and therefore are **diploid (2n)**. Germ

cells in the testes and ovaries are diploid; they have all 46 chromosomes, and during meiosis the diploid germ cells divide to produce **haploid (n)** gametes that have 23 chromosomes (Table 5.1). When a sperm and egg cell combine during fertilization, a diploid zygote is formed.

Stages of Mitosis

During mitosis, five distinct phases can be recognized: prophase, metaphase, anaphase, telophase, and cytokinesis (Figure 5.7). The first four phases involve division

TABLE 5.1

Cell Division

Cell Type	Cell Division	Purpose	Diploid or Haploid	Number of Chromosomes
Somatic cells	Mitosis	Producing two new identical cells	Diploid (2n)	46
Germ cells	Meiosis	Producing gametes	Diploid (2n)	46
Gametes (sperm or egg)	None	Fertilization	Haploid (n)	23

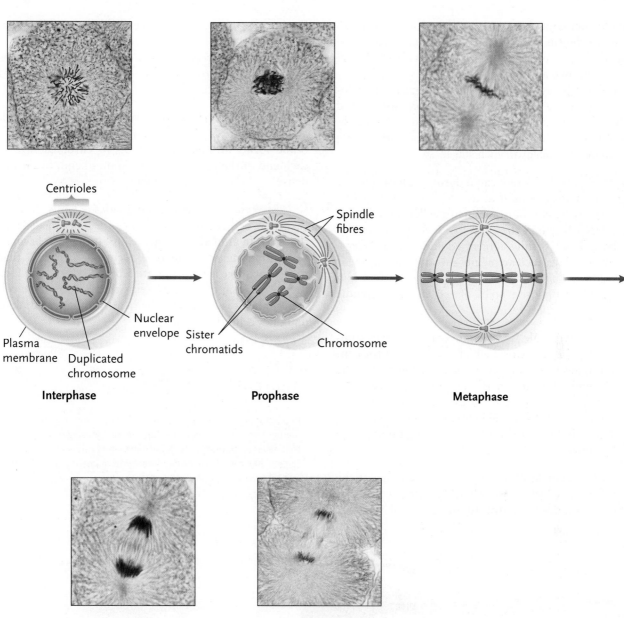

Figure 5.7 Stages of Mitosis

Articulate Graphics/illustrationOnLine.com

of the contents of the nucleus, primarily the chromosomes. In the final stage, called *cytokinesis*, the cytoplasm divides, and two new *identical* cells are produced; these are known as *daughter cells*.

During mitosis, the stages from prophase to telophase involve the separation of the chromosomes, and during the stage of cytokinesis the division of the cytoplasm occurs. The following sections describe the specific events that occur in each stage.

Prophase The DNA replicated during S phase and the sister chromatids remain attached at the centromere, and there are 92 chromosomes in the cell. Condensation of the DNA began at the end of G_2, and by the end of prophase, DNA is fully condensed and visible under a microscope. The nuclear membrane and the nucleolus break down. The centrioles that replicated in G_1 begin to migrate to opposite poles of the cell, and **spindle fibres** (microtubules) form.

Metaphase Spindle fibres connect to a protein complex called a **kinetochore**, which is located at the centromere region of each set of sister chromatids. The chromosomes line up in the centre of the cell: the **equatorial plane**. Spindle fibres consist of microtubules (Chapter 3), and the centrioles (also composed of microtubules) are involved in organizing the spindle fibres.

Anaphase Spindle fibres begin to break down at the centriole end, which causes the shortening spindles to pull the chromosomes toward the poles of the cell.

During this phase, an enzyme called *separase* is produced, which breaks down the *cohesin* proteins that hold the sister chromatids together, causing the centromeres to split, and then each sister chromatid can move to opposite ends of the cell.

Telophase Sister chromatids are at each end of the cell, and the spindle fibres break down. A new nuclear membrane forms around each set of 46 chromosomes at each end of the cell, the chromosomes begin to uncondense, and the nucleolus reforms.

Cytokinesis The cytoplasm divides equally so that each new cell receives a portion of the organelles and the fluid. In animal cells the membrane indents because of contracting **actin** microfilaments (Chapter 3) that form a **cleavage furrow** (Figure 5.8). ATP is required for the contraction of these proteins that constrict the cell membrane. Two new identical, somatic, diploid daughter cells are produced.

> **DID YOU KNOW?**
>
> Red blood cells live for approximately three to four months, and mature bone cells, osteocytes, can live for over 25 years. Cells that never go through mitosis include mature red blood cells, skeletal muscle cells, neurons, osteocytes, gametes, and the outermost layer of skin cells (stratum corneum); those cells are replaced by adult stem cells.

Contractile ring of microfilaments

David M. Phillips/Science Source

1 The furrow begins as an indentation running completely around the cell in the plane of the former spindle midpoint.

2 The furrow deepens by contraction of the microfilaments, like a drawstring tightening around the cell.

3 Furrowing continues until the daughter nuclei are enclosed in separate cells.

Figure 5.8 Cleavage Furrow

A cleavage furrow is formed when actin microfilaments in the cell contract to form two new cells.

CONCEPT REVIEW 5.3

1. Describe the main events that occur in each of the five stages of mitosis.

2. What is the role of the spindle fibres?

3. How do chromosomes become equally distributed?

4. Does it matter which sister chromatid moves into which new cell?

5. How many chromosomes are in each new daughter cell?

6. Give examples of some cell types that undergo mitosis.

5.4 CELL-CYCLE REGULATION AND CANCER

The cell cycle is controlled by **checkpoints** that ensure the full completion of one phase before advancing to the next phase. During the checkpoint stage of cell division, specific proteins become involved in searching the DNA for mutations. Some mutations may be obvious, for example, if two purines have combined, certain proteins recognize a bulge in the DNA helix. If these proteins do not recognize any DNA mutations, the cell cycle proceeds to the next stage. If they do find a mutation, the cell cycle becomes arrested at that stage. Then the mutation is either fixed, or the cell dies in a programmed process called **apoptosis**. Apoptosis is important for preventing damaged cells from dividing and perpetuating more mutations. Three principal checkpoints control the cycle in eukaryotes during G_1, G_2, and M (Figure 5.9). During the G_1 checkpoint the cell determines whether DNA replication can occur; if there are no DNA mutations, the cell progresses to S phase. During the G_2 checkpoint, the cell determines whether any DNA mutations occurred during DNA replication. Replication mistakes can be quite high, and the body has many DNA repair enzymes that help fix mistakes (Chapter 8). The final checkpoint occurs during metaphase of mitosis to ensure the cell cycle should be completed. (Figure 5.9).

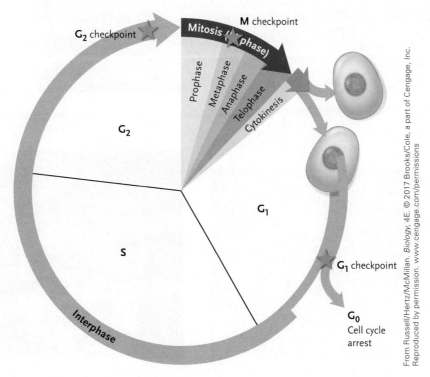

From Russell/Hertz/McMillan. *Biology*, 4E. © 2017 Brooks/Cole, a part of Cengage, Inc. Reproduced by permission. www.cengage.com/permissions

Figure 5.9 Cell-Cycle Checkpoints

The cell-cycle checkpoints are important for stopping the cell cycle when DNA mutations occur.

DNA Mutations That Go Unrecognized

Cancer occurs when mutated cells continue to divide and are not regulated by the cellular checkpoints (Figure 5.10). This process results in a growth of cells called a **tumour**. Sometimes an overgrowth of cells is not cancerous and is called a **benign tumour**. Since these cells are usually surrounded by a healthy layer of cells, they are known as **encapsulated**, and they do not spread to other areas of the body. Cancer cells generally have two or more of the following characteristics:

- Up-regulated genes, oncogenes, increase the rate of the cell cycle.
- They do not undergo cell death but become immortalized.
- They spread to other regions of the body—metastasis.

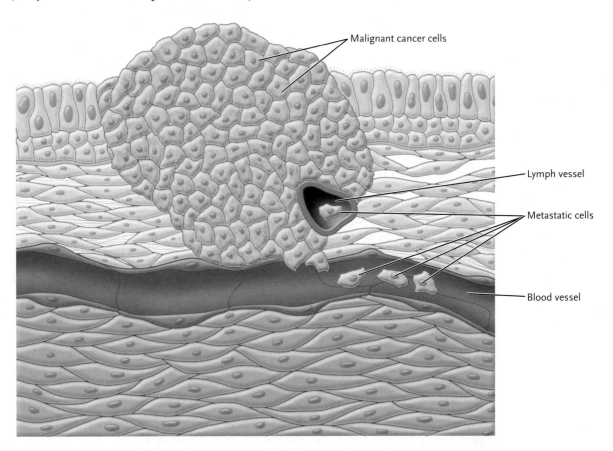

Malignant cancer cells

Lymph vessel

Metastatic cells

Blood vessel

Figure 5.10 Cancer Cells

Cancer cells are groups of cells that grow more rapidly than normal and form a tumour.

jxfzsy/Getty Images

Du Cane Medical Imaging Ltd./Science Source

- They cause the growth of new blood vessels so they can access more nutrients—angiogenesis.

The genes that normally regulate the rate of the cell cycle are called **proto-oncogenes** and tumour-suppressor genes. Proto-oncogenes code for proteins that increase the rate of the cell cycle during periods of growth or wound healing or pregnancy. Tumour-suppressor genes code for proteins that act at the three checkpoints and stop the cell cycle if DNA damage is detected. If proto-oncogenes or tumour-suppressor genes are themselves mutated, the rate of cell division cannot be regulated and cancer can occur. A mutated proto-oncogene that up-regulates cell division more than normal is an oncogene. The study of cancer is called **oncology**.

Causes of DNA Mutations

DNA mutations can be inherited or caused by environmental factors, such as radiation, viruses, chemicals. Errors during DNA replication occur quite frequently; however, most are fixed either by DNA polymerase or DNA repair enzymes (Chapter 8). Of all cancers, 5% are inherited gene mutations, and 95% are caused by environmental factors. Anything that damages DNA is called a **mutagen**. Anything that damages DNA and also causes cancer is called a **carcinogen**. Note that the following factors may be carcinogenic in different doses in different people; the overall health of an individual affects the capacity of cells to remove toxins and repair DNA.

Most Common Carcinogens

- Radiation—x-rays, nuclear power plant radiation
- Asbestos—not so common now that it is no longer used as insulation
- Dioxins—herbicides and pesticides, and many cleaning products
- Tobacco smoke
- Benzene—in gasoline, paint remover, and rubber, and also released by volcanoes
- Aflatoxin—fungus that grows on raw grain and nuts; reason why most nuts are roasted
- Formaldehyde—plastics, and also in aspartame
- Industrial smoke and pollution
- Viruses—three most common: HIV, HPV, and hepatitis
- Charred meat or burned toast—contains benzopyrene
- Acrylamide—can be found in starchy foods cooked at high temperatures, such as French fries or potato chips
- Pressure-treated wood

- Excessive UV light (but some sunlight is important for good health)
- Chronic inflammation—causes overgrowth of cells and angiogenesis (growth of blood vessels)

Cancer cells may develop if any of the following categories of genes are mutated: (1) genes that regulate the cell cycle, such as proto-oncogenes and tumour-suppressor genes; (2) DNA repair enzymes; (3) genes involved with angiogenesis; (4) genes that promote apoptosis; and (5) genes that regulate the immune response.

One very important tumour-suppressor gene is p53. It prevents cell division at the G_1 checkpoint if it detects any abnormal DNA before the DNA replication phase. Mutations of the p53 gene have been shown to play a significant role in the development of many cancers. The important functions of p53 include preventing the cell from moving into S phase, recruiting DNA repair enzymes to fix DNA mutations, and initiating apoptosis if the mutation can't be fixed (Figure 5.11). Note that p53 does not repair DNA mutations itself.

What happens to cancer cells that get passed the checkpoints? Our cells have multiple mechanisms for managing DNA mutations. We have many genes that code for proteins that either fix mutations—such as DNA repair enzymes—or initiate apoptosis if mutations cannot be fixed. But if cells with mutations continue to divide unregulated, our immune system can kill them. We have **natural killer cells**: white blood cells that specifically recognize cancer cells and can destroy them. Cytotoxic T cells can also kill cancer cells. Everyone develops some cancer cells all the time. However, if they develop faster than they can be destroyed, then malignant tumours can form and spread throughout the body. To minimize your risk of cancer it is important to avoid toxins as much as possible and to have a healthy immune system. Chronic stress inhibits the immune response; therefore, having healthy outlets for dealing with stress such as exercise, yoga, meditation, or counselling is very beneficial.

DID YOU KNOW?

Anti-cancer foods, teas, and spices contain important chemicals that regulate our immune response, decrease angiogenesis, or are antioxidants that prevent free radical molecules from damaging our DNA. Some great ideas to add to your diet include vitamin E foods (fish and nuts), green tea, lavender tea, blueberries, red grapes (also red wine), kale, turmeric, garlic, onion, pumpkin and pumpkin seeds, tomato, pineapple, coconut oil, olive oil, and dark chocolate.

Figure 5.11 Tumour-Suppressor Gene: P$_{53}$

P$_{53}$ is a tumour-suppressor gene that stops the cell cycle at the G$_1$ checkpoint when DNA mutations are detected.

CONCEPT REVIEW 5.4

1. What is apoptosis?
2. What is angiogenesis?
3. Describe the three checkpoints and how they are involved in cell-cycle regulation.
4. What is a carcinogen?
5. What types of factors can cause cancer? Give a few examples.

6. Explain the difference between a proto-oncogene, an oncogene, and a tumour-suppressor gene. Give one example of a tumour-suppressor gene.
7. What kinds of DNA mutations may be involved in cancer?
8. Describe how p53 prevents cancer cell development.
9. How does our immune system play a role in preventing cancer?

5.5 MEIOSIS

During mitosis, somatic cells divide in order to replace the body's old and damaged cells. During meiosis, haploid sperm or eggs (also called *gametes*) are generated from diploid germ cells so the organism can reproduce. The stages of meiosis are similar to mitosis because the DNA replicates and then the chromosomes become distributed into the new cells. The main difference between meiosis and mitosis is that meiosis involves two cell divisions and mitosis has only one. During meiosis haploid cells are produced with 23, instead of 46 chromosomes. During replication, a sperm and egg that each has 23 chromosomes form an organism with the normal diploid number of 46 (Figure 5.12).

Haploid egg   Haploid sperm

Maternal homologue

Paternal homologue

Diploid zygote

Articulate Graphics/IllustrationOnLine.com

Figure 5.12 Haploid Gametes

Haploid gametes have 23 chromosomes, one of each of the homologues.

A very important feature of meiosis is that the separation of the homologous chromosomes into the gametes is *random*, and therefore different combinations of homologues can be found in every gamete, which means that any combination of gametes that leads to fertilization will lead to a genetically unique individual. Two people can theoretically produce millions of different offspring. The first cell is called a **zygote,** and it undergoes mitosis many times to eventually produce an embryo. Genetic variation is an important characteristic of every population; this is discussed further in Chapters 6 and 7.

Unlike mitosis, meiosis involves two cell divisions: meiosis I and meiosis II. The difference between these divisions relates to which chromosomes separate into the new cells. During meiosis I the homologous pairs separate, and during meiosis II the sister chromatids separate. DNA replication occurs only once—during S phase before the first meiotic division.

Stages of Meiosis

Prophase I The first prophase involves all the same events that occur in prophase of mitosis. The DNA replicated during S phase and the sister chromatids remain attached at the centromere, and the cell has 92 chromosomes. Condensation of DNA occurs, the nuclear membrane and the nucleolus break down, the centrioles migrate to opposite poles of the cell, and spindle fibres (microtubules) form.

Metaphase I Spindle fibres connect to the kinetochore at the centromere region of each set of sister chromatids. The chromosomes line up at the equatorial plane. However, compared to what happens in mitosis, where all 46 chromosomes with their attached sister chromatid line up, in meiosis I it is the homologous pairs with their attached sister chromatids that line up

(Figure 5.13). This difference is important because during this first cell division the daughter cells are still haploid (have 46 chromosomes) because, although they have only one of each homologous chromosome, they are still connected to their sister chromatids, which also contain only *one* of each homologous pair.

Anaphase I Spindle fibres begin to break down at the centriole end, which causes the shortening spindles to pull the homologous chromosomes toward the poles of the cell, with sister chromatids still connected together.

Telophase I Homologous chromosomes are situated at each end of the cell, and the spindle fibres break down. A new nuclear membrane forms around each set of chromosomes at each end of the cell, the chromosomes begin to uncondense, and the nucleolus reforms.

Cytokinesis The cytoplasm becomes divided equally so that each new cell receives a portion of the organelles and fluid. Two haploid cells are produced, but because each homologous chromosome remains connected to its sister chromatid, there are 46 chromosomes in each cell. Therefore, one more cell division occurs so that each gamete contains only 23 chromosomes.

Prophase II DNA does *not* replicate before the next prophase. The cells produced from meiosis I move directly into meiosis II, so there is no second S phase. Condensation of the DNA occurs, the nuclear membrane and the nucleolus break down, centrioles migrate, and spindle fibres form.

Metaphase II Spindle fibres connect to the kinetochore, and the chromosomes line up at the equatorial plane, but now in meiosis II, the chromosomes with their sister chromatid all line up (Figure 5.13).

Anaphase II Spindle fibres begin to break down at the centriole end, which causes the sister chromatids to separate toward each end of the cell.

Telophase Sister chromatids are situated at each end of the cell, and the spindle fibres break down. A new nuclear membrane forms around each set of 23 chromosomes at each end of the cell, the chromosomes begin to uncondense, and the nucleolus reforms.

Cytokinesis Cytokinesis in meiosis I and meiosis II occur differently in males and females. In males, by the end of meiosis II, there are four sperm cells, each with 23 chromosomes. In females, each cell division occurs so that more of the cytoplasm goes into one cell and the other becomes a polar body that dissolves, leaving only one egg cell. To further complicate this process, females don't complete the last meiotic division unless fertilization occurs. Eggs stay arrested in metaphase II until ovulation, and do not complete the last stage of cell division if unfertilized.

MEIOSIS I

Crossing over

Homologous pair

Prophase I   Metaphase I   Anaphase I   Telophase I

MEIOSIS II

Prophase II   Metaphase II   Anaphase II   Telophase II and cytokinesis

Articulate Graphics/illustrationOnLine.com

Figure 5.13   Stages of Meiosis

Human Genetic Variation

Considering all the possible combinations of homologues that could separate during the first meiotic cell division into each new cell, there are two possible homologues for each cell, and since we have 23 pairs of chromosomes, there are 2^{23} possible combinations of chromosomes that can occur in each gamete from one individual. Therefore, each person can produce over 8 million possible different gametes. This scope of possible variation increases further because every individual is formed from two different gametes from two different people. The probability of two parents having two children with the same genetic material (just looking at random assortment of chromosomes) is one in 2.5 billion.

Crossing Over

Another element ensures that genetic variation predominates. There is not only random separation of homologues—also called **independent assortment** (Chapter 6)—but also a process called **crossing over**, which occurs during prophase I. When homologous chromosomes start to pair up at the beginning of meiosis, the homologous chromosomes overlap. Fragments of non-sister chromosomes break in the same place, and a section of each chromosome is swapped, resulting in a hybrid chromosome (Figure 5.14).

Meiosis has unique features that are not seen in mitosis. Meiosis has two cell divisions so that haploid gametes are produced, and meiosis has crossing over, which ensures that every individual is unique (except in the case of identical twins when the zygote splits to form two embryos that have identical DNA).

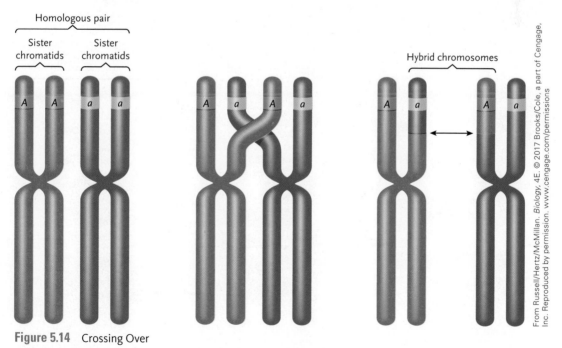

Figure 5.14 Crossing Over

Crossing over involves the transfer of sections of homologous chromosomes during prophase I and increases genetic variation.

From Russell/Hertz/McMillan. Biology, 4E. © 2017 Brooks/Cole, a part of Cengage, Inc. Reproduced by permission. www.cengage.com/permissions

CONCEPT REVIEW 5.5

1. What is the purpose of meiosis?

2. Which cells are haploid? Diploid? How many chromosomes are in each?

3. When do homologous chromosomes and sister chromatids separate during meiosis?

4. What is a gamete? Where are they formed in males and females?

5. Why is genetic variation important?

6. How does meiosis ensure that every individual has a unique combination of genes?

7. How is meiosis different in males and females?

8. What is crossing over, and why is it important?

CHAPTER REVIEW

Labelling Activity

Label or identify the following diagrams.

1. Label the first two diagrams and identify the process in each of the remaining diagrams.

Articulate Graphics/IllustrationOnLine.com

2. Identify the process in each diagram.

_____   _____   _____   _____

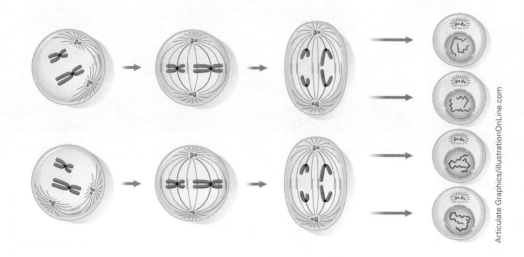

_____   _____   _____   _____

Answer to Application Question

The yellow fever mosquito has a diploid number of 6; therefore, it has 6 chromosomes in its somatic cells that undergo mitosis. A somatic cell in metaphase looks like this:

This cell will have 12 chromosomes after S phase, and each new daughter cell will have 6. After meiosis, this mosquito would produce gametes containing three chromosomes per gamete. After DNA replication, mistakes would be detected by the G_2 checkpoint.

Metaphase

6 Genetic Inheritance

Develop a passion for learning. If you do, you will never cease to grow.

Anthony J. D'Angelo

6.1 MENDEL'S THEORY OF INHERITANCE

Heredity refers to the transfer of traits from parent to offspring. Humans acquire 23 chromosomes from each parent, resulting in a total of 46 chromosomes that contain all 25 000 genes in every nucleated cell. Recall from Chapter 5 that each pair of chromosomes are homologous pairs, so they are *almost* exactly the same; whereas sister chromatids that result from DNA replication are *exactly* identical, assuming no mutations occurred during S phase (Figure 6.1).

Each gene codes for an RNA molecule that acts as the template for protein synthesis. The resulting proteins constitute a particular trait. Recall from Chapter 2 that proteins have many functions, including as membrane transporters such as a chloride channel, hemoglobin that carries oxygen in red blood cells, hormones that signal cells to perform various functions, and enzymes that catalyze reactions. The visible appearance or function related to a specific protein—such as brown hair pigmentation, or freckles, or a disease such as cystic fibrosis—is called a *phenotype*. Note that a phenotype can also be due to the combination of genes expressed; for example, eye colour is actually due to the combined expression of many genes.

The genes that code for a phenotype (trait) are the genotype. Each of the two copies of every gene, one from each parent, is called an *allele*. Remember that we have two copies of every chromosome, one from each parent. When the alleles are the same, they are homozygous; when they

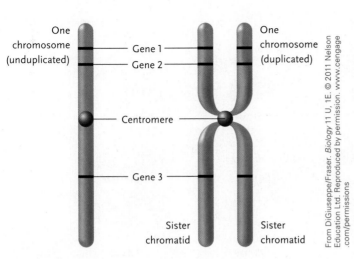

Figure 6.1 Sister Chromatids

Sister chromatids—formed during DNA replication—are exact copies of each other.

One chromosome (unduplicated) — Gene 1 — Gene 2 — Centromere — Gene 3 — One chromosome (duplicated) — Sister chromatid — Sister chromatid

TABLE 6.1

Genotype and Phenotype

Genotype	Phenotype
Homozygous dominant (AA)	Healthy
Heterozygous (Aa)	Healthy
Homozygous recessive (aa)	E.g., sickle cell anemia

are different, they are heterozygous. Sometimes one allele is dominant and another one is recessive. For example, in humans the disease allele that causes sickle cell anemia is recessive to the healthy allele. In this case, a person has the sickle cell anemia phenotype if they have both disease alleles, making their genotype homozygous recessive. The convention is to use a capital letter (e.g., *A*) to denote a dominant allele and a lower-case letter (e.g., *a*) to denote a recessive allele. Table 6.1 shows that three genotypes are possible because we have two alleles for each trait. Two phenotypes are possible for a given trait, they show either the dominant or the recessive trait.

In the 1800s, a monk by the name of Gregor Mendel studied the phenotypes of generations of pea plants. However, he did not know that organisms have two copies of each gene, or that DNA carries traits from parents to offspring. He called homozygous traits "true-breeding" and heterozygous traits "not true-breeding," and he discovered the inheritance patterns of dominant-recessive traits. He was fortunate to study peas that have many single-gene traits. In humans, most phenotypes are determined by multiple genes (Section 6.4). Some phenotypes of the pea plant are as follows (Figure 6.2):

- purple flowers or white flowers
- yellow seeds or green seeds
- wrinkled seeds or round seeds
- tall or short plants

Mendel cross-fertilized different pea plants with different traits and studied the frequency of the phenotypes in the offspring. The true-breeding plants always gave rise to offspring with the same traits, such as all purple flowers or all white flowers. When he crossed true-breeding purple-flowered pea plants with true-breeding white-flowered pea

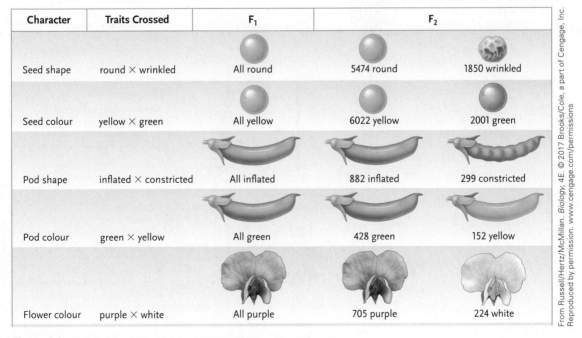

Character	Traits Crossed	F₁	F₂	
Seed shape	round × wrinkled	All round	5474 round	1850 wrinkled
Seed colour	yellow × green	All yellow	6022 yellow	2001 green
Pod shape	inflated × constricted	All inflated	882 inflated	299 constricted
Pod colour	green × yellow	All green	428 green	152 yellow
Flower colour	purple × white	All purple	705 purple	224 white

Figure 6.2 Examples of Dominant and Recessive Pea Plant Phenotypes

(Continued)

Character	Traits Crossed	F₁	F₂	
Flower position	axial (along stems) × terminal (at tips)	All axial	651 axial	207 terminal
Stem length	tall × dwarf	All tall	787 tall	277 dwarf

Figure 6.2 *(Continued)*

plants, the first generation, called F1, always had purple flowers. When he cross-fertilized that F1 generation, the next generation, F2, contained plants with white flowers again (Figure 6.3). He called the phenotype of the F1 generation the **dominant** trait, and the phenotype that reappeared in the F2 generation, the **recessive** trait. Crossing purple-flowered plants with white-flowered ones resulted in 100% with purple flowers—that is, dominant—in the F1 generation; crossing two purple-flowered plants from the F1 generation resulted in 75% with purple flowers and 25% with white flowers. Breeding many F1 generation pea plants always resulted in the same proportion of 3:1

(dominant:recessive) in the F2 generation. Now that we know the "true-breeding" plants are homozygous and the "not-true-breeding plants" are heterozygous, we can predict the phenotype ratios for any combination of genotypes by using a **Punnett square** (Section 6.2).

The possible genotypes and phenotypes for pea plant flower colour are as follows:

- PP (homozygous dominant)—purple
- Pp (heterozygous)—purple
- pp (homozygous recessive)—white

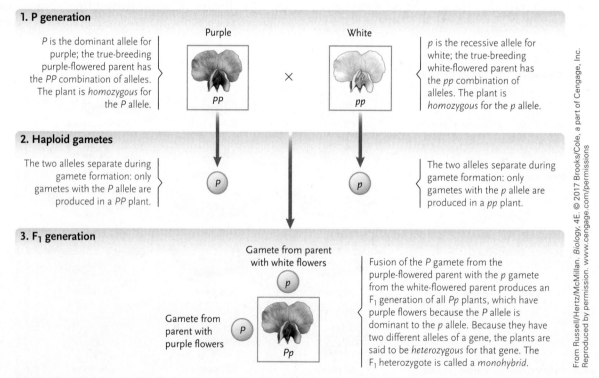

Figure 6.3 Inheritance Pattern

The inheritance pattern for breeding dominant-recessive, single gene traits in pea plants can be determined using a Punnett square.

From Russell/Hertz/McMillan. *Biology*, 4E. © 2017 Brooks/Cole, a part of Cengage, Inc. Reproduced by permission. www.cengage.com/permissions

(Continued)

4. F₁ × F₁ self

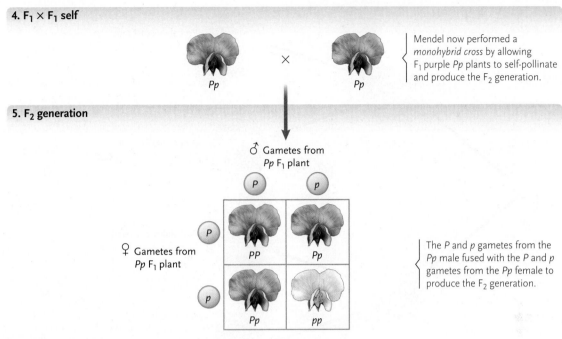

Mendel now performed a *monohybrid cross* by allowing F₁ purple *Pp* plants to self-pollinate and produce the F₂ generation.

5. F₂ generation

The *P* and *p* gametes from the *Pp* male fused with the *P* and *p* gametes from the *Pp* female to produce the F₂ generation.

Figure 6.3 *(Continued)*

CONCEPT REVIEW 6.1

1. Define the following: homozygous, heterozygous, dominant, recessive, allele, phenotype, and genotype.

2. Consider two pea plants that are heterozygous for flower colour, with purple and white being the possible alleles.

   a. What colour would the flowers be?

   b. What are the possible gametes they could form?

   c. What are the possible offspring they could produce?

   d. What is the probability that those pea plants would have offspring with white flowers?

6.2 PROBABILITY AND PUNNETT SQUARES

Probability refers to the likelihood that some event will occur. If we flip one coin, there are two possible outcomes; it could land on heads or tails. If we flip two coins there are four possible outcomes; two heads, one head and one tail, one tail and one head, or two tails (Figure 6.4). The probability of each possible outcome is as follows:

- 2 heads—¼ or 25%
- 1 head and 1 tail (in either order)—½ or 50%
- 2 tails—¼ or 25%

Considering genetic inheritance and each individual having two alleles, the possible outcomes are the same:

- 2 dominant—¼ or 25%
- 1 dominant and 1 recessive—½ or 50%
- 2 recessive—¼ or 25%

DID YOU KNOW?

Some human traits are passed on by dominant-recessive inheritance (Table 6.2). However, in these examples, other genes can act on these features. So these traits can also be considered polygenic, and many variations will be seen in the population. For example, face shape can vary from round, oval, oblong, to square, but oval shapes tend to be dominant over square shapes.

A Punnett square provides a method for determining the possible offspring genotypes and phenotypes from the alleles of each parent. If we look at only

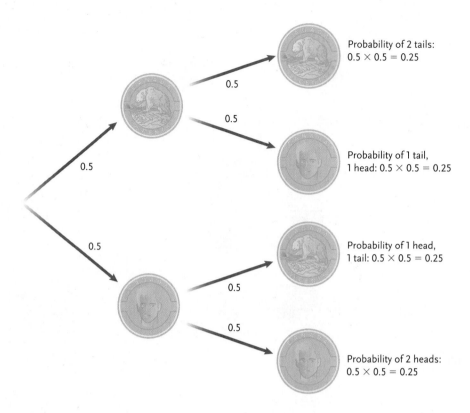

Figure 6.4 Probability Tree
A probability tree shows all of the possible outcomes from flipping a coin twice.

TABLE 6.2

Dominant-Recessive Human Traits

Trait	Dominant	Recessive
Shape of face	Oval	Square
Cleft in chin	No cleft	Cleft
Hairline	Widow peak	Straight
Eyebrow size	Broad	Slender
Eyebrow shape	Separated	Joined
Eyelash length	Long	Short
Earlobes	Free	Attached
Freckles	Freckles	No freckles
Tongue rolling	Rolling	Non-rolling
Hitchhiker's thumb	Straight	Hitchhiker
Little finger	Bent	Straight
Interlaced fingers	Left thumb over right	Right thumb over left

a single trait, this is considered a **monohybrid cross**. Since each person has two copies of each gene—alleles—on homologous chromosomes, and since homologous chromosomes separate during meiosis into different gametes (Chapter 5), we can determine the probability of each possible genotype and phenotype in the offspring (Figure 6.4): the possible genes in the gametes are shown *outside* the boxes, and the possible combinations of genes of the offspring are shown *inside* the boxes. For example (Figure 6.5), if a parent pea plant is heterozygous for flower colour, one allele is dominant and the other allele is recessive. When the homologous chromosomes separate during meiosis, it is possible for a gamete to have a dominant allele or a recessive allele. If both parents are heterozygous, the following possible offspring can arise:

- genotype ratio—1:2:1—homozygous dominant: heterozygous:homozygous recessive
- phenotype ratio—3:1—purple:white

Considering disease alleles, as in the case of cystic fibrosis, these alleles are most often recessive, which means a person with the disease would have both disease alleles (cc). A person who is heterozygous for the cystic fibrosis allele (Cc), does not have the disease because the dominant allele can still produce normal proteins, but this person is said to be a carrier of the cystic fibrosis (CF) allele.

Gametes from F₁ purple *Pp* plant

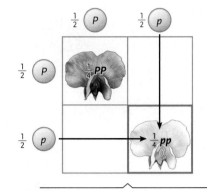

A. To produce an F₂ plant with the *PP* genotype, two *P* gametes must combine. The probability of selecting a *P* gamete from one F₁ parent is $\frac{1}{2}$, and the probability of selecting a *P* gamete from the other F₁ parent is also $\frac{1}{2}$. Using the product rule, the probability of producing purple-flowered *PP* plant from a *Pp* × *Pp* cross is $\frac{1}{2} \times \frac{1}{2} = \frac{1}{4}$.

B. To produce an F₂ plant with the *pp* genotype, two *p* gametes must combine. The probability of selecting a *p* gamete from one F₁ parent is $\frac{1}{2}$, and the probability of selecting a *p* gamete from the other F₁ parent is also $\frac{1}{2}$. Using the product rule, the probability of producing white-flowered *pp* plant from a *Pp* × *Pp* cross is $\frac{1}{2} \times \frac{1}{2} = \frac{1}{4}$.

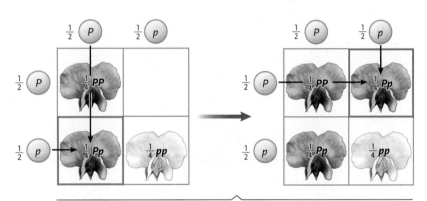

C. To produce an F₂ plant with the *Pp* genotype, a *P* gamete must combine with a *p* gamete. The cross *Pp* × *Pp* can produce *Pp* offspring in two different ways: (1) a *P* gamete from the first parent can combine with a *p* gamete from the second parent; or (2) a *p* gamete from the first parent can combine with a *P* gamete from the second parent. We apply the sum rule to obtain the combined probability: each of the ways to get *Pp* has an individual probability of $\frac{1}{4}$, so the probability of *Pp*, purple-flowered offspring is $\frac{1}{4} + \frac{1}{4} = \frac{1}{2}$.

Figure 6.5 Monohybrid Cross

This example shows a monohybrid cross between heterozygous pea plants and the possible genotypes and phenotypes of the offspring.

Testcross

Mendel devised the **testcross** to determine the genotype of individuals expressing the dominant phenotype. In pea plants, yellow peas are dominant over green peas, so a yellow pea could be homozygous dominant (YY) or it could be heterozygous (Yy). If a yellow pea plant is crossed with a homozygous recessive green pea plant, different outcomes will occur depending on whether the yellow plant is YY or Yy (Figure 6.6). If the yellow plant is homozygous dominant, the phenotype ratios of the offspring will be 100% yellow. If the yellow plant is heterozygous, the phenotypes ratios of the offspring will be 50% yellow and 50% green.

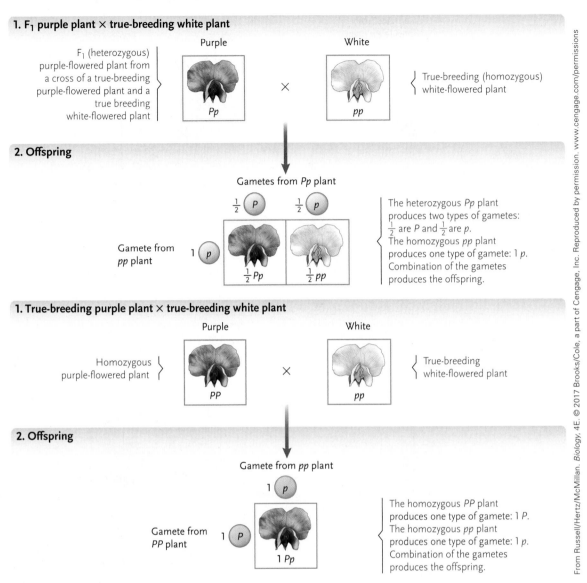

1. F₁ purple plant × true-breeding white plant

F₁ (heterozygous) purple-flowered plant from a cross of a true-breeding purple-flowered plant and a true breeding white-flowered plant

Purple Pp

×

White pp

True-breeding (homozygous) white-flowered plant

2. Offspring

Gametes from Pp plant

$\frac{1}{2}$ P $\frac{1}{2}$ p

Gamete from pp plant 1 p

$\frac{1}{2} Pp$ $\frac{1}{2} pp$

The heterozygous Pp plant produces two types of gametes: $\frac{1}{2}$ are P and $\frac{1}{2}$ are p. The homozygous pp plant produces one type of gamete: 1 p. Combination of the gametes produces the offspring.

1. True-breeding purple plant × true-breeding white plant

Homozygous purple-flowered plant

Purple PP

×

White pp

True-breeding white-flowered plant

2. Offspring

Gamete from pp plant

1 p

Gamete from PP plant 1 P

1 Pp

The homozygous PP plant produces one type of gamete: 1 P. The homozygous pp plant produces one type of gamete: 1 p. Combination of the gametes produces the offspring.

Figure 6.6 Testcross
A testcross is used to determine the genotype of an organism expressing the dominant phenotype.

CONCEPT REVIEW 6.2

1. Yellow peas are dominant over green peas. Given the case that a heterozygous yellow pea plant was crossed with a green pea plant, use a Punnett square to determine the probability of the offspring pea plant being the following:

   a. heterozygous
   b. yellow
   c. green
   d. homozygous dominant

2. A pea plant has purple flowers, but the genotype is not known. Explain how a testcross can be used to determine whether the purple plant is PP or Pp.

3. The ability to curl your tongue up on the sides is dominant over the inability to curl your tongue on the sides. A woman who can curl her tongue marries a man who cannot curl his tongue. Their first child cannot curl their tongue like their dad. What is the genotype of the mom? Dad? Child?

4. In rats, long tails are dominant over short tails. Can two long-tailed rats have short-tailed offspring?

5. Cystic fibrosis is a recessive disease. If a mother is a carrier of the CF allele (heterozygous) but does not have the disease, and the father is homozygous healthy, is it possible for these parents to have children with cystic fibrosis?

6.3 DIHYBRID CROSS

Mendel's **law of segregation** states that the two alleles of a trait separate from each other during the formation of gametes, so half the gametes will carry one copy and half will carry the other copy. This is because diploid germ-line cells form haploid gametes; each gamete has one of each homologous chromosome. Mendel's **law of independent assortment** states that how the homologous pairs separate into gametes is completely random, so any possible combination can occur.

If the genes for *two* different traits are found on different chromosomes, we can use a **dihybrid cross** to look at the possible genotype and phenotype outcomes. For example, in peas, the genes that code for round and yellow are dominant over the genes that code for wrinkled and green. If a homozygous round and yellow pea plant is crossed with a homozygous wrinkled and green pea plant, all the offspring will be heterozygous round and yellow peas (Figure 6.7). If those heterozygous offspring are crossed, all the possible genotypes and phenotypes can be determined using a Punnett square (Figure 6.7). As for a monohybrid cross, the possible gametes are written outside the squares, and the possible offspring are written inside the squares.

To determine the possible gametes when looking at two different traits, we can use the acronym FOIL (first, outer, inner, last). If a pea plant is heterozygous for round and yellow traits, the genotype is RrYy. The R and r are homologous pairs that separate randomly into different gametes during the first meiotic division; similarly, the Y and y are homologous chromosomes that separate randomly into different gametes during meiosis. Each gamete must include either the R or r AND a Y or y. Therefore, a diploid parent pea plant can form the following possible haploid gametes:

RrYy

First—RY

Outer—Ry

Inner—rY

Last—ry

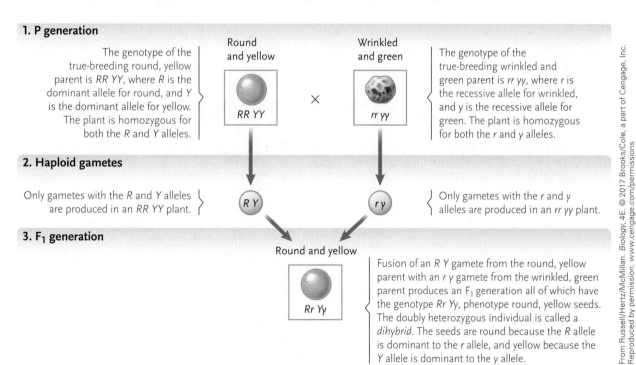

1. P generation

The genotype of the true-breeding round, yellow parent is *RR YY*, where *R* is the dominant allele for round, and *Y* is the dominant allele for yellow. The plant is homozygous for both the *R* and *Y* alleles.

Round and yellow — *RR YY*

×

Wrinkled and green — *rr yy*

The genotype of the true-breeding wrinkled and green parent is *rr yy*, where *r* is the recessive allele for wrinkled, and *y* is the recessive allele for green. The plant is homozygous for both the *r* and *y* alleles.

2. Haploid gametes

Only gametes with the *R* and *Y* alleles are produced in an *RR YY* plant.

R Y

r y

Only gametes with the *r* and *y* alleles are produced in an *rr yy* plant.

3. F₁ generation

Round and yellow — *Rr Yy*

Fusion of an *R Y* gamete from the round, yellow parent with an *r y* gamete from the wrinkled, green parent produces an F₁ generation all of which have the genotype *Rr Yy*, phenotype round, yellow seeds. The doubly heterozygous individual is called a *dihybrid*. The seeds are round because the *R* allele is dominant to the *r* allele, and yellow because the *Y* allele is dominant to the *y* allele.

From Russell/Hertz/McMillan. *Biology*, 4E. © 2017 Brooks/Cole, a part of Cengage, Inc. Reproduced by permission. www.cengage.com/permissions

Figure 6.7 Dihybrid Cross

A dihybrid cross is used to determine the probability of offspring acquiring two traits.

(Continued)

4. F$_1$ × F$_1$ self

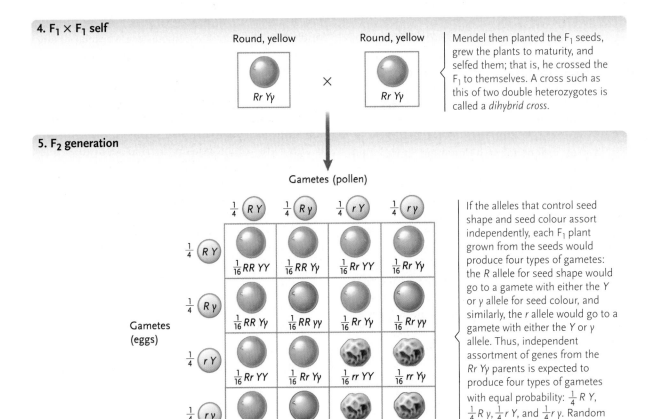

Round, yellow

Rr Yy

×

Round, yellow

Rr Yy

Mendel then planted the F$_1$ seeds, grew the plants to maturity, and selfed them; that is, he crossed the F$_1$ to themselves. A cross such as this of two double heterozygotes is called a *dihybrid cross*.

5. F$_2$ generation

Gametes (pollen)

$\frac{1}{4}$ RY $\frac{1}{4}$ Ry $\frac{1}{4}$ rY $\frac{1}{4}$ ry

Gametes (eggs)

$\frac{1}{4}$ RY

$\frac{1}{16}$ RR YY $\frac{1}{16}$ RR Yy $\frac{1}{16}$ Rr YY $\frac{1}{16}$ Rr Yy

$\frac{1}{4}$ Ry

$\frac{1}{16}$ RR Yy $\frac{1}{16}$ RR yy $\frac{1}{16}$ Rr Yy $\frac{1}{16}$ Rr yy

$\frac{1}{4}$ rY

$\frac{1}{16}$ Rr YY $\frac{1}{16}$ Rr Yy $\frac{1}{16}$ rr YY $\frac{1}{16}$ rr Yy

$\frac{1}{4}$ ry

$\frac{1}{16}$ Rr Yy $\frac{1}{16}$ Rr yy $\frac{1}{16}$ rr Yy $\frac{1}{16}$ rr yy

If the alleles that control seed shape and seed colour assort independently, each F$_1$ plant grown from the seeds would produce four types of gametes: the *R* allele for seed shape would go to a gamete with either the *Y* or *y* allele for seed colour, and similarly, the *r* allele would go to a gamete with either the *Y* or *y* allele. Thus, independent assortment of genes from the *Rr Yy* parents is expected to produce four types of gametes with equal probability: $\frac{1}{4}$ *R Y*, $\frac{1}{4}$ *R y*, $\frac{1}{4}$ *r Y*, and $\frac{1}{4}$ *r y*. Random fusion of the four different male gametes with the four different female gametes produces the F$_2$ generation.

Figure 6.7 *(Continued)*

CONCEPT REVIEW 6.3

1. Round and yellow are dominant over wrinkled and green peas. Given the case that one parent plant is homozygous for the round pea trait and heterozygous for the yellow trait, and the other parent plant is heterozygous for the round trait and is green, answer the following questions:

   a. What are the genotypes of the parent plants?
   b. What are the possible gametes those parent plants can form?
   c. What are the possible genotypes of the offspring?
   d. What are the possible phenotypes of the offspring?
   e. What are the phenotypic and genotypic ratios for the above cross?

2. Freckles and brown hair are dominant over no freckles and blond hair. Suppose mom has no freckles and blond hair, and dad is heterozygous for both traits. What is the probability that their children will have brown hair and freckles?

6.4 NON-MENDELIAN INHERITANCE

Polygenic Inheritance

Often a phenotype does not express a simple dominant or recessive inheritance pattern. Multiple genes determine most traits, and the phenotype varies from one extreme to another: for example, human height ranges from very short to very tall, and most people's height is somewhere in the middle (Figure 6.8). When a phenotype shows continuous variation in a population, it is known as **polygenic** because multiple genes play a role in that phenotype. Other examples of polygenic traits in humans include body weight, skin colour, eye colour, intelligence, and diseases such as cancer or schizophrenia.

Pleiotropic

When one allele causes multiple phenotypes in a population, it is considered **pleiotropic**. For example, cystic

A. **Students at Brigham Young University, arranged according to height**

B. **Actual distribution of individuals in the photo according to height**

Number of individuals in each height category

1 4 8 10 16 16 16 15 14 13 13 11 9 8 8 5 1 2

Shortest ◀——— Range of heights ———▶ Tallest

C. **Idealized bell-shaped curve for a population that displays continuous variation in a trait**

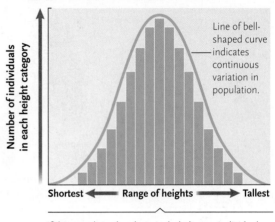

Line of bell-shaped curve indicates continuous variation in population.

Number of individuals in each height category

Shortest ◀——— Range of heights ———▶ Tallest

If the sample in the photo included more individuals, the distribution would more closely approach this ideal.

Figure 6.8 Polygenic Inheritance

Polygenic inheritance occurs when a single trait, such as body height, is determined by multiple genes, usually showing a continuous variation distribution in a population.

fibrosis (CF), an inherited gene mutation in the chloride channel gene, can cause a wide variety of symptoms in different people with that allele. Cystic fibrosis is a recessive disease, so a person must have both mutated alleles to have the disease. A person heterozygous for CF does not have cystic fibrosis. The symptoms of CF range from mild to severe and can affect the lungs, liver, pancreas, and sweat glands. Recessive alleles can be passed on from parents to offspring. If breeding occurs among closely related people or animals, there is a much higher chance that offspring will acquire two recessive alleles. Recall from Chapter 1 that genetic variation within a population of a given species is important for that population to be able to survive environmental changes (see also Chapter 7). When organisms within a population reproduce with organisms that have very different alleles (i.e., they are not closely related), genetic variation in the population increases.

Incomplete Dominance

The alleles for each trait are neither fully dominant nor fully recessive in heterozygotes. Sometimes a heterozygote has an intermediate trait in comparison to the homozygous dominant and the homozygous recessive (Figure 6.9). For example, if red flowers are incompletely dominant to white flowers, a heterozygote would be pink. The following three genotypes and three phenotypes are possible in an inheritance pattern of incomplete dominance:

- homozygous dominant (RR)—red
- heterozygous (Rr)—pink
- homozygous recessive (rr)—white

Codominance

A gene may have more than two alleles in a population, and sometimes in heterozygotes *both* alleles are expressed. For example, in humans, there are three possible alleles for blood type (A, B, or O), and each person can have any two alleles. If a heterozygote has an A and a B allele, both the A and B proteins will be expressed, giving a phenotype of blood type AB; this is an example of **codominance** (Figure 6.10). The ABO blood types (not including the Rh factor) represent genes that code for an enzyme that adds sugar groups

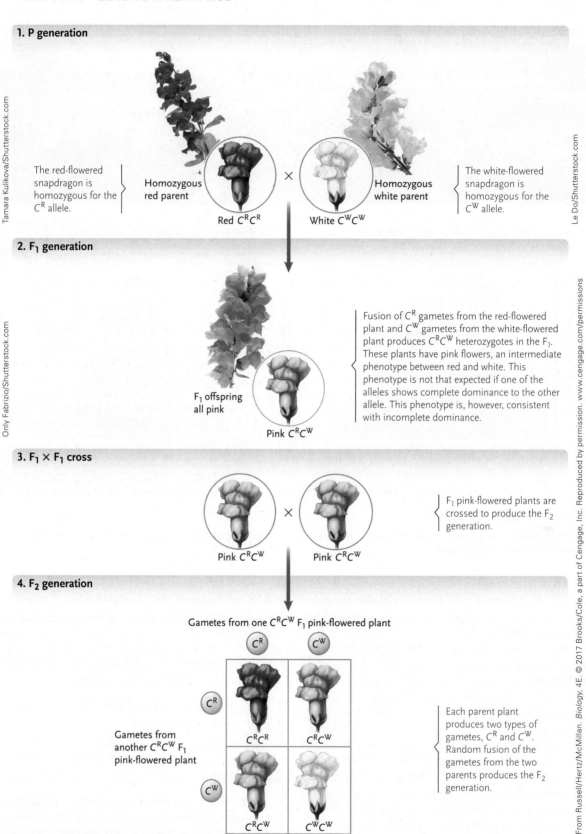

1. P generation

The red-flowered snapdragon is homozygous for the C^R allele.

Homozygous red parent × Homozygous white parent

Red $C^R C^R$ White $C^W C^W$

The white-flowered snapdragon is homozygous for the C^W allele.

2. F₁ generation

F₁ offspring all pink

Pink $C^R C^W$

Fusion of C^R gametes from the red-flowered plant and C^W gametes from the white-flowered plant produces $C^R C^W$ heterozygotes in the F₁. These plants have pink flowers, an intermediate phenotype between red and white. This phenotype is not that expected if one of the alleles shows complete dominance to the other allele. This phenotype is, however, consistent with incomplete dominance.

3. F₁ × F₁ cross

Pink $C^R C^W$ × Pink $C^R C^W$

F₁ pink-flowered plants are crossed to produce the F₂ generation.

4. F₂ generation

Gametes from one $C^R C^W$ F₁ pink-flowered plant

C^R C^W

Gametes from another $C^R C^W$ F₁ pink-flowered plant

C^R C^W

$C^R C^R$ $C^R C^W$
$C^R C^W$ $C^W C^W$

Each parent plant produces two types of gametes, C^R and C^W. Random fusion of the gametes from the two parents produces the F₂ generation.

Figure 6.9 Incomplete Dominance

Incomplete dominance occurs when the heterozygote phenotype is a blend of the dominant and recessive traits, such as pink flowers.

Figure 6.10 Codominance—Blood Types

Codominance occurs when the heterozygote expresses *both* traits; for example, both A and B alleles are present in type AB blood.

TABLE 6.3

Blood Type Genotypes and Phenotypes

Genotype	Phenotype
$I^A I^A$	Type A
$I^A i$	Type A
$I^B I^B$	Type B
$I^B i$	Type B
$I^A I^B$	Type AB
$i i$	Type O

Example: If mom has blood type A and dad has blood type B, is it possible for them to have a child that is type O? The answer is yes, because the parents could be heterozygous $I^A i$ and $I^B i$. If the gametes containing i alleles combined, then the baby can be ii, which is type O.

From Russell/Hertz/McMillan. *Biology*, 4E. © 2017 Brooks/Cole, a part of Cengage, Inc. Reproduced by permission. www.cengage.com/permissions

to proteins in the cell membranes of red blood cells. These glycoproteins act as cell recognition markers called *antigens*. The gene that encodes the enzyme, designated *I* (for immunoglobulin), has three alleles: I^A, I^B, which code for either antigen A or B, and *i*, which does not code for an antigen at all. Since antigens can stimulate an immune response (Chapter 23), it is important for people to receive the correct blood type during transfusions.

Different combinations of the three blood-type alleles produce four different possible phenotypes, or blood types (A, B, AB, and O) (Table 6.3). I^A and I^B are completely dominant over *i*, and I^A is codominant with I^B.

More than 40 other antigens are expressed on red blood cells, the most significant being Rh factor (antigen D). If this antigen is present the blood type is *positive*, if this antigen is not present the blood type is *negative*. Positive is dominant over negative because if one of the alleles expresses the antigen, that antigen will be present on the red blood cells. When an individual is exposed to an antigen that they themselves do not express, an immune reaction can occur and can be fatal.

DID YOU KNOW?

Chimpanzees have blood type A and are thought to have a common ancestor with the Cro-Magnon. Gorillas have blood type B and are thought to have a common ancestor with the Neanderthal.

DID YOU KNOW?

Some rare blood types can react with O negative blood (considered the "universal donor" blood type because it does not contain A, B, or Rh antigens) because of other antigens on the red blood cells. People who have this rare reaction to type O negative should carry medical information with them in case of an emergency.

DID YOU KNOW?

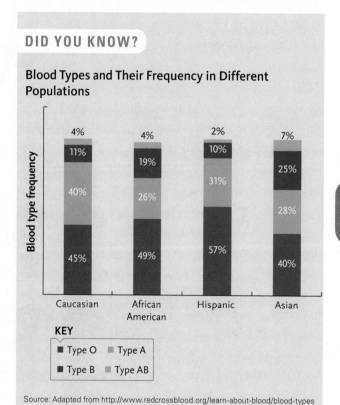

Blood Types and Their Frequency in Different Populations

Source: Adapted from http://www.redcrossblood.org/learn-about-blood/blood-types

Epistasis

Epistasis refers to the phenomenon where the effects of one gene are modified by one or more other genes; these are sometimes called **modifier genes**. For example, Labrador retrievers can have the phenotype black, brown, or yellow. For these dogs to be black or brown, they must have one or both dominant alleles for dark fur pigmentation (E); another allele will determine if that colour is black (BB or Bb) or brown (bb). A dog with EE or Ee will be black or brown, and a dog with ee will be yellow, regardless of the other colour alleles (Figure 6.11). A dog with the BBee or Bbee or bbee

A. **Black Labrador**

B. **Chocolate brown Labrador**

C. **Yellow Labrador**

D. **Black × yellow Labrador cross**

Homozygous parents:

Black $BB\,EE$ × Yellow $bb\,ee$

F_1 puppies:

Black $Bb\,Ee$

F_2 offspring from cross of two F_1 $Bb\,Ee$ dogs:

Gametes from one $Bb\,Ee$ F_1 dog:

Gametes from another $Bb\,Ee$ F_1 dog:

	BE	Be	bE	be
BE	$BB\,EE$	$BB\,Ee$	$Bb\,EE$	$Bb\,Ee$
Be	$BB\,Ee$	$BB\,ee$	$Bb\,Ee$	$Bb\,ee$
bE	$bB\,EE$	$Bb\,Ee$	$bb\,EE$	$bb\,Ee$
be	$Bb\,Ee$	$Bb\,ee$	$bb\,Ee$	$bb\,ee$

F_2 phenotypic ratio is 9 black:3 chocolate:4 yellow

Figure 6.11 Epistasis

Epistasis occurs when the phenotype of one gene masks the phenotype of another gene.

genotype will be yellow, because the epistasis (E) allele can mask the expression of the B or b allele.

Environmental Influences

Sometimes environmental factors determine how genes are expressed: for example, temperature, or the season, or the length of daylight. Hydrangea flowers have blue flowers if the soil is acidic and pink flowers if the soil is more alkaline (Figure 6.12). Another example is the arctic fox, which has white fur in the winter when it is cold and brown fur in the summer when it is warm. The sex of many reptiles depends on the temperature of the eggs; eggs incubated at temperatures higher than 30 degrees will be female, lower temperatures will be male. In humans, environmental factors, such as nutrition or

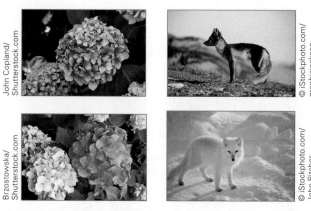

Figure 6.12 Environmental Factors

Environmental factors can influence the expression of certain genes; for example, the pH level in the soil alters the colour of hydrangeas, and the colour of arctic fox fur changes with the seasons.

exposure to infectious organisms, affect phenotypes such as height, weight, and immune function.

X-Linked Influences

Some inheritance patterns depend on which chromosomes the gene is located on. We have 23 pairs of homologous chromosomes: numbered 1 to 22 and also an X or Y. Even though X and Y are considered homologous, they do not have similar genes. Females have XX and males have XY. The Y chromosome contains the sex-determining factor that causes testosterone production and male development. The X chromosome contains many genes that are not related to sex characteristics, and those genes are called **X-linked**. For example, fruit flies have a mutant eye colour gene that causes white eyes instead of the normal red eyes (Figure 6.13), mostly in males. A female fruit fly with red eyes could have the genotype $X^R X^R$ or $X^R X^r$, with red eyes (R) being dominant over white eyes (r). A male fruit fly with red eyes has the genotype $X^R Y$, and with white eyes has the genotype $X^r Y$. It is more common for males to have white eyes because they only need one white-eye allele. It is possible for a female to have white eyes, but she must have both X chromosomes with the eye colour mutation ($X^r X^r$).

Human X-linked, recessive traits include hemophilia, which is a blood clotting disorder; red-green colour blindness; and Duchene muscular dystrophy. Each of

Figure 6.13 Drosophila Melanogaster

Drosophila melanogaster normally have red eyes, but sometimes males can have white eyes.

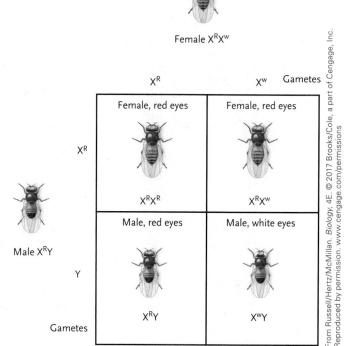

Figure 6.14 X-Linked Inheritance

X-linked inheritance is related to genes on the X chromosome. Males are more likely than females to express an X-linked mutation because males have only one X chromosome and females would need to have the mutation on both of their X chromosomes.

these traits are much more common in males than females because males have only one X chromosome and females need to inherit *two* mutated alleles to have the disease (Figure 6.14).

Linkage

Some genes tend to be frequently inherited together. When genes are located close together on the same chromosome, they are considered **linked** (Figure 6.15). The further two genes are away from each other on a chromosome, the likelihood of linkage decreases; this is due to the possibility that those genes may be separated by crossing-over during meiosis.

DID YOU KNOW?

Black and orange calico cats can only be female. The fur-colour gene is located on X chromosomes, so male calicos can only be black or orange. However, females have two X chromosomes, and the black and orange gene is codominant, so females can have a black and an orange allele. This is an example of epistasis, because a separate gene determines the pattern of fur-colour distribution, but only in females. This is also an example of a polygenic trait because another gene codes for white fur. The overall phenotype of fur colour and distribution is determined by multiple genes.

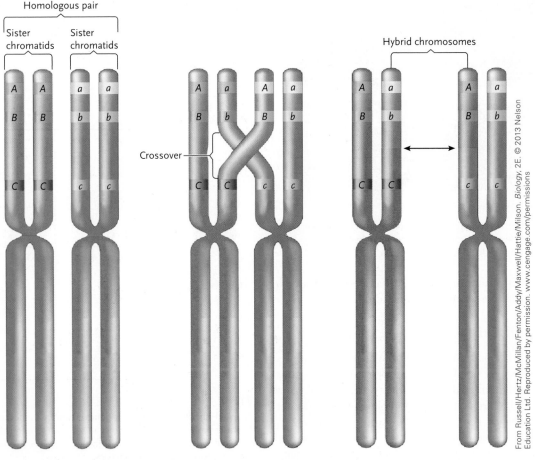

Figure 6.15 Linkage

Linkage occurs when two genes are located close together on a chromosome, and so they are much less likely to be separated during crossing over; the traits tend to be inherited together.

From Russell/Hertz/McMillan/Fenton/Addy/Maxwell/Hattie/Milson. *Biology*, 2E. © 2013 Nelson Education Ltd. Reproduced by permission. www.cengage.com/permissions

CONCEPT REVIEW 6.4

1. Define polygenic inheritance, and give three examples.

2. Define pleiotropic, and give one example.

3. If a bird's feather colour is incompletely dominant, and one allele causes red feathers and another allele gives blue feathers, what will a heterozygote look like? If two heterozygous birds mate, how many of their offspring will be blue?

4. If a mom has blood type A, and her baby has blood type O, which of the following is the dad: Larry with type O, Bill with type B, or Fred with type AB. What is mom's genotype? Explain how you know that.

5. Mom has type O blood, and dad is AB. What are the possible genotypes and phenotypes of their children?

6. What is epistasis? Explain how a dog can have the dominant black-fur gene but is a yellow lab?

7. Give two examples of environmental factors that affect gene expression.

8. Colour blindness is X-linked. If dad is colour blind and mom is a carrier, what is the probability that they will have a child with colour blindness?

9. What is linkage?

6.5 GENETIC DISORDERS

Errors can occur during meiosis. **Nondisjunction** refers to the failure of chromosomes to separate correctly during either meiosis I or meiosis II. Recall from Chapter 5 that during the first meiotic division, the homologous chromosomes separate, and during the second division, the sister chromatids separate. When one gamete has more or fewer chromosomes than it is supposed to, this is called **aneuploidy**: an abnormal chromosome number. If gametes with abnormal chromosome numbers are involved in fertilization, normal development is usually

Figure 6.16 Nondisjunction

Nondisjunction occurs when the chromosomes do not separate properly during meiosis and gametes have more or fewer chromosomes than normal.

not possible and early death occurs. However, if there is an extra chromosome 21 or, more rarely, an extra chromosome 22, development can continue (Figure 6.16). A child with an extra chromosome 21—**trisomy 21**—has **Down syndrome** and will have delayed development and mental impairment. Nondisjunction of the **sex chromosomes** can also lead to three viable genotypes (Table 6.4).

DID YOU KNOW?

The risk of a child being born with Down syndrome increases as the *mother's* age increases. All of the eggs in the ovaries were formed during embryonic development, and unless fertilization occurs the last meiotic cell division is not completed. As a woman's age increases, there is a higher chance that the last meiotic division will not happen normally. The risk of Down syndrome is approximately 2% or less when the mother is 35 years old or younger, and this increases to 10% at age 40, and 30% at age 45.

Single-Gene Hereditary Diseases

Hereditary mutations that do not cause death almost always result in recessive alleles. Recessive alleles are not eliminated from the population because usually they don't cause disease in heterozygous individuals,

TABLE 6.4

Nondisjunction of the Sex Chromosomes	
Genotype	**Phenotype**
XXX—Trisomy X	Female, tall stature, learning disabilities, limited fertility
XO—Turner syndrome	Female, short stature, sex organs don't mature at adolescence, may be infertile
XXY—Klinefelter syndrome	Male, severity varies, small gonads, breast development, may be infertile
XYY	Normal male

who have one recessive and one dominant allele. When a gene mutation occurs on chromosomes 1 to 22, it is **autosomal**; when it occurs on the X chromosome, it is X-linked. If individuals have the disease only when they have both mutated alleles, the disease is recessive. If individuals have the disease when they have only one mutated allele, the disease is dominant.

The following list describes seven examples of single-gene hereditary diseases.

• *Sickle cell anemia* is an autosomal recessive disease, where the **hemoglobin** gene is mutated, causing abnormal red blood cell formation and reduced capacity to carry oxygen (Figure 6.17). Sickle cells are stiff and sticky and tend to block blood vessels,

Figure 6.17 Sickle Cell Anemia

Sickle cell anemia is an autosomal recessive disorder caused by a mutation in the hemoglobin gene.

and they can damage many organs. Symptoms begin in children around four months old; with advancing medical treatment, life expectancy is now into the 40s or 50s.

- *Tay Sachs* is an autosomal recessive disease, where the HEXA gene that codes for a lysosomal enzyme is mutated, causing brain deterioration and leading to death most often around age four. Different HEXA gene mutations cause variations in the disease and populations that are affected. There are infantile, juvenile, and adult-onset variations. For example, a 4-base insertion in exon 11 causes the infantile form, which is found in Ashkenazi Jews and a population of Cajuns in Louisiana.

- *Cystic fibrosis* is an autosomal recessive disease caused by a mutation in the chloride channel membrane protein, called the *cystic fibrosis transmembrane conductance regulator (CFTR) gene*. The mutated protein cannot properly insert into the plasma membrane and, therefore, cannot regulate the movement of chloride and sodium ions across epithelial cells membranes, and this leads to an altered water balance that primarily affects the lungs, liver, pancreas, and sweat glands. Early symptoms include salty sweat, poor growth, and frequent respiratory infections, but many body systems can be affected (Figure 6.18).

- *Huntington's disease* is an autosomal dominant disease, where the CAG nucleotide repeats are inserted into the Huntington gene, which increases the number of glutamine amino acids incorporated into the protein. Huntington's is a neurodegenerative disease that affects mood, cognitive ability, muscle coordination, and involuntary muscle movements, called *chorea*. Huntington's primarily affects people of European descent.

- *Hemophilia* is a recessive X-linked disease that affects a clotting factor gene (either factor VIII or factor IX) and prevents the formation of fibrin, which is required to form a clot whenever a blood vessel

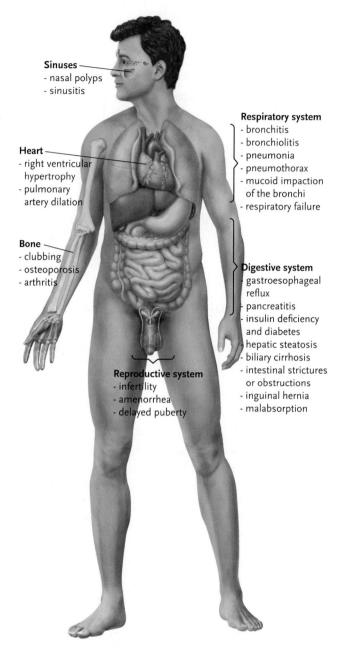

Figure 6.18 Cystic Fibrosis

Cystic fibrosis is an autosomal recessive disorder caused by a mutation in the chloride channel gene, called the *CFTR gene*, which stands for *cystic fibrosis transmembrane conductance regulator*.

has been damaged. People with hemophilia can have internal or external bleeding episodes that can be treated with infusions of the missing clotting factor.

- *Duchene muscular dystrophy* is a recessive X-linked disease caused by a mutation in the **dystrophin** gene, which holds myofibrils of muscle tissue together (Chapter 14). Without proper functioning of this protein, muscles degenerate and the disease progresses from muscle weakness to paralysis by age 10 to 12. Life expectancy is usually not longer than 25 years.

- *Familial hypercholesterolemia* is an autosomal dominant disease, where a mutation occurs in the **LDLR (low-density lipoprotein receptor)** gene. This gene is required for removing cholesterol from the bloodstream so cells can use it to make energy or the steroid hormones derived from cholesterol, such as testosterone or estrogen (Chapter 17). Those affected can have cardiovascular disease by age 30.

DID YOU KNOW?

Before 1985, donated blood was not tested for diseases such as HIV or hepatitis. It is estimated that 50% of hemophiliacs that received blood transfusions contracted HIV: approximately 10 000 people in North America.

CONCEPT REVIEW 6.5

1. Which chromosome is affected in Down syndrome?

2. Which sex chromosome nondisjunctions are viable?

3. List the seven most common single-gene hereditary diseases, the inheritance pattern, the gene affected, and the main phenotype for each.

7 Evolution

CHAPTER OUTLINE

Education's purpose is to change an empty mind to an open one.

Malcom Forbes

7.1 DARWIN'S THEORY OF EVOLUTION

Charles Darwin (Figure 7.1) published his book *On the Origin of Species by Means of Natural Selection* in 1859. Darwin's theory can be divided into two major components: microevolution and macroevolution. **Microevolution** refers to small changes within a single species, and **macroevolution** refers to the appearance of new species over thousands to millions of years.

Evolution is defined as the change in the inherited traits in a **population** of organisms over time. A single organism does not *evolve*, because individuals are born with the DNA they acquired through the process of reproduction. Individuals can *adapt* to some environmental changes, such as increasing red blood cell production at higher altitudes where oxygen concentrations are lower. Adaptation involves changing the *regulation* of the expression of certain genes. For example, the gene expression of erythropoietin—a hormone responsible for the production of red blood cells (Chapters 10, 17, and 21)—increases in people who move to high altitudes; this is because they need the higher level of red blood cell production due to the lower oxygen content in the atmosphere. The body continually adapts—changes gene expression—based on changes in the environment. So, the body adapts to meet different requirements: produces more erythropoietin if there is little oxygen in the environment, produces insulin when we eat, or makes growth hormone when we exercise.

Figure 7.1   Charles Darwin

Figure 7.2   Genetic Variation

Genetic variation is important for the survival of populations of organisms.

Evolution involves passing on different genes to off-spring, such as blood type A versus blood type O. Every living bacterium, protist, fungus, plant, and animal has their own unique combination of genetic material that will prove to be beneficial or deleterious to their survival and subsequent ability to reproduce. Organisms that survive will pass on their DNA to their offspring. Recall from Chapters 5 and 6 that during **meiosis** every gamete and, therefore, every zygote has a different combination of genes compared to their parents. Through sexual reproduction and **genetic variation** over many generations, new combinations of traits continually emerge within all populations (Figure 7.2).

One way that populations evolve is by means of **natural selection**. Darwin believed that all species over-reproduce, and only a small percentage of those born will survive. Herbert Spencer, a philosopher and biologist, coined the phrase "**survival of the fittest**" in 1864, after reading Darwin's *On the Origin of Species*. This term has persisted and is used frequently to describe the process of evolution: individuals born with the most beneficial traits for a *particular environment* will survive, reproduce, and pass on their DNA. The most obvious examples involve predator–prey relationships. The predators that are able to catch prey acquire the nutrients and good health for mating,

reproducing, and fighting off disease and competitors. Predators born with the DNA that gives them, for example, the ability to run fast, see in the dark, hear movement, or have a heightened olfactory sense for detecting prey will have an advantage over other individuals in that species and will more likely pass on their DNA to offspring (Figure 7.3). Likewise, prey that can evade predators, for example, by running quickly, or having a keen sense of hearing, or camouflage colouring will also be more likely to survive and pass on their DNA. Organisms born with genetic defects that prevent them from finding food, evading predators, competing for a mate, or being able to produce offspring will not pass on their DNA. In this way, the "fittest" organisms perpetuate certain genes in future generations.

Figure 7.3   Predator–Prey Relationships

Predator–prey relationships act as selection pressures and cause natural selection to occur in both the predators and the prey.

Survival of the fittest does not necessarily mean that the biggest, fastest organisms will be the most "fit." In many circumstances and depending on the environmental conditions, it may be more beneficial to be small, or nocturnal, or have camouflage colouring. For humans, and depending on the environment we live in, having access to healthcare, or clean water, or an education may be more important factors than being able to run fast or fight off predators.

Why don't organisms reach a point when they have the best combination of traits in relation to their environment and then just stay the same? Why does evolution continue to occur in all living things? Organisms live in continually changing circumstances, and **ecology**—the relationship between organisms and their environment—determines evolution. Since Earth first formed 4.5 billion years ago, conditions in every part of the planet have continued to change: such as changes in seasons, weather patterns, ice ages, forest fires, food availability, predators, and infectious organisms (Figure 7.4a). Each organism has different combinations of traits that are more or less beneficial in their particular environment. Therefore, depending on the type of environment change and the DNA of the organism, certain individuals are more likely to survive, reproduce, and pass on their DNA. This explains why genetic variation is important (Chapters 5 and 6).

Environmental changes do not necessarily impact large ecosystems. An environmental change can be as simple as putting an electric bug zapper in your backyard (Figure 7.4b). Most moths (and many other insects) have a gene that causes them to be drawn to light, but some have a mutation in this gene so that they are not drawn toward light (your bug zapper will kill only the moths that have this gene without the mutation). In this example, the environmental change is the addition of the bug zapper, and the fittest moths in your backyard

environment have the mutated version of the light-drawing gene; they are not attracted to light and, therefore, they will survive.

DID YOU KNOW?

Insects have a gene that draws them toward light because it is beneficial for them to be able to navigate at night when the moon is out. The light from the moon helps insects travel in a straight direction. Since we use artificial lights, insects get trapped when they fly around light bulbs at night, instead of following the moonlight.

Darwin's Theory of Natural Selection

Darwin's idea of how organisms change is based on the following steps:

1. All individuals have a unique combination of the genetic traits they acquired from their parents. All populations have genetic variation.

2. The organisms that survive pass on their DNA to their offspring, so closely related individuals have more similar DNA than do less closely related individuals.

3. All living organisms over-reproduce, and only some of those offspring will survive, depending on their DNA and the current environmental conditions (Figure 7.5).

4. The organisms that do survive and pass on their DNA provide the basis for future generations to also be able to survive.

5. Over time, populations of organisms change because of new combinations of traits in conjunction with changes in the environment.

A. B.

Figure 7.4 Environmental Change
Environmental changes can be large (A) or small (B).

Figure 7.5 Over-Reproduction
All organisms over-reproduce, which ensures that some will likely survive any given environment.

CONCEPT REVIEW 7.1

1. Define evolution.
2. What does *survival of the fittest* mean? Give an example.
3. What is the relationship between evolution and ecology?
4. Explain the major factors involved in natural selection.
5. Give some examples of environmental factors that affect evolution.
6. Give examples of some human inherited traits that you think would be beneficial for survival today.

7.2 MICROEVOLUTION

Microevolution is defined as the change in inherited traits in a population of organisms within a single species over time. For example, putting a bug zapper in your backyard will change the insect populations so that more individuals will carry the mutated gene and will not be drawn toward light. At the same time, other environmental changes—for example, the weather, a change in the bird population that feeds on these insects, or infection can affect the evolution of other traits in this insect population. Over very long periods of time many small changes can lead to large-scale changes that may give rise to a new species; this is macroevolution. A **species** is a group of organisms that can breed only with each other and produce viable offspring. Therefore, if many small changes in moth DNA occur over time so that they are no longer able to breed with the previous group of moths, they would be considered a new species of moth. There are currently 160 000 moth species.

ART Collection/Alamy Stock Photo

Figure 7.6 Jean Baptiste de Lamarck

DID YOU KNOW?

There are over 400 000 species of beetle, 5000 species of frog, and 400 species of primate. Primates first existed approximately 65 million years ago.

Darwin is not the only person who questioned where organisms come from and why so much diversity exists among organisms. The French scientist **Jean Baptiste de Lamarck** (Figure 7.6) presented the first ideas of evolution in 1809. The fossil record shows that today's giraffes are much taller than their ancestors, indicating that microevolution has occurred in the giraffe population over time. Darwin and Lamarck had different ideas about why these changes occurred.

Lamarck believed that organisms inherited characteristics that were *acquired* by their parents during their lifetime. Figure 7.7 illustrates Lamarck's view that giraffes had to continually stretch to reach food, which caused those individual giraffes to acquire slightly longer necks during their lifetime. Then, the offspring of those giraffes would be born with longer necks than the offspring of giraffes that didn't stretch to reach food: leading to the long-necked giraffes of today.

DID YOU KNOW?

Giraffes are the tallest living terrestrial mammal and can be 6 m tall (20 ft) and weigh 1600 kg (3500 lb).

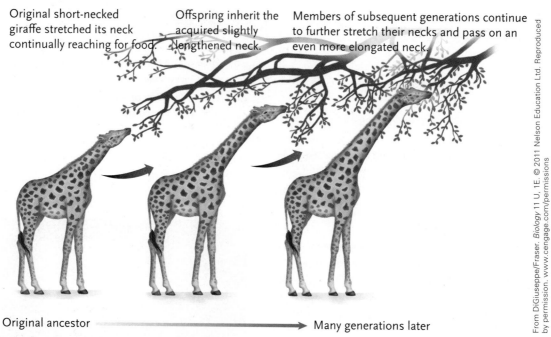

Original short-necked giraffe stretched its neck continually reaching for food.

Offspring inherit the acquired slightly lengthened neck.

Members of subsequent generations continue to further stretch their necks and pass on an even more elongated neck.

Original ancestor ⟶ Many generations later

From DiGiuseppe/Fraser. *Biology 11 U*, 1E. © 2011 Nelson Education Ltd. Reproduced by permission. www.cengage.com/permissions

Figure 7.7 Lamarck's Giraffes

Lamarck believed that giraffes evolved to have longer necks than their ancestors because they had to continually stretch to reach higher sources of food.

Darwin believed that giraffes are taller now because of natural selection. Several genes are involved in determining the height of a giraffe: similar to the multiple genes that determine height in humans (polygenic trait, Chapter 6). Genetic variation among giraffes allows for a range of heights within their population at any time. In ancestral giraffes, the tall giraffes were not as tall as they are now, but some of them were slightly taller than others at that time. These giraffes could reach more food higher in the trees, compared to the shorter giraffes that perhaps had to compete with other animals that ate the same food. If food supply was a continual selection pressure, over many years the tallest giraffes were the most likely to survive and pass on their "tall" genes to their offspring, leading to the very tall giraffes that exist today.

DID YOU KNOW?

Lamarck wasn't as far off as it may seem. Current evidence shows that behavioural changes in many organisms, including humans, can affect the regulation of genes in their children; this is called *epigenetics*. For example, rats that are not licked by their mother at birth will have an impaired stress response when they are older, not because of a change in their DNA sequences, but because of a change in how those stress response genes are expressed.

Importance of Microevolution to Humans

Microevolution occurs all the time in our ecosystems, and we interact with and influence many species of bacteria, fungi, plants, and animals. Since bacteria have lived everywhere on earth for over 3.5 billion years, humans co-evolved with bacteria, and 95% of bacteria are beneficial to humans. (Chapter 24).

An example of microevolution in bacteria involves **antibiotic resistance**: an extremely important health care concern. Bacteria can reproduce every 30 minutes in perfect conditions and, therefore, have a very rapid generation time. Bacteria undergo genetic changes based on changes in the environment, such as lack of a nutrient, or change in temperature or level of moisture. Since penicillin was discovered in 1928, bacterial microevolution has occurred at a rapid rate. And since that time, many beneficial **antibiotics** have been created to fight a myriad of bacterial infections in people and animals. However, antibiotics have been overused.

Antibiotics kill most, but not all, bacteria in any given population, making antibiotics a **selection pressure** that affects the microevolution of bacteria. Some bacteria have antibiotic **resistance** genes that allow them to survive in the presence of antibiotics; these bacteria are the fittest in an environment where antibiotics occur (Figure 7.8). Through natural selection, the fittest bacteria survive the antibiotic environment and pass on

Figure 7.8 Example of Microevolution

Bacteria that have antibiotic resistance genes will survive in an environment where antibiotics are present.

the resistant gene(s) to their bacterial offspring. Over the last few decades, more and more bacterial species that cause infection in humans and animals have become resistant to one or more antibiotic drugs. These bacterial populations have undergone microevolution in a relatively short period of time. Resistance genes have existed in bacterial populations for millions of years as a defence mechanism against fungus and other species that produce natural antibiotic molecules. Now that scientists have figured out how to isolate these molecules for medical use, the large amounts of antibiotics consumed by people and farm animals has brought about an increase in resistant bacterial populations. This happens because the non-resistant bacteria die off, but the resistant bacteria survive. This may become an extremely critical health care issue in the coming years. There are already several multi-resistant bacterial species that can cause fatal infections in humans: "super bugs" (Chapter 24).

DID YOU KNOW?

In 1928, Alexander Fleming isolated the first antibiotic ever used, penicillin, from the common indoor fungus found on salted food products, *Penicillium chrysogenum*.

Artificial Selection

People have an impact on the microevolution of many species, specifically domesticated animals and food plants. Human influence is a selection pressure that generates a variety of new species from the same ancestral species. As a result of this human selection pressure, microevolution leads to genetic variation that in the long term leads to macroevolution.

For example, all dogs from Chihuahuas to Great Danes to the grey wolf are the same species, *Canis lupus*. Domesticated pet dogs of all breeds are considered a subspecies of the grey wolf, *Canis lupus familiaris*. Being part of the same species means that

organisms can interbreed and produce viable, fertile offspring. The current enormous variation in breeds of dog is much more extreme than would ever naturally occur. This variation is due to **artificial selection**, whereby breeders choose which traits are most desirable and then breed for those traits: either in animals or in plants (Figure 7.9).

Humans began growing and artificially selecting food crops and domesticating animals approximately 8500 BC, the beginning of the agricultural revolution. Plants or animals with desirable characteristics were selected for reproduction, which caused rapid microevolutionary changes. Some examples of selected traits include cows that produce more milk; sheep with more wool; chickens that lay more eggs; horses that run faster; pigs that produce the most meat; seedless grapes or bananas, which can only be produced with human intervention. Also, broccoli, cauliflower, Brussel sprouts, cabbage, kale, and collard greens are all variations of the ancestral species *Brassica oleracea* (Figure 7.10). These vegetables exist because of the artificial selection of traits such as taste, appearance, colour, texture, size, and also type of leaf and, flower.

Courtesy of Garry Roscoe

Figure 7.9 Artificial Selection—Dogs

Humans have bred many varieties of dogs based on traits that were appealing. When humans determine which organisms should reproduce, this is called *artificial selection*.

A. B. C. D. E. F.

Figure 7.10 Artificial Selection—Vegetables

All our food plants have been artificially selected over thousands of years to give us the variety that we eat today. (A) broccoli, (B) Brussels sprouts, (C) cauliflower, (D) cabbage, (E) kale, and (F) ancestral wild mustard species.

CONCEPT REVIEW 7.2

1. Using bacteria and antibiotic resistance as an example, explain how microevolution occurs.

2. How do antibiotic-resistant bacteria become a selection pressure that affects humans?

3. What is the difference between microevolution and macroevolution?

4. How would Darwin and Lamarck each explain why modern-day giraffes have longer necks than ancestral giraffes?

5. How is natural selection different from artificial selection?

7.3 MACROEVOLUTION

Macroevolution refers to the production of new species that occurs through thousands or millions of years of microevolution.

Darwin's Finches

When Darwin travelled between South America and the Galapagos Islands, he observed the similarities and differences of several species of finch. An important difference between the species is the shape and size of the beak, which is highly adapted to different types of food in South America and the Galapagos. The finches in the two geographical locations are different enough to make them different species; however, they also have a number of similarities even though they live hundreds of kilometres apart. Darwin hypothesized that microevolution led to macroevolution, and this created all the finch species that exist today.

Two genes play a major role in the development of the shape and size of the beak: bone morphogenetic protein 4 (BMP4) and calmodulin. Changes in the timing and amount of expression of these genes can account for variations in beak shape and size in the population (Figure 7.11). Depending on the environment and food supply, certain finches will be more fit than others to eat certain foods: such as large, hard seeds; small seeds in small cracks; berries; worms; or insects. The theory of macroevolution is based on the idea that small changes in genes or gene expression over time—that is, microevolution—cause great changes over long periods of time, and this leads to the formation of new species.

Rate of Evolution

The general view is that evolution occurs over very long periods of time, perhaps millions of years. In different species, and depending on the environmental factors involved, macroevolution occurs at different rates. Two key theories explain the rate of evolution: gradualism, and punctuated equilibrium. **Gradualism** is a slow and gradual process whereby very small

Geospiza conirostris

Birds with long bills open cactus fruits to feed on the fleshy pulp.

Birds with intermediate bills may be favoured during nondrought years when many types of food are available.

Birds with deep bills strip bark from trees to locate insects.

© Krystyna Szulecka/Alamy Stock Photo

Figure 7.11 Macroevolution—Darwin's Finches
Darwin studied the geographical distribution of finches, which helped him develop his ideas about macroevolution.

Time

A.

B.

From DiGiuseppe/Fraser. *Biology* 11 U, 1E. © 2011 Nelson Education Ltd. Reproduced by permission. www.cengage.com/permissions

Figure 7.12 Gradualism and Punctuated Equilibrium
Two theories that relate to the rate of evolution are *gradualism* and *punctuated equilibrium*.

changes occur over millions of years. **Punctuated equilibrium** is a process whereby periods of no change are followed by periods of rapid change, as can be seen in the fossil record (Figure 7.12).

Some organisms hardly change over millions of years; for example, alligators have few noticeable changes throughout their existence of at least 200 million years (Figure 7.13).

Evidence for Macroevolution

Three main lines of evidence suggest that existing species came from a common ancestor:

(1) fossils, (2) homologous anatomy, and (3) homologous molecules.

© Bmoore81/Dreamstime.com

Figure 7.13 Alligator
Alligators have not changed significantly over 200 million years.

Fossils Fossils are considered the most direct evidence of macroevolution. Any part of an organism that contains minerals that are not degraded by micro-organisms can form rock; this includes bones, feathers, hooves, scales, horns, teeth, and shells. Also, an organism that is encased in tree sap (amber) can also be preserved: this idea was used in the film *Jurassic Park*. Fossilization is rare and occurs only under certain circumstances, which is why the fossil record is incomplete. Fossils show similarities between organisms over time.

Using the example of primate skulls, Figure 7.14 shows a range of skulls from modern chimpanzees, to primitive human primates that existed millions of years ago, and to current modern human skulls. Viewing the skull shape, you can see a trend in skull change that could indicate that modern humans came from an ancestor in common with the modern chimpanzee. Figure 7.15 shows some other famous fossil examples. Figure 7.16 shows the approximate order of the existence of primates. Figure 7.17 shows the general time periods when some of the main hominid species existed on earth. Note that *Homo sapiens* have been on earth for only 200 000 years and that includes the time period of the earlier Cro-Magnons. "Modern humans" with our current anatomical structures, have been on earth for approximately 50 000 years. Also note that Neanderthals and *Homo sapiens* coexisted for a few hundred thousand years; Neanderthals became extinct around the end of the last ice age, 20 000 years ago, and now we represent the only remaining species of the Homo genus.

Homologous Anatomy Recall from Chapter 5 that there are two copies of each chromosome, one from each parent. These chromosomes are very similar but not exactly the same because different alleles come from each parent; therefore, these chromosomes are called *homologous*. Evolutionists use the term **homologous structures** to compare anatomy between different

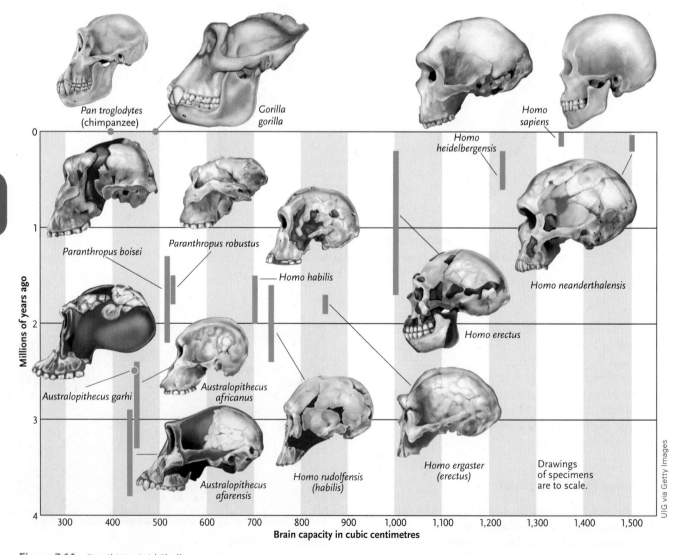

Figure 7.14 Fossil Hominid Skulls

Figure 7.15 Examples of Fossils

have **pharyngeal pouches**. These differentiate into gills in fish but disappear in mammals at a later stage in their embryonic development. All vertebrate embryos have a bony tail, which disappears in many mammals. Humans even have fur—called **lanugo**—at certain stages of embryonic development. Comparing the anatomical structures of a variety of vertebrates, we can see that most have the same basic bone structure, but with some differences in size and shape. Figure 7.18 shows examples of homologous structures in numerous vertebrate animals. Humans, cats, whales, bats, horses, and many other species all have a humerus, radius, ulna, carpals, metacarpals, and phalanges; they all have the same bones but with modifications.

Homologous Molecules To prove that macroevolution occurred, caused the formation of all species on earth today, and continues to occur, researchers use the evidence that the primary molecules of all living things are homologous. Every organism contains DNA,

organisms and to show that similarities indicate a common ancestor. Numerous similarities between different organisms appear at various times during embryonic development. For example, all vertebrate embryos

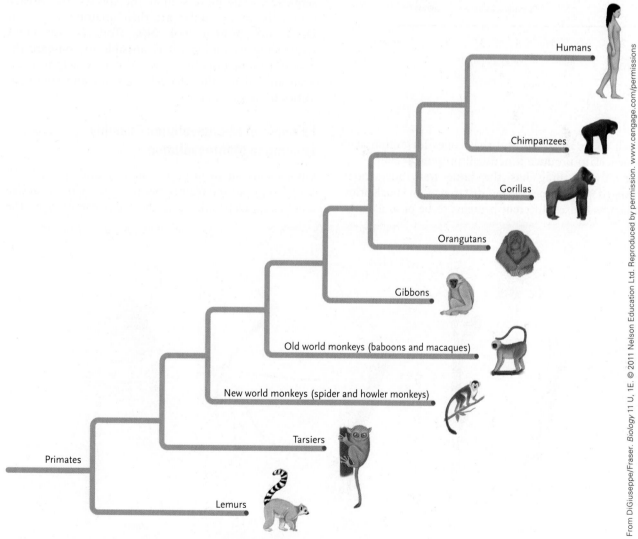

Figure 7.16 Approximate Order of the Establishment of Various Primate Species

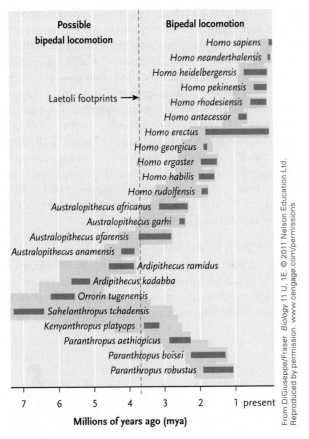

Figure 7.17 Human Evolution Timeline

the human body, different arrangements of nucleotides code for insulin, or collagen, or hemoglobin, or any of the 100 000 different proteins that humans produce. In finches, birds with bigger beaks have increased expression of the gene responsible for bone morphogenetic protein 4 (BMP4).

Many common genes exist in many organisms, and the closer relationship between species the more similarities exist. Scientists use mice to research many human diseases because mice possess many of the same genes and proteins that humans do; the genes and proteins might be slightly different but have the same function. For example, mice and humans both have a gene that codes for a cytochrome C protein that is required for the production of ATP. Figure 7.19 shows the homologous DNA sequence for the beginning of the cytochrome C gene in mice and humans. The sequences are almost identical, but there are some small differences. However, a comparison between the chimpanzee cytochrome C gene and that of humans shows more similarity than the comparison between the human and the mouse cytochrome C gene sequences. The more similar the species are to each other, the more similar are their anatomy and their DNA and protein molecules. Today, with DNA sequencing technology, it is possible to compare the genes of many organisms, and this may help to more accurately determine the relationships and commonalities between organisms.

Example of Microevolution Possibly Leading to Macroevolution

Guppies are small brightly coloured and spotted fish often kept in aquariums because of their attractive colouring. Depending on their environment in the

which codes for RNA and that codes for amino acid sequences to produce functional proteins (Chapter 2). Every living thing has the same four nucleotides arranged into genes. The difference in nucleotide sequences causes different proteins to be produced. In

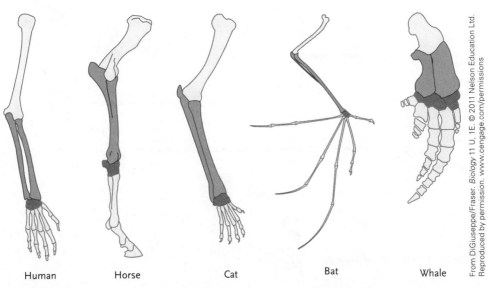

Figure 7.18 Examples of Homologous Structures

| Human DNA | GTT | GAG | AAA | GGC | AAG | AAG | ATT | TTT | ATT | ATG | AAG | TGT | TCC | GAG | TGC | CAC | ACC |
| Mouse DNA | GTT | GA_A_ | AAA | GGC | AAG | AAG | ATT | TTT | GTT | CAG | AAG | TGT | GCC | GAG | TGC | CAC | ACT |

| Human AA | Val | Glu | Lys | Gly | Lys | Lys | Ile | Phe | Ile | Met | Lys | Cys | Ser | Gln | Cys | His | Thr |
| Mouse AA | Val | Glu | Lys | Gly | Lys | Lys | Ile | Phe | Val | Gln | Lys | Cys | Ala | Gln | Cys | His | Thr |

Figure 7.19 Homologous Molecules
A comparison of a human and mouse gene sequence shows homologous molecules.

wild, populations of guppies show altered phenotypes over short periods of time. Microevolution involves the change in allele frequencies within a population over time, where organisms with the most beneficial traits for that environment survive and reproduce, passing along their alleles to their offspring. In one study, a guppy population was divided so that one group remained in a predator-free environment and the other group existed with a predator, such as pike. As a result, the allele frequencies and phenotypes of those populations changed within six months to a year. The types of changes observed for both populations are as follows:

1. Guppies in the predator-free environment
   - Brightly coloured and spotted males attracted females and were more likely to reproduce, so the number of guppies with spots and bright colours increased.
   - Reproduction age was slightly later than reproductive age of guppies living with predators.
   - Swimming speed was lower.

2. Guppies in the high-predation environment
   - Males with fewer spots and camouflage colouring were less likely to get eaten by the predator, so they survived to reproduce.
   - Due to predation stress, age of reproduction decreased, and body size was smaller.
   - Swimming speed increased.

The guppy population changed over time in the environment with the predator through natural selection; it changed from a brightly coloured and spotted phenotype to a camouflage phenotype. This indicates directional selection because the guppies with camouflage (one extreme) were more fit than the brightly coloured ones (the other extreme), or the intermediates. Guppies in the non-predation environment propagated the brightly coloured phenotype because females preferred those males for mating: that is, non-random mating.

If these guppy populations remained separate and DNA mutations or migration and mixing of alleles from other populations occurred, and they had to continually adapt to different environments—such as the addition of more predators, or disease, or different food availability—eventually these populations would be so different that they would not be able to breed with each other and would be considered different species: an example of macroevolution.

Factors That Affect Evolution

Several factors are involved in altering allele frequencies over time besides natural and artificial selection, such as DNA mutations, migration, non-random mating, and genetic drift.

Application

As you read through the rest of this chapter, consider that the human population is not the same today as it was 100 or 1000 years ago. As you read about the factors that affect evolution, make a list of how those factors affect human evolution.

Types of Selection Artificial selection occurs when people choose the traits in an animal or plant that they deem desirable: such as good-natured dogs, fast race horses, large tasty tomatoes.

Natural selection refers to the process where the organisms that are most fit for an environment survive and reproduce and pass on their DNA to their offspring. There are three types natural selection: stabilizing, disruptive, and directional (Figure 7.20).

Stabilizing selection occurs when the extremes of a phenotype are eliminated from a population. It is impossible to say for certain that a specific individual will have a better or worse chance of survival because many factors play a role; however, some traits that on average benefit survival in a whole population can usually be predicted. In stabilizing selection, genetic variation decreases as one specific trait is selected for. Human birth weight provides an example: a normal distribution of birth weights varies from very small to very large, with most babies being somewhere in the middle. Very small babies are less likely to survive because they may be underdeveloped, less likely to feed properly, and lose body heat easily. Very large babies are less likely to survive the birthing process. Babies born of intermediate size are the most likely to survive and pass on their intermediate-size genes.

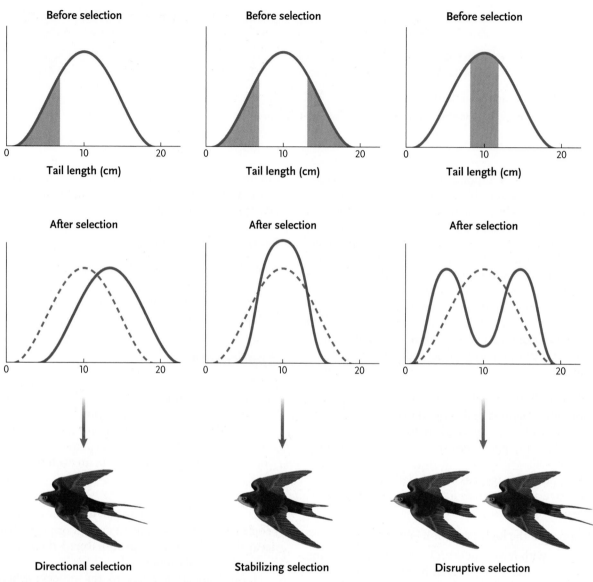

Figure 7.20 Natural Selection

Natural selection can be divided into three categories: stabilizing, disruptive, and directional. The dotted red line indicates the original population, and the blue solid line shows the change in the population after the particular selection pressure has occurred. Note the different tail length of each bird, based on the different selection types.

Sickle cell anemia is an example of stabilizing selection. It is an inherited disease, caused by a mutation in the hemoglobin gene such that the hemoglobin protein doesn't fold properly and is much less efficient at carrying oxygen (Chapter 6). Individuals need both mutated alleles to have the disease. If people who have the disease have a shortened lifespan, why is the sickle cell mutation still prevalent in certain populations? The possible genotypes in a population are as follows:

- homozygous dominant—two healthy hemoglobin alleles, so no sickle cell anemia occurs

- heterozygous—one healthy allele and one mutated allele, so no sickle cell anemia occurs

- homozygous recessive—two mutated alleles, so sickle cell anemia occurs

Recall that individuals considered more or less fit are the ones that survive to reproduce and pass on their DNA to their offspring, and this depends on the environment they live in. In Africa, the sickle cell anemia allele is much more prevalent than in North America because certain environmental factors play a role in each geographical location. In North America, the sickle cell allele has no benefit, so homozygous recessive individuals have a shorter lifespan, and people without the disease are more fit, and the allele continues to decrease in the population. However, in Africa, the sickle cell allele continues to be prevalent in the population because heterozygous individuals are less likely to die from malaria. Malaria is a *Plasmodium* parasite that infects red blood cells, and it is a leading cause of death in Africa. In Africa, homozygous dominant individuals without a sickle cell allele are more likely to die from malaria, and homozygous recessive individuals are more likely to die from sickle cell anemia. Therefore, the heterozygous individuals with one healthy allele and one mutated allele are the most "fit" in that environment because they do not have the sickle cell symptoms and they are less likely to die from malaria. This stabilizing selection in Africa acts on the sickle cell allele; this is also referred to as the **heterozygote advantage**. In central Africa the frequency of the sickle cell allele is 0.12 (12%) compared to 0.1 (1%) in North America.

Disruptive selection, the opposite of stabilizing selection, favours the extremes of a trait, and the intermediate trait is eliminated. This type of selection forms two distinct groups within the population, and it may be responsible for the early changes involved in macroevolution where, eventually, the two groups become different species. For example, the Galapagos finches that Darwin studied had a range of beak sizes and shapes in different environments, and these were adapted for different types of food. Darwin believed the finches came from a common ancestor population. Perhaps in one environment the food supply was such that birds could survive if they had very large beaks to eat large hard seeds or if they had very small beaks that could reach food in small cracks. The birds with intermediate beak sizes would have been less adapted to find food in that environment, leading to a population of finches with two distinct beak sizes.

Directional selection favours one extreme trait in a population so that the opposite extreme is eliminated. Recall the example from Section 7.2. In an environment with many backyard electric bug zappers, insects with a mutation in the gene that draws them toward light are more fit than insects without this mutation. Eventually, the population of insects in that area will be altered so that the allele frequency of the mutant gene will be more prevalent. Another example is antibiotic resistance in bacterial populations. In a population of bacteria that has a range of resistance to antibiotics, only the most resistant bacteria will survive during exposure to those drugs, leading to future generations with greatly increased expression of antibiotic resistance traits (Figure 7.8).

DNA Mutations Most DNA mutations are not beneficial, but they are considered the ultimate source of genetic variation in every population. All living organisms have genetic material made up of specific sequences of nucleotides that form genes that code for proteins (Chapter 2). Some DNA mutations have no effect on the structure or function of a particular protein (Chapter 8), but some mutations cause disease (Chapter 6), such as the mutation in the hemoglobin gene that causes sickle cell anemia, or the mutation in the chloride channel gene that causes cystic fibrosis. It is thought that some DNA mutations can cause the formation of a new allele that is a new protein that does not cause disease. An example is the evolution of the human blood cell types. Blood type A and B have existed throughout the existence of primate species, and continue to exist in humans. When parents of each blood type have children, sometimes that child has an A and a B allele, giving them blood type AB (Chapter 6). It is thought that type O came about later than type A and B as a mutation in the type A allele. So a DNA mutation led to the formation of a new allele and therefore a new phenotype that is not deleterious. It is also important to note that any DNA mutations that occur in somatic cells during an individual's life—for example, due to chemical exposure or any kind of mutagen—will not be passed onto their offspring unless it is also found in the gametes.

DID YOU KNOW?

Chimpanzees have only blood type A, and gorillas have only blood type B. Some researchers believe that chimpanzees and Cro-Magnons have a common ancestor that gave rise to Caucasians, and that gorillas and Neanderthals have a common ancestor that gave rise to Asians and Africans. Interestingly, more Caucasians have blood type A, and more Asians and Africans have blood type B. Type O blood is more common among First Nations.

Migration The movement of individuals to new locations is known as *migration*. This movement affects the allele frequencies of many populations. **Gene flow** is the transfer of alleles from one population to another, which occurs when groups of people or animals migrate and join other populations. Migration can also increase genetic variation in a population. For example, with the current ease of transportation around the globe, people migrate to different countries to live and have families. Consider a simple phenotype, skin colour: increased migration of people from different countries has dramatically increased the genetic variation and frequency of various skin-colour alleles in North America.

Figure 7.21 Blue-Footed Booby

The blue-footed booby has blue feet and a special dance that it uses to attract females.

Non-random Mating Random mating occurs when there are no preferred genes for reproduction: for example, pollination. Flowering plants pollinate nearby flowers that are the same species. In the animal kingdom, however, random mating rarely occurs. Among animals, genotypes are usually highly linked to reproductive behaviours, such as dancing rituals (blue-footed booby, Figure 7.21); strength (many animal males fight for a female); bird songs; feather colour; pheromone scent. Consider what preferences and therefore what genes may be involved in human mating. For example, people usually select a mate non-randomly and usually for observable traits. Cultural and social values play a role in mate selection, and people tend to prefer mates that have similar skin colour, stature, and personality.

Genetic Drift Genetic drift is the process of random changes in allele frequencies that are not due to an the fitness of an organism, but are due to major environmental impacts: for example, a hurricane that kills many organisms in a population. This is also called the **bottleneck effect**. The individuals that survive and continue to pass on their DNA may not necessarily be the ones with the most beneficial traits for the next generation. Also, the small population that is left may not be representative of the original population (Figure 7.22).

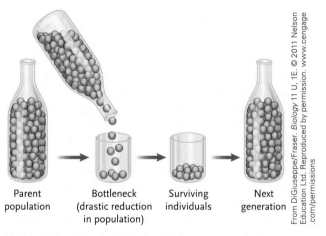

Figure 7.22 Bottleneck Effect

Environmental changes can cause a bottleneck effect, which alters the frequency of alleles in a population after many organisms in that population die.

CONCEPT REVIEW 7.3

1. Using Darwin's finches as an example, explain how macroevolution occurs.

2. What is the difference between gradualism and punctuated equilibrium?

3. Name the three main methods used as evidence of macroevolution.

4. Humans are more closely related to bats than to birds. How can anatomical structures or molecular sequences be used to determine if this is true or not?

5. What is a selection pressure? Give an example of a selection pressure that affects bacteria.

6. What are three types of natural selection that can affect allele frequencies? Give an example of each.

7. What type of selection is occurring with the sickle cell anemia alleles in Africa?

8. What is a heterozygote advantage?

9. What is the selection pressure acting on the guppies?

10. What factors cause allele frequency changes in populations?

11. Explain why the bottleneck effect does not select for the fittest individuals.

12. What is the source of new alleles in a population?

7.4 POPULATION GENETICS

Population genetics is the study of the frequency of **alleles** in a population. Every organism has at least two copies of every gene, one from each parent—although plants can sometimes have more than two copies. Each copy of the gene is not necessarily the same; for example, a pea plant can have a purple flower allele and a white flower allele, or it can have two alleles for the same colour of flower. When the two alleles are the same, the genotype is **homozygous,** and when they are different, the genotype is **heterozygous** (Chapter 6). Recall that homozygous or heterozygous refers to the pea plant's **genotype**, and having the trait of purple or white flowers describes the **phenotype**.

Within a population as a whole, there are a certain number of dominant and recessive alleles, and this is known as the **gene pool**. It is sometimes assumed that dominant traits will eventually predominate in a population and recessive traits will disappear, but this does not happen in a stable, non-evolving population. Two scientists, G. H. Hardy and Wilhelm Weinberg studied allele frequencies in stable populations and determined that the amount of dominant and recessive alleles stays the same over several generations; this is known as the *Hardy–Weinberg principle*. For a population to be considered stable or in equilibrium, the following assumptions apply:

1. The population has to be large. Small populations can have random changes in **allele frequency**.

2. Mating has to be random. The population will not be stable if certain phenotypes determine mating behaviour and evolution occurs.

3. There are no DNA mutations. Mutations can create new alleles that alter the gene pool.

4. There is no migration into or out of the population. If alleles are entering or leaving the population the allele frequencies change and evolution occurs.

5. There is no natural selection, that is, no selection pressure on a specific trait.

6. There are no environmental changes that cause random changes in allele frequencies, which is called *genetic drift*.

The **Hardy–Weinberg equilibrium** refers to populations where the above assumptions are true—the population is stable, not evolving, and allele frequencies stay the same over successive generations. The following example illustrates what can be determined about a given population based on the Hardy–Weinberg equilibrium.

Example 1
Suppose 100 mice live in a field, and they have black or white fur. In a stable population, fur colour does not influence mating, no mice are entering or leaving the field, there are no DNA mutations, no selection pressures such as predators, and no environmental changes. If black fur is dominant, the mice with two black fur alleles (homozygous dominant) and the heterozygous mice will have black fur; only homozygous recessive mice will have white fur. How many total fur-colour alleles exist in that field of mice? If there are 100 mice and each mouse has two alleles, then 200 fur-colour alleles exist in that population. If we suppose that exactly half of those alleles code for white fur and half for black fur, how many mice in that population would have white fur? Did you guess 50%? Did you guess 25%? If you guessed 25% you would be correct. This is because heterozygous mice have a white fur allele, but they have a black fur phenotype. This is explained with the following two equations.

The letters p and q represent the dominant and recessive alleles:

p = dominant allele, black fur gene

q = recessive allele, white fur gene

Every mouse has two copies of every gene. Table 7.1 shows the possible genotypes and phenotypes in this population of mice

In the total mouse population, the sum of the dominant and recessive alleles is 100% of the population (assuming that the trait is dominant-recessive inheritance only).

$p + q = 1$ (all the dominant alleles + all the recessive alleles = 100%)

In the total mouse population, the sum of all the genotypes is 100% of the population.

$p^2 + 2pq + q^2 = 1$ (homozygous dominant + heterozygous + homozygous recessive = 1)

Continuing with the fur-colour example, if half the alleles in the mouse population are black, $p = 0.5$ and, therefore, $q = 0.5$.

$p + q = 1$

$0.5 + 0.5 = 1$

How many mice are white?

$p^2 + 2pq + q^2 = 1$

$(0.5)^2 + 2(0.5)(0.5) + (0.5)^2 = 1$

$0.25 + 0.5 + 0.25 = 1$

Since q^2 represents the white mice, and q^2 equals 0.25, then 25% of the mouse population is white. According to the Hardy–Weinberg equilibrium, the frequency of alleles would not be changing over time; therefore, the mouse population is not evolving.

TABLE 7.1

Genotypes and Phenotypes in Population of Mice (Example 1)

Genotype	Phenotype
pp or p^2 (homozygous dominant)	Black fur
pq (heterozygous)	Black fur
qq or q^2 (homozygous recessive)	White fur

Example 2

What is the expected number of mice with black fur in a population where the frequency of the black fur allele is 0.6 (60% of all alleles in the population are for black fur), and black fur is dominant over white fur?

We know that $p + q = 1$, and since black fur is dominant, p must equal 0.6 and, therefore, q must equal 0.4. We also know that the possible genotypes for mice with black fur are homozygous dominant (p^2) *and* heterozygous ($2pq$). Therefore, the number of black mice in this population would be equal to $p^2 + 2pq$, which is $0.36 + 0.48 = 0.84$. Therefore, 84% of this population of mice will have black fur, and 16% will have white fur.

Example 3

Determine the frequency of the white fur allele in a population of mice if 9% of the population has white fur.

We know that $p^2 + 2pq + q^2 = 1$, and q^2 represents homozygous recessive white mice, which is 9% of the population, so $q^2 = 0.09$. To determine the frequency of the white fur allele, we need to solve for q:

$q^2 = 0.09$

$q = $ square root of 0.09

$q = 0.3$, or 30%

Therefore, 30% of the alleles in this population are the white fur allele.

DID YOU KNOW?

Cystic fibrosis is a genetically inherited mutation in the gene that codes for cell membrane chloride ion channels. When this gene is mutated the protein channel does not form properly, which causes disease in individuals that are homozygous recessive for the mutation. In North America, the frequency of the mutant allele is 0.025, which means that 2.5% of the alleles for cell membrane chloride channels are mutated, and 97.5% of those alleles are normal.

CONCEPT REVIEW 7.4

1. What factors maintain Hardy–Weinberg equilibrium?

2. Consider the following: Hardy–Weinberg equilibrium occurs in a population of rats, and long-tailed rats are dominant over short-tailed rats, and the dominant allele is present in the population at a frequency of 0.5. How many short-tailed rats are in the population? How many long-tailed rats are in the population?

CHAPTER REVIEW

Answer to Application Question

Factors that affect human evolution include the following:

- Natural selection. Humans inherit genes that affect their ability to reproduce and to adapt to their environment. Some examples include being physically capable of reproduction, ability to survive infections or extreme weather changes, and ability to learn, work, and socialize.

- DNA mutations. Most DNA mutations are detrimental and cause deformities or disease and, therefore, most often decreases the ability to reproduce. DNA mutations are the driving force behind evolution; mutations lead to increased genetic variation, and this is important for the survival of any population.

- Migration. As people increasingly travel and migrate around the world, reproduction between people of different backgrounds leads to offspring with DNA representing a diversity of physical characteristics.

- Non-random mating. Humans have preferences for partners with certain phenotypes. People of various cultures have different preferences such as physical appearance, intelligence, and personality traits such as honesty or compassion.

- Genetic drift. Human populations are affected by random events such as tsunamis, earthquakes, and plane crashes.

8 DNA Structure and Function

Andrea Danti/Shutterstock.com

8.1 DNA STRUCTURE

Composition of DNA

DNA is a macromolecule—a polymer—made up of building blocks—monomers—called **nucleotides** (Chapter 2). Every nucleotide consists of a phosphate molecule, a five-carbon sugar, and a nitrogenous base. DNA is made up of only four different nucleotides: adenine (A), guanine (G), thymine (T), and cytosine (C). A fifth nucleotide, called *uracil*, is found only in RNA. Adenine and guanine are purines, which are double-ringed molecules, and thymine, cytosine, and uracil are single-ringed molecules called *pyrimidines* (Figure 8.1). In the DNA molecule, a purine always binds with a pyrimidine, which is why DNA has a consistent symmetrical shape. Adenine always binds with thymine, and cytosine always binds with guanine.

Learning is not attained by chance, it must be sought for with ardor and diligence.

Abigail Adams

DID YOU KNOW?

Humans have 46 DNA macromolecules (chromosomes) in every cell. Chimps and potatoes have 48, chickens and dogs have 78, ants have 2, and shrimp have 254.

Figure 8.1 Four Nucleotides in DNA

DNA consists of sequences of four nucleotides: adenine, guanine, thymine, and cytosine.

Figure 8.2 Ribose and Deoxyribose

Ribose and deoxyribose sugars in nucleic acids have five carbon atoms, numbered C1–C5.

hydrogen atoms attached to the second carbon, whereas ribose has a hydroxyl group attached to the second carbon.

Double-Helix Structure

DNA is composed of two strands that naturally twist into a helical shape, the DNA double helix (Figure 8.3). This structure was discovered in 1953 through the combined research efforts of Rosalind Franklin, Francis Crick, and James Watson at Cambridge University, England. The DNA backbone consists of alternating sugar-phosphate molecules that form covalent bonds called *phosphodiester bonds*. The sugar is connected to the purine or pyrimidine by a glycosidic bond, and the bases are connected in pairs by hydrogen bonds. DNA has two complementary strands, and they connect with each other by aligning in opposite directions; this means the two DNA strands are **antiparallel**. Phosphate molecules bind to either the fifth or third carbon on the deoxyribose, which gives direction to the DNA strand. When the phosphate binds to the fifth carbon, which is called the 5′ (5 prime) *end* of one DNA strand, the other end has a free OH group on the third carbon, which becomes the 3′ (3 prime) end.

Nucleosides and Nucleotides

Nucleosides are composed of a purine or pyrimidine base attached to either ribose or deoxyribose (Figure 8.4). A nucleotide is a nucleoside with a phosphate group. When DNA replicates, nucleoside triphosphates are added to the growing DNA strand, and two of the phosphates are removed. ATP (adenosine triphosphate) is a nucleoside triphosphate. Recall from Chapter 4 that ATP is produced in all cells and is the primary source of energy for all cell functions. GTP can also be used as energy in cells. Both ATP and GTP provide energy through the loss of one or more phosphate groups.

Our diet is the source of the nucleosides that are made into nucleotides for DNA replication or the transcription of RNA molecules. The liver can synthesize nucleosides, but our main source is the foods we

Comparison of Sugar Molecule in DNA and RNA

The sugar molecule in a DNA nucleotide is different from the sugar molecule in RNA nucleotides. DNA has the five-carbon sugar called **deoxyribose**, and RNA has the five-carbon sugar called **ribose** (Figure 8.2). The carbon atoms of each sugar are numbered 1, 2, 3, 4, 5 starting with the first carbon within the ring; this carbon atom binds to the nitrogenous base. Carbon 3 and 5 bind with the phosphate groups that make up the sugar-phosphate backbone. Deoxyribose has two

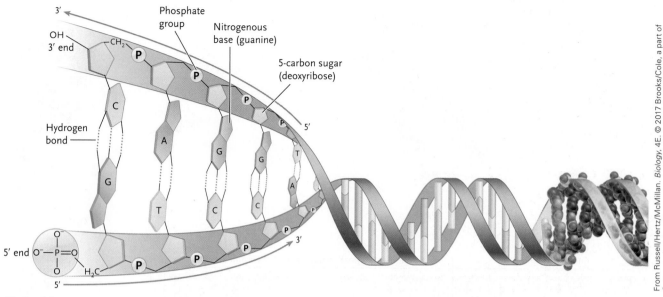

Figure 8.3   Double Helix Structure of DNA

Figure 8.4   Base, Nucleoside, and Nucleotide

Nucleotides are built from nucleosides (sugar and nitrogenous base) and a phosphate group.

eat: anything that is plant or animal and not processed. All plant and animal cells contain DNA and RNA, which get broken down in our digestive tract with specific enzymes (Chapter 20) and then used to form nucleotides.

Telomeres

All eukaryotic chromosomes have special repeated DNA sequences at the ends of each chromosome: **telomeres** (Figure 8.5). The human telomere repeat is 5′ TTAGGG 3′, and it is 5 to 10 kb (kilobases) long. Telomere sequences protect the genes on the ends of chromosomes from deteriorating during DNA replication. Each time the cell enters S-phase and DNA replication occurs, the very first 5′ end of each

chromosome is not replicated. The beginning of the DNA replication process involves the addition of primers, which are later removed and replaced with DNA (see Section 8.2). All the primers along the chromosome can be replaced with DNA, except the first 5′ end. Therefore, with each successive replication, the chromosomes become slightly shorter.

The body's cells have a specific lifespan, mainly because of this shortening process, and this is part of the reason why we age as a whole organism. If telomeres were not present the genes at the ends of the chromosomes would degrade and cause cell dysfunction. When the telomeres shorten extensively after many cell divisions, the cell dies or becomes **senescent** and loses the ability to divide. New cells arise from the adult stem cells in many tissues.

From Russell/Hertz/McMillan/Fenton/Addy/Maxwell/Hattie/Milson. *Biology*, 2E. © 2013 Nelson Education Ltd. Reproduced by permission. www.cengage.com/permissions

Figure 8.5 Telomeres

All eukaryotic chromosomes have telomere sequences at the end of each chromosome, and they shorten with each DNA replication.

DID YOU KNOW?

Different organisms have different rates of aging. A mouse lives for approximately 3 years; humans, 80 years; tortoises, 160 years; and black coral, 4000 years.

Absence of Telomeres in Prokaryotes

Most bacterial species do not have telomeres because their DNA molecules are circular. Bacteria can also divide indefinitely and don't have a specified lifespan, unless environmental factors cause them to die.

CONCEPT REVIEW 8.1

1. What are the five nucleotides that make up nucleic acid macromolecules?

2. Which of the nucleotides are purines, and which are pyrimidines?

3. Which nucleotides always combine in the DNA molecule?

4. What three types of bonds occur in the DNA molecule?

5. What is the difference between a nucleoside and a nucleotide?

6. How do our cells acquire nucleosides?

7. What kind of structure does DNA form?

8. How are DNA and RNA different?

9. Explain how the carbons of the sugar molecules are named and why this is important.

10. What is the difference between ribose and deoxyribose?

11. What is the energy molecule of the cell?

12. What are telomeres? Why are they important?

13. Do prokaryotes have telomeres? Why or why not?

8.2 DNA REPLICATION

DNA Semi-conservative Replication

DNA replicates according to a method known as **semi-conservative replication**: the original strands of DNA separate and act as a template for two new strands of DNA (Figure 8.6). The DNA in every living organism replicates during the S-phase of the cell cycle before mitosis (see Chapter 5).

In prokaryotes, DNA replication begins at a specific location in the genome, called the origin of replication (Figure 8.7). Eukaryotes have a lot more DNA, and so each chromosome usually has multiple sites of origin of replication.

In all living organisms, several enzymes perform specific functions that enable the replication process and ensure that two complete copies of the DNA are produced. **Helicase** is the enzyme that breaks the hydrogen bonds between the bases of the complementary strands of DNA, starting at the origin of replication. Helicase gets its name because DNA has a double helix structure. When the DNA begins to separate and the helicase moves along the DNA molecule separating the two strands, the open portion is called the **replication fork**. DNA can become overwound ahead of the replication fork and therefore, an enzyme called topoisomerase is used to cut the DNA backbone to relieve the supercoiling. **Single-strand binding proteins** stabilize the single-strand region of the DNA molecule, and the DNA polymerase enzyme adds the

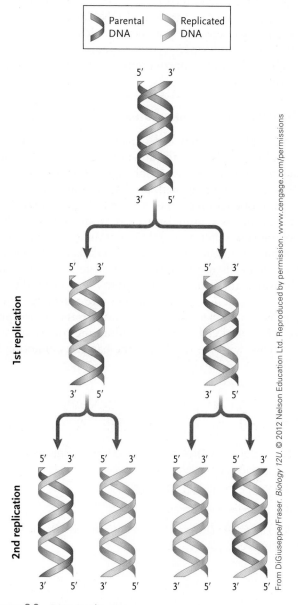

Figure 8.6 DNA Replication

Each original DNA strand separates during DNA replication and acts as a template for the new strand. The process is called *semi-conservative replication* because the new strands of DNA are made up of one original strand and one new strand.

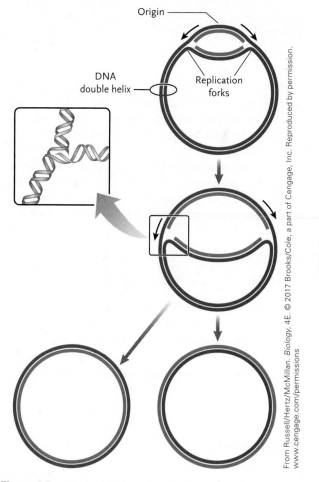

Figure 8.7 Bacterial DNA Replication

Bacterial DNA replication is also semi-conservative; the process starts at one point of origin, and the DNA is replicated in both directions at once.

complementary nucleotides to each single strand. Adenine always pairs with thymine, and cytosine always pairs with guanine. It is possible for all living things to replicate DNA almost perfectly because a purine always binds with a pyrimidine. Some incorrect nucleotides may be added and lead to DNA mutations (discussed further in Section 8.3).

Polymerase can add **nucleoside triphosphates** only to the 3′ end of a nucleotide on the *new* strand. Then two phosphate groups are removed, which leaves a nucleotide: sugar, phosphate, and base (Figure 8.8). The problem arises when the replication fork begins and there is no 3′ hydroxyl group yet available. This problem is solved when the enzyme **primase** adds to the open strand of DNA a small RNA primer: a short sequence of nucleotides that has a free 3′ hydroxyl group.

DID YOU KNOW?

E. coli bacteria have 4.7 million nucleotides on their single chromosome, which can be replicated within 40 minutes (1000 bases per second). An average human chromosome contains approximately 150 million nucleotides, but due to the multiple origins of replication, an entire genome can be replicated in about one hour.

After an RNA primer has been added, DNA polymerase copies the DNA by adding the complementary nucleotides to the template strand. Because the two strands of DNA are antiparallel and helicase moves only in one direction and polymerase can add only to the 3′ end, the two strands of DNA are copied in different ways. The two template strands are called a *leading strand* and a *lagging strand*. On the leading strand, only one RNA primer is required, and polymerase can follow the helicase and add complementary nucleotides in the 5′ to 3′ direction (on the new strand). The lagging strand, however, needs to replicate in fragments; these are called *Okazaki fragments*, named after the scientist who discovered them. These fragments require new RNA primers so that the polymerase can still add nucleotides in the 5′ to 3′ direction (Figure 8.9).

As the replicated DNA is produced, the RNA primers are removed by DNA polymerase, and DNA nucleotides replace the RNA primer. An enzyme called **ligase** forms phosphodiester bonds to seal the gaps. As the replication fork moves along the chromosome, the double helix structure re-forms naturally. Eventually, in eukaryotic cells, the replication forks meet, and two new strands of DNA are produced. In prokaryotic cells, the polymerase adds approximately 1000 nucleotides per second. In eukaryotic cells, the polymerase adds approximately 50 nucleotides per second.

Supercoiling in Prokaryotes

DNA replication enzymes in bacteria are the same as in eukaryotic cells: helicase unwinds the DNA, and polymerase adds nucleoside triphosphates in the 5′ to 3′ direction, beginning with a primer. However, because prokaryotic chromosomes are small and circular, only one primer is required on each strand initially; then polymerase replicates DNA continuously on both strands, without the need of Okazaki fragments. Bacterial cells use topoisomerase to relieve supercoils. In viruses, topoisomerase is used to integrate viral DNA into the genome of the cell they are infecting. Some antibiotics, such as the fluoroquinolone antibiotics, prevent bacterial cell growth by inhibiting topoisomerase enzymes, with the result that the bacteria cannot replicate DNA and divide.

CONCEPT REVIEW 8.2

1. What is semi-conservative replication?
2. Make a chart to show the main functions of the following enzymes and molecules involved in DNA replication: helicase, single-strand binding protein, primase, polymerase (two functions), and ligase.
3. Explain how DNA replication differs on the leading strand and the lagging strand.
4. Why do we need primase?
5. What is topoisomerase?
6. How does DNA replication differ between prokaryotes and eukaryotes?

Application

As you read through the rest of this chapter, consider the following question. Viruses can cause DNA mutations that lead to cancer. Viruses inject strands of DNA into the cell during infection. What type of DNA mutation can this cause—a point mutation or a recombination mutation? Look at Figure 8.11, and then describe how this type of mutation can cause cancer.

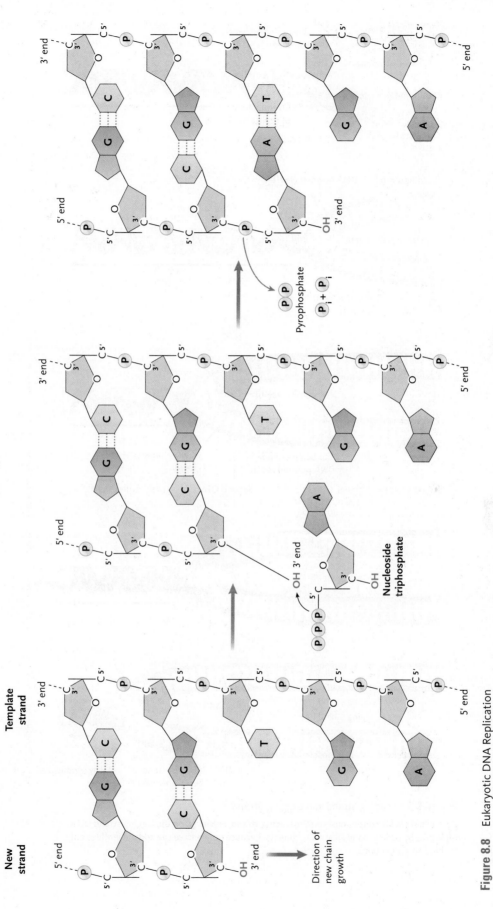

Figure 8.8 Eukaryotic DNA Replication

Nucleoside triphosphates are added by polymerase to the 39 OH group of an existing nucleotide.

Figure 8.9 Leading Strand and Lagging Strand

DNA can replicate continuously on the leading strand. However, on the lagging strand DNA replicates by adding many Okazaki fragments, because polymerase can add nucleotides only in the 5′ to 3′ direction.

8.3 DNA MUTATIONS

DNA mutations are the primary source of genetic variation and the creation of new alleles within a population. Some gene sequences are highly conserved between species because of their importance to survival. For example, the gene responsible for the sodium-potassium pump is found in every living organism, as are genes that produce the proteins required for glycolysis—the production of ATP in the absence of oxygen. DNA mutations are produced through errors in replication, exposure to environmental chemicals, radiation, or some viruses. Whatever way they occur, mutations may be detrimental to the survival of the organism, beneficial, or have no effect at all (silent mutations).

Errors in Replication

Errors in replication occur approximately once in every 1 million nucleotides. These mutations would accumulate very quickly if they were not fixed. Not all DNA mutations are fixed, and this leads to many illnesses (Figure 8.10). During replication, polymerase adds complementary nucleotides according to the sequence of nucleotides on the template strand. However, the following errors can result from the action of polymerase: (1) **substitution**, or mismatch, when the wrong nucleotide is added; (2) **insertion**, or addition, when an extra nucleotide is added; and (3) **deletion**, when a nucleotide is skipped. Errors in the DNA sequence lead to errors in the amino acid sequence of proteins, which can affect the function of the particular protein.

Categories of Mutations

Point mutations involve either one or several nucleotides and can be substitutions, insertions, or deletions. A **frameshift mutation** occurs when an insertion or deletion mutation occurs, and the whole gene sequence is shifted. This type of mutation is very harmful because the amino acid sequence of a protein is determined by every three nucleotides of an RNA molecule, and the production of RNA is based on the DNA sequence; therefore, an entire protein is affected. **Recombination mutations** involve the moving of large sequences of DNA from one location to another within the genome. **Transposable elements** (TEs), also called *transposons* or *jumping genes*, refer to sequences of DNA that move or transpose themselves to another chromosome within a cell. The mechanism of transposition can be either "copy and paste" or "cut and paste." All living organisms contain TEs, and these play a significant role in phenotypic variation and evolution, but they can also cause harmful mutations if genes are damaged. When a mutagen causes a break in the sugar-phosphate backbone, the cell tries to repair this by adding the resulting fragment to another piece

Figure 8.10 DNA Mutations

DNA mutations can lead to the production of non-functional proteins.

From DiGiuseppe/Fraser. *Biology 12U*, 1E. © 2012 Nelson Education Ltd. Reproduced by permission. www.cengage.com/permissions

of DNA. Then some segments of DNA move to entirely different chromosomes. If this occurs in germ-line cells during gametogenesis, the homologous chromosomes may not line up properly, and meiosis cannot continue.

Insertional inactivation occurs when a transposable element inserts a sequence of DNA into the middle of a gene, and so that gene can no longer produce a functional protein (see Chapter 10). This can happen in a number of ways: (1) a TE may insert a sequence of DNA into the regulatory region upstream of the beginning of a gene sequence—sometimes hundreds of nucleotides before the beginning of the actual gene—and alter how often a particular protein is transcribed and translated; (2) a TE may turn on or off that gene in the wrong cell type; and (3) a TE that inserts into a non-coding region

DNA segment or transposon

Normal gene and regulatory sequences: Gene is expressed in appropriate cells and transcribed to functional protein product.

Normal gene

Regulatory sequences

1 Mutated gene, normal regulatory sequences: Gene is expressed in appropriate cells but is transcribed to non-functional protein product.

2 Normal gene, mutated regulatory sequences: Functional protein product is transcribed but not in appropriate cells or in appropriate amounts.

3 No mutations in gene or regulatory sequences: Silent mutation.

Figure 8.11 Transposition Mutations

Transposition or recombination can result in the production of a non-functional protein, an altered regulation of protein expression, or no noticeable effect.

of the chromosome may have no noticeable effect at all. (See Figure 8.11.)

Mutations can occur in somatic cells or gametes (Chapters 5 and 6). Somatic cell mutations are not passed on to future generations but can be detrimental and possibly lethal to the individual. Mutations that occur in gametes are passed on to offspring, and they are a source of genetic variation between individuals (Figure 8.12). DNA mutations can affect morphological

Wild type	Bar eyes	Rotated abdomen
Bithorax	Dichaete	Rudimentary wings
Vestigial wings	Cut wings	Curly wings

Figure 8.12 Non-detrimental DNA Mutations

Non-detrimental DNA mutations are the primary source of new alleles in a population, for example, morphological differences in bees.

traits, development, biochemical processes, behaviour, or the regulation of expression of other genes. Mutations in gametes passed on to future generations can be dominant or recessive, autosomal, or X-linked (Chapter 6).

Cellular Response to DNA Mutations

The enzyme polymerase recognizes some mistakes during replication through a process known as **proofreading**. When polymerase adds the wrong nucleotide, it can stop replicating, reverse direction (move in the 3′ to 5′ direction), and remove incorrect nucleotides using another enzyme function called **exonuclease activity**. Then replication continues in the 5′ to 3′ direction.

Humans have approximately 130 different genes that code for **DNA repair enzymes**. DNA mutations within a DNA repair enzyme gene would be significantly damaging. DNA repair enzymes can locate mutations that polymerase missed during replication. Any purine-purine combinations create a bulge in the DNA molecule, and pyrimidine-pyrimidine combinations create an indent; repair enzymes can recognize both of these combinations. Also, during DNA replication some methyl groups (CH_3) are added to some of the cysteine bases or adenine bases; in this way the new strand is distinguished from the strand of the original molecule. DNA repair enzymes, however, change only the non-methylated strand, which means they would fix the correct strand.

If mutations continue to persist in the DNA molecule, tumour suppressors function to prevent cell division (recall the cell-cycle checkpoints in G_1, G_2, and M). Also DNA repair enzymes can become recruited to fix

© Kage-Mikrofotografie/First Light

Figure 8.13 Cancer and Natural Killer Cells
Natural killer cells can kill cancer cells.

mutations. When DNA mutations cannot be corrected, p53 induces apoptosis (programmed cell death) so that cells with mutations do not continue to divide and propagate those mutations. If apoptosis does not occur and the cell continues to divide with mutations, the body uses its last mechanism, the immune system, to kill damaged cells (Chapter 23). **Natural killer cells** are a type of white blood cell that recognize and kill cancer cells that are formed in our body (Figure 8.13). If mutations continue to exist, disease may occur. As we age, the accumulated amount of DNA mutations increases, as does the risk of cancer and other illnesses.

Causes and Consequences of Mutations

Mutagens and carcinogens, such as radiation and chemicals, are the most common causes of mutations (see Chapter 5 regarding cancer-causing mutations). The other causes of mutation are mistakes during DNA replication, transposition of genetic material, and viruses.

Mutations in viruses are responsible for diseases such as the following:

• Human papilloma virus (HPV)—genital warts—can cause cervical cancer.

• Human immunodeficiency virus (HIV)—AIDS—can cause Kaposi sarcoma (abnormal growth of blood vessels).

• Hepatitis B or C—liver infection—can cause liver cancer.

• Epstein–Barr virus (EBV)—mononucleosis—can cause lymphoma (rare).

Genetic diseases caused by inherited DNA mutations account for approximately 10% of all diseases (Chapter 6). The following list describes the type of DNA mutation inherited in some common genetic diseases:

• Cystic fibrosis—due to one mismatch base in the gene that codes for chloride channels

• Huntington's disease—due to insertion of multiple CAG repeats in a gene on chromosome 4

• Sickle cell anemia—due to one mismatch in the hemoglobin gene

• Cancer—usually due to two or more mutations in genes that code for repair enzymes, or genes that affect the cell cycle, oncogenes, or tumour-suppressor genes

• Phenylketonuria (PKU)—due to point mutation in liver enzyme gene that causes brain damage; babies tested for this at birth

• Nonpolyposis colorectal cancer—autosomal dominant—due to repeated CA mutation in DNA repair enzyme gene expressed in cells of the large intestine.

CONCEPT REVIEW 8.3

1. What is proofreading?

2. How do our cells cope with DNA mutations?

3. Explain the differences between the following types of DNA mutations: substitution (mismatch), addition (insertion), deletion, and recombination (transposition).

4. What is a frameshift mutation?

5. What are the consequences of DNA mutations that occur in somatic cells or gametes?

6. How do mutations contribute to genetic variation?

7. What is insertional inactivation?

8. What are transposable elements? Do they always cause DNA mutations?

9. How do we acquire DNA mutations?

10. Make a list of the most common types of carcinogens.

11. What are some common viruses that can cause cancer through DNA mutations?

12. Make a list of some common genetic diseases caused by the inheritance of DNA mutations, and state what type of DNA mutation is involved.

EFFECTS OF AGING

DNA aging is based on the shortening of the protective telomeres at the end of each chromosome. The length of our telomeres represents our biological age. When telomeres become too short, the cells can no longer reproduce (mitosis), and tissues begin to degenerate. Certain cell types, such as the skin, the epithelial lining of GI tract, and immune cells, are exposed to many factors that cause them to replicate more often than other cell types. These cell types are, therefore, more prone to the effects of aging.

An enzyme called **telomerase** is present in stem cells such as those in the bone marrow that give rise to new blood cells, and germ-line cells in the gonads that give rise to gametes. Telomerase functions to replace lost telomere sequences. If inducing telomerase expression in cells prevents telomere shortening and, therefore, cell death, why don't we use telomerase as the "fountain of youth"? Researchers have successfully prevented the shortening of telomeres in mice; however, this process is directly linked to the development of cancer. If cells do not die or decrease cell division after a period of time—depending on the cell type—many mutations would accumulate in that cell, depending on the amount of exposure to mutagens and carcinogens. Research has shown the risk of cancer development increases when cell death is prevented. In fact, cancer cells begin to express telomerase, which explains why late-stage cancer cells do not die; they become immortalized and can spread throughout the body.

Recent research indicates that our health can affect telomere length. In a study of men with prostate cancer, telomere length was measured after five years during which the men increased their consumption of plant foods, exercised 30 minutes each day for six days a week, and added stress-reduction strategies, such as yoga or meditation. The group that incorporated the lifestyle changes had 10% longer telomeres, indicating that healthy lifestyle affected telomere length and lifespan by slowing the rate of telomere degradation.

CHAPTER REVIEW

Answer to Application Question

When viruses that cause cancer infect human cells, they insert their DNA into the cell, and that DNA can be incorporated into the human DNA. This is a case of recombination mutation because it involves a large sequence of DNA. When virus DNA is inserted, this can sometimes cause insertional inactivation, which means the viral DNA interrupts a human gene and inactivates it. If the affected gene is one that is involved in the cell cycle, such as a tumour-suppressor gene, this can lead to cancer. If the viral DNA inserts into a regulatory region of a gene, this can alter the regulation of the expression of that gene—with the result that too much or too little of the gene may be expressed—and can lead to cancer. If the viral DNA inserts into a region of the chromosome that does not contain genes or regulatory regions, the viral infection will not cause cancer. For example, some strains of human papilloma virus can cause cervical cancer, but not every person infected with HPV will have cancer.

9 Gene Expression and Regulation

9.1 CENTRAL DOGMA OF GENE EXPRESSION

Recall from Chapters 2 and 8 that each DNA macromolecule is a chromosome that's made up of sequences of the nucleotides: adenine, cytosine, guanine, and thymine (Figure 9.1). DNA is double stranded, and complementary nucleotides on each strand always bind so that adenine binds to thymine and cytosine binds to guanine; these are called **base pairs**. Sequences of nucleotides make up genes; for example, the human insulin gene sequence involves 4044 nucleotides on chromosome number 11. Only one strand of the DNA codes for the gene: known as the **coding strand** or **sense strand**. The complementary DNA strand is called the **template strand** or **antisense strand** (Figure 9.2). The coding strand has the *same* sequence as the mRNA molecule (except that uracil takes the place of thymine in the RNA). We have two copies of every gene—alleles—on each homologous chromosome (Chapter 6), one allele from each parent.

Genes code for messenger RNA (mRNA) sequences that consist of nucleotides (Chapter 2), and this process is called **transcription**. A mature mRNA molecule moves out of the nucleus and through nuclear pores to a ribosome, where protein synthesis (translation) occurs (Chapter 3). The flow of information from gene to RNA to protein is

The object of education is to prepare the young to educate themselves throughout their lives.

Robert M. Hutchins

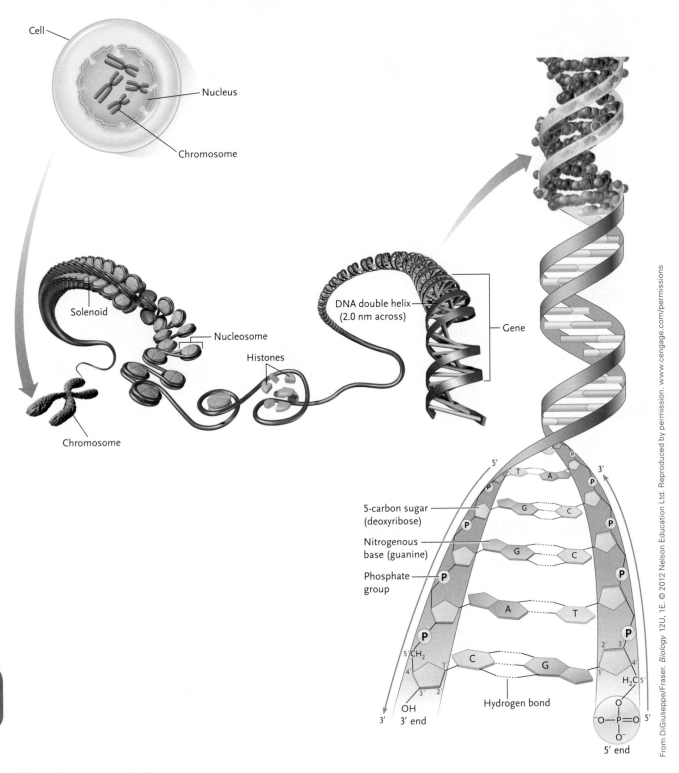

Figure 9.1 Structure of a Chromosome

what is known as the **central dogma of gene expression** (Figure 9.3). Although every cell nucleus contains all 46 chromosomes, which consist of the approximately 25 000 genes that make up the total human genome, only certain genes are transcribed and translated in certain cells, at specific times and in specific amounts. This process is called **gene expression**. The regulation of gene expression is covered in Section 9.4.

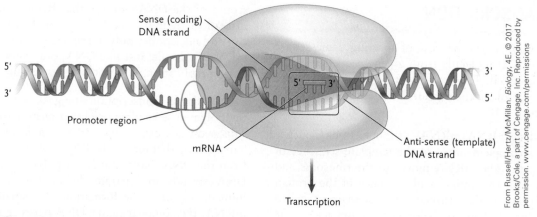

Figure 9.2 Sense and Antisense Strands of DNA

The sense strand is also called the *coding strand*, and the antisense strand is also called the *template strand*.

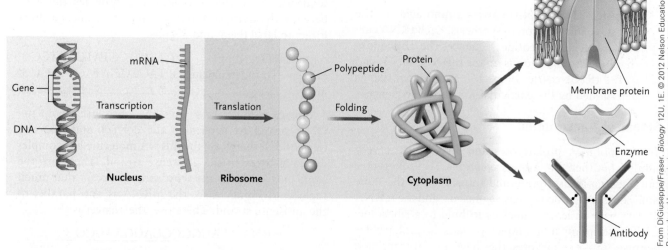

Figure 9.3 The Central Dogma of Gene Expression

The central dogma of gene expression is that DNA codes for RNA, which codes for proteins.

CONCEPT REVIEW 9.1

1. Which nucleotides form complementary base pairs in DNA?

2. Explain what the central dogma of gene expression is.

3. What is the difference between the sense strand and the antisense strand of DNA?

4. How many genes are found in the human genome?

Application

As you read through the rest of this chapter, consider the following. Use the genetic code to figure out the amino acid sequence for the following gene sequence.

sense strand: 5′ ATGGCCTATGAATCG 3′
antisense strand: 3′ TACCGGATACTTAGC 5′

What would happen to that protein sequence if during replication an error occurred that changed the first G to a C instead?

9.2 TRANSCRIPTION

Transcription is the process of forming a messenger RNA (mRNA) molecule from a gene sequence in the nucleus of a cell. Three different types of RNA molecules play a role in the production of a functional protein (Figure 9.4).

1. **Messenger RNA (mRNA)** is the nucleotide sequence that's complementary to the template sequence of the gene. The mRNA moves to the ribosome and codes for the amino acid sequence of the protein (translation). Each three-nucleotide segment, called a **codon**, determines the order of amino acids.

2. **Ribosomal RNA (rRNA)** makes up part of the ribosome and plays an important role in ensuring the mRNA can bind to the ribosome. Recall from Chapter 3 that transcription of rRNA molecules occurs in the nucleolus.

3. **Transfer RNA (tRNA)** carries amino acids to the ribosome during protein synthesis. Each tRNA carries a special nucleotide sequence, the **anticodon**, that binds to the codons on the mRNA. The 3′ end of the tRNA carries a specific amino acid that is determined by the particular anticodon sequence.

Stages of Transcription

Transcription in prokaryotes and eukaryotes is very similar, but there are a few exceptions: (1) Prokaryotes can undergo transcription and translation at the same time because they do not have a nucleus, and the mRNA doesn't have to leave a nucleus to bind to a ribosome; (2) Transcription is a simpler process in prokaryotes because, unlike eukaryotes, they have no non-translated regions—**introns**. Prokaryotes and eukaryotes also have different mechanisms for regulating gene expression (Section 9.4). Transcription in both prokaryotes and eukaryotes has three stages: initiation, elongation, and termination.

Initiation Transcription begins when the enzyme **RNA polymerase** binds to a **promoter region** of the DNA upstream from the gene that is transcribed (Figure 9.5). Promoters are nucleotide sequences that have many variations. The most common bacterial promoter sequences are TATTAA or CAAT, which are referred to as a "TATA box" or a "CAT box." Promoters act as **regulatory regions** of the DNA; they are not transcribed but help the polymerase locate the beginning of a gene. The first three nucleotides (first codon) of almost all mRNA molecules is **AUG**. Therefore, the *start* site of the gene begins with ATG on the coding strand.

Elongation Starting with the first nucleotide, RNA polymerase adds complementary nucleotides to the template strand of the gene so that the mRNA sequence occurs in the same order of nucleotides as the coding

strand of the DNA—except that RNA contains uracil instead of thymine (see Figure 9.4). RNA polymerase can add nucleotides only in the 5′ to 3′ direction on the mRNA molecule: just like DNA, polymerase functions during replication. RNA polymerase requires ATP.

Termination Transcription of a gene is complete when the polymerase enzyme reaches a **terminator sequence**. This sequence causes the mRNA molecule to form a hairpin loop that makes the RNA polymerase dissociate from the DNA. Note that RNA polymerase causes the DNA strands to separate during transcription, and immediately after the formation of a small section of mRNA, the complementary DNA bases re-form. Upon completion of transcription, the DNA is intact, and the mRNA molecule is ready for translation.

mRNA Sequencing Exercise Examine the following DNA sequence, and figure out the corresponding mRNA sequence. Note that this sequence provides the gene between the start site and the terminator sequence; don't forget to label the 5′ and 3′ ends.

sense strand (coding): 5′ ATGGCCTATGAATCG 3′
antisense strand (template): 3′ TACCGGATACTTAGC 5′
mRNA: _____

Answer: Polymerase uses the antisense (template) DNA strand to determine the correct and opposite nucleotides; therefore, the mRNA molecule has complementary bases to the antisense strand. These are the same nucleotides as the sense strand, except that uracil replaces thymine. Also, the mRNA is anti-parallel to the antisense strand. Therefore, the answer is

mRNA: 5′ AUGGCCUAUGAAUCG 3′

RNA Processing

In eukaryotes, genes contain sequences of DNA, the introns, that do not translate into part of the amino acid sequence. The regions that do become the code for the protein are called **exons** (Figure 9.6). In eukaryotes, transcription produces a pre-mRNA molecule that must be processed so that a mature mRNA molecule is ready for translation. This process, called **splicing**, involves removing the introns and adding a **5′ cap** and a **3′ tail**. A protein-RNA complex called a **spliceosome** removes the introns and forms new bonds between the exons. The 5′ cap is a methylated GTP (guanine triphosphate), and it enables the mRNA to efficiently begin translation at the ribosome. A process called *polyadenylation* creates a length of several hundred adenine nucleotides that attach to the 3′ end of the mRNA, called the *poly A tail*. This tail is required to protect the mRNA from rapid enzymatic degradation. RNA processing occurs in the nucleus after transcription, and then the mature mRNA molecule moves out of the nucleus to a ribosome in the cytoplasm or onto the rough ER for translation.

From Russell/Hertz/McMillan. *Biology*, 4E. © 2017 Brooks/Cole, a part of Cengage, Inc. Reproduced by permission. www.cengage.com/permissions

Figure 9.4 Types of RNA Molecules
(A) mRNA, (B) rRNA, and (C) tRNA

Figure 9.5 Transcription

Transcription occurs in the nucleus. RNA polymerase unwinds the DNA and adds complementary nucleotides to the template strand. After transcription, the RNA leaves the nucleus, and the DNA rewinds.

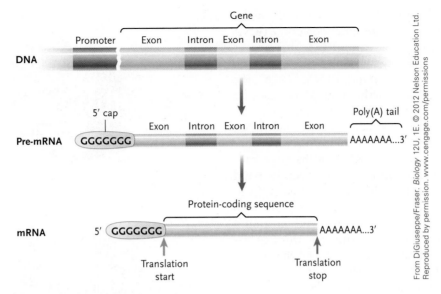

Figure 9.6 RNA Processing

RNA processing involves splicing out the introns and adding the 5' cap and the 3' tail.

From DiGiuseppe/Fraser. *Biology* 12U, 1E. © 2012 Nelson Education Ltd. Reproduced by permission. www.cengage.com/permissions

CONCEPT REVIEW 9.2

1. How does transcription differ in prokaryotes and eukaryotes?

2. Where does transcription take place in human cells?

3. What is the role of RNA polymerase?

4. How does RNA polymerase "know" where the beginning of a gene is located?

5. What is almost always the first codon in an mRNA molecule?

6. Which strand of DNA does the polymerase use to form the mRNA molecule? In which direction does polymerase transcribe mRNA?

7. What prevents the RNA polymerase from transcribing nucleotides that are not part of the gene sequence? (What stops transcription?)

8. Explain what occurs during RNA processing and why each process is important.

9. Extra study: compare the similarities and differences between DNA replication (Chapter 8) and transcription.

10. What mRNA sequence corresponds to the following sequences?

    a. sense strand: 5' TACGGCTATC 3'
    b. antisense strand: 3' ATGCCGATAG 5'

9.3 TRANSLATION

Translation is the process of synthesizing a protein from an mRNA sequence on a ribosome. Recall from Chapter 2 that the human body produces over 100 000 different proteins that have various functions in the body. This section describes how an mRNA molecule becomes a functional protein from a gene inherited from our parents.

Composition of Ribosomes

Ribosomes are made up of a small subunit and a large subunit, both of which contain ribosomal RNA and proteins. The mRNA binds to the small subunit, and the tRNAs bind to the large subunit (Figure 9.7). The

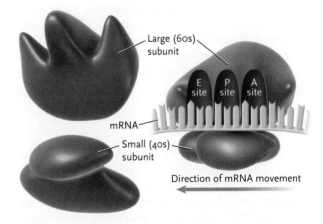

Figure 9.7 Structure of a Ribosome

Ribosomes consist of ribosomal RNA and proteins.

large subunit has three binding sites for tRNA molecules: the A, P, and E sites. The anticodon of the tRNA must match the codon of the mRNA that moves between the small and large subunits. Since the first codon, located at the 5′ end of the mRNA, is **AUG**, the first amino acid is always **methionine** (Met). The first tRNA carrying the Met amino acid binds to the P site, where it pairs with the first codon. The second tRNA then binds to the A site, pairing with the second codon. The enzymatic activity of the RNA in the large subunit catalyzes the formation of the peptide bond between the amino acids in the P site. Because RNA in ribosomes has enzyme activity, ribosomes are also called **ribozymes**. The first tRNA, now with no amino acid attached, moves into the E site. The third tRNA binds to the third codon in the A site, and the second peptide bond forms between the second and third amino acid. This process continues—each new tRNA entering the A site, the peptide bond forming in the P site, and the empty tRNA moving into the E site (Figure 9.8)—until a *stop* codon is reached. Three codons (UAA, UAG, and UGA) do not code for any amino acid. Since no tRNA molecules bind to the A site when a stop codon is present, the ribosome dissociates, and translation ends. Similar to transcription, the process of translation requires energy. An empty tRNA becomes recycled when it leaves the ribosome, and **activating enzymes** in the cytoplasm attach the correct amino acid to the correct tRNA molecules (Figure 9.9).

As translation takes place, the primary protein structure begins to form (see Chapter 2), and immediately the secondary structures (alpha helices and beta sheets) begin to form due to hydrophobic and hydrophilic interactions between the functional groups of the amino acids. The full tertiary structure is complete when translation has finished. Proteins that have been translated on the ribosomes in the cytoplasm are released and can immediately function in the cell. Proteins produced on ribosomes on the rough ER are packaged into **vesicles** that pinch off and move to the **Golgi bodies** for further processing (see Chapter 3), called **post-translational modification**. Within the Golgi, certain proteins combine with lipids or sugars to produce **lipoproteins** (e.g., LDLs made in the liver), or **glycoproteins** (e.g., blood type markers, A, B, and AB).

Use of the Genetic Code

Each mRNA codon consists of three nucleotides. The sequence of codons determines the amino acid sequence. Therefore, we can predict the amino acid sequence of any protein from the gene, because each tRNA always carries a specific amino acid, depending on its anticodon. A chart called the **genetic code** can be used to figure out an amino acid based on the mRNA codon

1 An aminoacyl-tRNA binds the A site.

2 Peptidyl transferase cleaves the amino acid from the P site tRNA and bonds it to the amino acid on the A site tRNA.

3 The ribosome moves along the mRNA to the next codon, thereby bringing the tRNA with the growing polypeptide to the P site and moving the empty tRNA to the E site.

Figure 9.8 Translation

Translation of mRNA occurs on ribosomes to synthesize proteins.

(Figure 9.10). There are 64 possible codons—each of the three nucleotides in a codon can be one of the four possible nucleotides. However, there are only 20 amino acids, so usually more than one codon matches each amino acid.

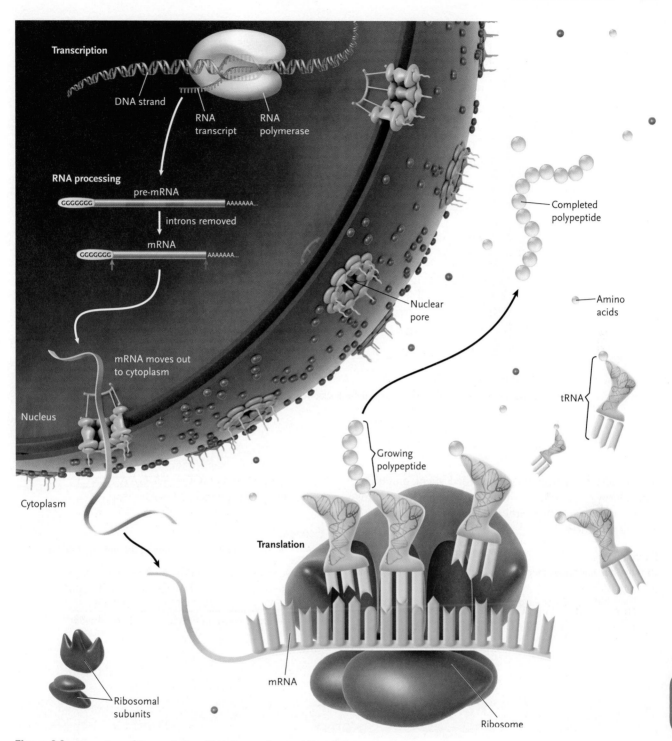

Figure 9.9 Overview of Transcription, RNA Processing, and Translation

Use the genetic code in Figure 9.10 to find the amino acid that coincides with the ATG codon. Find the *A* on the left-hand side, and follow that line across until you see the second letter as *U*. Then you'll see the AUG codon is Met: methionine. Remember that a normal gene codes for hundreds of amino acids. Note: a chart of the amino acid short forms can be found in Appendix A.

Effect of Mutations on the Protein Sequence

Recall from Chapters 5 and 6 that mutations in a gene sequence can alter a protein, making it non-functional: for example, alterations in the cystic fibrosis chloride channel protein, or in the formation of oncogenes that can cause an unregulated increase in cell division. Some changes in nucleotide sequence may have no effect on the

Second base of codon

		U	C	A	G	
First base of codon	**U**	UUU }Phe UUC UUA }Leu UUG	UCU UCC }Ser UCA UCG	UAU }Tyr UAC UAA UAG	UGU }Cys UGC UGA UGG Trp	U C A G
	C	CUU CUC }Leu CUA CUG	CCU CCC }Pro CCA CCG	CAU }His CAC CAA }Gln CAG	CGU CGC }Arg CGA CGG	U C A G
	A	AUU AUC }Ile AUA AUG Met	ACU ACC }Thr ACA ACG	AAU }Asn AAC AAA }Lys AAG	AGU }Ser AGC AGA }Arg AGG	U C A G
	G	GUU GUC }Val GUA GUG	GCU GCC }Ala GCA GCG	GAU }Asp GAC GAA }Glu GAG	GGU GGC }Gly GGA GGG	U C A G

Third base of codon

Figure 9.10 The Genetic Code

The genetic code is determined by the three-nucleotide sequences called *codons* that code for specific amino acids.

amino acid sequence; they are called **silent mutations**. The following examples show a single nucleotide mismatch; these can cause either a detrimental mutation or a silent mutation.

Example A

Detrimental mutation causing a truncated and non-functional protein

sense strand: 5′ ATGGCCTAAGAATCG 3′
antisense strand: 3′ TACCGGATTCTTAGC 5′

mRNA: 5′ AUG GCC UAA GAA UCG 3′
amino acid sequence: Met-Ala-Stop

Example B

Silent mutation causing no change in the amino acid sequence

sense strand: 5′ ATGGCTTATGAATCG 3′
antisense strand: 3′ TACCGAATACTTAGC 5′
mRNA: 5′ AUG GCA UAU GAA UCG 3′
amino acid sequence: Met-Ala-Tyr-Glu-Ser

9

CONCEPT REVIEW 9.3

1. What are ribosomes composed of?
2. How are mRNA, rRNA, and tRNA molecules involved in translation?
3. What happens in the A, P, and E sites of the ribosome?
4. Why are ribosomes also called *ribozymes*?
5. What is the role of activating enzymes?
6. Briefly explain how translation works.

7. What happens when a stop codon enters the A site?
8. When do proteins fold into tertiary structure?
9. How can proteins be modified after translation? Give two examples.
10. Use the genetic code to determine the protein sequence from the following mRNA:

    mRNA: 5′ CUC GAU ACC GAG UAG 3′
11. What is a silent mutation?

9.4 REGULATION OF GENE EXPRESSION

Every cell must be able to regulate when and how much protein is made from each gene that is expressed in its particular cell type. When we eat food, our pancreas beta islet cells (Chapter 17) detect an increase in blood sugar and then respond by producing insulin. Insulin regulates the transcription of the cell's glucose transporters (GLUT proteins) that then take up the blood sugar.

Although every cell that contains a nucleus contains all 46 chromosomes and all 25 000 genes, only certain genes are transcribed in certain cell types. Pancreas beta islet cells are the only cells that produce insulin, only red blood cells produce hemoglobin, and only muscle cells produce dystrophin. However, all cells produce the enzymes involved in ATP production. The regulation of gene expression is the essence of homeostasis (Chapter 1). When our body temperature decreases, receptors detect the temperature change and signal the hypothalamus integrating centre. Then the hypothalamus responds by signalling the skeletal muscles to shiver to create heat. All of this occurs because of regulated gene expression. Hundreds of genes are involved in this simple example: those involved in the production of neurotransmitters, ion channels, and synaptic vesicles and those involved in muscle contraction and the production of ATP and therefore heat. The regulation of gene expression can occur at the level of transcription, RNA processing, RNA degradation, or translation.

Prokaryotic Gene Expression

The *lac* operon in *Escherichia coli (E. coli)* bacteria was one of the first systems of gene regulation that scientists studied. The *lac* operon is a set of genes involved in *lactose* metabolism. The bacterial cell does not use energy to produce proteins that are not required, so the *lac* operon genes become transcribed and translated only when lactose is present and glucose is not. (*E. coli* preferentially metabolize glucose rather than lactose if both are present.) In this case, gene regulation occurs at the level of transcription using a combination of an **activator** and a **repressor**.

It is important to understand the differences between regulatory DNA sequences and the proteins that bind to them (Figure 9.11). Figure 9.11 shows that in *E. coli* the three genes required for lactose metabolism are clustered in an **operon** so that all three are transcribed and translated together whenever needed; this clustering does not occur in eukaryotes. The regions of the regulatory DNA sequences are not transcribed, but they are binding sites for proteins that either initiate or prevent transcription. The **CAP-binding protein** (catabolite activator protein) is the **activator** that initiates transcription by binding to the CAP-binding sequence. The **repressor** prevents transcription by binding to the operator sequence, and the RNA polymerase binds to the promoter region.

Transcription will occur or not occur depending on the presence of lactose and glucose. Transcription occurs only if lactose *is* present and glucose is *not* present. When lactose is present, some molecules bind to the repressor and remove it from the operator sequence; therefore, the lactose acts as the **inducer**. In the absence of glucose, the CAP activator binds to the CAP-binding site and directs the RNA polymerase to the promoter (Figure 9.12). So transcription occurs when the repressor is removed by a molecule of lactose and CAP binds and directs the RNA polymerase to the promoter. When glucose is present, the CAP activator does not bind, and the lactose metabolizing genes are not transcribed. Instead, genes involved in the utilizing of glucose will be transcribed.

Eukaryotic Gene Expression

In eukaryotes, the regulation of gene expression is much more complex than that in prokaryotes because of the condensed structure of their DNA (Figure 9.1). Different regions of each chromosome are condensed to a different degree, depending on the regions of DNA transcribed by a particular cell type. Four of the mechanisms involved

Figure 9.11 The *Lac* Operon

The *lac* operon is found in *E. coli* bacteria and contains three genes involved in lactose metabolism, all regulated by one promoter.

From DiGiuseppe/Fraser, *Biology 12U*, 1E. © 2012 Nelson Education Ltd. Reproduced by permission. www.cengage.com/permissions

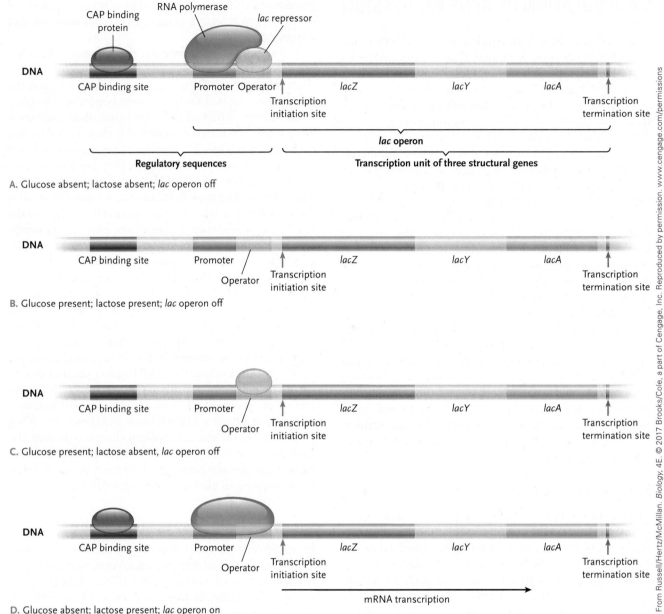

Figure 9.12 Overview of Regulation of the Transcription of the *Lac* Operon

A. Glucose absent; lactose absent; *lac* operon off

B. Glucose present; lactose present; *lac* operon off

C. Glucose present; lactose absent, *lac* operon off

D. Glucose absent; lactose present; *lac* operon on

in regulation of gene expression are described below: methylation, RNA processing, mRNA degradation, and transcription factors.

1. **Methylation.** The prevention of gene transcription by the addition of methyl groups ($-CH_3$) is called **gene silencing.** Methylation of cytosine nucleotides most commonly occurs in the tightly condensed regions of DNA where the genes do not become transcribed in certain cell types. For example, all cells except the red blood cells have the hemoglobin gene region tightly condensed and methylated to prevent transcription. Methylation also occurs during DNA replication (Chapter 8, Section 3). Abnormal methylation is thought to play a role in cancer.

2. **RNA processing.** A mature mRNA molecule forms after introns have been removed and exons have been spliced together during RNA processing, but this does not have to occur in the exact same manner each time the gene is transcribed. In a process called **alternate splicing** or **exon shuffling,** certain exons become part of the mature mRNA at certain times, and for other exons this happens at other times. The particular function of a protein depends on what exons are involved in forming the mature mRNA. This mechanism makes it possible for over 100 000 different proteins to be produced from only 25 000 genes.

3. **mRNA degradation.** Various mRNA transcripts have a specified lifespan, and mRNA molecules can

Figure 9.13 Transcription Factors

Transcription factors regulate eukaryotic gene expression.

From Russell/Hertz/McMillan. *Biology*, 4E. © 2017 Brooks/Cole, a part of Cengage, Inc. Reproduced by permission. www.cengage.com/permissions

Figure 9.14 Transcription of the Insulin Gene

Transcription factors are recruited to the enhancer region of the insulin gene when glucose levels increase.

undergo translation as long as they are not broken down. Some mRNA molecules are broken down immediately, but others may stay in the cytoplasm for several hours. The lifespan of each mRNA molecule is determined by the gene.

4. **Transcription factors.** Various transcription factors can bind to regulatory regions of the DNA to promote transcription of certain genes. This is the most common mechanism for regulating gene expression in eukaryotes. All hormones, both water soluble and fat soluble, are signalling molecules that regulate the transcription of other genes. Transcription factors are regulated by cellular signals as well as environmental stimuli. For example, glucose levels in the blood are detected by cells in the pancreas. Glucose entering the cell stimulates three transcription factors to bind to the enhancer region of the insulin; this promotes

RNA polymerase to bind to the promoter and begin transcription (Figure 9.14). It is essential that our pancreas produces insulin when blood sugar levels increase after eating so that cells can take up the glucose and use it for either production of ATP or fat.

DID YOU KNOW?

Thinking can affect gene expression! Close your eyes and think of something stressful or scary. Your thoughts just caused the transcription and translation of many molecules involved in the stress response. Conversely, meditation works by slowing thoughts and induces expression of genes involved in rest and relaxation. Meditation is becoming a widely popular adjunctive treatment for many illnesses. It has also been shown that optimistic people live longer on average than pessimistic people.

6

CONCEPT REVIEW 9.4

1. How is the regulation of gene expression related to homeostasis? Give one example.

2. In prokaryotes, what are two types of regulatory proteins that affect transcription?

3. What is the *lac* operon?

4. What is the regulatory region of DNA that the repressor, CAP activator, and RNA polymerase each bind to?

5. How is the repressor removed from the DNA?

6. Under what conditions is the *lac* operon transcribed?

7. If glucose and lactose are both present, are the activator and/or repressor bound or not? Does transcription occur? Why or why not?

8. List and briefly describe the main mechanisms of gene regulation in eukaryotes. Which is the most common?

9. What is gene silencing?

10. What is alternate splicing, and why is it important?

CHAPTER REVIEW

Labelling Activity

In each example, state whether glucose and/or lactose is present and whether transcription occurs.

1. a Label the following diagram, and circle Yes or No for each of the following.

   Glucose present: Yes   No

   Lactose present: Yes   No

   Transcription occurs: Yes   No

   b Circle Yes or No.

   Glucose present: Yes   No

   Lactose present: Yes   No

   Transcription occurs: Yes   No

   c Circle Yes or No.

   Glucose present: Yes   No

   Lactose present: Yes   No

   Transcription occurs: Yes   No

   d Circle Yes or No.

   Glucose present: Yes   No

   Lactose present: Yes   No

   Transcription occurs: Yes   No

Answer to Application Question

RNA polymerase uses the antisense strand as the template and adds complementary bases.

sense strand: 5′ ATGGCCTATGAATCG 3′

antisense strand: 3′ TACCGGATACTTAGC 5′

Therefore, the RNA transcript is 5′AUG GCC UAU GAA UCG 3′.

Note that the RNA transcript is the same sequence as the sense strand, except that there is uracil instead of thymine.

Using the genetic code (Figure 9.10) and the RNA transcript, and starting from the 5′ end, the amino acid sequence is: Met-Ala-Try-Glu-Ser.

If the first G was changed to a C, then the first codon would be AUC, which would code for Ile instead of Met. Since most transcripts begin with Met (methionine) this protein would not be translated.

6

10 Biotechnology

Education is the passport to the future, for tomorrow belongs to those who prepare for it today.

Malcom X

10.1 OVERVIEW OF BIOTECHNOLOGY

Biotechnology is the use of living organisms to produce biological products. For example, bacteria are used to produce cheese, yeast is used to make alcohol, and a fungus is used to produce penicillin. In the last few decades, it has become possible to move specific sequences of DNA from one organism to another; this process is called **recombinant DNA technology, genetic engineering,** or **gene splicing.** DNA sequences, or genes, can be cut at specific sites in the genome of any organism and can be inserted into the genome of any other organism. A *hybrid* or **recombinant DNA** molecule contains nucleotide sequences from two different organisms. For example, human DNA sequences, such as the insulin gene, can be inserted into a bacterial plasmid DNA molecule to produce a human-bacteria recombinant DNA molecule.

A **genome** consists of all of the genes in an organism. A **proteome** is all of the proteins produced from the DNA in an organism. **Genetic modification** refers to any kind of alteration in the genome, including the deleterious effects of mutations caused by radiation or chemicals. Humans have been altering DNA in plants and animals gradually for over 200 000 years by selective breeding (Chapter 7). It is now possible to insert specific DNA sequences into organisms such as plants or animals; the resulting organisms are called **genetically modified organisms (GMOs).**

© iStockphoto.com/DNY59

Overview of Recombinant DNA Technology

To create recombinant DNA molecules the following essential "ingredients" are required:

1. A gene of interest
2. A **restriction endonuclease** enzyme (a restriction enzyme) that cuts DNA at specific sequences
3. A **vector** that carries the inserted gene; most commonly a bacterial plasmid, but also other types of vectors (Section 10.3)
   a. Vectors must be able to replicate so that the inserted gene is copied.
   b. Vectors must be able to be transcribed and translated to produce a functional protein.
4. Ligase, an enzyme that's used to bind DNA together
5. A **host cell,** usually bacteria, to take up the recombinant plasmid; will replicate and copy the DNA, as well as transcribe and translate the inserted gene

Human insulin was the first genetically engineered medication, produced in 1982. The human insulin gene can be inserted into a bacterial plasmid; the bacteria replicate and transcribe the human insulin gene and then synthesize the protein, which can be extracted and used to treat people with type 1 diabetes.

DID YOU KNOW?

Papaya trees are genetically engineered to contain a ringspot virus coat protein so that the trees react by producing antiviral chemicals. This prevents the trees from being infected by actual ringspot viruses. The virus coat protein is not expressed in the fruit, so it doesn't harm humans.

CONCEPT REVIEW 10.1

1. What is recombinant DNA technology?
2. What is recombinant DNA?
3. What is a genetically modified organism? Give an example.
4. What are the primary ingredients required for recombinant DNA technology?

10.2 RESTRICTION ENZYMES

Restriction enzymes are special enzymes that cut DNA phosphodiester bonds at very specific symmetrical sequences called **palindromes,** usually four to eight nucleotides long. A palindrome sequence can be read the same way in both directions, such as the word RADAR. The following is an example of a DNA palindrome:

5′ GAATTC 3′
3′ CTTAAG 5′

On both strands of the DNA, the sequence is read the same in the 5′ to 3′ direction. Restriction enzymes do not require ATP to function.

Restriction enzymes have been isolated from many species of bacteria. Bacteria produce many types of restriction enzymes that function to prevent viral infections caused by **bacteriophages:** viruses that infect bacteria. The restriction enzymes cut the DNA molecules and, therefore, prevent viral replication. Each restriction enzyme always recognizes and cuts the DNA at exactly the same sequence. Figure 10.1 shows the enzyme *Eco*RI, isolated from *E. coli* bacteria, always cuts 5′ GAATTC 3′ sequences. Bacteria protect their own DNA from these enzymes by using **methylase enzymes** to block those recognition sites on their chromosome with methyl groups (–CH3).

When restriction enzymes cut DNA, they don't necessarily cut it exactly in half (Figure 10.1). The DNA is

Figure 10.1 Restriction Enzymes
Restriction enzymes cut specific palindrome sequences.

usually cut so there are overhanging ends of non-paired nucleotides: called **sticky ends**. This is a very useful feature that allows the DNA cut with a restriction enzyme to match up (**anneal**) with other DNA that has been cut with the same enzyme. The enzyme DNA ligase is used to re-form the phosphodiester bonds. Over 3000 restriction enzymes have been studied, and several hundred of these are available for genetic research. Each restriction enzyme cuts different palindrome sequences. The following shows the palindrome sequences related to three of the restriction enzymes:

> *Eco*RI   5′.......GAATTC.......3′
>   3′.......CTTAAG.......5′
>
> *Bam*HI   5′.......GGATCC.......3′
>   3′.......CCTAGG.......5′

> *Hind*III   5′......AAGCTT.......3′
>   3′......TTCGAA.......5′

Usefulness of Restriction Enzymes

Restriction enzymes can be used to cut DNA from two different organisms—such as bacterial DNA and human DNA—to produce recombinant DNA molecules. For example, the human insulin gene can be cut with a restriction enzyme and then inserted into a bacterial plasmid that has been cut with the *same* restriction enzyme. Once the human insulin gene anneals with the bacterial plasmid DNA, that recombinant DNA molecule can be taken up by bacterial cells; there, the human insulin protein will be expressed by the bacteria.

CONCEPT REVIEW 10.2

1. What are palindromes?
2. Where do restriction enzymes come from?
3. How do bacteria protect themselves from cutting their own DNA?
4. What bonds are broken by restriction enzymes?
5. Why are sticky ends important?
6. Why are restriction enzymes useful to genetic researchers?

10.3 VECTORS

A vector, also called a *vehicle*, is a DNA molecule that can carry extra pieces of DNA. Among the many different types of vectors, the most common ones are plasmids, bacteriophages, cosmids, and YACs. All vectors must have certain characteristics for them to be useful to genetic scientists.

- They must be able to **replicate** within the host cell.
- They must contain **restriction enzyme cut sites**.
- They should carry a **selectable marker** to allow for identification of the host cells that contain the recombinant DNA.
- The vector should be easy to recover from the host cells that contain it.
- The inserted gene must be expressed as a protein.

Plasmids

Plasmids are small pieces of bacterial cell DNA that are separate from the main chromosome (Figure 10.2). Plasmids usually contain around 2000 to 5000 nucleotides and are self-replicating once inside bacterial host cells because they have an origin of replication.

Figure 10.2   Bacterial Plasmid

A bacterial plasmid is a small, circular, extra piece of DNA found in some bacteria.

Figure 10.3 shows an example of a plasmid. Plasmids are smaller than other vectors and can hold only small pieces of DNA, but they are usually sufficient for replicating single gene sequences. Bacteria are easy to grow and easily take up plasmids by a process called **transformation**. Many plasmids have been engineered to contain all of the required characteristics listed above. A plasmid contains at least one selectable marker that is usually an antibiotic resistance gene. It also contains many restriction enzyme cut sites, which are typically placed into a **multiple cloning site** (MCS) that is sometimes found within another selectable marker, such as the *lacZ* gene.

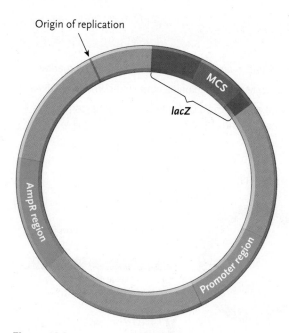

Origin of replication

MCS

lacZ

AmpR region

Promoter region

Figure 10.3 Important Features of a Plasmid

The multiple cloning site (MCS), a promoter, the origin of replication, and a selectable marker are important features of a plasmid.

anyaivanova/Shutterstock.com

Figure 10.4 Blue and White Bacterial Colonies

Blue bacterial colonies do not contain recombinant DNA; white bacterial colonies do contain recombinant DNA.

Use of Selectable Markers

Genetically modified bacteria are grown in the presence of the antibiotic that the plasmid is resistant to, such as ampicillin. It is very important for plasmids to have an antibiotic resistance gene for two reasons:

1. Only bacteria that contain plasmids will grow.
2. Antibiotics in the growth medium will kill any contaminating bacteria.

LacZ is one of the genes in the *E.coli* bacterial lac operon (Chapter 9), and it is used as a selectable marker. *LacZ* is one of the proteins produced by *E.coli* that are involved in breaking down lactose (Chapter 9). When bacteria are grown in the presence of a molecule that is similar to lactose, called X-galactosidase (X-gal), the bacterial colonies turn blue, making this an excellent selectable marker (Figure 10.4). When bacteria are grown on a plate containing all nutrients required for bacterial growth, including an antibiotic, some of those bacteria may contain plasmids that do not contain a piece of recombinant DNA. In Figure 10.3, notice how the *lacZ* gene is located within the MCS. The MCS contains all of the restriction enzyme cut sites. When the plasmid is cut with a restriction enzyme and then another piece of DNA is inserted into that region, the *lacZ* gene becomes inactivated: a process known as *insertional inactivation*. If *lacZ* is inactivated, the bacteria do not turn blue in the presence of X-gal. Therefore, only the white bacterial colonies will contain recombinant DNA (Figure 10.4).

Other Useful Vectors

Bacteriophages are viruses that infect bacteria. The lambda phage is widely used in recombinant DNA technologies. All of the lambda phage genes are known and sequenced. The genes involved in the lysogenic phase can be removed and replaced with foreign DNA without affecting the ability of the phage to infect bacteria and replicate. The phage containing recombinant DNA can then infect bacterial cells: a process called **transduction**.

Cosmids are laboratory-constructed vectors that contain the *cos* site from the lambda phage, which is required for packaging chromosomes into virus particles. They contain plasmid features such as antibiotic resistance and an origin of replication. They are used to carry much larger sequences of DNA than plasmids, and then they can be introduced to bacterial cells by transformation, in the same way as plasmids.

Yeast artificial chromosomes (YACs) are beneficial because these vectors can be grown in yeast cells, which are eukaryotic. Since prokaryotes lack the ability to process RNA, a eukaryotic system is very beneficial. YACs have an origin of replication; a centromere, so that DNA can be distributed during mitosis; restriction enzyme cut sites; and yeast-specific selectable markers. Yeast cells are easily grown and have been used to express many mammalian genes.

10

CONCEPT REVIEW 10.3

1. What are the important features of a vector?
2. Describe the important components of a plasmid.
3. What is the role of selectable markers?
4. How is the *lacZ* gene used as a selectable marker?
5. Do white or blue colonies contain recombinant DNA?
6. What are the main benefits of using a bacteriophage, a cosmid, and a YAC?

10.4 CLONING A GENE

Using a plasmid vector as an example, we can look at the process of **gene cloning**: making many copies of a specific gene with bacteria. If the plasmid is an expression vector, then the gene can be transcribed and translated into a protein. A gene of interest, such as the human insulin gene, can be cut from the human chromosome and ligated into a bacterial plasmid that is cut with the same restriction enzyme.

Human DNA can be isolated from any human cell that contains a nucleus, because all 46 human chromosomes occur in every nucleated cell (mature red blood cells do not contain a nucleus). Since the human genome project was completed in 2003, all three billion nucleotides have been sequenced. Therefore, the human insulin gene sequence is known, along with the restriction enzyme cut sites that are within and outside the gene. It is possible now for scientists to order the insulin gene sequence, rather than having to isolate the gene from an entire human genome. However, scientists need to be able to figure out what restriction enzymes to use so that the gene will remain intact, that is, without cutting the gene into pieces.

You can find a **restriction map** (Figure 10.5) for any gene through NCBI, the National Center for Biotechnology Information. You can also use restriction mapping software—capable of doing virtual digests with various restriction enzymes—to determine possible outcomes. The example in Figure 10.5 illustrates that you would choose *Eco*Ri or *Hind*II, used with *Bam*HI, to safely extract the insulin gene, not *Sma*I or *Hind*III, which would cut the gene into pieces.

Once the restriction enzymes are chosen, they are used to cut the bacterial plasmid DNA. The cut plasmid DNA and the cut human DNA can be mixed together with DNA ligase, to form recombinant DNA (Figure 10.6).

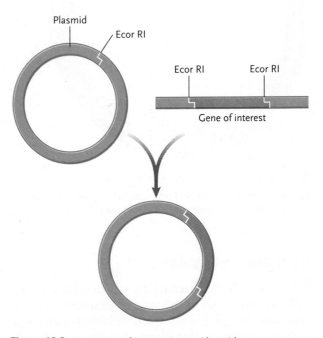

Figure 10.6 Insertion of a Gene into a Plasmid

Insertion of a gene into a plasmid involves cutting the gene and the plasmid with the same restriction enzyme, and then ligating the pieces of DNA together with the ligase enzyme.

The plasmid needs be taken up by bacteria through a process called *transformation*. Simply by mixing the plasmids and bacterial cells together and incubating them at 37°C for two minutes (using a specific strain of *E. coli* bacteria), the bacterial cells will take up the plasmid DNA. The transformed bacteria can be grown overnight to make millions of bacteria (Figure 10.7).

Screening Colonies

The transformed bacteria can be grown overnight on plates containing a growth medium, an antibiotic that the plasmid has a resistance gene for, and X-gal.

Screening Using Antibiotic Resistance Gene Any bacteria that have not taken up a plasmid cannot grow on the antibiotic medium because the antibiotic resistance gene is on the plasmid.

Screening Using the *LacZ* Gene Any bacteria that have a plasmid that did not take up the gene of interest

Figure 10.5 Restriction Map

A restriction map is used to locate restriction enzyme cut sites.

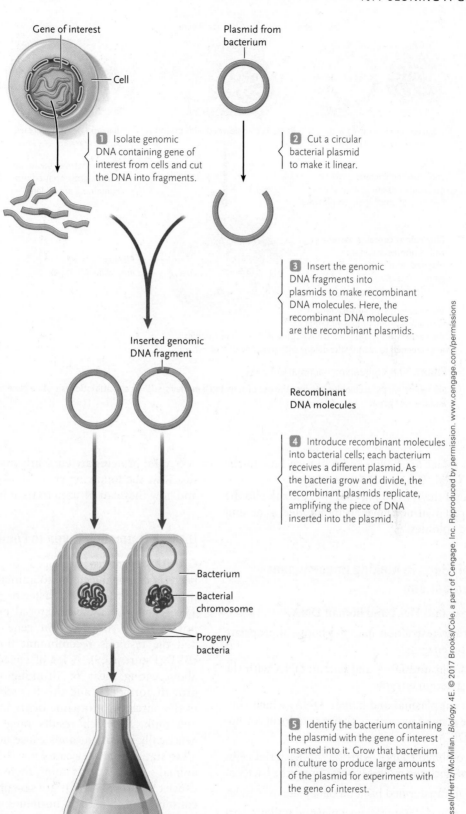

Gene of interest

Cell

Plasmid from bacterium

1 Isolate genomic DNA containing gene of interest from cells and cut the DNA into fragments.

2 Cut a circular bacterial plasmid to make it linear.

3 Insert the genomic DNA fragments into plasmids to make recombinant DNA molecules. Here, the recombinant DNA molecules are the recombinant plasmids.

Inserted genomic DNA fragment

Recombinant DNA molecules

4 Introduce recombinant molecules into bacterial cells; each bacterium receives a different plasmid. As the bacteria grow and divide, the recombinant plasmids replicate, amplifying the piece of DNA inserted into the plasmid.

Bacterium

Bacterial chromosome

Progeny bacteria

5 Identify the bacterium containing the plasmid with the gene of interest inserted into it. Grow that bacterium in culture to produce large amounts of the plasmid for experiments with the gene of interest.

From Russell/Hertz/McMillan. *Biology*, 4E. © 2017 Brooks/Cole, a part of Cengage, Inc. Reproduced by permission. www.cengage.com/permissions

10

Figure 10.7 Cloning a Gene

Cloning a gene involves inserting a gene into a plasmid, and then the bacteria transform as they take up that plasmid and grow and replicate the DNA.

lac Z gene   *lac Z* gene

Bacteria transformed with
recombinant plasmids

Bacteria transformed with plasmids

Bacteria not transformed
with a plasmid

Selection:
Transformed bacteria grow on
medium containing ampicillin
because of *amp*^R gene on plasmid.

Untransformed bacterium
cannot grow on medium
containing ampicillin.

Screening:
Blue colony contains bacteria
with a nonrecombinant
plasmid; that is, the *lacZ*^+
gene is intact.

Plate containing
ampicillin and X-gal

White colony contains bacteria
with a recombinant plasmid, that is,
the vector with an inserted DNA fragment. The white colonies
are screened to identify the colony with the gene of interest.

Figure 10.8 Screening Bacterial Colonies

Screening bacterial colonies involves the use of selectable markers, such as an antibiotic resistance gene
and the *lacZ* gene.

will have an intact *lacZ* gene and will turn blue in the presence of X-gal (Figures 10.4 and 10.8).

Any bacteria that have a plasmid that did take up the gene of interest will not have an intact *lacZ* gene and will be white colonies.

Summary of Steps in Making Recombinant Protein Medications

1. Isolate plasmid DNA and human DNA.

2. Look at the restriction map to choose appropriate restriction enzymes

3. Cut both plasmid DNA and human DNA with the same restriction enzyme.

4. Mix the cut plasmid and human DNA, which contains the gene of interest, with ligase to form recombinant DNA.

5. Mix recombinant DNA with *E. coli* bacterial cells, and heat-shock the mixture to transform the bacteria.

6. Grow the transformed bacteria on a plate overnight.

7. Bacteria that did not take up a plasmid will not survive the antibiotics.

8. Choose white colonies, they will have plasmids that contain the gene of interest.

9. Those bacteria are now capable of transcribing and translating the gene of interest, and the protein can be isolated.

Several proteins are currently made using bacteria in this way, the following chart shows the proteins made and how they can be used to treat humans.

Growth Hormones Given to Food Animals

Growth hormones produced by genetically altered bacteria can be injected into animals such as cows and pigs to make them grow faster or produce more milk. The genetically altered bacterial plasmids contain the bovine (cow) somatotropin gene (growth hormone), and the result is **recombinant bovine somatotropin, rBST** (Figure 10.9). A lot of concern has been raised about eating meat or drinking milk that contains growth hormones, and this has stimulated an increase in the purchase of organic or traditionally raised meat and milk. Research results range from stating that genetically modified foods cause no harmful effects at all to stating that they cause massive tumour growth in animal studies. At this point, there is no conclusive evidence that proves GMOs are safe or not safe for human consumption, but rBST hormone is not used in dairy cows in Europe or Canada; it is, however, used in the United States. Cows that are given growth hormone have decreased fertility and increased infections that require antibiotic treatment. Milk containing growth hormones has been implicated in the earlier onset of puberty in girls since the 1980s. Drinking milk that contains rBST can stimulate other hormones in humans,

Somatotropin gene

Cow cell

Plasmid from bacterium

1 Cut the DNA into fragments using restriction enzymes; one fragment should include the target gene.

2 Use an artificial engineered plasmid that is antibiotic resistant and has a multiple cloning site. Cut the circular plasmid to make linear.

3 Insert a DNA fragment into each plasmid to make recombinant DNA. The recombinant DNA molecule is the recombinant plasmid.

Inserted DNA fragment

Recombinant DNA molecules

4 Transform bacterial cells with recombinant DNA plasmids. Each bacterium receives a different plasmid. As the bacteria divide, the recombinant plasmids replicate, amplifying the piece of DNA inserted into the plasmid.

Somatotropin gene

Bacterium

Bacterial chromosome

Progeny bacteria

5 Identify the bacterium containing the plasmid with the somatotropin gene inserted into it, and multiply this bacterium in fermentation tanks.

6 Somatotropin is isolated and purified.

7 Somatotropin is administered to cows to increase milk production.

From Russell/Hertz/McMillan. *Biology*, 4E. © 2017 Brooks/Cole, a part of Cengage, Inc. Reproduced by permission. www.cengage.com/permissions

Figure 10.9 Production of Recombinant DNA

The production of recombinant DNA can be used to grow specific proteins within transformed bacteria; the protein can then be isolated and given to humans or animals.

such as **insulin-like growth hormone (IGF)**, which is another type of growth factor that can increase cancer cell growth, and this is why it is not approved for use in Canadian dairy farming. See Table 10.1.

TABLE 10.1

Protein Medications Made by Bacteria to Treat Human Conditions	
Protein Medication	**Condition Treated**
Insulin	Type 1 diabetes
Erythropoietin	Anemia
Growth hormone	Dwarfism
Anticoagulants	Blood clotting after stroke or heart attack
Clotting factors	Hemophilia
Interferons	Some autoimmune diseases

DID YOU KNOW?

Scientists have created bacteria that do not produce lactic acid, and these organisms could be used to colonize people's mouths and therefore prevent cavities. Cavities are formed by the lactic acid produced by *Streptococcus mutans* bacteria in the mouth, which digest sugars and form lactic acid. If those bacteria didn't produce lactic acid, we would not form cavities. In any case, we can also prevent cavities by eating less sugar.

DID YOU KNOW?

Genetically modified (transgenic) bacteria are used to produce toxic chemicals that kill insects, and therefore are used as pesticides on crop foods. Other transgenic bacteria have been created to take up mercury and help decrease pollution.

CONCEPT REVIEW 10.4

1. What is gene cloning?
2. What is transformation?
3. Explain how a restriction map is useful.
4. How can you determine which bacterial colonies have the gene of interest?
5. List the steps required for bacteria to produce a medication.
6. What are some examples of medications made by bacteria? What conditions do they treat?

10.5 OTHER DNA TECHNIQUES

Gel Electrophoresis

Gel electrophoresis is a method that uses an electrical current to separate strands of DNA based on the length of the DNA sequence. A type of plant carbohydrate called *agarose* is used to form a tiny mesh-like gel. The phosphate molecules in nucleotides are negatively charged, so DNA moves toward the positive end of the gel. The DNA molecules have to squeeze through the gel as the electrical current attracts the negative charges. The small fragments can move through the gel faster than the longer fragments (Figure 10.10). Gel electrophoresis is beneficial for determining if the correct length of fragment has been inserted into a plasmid. The DNA is run on the gel along with a series of known fragments ranging in size from 300 to 2000 base pairs; this is called a *ladder*.

PCR

Polymerase chain reaction (PCR) is a technique to generate multiple copies of a specific DNA sequence (Figure 10.11). A specific primer must be used: a single-stranded fragment of approximately 20 nucleotides that match up with one strand of the gene of interest. Since DNA is double stranded, a primer is needed for each end of the DNA, that is, a 5′ primer and a 3′ primer. The primers must match the correct end of the DNA, since DNA polymerase can replicate DNA only in the 5′ to 3′ direction (Chapter 8). Using PCR to amplify a gene is a much faster and more reliable technique than using bacteria because the bacteria can produce point mutations during DNA replication.

Main Ingredients Required for PCR

- Template DNA—can come from a plasmid, or a blood sample, or any DNA that contains the gene you want to amplify
- Primers—complementary DNA sequences that will bind to the DNA sequence you want to amplify
- dNTPs (nucleotide triphosphates)—the (A, T, C, and G) nucleotides needed to make the new DNA
- Taq polymerase—polymerase enzyme from a strain of bacteria (*Thermus aquaticus*) that lives at very high temperatures and therefore can withstand the increase in temperature needed to separate the DNA strands without itself being denatured.

Figure 10.10 Gel Electrophoresis

Gel electrophoresis is used to separate strands of DNA based on the length of the DNA fragments. Small strands of DNA move through the gel faster than large strands.

Figure 10.11 Polymerase Chain Reaction (PCR)

PCR is used to produce millions of copies of a DNA sequence.

Three Steps Involved in PCR

1. **Denaturation.** At increased temperature, DNA strands separate by breaking the hydrogen bonds that hold the bases together.

2. **Primer annealing.** 5′ and 3′ primers bind to the single-stranded DNA.

3. **Primer extension.** DNA polymerase replicates the DNA using the primers as a starting point.

Using PCR, researchers can amplify very small samples of DNA and then use them for testing: for example, when looking for an inherited genetic mutation, or for determining paternity, or for identifying a specific individual (Figure 10.12). PCR is used to amplify specific regions of DNA where known genetic markers are linked to hereditary diseases, and this makes it a valuable tool for genetic testing.

DNA Fingerprinting

Everyone has repeated DNA sequences in their introns. These repeated sequences are called **variable number tandem repeats (VNTRs)**. Everyone has the same sequences but different numbers of repeats of those sequences. Here is an example of a human VNTR:

> GTCGGAATCG GTCGGAATCG GTCGGAATCG GTCGGAATCG

One person may have 10 copies of this sequence, and someone else may have 23. Consider how this information could be used. Imagine a crime scene where blood samples are taken from the victim, the crime scene, and the suspects. The DNA from the samples is digested with restriction enzymes, and then the DNA fragments are run on a gel to separate the varying *sizes* of DNA. Since each person has a different restriction pattern, the variation in lengths of their DNA shows up, except in identical twins (Figure 10.12).

DID YOU KNOW?

Closely related individuals have common VNTR sequences, but there are still enough differences to distinguish each individual. Since a child has approximately half of their VNTR sequences in common with each parent, they have their own unique pattern.

Figure 10.12 DNA Fingerprinting

DNA fingerprinting can be used to identify people because everyone except identical twins has their own unique DNA fingerprint. Suspect 2 has the same DNA pattern as the DNA evidence.

When multiple probes are used on multiple VNTRs, it is as reliable as traditional fingerprinting; everyone has their own specific pattern. DNA fingerprinting has been a reliable resource in criminal court since 1987.

Gene Therapy

Gene therapy refers to the process of inserting a gene into an individual's cells to replace a mutated gene. The goal is to restore a cell's normal function with the healthy gene. So gene therapy can be used for single-gene, inherited diseases such as cystic fibrosis or muscular dystrophy. Despite some great advances in this area, it remains difficult to insert a particular gene into the correct location in the genome in the correct cell type. Viruses are often used to target a specific cell type, since viruses bind only to specific receptors (Chapter 24), and therefore, only the targeted cell type will receive the gene. However, when viruses are used to inject DNA into cells, the DNA may be integrated randomly into the genome, not specifically; as a result, the normal regulation of the gene's process of transcription can become altered. A more common variation of gene therapy is called *antisense therapy*, where a homologous strand of DNA or RNA is used to bind to its complementary sequence and inactivate it. This technique is most commonly used to treat some cancers and has shown some success in blocking viral infections such as cytomegalovirus.

CONCEPT REVIEW 10.5

1. What is gel electrophoresis? What is the main use for it?

2. Why is PCR useful?

3. What are the main ingredients and steps in PCR?

4. What are VNTRs? How can we make use of them?

5. What is gene therapy?

6. Why is gene therapy frequently not successful?

10.6 GENETICALLY MODIFIED ORGANISMS—GMOs

Genetically modified organisms (GMOs) include not only bacteria that contain human genes or genes from any other organism but also entire plants or animals that contain extra genes or are modified so that certain genes are turned off. Following are some examples of GMOs (see also Figure 10.13):

1. A gene that codes for a blue pigment can be added to plant cells, creating blue roses or blue strawberries.

2. A gene that causes fruit to produce ethylene gas, which causes fruit to ripen, can be inhibited with antisense RNA, and so the fruit doesn't ripen and rot during shipping; ethylene gas sprayed on the tomatoes at the store causes the tomatoes to ripen—known as *flavr savr* tomatoes.

3. Frogs engineered to be transparent are used to study the living frog.

4. Fluorescent genes from jellyfish can be injected into mouse and cat embryonic cells; Mr. Green Genes was the first glow-in-the-dark cat.

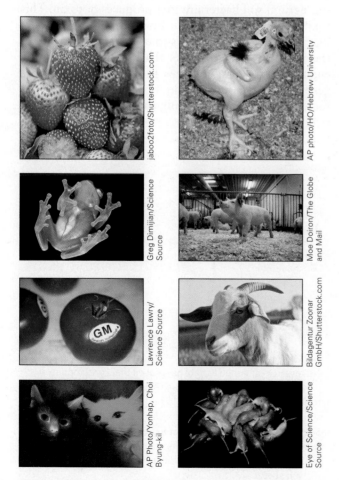

Figure 10.13 Examples of Genetically Modified Organisms

5. Chickens that don't grow feathers have been created so they don't have to be plucked.

The list of possible GMOs increases exponentially every year. Since all organisms are made of DNA—the same ATCG nucleotides—any DNA molecules can be combined to produce GMOs.

Application

As you read through the following section on genetically modified animals that describes the production of BioSteel, the information provided to determine how you could make an "enviro pig"—a pig that generates environmentally friendly manure. Enviro pigs have an extra gene called *phytase* that is expressed in their salivary glands and digests phosphates.

Genetically Modified Animals

Genetically modified (also called *transgenic*) animals can be produced by inserting a recombinant plasmid into a germ-line cell population (Figure 10.14). Instead of transforming bacteria with the recombinant plasmid to produce a protein, the plasmid can be introduced to stem cells that can be implanted into the blastocyst of a mouse, or cat, or goat, etc. Once the new gene is incorporated into all of the embryonic cells, that gene will be expressed in the organism. *Where* the gene is expressed depends on which promoter the gene is attached to because that will affect how transcription is regulated.

Simplified Steps to Make a Transgenic Goat Produce Spider Silk (BioSteel)

1. Isolate the spider silk gene.

2. Ligate the gene into a plasmid that contains the promoter sequence involved in the regulation of milk production in goats.

3. Microinject the recombinant plasmid into a goat egg cell, and let it grow to blastocyst stage.

4. Implant the blastocyst into a pseudopregnant goat—a female goat that has been given hormone injections to cause the uterus to be prepared for implantation.

5. The offspring of the goat will have the spider silk gene connected to the promoter region involved in lactation. When lactation occurs in that goat, the milk will contain spider silk that can be isolated and used to build many very strong substances. The first product ever made from BioSteel was a bullet-proof vest.

Since animals are eukaryotic and have the cellular machinery to process RNA and have Golgi bodies

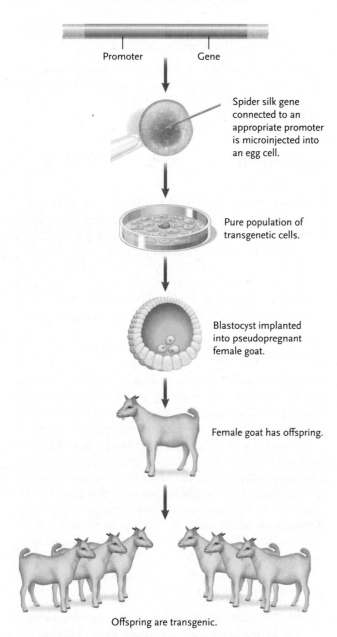

Promoter   Gene

Spider silk gene connected to an appropriate promoter is microinjected into an egg cell.

Pure population of transgenetic cells.

Blastocyst implanted into pseudopregnant female goat.

Female goat has offspring.

Offspring are transgenic.

Figure 10.14   Production of a Recombinant Organism
The production of a recombinant organism involves the introduction of a new gene into embryonic stem cells.

that modify proteins, they can be more reliable than bacteria for producing complete functional proteins. For example, ATryn, a drug produced in the milk of goats, is used to treat hereditary antithrombin deficiency.

Animals can also be genetically engineered to *not* produce a specific protein. For example, genes are deleted from the genomes of mice—known as **knockout mice**—so researchers can figure out the functions of those genes and what would happen to people with mutations in those genes. Several thousand single-gene or double-gene knockout mice exist for research purposes, and many are used to model human diseases, such as cancer, Parkinson's, diabetes, arthritis, and

heart disease; these mice have been very valuable in finding treatments for many diseases.

Genetically Modified Plants

Many food crops are engineered to contain genes that give plants particular properties, such as the following:

- Resistance to herbicides called Roundup Ready®. Herbicides can be sprayed on whole crops, and only the weeds die.
- Ability to produce their own pesticides. This eliminates the need to spray for insects.
- Resistance to cold temperatures. This lengthens the growing season.
- Ability to prevent ripening. This enables long-distance transport without loss from rotting.
- Added nutrients. Golden rice contains a gene for beta-carotene, which is a precursor for vitamin A production.

One prevalent method of food plant engineering involves integrating a bacterial gene into a plant that codes for a protein that is toxic to insects, thereby producing insect-resistant crops. *Bacillus thuringiensis* (bt) is a natural soil bacterium that produces bt toxin. This toxin kills insects that eat it by creating pores in their digestive tracts, which leads to cell death and death of the insect. The crop plants that are now genetically engineered to contain either bt toxin or some other genetic modification include tobacco, potatoes, tomatoes, soybeans, corn, broccoli, canola, flax, sugar beet, rice, papaya, and cotton.

DID YOU KNOW?

Approximately 70–85% of processed foods contain genetically modified ingredients, mainly because many processed foods contain some form of corn starch or soybean oil.

DID YOU KNOW?

No research published to date has proven that consumption of genetically engineered foods cause harm. One study attempted to show that GMO corn caused large tumours in rats, but that study had major flaws, and the conclusions drawn are incorrect. See Forbes online references that explain why the rat tumour research is flawed: http://www.forbes.com/sites/stevensalzberg/2012/09/24/does-genetically-modified-corn-cause-cancer-a-flawed-study/

Implications of Genetically Engineered Plants

- Insects become resistant to the bt toxin. Recall from Chapter 7 that microevolution occurs in every population. There are always some organisms that resist drugs, or antibiotics, or toxins. In fact, the first resistant organism was found in 2009: a bollworm that infected cotton plants. As a result, genetic engineers made cotton plants that contained multiple types of bt toxin-producing genes.

- Changes in ecosystems occur if even one population of organisms is altered. In many cases secondary pests become a problem, and the crops need to be sprayed with pesticides anyway.

- The engineered gene can be transferred into other closely related species. Many plants cross-pollinate with other similar species, and a wide variety of weed plants are similar to food plants. If the bt toxin gene crosses into other species, it could have an effect on that ecology.

- The implications of genetically engineered foods on human health are not known. No reliable, long-term studies have been done with humans. The bt toxin gene is expressed only by the parts of the plant that humans don't eat, such as the stem or leaves, but it is found in all of the plant's cells, including the parts used for food. Roundup Ready crops are sprayed with toxic herbicides that do not wash off the plants: a topic of considerable debate by health experts.

CONCEPT REVIEW 10.6

1. Give some examples of genetically engineered organisms that have been made.

2. In a genetically engineered animal, how is the expression of the gene controlled? For example, describe the difference between goats that produce spider silk in milk and mice that express GFP in their paws.

3. What is the advantage of using transgenic animals rather than bacteria to produce medications?

4. List the steps involved in making a transgenic goat that can produce spider silk in its milk.

5. Why are knockout mice so valuable to researchers?

6. What are the main ways that plants are genetically engineered?

7. What is bt toxin and how is it used to genetically engineer plants?

CHAPTER REVIEW

Answer to Application Question

- Isolate the phytase gene.
- Ligate the gene into a plasmid that contains the promoter sequence involved in the production of saliva.
- Microinject the recombinant plasmid into a pig egg cell, and let it grow to blastocyst stage.
- Implant the blastocyst into a pseudopregnant pig. The offspring of the pig will have the phytase gene connected to the promoter region involved in salivation, allowing that gene to digest phosphates.

10

Christopher Meade/Shutterstock.com

11 Tissues

Any fool can know. The point is to understand.

Albert Einstein

A tissue is a combination of cells and proteins that function together in a specialized role. **Histology** is the study of tissues. Four main types of tissues are found throughout the body and support the function of our organ systems: **epithelial tissue**, connective tissue, muscular tissue, and nervous tissue.

Application

As you read through this chapter, consider the following. The small intestine is an organ that contains all four types of tissues. What kind of epithelial cells line the lumen of the small intestine? Which type of muscle tissue is in the small intestine (skeletal, smooth, or cardiac)? Which branch of the nervous system innervates the digestive tract (somatic or enteric)? Which types of connective tissue are found in the small intestine (loose, dense, cartilage, bone, blood)?

11.1 EPITHELIAL TISSUE

Epithelial tissue has two major functions: (1) covering and lining the outside of the body and the inside of hollow organs, body cavities, and ducts; and (2) forming glands. The first function protects organs and structures, and allows molecules and nutrients to move from one location to another by processes such as diffusion, filtration, secretion, and absorption. Recall from Chapter 3 that simple diffusion refers to the movement of small molecules across a cell membrane, and facilitated diffusion (facilitated transport) occurs when a membrane protein is required for an ion, such as a sodium ion, to cross the cell membrane. **Filtration** occurs in the glomerulus of the kidney when substances move from the blood into the Bowman's capsule (Chapter 21). **Secretion** is the movement of chemicals or molecules from a cell, usually into the bloodstream or into the lumen of a hollow organ, such as the small intestine or the kidney nephron. **Absorption** that involves epithelial cells is the movement of chemicals or molecules into the bloodstream, such as from the lumen of the digestive tract, or through the skin. In the kidney, molecules are reabsorbed through nephron epithelial cells from the lumen into the surrounding capillaries.

Glandular epithelial tissue is found in both exocrine glands and endocrine glands. Exocrine glands secrete substances into ducts that empty into a hollow organ, such as gastric glands in the stomach, or onto the surface of the skin, such as sebaceous oil glands (Chapter 12). Endocrine glands secrete hormones into the bloodstream, such as the thyroid gland secreting thyroid hormones (Chapter 17).

Classification of Epithelial Cells

Epithelial cells are categorized based on their shape and the number of layers (Figure 11.1). They can have the following shapes: **squamous** (flattened), **cuboidal** (cube), **columnar** (rectangular, or column-like). Some epithelial cells that change shape are called **transitional epithelial cells**, such as bladder cells that change from cuboidal, when the bladder is empty, to squamous, when the bladder is full. *Simple* epithelial cells are a single layer, whereas those that are **stratified** occur in multiple layers. Some epithelial cells appear to be stratified when in fact they are a single layer: these cells are called **pseudostratified** because some cells do not reach the surface and so their nuclei appear layered. Some epithelial cells can have extra surface structures: such as **microvilli** to increase surface area for absorption of nutrients in the small intestine (Chapter 20), or **cilia** in the upper airways that help move mucus up and out of the respiratory tract (Chapter 19). See Table 11.1.

Characteristics of Epithelial Cells

Since epithelial cells cover and line the surfaces and cavities of the body or glands, each side of the cell has different functions. Their specific function depends

Classifying epithelium:
Cell shape

Transitional epithelial   Squamous   Cuboidal   Columnar

Arrangement of layers

Apical surface
Basal layer
Basement membrane

Simple   Pseudostratified   Stratified

Figure 11.1   Epithelial Tissue
Epithelial tissue is classified by the shape of the cells and the arrangement of the layers.

on whether the side of the cell is facing the surface or the cavity (**apical** side), or if it is in contact with the **basement membrane** (**basal** side)—which is a protein region composed mainly of laminin and integrins that connect the epithelial cells to the connective tissue (Section 11.2). Lateral surfaces of epithelial cells attach to adjacent cells. Epithelial tissue is **avascular**, meaning that it does not have any blood vessels, and so old cells can be shed without causing bleeding. Nutrients and oxygen diffuse into these cells from the underlying connective tissue. Epithelial cells are also highly connected by cellular junctions, such as gap junctions, tight junctions, adherens junctions, and desmosomes (Chapter 3). This means they are strongly held together and can transfer signalling molecules that allow the cells to act as a unit, that is, as a tissue. Because of the location of epithelial tissue and its role in protecting other structures, these cells undergo the most rapid cell cycle in the body, and new cells are formed daily.

DID YOU KNOW?

A pap test is used to look at a sample of non-keratinized, stratified, squamous epithelial cells taken from the cervix to determine if cells are normal, precancerous, or cancerous. The cells are examined microscopically for shape or density changes. The most common type of cervical cancer is squamous cell carcinoma. Prognosis is very good if the cancer is detected early.

11

TABLE 11.1

Classification of Epithelial Tissue

Simple Squamous

Squamous cell

Apical surface
Basal surface
Basement membrane

From Russell/Hertz/McMillan. *Biology*, 4E. © 2017 Brooks/Cole, a part of Cengage, Inc. Reproduced by permission. www.cengage.com/permissions.

Ed Reschke/Photolibrary/Getty Images

Functions

Single layer of flattened cells allows for diffusion, osmosis, filtration (glomeruli of kidneys), and secretion (surfactant in the alveoli).

Location

These cells line heart, blood vessels, lymphatic vessels, alveoli in lungs, and glomeruli in the kidney. It forms part of any serous membrane that covers internal organs, such as the heart (pericardium), lungs (pleura), and abdominal organs (peritoneum).

Simple Cuboidal

Apical surface

Squamous cell
Basal surface
Basement membrane

From Russell/Hertz/McMillan. *Biology*, 4E. © 2017 Brooks/Cole, a part of Cengage, Inc. Reproduced by permission. www.cengage.com/permissions.

Ed Reschke/Photolibrary/Getty Images

Functions

Single layer of cuboidal cells allows for secretion and absorption.

Location

These cells are found primarily in the ducts of exocrine glands, the lumen of the nephron in the kidney, the surface of the ovaries, and parts of the thyroid gland.

Simple Columnar

Cilia
Apical surface
Goblet cell
Ciliated columnar cell
Basolateral surface
Basement membrane
Underlying tissue

Jose Luis Calvo/Shutterstock.com

Functions

Ciliated, simple, columnar epithelial cells function to *move* substances, such as foreign particles trapped by mucus in the upper respiratory tract, or to move the egg along the fallopian tubes. These cells secrete mucus.

Location

These ciliated cells are found in the upper respiratory tract, sinuses, fallopian tubes, and uterus. (Cilia are hair-like projections that move in wavelike patterns.)

Apical surface (location of microvilli)
Non-ciliated columnar cell
Basal surface
Basement membrane

From Russell/Hertz/McMillan. *Biology*, 4E. © 2017 Brooks/Cole, a part of Cengage, Inc. Reproduced by permission. www.cengage.com/permissions.

Jose Luis Calvo/Shutterstock.com

Non-ciliated, simple, columnar epithelial cells usually contain specialized extensions called *microvilli* that are cell extensions that allow for absorption of nutrients. This tissue contains goblet cells (considered unicellular glands) that secrete mucus.

These non-ciliated cells are found primarily in digestive tract (stomach to anus), and also in the gallbladder. They secrete mucus that protects the lining of the digestive tract.

Pseudostratified Columnar

Apical surface — Cilia
Goblet cell
Ciliated pseudostratified columnar cell
Basal cell
Basolateral surface
Basement membrane
Underlying tissue

Jose Luis Calvo/Shutterstock.com

Functions	Location
Ciliated, pseudostratified, columnar epithelial cells function to secrete mucus, trap foreign particles, and move substances along the surface of the tissue. Notice how the cells appear to be stratified, but they are actually in a single layer with some cells not extending to the surface (*pseudo* means "false"). Non-ciliated cells function in absorption and secretion.	These cells are found in similar locations as ciliated, simple, columnar cells— primarily the respiratory tract. But non-ciliated, pseudostratified epithelial cells can be found in the epididymis and many glands.

Stratified Squamous

Flattened squamous cell
Apical surface
Squamous cell
Basolateral surface
Basement membrane
Underlying tissue

Biophoto Associates/ Science Source

Functions	Location
Stratified squamous is the most abundant of the stratified tissue types. The multiple layers function to protect the underlying tissues from abrasion and from microorganisms. Note that the apical cells are squamous, and the deeper basal cells are cuboidal.	**Keratinized,** stratified epithelial cells are found in the skin; the outermost layers contain keratin protein in dead cells to provide a tough barrier (Chapter 12). Non-keratinized, stratified epithelial cells are located in the mouth, tongue, esophagus, pharynx, and vagina.

(Continued)

11

11

TABLE 11.1

Classification of Epithelial Tissue (Continued)

Stratified Cuboidal

Apical surface
Cuboidal cells
Basolateral surface
Basement membrane
Underlying tissue

Alvin Telser/Science Source

Functions	Location
Stratified, cuboidal, epithelial cells can be two or more layers, and they function to protect underlying tissues and have some secretion and absorption properties.	These cells are found mainly in esophageal ducts, sweat glands, mammary glands, and salivary glands.

Stratified Columnar

Apical surface
Columnar cells
Basolateral surface
Basement membrane
Underlying tissue

Science Vu, Visuals Unlimited/Science Photo Library

Functions	Location
This less common cell type, functions in protection and secretion.	These cells are found in the ducts of some exocrine glands, such as esophageal glands, and some parts of the urethra.

Transitional

Apical surface
Epithelial cells
Basolateral surface
Basement membrane
Underlying tissue

© M I (Spike) Walker/ Alamy Stock Photo

Functions	Location
Having characteristics of both squamous and cuboidal, transitional epithelial cells allow for a hollow organ to distend and stretch. Cell shape is squamous when stretched and cuboidal when relaxed.	These cells are found in the bladder, ureters, and urethra.

CONCEPT REVIEW 11.1

1. What are the main functions of epithelial cells?

2. How are epithelial cells classified?

3. Briefly describe the location and function of each type of epithelial tissue.

11.2 CONNECTIVE TISSUE

Connective tissue is the most abundant and diverse tissue in the body. It is composed of combinations of different **cell types** and **extracellular matrix**. The matrix consists of fibrous proteins and molecules that make up the **ground substance**, which is a gel-like substance composed of water and non-fibrous proteins (Figure 11.2). Cell types include fibroblasts, adipocytes, and reticular cells. Connective tissue can also contain immune cells, such as mast cells, macrophages, and lymphocytes. The immune cells play a role in the inflammatory response and wound healing. Mast cells produce **histamine** when there is tissue injury, macrophages engulf pathogens and dead cells, and lymphocytes are involved in the adaptive immune response (Chapter 23).

Overview of Connective Tissue Functions

- **Loose connective tissue**—found in almost every body structure—connects epithelial tissue to muscle layers (skeletal or smooth), gives tissues strength and elasticity, stores fat (adipose tissue), and acts as a filter in spleen and lymph nodes (loose reticular tissue).

- **Dense connective tissue** forms strong attachments between structures, such as ligaments connecting bones within a joint, and allows organs to stretch and recoil, such as within arteries or the lungs.

- **Cartilage**—a very strong tissue—acts as a smooth surface for joint movement and gives particular structures their shape, such as the ear.

- **Bone** forms the skeleton, stores calcium and other minerals, produces red and white blood cells, stores some fat, and protects internal organs.

- **Blood** transports oxygen, nutrients, waste products, and hormones throughout the body.

Cell Types in Connective Tissues

Fibroblasts make up the most abundant connective tissue cell type. They produce the collagen, elastin, and reticular fibres, as well as ground substance that make up the extracellular matrix. Fibroblasts are very important in the healing of wounds; when tissues are damaged, fibroblasts increase the synthesis of these substances to bind tissues together.

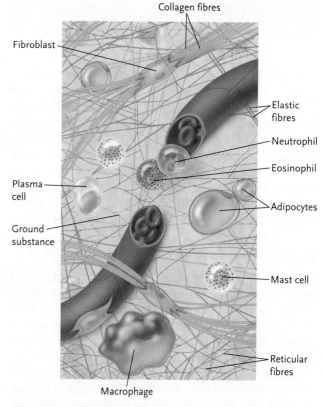

Figure 11.2 Connective Tissues

Connective tissues contain various cell types, protein fibres, and ground substance.

Adipocytes are cells that store fat as triglycerides. Any extra calories we eat—whether in the form of protein, carbohydrate, or fat—that are not used to build cell structures or to produce energy become converted into fat and stored in adipocytes in the subcutaneous layer of the skin, which is loose adipose connective tissue.

DID YOU KNOW?

A scar is formed from increased collagen production by fibroblasts when the skin undergoes significant damage and the epithelial cells of the skin cannot be completely replaced. Scar tissue is significantly stronger but less stretchable than surrounding skin cells.

Reticular cells are found in loose reticular connective tissue that produce the reticular fibres involved in filtering pathogens and old red blood cells in lymph nodes, the spleen, and the liver. **Chondrocytes** are the cells that make up cartilage connective tissue. **Osteocytes** are the cells that make up bone connective tissue. **Blood cells** include red blood cells that carry oxygen and some carbon dioxide, and the various white blood cells involved in the immune response (Chapter 23). See Table 11.2.

Types of Protein Fibres in Connective Tissue

Collagen, the most abundant protein in the body, makes tissues very strong. The more collagen fibres in a tissue, the stronger that tissue is. Tendons have a very high amount of collagen and are very strong because they connect muscles to bones and must withstand high pressure when muscles contract and pull on bones.

Elastin is a protein that gives tissues the ability to stretch and recoil back to their original shape. The connective tissue layers beneath the epithelial cells of the skin have a high amount of elastin.

Reticular fibres are produced by reticular cells and are a type of thin collagen fibre surrounded by glycoproteins. These fibres form a framework for many tissues and act as a filter in the spleen and lymph nodes. Reticular fibres are also found in blood vessels; they surround fat cells, nerves, and muscles; and they make up one of the two layers of basement membranes (composed of basal lamina and the reticular lamina).

Composition of Ground Substance

Ground substance is the fluid part of the connective tissue that supports the functions of cells and fibres. Oxygen, nutrients, and hormones move from the blood vessels through ground substance to all the cells in the body. Ground substance contains water and combinations of proteins and carbohydrate chains, called **proteoglycans**. Some of the most important examples include glucosamine sulfate, hyaluronic acid, and chondroitin sulfate.

Glucosamine sulfate makes up a significant proportion of the ground substance in cartilage, tendons, and ligaments, and it is found in the fluid surrounding joints. **Chondroitin sulfate** is an important component of the fluid surrounding joints and within cartilage; it allows the joints to withstand compression.

Hyaluronic acid is found in most types of connective tissue and plays a role in cell proliferation and migration. Immune cells and bacteria produce the **hyaluronidase** enzyme that breaks down hyaluronic acid, and this lets the immune cells and bacteria move easily through connective tissues. It is thought that cancer cells also produce this enzyme, which allows cancer cells to spread (metastasis).

TABLE 11.2

Classification of Connective Tissue

Loose Areolar

Functions	Location
This is the most widely distributed connective tissue, and it functions to strengthen, support, and provide elasticity to every body structure.	Found around all epithelial tissues, such as the subcutaneous layer and dermis layer of the skin, mucus membranes, blood vessels, nerves, and all organs.

Patrick J. Lynch/
Science Source

Adipocyte Reticular Fibroblast Ground
           fibres   substance

Elastic
fibres

Mast
cell

Plasma cell Collagen fibres

Loose Adipose

Functions	Location
Adipose tissue, which stores fat as triglycerides, can be used to produce energy (ATP), insulate the body, and protect and cushion organs.	Found wherever there is areolar connective tissue—in the subcutaneous layer of the skin and around blood vessels, nerves, and organs.

Courtesy of Wendi Roscoe

Nucleus

Adipocyte

Fat storage

Blood
vessel

(Continued)

11

TABLE 11.2

Classification of Connective Tissue (*Continued*)

Loose Reticular	Functions	Location
 Reticular fibres Reticular cells Biophoto Associates/Science Source	This type provides structural support, forms part of basement membranes that surround organs, blood vessels, muscles, and acts as a filter to remove old red blood cells in the spleen and pathogens in the spleen and lymph nodes.	Found in liver, spleen, lymph nodes, bone marrow, blood vessels, all organs, and smooth muscle

Dense Regular	Functions	Location
 Collagen fibre Fibroblast cell Ed Reschke/Photolibrary/Getty Images	This type provides very strong attachments for connecting muscle to bone (tendons) and connecting bone to bone (ligaments). Dense, regular, connective tissue contains a high number of parallel strands of collagen, making it very strong. This tissue does not contain blood vessels, so injuries heal very slowly.	Found in tendons, ligaments, and aponeuroses (the sheet-like tendons that connect muscle to muscle) (Chapter 14)

Dense Irregular

Collagen fibre   Elastic fibre   Fibroblast

Biophoto Associates/
Science Source

Functions	Location
This type consists primarily of collagen fibres, similar to dense, regular tissue but the fibres are not parallel; they are arranged randomly and form a strong sheet-like connective tissue that provides strength when pulled in many directions.	Makes up the fascia that surrounds muscles, organs—such as the heart (pericardium)—digestive tract, bones (periosteum), and that forms part of the lower dermis region of the skin

Dense Elastic

Elastic fibre   Fibroblast

Biophoto Associates/
Science Source

Functions	Location
This tissue allows a high amount of stretch and recoil.	Found in structures that undergo stretching forces, such as arteries, lungs, trachea, bronchial tubes, and vocal cords

(Continued)

11

TABLE 11.2

Classification of Connective Tissue (Continued)

Hyaline Cartilage	Functions	Location
 Lacuna with chondrocytes inside — Ground substance Source: M. I. Walker/Science Source	This tissue provides a strong, smooth surface for joint movement. Cartilage does not contain any blood vessels. Hyaline cartilage is the weakest of the three types and can fracture easily.	Found at the ends of long bones, ribs, nose, trachea, larynx, bronchi, and comprises most of the fetal skeleton

Elastic Cartilage	Functions	Location
 Lacuna with chondrocytes inside — Elastic fibre — Ground substance Herve Conge, ISM/ Science Photo Library	This tissue provides strength as well as the elasticity and shape of certain structures.	Found in the epiglottis, external ear, and the Eustachian tubes (tubes that connect the inner ear to the throat)

Fibrocartilage

Lacuna   Chondrocytes   Collagen fibre

Biophoto Associates/
Science Source

Functions	Location
This is the strongest type of cartilage, containing thick collagen fibres that connect certain structures together and act as shock absorbers.	Found in the intervertebral discs, pubic symphysis, and knee joint

Bone

Artery
Nerve
Vein
Ground substance

Lacunae containing osteocytes

Herve Conge, ISM/
Science Photo Library

Functions	Location
Bones move when muscles contract and produce movement. Bones are important in mineral homeostasis and provide a reservoir for calcium and phosphorus, and they protect internal organs. Red marrow produces red and white blood cells; yellow marrow stores some fat.	Includes long bones, skull, ribs, vertebrae, and pelvis (Chapter 13)

Blood

White blood cell
Plasma
Red blood cells
Platelets

Biophoto Associates/
Science Photo Library

Functions	Location
Blood transports oxygen, nutrients, hormones, and waste products throughout the body. Red blood cells transport oxygen and some carbon dioxide (Chapters 18 and 19). White blood cells include a variety of immune cells (Chapter 23). Platelets are important for blood clotting.	Found within blood vessels, and also where red and white blood cells and platelets are produced in red bone marrow

11

1. Name the three main components of connective tissue.
2. What are the main functions of each type of connective tissue?
3. Describe the main functions of each cell type found in connective tissue.
4. What kinds of protein fibres are found in connective tissue, and why are they important?
5. Why is ground substance important?
6. Describe the main function and locations of each type of connective tissue.

11.3 MUSCLE TISSUE

The function of muscle tissue is to contract and generate force. When muscle fibres contract they produce movement and heat. Muscle fibres can contract without producing movement—called *muscle tone*—which is important for maintaining body posture and protecting internal organ structures. Specialized muscle tissue, **sphincters**, regulates the movement of substances within the body: for example, in diverting blood flow into certain capillary beds, and the movement of food through the digestive track.

Muscle tissue can be classified into three categories: skeletal, smooth, and cardiac. Each type of muscle tissue has distinctive properties that allow it to function optimally (see Figure 11.3).

Skeletal Muscle

Skeletal muscle—the muscle tissue that is connected to our bones (skeleton)—allows our limbs to move. Muscles are connected to two different bones across a joint so that contraction will cause movement of the bones. For example, to bend the arm at the elbow, the brachialis muscle is connected to the humerus (upper arm bone), crosses the elbow, and connects to the ulna (one of the forearm bones). Mostly, skeletal muscle cells (also called *fibres*) contract as a result of something we decide to do. In this case, skeletal muscle contraction is considered **voluntary**. For example, when you pick up something from the ground, contraction of the necessary skeletal muscles primarily involves the motor cortex of the brain's frontal lobe. Sometimes certain skeletal muscles contract involuntarily; these are **reflexes**, such as pulling our hand away from a hot stove.

Skeletal muscle cells are **multinucleated** because during their development, many muscle cells merge together. Skeletal muscle fibres are **striated** due to a light and dark banding pattern that's visible under a microscope. The individual units, called *sarcomeres*, line up evenly, unlike smooth muscle. Skeletal muscle fibres can be subdivided into those that are fast twitch, slow twitch, and intermediate, giving various muscle groups their different capacity for endurance or power movement (Chapter 14). (See Figure 11.3a.)

Smooth Muscle

Smooth muscle is found within the walls of hollow structures, such as blood vessels, lymphatic vessels, stomach, intestines, airways, fallopian tubes, ureters, bladder, uterus, vas deferens, erector pili muscles in the skin, and the iris of the eye. Smooth muscle contraction and relaxation is controlled by the **autonomic nervous system** and is therefore considered **involuntary movement** that is unconsciously controlled. Different smooth muscles contract or relax when stimulated by either the sympathetic ("fight or flight") or parasympathetic ("rest and digest") components of the nervous system (see Chapter 15).

Smooth muscle cells each contain one single nucleus, and they are **non-striated** because they overlap unevenly, so there are no visible banding patterns. Smooth muscle cells are much smaller and contract very slowly, with much less force compared to skeletal and cardiac muscle cells. There are two kinds of smooth muscle cells, single unit and multiunit. **Single-unit smooth muscle** occurs in sheets that surround visceral organs, where many cells contract simultaneously: for example, as found in the stomach, intestines, uterus, and bladder. **Multiunit smooth muscle** consists of many cells contracting individually, which gives more precise control: for example, as found in blood vessels, the eye, erector pili, and airways (see Figure 11.3b).

Cardiac Muscle

Cardiac muscle, found only in the walls of the heart, contracts in order to circulate blood throughout the body. Cardiac muscle has some features similar to skeletal muscle and some features similar to smooth muscle. Cardiac muscle fibres are involuntary in the sense that their contraction is not under conscious control. Cardiac muscle is **autorhythmic**, meaning that specialized cells

Figure 11.3 Muscle Cell Types
There are three types of muscle cells: (A) skeletal, (B) smooth, and (C) cardiac.

From Russell/Hertz/McMillan. *Biology*, 4E. © 2017 Brooks/Cole, a part of Cengage, Inc. Reproduced by permission. www.cengage.com/permissions

Eric V. Grave/Science Source

Biophoto Associates/Science Source

Biophoto Associates/Science Source

in the heart called **pacemaker cells** control the rate of contraction. However, the autonomic nervous system can influence the heart rate. Cardiac muscle fibres are striated like skeletal muscle, but have one to three nuclei per cell. Cardiac muscle fibres are branched and connected by **intercalated discs**, which contain gap junctions that allow signalling molecules to quickly pass between cells (see Chapter 14). This very unique feature of cardiac muscle is required for controlled and simultaneous contraction of the chambers of the heart so that blood circulates efficiently throughout the body (see Figure 11.3c).

CONCEPT REVIEW 11.3

1. What are the functions of muscle tissue?
2. What are the three types of muscle tissue?
3. Make a study chart that summarizes the characteristics of each type of muscle.
4. Where is smooth muscle located?
5. What are the two types of smooth muscle, and where are they located?
6. What are the special features of cardiac muscle tissue?
7. Why are intercalated discs an important component of cardiac muscle cells?

11.4 NERVOUS TISSUE

The function of nervous tissue is to detect and transmit sensations, interpret environmental stimuli, and respond to stimuli by signalling other **neurons**, muscles (skeletal, smooth, or cardiac), or glands (endocrine or exocrine).

Nervous tissue is composed of the neurons that send and receive signals (Figure 11.4), and **neuroglia**, which are cells that support the neurons in various ways.

There are three general categories of neurons: multipolar, bipolar, and unipolar (Figure 11.5). Each type of neuron is found in different areas of the nervous system.

11

Motor neurons, also called **multipolar neurons**, transmit information from the CNS to muscles or glands (Chapter 15). **Sensory neurons**, also called **unipolar neurons**, transmit information from sensory receptors to the central nervous system (brain and spinal cord); **bipolar neurons** are found in the olfactory sensory system and the retina (Chapter 16).

Neuroglia are the cells that support the neurons and ensure proper functioning of the central (CNS) and peripheral (PNS) nervous systems. **Oligodendrocytes** (CNS) and **Schwann cells** (PNS) produce the protective myelin covering of neurons as found in CNS white matter and on long sensory and motor neurons. **Astrocytes** (CNS) and **satellite cells** (PNS) maintain ion concentrations surrounding neurons, and astrocytes form the blood-brain barrier. **Microglia** are resident immune cells in the brain and spinal cord. **Ependymal cells** produce cerebral spinal fluid.

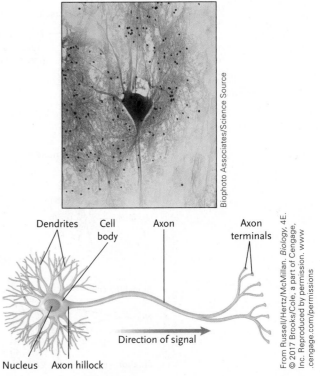

Figure 11.4   Structure of a Typical Neuron

A typical neuron is composed of dendrites, a cell body, an axon, and axon terminals.

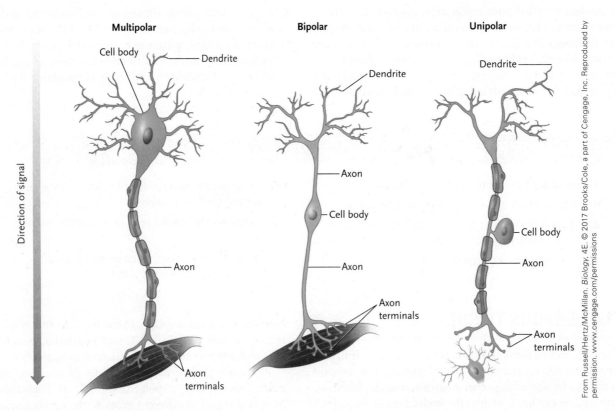

Figure 11.5   Types of Neurons

The three types of neurons are multipolar, bipolar, and unipolar.

CONCEPT REVIEW 11.4

1. What are the functions of nervous tissue?
2. What types of cells can neurons signal?
3. What cell types are found in nervous tissue?

4. Name the three types of neurons and where they are located.
5. Briefly, name the types of neuroglia and their main functions.

CHAPTER REVIEW

Labelling Activity

Label the following diagrams.

1.

2.

11

Identify the following neurons.

3.

a.

_____

b.

_____

c.

_____

Identify the following tissues.

4.

_____

5.

_____

6.

_____

7.

_____

8.

_____

9.

_____

10.

_____

11.

_____

12.

_____

13.

Patrick J. Lynch/Science Source

14.

Courtesy of Wendi Roscoe

15.

Biophoto Associates/Science Source

16.

Ed Reschke/Photolibrary/Getty Images

17.

Biophoto Associates/Science Source

18.

Biophoto Associates/Science Source

19.

M. I. Walker/Science Source

20.

Herve Conge, ISM/Science Photo Library

21.

Biophoto Associates/Science Source

22.

Herve Conge, ISM/Science Photo Library

23.

Biophoto Associates/Science Photo Library

24.

Eric V. Grave/Science Source

25.

Biophoto Associates/Science Source

26.

Biophoto Associates/Science Source

Answer to Application Question

Answer to Application question
The small intestine is lined with simple, columnar epithelial cells. The small intestine contains smooth muscle, which is controlled by the autonomic nervous system (ANS). The enteric nervous system, a branch of the parasympathetic nervous system (a division of the ANS) innervates the digestive tract. The following types of connective tissue are found in the small intestine:

- Loose areolar, in the mucus membranes, blood vessels, and nerves

- Loose adipose, in the visceral fat surrounding organs
- Loose reticular, found in all organs, smooth muscle, and blood vessels
- Dense irregular, found in the fascia surrounding the digestive tract
- Blood, in the capillaries within the digestive tract where nutrients are absorbed

11

12 The Integumentary System

Alex Luengo/Shutterstock.com

CHAPTER OUTLINE

Live as if you were to die tomorrow. Learn as if you were to live forever.

Mahatma Gandhi

The **integumentary system** consists of skin, glands, hair, and nails. The **skin** is the largest organ in the body, making up about 12% of our total body weight. In other animals the integumentary system includes fur, feathers, scales, claws, and hooves.

12.1 FUNCTIONS OF THE INTEGUMENTARY SYSTEM

- Protection. Our skin protects our internal organs from infections, physical damage, dehydration, and environmental hazards such as chemicals and UV light. The skin is the first barrier to pathogens (bacteria, viruses, fungi, or parasites) that we come in contact with. Sebaceous glands secrete oil called *sebum*, which lubricates the outer layers of skin (epithelial cells), helps skin act as a waterproof barrier, and provides antimicrobial properties (see Section 12.3).

- Protection provided by the hair and nails. Humans have hair on most surfaces of the skin, except on the palms of the hands and soles of the feet (thick skin regions). Having more hair on the head provides protection from UV damage, and eyebrows and eyelashes help protect the eyes from damage by foreign particles. For example, certain reflexes cause blinking when objects come close to our eyes, and the hairs of the eyebrows and lashes help trap small particles. Differences in hair distribution patterns are related to many factors, including genetics, hormones, and age. Nails are thick layers of dead keratinocytes that help us grip objects. Our nails are not as strong or sharp as claws in other animals, which need them for survival, such as in climbing to find food or escaping predators.

- Regulation of body temperature. The deepest layer of the skin, the hypodermis (also called the *subcutaneous* layer), contains fat cells (adipose tissue) that help the body retain body heat. Some layers of the skin (dermis and hypodermis) have blood vessels that constrict in cold temperatures to keep heat within the body, or dilate in hot temperatures to promote heat loss. When blood vessels constrict and there is less blood flow near the cold body surface, less body heat escapes. By contrast, sweat glands in the dermis layer increase heat loss from the body through evaporation of sweat (water and ions). As water is pulled away from the body during evaporation, heat also leaves the body.

- Vitamin D synthesis. Cholesterol molecules in the lower layers of the epidermis of the skin are converted into a vitamin D precursor by sunlight, which the liver and kidneys then convert into active vitamin D, called *calcitriol* (Chapter 21). Although vitamin D can also be acquired through the diet, some daily sunlight exposure is very important. Foods that contain vitamin D include fatty fish, like salmon, liver, egg yolk, and fortified foods. Vitamin D plays many important roles in bone mineral regulation, hormone synthesis, and immune system regulation (see Chapters 17 and 23).

- Sensation. Our skin contains several different sensory receptors that allow us to be aware of changes in temperature, pain, touch, vibration, and pressure (Chapter 16).

- Excretion. Some waste products, including salts and urea, are removed from the body through the sweat glands. The skin, lungs, liver, and kidneys are all involved in breaking down or removing waste products from the body (Chapter 21).

- Absorption. The skin is highly permeable and many fat-soluble substances can be absorbed through the dermal layers, glands, or hair follicles into the bloodstream. Medications can be absorbed into the bloodstream through dermal patches, such as nicotine patches or hormone replacement therapy patches. A skin patch is an effective way to provide lower doses of a drug for a longer period of time, compared to an oral or injectable mode of delivery.

DID YOU KNOW?

Vitamin D has been shown to play a role in preventing seasonal affective disorder (SAD), or winter depression. SAD is correlated with decreased sunlight exposure, which stimulates the pineal gland that is important for regulating sleep-wake cycles and mood. Vitamin D is involved in the production of mood neurotransmitters such as serotonin and dopamine. A combination of low light and low vitamin D increases the risk of developing SAD. Treatments include light therapy and vitamin D supplementation.

CONCEPT REVIEW 12.1

1. List the components of the integumentary system.
2. List the main functions of the integumentary system.
3. Why is vitamin D important?
4. How does the skin help us regulate body temperature?

12.2 EPIDERMIS, DERMIS, AND HYPODERMIS

The skin is made up of an **epidermis** and a **dermis**, which is attached to a subcutaneous layer called the *hypodermis* (Figure 12.1). The epidermis is the outermost region and is made up of either four layers of cells (thin skin) or five layers of cells (thick skin), depending on the location in the body. Thick skin is located on the palms of the hands and soles of the feet. The dermis is divided into a **papillary layer** and a **reticular layer**, which is primarily areolar and dense connective tissue.

The subcutaneous layer contains adipose tissue and is the main area for storing fat, which provides both stored energy and insulation for helping maintain body temperature.

Epidermis

The epidermis is made of stratified, squamous epithelial cells (mostly keratinocytes), and it is avascular (Chapter 11). Nutrients and oxygen must diffuse from capillaries in the dermis into the epidermis. The deepest layer of the epidermis where it contacts the dermis is the **stratum basale**, which consists of a single row of

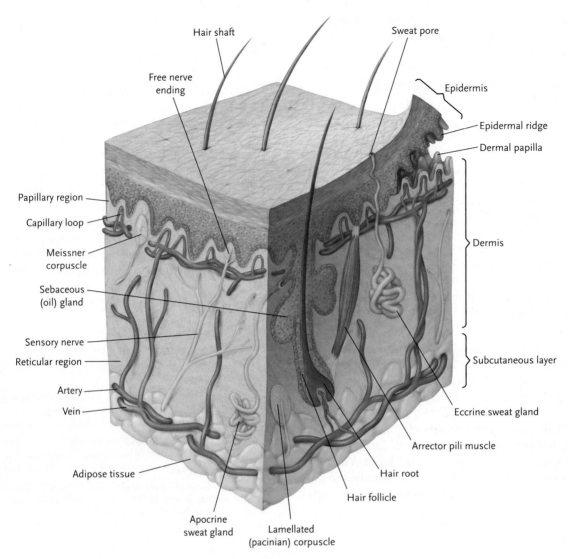

Figure 12.1 Integumentary System
The integumentary system is composed of the epidermis, dermis, and subcutaneous layers.

keratinocytes that divide very rapidly, some Merkel cells, and some melanocytes (Figure 12.2). This basal layer contains the most rapidly dividing cells in the whole body, and it is where all epidermis cells are produced. **Merkel cells** are specialized mechanoreceptors that detect touch. They are closely associated with sensory neurons that send information to the somatosensory cortex of the brain where sensory stimuli are interpreted (Chapter 16). **Melanocytes** are the cells that produce **melanin** when exposed to UV light; they are the brown pigmentation associated with a tan. Melanin is produced by melanocytes from the amino acid tyrosine, and it acts as a photo-protectant because it can absorb UV energy, which then protects the DNA in skin cells from UV damage.

Newly replicated keratinocytes move up into the next layer and make up the greatest proportion of the epidermis: the **stratum spinosum**. This second layer is composed primarily of keratinocytes but also contains **Langerhans cells**. Langerhans cells are a type of immune cell that helps prevent infections by phagocytosis of infectious pathogens (Chapters 3 and 23). Some mitosis does occur in this layer, but not to the extent of the cell division that occurs in the basal layer. In this layer, keratinocytes begin to produce keratin, which is a strong, fibrous, structural protein that forms into intermediate filaments (Chapter 3) and makes the skin tough and strong; it also occurs in large amounts in hair and nails. Keratin is found in abundance in animal claws, scales, hooves, shells, and horns. A **callus** is formed when we use an area of skin repetitively. This happens because the keratinocytes increase the synthesis of the tough keratin protein that hardens the skin and protects it from excessive damage. A callus also has stratum corneum layers.

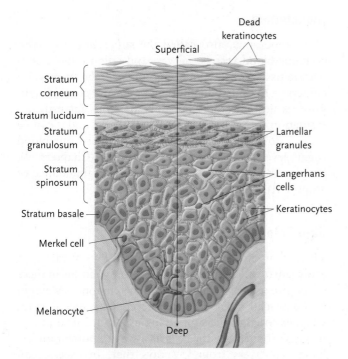

Figure 12.2 Epidermis

The epidermis is composed of layers of keratinocytes from the stratum basale to the stratum corneum. The epidermis also contains melanocytes, Langerhans cells, and Merkel cells.

DID YOU KNOW?

Some infectious fungal organisms, such as in athlete's foot (ringworm), feed on the keratin proteins in our skin.

As keratinocytes continue to migrate to the next layer, the **stratum granulosum, lamellar granules** are released and keratin production increases. Lamellar granules are the organelles in keratinocytes. They merge with the cell membrane and release lipids, enzymes, and proteins into the extracellular space, which acts as a protective waterproof barrier, and they also help prevent infection by breaking down bacterial cell walls (Chapter 23). Lamellar granules are also found in certain cells of the lung, and they secrete surfactant (Chapter 19).

In areas of the body that have thicker skin, such as the soles of the feet and the palms of the hands, an extra layer of a few rows of dead keratinocytes form the **stratum lucidum**. This layer consists of three to five layers of dead keratinocytes that have high amounts of keratin; this is necessary to protect these areas that endure much more physical stress than other skin areas.

The final layer of the epidermis is the **stratum corneum**, which consists of approximately 30 layers of dead keratinocytes that are completely filled with keratin. The stratum corneum cells are continually being sloughed off our bodies and replaced with cells migrating up from the deeper layers.

Dermis

The dermis consists of strong and flexible connective tissue that allows the skin to move and stretch and return to its original shape. The dermis is divided into two sections—the papillary layer, which binds the dermis to the epidermis, and the deeper reticular layer, which connects to the underlying subcutaneous region.

The papillary layer is made up of loose areolar connective tissue that contains fibroblasts, collagen, reticular fibres, and elastin fibres (Chapter 11). The papillary layer of the dermis projects into the epidermal layer. This is called **dermal papillae**, which increases the surface area for the diffusion of oxygen and nutrients into the avascular epidermis. On the palms of the hands and soles of the feet, the dermal papillae lie on top of **dermal ridges** and cause the epidermis to also form **epidermal ridges**; these form our fingerprints and footprints (Figure 12.3). The pattern of the epidermal ridges is genetically determined, and everyone has a different pattern.

The reticular layer of the dermis consists of dense, irregular, connective tissue that contains many elastin and collagen fibres and fibroblasts in an irregular pattern that gives the skin strength and the ability to withstand pulling forces and to recoil from stretch.

DID YOU KNOW?

Stretch marks are a type of scar formed on the skin when growth occurs rapidly, such as during pregnancy. The skin has to grow so quickly that the dermis layer tears and new fibroblast cells fill in the torn region. Because the fibroblasts have to fill in the gaps so quickly, they don't produce the normal amount of collagen and elastin, and this causes a groove in the skin. Interestingly, young pregnant women have a higher risk than older women for developing severe stretch marks; teenagers have the highest risk.

Middle Temple Library/Science Photo Library

Figure 12.3 Fingerprints

Fingerprints are formed from epidermal ridges.

12

Many structures are found within the dermis layer, including blood vessels, hair follicles, sebaceous glands, sweat glands, neurons, sensory receptors, and the arrector pili muscles attached to every hair follicle.

Merkel cells are touch receptors located in the epidermis, but there are many other sensory receptors in the dermis. **Pacinian corpuscles** detect pressure, nerve endings (nociceptors) detect pain, thermoreceptors detect temperature, and **Meissner corpuscles** detect touch and vibration. It is important to note that sensory receptors are not the same as membrane protein receptors (Chapter 3). Membrane receptors bind to specific molecules, such as a hormone or neurotransmitter, but a sensory receptor can be a whole cell, such as Merkel cells and Pacinian corpuscles.

Arrector pili muscles contract and cause hairs to stand up (**piloerection**)—"goose bumps"—when the sympathetic nervous system is stimulated by fear or cold. The evolutionary significance of this response is much more useful in animals with fur; it makes them appear larger to try to avoid confrontation and helps them maintain body temperature. Since humans do not have fur, piloerection doesn't do much to prevent heat loss or scare off rivals and is therefore considered a vestigial structure (a remnant of evolution that is no longer useful).

DID YOU KNOW?

Tattoo ink is injected into the dermal layer of the skin. Cell division is much slower in this region, and most of the dermis is connective tissue, so the ink will stay permanently. The epidermis is clear, so the ink in the dermis is visible through the epidermis.

Hypodermis

The hypodermis, also called the *subcutaneous layer*, contains blood vessels, nerves, and adipose loose connective tissue. The hypodermis connects the dermis and epidermis to underlying muscle tissue and contains adipose tissue that stores fat. Subcutaneous fat provides insulation that helps the body retain heat, and it can be broken down and used to make ATP when our blood sugar level decreases between meals (Chapters 4). People who are overweight have an overabundance of adipose tissue under their skin.

Skin Colour

The colour of our skin is determined by several genes (polygenetic inheritance, Chapter 6). Variations in these genes give us a vast range of human skin tones. Melanin is the black-brown pigment secreted by melanocytes when exposed to the sun, giving us a tan. Melanin pigment is produced by the cells as a way to protect underlying cell layers from UV rays that, in excess, can damage DNA. Hemoglobin is the oxygen-carrying molecule in red blood cells, so when more oxygen is bound to hemoglobin, this becomes visible through the skin, which appears redder. Also, when blood vessels in the skin dilate, skin colour becomes pink or redder. **Carotene** is the orange-yellowish pigment found in many foods we eat, such as carrots and egg yolk, and some is stored in the skin. Eating excessive amounts of carrots can make the skin to look orange; this is called **carotenemia** and is not harmful. Carotene is an important pigment to have in our diet because it is the precursor for **vitamin A**, which is essential for healthy skin and vision.

CONCEPT REVIEW 12.2

1. Name the layers of the epidermis.
2. What types of cells are found in the different layers of the epidermis?
3. Describe the function of the various cell types.
4. Describe the function of melanin, keratin, and lamellar granules.
5. What are the layers and components of the dermis?
6. What types of sensory receptors are found in the epidermis and dermis? What type of sensation does each detect?
7. What are the main functions of the hypodermis?
8. Name three molecules that affect skin colour?

12.3 GLANDS

Several types of glands are located in the integumentary system: **sudoriferous glands** (sweat glands), **sebaceous glands** (oil glands), **ceruminous glands** (produce ear wax), and **mammary glands** (in females produce milk).

The two types of sudoriferous glands are eccrine and apocrine (Figure 12.4).

Eccrine glands are found throughout the body, particularly on the forehead, palms, and soles. Eccrine glands are located in the deep reticular region of the dermis, and they have a duct leading to a pore on the pheromones, molecules that are important for attracting a mate in most animal species.

Sebaceous glands that produce sebum and surround each hair follicle in all regions of the body are not found on the palms or soles. Sebum primarily consists of fatty acids, triglycerides, and **squalene**, which is similar in structure to long-chain fatty acids. Sebum lubricates the skin and hair, and it contains important molecules, such as antioxidants, pheromones, and antimicrobial substances that help protect us from infection. Acne is caused in part by an increased production of sebum that is not released from the gland and by an increase in bacteria that ingest sebum and initiate a small inflammatory response. During puberty, increased sebum production coincides with increased testosterone or estrogen production in males and females, respectively. Acne medication kills the bacteria that cause the inflammatory response.

Figure 12.4 Glands of the Skin

The skin contains sebaceous oil glands and two types of sudoriferous sweat glands: eccrine and apocrine.

surface of the epidermis. Eccrine sweat production is important for body temperature regulation.

Apocrine glands are located primarily in the axilla (armpit), the areola of the nipples, and the genital regions. These glands are located deep in the reticular dermis or in the hypodermis. Sweat secreted from these glands is thicker because it contains fatty acids in addition to water and ions. When the fatty acids are metabolized by the bacteria that reside on the skin, the fatty acids are broken down into **butyric acid**, which has an odour. Apocrine glands begin to function during puberty. Apocrine gland sweat is thought to function as

DID YOU KNOW?

A thick layer of sebaceous oil called *vernix* covers a baby during its embryonic development. This layer helps protect the baby's skin while surrounded by amniotic fluid, and sometimes babies are born with some vernix still on their skin.

DID YOU KNOW?

The only difference between a pimple and a blackhead is exposure to oxygen. Blackheads turn dark because the sebum is oxidized when exposed to air. Pimples are not dark because the sebum stays under the skin within the gland.

Ceruminous glands located in the dermis of the ear canal produce ear wax. Ear wax is a substance called **cerumen**, which is very similar in composition to sebum, but it is thicker and coats the inside of the external ear canal to trap foreign particles. It also acts as water barrier and inhibits the growth of bacteria.

Mammary glands are located in breast skin tissue and are specialized to produce milk during lactation (see Chapters 17 and 22).

CONCEPT REVIEW 12.3

1. What are the two types of sudoriferous glands? Where are they found? What is their function?

2. What is the location and function of the sebaceous glands?

3. What causes acne?

4. What is the function of ceruminous glands and mammary glands?

Application

As you read through the rest of this chapter, consider this scenario. While riding his bike, Bert hit a hole in the road; this caused him to fall and slide on the gravel shoulder. Bert badly scraped the whole right side of his body. Looking at his leg, he not only saw a lot of blood but also some muscle exposed beneath his skin. Which of the three layers of skin were likely affected? Similar to people with severe burns, what two things should Bert be aware of?

12.4 BURNS AND SKIN CANCER

Burns

Burns can range from very mild to very severe depending on the layers of the skin that are affected. A **superficial burn** (first-degree burn) involves the epidermis, such as sunburn. Peeling that occurs after sunburn involves the loss of epidermal tissue only. A **partial-thickness burn** (second-degree burn) damages the epidermis and the papillary region of the dermis, and blisters are visible. A **full-thickness burn** (third- and fourth-degree burns) results in damage to both the epidermis and dermis of the skin, and often includes underlying tissues, such as the subcutaneous layer and possibly muscle or bone tissue (Figure 12.5). Full-thickness burns require skin grafting in order for the skin to heal because there are no epithelial or dermal cells left for regeneration. Cells of the epidermis can regenerate quite easily through cell division of the undamaged healthy cells. The dermis can regenerate to some extent, but any burns that include the reticular dermis or the subcutaneous layer usually form scar tissue and possibly permanent damage.

The primary cause of death from extensive burns is loss of body fluids; the second major cause of death is infection. Both the severity of the burn and the extent of the body surface affected are important factors in determining the risk of fluid loss and infection. One method of determining the percentage of surface area of the skin affected by burns is to use the "rule of nines": certain body parts make up approximately 9% of the total surface area of the skin (Figure 12.6). If more than 15–20% of the body has deep or full-thickness burns, the person may go into shock and require intravenous to replace body fluids. The percentages for each area differ in infants, children, and people who are obese because the surface area ratios differ. The surface areas in Figure 12.6 are based on an average-weight adult.

> ### DID YOU KNOW?
>
> Burns are not only caused by heat or fire. People can have significant burns from electricity, chemicals, or radiation (such as a sunburn).

Skin Cancer

The three most common types of skin cancer are basal cell carcinoma, squamous cell carcinoma, and melanoma (Figure 12.7). **Basal cell carcinoma** originates in the basal cells of the stratum basale of the epidermis. Recall that it is this layer of the epidermis that undergoes very rapid cell division. Basal cell carcinoma is the most common type of skin cancer and is caused by excessive sunlight exposure. It most commonly occurs on the face or areas exposed to excessive sunlight, and it appears as a raised, smooth bump that can sometimes ulcerate. This type of cancer rarely spreads and is easily treated.

Squamous cell carcinoma originates in the keratinocytes of the stratum spinosum layer. Recall that these cells make up the largest portion of the epidermis and have some cell division, but not as rapid as the basal layer. This type of cancer is also caused by excessive sun exposure and can spread quickly if it is not removed, but it is still easily treatable. Squamous cell carcinoma usually looks like a red, raised, scaly patch on the skin and can become quite large if not treated. These types of cancers can bleed and ulcerate.

Melanoma is the uncontrolled growth of the pigment-producing cells, melanocytes. Melanoma is the most dangerous type of skin cancer, often arising from moles; if left untreated they can be fatal. Melanoma is the rarest type of skin cancer but has the highest mortality rate because it can spread to many other tissues, making treatment much more difficult. A common method to determine possible melanoma is the mnemonic ABCDE, where A = asymmetrical, B = borders are irregular, C = colour is dark brown or black but not consistent, D = diameter is larger than 6 mm, and E = evolves over time. Harmless moles are generally round with smooth edges, all the same colour, and do not change over time. The cause of melanoma is actually 75% more related to ethnicity and pigmentation than to sun exposure.

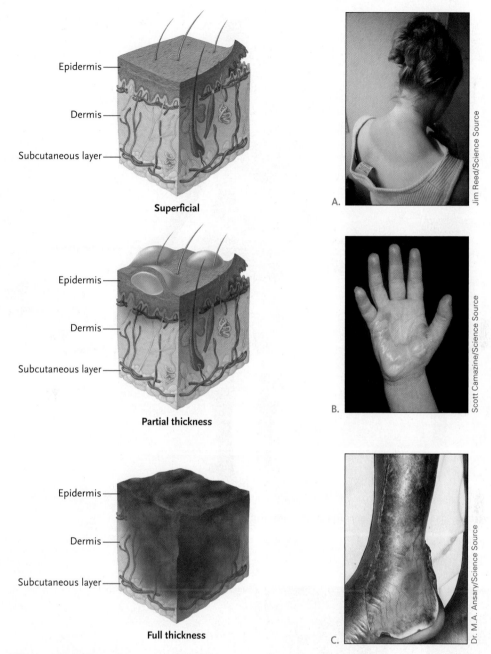

Superficial

Partial thickness

Full thickness

A. Jim Reed/Science Source

B. Scott Camazine/Science Source

C. Dr. M. A. Ansary/Science Source

Figure 12.5 Burns

Burns are categorized based on the layer of skin that is damaged: superficial burns affect the epidermis, partial-thickness burns affect the dermis, and full-thickness burns affect the subcutaneous layer.

DID YOU KNOW?

Sunlight is not the only cause of skin cancer. Other factors that play a significant role in skin cancer development include smoking; chronic inflammation (true of all cancers) or continual wound healing in the same location; being immunosuppressed or using immunosuppressant medication; and chemicals found in cosmetics, lotions, and sunscreens that can contain harmful chemicals that induce production of free radicals. Warts caused by the human papilloma virus (HPV)—of which there are over 100 different types—can also increase the risk of squamous cell carcinoma.

DID YOU KNOW?

According to the Public Health Agency of Canada, the incidence of melanoma in Canada is 11.5 cases per 100 000 people in the population, and the five-year survival rate is 90%. As a comparison, breast cancer affects 96 people per 100 000 and has a five-year survival rate of 83%.

12

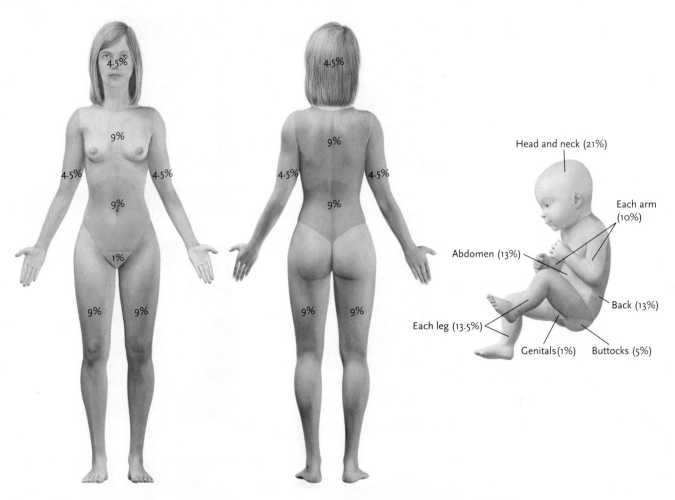

Figure 12.6 Rule of Nines

The rule of nines can be used to determine the approximate percentage of an area of skin that is burned.

Figure 12.7 Types of Skin Cancer

(A) Basal cell, (B) squamous cell, and (C) melanoma

DID YOU KNOW?

Fat-soluble vitamins such as A and E are found in seafoods such as fish and shellfish (including oysters, snails, and clams); healthy oils such as cod liver oil and olive oil; and in eggs, nuts, and dairy products. Vitamin A can also be produced in our body by converting the beta-carotene found in orange foods, such as carrots, squash, pumpkin, and sweet potatoes.

CONCEPT REVIEW 12.4

1. Describe the three types of burns and which skin layers of the skin are affected for each.

2. Name the three main types of skin cancer and the cells that are affected.

3. What are the risks with each type of skin cancer? How could you determine possible melanoma?

EFFECTS OF AGING

The most common effects of aging on the skin include wrinkles, sagging skin, age spots, greater visibility of blood vessels, skin tags, and increased dryness. These effects are related to genetic and environmental factors and can begin to be noticeable in a person's late 30s and early 40s.

When we age, the skin becomes thinner and can sag and wrinkle because of decreased production of connective tissue proteins, mainly collagen and elastin. Collagen keeps the skin tough, and elastin allows the skin to go back to its normal position after being stretched. Sebaceous glands become less active, and therefore the skin becomes dry and more likely to crack, itch, and wrinkle. Wrinkles occur in areas where the skin is often creased from smiling or frowning. If you make a big smile or frown in the mirror, you can predict where you will end up with wrinkles later in life. Keeping skin moisturized can help slow the wrinkling process. With aging, the melanocytes decrease in number but increase in size, and they contribute to age spots that become more noticeable with excessive sun exposure. The amount of subcutaneous fat decreases, which leads to sagging skin, especially in the face and neck region. Losing a lot of weight when older causes more sagging of the skin because the skin becomes less able to repair and regenerate.

Certain factors can increase the rate of the effects of aging on the skin: excessive sun exposure, including tanning beds; increased exposure to toxins, such as cigarette smoke and pollution; lack of sleep; lack of exercise; excess stress; and poor nutrition. For example, smoking prevents proper collagen formation, adds toxins to the body that damage cells and DNA, and causes vasoconstriction and therefore decreased blood flow to the skin. Burning of the skin from excess sun exposure damages the epidermis, prevents sebum production, damages DNA, and prevents protein synthesis (of collagen and elastin), and causes inflammatory reactions that damage cells.

Some hormones help maintain the skin and the connective tissue proteins, including **growth hormone**, insulin-like growth factor (IGF), and the sex hormones (Chapter 17). When we get older, growth hormone production decreases. Exercise helps ensure the production of some growth hormone and other **androgens** in males and females, and this helps maintain cell regeneration processes and protein synthesis. When women go through menopause, the growth-promoting effects of **estrogen** are lost, and exercise becomes much more important to slow the effects of aging. Exercise also has many other health benefits (Chapter 14). Stress hormones, particularly cortisol, increases blood sugar that contributes to glycation (Chapter 3, Effects of Aging). Stress increases the production of free radicals in the body that damage cells. Stress hormones also reroute blood flow away from the skin so it receives less nutrients. Increasing blood flow to the skin, such as during exercise or by sitting in a sauna helps bring more nutrients to the skin.

The most important vitamins that contribute to skin health are vitamins C, A, and E. Vitamin C (ascorbic acid) is required for collagen production in our connective tissues and is an important antioxidant. Therefore, eating a diet high in foods such as fresh fruits and vegetables is very beneficial. Vitamin E is a powerful antioxidant secreted in sebum, and it plays an important role in protection from UV light. Vitamin A has many functions and, with respect to the skin, it is important for epithelial cell growth. A form of vitamin A is *retinoic acid*, which is an important growth factor and is added to many anti-aging skin creams. Eating vitamin A-rich foods in addition to using good skin creams containing retinoic acid can reduce the signs of aging (fat-soluble vitamins can easily be absorbed through the skin). It is important to note that if you use a skin cream that contains retinoic acid you need to stay away from UV light because the retinoic acid will oxidize and can cause burning.

CHAPTER REVIEW

Medical Terminology

Term	Definition	Example
acanth/o	thorny, spiny	acanthoma
adip(o)	relating to fat	adipocyte
alb-	white or pale	albino
brom/o	odour	bromoderma
cauter/o	burn, burning	cauterize
chrom/o	colour	chromophore
cili-	eyelid or eyelashes	cilia
cry/o	cold	cryotherapy
cutane/o	skin	subcutaneous
derm/o	skin	dermatologist
erythem/o	flushed, redness	erythema
eschar/o	scab	escharotomy
graph/o	writing	electrocardiograph
hidr/o	sweat	hyperhidrosis
ichthy/o	fish	ichthyosis
iod/o	iodine	iododerma
kerat/o	horny tissue, cornea	keratolysis
lepid/o	flakes, scales	lepidosis
onych/o	nail	onychomycosis
pachy-	thick	pachydermatocele
papul/o	papule, pimple	papulopustular
perspir/o	breathe through	perspiration
phyt/o	plant	phytophotodermatitis
pil/o	hair	pilomotor
pseudo-	false	pseudomonas
psor/o	itching	psoriasis
py/o	pus	pyodermatitis
seb/o	sebum	seborrhea
strata	layer	stratified
trich/o	hair	hypertrichosis
ul/o	scar, scarring	uloid
ungu/o	nail	subungual
verruc/i	wart	verrucosis
xer/o	dry	xeroderma

Labelling Activity

Label the following diagrams.

1.

2.

Answer to Application Question

Since his muscle tissue is visible, Bert has damaged all three layers of skin: the epidermis, dermis, and subcutaneous layer. When his skin heals he will have scarring. Considering the similarity of his skin damage with a severe burn, Bert should be aware that the risk of infection and water loss increases when there is extreme skin loss.

13 The Skeletal System

CHAPTER OUTLINE

That is what learning is. You suddenly understand something you've understood all your life, but in a new way.

Doris Lessing

13.1 OVERVIEW OF THE SKELETAL SYSTEM

The skeletal system is made up of the axial skeleton and the appendicular skeleton (Figure 13.1). The axial skeleton includes the skull, vertebrae, sternum, and ribs. The appendicular skeleton includes the shoulders, pelvis, and limbs. The adult skeleton has 206 bones: 80 in the axial skeleton and 126 in the appendicular.

The skeletal system performs the following functions:

- Providing a structural framework for the body
- Providing movement because of attachments of skeletal muscles
- Protecting internal organs, such as the brain, spinal cord, heart, and lungs
- Storing minerals such as calcium, magnesium, and phosphorus, and being involved in the regulation of blood levels of those minerals
- Producing red and white blood cells and platelets in the red bone marrow by a process called *hematopoiesis*
- Storing fat as triglycerides in the yellow bone marrow

Figure 13.1 Axial Skeleton and Appendicular Skeleton

The axial skeleton (green) consists of the skull, vertebrae, chest, and ribs; the appendicular skeleton (red) consists of the shoulders, pelvis, and upper and lower limbs.

- Facilitating hearing by means of the very small bones in the inner ear that vibrate and transmit sound to the auditory nerve, which signal the brain and allow us to hear and interpret sounds

DID YOU KNOW?

Exercise can help prevent type-2 diabetes. Exercise that puts stress on the bones stimulates bone growth, specifically osteoblasts, which are cells that build up bone density. Stimulated osteoblasts secrete a hormone called *osteocalcin*, which signals fat cells to produce adiponectin, which then increases cellular sensitivity to insulin. (Type-2 diabetes is caused by insulin resistance; see Chapter 17.)

CONCEPT REVIEW 13.1

1. How many bones are in the adult body?
2. What major bones make up the axial skeleton and the appendicular skeleton?
3. What are the main functions of the skeletal system?

13.2 BONE TISSUE

Bones are an organ composed of *mineralized* bone connective tissue (Chapter 11), as well as bone marrow, epithelial tissue, blood vessels, and nerves. Bone tissue is made up of cells and extracellular matrix proteins such as collagen, and minerals such as calcium and phosphorus (Figure 13.2).

Bone Cells

There are three main types of bone cells. **Osteoblasts** are the cells that build up the bone tissue. First, osteoblasts produce the collagen scaffolding, and then they take up minerals from the blood and deposit them in the extracellular matrix. Osteoblasts are very important for bone growth and repair. Once the bone tissue is formed around the osteoblast, it becomes a mature osteocyte. Osteocytes are the mature "maintenance" cells that take up nutrients, get rid of waste, produce ATP, and communicate with other bone cells.

Figure 13.2 Bone Tissue

Bone tissue is made up of cells, collagen fibres, and minerals.

Osteoclasts are the cells that break down the mineral portion of the extracellular matrix to release calcium, phosphate, and magnesium into the bloodstream; then these minerals are used by other cells, such as muscles or neurons. Bone breakdown is a normal and very important process for bone remodelling and is regulated by hormones, such as **calcitonin** and parathyroid hormone (Chapter 17).

Extracellular Matrix

The extracellular matrix of bone tissue consists of water, collagen protein fibres, and minerals. Although bone is very dense tissue, water makes up approximately 25% of the extracellular matrix. Collagen is a very strong but flexible protein, composing another 25% of the bone tissue and allowing bones to have some *flexibility*.

Minerals

The minerals make up approximately half of the bone matrix and give the bones their density and *strength*.

Types of Bone Tissue

There are two types of bone: compact bone and spongy bone (Figure 13.3).

Compact Bone Compact bone is tightly arranged into units called **osteons**. Each osteon has a **central canal** (also

called a *Haversian canal*) containing blood vessels and lymphatic vessels. Each central canal is connected to other central canals by **perforating canals**, which also connect blood and lymph vessels with the larger vessels surrounding the outer regions of the bone. The rest of the osteon is bone tissue made up of osteocytes found within small spaces called **lacunae**, and concentric rings of mineralized extracellular matrix; each ring is called a **lamella**. Each lacuna is connected—by very small channels called **canaliculi**—to other lacunae so that cells within the bone can communicate and nutrients and oxygen can reach all cells.

Spongy Bone Spongy bone (also called *trabecular bone*) is made up of **trabeculae** that consist of osteocytes in lacunae surrounded by mineralized extracellular matrix; but it is not arranged into osteons like compact bone. The space around the trabeculae in spongy bone is filled with **red bone marrow**, where hematopoiesis occurs (production of blood cells). Since spongy bone has more space between the trabeculae, it is lighter in

Figure 13.3 Types of Bones

There are two types of bone: compact and spongy.

weight. Spongy bone is found in short, flat, and irregular bones and also in the ends of long bones, and it surrounds the hollow medullary cavity (Figure 13.3).

Classification of Bones

Every bone in the body can be generally classified into one of five categories: long, short, flat, irregular, and sesamoid (Figure 13.4). The **long bones** are found in the arms, legs, fingers, and toes. **Short bones** are in the wrists and ankles. **Flat bones** are in the sternum, skull, ribs, and shoulder blades. **Irregular bones** include the vertebrae, some bones of the face and skull (such as the ethmoid and sphenoid bones), and the pelvis. **Sesamoid bones** are embedded within a tendon and cross over a joint; they provide strength to the tendon. The sesamoid bones in the body are the patella, two bones in the hand, the pisiform bone in the wrist, and one bone in the foot that connects to the big toe.

Structure of a Typical Long Bone

All bones are covered by a layer of dense, irregular, connective tissue called the **periosteum** (Chapter 11). Under the periosteum and surrounding the bone tissue are blood vessels, nerves, and lymphatic vessels that infiltrate the entire bone tissue through Haversian and perforating canals. Bones are highly vascular. The periosteum also provides an attachment point for tendons and ligaments. The bone precursor cells that become osteoblasts lie directly under the periosteum and form the compact bone. These cells contribute only to bone width (increases with weight-bearing exercises), and they are important in repair of fractures, called *intramembranous ossification*.

Each end of a long bone (Figure 13.5) is called an *epiphysis*; the superior end is the proximal epiphysis and the inferior end is the distal epiphysis. (See Chapter 1 for directional terms.) Each epiphysis is covered by a

13

Femur

Parietal

Patella

Flat bone

Sesamoid bone

Carpal

Vertebra

Irregular bone

Long bone

Short bone

Figure 13.4 Classification of Bones

Bones can be classified as long, short, flat, irregular, and sesamoid.

13

Articular cartilage

Spongy bone

Epiphyseal line

Red bone marrow

Compact bone

Artery

Endosteum

Medullary cavity

Periosteum

Proximal epiphysis

Metaphysis

Diaphysis

Metaphysis

Distal epiphysis

Articular cartilage

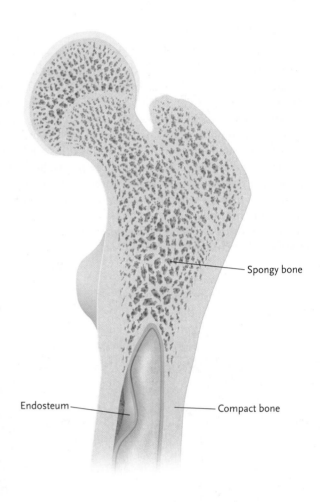

Spongy bone

Endosteum

Compact bone

Figure 13.5   Long Bone Structure

protective layer of articular cartilage. This type of cartilage articulates or connects to another bone. It is hyaline cartilage connective tissue (Chapter 11). The epiphyses contain spongy bone with red bone marrow where hematopoiesis occurs.

A region next to each epiphysis called the *metaphysis* is the region of the bone that grows during development—known as *endochondral ossification*. During bone development, chondrocytes in a layer of hyaline cartilage called the *epiphyseal plate* produce extracellular matrix, and then mineralization occurs, which allows the bone to lengthen. Once growth is complete, the epiphyseal plate is replaced by bone, and is called the *epiphyseal line*.

DID YOU KNOW?

Exercise increases bone strength and density because the mechanical stress stimulates the production of growth hormone that stimulates osteoblasts to increase production of the extracellular matrix collagen proteins and mineralization. Without some weight-bearing exercise, the bone doesn't remodel and becomes weak, increasing the risk of injury. It has been shown that astronauts can lose as much as 1% of their bone mass (and muscle mass) per week when in space without gravity.

The mid-region or shaft of the long bone is called the *diaphysis*. In the diaphysis, under the periosteum, is the compact bone. Within the centre of the shaft is a hollow cavity called the *medullary cavity*. This cavity is lined with dense, irregular connective tissue, called the *endosteum*: like the periosteum, this tissue contains bone precursor cells. The medullary cavity is highly vascularized and contains yellow bone marrow, which is a storage site for triglycerides.

CONCEPT REVIEW 13.2

1. What is the difference between a bone and bone tissue?
2. What is the role of the three major cells types?
3. What components of bone tissue provide strength and flexibility?
4. Name the five general categories of bone types, and give an example of each.
5. Label a blank diagram of a long bone.

13.3 AXIAL SKELETON

Skeletal System Terminology

- Process—projection or bump
- Ramus—extension of a bone making an angle
- Trochanter—large, rough projection; only in femur
- Tuberosity—small, rough projection
- Tubercle—small, rounded projection
- Crest—prominent ridge
- Line—low ridge
- Spine—pointed process
- Head—articular end of an epiphyses, separated from the shaft by the neck
- Neck—marrow connection between the epiphysis and diaphysis
- Condyle—smooth, rounded articular process
- Trochlea—smooth, grooved articular process
- Facet—small, flat articular surface
- Fossa—depression in or on a bone; hollow area
- Foramen—rounded passageway for blood vessels and/or nerves
- Canal—large passageway through bone
- Fissure—elongated cleft-like opening between adjacent parts of bones; where blood vessels pass through
- Sinus/antrum—chamber within a bone

The axial skeleton consists of the 8 cranial bones, 14 facial bones, 1 hyoid, 6 auditory ossicles, 26 vertebrae, 1 sternum, and 24 ribs.

Cranial and Facial Bones

Twenty-two **craniofacial bones** provide significant protection for the brain, eyes, and inner ear. The cranial bones include the frontal bone, two temporal bones, two parietal bones, and the occipital bone: the names of all these bones correspond to particular aspects of the cerebral cortex (Chapter 15). There are also the cranial bones behind the eye sockets, the sphenoid bone, and the ethmoid bone (Figures 13.6–13.11).

The cranial bones are fused together after birth and form **sutures**, which are a type of joint, categorized as fibrous or **synarthrosis** because they are not movable (Section 13.7). The names and locations of the four main sutures of the cranial bones are as follows:

1. **Coronal suture**—between the frontal bone and the anterior portion of the parietal bones
2. **Sagittal suture**—between the medial edges of the parietal bones
3. **Lambdoid suture**—between the occipital bone and the posterior edges of the parietal bones
4. **Squamous sutures**—one between the temporal bone and the lateral parietal bone, and the second at the edges of the occipital and sphenoid bones on each side of the skull

The 14 facial bones are all in pairs, except the **mandible** and the **vomer** (Figures 13.6–13.11). Note that the vomer can be seen in Figure 13.10. The other 12 facial bones include two nasal bones, two **maxilla bones**, two **zygomatic bones**, two **lacrimal bones**, two **palatine bones**, and two **inferior nasal conchae**.

The **frontal bone** forms the forehead, superior eye orbits, and most of the cranial floor, and it contains two frontal sinuses. The superior edge of the orbital is called the *supraorbital margin*. The supraorbital notch allows blood vessels and the nerve to pass

13

through. During infancy, the frontal bone is separated down the midline (called the *anterior fontanel*) but fuses during development forming the frontal suture or frontal squama.

The supraorbital ridge is the brow ridge where eyebrows grow (Figure 13.6).

The **parietal bones** form the lateral and superior portions of the skull.

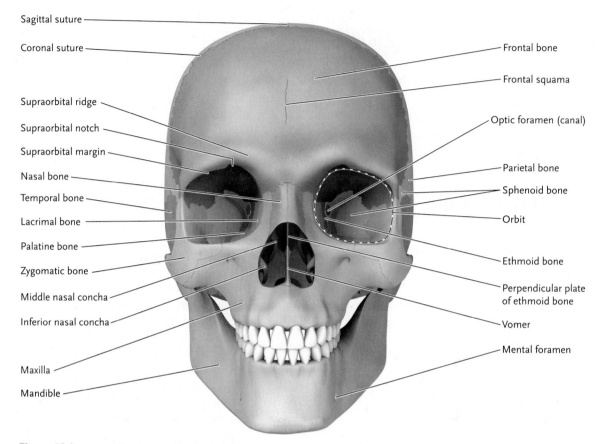

Sagittal suture
Coronal suture
Supraorbital ridge
Supraorbital notch
Supraorbital margin
Nasal bone
Temporal bone
Lacrimal bone
Palatine bone
Zygomatic bone
Middle nasal concha
Inferior nasal concha
Maxilla
Mandible

Frontal bone
Frontal squama
Optic foramen (canal)
Parietal bone
Sphenoid bone
Orbit
Ethmoid bone
Perpendicular plate of ethmoid bone
Vomer
Mental foramen

Figure 13.6 Craniofacial Bones: Frontal View

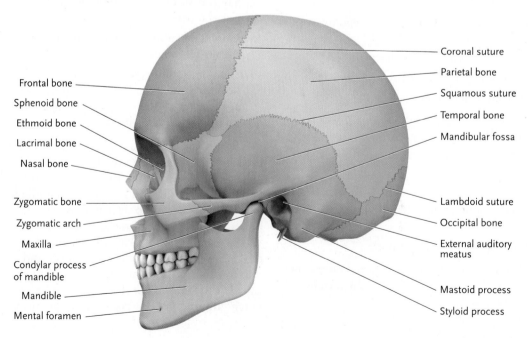

Frontal bone
Sphenoid bone
Ethmoid bone
Lacrimal bone
Nasal bone
Zygomatic bone
Zygomatic arch
Maxilla
Condylar process of mandible
Mandible
Mental foramen

Coronal suture
Parietal bone
Squamous suture
Temporal bone
Mandibular fossa
Lambdoid suture
Occipital bone
External auditory meatus
Mastoid process
Styloid process

Figure 13.7 Craniofacial Bones: Lateral View

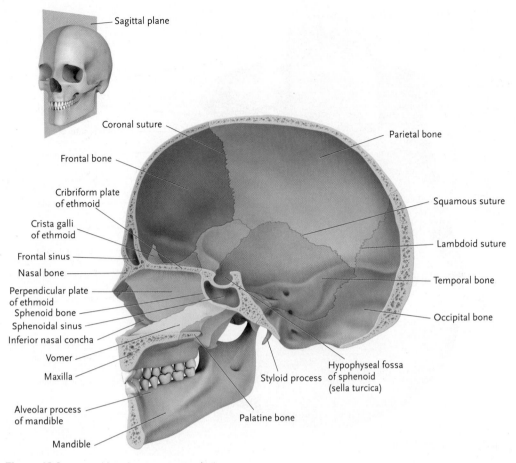

Figure 13.8 Craniofacial Bones: Sagittal View

Figure 13.9 Craniofacial Bones: Superior View

13

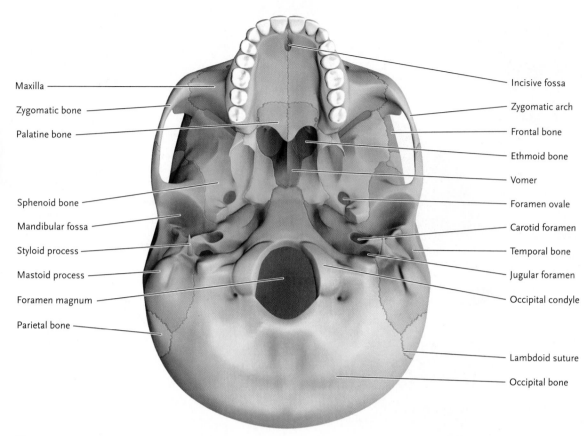

Maxilla

Zygomatic bone

Palatine bone

Sphenoid bone

Mandibular fossa

Styloid process

Mastoid process

Foramen magnum

Parietal bone

Incisive fossa

Zygomatic arch

Frontal bone

Ethmoid bone

Vomer

Foramen ovale

Carotid foramen

Temporal bone

Jugular foramen

Occipital condyle

Lambdoid suture

Occipital bone

Figure 13.10   Craniofacial Bones: Inferior View

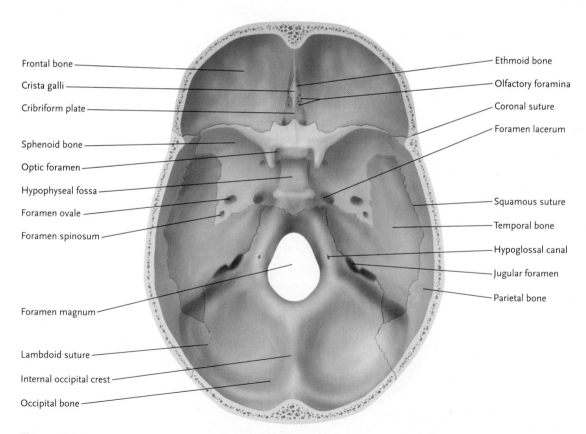

Frontal bone

Crista galli

Cribriform plate

Sphenoid bone

Optic foramen

Hypophyseal fossa

Foramen ovale

Foramen spinosum

Foramen magnum

Lambdoid suture

Internal occipital crest

Occipital bone

Ethmoid bone

Olfactory foramina

Coronal suture

Foramen lacerum

Squamous suture

Temporal bone

Hypoglossal canal

Jugular foramen

Parietal bone

Figure 13.11   Craniofacial Bones: Inferior-Transverse View

The **temporal bones** form the lateral portions of the skull and contain the **external auditory meatus**, which is where sound waves enter the inner ear. The temporal bone also forms the **zygomatic arch** and connects to the zygomatic bone. The **condylar process** of the mandible sits in a groove in the temporal bone, called the **mandibular fossa**; this forms the **temporo-mandibular joint (TMJ)**. The temporal bone has a process extending posteriorly—called the **styloid process**—and a small opening called the **carotid foramen**, where the carotid artery passes through the bone (Chapter 18).

DID YOU KNOW?

Myofacial pain is most commonly caused by an improper alignment of the condylar process within the mandibular fossa, which causes friction and inflammation, and is often accompanied by clicking or popping sounds when the jaw is opened. This is most often caused by imbalanced contraction of the chewing muscles because many people grind their teeth when they have stress. Both chiropractors and massage therapists can help relax the muscles and align the jaw to minimize the problem.

The **occipital bone** forms the posterior skull and most of the base of the skull. It contains the largest foramen, called the **foramen magnum**, and is the opening that the spinal cord passes through. If you feel the back and base of your skull, you will feel two large processes; those are called the **occipital condyles**, and they sit on the first vertebra and allow the head to move.

The **sphenoid bone** forms the middle base of the skull and is behind the eye sockets. The sphenoid bone contains sinuses, and also has a structure called the **sella turcica**, which protects the pituitary gland. The sphenoid bone contains the **foramen ovale**, which is the opening for the mandibular nerve. The **optic foramen** is located in the sphenoid bone behind each eye and is where the optic nerve passes through.

The **ethmoid bone** forms the anterior portion of the base of the skull (the entire base consists of the occipital bone, the sphenoid bone, the vomer, and the ethmoid bone). The ethmoid bone forms the **nasal septum**, and the **cribriform plate** forms the roof of the nasal cavity. The olfactory nerves pass through the **olfactory foramina**. The ethmoid bone also forms the superior and middle nasal conchae; the inferior nasal conchae are facial bones.

All of the **paranasal sinuses** are within the frontal bone, sphenoid bone, ethmoid bone, and maxilla bones (Figure 13.12). The sinuses are lined with mucus

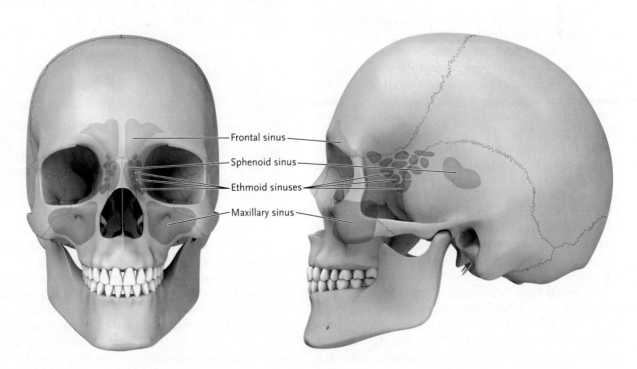

Frontal sinus
Sphenoid sinus
Ethmoid sinuses
Maxillary sinus

Figure 13.12 Paranasal Sinuses

Hollow cavities in the craniofacial bones are paranasal sinuses, located in the frontal, ethmoid, sphenoid, and maxilla bones.

membranes composed of epithelial tissue that is highly vascularized to warm the incoming air, and the mucus moisturizes the air before it enters the lungs. The sinuses also decrease the weight of the skull and give resonance to the voice. Since everyone has a unique structure and size of craniofacial bones and sinuses, everyone has a unique voice.

The vomer forms a small portion of the base of the skull that articulates with the maxilla and palatine bones, and is one of the components of the nasal septum. The mandible, which forms the lower jawbone, is the largest and strongest facial bone. The mandible articulates with the temporal bone, forming the TMJ joint, and the condylar process on the mandible articulates with the mandibular process of the temporal bone. The mandible contains **alveolar processes and sockets** that house the lower teeth.

The nasal bones form the top-centre portion of the nose, superior to the movable hyaline cartilage that forms most of the shape of our nose.

DID YOU KNOW?

The mandible has two openings called *mental foramina*, where a branch of the trigeminal nerve enters (Chapter 15). This nerve is used by dentists to "freeze" the lower jaw for dental work.

The maxilla bones form the upper jawbone and, like the mandible, have alveolar processes and sockets that house the upper teeth. The maxilla bones articulate with every other facial bone, except the mandible, and they have sinuses that empty into the nasal cavity. They also form most of the upper hard palate of the mouth: with the **palatine bones** forming the posterior region of the hard palate.

The zygomatic bones form the cheekbones, and they articulate with the zygomatic arch of the temporal bones. The lacrimal bones—the smallest facial bones— form the inner part of the medial portion of the eye sockets. The inferior nasal conchae form the inferior portion of the nasal cavity (the ethmoid bone forms the middle and superior nasal conchae). Air circulates through the nasal cavity and the sinuses before it enters the lungs.

The **hyoid bone** is a unique U-shaped bone that is connected to the styloid processes of the temporal bone by ligaments, but it does not form a joint with any other bone. The hyoid is located just inferior to the mandible and is an attachment site for the tongue and neck muscles.

Infant Skull

When babies are born, the bones of the skull are not fully fused; this allows the infant skull to squeeze through the mother's vaginal canal and allows space for the skull bones to grow. Newborns have two very important "soft spots" called *fontanels*. The **anterior fontanel**, located between the frontal bones and anterior edge of the parietal bones, can be palpated for up to two years after birth; this fontanel takes the longest to close (Figure 13.13). The **posterior fontanel** is located between the occipital bone and the posterior edge of the parietal bones. The two **anterolateral fontanels** are located on each side of the head: superior and anterior to the temporal bone and posterior to the sphenoid bone. Two **posterolateral fontanels** are located on each side of the head, posterior to the temporal bone where the temporal, occipital, and parietal bones meet.

Vertebrae

Babies have 33 vertebrae, but adults have only 26 because the **sacrum** and **coccyx vertebrae** fuse. Specific types of vertebrae are located in specific regions of the spinal column: cervical (seven), thoracic (twelve), lumbar (five), sacrum (five fused bones), coccyx (four fused bones). The vertebrae protect the spinal cord, articulate with thoracic ribs, and play a role in movement because of the many muscles that are attached to them (Figure 13.14). The curve of the spinal column allows for increased strength and movement, as well as absorption of shock during physical activities such as jumping or running.

DID YOU KNOW?

The cervical curve forms at approximately three months of age, when the baby begins to hold its head up, and the curve of the lower back develops once the child begins learning to walk, when gravity and muscle contractions pull the spinal column into the curved position.

Vertebrae in each region of the spinal column differ in structure and size but have some similarities. The similar components include the body, foramina, articular facets, and processes. In between each vertebra is an **intervertebral disc** (Figure 13.15), which is composed of fibrocartilage (Chapter 11) that is very important for shock absorption and protection of the vertebrae. The body supports the weight of the spinal column, and the largest vertebrae, the **lumbar**

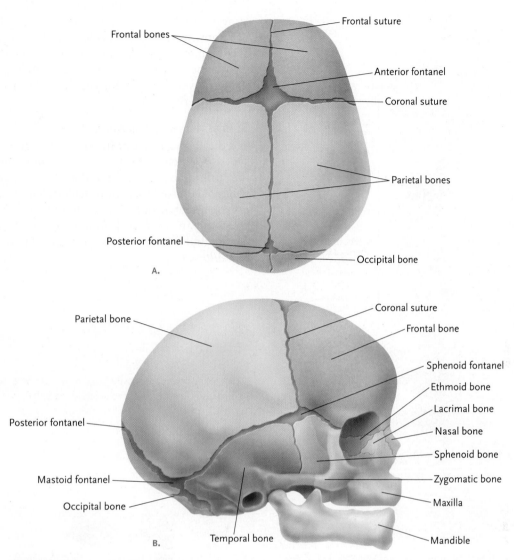

Frontal bones
Frontal suture
Anterior fontanel
Coronal suture
Parietal bones
Posterior fontanel
Occipital bone

A.

Parietal bone
Coronal suture
Frontal bone
Sphenoid fontanel
Ethmoid bone
Lacrimal bone
Posterior fontanel
Nasal bone
Sphenoid bone
Mastoid fontanel
Zygomatic bone
Occipital bone
Maxilla
Mandible
Temporal bone

B.

Figure 13.13 Infant Skull
The infant skull has fontanels where the skull bones have not yet fused.

vertebrae, bear the most body weight and attach to large back muscles.

A **foramen** (plural, foramina) is an opening within the vertebrae. All vertebrae have a **vertebral foramen**, which is an opening for the spinal cord. The cervical vertebrae also have transverse foramen on either side of the body of the vertebrae, and these protect the blood vessels and nerves of the neck. All vertebrae have an intervertebral foramen, which is the opening for the spinal nerves.

Superior and inferior articular facets connect to vertebrae above and below and have a hyaline cartilage surface to prevent friction and allow smooth movement. The **spinous processes** extend out posteriorly and are the ridges you can feel along your spinal column; the **transverse processes** extend out

transversely. Both spinous and transverse processes provide attachment sites for deep and superficial muscles of the back (Chapter 14).

Cervical There are seven cervical vertebrae (C1–C7). The C1, also called the **atlas**, has superior articular facets that connect with the occipital condyles of the occipital bone and bear the weight of the skull. Due to the shape of the articular facets, the only movement provided at this joint is nodding the head forward and backward. The atlas does not have a vertebral body or spinous process. Movement side to side is due to the C2, also called the **axis**, which contains a unique structure—the **dens**—that projects into the atlas where a vertebral body would be located, and allows for rotation of the head

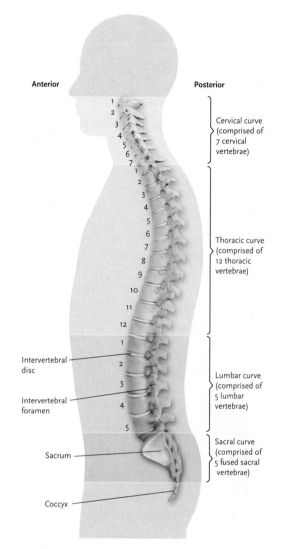

Anterior   Posterior

Cervical curve (comprised of 7 cervical vertebrae)

Thoracic curve (comprised of 12 thoracic vertebrae)

Intervertebral disc

Intervertebral foramen

Lumbar curve (comprised of 5 lumbar vertebrae)

Sacral curve (comprised of 5 fused sacral vertebrae)

Sacrum

Coccyx

Figure 13.14 Vertebral Column

The vertebral column consists of seven cervical, twelve thoracic, five lumbar vertebrae, and also the sacrum and coccyx.

(Figure 13.16). Also unique to the cervical vertebrae are the bifurcated spinous processes (Figure 13.16c). The C7 is also called the *vertebra prominens* because of its long spinous process, which is the large palpable bone at the base of the neck.

Thoracic Twelve **thoracic vertebrae** (T1–T12) connect with 12 pairs of ribs. Thoracic vertebrae are larger and stronger than the cervical vertebrae. They have long, downward-sloping spinous processes that minimize movement in the thoracic region to protect the delicate lungs. Thoracic vertebrae also have two extra articular processes for rib attachment, and this limits movement of the thoracic vertebrae. All articular processes have hyaline cartilage to decrease friction (Figure 13.17).

Lumbar The five lumbar vertebrae (L1–L5) are the largest vertebrae; they have the strongest bodies and bear more weight than the other vertebrae (Figure 13.18). The spinous processes are shorter, which allows for greater movement of the back in the lumbar region, and they are well adapted for the attachment of the large muscles of the back.

Sacral The adult sacrum consists of five fused bones, and the coccyx has four fused bones (Figure 13.19). The superior articular facet of the sacrum connects with the fifth lumbar vertebra and also articulates with the pelvis. The fused sacrum provides a large, strong, supportive attachment for the spine and the pelvis and contains a pair of sacral foramina that accommodate nerves and blood vessels. The coccyx is known as the *tailbone*.

DID YOU KNOW?

Your sacral bones are mostly fused by the age of 16, but are not completely fused until approximately age 30.

A.

Vertebral body

Intervertebral disc

Intervertebral foramen

Spinous process

B.

Transverse process

Vertebral foramen

Vertebral body

Figure 13.15 Anatomy of Vertebrae

(A) Lateral view and (B) superior view

(A) Courtesy of Mark Nielsen, AnatBooks, LTD

13

A. Atlas, superior view

Posterior arch

Groove for vertebral artery and first cervical spinal nerve

Posterior

Vertebral foramen

Transverse foramen

Superior articular facet

Anterior

B. Axis, superior view

Lamina

Transverse process

Dens

C. Typical cervical vertebra, superior view

Posterior

Vertebral foramen

Transverse foramen

Transverse process

Vertebral body

Lamina

Superior articular facet

Pedicle

Anterior

D. Cervical vertebrae, posterior view

Superior articular facet on atlas

Dens of axis

Atlas

Axis

Bifurcated spinous processes

Figure 13.16 Cervical Vertebrae

(A) Atlas, (B) axis, (C) typical cervical vertebrae, and (D) posterior view

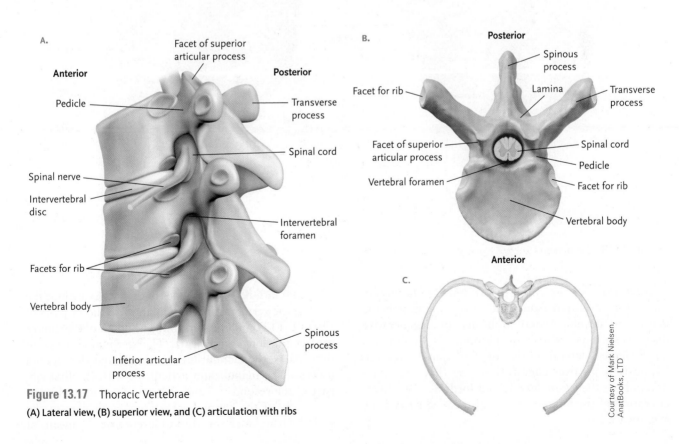

A.

Anterior

Pedicle

Spinal nerve

Intervertebral disc

Facets for rib

Vertebral body

Inferior articular process

Facet of superior articular process

Posterior

Transverse process

Spinal cord

Intervertebral foramen

Spinous process

B.

Posterior

Spinous process

Lamina

Transverse process

Facet for rib

Facet of superior articular process

Vertebral foramen

Spinal cord

Pedicle

Facet for rib

Vertebral body

Anterior

C.

Figure 13.17 Thoracic Vertebrae

(A) Lateral view, (B) superior view, and (C) articulation with ribs

Courtesy of Mark Nielsen, AnatBooks, LTD

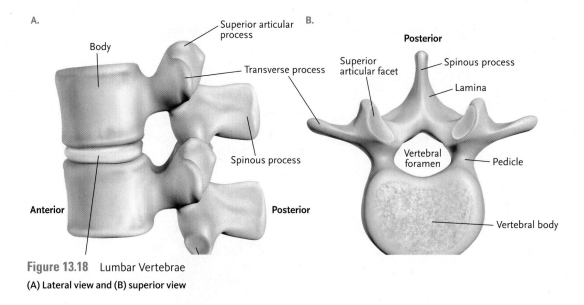

A.

Body

Superior articular process

Transverse process

Spinous process

Anterior

Posterior

B.

Posterior

Superior articular facet

Spinous process

Lamina

Vertebral foramen

Pedicle

Vertebral body

Figure 13.18 Lumbar Vertebrae
(A) Lateral view and (B) superior view

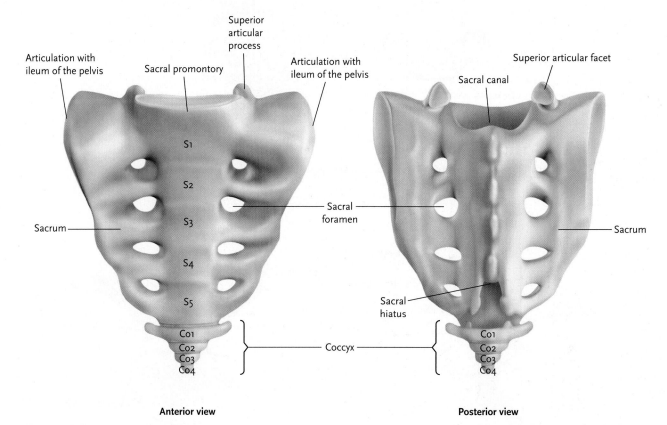

Superior articular process

Articulation with ileum of the pelvis

Sacral promontory

Articulation with ileum of the pelvis

Superior articular facet

Sacral canal

S1

S2

S3

S4

S5

Co1
Co2
Co3
Co4

Sacrum

Sacral foramen

Sacral hiatus

Co1
Co2
Co3
Co4

Coccyx

Sacrum

Anterior view

Posterior view

Figure 13.19 Sacrum and Coccyx

Alterations in the curve of the spinal column can occur for several reasons, the most common being genetic or developmental problems, or degenerative diseases such as arthritis or osteoporosis. Scoliosis is an S-shaped lateral curve of the spinal column; **kyphosis** is an over-curvature of the thoracic vertebrae, also called *hunchback*; and **lordosis** is an over-curvature of the lumbar region, also called *sway back* (Figure 13.20).

Sternum and Ribs The bones of the thorax include the 12 thoracic vertebrae that articulate with the 12 pairs of ribs (T1–T12), and the sternum. The bones of the thorax protect the lungs and heart. The **sternum** consists of three parts: the manubrium, the body, and the xiphoid process. The **manubrium** articulates with the first rib, part of the second rib, and the clavicles (Figure 13.21). The body of the sternum articulates with part of the second rib and with the third to seventh ribs by means of

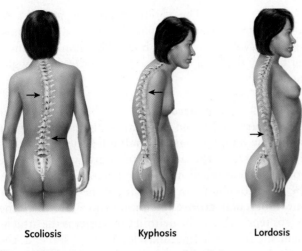

Figure 13.20 Abnormal Curvatures of the Spine

Abnormal curvatures of the spine include scoliosis, kyphosis, and lordosis.

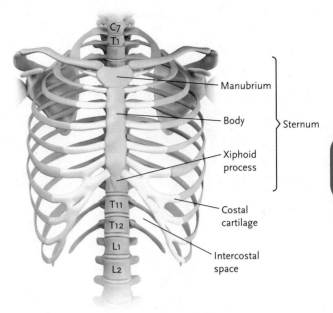

Figure 13.21 Sternum

The sternum consists of the manubrium, body, and xiphoid process. Hyaline cartilage connects the ribs to the sternum.

costal (hyaline) cartilage. The *true ribs* are ribs 1 to 7 that articulate directly with either the manubrium or the sternum. Ribs 8 to 12 are the *false ribs* because they either connect to the sternum indirectly through the cartilage of the seventh rib (rib 8–10) or they do not connect to the sternum at all (ribs 11–12, also called *floating ribs*). The **xiphoid process** is a small triangular piece of hyaline cartilage directly below and slightly deeper than the **sternal notch**, and it doesn't ossify into bone tissue until approximately adulthood. The sternal notch can be palpated where the seventh ribs attach to the lower region of the sternum.

DID YOU KNOW?

The sternal notch is used as a landmark for ensuring that chest compressions during CPR are done above the xiphoid process. If chest compressions are given too low on the sternum, the xiphoid process can break off and its sharp edges can lacerate the lungs.

CONCEPT REVIEW 13.3

1. What are the sutures of the skull, and where are they located?
2. Name the bones that form the temporomandibular joint.
3. What are fontanels?
4. Name three abnormal curvatures of the spine.
5. Name the cranial bones that articulate with the frontal bone.
6. Name the bones that contain sinuses.
7. Which bone contains the foramen magnum?
8. Which bone contains the optic foramen?
9. Which bone protects the pituitary gland?
10. How many vertebrae are in each region of the spine?
11. Which vertebrae are the largest and strongest?
12. Name the seven processes in a vertebra.
13. Which ribs are considered "floating ribs"?
14. Name the three parts of the sternum.

13.4 APPENDICULAR SKELETON

The appendicular skeleton has a total of 126 bones, which include the upper and lower limbs, the shoulder blades (scapula and clavicles), and the hips (pelvis).

Upper Limb

The upper arms are connected to the body through the scapula and clavicles, which articulate with the thorax (Figure 13.22).

13

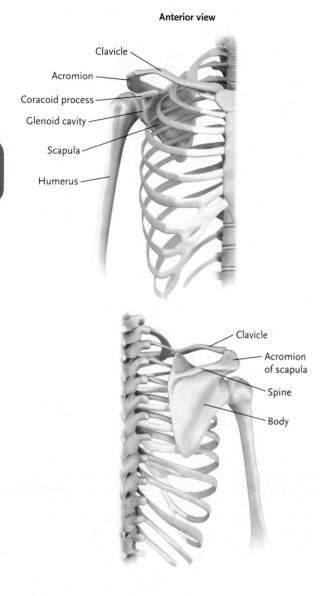

Anterior view

Clavicle
Acromion
Coracoid process
Glenoid cavity
Scapula
Humerus

Clavicle
Acromion of scapula
Spine
Body

Posterior view

Figure 13.22 Shoulder
The shoulder girdle is made up of the scapula and clavicle bones.

Scapula The **scapula** bones are the shoulder blades and are flat bones with a slightly concave anterior surface that articulates with the upper posterior ribs. The posterior scapula has a ridge called the *spine* that ends laterally at the **acromion**. Anteriorly, the scapula has a protrusion called the *coracoid process*. The acromion and the **coracoid process** form the edges of the "socket" of the shoulder joint, called the **glenoid cavity**, which is where the head of the humerus sits.

Clavicles The **clavicles** are the collar bones, and they articulate laterally with the acromion and medially with the superior surface of the manubrium of the sternum. It is common for children to break their clavicles because falling sideways onto the upper arm transfers force to the body through the clavicle.

Shoulder Joint The shoulder joint is a ball-and-socket joint (Section 13.7) like the hip, and is the most highly movable joint in the body. The shoulder joint is formed by the head of the humerus in the shallow glenoid cavity.

Humerus The **humerus** is the only bone in the upper arm (Figure 13.23) and is the largest bone of the upper limb. The humerus has an **anatomical neck**, which is the area where the articulating cartilage meets the bone, and a **surgical neck**—the narrow region that is the most commonly broken part of the humerus. The **intertubercular groove** (bicipital groove) is where the biceps tendon lies. The **deltoid tuberosity** is the attachment site for the deltoid muscle. At the distal end, the **capitulum** articulates with the proximal **radius**, and the **trochlea** articulates with the proximal **ulna**. Just proximal to the capitulum and trochlea are the **radial fossa** and **coronoid fossa**. These indentations allow the head of the radius and the coronoid process of the ulna to move together, which increases the range of motion when the elbow is bent. On the posterior, distal region of the humerus there is a large indentation called the **olecranon fossa**, which allows room for the **olecranon** (elbow) of the ulna when the arm is straight. This fossa allows the arm to be straight, but prevents the elbow from hyperextending. The medial and lateral bone processes that can be felt at the elbow are the **medial and lateral epicondyles** of the distal humerus.

Forearm The forearm consists of two bones, the radius and the ulna (Figure 13.24). The head of the radius and the coronoid process of the ulna articulate with the humerus. The head of the radius also articulates with the ulna at the **radial notch**, forming the **radioulnar joint**, which allows the forearm to rotate (Section 13.7). In between the radius and ulna is the **interosseous membrane**, composed of dense, irregular connective tissue. At the distal end, the radius and ulna articulate with the carpals at the wrist. The medial and lateral palpable bones in the wrist are the styloid process of the radius (lateral in anatomical position, which is near the thumb), and the styloid process of the ulna (medial, near the little finger).

Wrist and Hand The wrist bones are formed from eight **carpals**, which are short bones that form a small concave region in the front of the wrist, called the **carpal tunnel**. Nine tendons that flex the fingers and the median nerve that causes flexion of the thumb, index finger, middle finger, and half of the ring finger, enter through the carpal tunnel. The bones of the hand are formed by the five metacarpals. The fingers are formed by 14 phalanges. Each finger has a proximal, medial, and distal phalange, whereas the thumb has only two.

13

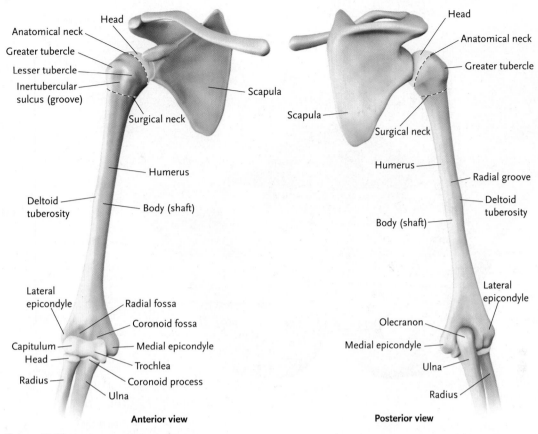

Figure 13.23   Anatomy of the Upper Arm

Anterior view (left diagram labels): Head, Anatomical neck, Greater tubercle, Lesser tubercle, Inertubercular sulcus (groove), Surgical neck, Scapula, Humerus, Deltoid tuberosity, Body (shaft), Lateral epicondyle, Radial fossa, Coronoid fossa, Capitulum, Medial epicondyle, Head, Trochlea, Radius, Coronoid process, Ulna

Anterior view

Posterior view (right diagram labels): Head, Anatomical neck, Greater tubercle, Scapula, Surgical neck, Humerus, Radial groove, Deltoid tuberosity, Body (shaft), Lateral epicondyle, Olecranon, Medial epicondyle, Ulna, Radius

Posterior view

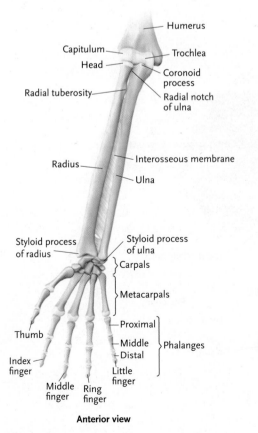

Figure 13.24   Anatomy of Forearm and Hand

Labels: Humerus, Capitulum, Trochlea, Head, Coronoid process, Radial tuberosity, Radial notch of ulna, Radius, Interosseous membrane, Ulna, Styloid process of radius, Styloid process of ulna, Carpals, Metacarpals, Thumb, Proximal, Middle, Distal, Phalanges, Index finger, Middle finger, Ring finger, Little finger

Anterior view

DID YOU KNOW?

Carpal tunnel syndrome is a common, painful condition of the hand and arm caused by repetitive and prolonged overuse of the hands and fingers, which causes inflammation. The swelling causes the carpal tunnel to become narrowed, and this puts pressure on the median nerve, causing pain, weakness, numbness, and tingling in the hand and forearm. Non-surgical treatment of carpal tunnel syndrome involves reducing repetitive movements and allowing rest and time for the inflammation to decrease. Ensuring proper body positioning and wearing a brace may help.

Lower Limb

Pelvis The pelvis is formed by three components of the hip bones: the ilium, pubis, and ischium on each side (Figure 13.25). The **ilium** is the large superior region, and the palpable hip bones are the **iliac crest**. The right and left pubis bones articulate anteriorly with the **pubic symphysis** made of fibrocartilage: a cartilaginous joint (Section 13.7). The ilium bones articulate with the sacrum posteriorly, forming the sacroiliac joint. When we sit down, the bones we sit on are the ischium bones, specifically the **ischial tuberosities**, which are the protruding processes. The opening in the pelvis formed by the sacrum, ilium bones, and pubis bones is the **pelvic brim**,

13

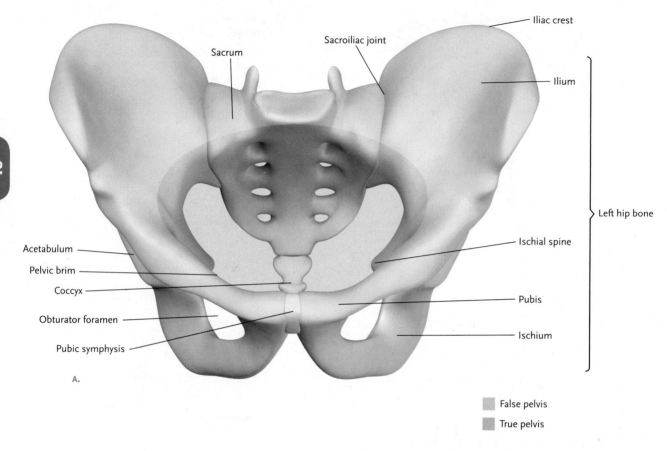

A.

False pelvis

True pelvis

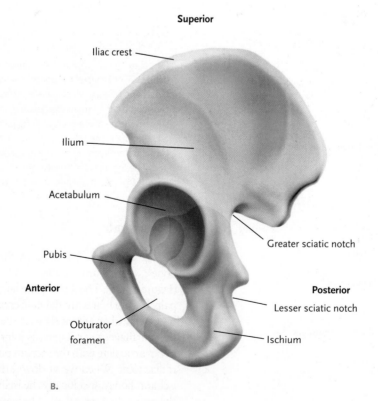

B.

Figure 13.25 Pelvis
(A) Anterior view and (B) lateral view

and this forms the top of the **true pelvis**, which protects the internal, lower abdominal organs. The true pelvis is wider in females to allow for childbirth. The false pelvis is the region between the top of the iliac crests to the top of the pelvic brim. Within the interior of the true pelvis are medial protrusions from the ischium, called the **ischial spines**; these mark the narrowest part of the pelvic opening.

The openings in the ischium bones are the **obturator foramina**, which allow nerves and blood vessels and lymphatic vessels to pass through. These are the largest foramen in the body. All three pelvic bones meet in a deep socket called the **acetabulum**, which is where the head of the femur articulates and forms the hip joint. The iliac bone has a large indentation posteriorly, and this is called the **greater sciatic notch**; this is where the **sciatic nerve** travels down the posterior lower limb (Chapter 15).

Femur The head of the **femur** in the thigh articulates with the acetabulum in the pelvis, forming the hip joint. Like the shoulder, this is a ball-and-socket joint with a large range of motion (Figure 13.26). The femur is the longest and largest bone in the body. The narrow neck of the femur is the most commonly broken region in hip fractures. The femur has a palpable, lateral projection, called the **greater trochanter**, and a medial lesser trochanter, and both of these provide sites for muscle attachment. The **gluteal tuberosity**, the attachment site for the gluteal muscles, is located on the proximal, posterior femur. The large

medial and lateral knee bones are the medial and **lateral condyles** of the femur. Connecting to the distal anterior femur is the **patella**, also called the *knee cap*, which is a sesamoid bone connecting the quadriceps tendon and the patellar tendon that attaches the quadriceps to the tibia. The patella helps protect the knee joint. The **intercondylar fossa** is the indentation at the distal posterior femur.

Lower Leg The distal femur articulates with the superior **tibia**, which is the larger of the bones in the lower leg. The bump on the proximal anterior tibia below the patella is the **tibial tuberosity** and is the attachment site for the **patellar ligament**, which is a continuation of the quadriceps tendon. The **fibula** is the smaller lateral bone in the lower leg that articulates with the proximal tibia, but does not articulate with the femur (Figure 13.27). Like the forearm, there is connective tissue—the interosseous membrane—between the tibia and fibula. At the distal end, the fibula and tibia articulate at the fibular notch, forming the **tibiofibular joint** (Section 13.7). The protruding medial and lateral ankle bones are the lateral malleolus of the fibula and the medial malleolus of the tibia.

Ankle and Foot The ankle bones consist of seven tarsal bones. The foot consists of five metatarsals, and the toes are named phalanges, just like the fingers. All toes except the big toe have a proximal, medial, and distal phalange; the big toe has two phalanges.

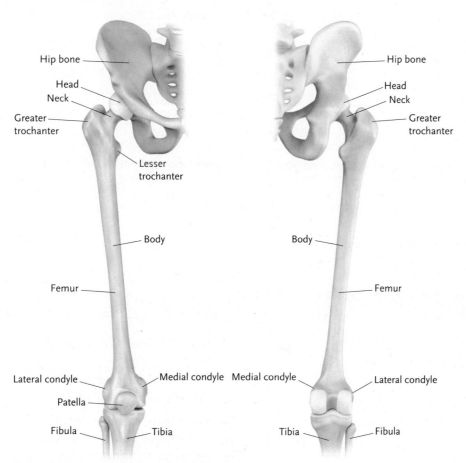

Figure 13.26 Anatomy of the Femur

13

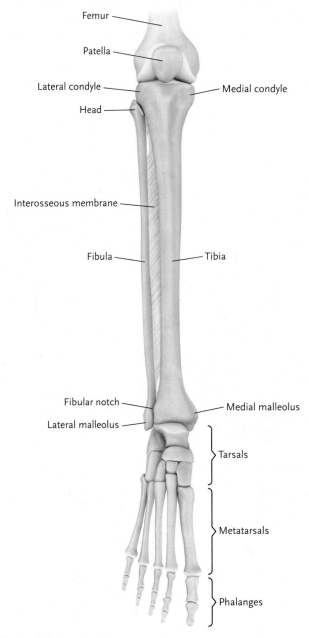

Femur

Patella

Lateral condyle

Medial condyle

Head

Interosseous membrane

Fibula

Tibia

Fibular notch

Medial malleolus

Lateral malleolus

Tarsals

Metatarsals

Phalanges

Figure 13.27 Anatomy of Lower Leg and Foot

See Figure 13.28 for an illustration of the full appendicular skeleton.

CONCEPT REVIEW 13.4

1. List the bones in one upper limb.
2. List the bones in one lower limb.
3. Which pelvic bones do we sit on?
4. What region makes up the true pelvis?
5. What is the function of the patella?
6. Where on the humerus do the radius and ulna articulate?

7. Name the medial and lateral protrusions of the wrist.
8. Which bone is the olecranon?
9. Which bones make up the wrist?
10. Which bone forms the glenoid cavity?
11. Which bones form the acetabulum?
12. Where are the medial and lateral condyles?

13

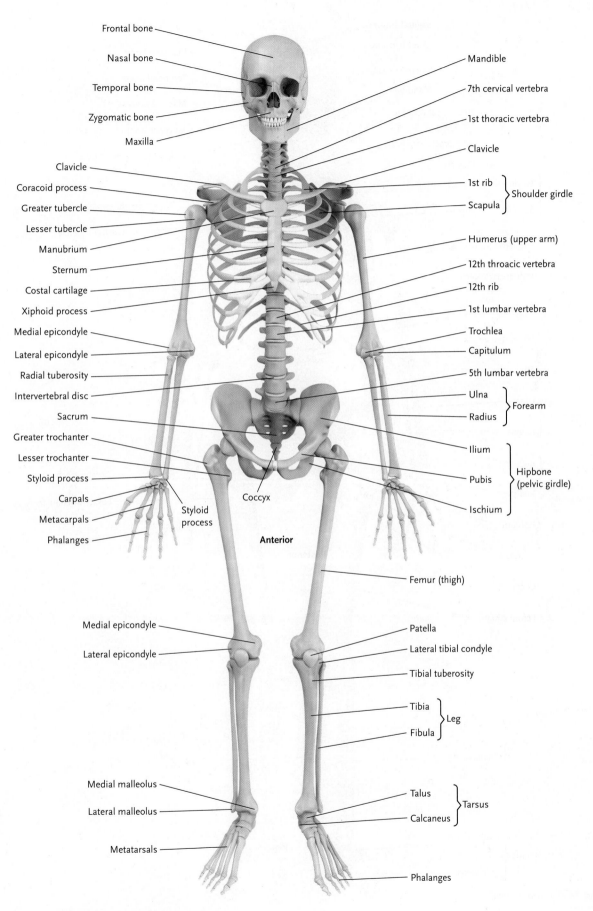

Frontal bone
Nasal bone
Temporal bone
Zygomatic bone
Maxilla

Mandible
7th cervical vertebra
1st thoracic vertebra
Clavicle

Clavicle
Coracoid process
Greater tubercle
Lesser tubercle
Manubrium
Sternum
Costal cartilage
Xiphoid process
Medial epicondyle
Lateral epicondyle
Radial tuberosity
Intervertebral disc
Sacrum
Greater trochanter
Lesser trochanter
Styloid process
Carpals
Metacarpals
Phalanges

1st rib
Scapula
} Shoulder girdle

Humerus (upper arm)
12th throacic vertebra
12th rib
1st lumbar vertebra
Trochlea
Capitulum
5th lumbar vertebra
Ulna
Radius
} Forearm

Ilium
Pubis
Ischium
} Hipbone (pelvic girdle)

Styloid process
Coccyx

Anterior

Femur (thigh)

Medial epicondyle
Lateral epicondyle

Patella
Lateral tibial condyle
Tibial tuberosity

Tibia
Fibula
} Leg

Medial malleolus
Lateral malleolus

Talus
Calcaneus
} Tarsus

Metatarsals

Phalanges

Figure 13.28 The Human Skeleton

(Continued)

13

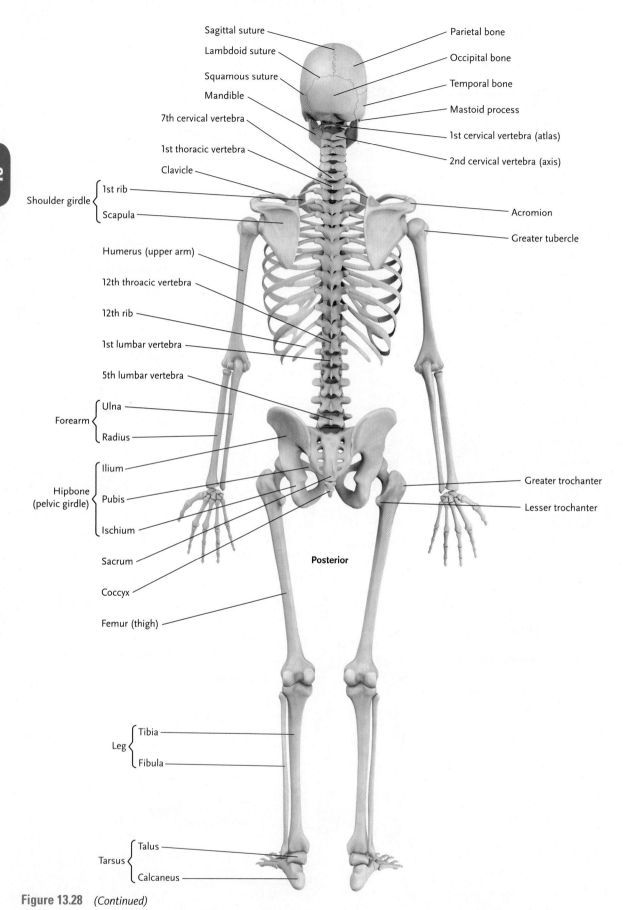

Sagittal suture
Lambdoid suture
Squamous suture
Mandible
7th cervical vertebra
1st thoracic vertebra
Clavicle
Shoulder girdle {
1st rib
Scapula
Humerus (upper arm)
12th throacic vertebra
12th rib
1st lumbar vertebra
5th lumbar vertebra
Forearm {
Ulna
Radius
Hipbone (pelvic girdle) {
Ilium
Pubis
Ischium
Sacrum
Coccyx
Femur (thigh)

Parietal bone
Occipital bone
Temporal bone
Mastoid process
1st cervical vertebra (atlas)
2nd cervical vertebra (axis)
Acromion
Greater tubercle
Greater trochanter
Lesser trochanter

Posterior

Leg {
Tibia
Fibula

Tarsus {
Talus
Calcaneus

Figure 13.28 *(Continued)*

13.5 MINERAL HOMEOSTASIS

The hormones involved in regulating bone growth and remodelling include **parathyroid hormone (PTH)**, calcitonin, growth hormone, insulin-like growth factors, thyroid hormones, insulin, and sex hormones (estrogen and testosterone) (Chapter 17).

PTH is released by the parathyroid gland in response to low blood calcium. PTH stimulates the osteoclasts to break down bone extracellular matrix and release calcium into the bloodstream. PTH also causes the kidneys to reabsorb Ca^{++} and Mg^{++} (Chapter 21), and stimulates the kidneys to produce **calcitriol** (active form of vitamin D) that causes the small intestine to absorb more calcium from the diet (Chapter 20).

Calcitonin is released from the thyroid gland and prevents the osteoclasts from breaking down bone extracellular matrix. This hormone plays only a small role in the regulation of calcium.

Growth hormone (GH) is produced by the anterior pituitary gland in response to stimulation from the hypothalamus during exercise in adults or growth spurts in adolescents. GH stimulates the growth of many tissues, including bone and cartilage, by stimulating the production of **insulin-like growth factors (IGFs)**.

IGFs are secreted mainly by the liver, bone, and muscle cells in response to growth hormone and, in bone, stimulate osteoblasts to increase bone mineralization and the production of collagen fibres in the bone extracellular matrix.

Thyroid hormones (T3 and T4) are produced by the thyroid gland and play an important role in metabolism in many tissues (Chapter 17), including bones where T3 and T4 stimulate osteoblasts.

Insulin, secreted by the pancreas when we eat food, is a general signal to many cells of the body to take up and use or store nutrients that are circulating in the bloodstream. This movement of nutrients also occurs in bones, where extra circulating calcium is taken up by osteoblasts and incorporated into the bone matrix.

Sex hormones are produced by the gonads (ovary or testes) and stimulate osteoblasts to increase bone growth. These hormones play a significant role through growth spurts during puberty and continue to stimulate bone remodelling during adulthood. Sex hormones also cause the epiphyseal plate to mineralize into the epiphyseal line when bone length is complete.

DID YOU KNOW?

Women in menopause have a higher risk of osteoporosis due to the loss of the bone growth, which promotes the stimulation of estrogen. It is very important for menopausal women to continue to exercise to ensure the production of growth hormone and adrenal androgens to prevent loss of bone density.

The key minerals required for bone health are calcium, magnesium, and phosphorus because they make up the majority of the mineral component of the extracellular matrix. Magnesium is required as a coenzyme in every tissue in the body.

Vitamin requirements for healthy bones include vitamins D, A, and C. **Vitamin D** is produced from cholesterol molecules in the skin when we are exposed to sunlight, and it is converted into calcitriol by the kidneys. Vitamin D is required for the absorption of calcium from food in the small intestine. Without vitamin D, any calcium in food stays in the small intestine and is excreted.

Vitamin A is required for osteoblast function; however, an excessive amount of vitamin A has been shown to stimulate osteoclasts and have a detrimental effect on bone health. It is highly unlikely that anyone would take in too much vitamin A from dietary sources alone, but this can occur by frequent consumption of supplements.

Vitamin C is required for the movement of collagen from cells where it is produced and into the extracellular matrix.

DID YOU KNOW?

Vitamin D deficiency in children causes rickets, which causes malformation of the bones, most commonly leading to bowing femurs, greenstick fractures, lordosis, and dental problems.

CONCEPT REVIEW 13.5

1. Which hormones increase bone growth, and which hormones cause bone to be broken down?
2. Which type of bone cells are stimulated by hormones that promote growth?
3. Which hormones target osteoclasts?
4. What are the key minerals required for bone growth?
5. What is the role of vitamins A, C, and D in bone health?

13

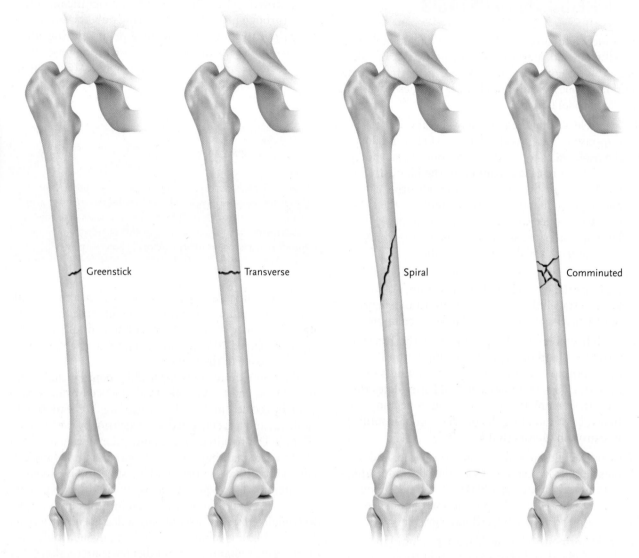

Greenstick   Transverse   Spiral   Comminuted

Figure 13.29 Bone Fractures
The categories of bone fractures are named according to how the bone breaks.

13.6 FRACTURES AND OSTEOPOROSIS

A fracture is any type of break in a bone (Figure 13.29). The most commonly broken bones are the long bones in the upper and lower limbs; this includes hip fractures because they usually involve the head of the femur, not the pelvis. Types of fractures include the following: greenstick, transverse, comminuted, spiral, stress, or open.

Fractures that do not break the skin are called *closed*, and if the bone protrudes through the skin the break is called *open*. **Greenstick fractures** are more common in children whose bones are very flexible. Like when you break a green tree branch, the bone cracks on one side but only bends on the other side. Adults tend to have fractures that completely break the bone. A **transverse fracture** is a simple break across the transverse section of the bone, a twisted break is called a **spiral fracture**,

and crushed bone is called **comminuted fracture. Stress fractures** that cause the bone to crack but not break apart are common in athletes that jump or run on hard surfaces. Since bones are highly vascularized, a break can cause blood vessels to be broken and blood to fill up the area between the periosteum and the bone. This is called a *hematoma*, which is very painful.

When fractures heal, phagocytic immune cells (Chapter 23) move in to clean up the dead cells. Chondrocytes form **fibrocartilage** (Chapter 11) in the broken region and join the broken sections of bone. Then osteoblasts begin to lay down the new bone matrix of collagen and mineralize the newly forming bone: first spongy bone, which is then converted into compact bone. Osteoclasts are also highly involved in remodelling the new compact bone to replace the spongy bone in the right areas.

Osteoporosis

Bones need to be under mechanical stress to maintain density and strength. This is mostly accomplished by the hormonal influences mentioned above, including exercise. The combination of hormones (mainly thyroid hormones, growth hormones, IGFs, insulin, and sex hormones) as well as vitamins and minerals from the diet, ensure that osteoblasts continue to build up the extracellular matrix of bone tissue. When one or more of these hormones or nutrients are not available, the osteoclasts break down bone faster than the osteoblasts can build it up. A loss of bone density leads to a loss of tensile strength, and this process is called *osteoporosis* (Figure 13.30).

In adults, the primary factors are diet and exercise. Calcium levels in the blood must be maintained because calcium is required for the proper functioning of the nervous system (Chapter 15) and the muscles (Chapter 14). Exercise must include gravity and mechanical stress, such as jogging or running, jumping, weight lifting, or most other physical activities. Swimming does not produce enough force on the bones to increase bone density. Osteoporosis tends to affect primarily the bones of the

Professor Pietro M. Motta/Science Source

Figure 13.30 Osteoporosis

Osteoporosis is a progressive disease caused by ongoing loss of mineralization of the bones, and it leads to an increased risk of bone fractures.

wrist, shoulder, hip, and vertebrae. It mostly affects elderly people, women in menopause, and people who do not or cannot engage in physical activity. Among people age 50 and over, 80% of hip fractures are due to osteoporosis.

CONCEPT REVIEW 13.6

1. Name the different types of fractures. Which type is most common in children?

2. Which cell types are involved in healing fractures?

3. What are the primary factors involved in osteoporosis?

4. Why is osteoporosis more common in menopausal women?

Application

As you read Section 13.7, consider Fred, a soccer player, who is about to kick the ball. Which bones in the leg does he use to kick the soccer ball? Which joints are involved? What type of joints are they? What type of movement occurs in each joint?

13.7 JOINTS AND MOVEMENT

A **joint** is an articulation of either two or more bones, cartilage and bone, or teeth and bone. Different types of joints have different structures and functions. Structurally, joints are classified as fibrous, cartilaginous, or synovial. Functionally, they are classified as syndesmosis (no movement), amphiarthrosis (slight movement), and diarthrosis (highly movable) (Figure 13.31).

Fibrous joints can be either immovable (syndesmosis) or only slightly movable (amphiarthrosis) and are connected by dense, irregular connective tissue. Skull sutures are immovable fibrous joints and also **syndesmosis** joints because they do not move. The joint between a tooth and the socket of the alveolar process, the *dentoalveolar joint*, which is connected by the *periodontal ligament*, is immovable and also called a **gomphosis joint**. Although

this joint is considered immovable, there is enough space between the tooth root and the alveolar bone process to allow some movement; this is why teeth can be straightened with orthodontic braces. The connection between the tibia and fibula by the interosseous ligament (also called the *anterior tibiofibular ligament*), as well as between the radius and ulna (also connected by an interosseous membrane), is **amphiarthrosis** because it is slightly movable.

Cartilaginous joints are connected by fibrocartilage or hyaline cartilage, rather than connective tissue, and are tightly connected with little or no movement (syndesmosis). Examples of cartilaginous joints are the connection between the costal (hyaline) cartilage of ribs and the sternum, the connection between the pubic symphysis (fibrocartilage) and the pelvic bones, the epiphyseal plates in the ends of the long bones

Figure 13.31 The Body's Joints
The joints in the body are categorized based on their structure—fibrous, cartilaginous, or synovial—or their function—syndesmosis, amphiarthrosis, or diarthrosis.

Figure 13.32 Synovial Joints

Synovial joints are highly movable and contain a capsule filled with synovial fluid between articulating bones to prevent friction during movement.

Types of Synovial Joints

Planar joint. The bones in the joint move in one plane back and forth or side to side and may allow rotation. The articulating surfaces are flat, and there is minimal movement. Examples of planar joints are the carpals in the wrists, the tarsals in the ankles, and the acromioclavicular joint (Figure 13.33a).

Hinge joint. The bones in the joint move only in one plane, like a door on a hinge. Hinge joints are connected by strong ligaments. Examples of hinge joints are the elbow, knee, ankle, and interphalangeal joints of the fingers and toes (Figure 13.33b).

Pivot joint. This joint occurs where one bone in the joint fits into a curve or ring of another bone and allows for rotational movement. Examples of pivot joints include the radioulnar joints that allow rotation of the forearm, and the atlanto-axial joint between the first and second cervical vertebrae, which allows the head to turn side to side (Figure 13.33c).

Condylar joint. Also called *condyloid*, this joint occurs where one bone of the joint has a concave surface that fits into the convex region of the other bone, allowing movement around two axes. One example is the metacarpophalangeal joint in our fingers, where the proximal phalanges connect with the metacarpals in the hand, which allows the fingers to rotate and also move back and forth and side to side. The wrist is also a condylar joint (Figure 13.33d).

Saddle joint. This is similar to the condylar joint, except that the bones fit together like a person sitting on a saddle, and it allows movement back and forth and side to side. The main example of a saddle joint is the thumb, where the proximal phalange connects with a metacarpal (Figure 13.33e).

Ball-and-socket joint. One bone has a ball-shaped surface that fits into a concave depression on the other bone. These joints are the most highly movable of all synovial joints. Examples include the shoulder and hip joints (Figure 13.33f).

before they ossify, the connection between the manubrium and the body of the sternum, and the connection between the intervertebral (fibrocartilage) discs and the vertebrae.

A **synovial joint**, also called **diarthrosis**, is highly movable. Synovial joints have a joint cavity called the **synovial cavity**. The synovial cavity contains synovial fluid, and a synovial membrane encloses the ends of the connecting bones in a fibrous capsule (Figure 13.32).

Synovial fluid is secreted by the cells in the synovial membrane (made up of epithelial and connective tissue). The fluid consists of plasma from the nearby blood vessels as well as hyaluronic acid, which increases the viscosity of the fluid. The purpose of synovial fluid is to provide nutrients, prevent friction between the ends of the connecting bones, and to absorb shock. Exercise and movement stimulate the production of synovial fluid. The bones in synovial joints are held together by **ligaments** (dense, irregular connective tissue), and muscles are connected to bones by **tendons**, which also consist of dense, irregular connective tissue.

Synovial joints often contain **bursa sacs**, which are small sacs filled with synovial fluid and lined with a synovial membrane. Located outside the synovial cavity, bursa sacs add extra protection and cushion between bones and tendons or muscles.

13

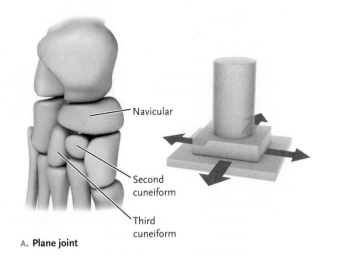

Navicular

Second cuneiform

Third cuneiform

A. **Plane joint**

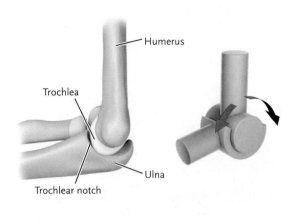

Humerus

Trochlea

Ulna

Trochlear notch

B. **Hinge joint**

Humerus

Annular ligament

Ulna

Radius

C. **Pivot joint**

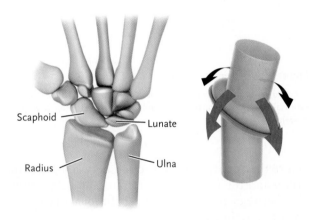

Scaphoid

Lunate

Radius

Ulna

D. **Condyloid joint**

Metacarpal of thumb

Trapezium

Radius

E. **Saddle joint**

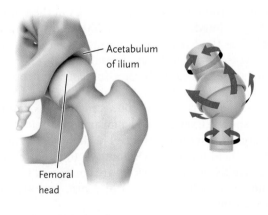

Acetabulum of ilium

Femoral head

F. **Ball-and-socket joint**

Figure 13.33 Types of Synovial Joints

(A) Plane, (B) hinge, (C) pivot, (D) condylar, (E) saddle, and (F) ball-and-socket

Types of Movement

Gliding movement occurs when nearly flat surfaces of bones move back and forth and side to side. This type of movement occurs in planar joints (Figure 13.34). **Rotation** is a type of movement that allows a bone to move around its longitudinal axis; this occurs in pivot joints (Figure 13.34). **Angular** movement occurs when the angle between bones increases or decreases. It occurs in hinge, condylar, saddle, and ball-and-socket joints. Specific angular movements that occur in these joints include flexion, extension, hyperextension, abduction, adduction, and circumduction (Figure 13.34).

At certain specific joints, other types of movements occur, including elevation, depression, protraction, retraction, inversion, eversion, dorsiflexion, plantar flexion, supination, pronation, and opposition (Figure 13.34).

Figure 13.34 Types of Synovial Joint Movement

(Continued)

13

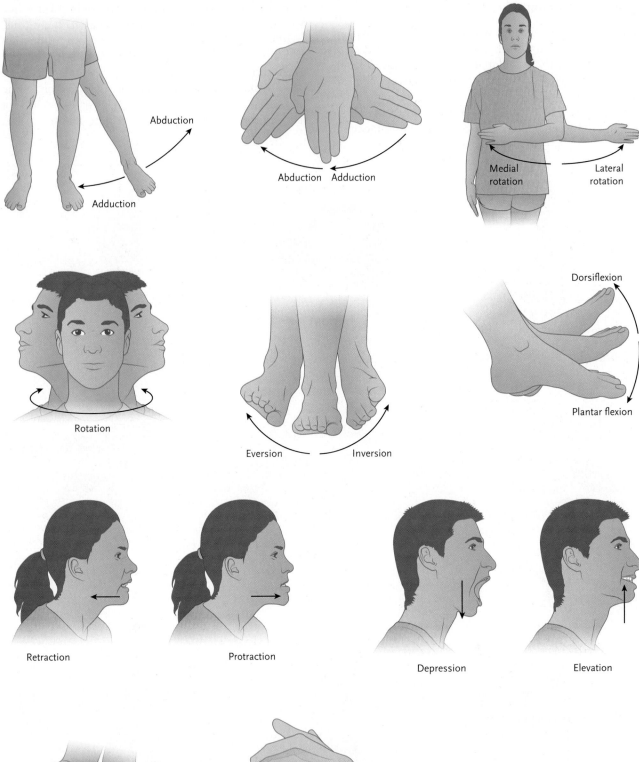

Abduction

Adduction

Abduction   Adduction

Medial rotation   Lateral rotation

Rotation

Eversion   Inversion

Dorsiflexion

Plantar flexion

Retraction   Protraction   Depression   Elevation

Palm posterior

Palm anterior

Supination   Pronation

Opposition

Figure 13.34 *(Continued)*

CONCEPT REVIEW 13.7

1. Explain how joints can be categorized by structure or function.
2. What are some examples of joints that are fibrous, cartilaginous, and synovial?
3. What are some examples of joints that are syndesmosis, amphiarthrosis, and diarthrosis?
4. What is the function of synovial fluid?
5. What are bursa sacs?

6. Label a diagram of a typical synovial joint.
7. What are the six types of synovial joints? Give an example of each.
8. What types of movements can occur in ball-and-socket joints?
9. What is an example of a joint that rotates?
10. What is the difference between pronation and supination?

EFFECTS OF AGING

As we age the skeleton changes with respect to posture, gait, joint mobility, and bone composition. The muscles have a major impact on the skeleton: muscle strength affects posture and gait, the force of muscle contractions on the bones affects the production of hormones that stimulate bone mineralization, and mobility affects the joints.

The most common bone condition that occurs with aging is loss of bone density, which is due to loss of minerals such as calcium and phosphorus. Mineralization and stimulation of the osteoblasts occurs when force is put on the bones, such as during exercise. As people age they tend to become less active, and therefore hormones, such as parathyroid hormone, stimulate osteoclasts to release minerals from the bones into the bloodstream to be used by the nervous system and muscles. Continuing to be active in later life is important for maintaining bone composition.

In the spine, there are discs that act as a cushion between each vertebra. As we age the discs gradually lose fluid and become thinner, making us a little shorter. We actually also become a little bit shorter at the end of the day, compared to the beginning of the day, because when we sleep there is no pressure is on the spinal discs, and they refill with fluid. Also, as we age foot arches can become less pronounced, which can slightly affect our height.

In the joints, less fluid is produced in the synovial sacs, which causes the epiphyseal cartilage to rub together, creating friction, and may form bone spurs that make movement more painful. Minerals can deposit around some joints (most commonly in the shoulder), decreasing mobility. Continuing to be mobile in older age is important for stimulating the production of synovial fluid. With a loss of mobility—for example, due to a broken hip, likely caused by brittle bones from mineral loss—all of these aging factors will occur more rapidly.

CHAPTER REVIEW

Medical Terminology

Term	Definition	Example
acr/o	extremity, topmost	acromion
arthr/o	joint	arthroscopy
articul/o	joint	articulation
brachi/o	arm	brachiocephalic
burs/o	bursa	bursitis
calcane/o	heel	calcaneodynia
carp/o	wrist	carpoptosis
cephal/o	head	cephaledema
cervic/o	neck, cervix	cervical spine
chondr/o	cartilage	chondrocyte
-clast	break, breaking	osteoclast

13

Term	Definition	Example
cleid/o	clavicle	cleidorrhexis
coccyg/o	coccyx	coccygodynia
condlye	smooth articular surface	occipital condyle
cost/o	rib	costosternal
cox/o	hip	coxarthrosis
crani/o	skull	cranioclast
cubit/o	elbow, forearm	genucubital
-dactyl(o)-	pertaining to a finger or toe	polydactyly
femor/o	femur	ischiofemoral
fibul/o	fibula	fibulocalcaneal
foramen	opening	foramen ovale
fossa	hollow area	fossa ovalis
hamus	hook	hamate
humer/o	humerus	humeroradial
ileo-	ileum	ileocecal
ili/o	ilium	iliolumbar
ischi/o	ischium	ischiodynia
itchy/o	erect, straight	ithylordosis
-itis	inflammation	hepatitis
kyph/o	humpback	kyphoscolilosis
lip/o	fat	liposuction
lord/o	curvature, bending	lordoscoliosis
maxill/o	maxilla	maxillotomy
mega-, megalo-	enlargement	acromegaly
-megaly	enlargement	dactylomegaly
-oid	resembling	osteoid
oste/o	bone	osteochondroma
pan-	all	panarthritis
patell/o	patella	patellofemoral
ped/o	foot, child	pedal
pelv/i	pelvis	pelvimeter
phalang/o	phalanges	phalangitis
-physis	growth, growing	diaphysis
pod/o	foot	podiatrist
pub/o	pubis	pubovesical
sacr/o	sacrum	sacrocoxalgia
scapul/o	scapula	scapulopexy
scoli/o	crooked, twisted	scoliorachitic
skelet/o	skeleton	skeletogenous
spin/o	spinal cord, spine	spinocerebellar
spondyl/o	vertebrae	spondylopyosis
stern/o	sternum	sternocostal
synov/o	synovia, synovial membrane	synovectomy
tal/o	talus	talofibular

(Continued)

TERM	DEFINITION	EXAMPLE
tars/o	tarsus, edge of eyelid	tarsoclasis
tempor/o	the temples	temporal bone
tibi/o	tibia	tibiotarsal
xiph/o	sword-shaped, xiphoid	xiphoid process
zyg/o	union, junction	zygomatic bone

13

Labelling Activity

Label the following diagrams.

1. a.   b.   c.

d.   e.

13

2.

3.

4.

5.

13

13

6.

7.

8.   a.

b.

13

9.

Anterior   Posterior

10.

Courtesy of Mark Nielsen, AnatBooks, LTD

11.   a.   b.

12.

a.

b.

13.

14.

15.

a.

b.

16.

13

17.   a.   b.

18.

19.

20.

13

13

Answer to Application Question

- The bones involved are the pelvis, femur, tibia, fibula, ankle and foot bones.
- The joints involved are all synovial joints:
  a. Hip—ball and socket
  b. Knee—hinge
  c. Ankle—condyloid
- Movements of each joint are as follows:
  d. Hip. During the movement when the kicking leg is pulling backward, the hip is

hyperextending; during the forward kicking motion, the hip is flexing.

e. Knee. During the movement of the leg pulling backward, the knee is flexing; during the forward kicking motion, the knee is extending.

f. Ankle. During the leg's backward motion, the foot is plantar flexing; during the forward kicking motion, the foot is dorsiflexing.

14 The Muscular System

CHAPTER OUTLINE

The beautiful thing about learning is that nobody can take it away from you.

B.B. King

14.1 OVERVIEW OF THE MUSCULAR SYSTEM

There are three types of muscle tissue: cardiac, smooth, and skeletal (Chapter 11). This chapter focuses on the superficial skeletal muscles and the major muscle groups. Each type of muscle tissue has different functions, including the following:

- Producing body movements
- Stabilizing body positions and posture
- Causing movement of internal organs, such as the smooth muscles of the blood vessels, digestive tract, fallopian tubes, bronchial tubes, and ureters

- Regulating capillary blood flow and movement of contents through the digestive tract by means of sphincters that are smooth or skeletal muscle
- Producing heat, as a by-product of ATP synthesis, with muscles producing the most ATP of any body tissue and playing an important role in body temperature homeostasis

DID YOU KNOW?

The strongest muscles in the body have the largest cross-sectional area and can exert the most force; these include the rectus femoris and the gluteus maximus. Shorter muscles with a shorter lever arm, such as the masseter (jaw muscle), can exert tremendous force. The heart muscle performs the largest quantity of contraction force in a lifetime.

CONCEPT REVIEW 14.1

1. What are the three types of muscle tissue?

2. What are the functions of muscles?

14.2 MUSCLE TISSUE

Table 14.1 summarizes the main features of each type of muscle tissue: cardiac, smooth, and skeletal.

Skeletal Muscle Tissue

There are approximately 650 skeletal muscles in the human body. Skeletal muscles are organs, and they are composed of all four tissue types: muscle fibres, nerves,

TABLE 14.1

Types of Muscle Tissue

Muscle Feature	Cardiac	Smooth	Skeletal
Appearance			
Location	Heart	Blood vessels, digestive tract, bronchial tubes, some sphincters, iris	Connected to bones or skin, some sphincters
Nervous system control	Involuntary Autonomic nervous system	Involuntary Autonomic nervous system	Voluntary Somatic nervous system
Cell structure	Branched, connected with intercalated discs containing desmosomes and gap junctions	Small cells with tapered ends; can function as a single cell or multi-unit group of cells connected by gap junctions	Long, parallel cells that function as small or large motor units
Nuclei	One to three nuclei per cell	One nucleus per cell	Multinucleated
Striated	Yes	No	Yes
Diameter of fibres	Intermediate	Small	Large, intermediate, and small
Speed of contraction	Intermediate	Very slow	Fast, intermediate, and slow

epithelial tissue, and connective tissue. Muscles are highly vascular, so blood can carry oxygen and nutrients to the muscle cells, also called **muscle fibres**, for ATP production. Every single muscle fibre is in contact with a neuron that stimulates contraction.

Skeletal muscle tissue is attached to bones by tendons and, when contracted, causes bones to move. Tendons consist of dense, regular connective tissue. Three dense, irregular connective tissue layers surround muscles: the epimysium, perimysium, and endomysium (Figure 14.1). The epimysium is the outermost covering below the **fascia**, which is also dense, irregular connective tissue that supports and surrounds all muscles and connects with the subcutaneous layer of the skin (Chapter 12). Each muscle is composed of bundles of muscle fibres called **fascicles**, and these are surrounded by the **perimysium**. Each individual muscle fibre is surrounded by the **endomysium**. All three layers of connective tissue meet beyond the muscle tissue at each end and are continuous with the tendons that connect the muscle to the bone. Within the fascicles and surrounding

every muscle fibre are blood vessels, lymphatic vessels, and neurons.

Skeletal muscle fibres (cells) contain all of the organelles of any typical cell type, except that these cells are very long, multinucleated cells. The prefix *sarco* means flesh and is associated with many terms regarding muscles. The cytoplasm is called **sarcoplasm**. The muscle fibre membrane is called the **sarcolemma**, and it is different from other cell membranes because it forms tunnel-like extensions throughout the cell called **transverse tubules (T-tubules)** (Section 14.4). Muscle cells contain mitochondria for producing ATP, ribosomes for protein synthesis, and also rough and smooth endoplasmic reticulum (Chapter 3). The smooth endoplasmic reticulum in muscles cells is called **sarcoplasmic reticulum** and has the special feature of storing calcium ions that are involved in muscle contraction.

In addition to the cellular organelles, muscle cells have an extensive arrangement of contractile and **regulatory proteins**, which are arranged into small bundles called **myofibrils** (Figure 14.1). Each myofibril is

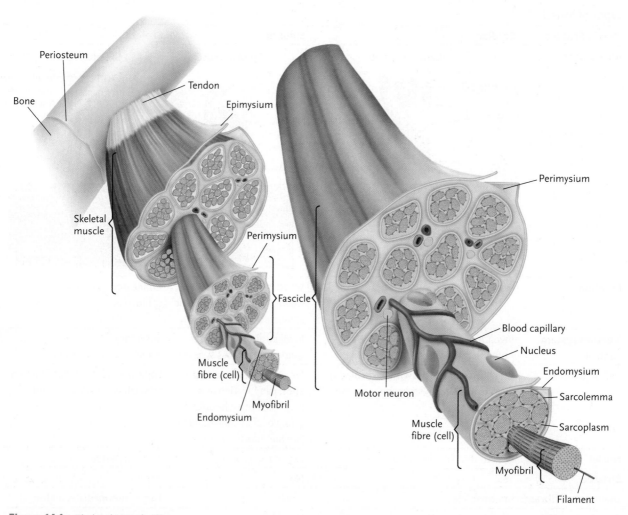

Figure 14.1 Skeletal Muscle Tissue

Skeletal muscle tissue is connected to bone by tendons. It is made up of bundles called *fascicles*, which consist of muscle fibres composed of smaller bundles called *myofibrils*.

composed of thick and thin filaments, which consist of the contractile and regulatory proteins. The **thick filaments** consist of **myosin** proteins that have a movable head region that causes the contraction. The **thin filaments** consist of **actin** proteins, which myosin binds to during contraction, as well as the two main regulatory proteins, **troponin** and **tropomyosin**. The thick and thin filaments overlap in highly organized structures called **sarcomeres**, which are considered the smallest single contractile unit; each muscle consists of millions of sarcomeres. The details of how sarcomeres contract is covered in Section 14.4.

CONCEPT REVIEW 14.2

1. What structure connects muscle to bone?
2. What is fascia?
3. What are the three layers of connective tissue in muscles?

4. List the structures of a muscle from smallest to largest, starting with thick and thin filaments, and ending with epimysium.

Application

As you read through the rest of this chapter, consider this scenario. Dave is going to do a squat workout, five sets of six to eight repetitions per set. List the lower body muscles that are involved in doing a squat, as shown in the accompanying illustration. What type of exercise is this workout, aerobic or anaerobic? What are the primary muscle fibre types used for this kind of workout? What are the primary sources for the production of ATP during this workout? Would creatine be a useful supplement for this type of exercise?

14.3 MUSCLES OF THE BODY

Muscles are named based on their location, orientation to the body, size, shape, and/or action (see Table 14.2).

Movement Produced by Skeletal Muscles

The end of the muscle that attaches to the stationary bone is called the **origin**. The end of the muscle attached to the movable bone is called the **insertion** (Figure 14.2). For example, when you contract your biceps muscle, the forearm moves toward the shoulder, so the end of the biceps that attaches to the scapula is the origin, and the insertion attaches to the radius (Chapter 13). The middle of the muscle is called the **belly**, and it is largest during a maximal contraction.

When we do complex movements, multiple muscles contract at the same time. The main muscle causing the contraction is called the **agonist**. In the example of doing a biceps curl, the biceps and underlying brachialis are the agonists. Since those two muscles work together during a curl (flexion of the arm at the elbow), they are **synergistic**. The muscle on the opposite side of the upper arm, the triceps, must relax and stretch when the biceps are contracting and shortening. Since the triceps cause the arm to extend—which is *opposite* to flexion during the curl—the triceps are the **antagonist**. We also have

TABLE 14.2

Muscle Names and Their Definition

Muscle Name	Definition
Maximus	Largest
Minimus	Smallest
Major	Larger
Minor	Smaller
Longus	Longer
Rectus	Parallel to midline
Transverse	Perpendicular to midline
Oblique	Diagonal to midline
Flexor	Decreases joint angle
Extensor	Increases joint angle
Abductor	Moves away from midline
Adductor	Moves toward midline
Deltoid	Triangular shape
Trapezius	Trapezoid shape
Biceps	Has two origins
Triceps	Has three origins
Quadriceps	Has four origins

14

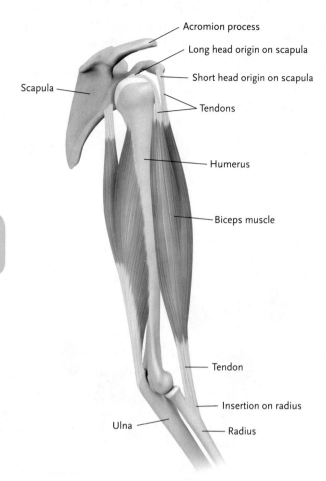

Acromion process
Long head origin on scapula
Short head origin on scapula
Scapula
Tendons
Humerus
Biceps muscle
Tendon
Insertion on radius
Ulna
Radius

Figure 14.2 Muscle Insertion and Origin
The insertion of a muscle is connected to the moving bone, and the origin is connected to the unmoving bone.

muscles called **fixators** that stabilize the origin of the prime mover so that it can act more efficiently.

Muscle movements can also be categorized based on the force of contraction throughout the movement.

Isotonic contraction causes the muscle to shorten with even tension throughout a particular movement, such as using a machine to do a bicep curl so that the level of difficulty is the same throughout the movement. Using a dumbbell to do a bicep curl is tougher when the bicep is extended and gradually becomes easier as the muscle shortens. When the velocity remains the same but the tension varies, this is called an **isokinetic contraction**.

Isotonic and isokinetic contractions can occur in two ways. If the muscle is shortening during the contraction, it is a **concentric contraction**. If the muscle is lengthening, such as slowly lowering the dumbbell and straightening your arm, it is an **eccentric contraction**. It is also possible to have a contraction without any movement. For example, if you hang from a bar with biceps contracted so that your eyes stay at the level of the bar (called a *flexed-arm hang*), the biceps are contracted and hold you in place without shortening or lengthening; this is called **isometric contraction**.

The rest of this section describes individual muscles and some groups of muscles, specifically, origin and insertion points, and how certain muscles contract to produce movement. It is important to review Chapter 13, Section 13.7 (Joints and Movement) before continuing with this chapter. See Figures 14.3 to 14.8 and Tables 14.3 to 14.5.

DID YOU KNOW?

Your muscles are approximately 40% stronger during eccentric rather than concentric contractions. If you add some eccentric contraction exercises (called *negatives*) into your regular concentric exercise routine, you will gain more strength if you do concentric contraction exercises alone.

Muscles of the face and neck are numerous and small, so we have the ability to make many facial expressions, chew, talk, swallow, squint, blink, sniff, move the eyes, and move the head. Facial expressions can be made because the insertion for many muscles is not on a bone but on the skin of the face. Depending on which facial muscles contract and how much they contract, humans can make over 10 000 different expressions.

DID YOU KNOW?

The limbic system, our primitive brain centre involved in emotions, is highly active when people are recognizing facial expressions and emotions of others. This capacity relates to our system of non-verbal communication, or body language.

The **pectoralis** major and minor muscles are the main chest muscles that allow pushing movements, such as push-ups, bench press, and anything involved in moving the arms forward. During pushing movements, other muscles contribute to the movement, including the triceps and anterior deltoids.

The abdominal muscles are considered **core muscles**, and their main role is to *stabilize* the body during movements involving multiple muscles, such as running, lifting, squatting, jumping, pushing, and pulling. The abdominals also help us bend forward, such as during sit-ups or raising legs up. The erector spinae muscles, the deep muscles of the back, are also core muscles that help us maintain posture and stabilize movements involving twisting or bending.

The **diaphragm** is the primary muscle involved in breathing. It moves downward in the thoracic cavity during contraction, causing an increase in thoracic volume, and this causes a brief decrease in pressure, which pulls air into the lungs, quickly equalizing the pressure (Chapter 19). During exhalation, the diaphragm

Figure 14.3   Major Superficial Muscles of the Body: Anterior View

Occipitofrontalis

Orbicularis oculi

Masseter

Sternocleidomastoid

Trapezius

Pectoralis major

Latissimus dorsi

Rectus abdominis

External oblique

Tensor fasciae latae

Psoas major

Pectineus

Vastus lateralis

Rectus femoris

Vastus medialis

Gastrocnemius

Soleus

Temporalis

Orbicularis oris

Platysma

Deltoid

Serratus anterior

Biceps brachii

Brachialis

Triceps brachii

Brachioradialis

Flexor carpi radialis

Sartorius

Adductor magnus

Gracilis

Tibialis anterior

Fibularis longus

14

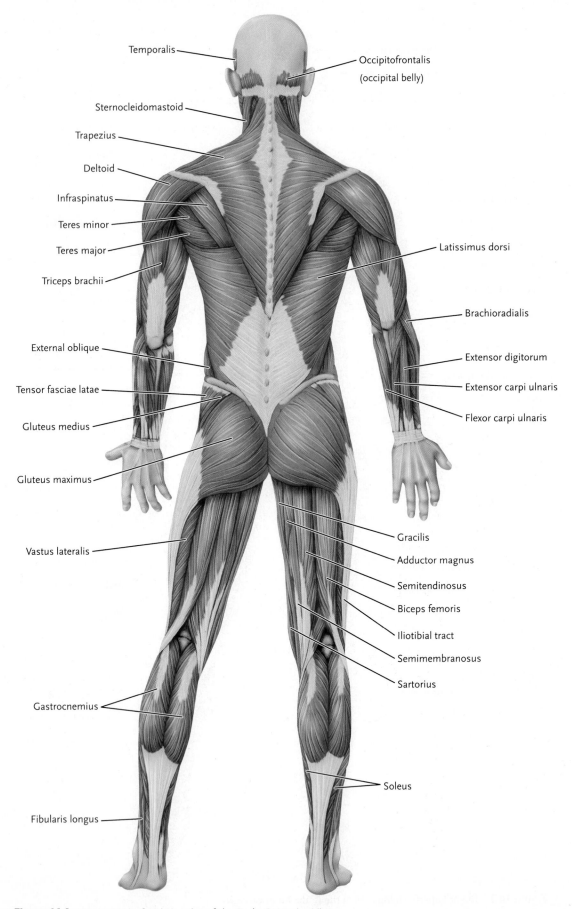

Temporalis

Occipitofrontalis
(occipital belly)

Sternocleidomastoid

Trapezius

Deltoid

Infraspinatus

Teres minor

Teres major

Triceps brachii

Latissimus dorsi

Brachioradialis

External oblique

Extensor digitorum

Tensor fasciae latae

Extensor carpi ulnaris

Gluteus medius

Flexor carpi ulnaris

Gluteus maximus

Gracilis

Adductor magnus

Vastus lateralis

Semitendinosus

Biceps femoris

Iliotibial tract

Semimembranosus

Sartorius

Gastrocnemius

Soleus

Fibularis longus

Figure 14.4 Major Superficial Muscles of the Body: Posterior View

TABLE 14.3

Muscles of the Head and Neck

Muscle	Origin	Insertion	Action
Occipitofrontalis Frontal belly	Epicranial aponeurosis	Skin superior to orbit	Raises eyebrows
Occipitofrontalis Occipital belly	Occipital and temporal bone	Epicranial aponeurosis	Draws scalp backward
Orbicularis oculi	Wall of orbit	Eyelid	Closes eye
Orbicularis oris	Muscles around mouth	Skin around mouth	Protrudes lips (as in kissing)
Zygomaticus major	Zygomatic bone	Skin at corners of mouth	Smiling
Platysma	Fascia over deltoid	Mandible, skin of lower face	Depresses mandible
Masseter	Zygomatic arch	Mandible	Elevates mandible (chewing)
Temporalis	Temporal bone	Mandible	Elevates and retracts mandible (chewing)
Sternocleidomastoid	Manubrium and clavicle	Occipital bone and mastoid process	Flexes and rotates head
Trapezius	Occipital bone and thoracic vertebrae	Clavicle and scapula (acromion anteriorly, and spine posteriorly)	Extends head and elevates scapula (shrugging) and pulls arms back, such as a seated row movement

14

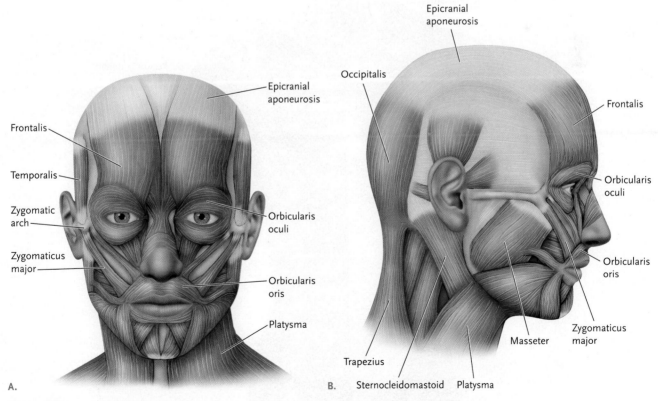

Figure 14.5   Head and Neck

(A) Frontal view and (B) lateral view

TABLE 14.4

Muscles of Anterior Trunk

Muscle	Origin	Insertion	Action
Pectoralis Major	Medial clavicle and sternum	Humerus	Adduction of the humerus, such as push-ups or moving straight arm toward midline; medial rotation of shoulder, such as arm wrestling
Pectoralis Minor	Ribs 3–5	Coracoid process of the scapula	Elevates ribs during forced inspiration, and pulls scapula forward and down
Rectus Abdominus	Pubic symphysis	Cartilage of ribs 5–7 and xiphoid process	Flexes vertebral column, aids in defecation, urination, forced expiration, and childbirth
External Oblique	Ribs 5–12	Linea alba	Flexion and rotation of vertebral column
Internal Oblique	Ilium	Cartilage of ribs 9–12	Flexion and rotation of vertebral column
Transverse Abdominus	Ilium, lumbar fascia, and ribs 6–12	Xiphoid process, linea alba, and pubis	Compresses abdomen and stabilizes trunk

14

Figure 14.6 Muscles of the Trunk

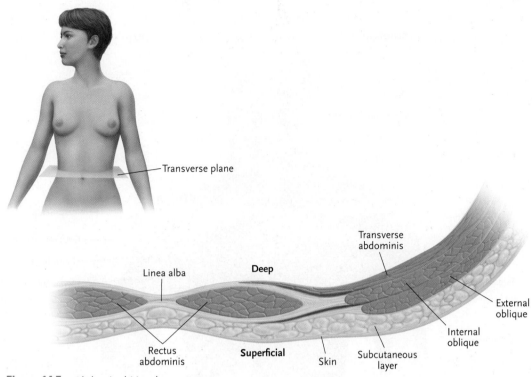

Figure 14.7   Abdominal Muscles

Transverse section

TABLE 14.5

Muscles Used for Breathing

Muscle	Origin	Insertion	Action
External intercostals	Rib above	Rib below	Increases thoracic volume during normal inspiration, involved in relaxation during normal expiration
Internal intercostals	Rib below	Rib above	Depresses ribs (forced expiration)
Diaphragm	Xiphoid process, cartilage of ribs 6–12, and lumbar vertebrae	Central tendon	Causes flattening during contraction, increases thoracic volume during normal inspiration

relaxes, moves upward, and lung volume decreases, which increases the pressure and causes the air to move out of the lungs. The synergistic **external intercostal muscles** help expand the rib cage during inspiration.

During exercise, more oxygen is required for muscle cells to produce ATP and more carbon dioxide is produced; therefore, breathing rate increases to match demand, and additional muscles play a role in breathing. During forced inspiration, the sternocleidomastoid contracts to help elevate the sternum and increase thoracic cavity volume. During a severe asthma attack, it is also possible to contract pectoralis muscles as well as many other deep muscles of the trunk. During forced expiration, the **internal intercostal muscles** and **abdominal muscles** contract to help push air out of the lungs.

The **rotator cuff** muscles include the supraspinatus, infraspinatus, teres minor, and subscapularis, which function to stabilize the shoulder joint and to produce movement. The shoulder joint is the most movable synovial joint in the whole body.

The **deltoid** muscle has an anterior, lateral, and posterior head that allows for different movements. Note that the insertion of the deltoid is the deltoid tuberosity, which is midway along the humerus, giving the deltoid muscles good leverage for movement; this makes the deltoids very strong muscles. Moving the arm up and forward (forward flexion) uses the anterior deltoid; moving the arm up sideways (lateral flexion) uses the lateral deltoid; and moving the humerus backward, as in any kind of pulling motion, uses the posterior deltoid. Pulling movements also use many other muscles, including the lower trapezius, teres major, rhomboids (deep to trapezius), as well as accessory muscles such as biceps and muscles of the forearm (Table 14.6).

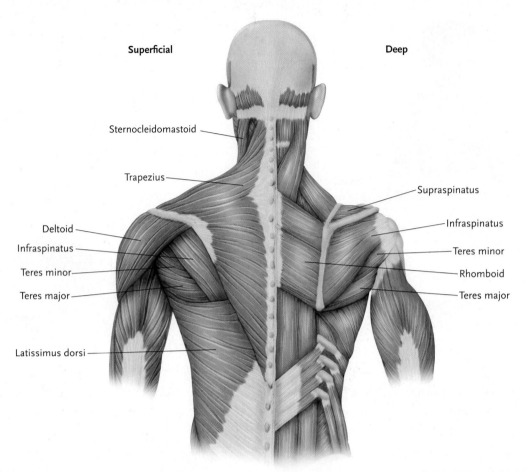

Superficial   Deep

Sternocleidomastoid

Trapezius

Deltoid
Infraspinatus
Teres minor
Teres major

Latissimus dorsi

Supraspinatus

Infraspinatus

Teres minor

Rhomboid

Teres major

Figure 14.8   Posterior Trunk Muscles

TABLE 14.6

Muscles of Posterior Trunk and Shoulder

Muscle	Origin	Insertion	Action
Supraspinatus	Supraspinous fossa of scapula	Superior humerus	Abduction of arm at shoulder
Infraspinatus	Infraspinous fossa of scapula	Superior, lateral humerus	Main external rotator
Subscapularis	Anterior scapula at medial border	Medial head of humerus	Medially rotates the humerus
Teres minor	Dorsal, lateral scapula	Lateral humerus	Mainly functions to hold humerus in the glenoid cavity
Teres major	Lower than the teres minor on the dorsal, lateral scapula	Lower than teres minor on anterior, lateral humerus	Extension and adduction of arm, such as chin-up, as well as medial rotation
Latissimus dorsi	Spines of ribs T7–L5, sacrum, ilium	Upper medial humerus	Extension and adduction of arm, such as wide grip, overhand pull-up, as well as internal rotation of the humerus
Deltoid	Clavicle and scapula (spine)	Humerus (deltoid tuberosity)	Abduction, lateral and anterior flexion and extension, rotation of arm, and circumduction

The **latissimus dorsi**, the largest back muscle, connects to many ribs as well as the pelvis. Note that it inserts lower and more medial on the humerus than the teres major. The teres major and latissimus dorsi are synergistic and are both involved in moving the humerus down from overhead, whether the arm is forward or out to the side.

Deep Muscles of the Back

The **erector spinae** are bundles of muscles that extend the length of the vertebral column. They are categorized into three main groups: spinalis, longissimus, and iliocostalis. The **spinalis** muscles originate and insert on the vertebral spinous processes. **Longissimus**

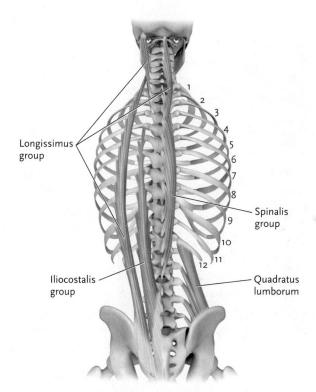

Longissimus group

Spinalis group

Iliocostalis group

Quadratus lumborum

Figure 14.9 Deep Muscles of the Back

muscles attach to the sacrum and the vertebral transverse processes. The **iliocostalis** muscles attach primarily to ribs (Figure 14.9). The function of the erector spinae muscles is primarily to maintain body postures and the extension of the vertebral column. We use our erector spinae muscles to lift objects up off the floor, but because these are small muscle bundles attached with small tendon insertions, they are easily injured with sudden heavy lifting. These muscles are part of the core muscles used for stabilization and also used synergistically with other large muscles during lifting movements: such as the gluteus maximus, hamstrings, and upper back muscles. It is important when lifting a heavy object to include these large muscles; this is done simply by bending the knees and using the legs as well as the back muscles. It is also important to keep erector spinae muscles strong with exercise (Section 14.5).

DID YOU KNOW?

Lower back pain affects 80% of the population at some point in a person's life. Most frequently, lower back pain is due to muscle sprains, strains, tears, or weakness of the erector spinae muscles; these conditions are completely preventable with strength exercises and proper technique when lifting.

TABLE 14.7

Arm Muscles

Muscle	Origin	Insertion	Action
Biceps brachii	Long head— superior coracoid process Short head— anterior coracoid process	Radius	Flexion and supination of the forearm, flexion of arm at shoulder
Brachialis	Lower humerus, near the deltoid tuberosity	Ulna	Flexion of forearm
Triceps brachii	Lateral posterior scapula	Posterior ulna	Extension of forearm

The **biceps** and **brachialis** muscles work together to flex the forearm; however, only the biceps supinates the forearm, because it inserts on the radius (Table 14.7). Both the biceps and brachialis are involved in pulling movements that include bending the elbow, such as chin-ups. The **triceps** are the agonist for extension of the arms and the antagonist to flexion. The forearm muscles cross the wrist joint, and long tendons insert onto the bones of the fingers. In the anatomical position, the anterior forearm muscles are the **flexors**, and the posterior forearm muscles are the **extensors** (Figure 14.10).

The **hip flexors (iliopsoas)** are the **psoas major** and **iliacus** muscles. Both of these function to flex and laterally rotate the thigh: for example, when pulling your leg toward your body or your body toward your legs, as in doing leg raises or sit-ups (Table 14.8).

DID YOU KNOW?

Shortened hip flexors (iliopsoas) often result from sitting or riding a bicycle for extended periods. This causes a common form of lower back pain during sleep when lying prone. This occurs because the extended leg elongates the hip flexors, causing them to pull on the lumbar vertebrae—their point of origin—thereby moving the pelvis into an anterior tilt. The pulling action on the pelvis creates pain, which is relieved when the knees are raised and the feet are brought toward the buttocks, that is, as the hip flexors are shortened. It is important to stretch the hip flexors if you sit or cycle a lot. You can stretch hip flexors by bending the knee and pulling your foot toward your buttocks.

The quadriceps are a group of four muscles. The largest is the **rectus femoris** (Figure 14.11), which is anterior to the **vastus intermedius**. The **vastus medialis**

14

14

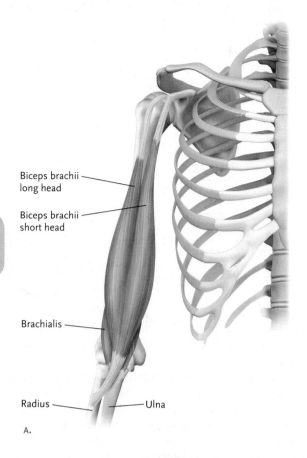

Biceps brachii
long head

Biceps brachii
short head

Brachialis

Radius

Ulna

A.

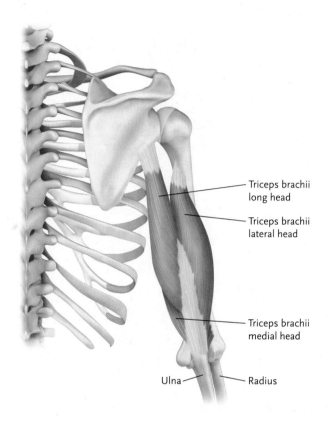

Triceps brachii
long head

Triceps brachii
lateral head

Triceps brachii
medial head

Ulna

Radius

B.

Flexors

C.

Extensors

D.

Figure 14.10   Muscles of the Arm

(A) Anterior arm, (B) posterior arm, (C) flexors, and (D) extensors

TABLE 14.8

Anterior Thigh Muscles

Muscle	Origin	Insertion	Action
Psoas major	Lumbar vertebrae	Femur	Hip flexion and lateral rotation
Iliacus	Ilium	Femur	Hip flexion and lateral rotation
Rectus femoris	Anterior iliac spine	Tibia (tibial tuberosity) via the patellar ligament	Extends the knee and flexes the hip
Vastus lateralis	Proximal femur	Tibia (tibial tuberosity) via the patellar ligament	Extends knee
Vastus medialis	Proximal femur	Tibia (tibial tuberosity) via the patellar ligament	Extends knee and adducts thigh
Vastus intermedius	Anterior lateral femur	Tibia (tibial tuberosity) via the patellar ligament	Extends knee
Sartorius	Ilium	Tibia	Flexes hip, abducts and laterally rotates thigh
Adductor longus	Pubis	Femur	Adducts and laterally rotates femur

Psoas major
Iliacus } Hip flexors

Sacrum
Inguinal ligament
Iliotibial band
Sartorius
Adductor longus
Rectus femoris (cut)
Vastus lateralis } Quadriceps
Vastus intermedius
Vastus medialis

Rectus femoris tendon
Patella
Patellar ligament

Figure 14.11 Anterior Leg Muscles

patella, and the patellar ligament attaches to the tibial tuberosity of the tibia. Since the quadriceps are all inserting on the tibia and crossing the knee, the primary movement is extension of the knee, like kicking a ball. The quadriceps are involved in every leg movement, including walking, running, jumping, squatting, kicking, as well as full-body pushing or pulling. For example, during a deep knee bend or squat, as you slowly sink down, the quads are contracting eccentrically because the knee is bending and the quads are lengthening. When you stand up, the most difficult part of the movement, the quads are contracting concentrically, with the hamstrings and gluteal muscles acting synergistically. If you were to do a squat with weight, you could lift more weight on the eccentric contraction—the negative part of the movement—than on the concentric contraction, standing up.

The **sartorius** muscle is important for the extension of the knee and lateral rotation of the leg. For example, when you are going to cross your legs, you lift your leg up, then move the knee out and the foot in. Several muscles, such as the **adductor longus**, help to adduct the leg in toward the midline.

is the muscle on the medial distal thigh near the knee, and the **vastus lateralis** is the large, long, lateral muscle of the thigh. The quadriceps all meet at the quadriceps tendon just above the knee; that tendon covers the

DID YOU KNOW?

Pushing a sled is considered one of the best full-body exercises for building strength and cardiovascular fitness. The muscles used include the forearm flexors; triceps; pectoralis major and minor; deltoids and rotator cuff muscles; lats, as stabilizers if arms are straight and shoulders are down; other upper back muscles, also as stabilizers; abdominals and erector spinae, as core stabilizers; hip flexors; and all leg muscles.

The **gluteus maximus** (also called *glutes*) is the largest and strongest muscle in the body and is involved in all lower body movements. The gluteus maximus causes the legs to extend and hyperextend (Figure 14.12 and Table 14.9). The **hamstrings** primarily function to flex the knee, but since they attach to the pelvis, they also play a role in extending and hyperextending the thigh at the hip joint. During walking or running, as we push off the ground and the leg moves backward, the gluteus maximus (agonist) and the hamstrings (synergists) contract

to push the leg back. Bending the knee of the back leg primarily uses the hamstrings. The forward leg uses hip flexors to pull the knee up and the quadriceps when landing on the front leg and pushing off; the movement quickly switches to the glutes and hamstrings as that leg moves back.

DID YOU KNOW?

You do not have to use as much posterior muscle strength to run on a treadmill compared to running on a road or track. Since the treadmill is moving, it pulls your leg back; however, the quadriceps have to work as much as usual to pull the leg forward and extend the leg. So the strength of the anterior and posterior muscles can become imbalanced in people who run only on treadmills. For those who run on a treadmill during the winter it is also common for them to pull a hamstring in the spring when they start running outside again; this occurs because the hamstrings have lost some strength compared to the quadriceps.

The lower leg muscles are used mainly to move the foot. The two largest posterior muscles are the soleus and the gastrocnemius. The **soleus** is deep to the flatter gastrocnemius and is the main muscle used to plantar flex the foot, as in standing on your toes, or pointing your toes. The origin for the **gastrocnemius** is on the distal femur, so it is also involved in flexing the knee with the hamstrings. When the knee is bent, the gastrocnemius has less tension and is less involved in plantar flexion. A seated calf-raise exercise targets the soleus, whereas a standing calf-raise targets the gastrocnemius (Table 14.10).

The main muscle in the anterior leg that dorsiflexes the foot, as in lifting your toes when you walk, is the tibialis anterior. This muscle can cramp and cause pain and inflammation when someone first begins jogging or running; this condition is called *shin splints* (Figure 14.13).

Figure labels:
- Gluteus medius
- Gluteus maximus
- Iliotibial band
- Semitendinosus
- Semimembranosus
- Long head of biceps femoris
- Short head of biceps femoris
- Hamstrings

Figure 14.12 Muscles of the Posterior Thigh

TABLE 14.9

Posterior Thigh Muscles

Muscle	Origin	Insertion	Action
Gluteus maximus	Posterior ilium, sacrum, and coccyx	Iliotibial tract and femur	Extends, hyperextends, and laterally rotates thigh
Gluteus medius	Ilium	Femur	Abducts and medially rotates thigh
Semitendinosus	Ischium	Medial tibia	Flexes knee and extends thigh
Biceps femoris	Ischium	Lateral tibia	Flexes knee and extends thigh
Semimembranosus	Ischium	Medial tibia	Flexes knee and extends thigh

TABLE 14.10

Lower Leg Muscles

Muscle	Origin	Insertion	Action
Soleus	Proximal tibia and fibula	Calcaneus via calcaneal (Achilles) tendon	Plantar flexion of foot
Gastrocnemius	Distal femur	Calcaneus via calcaneal (Achilles) tendon	Plantar flexion of foot and flexing of knee
Tibialis anterior	Proximal tibia	First metatarsal	Dorsiflexion of foot

Figure 14.13   Lower Leg

CONCEPT REVIEW 14.3

1. Name the insertion, origin, and action of each major muscle.

2. Describe how the muscles of the thigh and leg work together during walking or running.

14.4 MUSCLE CONTRACTION AND THE NEUROMUSCULAR JUNCTION

Muscle Cell Structure

Skeletal muscles cells (fibres) are surrounded by a covering of connective tissue called the *endomysium*. The cell membrane, called the **sarcolemma**, folds into the cell, forming the transverse tubules (T-tubules), so that action potentials from the nervous system can penetrate the interior of the cells to cause contraction. Each muscle cell contains bundles of filaments called *myofibrils* that contain thick and thin filaments that contract (Figure 14.14). Myofibrils are surrounded by mitochondria that produce ATP, the T-tubules, and the sarcoplasmic reticulum, which contains calcium ions required for contraction.

A single sarcomere, the smallest contractile unit, is composed of thick and thin filaments (Figure 14.15). The thick filaments consist of myosin proteins that look like bundles of golf clubs wrapped together. The thin filaments consist of actin, which looks like two strings of beads twisted together. Myosin contains **contractile proteins**. The thin filaments also include the regulatory proteins, troponin and tropomyosin. The protein of each myosin head (like the head of a golf club) has two binding sites: one binds to the actin proteins of the thin filament, and the other binds an ATP molecule. Energy in the form of ATP is required for every myosin head protein to bind to actin to form a cross-bridge. The thick and thin filaments overlap each other in a very organized pattern, forming light and dark bands that are visible under a microscope; these bands are called **striations**.

Sarcomere

The banding pattern of each sarcomere has specific labels and contains the specific components of thick and thin filaments (Figure 14.16). The ends of the sarcomere are the **Z discs**, which are structural proteins that hold the thick and thin filaments together and parallel

14

Figure 14.14 Skeletal Muscle Cells

Skeletal muscle cells are composed of bundles of myofibrils that are made up of thick and thin filaments that contract.

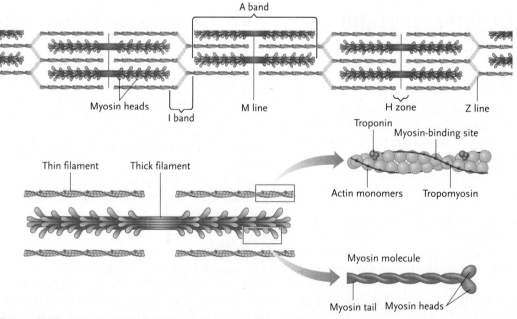

Figure 14.15 Thick and Thin Filaments

Thick and thin filaments are composed of actin and myosin.

Figure 14.16 Sarcomere

A sarcomere is the smallest contractile unit.

to each other. In the very centre of the sarcomere is another structural protein that forms the **M line**. The central region that contains only myosin is called the **H zone**. The entire region of the myosin proteins, including the overlapping thin filaments, is the **A band**. Between the A band and the edge of the A band of the next sarcomere—containing only thin filaments and the Z disc—is the **I band**.

Sliding Filament Theory

When the muscles contract, the myosin heads bind to actin, then pivot and pull the actin toward the M line. Each myosin head pivots toward the M line, pulling the actin of the thin filament toward the M line so that the thin filaments slide past the thick filaments; this is called the **sliding filament theory**. As the muscle contracts and the actin slides across the myosin, there are changes in the sarcomere (see Figure 14.17). The A band does not change because the myosin proteins are not contracting or shortening; they are pulling the thin filaments inward. The I bands and the H zone decrease, and the Z discs move closer together.

Regulatory Proteins

The main regulatory proteins are troponin and tropomyosin (Figure 14.18). The myosin-binding sites on the actin proteins are blocked by tropomyosin when there is no stimulation by the nervous system. So skeletal muscle contraction can occur only during voluntary movement: that is, when we use our frontal lobe and motor cortex to move and contract certain skeletal muscles. When a neuron stimulates a muscle fibre, it causes calcium to be released from the sarcoplasmic reticulum. The calcium ions then bind to troponin, which is bound to tropomyosin, which causes the tropomyosin to roll off the myosin-binding sites on the actin (Figure 14.19). After the calcium is released, it stays in the cytoplasm only as long as there is stimulation by a neuron. Once the stimulation stops, the calcium is pumped back into the sarcoplasmic reticulum by an active transport membrane protein; therefore, ATP is required. As the

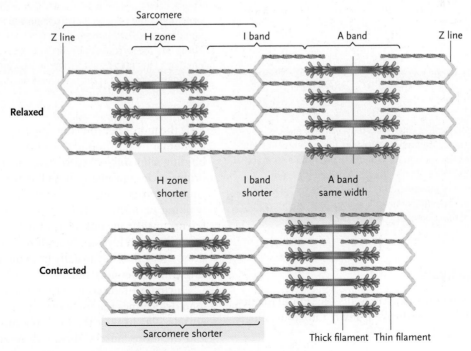

Figure 14.17 Changes in the Sarcomere during Contraction

During muscle contraction the H zone and I bands get smaller, but the A band does not.

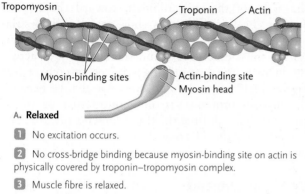

A. Relaxed

1 No excitation occurs.

2 No cross-bridge binding because myosin-binding site on actin is physically covered by troponin–tropomyosin complex.

3 Muscle fibre is relaxed.

B. Excited

1 Muscle fibre is excited and Ca²⁺ is released.

2 Released Ca²⁺ binds with troponin, pulling troponin–tropomyosin complex aside to expose cross-bridge binding site.

3 Myosin binds to the myosin-binding site on the actin.

4 Binding of actin and myosin cross-bridge triggers power stroke that pulls thin filament inward during contraction.

Figure 14.18 Role of Regulatory Proteins

Regulatory proteins bind calcium and allow contraction to occur only after an action potential has occurred.

cytoplasmic calcium level drops during relaxation, the tropomyosin moves back to cover the myosin-binding sites on actin.

DID YOU KNOW?

A decrease in magnesium (Mg²⁺) is often a cause of muscle cramps or twitching. Mg²⁺ helps to balance the movement of calcium ions across the membrane in muscle cells. After a contraction, Mg²⁺ enters the cell as Ca²⁺ leaves. A deficiency in Mg²⁺ prevents the normal movement of Ca²⁺ out of the cell, which leads to continued binding of myosin to actin, even in the absence of an action potential.

Cross-Bridge Cycle

How the myosin binds to actin, pulls the thin filaments, and is released can be broken down into four simple steps (Figure 14.19).

1. Myosin is unbound from the actin and contains an ATP molecule in the ATP-binding site on the myosin head. The ATP hydrolyzes, the ADP and phosphate stay bound, and the action of bond-breaking in the ATP energizes the myosin by changing the myosin head to tilt from its normal conformational shape.

2. The phosphate molecule leaves the myosin, and the energized myosin binds to the myosin-binding site on the actin proteins, as long as calcium ions are present to bind troponin and move the tropomyosin off the binding sites. The binding of myosin to actin forms the **cross-bridge**.

3. The ADP molecule leaves the myosin, the myosin head tilts back to its normal shape, and because it is bound to actin, the myosin moves the actin toward the M line; this is called the **power stroke**. After the power stroke, the myosin remains bound to the actin.

4. A new ATP molecule binds to the myosin head, which allows the myosin to be released from the actin. Then the cycle repeats, starting with ATP hydrolysis in step 1.

DID YOU KNOW?

When a person dies the body becomes completely rigid: a state called *rigor mortis*. This occurs because ATP is no longer produced after death, so the amount of ATP decreases until all the myosin heads remain bound to actin. Once the cells start to break down, calcium is released from the sarcoplasmic reticulum and floods the cells, binding troponin and allowing myosin to bind to actin. Furthermore, there is no ATP to pump the calcium back into the sarcoplasmic reticulum. It takes the body approximately 12 hours to be in full rigor mortis, and then gradually the stiffness decreases as the cells break down.

If the motor neuron continues to stimulate the muscle, calcium ions stay bound to troponin, leaving the myosin-binding sites unblocked by tropomyosin, and the cross-bridge cycle continues—if ATP is available—until muscle contraction is complete. You can visualize a millipede walking with different legs touching the ground at different times as the millipede moves forward. The process is similar with myosin. Not all myosin heads bind at the same time, and they usually bind multiple times during a single muscle contraction.

During a single contraction, such as a bicep curl, the **load** determines the **force** of contraction required by the muscle. The heavier the load, the more force is required and the more muscle fibres will be recruited, and more

Figure 14.19   Cross-Bridge Cycle

The cross-bridge cycle occurs when ATP hydrolyzes and causes myosin to bind to actin and move the actin toward the centre of the sarcomere.

myosin heads within each muscle fibre will be involved in the contraction. To lift a glass of water, not very many muscle fibres or actin-myosin connections are required, but to lift a heavy dumbbell many more muscle fibres and myosin heads need to be involved in the contraction.

During an eccentric contraction, such as lowering yourself slowly from a chin-up, the myosin heads bind to actin, but each time they are released they bind to an actin protein that is closer to the Z disc rather than to the M line; this is due to the sarcomeres being pulled into their lengthened state. It is easier to move the same weight using eccentric movements compared to concentric movements because the myosin head does not have to "pull" on the actin. It simply releases and binds to another site further from the M line.

Muscle Tone

When muscles are relaxed, there are always some myosin heads binding and releasing actin, so muscles are never 100% relaxed. This small amount of binding keeps muscles somewhat firm, and this is called **muscle tone**. Muscle tone is important for preventing our muscles from overstretching, and it helps maintain posture and balance when we are sitting or moving.

Neuromuscular Junction

Skeletal muscle contraction is voluntary, although many complex movements, such as walking or swimming, involve muscle contractions that we do not have to specifically think about; they become learned motor patterns that do not involve conscious thought. For the sake of simplicity, skeletal fibres are considered under voluntary control. Motor movements are initiated from the **motor cortex** of the brain, which connects to motor neurons in the spinal cord that allow action potentials to travel through spinal nerves to all skeletal muscle fibres (Chapter 15).

The neuromuscular junction (NMJ) is the **synapse**, which is the area where a motor neuron connects to a skeletal muscle fibre. A motor neuron may contact multiple muscle fibres (possibly up to 3000 fibres in a large motor unit) and multiple areas of the same muscle fibre because a neuron branches into **axon terminals**. Axon terminals contain **synaptic end bulbs**, which contain vesicles containing **neurotransmitters**. **Acetylcholine (Ach)** is the *only* neurotransmitter in motor neurons that leads to skeletal muscle contraction. The neurotransmitters cross the synaptic cleft—the space between the axon terminal and the sarcolemma—and bind to the specific receptors on the sarcolemma, thus stimulating an electrical signal in the muscle cell (Figure 14.20).

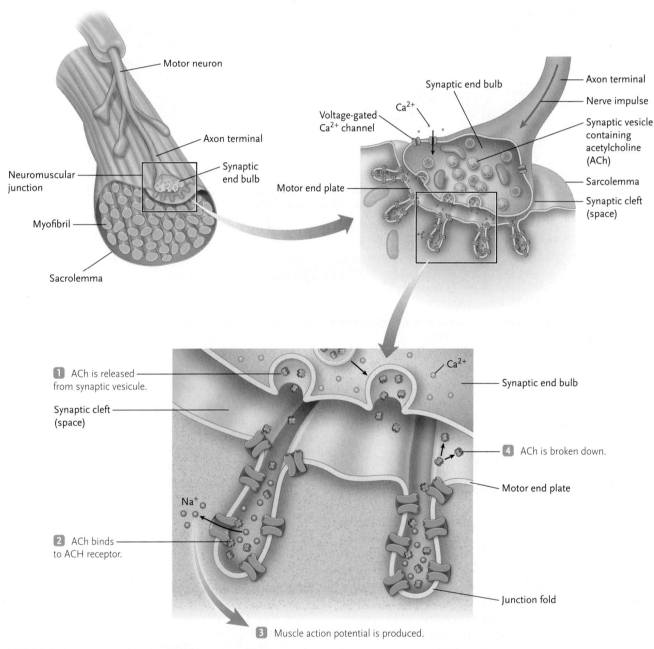

Figure 14.20 Neuromuscular Junction

The neuromuscular junction is the connection between a neuron and a muscle cell.

The electrical signal that travels down the motor neuron, from the cell body in the spinal cord, along the axon, to the axon terminals, is called an **action potential** (discussed in detail in Chapter 15). Simply, the action potential is the movement of sodium ions across the neuron membrane. As sodium ions cross the membrane in one section of the axon, they signal sodium ions to cross the next section of the axon, until sodium ions enter the axon at the axon terminal. At the axon terminal, the action potential (sodium ions moving into the cell) triggers the opening of **voltage-gated** calcium ion membrane channels. *Voltage-gated* means membrane channels open because of a change in the charge across the membrane, due to the Na+ entering the cell. When the Ca^{2+} channels open, Ca^{2+} can enter the synaptic bulb and trigger the release of ACh neurotransmitters. The vesicles containing the ACh fuse with the axon membrane, and the neurotransmitters are released by exocytosis into the synaptic cleft.

ACh then diffuses across the synaptic cleft and binds to specific receptors on the sarcolemma, causing **ligand-gated** sodium ion channels to open on the muscle membrane, which causes an action potential, just like in the neuron. Sodium ions enter the muscle cell sarcoplasm all along the T-tubules, which are located directly beside the sarcoplasmic reticulum. The action potential in the T-tubules causes the release of the calcium ions from the sarcoplasmic reticulum. The Ca^{2+} binds to troponin, moving the tropomyosin, and the cross-bridge cycle can begin. Back in the synaptic cleft, ACh is broken down by an enzyme called **acetylcholinesterase**. If the muscle needed to continue contraction, the motor neuron would release more Ach.

Steps of Skeletal Muscle Contraction from Nervous System Stimulation to Relaxation

1. Action potential travels down motor neuron to axon terminal (Figure 14.21).
2. Action potential triggers movement of Ca^{2+} into the axon terminal.
3. Ca^{2+} triggers release of ACh neurotransmitters into synaptic cleft.
4. ACh binds to specific receptors on the sarcolemma that trigger the opening of Na^+ channels.
5. Action potential occurs in muscle cell membrane as Na^+ crosses the membrane from the T-tubules and enters the sarcoplasm.
6. The action potential in the sarcolemma causes the sarcoplasmic reticulum to release Ca^{2+} into the sarcoplasm.
7. Ca^{2+} binds to troponin, which moves tropomyosin off the myosin-binding sites on the actin proteins.
8. ATP is bound to myosin in the relaxed state. Once an action potential occurs and the myosin-binding sites on the actin are available, ATP hydrolyzes, releasing the phosphate group, and myosin binds to actin (cross-bridge formation).
9. ADP leaves the myosin and causes myosin to tilt back into its normal position, which causes the actin to slide across the myosin (power stroke), and sarcomeres shorten.
10. A new ATP molecule binds to myosin, causing the release of myosin from actin. The cycle continues as long as a motor neuron stimulates the muscle cell; the muscle relaxes when there are no further action potentials and calcium has been pumped back into the sarcoplasmic reticulum.

DID YOU KNOW?

The choline used to make acetylcholine is acquired when we eat plant and animal cell membranes. Recall from Chapter 2 that the polar head of many phospholipids that make up cell membranes contains choline. The foods that have very high amounts of choline include seafood, eggs, chicken, beef, peanuts, flaxseeds, cauliflower, and spinach. Choline is also required to convert homocysteine to the amino acid methionine; a deficiency of choline leads to increased homocysteine levels, which are correlated with cardiovascular disease (Chapter 18, Section 12).

14

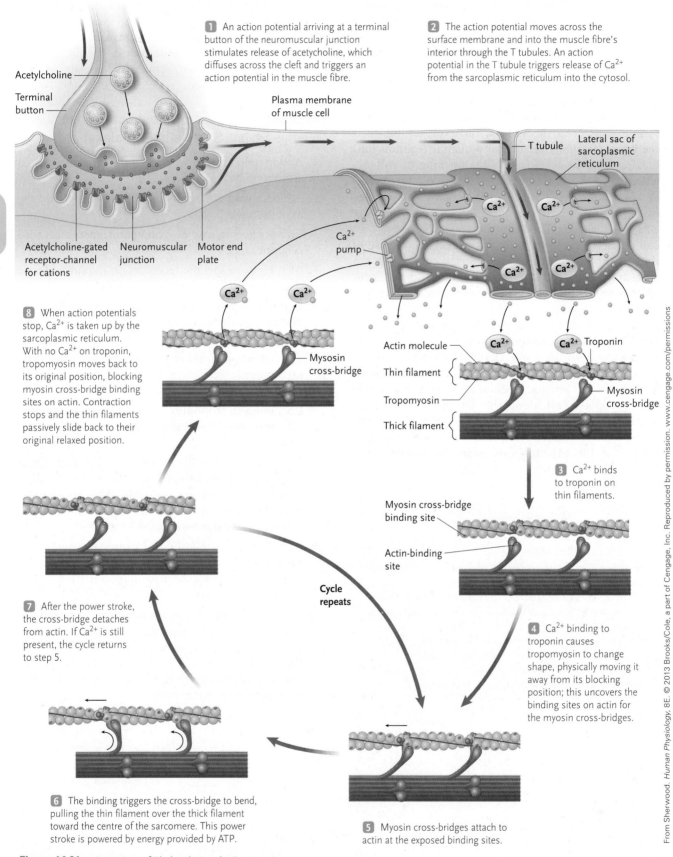

1 An action potential arriving at a terminal button of the neuromuscular junction stimulates release of acetycholine, which diffuses across the cleft and triggers an action potential in the muscle fibre.

2 The action potential moves across the surface membrane and into the muscle fibre's interior through the T tubules. An action potential in the T tubule triggers release of Ca^{2+} from the sarcoplasmic reticulum into the cytosol.

Acetylcholine

Terminal button

Plasma membrane of muscle cell

T tubule

Lateral sac of sarcoplasmic reticulum

Ca^{2+}

Ca^{2+}

Acetylcholine-gated receptor-channel for cations

Neuromuscular junction

Motor end plate

Ca^{2+} pump

Ca^{2+}

Ca^{2+}

Ca^{2+}

Ca^{2+}

8 When action potentials stop, Ca^{2+} is taken up by the sarcoplasmic reticulum. With no Ca^{2+} on troponin, tropomyosin moves back to its original position, blocking myosin cross-bridge binding sites on actin. Contraction stops and the thin filaments passively slide back to their original relaxed position.

Ca^{2+}

Ca^{2+}

Mysosin cross-bridge

Actin molecule

Ca^{2+}

Ca^{2+}

Troponin

Thin filament

Tropomyosin

Mysosin cross-bridge

Thick filament

3 Ca^{2+} binds to troponin on thin filaments.

Myosin cross-bridge binding site

Actin-binding site

Cycle repeats

7 After the power stroke, the cross-bridge detaches from actin. If Ca^{2+} is still present, the cycle returns to step 5.

4 Ca^{2+} binding to troponin causes tropomyosin to change shape, physically moving it away from its blocking position; this uncovers the binding sites on actin for the myosin cross-bridges.

6 The binding triggers the cross-bridge to bend, pulling the thin filament over the thick filament toward the centre of the sarcomere. This power stroke is powered by energy provided by ATP.

5 Myosin cross-bridges attach to actin at the exposed binding sites.

Figure 14.21 Overview of Skeletal Muscle Contraction

The process of skeletal muscle contraction, from neuron signalling to contraction

1. What are the two contractile proteins in muscle cells?

2. What are the two regulatory proteins? Which one binds calcium, and which one blocks the myosin-binding sites?

3. Describe the sliding filament theory.

4. Which regions of the sarcomere shorten during contraction?

5. Where is calcium stored in muscle cells?

6. What causes calcium to be released?

7. Name the four steps of the cross-bridge cycle.

8. Name two reasons why muscle cells use so much ATP.

9. What causes the number of muscle fibres involved in a contraction to increase?

10. Name three functions of muscle tone.

11. What ion causes action potentials in neurons? In muscle cells?

12. Describe how calcium plays a role in the neuron.

13. What neurotransmitter triggers the action potential in the muscle fibre?

14. What happens to the neurotransmitter after the action potential?

15. List the order of events that occur during muscle contraction, beginning with stimulation by the motor neuron.

14.5 TYPES OF SKELETAL MUSCLE FIBRES

There are three main types of skeletal muscle fibres: (1) slow oxidative, also called *slow-twitch fibres*; (2) fast oxidative-glycolytic, also called *intermediate fibres*; and (3) fast glycolytic, also called *fast-twitch fibres*.

Each fibre type has special characteristics that allow muscles to function in either endurance activities, intermediate-intensity activities, or high-intensity, short-duration activities. All the specific muscles discussed in Section 14.3 contain some ratio of all three fibre types; some muscles contain more of the slow oxidative, and others contain more of the fast glycolytic. The different ratios depend on the muscle, genetics, and types of training.

Slow Oxidative Fibres

Slow-twitch fibres have the smallest diameter, so they contain fewer myofibrils per cell than other fibre types. They are adapted for long periods of continuous, low-intensity contractions and are the most resistant to fatigue. They are called *oxidative* because they primarily use aerobic respiration to produce ATP. Muscles with more slow-twitch fibres are a deep red colour because they contain many blood vessels that carry the required oxygen and glucose and also a lot of **myoglobin**, which is red like hemoglobin when carrying oxygen. Since these fibres use aerobic respiration, in comparison to other fibre types they have the most mitochondria. Slow-twitch fibres store little glycogen and do not store creatine phosphate (discussed in Section 14.6); therefore, they develop tension more slowly than the fast fibres. Consider chickens that walk around using their leg muscles for long periods of time: the legs have dark meat because leg muscles consist of the slow-twitch fibres (Figure 14.22a). Chicken legs—the dark meat—also tends to be juicier

than white meat because slow-twitch muscles store fat, which is an important fuel supply for ATP production.

Fast Glycolytic Fibres

Fast-twitch fibres have the largest diameter, contain the most myofibrils per cell, and are the strongest of all the fibre types. They are adapted for short bursts of high-intensity activity, such as sprinting or lifting weights, and they fatigue quickly. Fast-twitch fibres are adapted for rapid movements, with periods of rest in between bursts of activity. For example, when bike riding, you can sprint uphill for approximately 30 seconds, then coast or peddle more slowly to catch your breath, and then sprint again once your muscles have recovered and produced more ATP. Or, when weight lifting, you can do repetitions and sets so that the muscles recover between lifts. Fast-twitch fibres primarily use glycolysis for ATP production; therefore, they do not need as much oxygen. They have fewer blood vessels, no myoglobin, and appear white in colour. They store the creatine phosphate that is used in the initial bursts of activity, and they have a large glycogen supply. Picture a chicken again: the chicken breast primarily consists of fast glycolytic fibres because a chicken uses the breast muscle only to flap its wings in short bursts for very short periods of time. Exercising fast glycolytic fibres causes increased muscle mass because the muscle fibres build up more myofibrils in each cell. Having more myofibrils means more ATP is required to fuel those muscles, and therefore, they burn more calories (Figure 14.22c).

Fast Oxidative-Glycolytic Fibres

Intermediate fibres have qualities similar to both slow oxidative and fast glycolytic. They are adapted to use glycolysis as well as oxidative phosphorylation for ATP production, they are involved in moderate- to somewhat high-intensity activities, and they are moderately resistant to fatigue. We

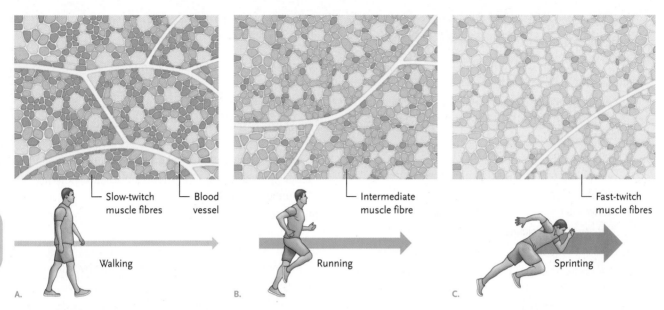

Figure 14.22 Muscle Fibre Types

(A) Slow-twitch fibres have the most myoglobin, blood vessels, and mitochondria; the least glycogen; and no creatine phosphate. (B) Intermediate fibres have some myoglobin, blood vessels, mitochondria, glycogen, and creatine phosphate. (C) Fast-twitch fibres have the least myoglobin, blood vessels, mitochondria; and the most glycogen and creatine phosphate.

use intermediate fibres (along with slow fibres) for most of our daily movement, such as carrying groceries, climbing stairs, jogging, biking, dancing, and swimming. These fibres have an intermediate diameter; contain many mitochondria, blood vessels, and myoglobin; and appear red in colour. Intermediate fibres can use both aerobic and anaerobic respiration for ATP production (Figure 14.22).

Muscle Types and Their Fibre Speed

Most skeletal muscles have combinations of all three types of fibres. Approximately 50% of our muscle fibres are slow oxidative, and the other 50% is a combination of fast glycolytic and fast oxidative-glycolytic. Different people have different proportions of each type of fibres based on their genetics and their primary types of activity. Many researchers now believe that fibres can be converted to slow- or fast-twitch fibres, either to increase blood vessel density and increase mitochondria or to increase fibre diameter and

utilize glycogen, but it is not known to what extent this can occur. By thinking about the types of sports or activities you like to engage in, you can probably determine whether your muscles are predominantly slow fibres or fast fibres. Endurance athletes have more slow-twitch fibres, and power-lifters and sprinters have more fast-twitch fibres.

Muscles that contain more slow oxidative fibres include postural muscles, erector spinae, abdominals, calves, forearm flexors and extensors, and deep upper back muscles. Also, the muscles of the face used for facial expressions and chewing contain more slow fibres. Muscles that tend to have a large number of fast glycolytic fibres include the deltoids, gluteus maximus, quadriceps (depending on the person and type of exercise), chest, latissimus dorsi, teres major, biceps, and triceps. The quadriceps and hamstrings have a lot of intermediate fibres and are adapted for walking long distances for hours, as well as for sprinting; so the exact proportions of muscle fibre types depend on the individual.

CONCEPT REVIEW 14.5

1. Make a chart comparing the main features of slow oxidative fibres, fast glycolytic fibres, and fast oxidative-glycolytic fibres.

2. From the following list of activities, determine which muscle fibre type would be primarily involved. Note some activities may include more than one fibre type.

   a. squatting 50 lb
   b. jogging 1 hr
   c. running 1500 m
   d. sitting in a chair with good posture
   e. typing
   f. eating
   g. downhill skiing
   h. walking up a flight of stairs
   i. running long jump

14.6 MUSCLE METABOLISM AND EXERCISE

Recall from Chapter 4 that ATP can be produced by aerobic respiration, with oxygen, or by anaerobic respiration, without oxygen. Muscles primarily use glucose as an initial fuel source, but they can also use lactic acid, ketones created from breaking down fat (Chapter 20), fatty acids, or amino acids. Different muscle fibre types use different fuel sources.

Slow-twitch fibres contain more myoglobin—which binds oxygen—and mitochondria for aerobic respiration, and can use glucose, fat, and protein to produce ATP. Aerobic activities include rhythmic movements such as walking, hiking, jogging, swimming, cross-country skiing, cycling, rowing, and kayaking. These exercises primarily use slow-twitch fibres; so breathing should be easy, and the activity should be sustainable for one or more hours. As intensity increases, breathing rate increases, and more intermediate fibres begin to be used: for example, when jogging uphill. When intensity increases significantly, such as running up a steep hill, breathing becomes quite difficult, and muscles begin to have a burning sensation; at this point fast-twitch fibres and anaerobic respiration kicks in. It is important to know that fitness level—in terms of how much intensity of exercise it takes to put someone into anaerobic respiration—makes a big difference. Jogging may be an aerobic activity for people with a good level of fitness and may be an anaerobic activity for others who are less fit.

Fast-twitch fibres are adapted to anaerobic respiration and break down glycogen into glucose to produce ATP, with the by-product of lactic acid. Fast-twitch fibres can also utilize creatine phosphate for energy in the initial part of each burst of activity. Anaerobic exercises include weight lifting, sprinting, soccer, and hockey, where bursts of high-intensity exercise are required, and also include high-intensity interval training, such as running or cycling uphill for short bursts. The higher the intensity, the quicker the muscles will fatigue, and rest is required before another burst is possible.

Methods of ATP Production

1. Creatine phosphate. Creatine is made from amino acids in the liver and is transported to the muscles. In the muscle cells, creatine is phosphorylated by an enzyme called **creatine kinase**. The phosphorylated creatine, which is stored in small amounts in the muscle, is used initially as a rapid supply of energy. Creatine phosphate can also transfer its phosphate group to ADP to form ATP, and this gives the cell enough energy for intense activity for approximately two to seven seconds, depending on the intensity of the activity (Figure 14.23). This brief period is enough time for the muscle cells to begin generating more ATP through aerobic or anaerobic mechanisms. When the intense exercise is complete, any excess ATP that was produced can be used to re-form

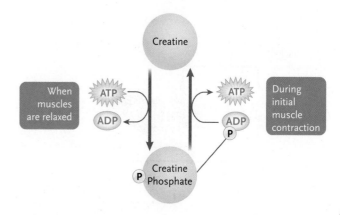

Figure 14.23 Creatine Phosphate

Creatine phosphate is the first energy used by muscle cells at the beginning of exercise.

phosphorylated creatine, which can be used during the next bout of exercise. This is a protective mechanism for muscle cells; it ensures there is always a supply of energy, without a lag period, while ATP is being produced through glycolysis or oxidative phosphorylation (Chapter 4). Creatine phosphate can be taken as a supplement by athletes, but it only benefits athletes in high-intensity sports such as sprinting.

2. Aerobic respiration. This method of ATP production includes glycolysis, pyruvate oxidation, Krebs cycle, and the electron transport chain (Chapter 4). Glucose is the starting fuel that undergoes glycolysis to produce pyruvate, which is then oxidized to produce the acetyl-CoA that enters the Krebs cycle. During all of these processes the cell accumulates important molecules of NADH that are used by the electron transport chain to produce 36 ATP molecules from every glucose that starts glycolysis. This process—called *oxidative phosphorylation*—requires oxygen and is the primary mechanism of ATP production in slow-twitch muscle fibres (Figure 14.24).

3. Glycogenolysis. This process breaks down glycogen to produce glucose. Fast-twitch muscle fibres store glycogen and can also break it down for a quick supply of energy for ATP production. The liver will also break down glycogen to maintain blood sugar levels during aerobic exercise so that glucose is always available for the slow-twitch fibres.

4. Lipolysis. This is one way that fat is broken down. The fat stored in slow-twitch muscle fibres is broken down by beta-oxidation (Chapter 4), producing two-carbon units that can be converted into acetyl-CoA and can enter the Krebs cycle. This method is highly used by slow-twitch fibres during aerobic exercise and one reason why slow-twitch fibres can sustain activity for a long period of time; recall from Chapter 4 that fatty acids can produce much higher amounts of ATP than a glucose molecule. Lipolysis also occurs after anaerobic exercise, so we do burn fat indirectly with high-intensity exercise.

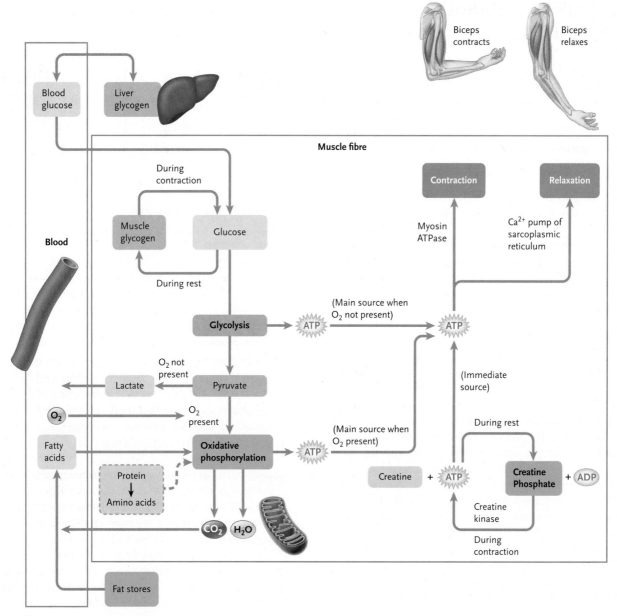

Figure 14.24 Aerobic and Anaerobic Respiration

Aerobic respiration utilizes oxygen, and anaerobic respiration involves the production of ATP in the absence of oxygen.

5. Gluconeogenesis. The is a second way that fat is broken down, specifically by adipose tissue and the liver. The liver converts fat into glucose, and this is important for maintaining blood sugar levels after glycogen has been mostly depleted, which can occur within approximately 20 to 30 minutes of aerobic exercise. Gluconeogenesis also occurs after anaerobic exercise, because the liver replenishes the used-up glycogen.

6. Lactic acid production. Lactic acid is produced during anaerobic metabolism, when oxygen is low and pyruvate from glycolysis is converted into lactic acid. This process is primarily used by fast-twitch fibres and can be produced within 10 to 20 seconds of very high-intensity exercise. The liver can convert lactic acid back into glucose, which can then be used to make more energy. Lactic acid can also be converted back

into pyruvate—which can be used by the cell for aerobic respiration—once oxygen is available, blood flow increases, or the intensity of the exercise decreases.

7. Ketone production. The liver produces ketones when it breaks down fat during periods of intense gluconeogenesis; this occurs in conditions such as a low-carbohydrate diet, periods of fasting, or during prolonged aerobic exercise. Ketones can be used to produce ATP because they can be converted by slow-twitch muscles cells into acetyl-CoA that can directly enter the Krebs cycle. Note that ketones are not produced during beta-oxidation of fatty acids, only through gluconeogenesis in the liver. Ketones can also cross the blood-brain barrier and can be used by the brain for ATP production.

8. Protein metabolism. Protein is the last resort for fuel for muscle cells to produce ATP. Muscle

protein can be broken down into amino acids during very prolonged exercise or during prolonged fasting. Amino acids can enter the oxidative pathway at various points, such as being converted to pyruvate or entering the Krebs cycle (Chapter 4).

DID YOU KNOW?

Lactic acid is produced in fast-twitch muscle fibres during very-high-intensity exercise, such as 20-to-30-second bursts as occurs in sprinting. That lactic acid affects the pH of the muscle, and a buildup causes the muscle to fatigue. The low pH can slow the recovery of those muscle fibres after a workout, unless the lactic acid is removed. How do we remove lactic acid from our muscles more efficiently? By using active recovery, meaning to continue to move around or walk for 15 minutes after a high-intensity workout. The continued blood flow through the muscles during active recovery helps to circulate the lactic acid out of the muscles to the liver, where it can be broken down.

Exercise and Weight Loss

The key to losing weight is consuming fewer calories and burning more by making more ATP. Producing ATP, whether through aerobic or anaerobic means is how we burn calories (Chapter 2). All cells use ATP. Just to fuel the sodium-potassium pumps in all cells of the body we use close to one-third of our daily required calories. We also use ATP to think, to digest food, to breathe, to fight infections, to pump blood, and to move around using our skeletal muscles. The number of calories we need to consume to maintain our cellular functions without gaining or losing weight is called our **basal metabolic rate (BMR)**. If we want to lose weight, stored as fat in our subcutaneous adipose tissue and in the abdominal cavity surrounding our organs, then we need to break down the stored fat and use it to make ATP; this can be accomplished by eating fewer calories or by increasing the need for ATP through exercise. If we consume more calories than required to maintain our BMR any excess nutrients will be stored as fat.

Since muscle cells use by far the most ATP compared to any other tissue, and increasing the amount of muscle contraction requires more ATP, we need to exercise. All three types of muscle fibres use ATP and burn calories. Slow-twitch fibres burn calories at a slower rate than fast-twitch fibres, but you can use slow-twitch fibres for a much longer period of time. Using fast-twitch fibres during high-intensity, short-duration exercises means more recovery time is required, and muscle cells need to repair, but high intensity exercise burns a large number of calories. The repair process can last 24 hours or longer after a workout, and building new myofibrils requires a lot of ATP. It depends on which type of activities you prefer. If you enjoy the activity, you are more likely to stick with it. The most important thing is to be active all the time: include some high-intensity activities to build speed and strength, and also low-intensity activities, such as walking, carrying groceries, taking the stairs, playing golf or tennis, or swimming. Any kind of activity causes muscles to contract; this requires ATP, so the more active you are, the more you will increase your metabolic rate.

Factors Affecting Weight Loss

Is weight loss really just that simple: eat less and be more active? The simple answer is yes, but there are several factors that play a significant role.

- People who weigh more burn more calories per day because they have more cells that require ATP.
- People with more muscle mass burn more calories because muscles use the most ATP.
- Males generally burn more calories than females, primarily because they usually have more muscle mass.
- Younger people burn more calories than older people. A teenage male who is very active in sports may require 3500 Cal per day, whereas a small elderly woman who does not exercise may only require 1200 Cal per day.

There are also physiological factors that affect BMR. **Lipoprotein lipase** (LPL) is a group of enzymes that break down triglyceride fats into components that can be used by cells to make ATP. LPLs are expressed primarily in adipose tissue, small intestine, heart, and skeletal muscles. Increased LPL is correlated with increased use of fats for ATP production. The hormones **glucagon** and **epinephrine** stimulate the production of LPL during fasting (Chapter 17), and it is also increased in growing children and teens. LPL expression decreases in sedentary people, making it more difficult to use fat for energy. Research has shown that people who exercise for 30 minutes, three times per week (the most common advice), but spend most of the rest of their time sitting, will have decreased LPL levels and will not use stored fat for energy. To summarize, yes, weight loss can be as simple as eating less and being active. Eating less means having periods of time between meals when you are not eating, and this stimulates hormones that break down fat. Being active means being active *throughout* the day, not just three times each week.

Health Benefits of Exercise

Exercise, especially exercise that uses all three fibre types, has many health benefits, including the following:

- Exercise increases bone density and prevents osteoporosis.
- Exercise strengthens muscles, which decreases the risk of injury during activities such as shovelling snow.
- Exercises develops balance and coordination and therefore decreases the potential for falls and the breaking of bones.
- Exercise increases the production of good mood hormones, helps decrease stress, and possibly prevents or reverses mild depression.

- Exercise helps decrease inflammation in the body and therefore helps reduce the risk of cardiovascular disease.
- Since muscles use blood sugar for energy, exercise helps to stabilize blood sugar levels and reduces insulin resistance, thereby reducing the risk of type 2 diabetes.
- Exercise stimulates the production of growth hormone and DHEA, which increases muscle growth and also helps regenerate all cells in the body.

CONCEPT REVIEW 14.6

1. What is the difference between aerobic and anaerobic respiration?
2. How is creatine phosphate used as an energy source for muscle cells?
3. How much ATP is produced from anaerobic respiration?
4. How much ATP is produced from aerobic respiration?

5. Which method of ATP production would allow for long-duration activities, such as jogging 5–10 km?
6. What are the two factors involved in maintaining a healthy body weight?
7. Name some benefits of exercise other than weight loss.

EFFECTS OF AGING

As we age the primary factor affecting the muscles is loss of strength and muscle mass. This can be slowed by continuing to exercise using all muscle fibre types. Some of the most common effects of aging on muscles are as follows:

- Muscle fibre diameter in fast-twitch fibres decreases as we age.
- Recovery time after exercise increases. Recovery from muscle tissue damage that occurs during exercise can take longer to heal. Older people may need to space out the timing of their higher-intensity exercise.
- Slow-twitch fibres do not change very much if an older person continues to do aerobic activities.
- Factors that affect the aging endocrine system also affect muscle metabolism; for example, less glucagon production

affects the ability of the liver to break down glycogen. Generally, testosterone, estrogen, and growth hormone decrease as we age, and this affects muscle recovery. If muscles cannot recover and regenerate proteins, muscles will lose strength.
- Muscle strength in fast-twitch fibres decrease due to a decrease in the number of muscle fibres numbers and the size of muscle fibres.
- The number of motor neuron axon terminals decreases as we age, which affects the electrical stimulation of skeletal muscles; without neuronal stimulation, muscle contraction cannot occur. If muscles are not stimulated to contract, they deteriorate.
- The loss of muscle strength, balance, and coordination can lead to an increase in injuries from falls.

CHAPTER REVIEW

Medical Terminology

TERM	DEFINITION	EXAMPLE
-asthenia, ashten/o	weakness	myasthenia gravis
aux/o	growth, acceleration	auxotrophy
bi-	double, twice	biceps
-chalasia	relaxation	achalasia
dynam/o	power, strength	dynamogenesis
erg/o	work	ergometer
fasci/o	fascia	fascia
fibr/o	fibre, fibrous	fibroblast
flex/o	bend	flexor
kinesi/o, kinesia, -kinetic	movement	kinesthesia

TERM	DEFINITION	EXAMPLE
lei/o	smooth	leiomyoma
ligament/o	ligament	ligamentopexy
-lysis	dissolution, breakdown	hydrolysis
muscul/o	muscle	musculoskeletal
my/o, myos/o	muscle	myorrhexis
orbis-	circle	orbicularis
pale/o	old	paleokinetic
pyg/o	buttocks	pygalgia
rectus	straight	rectus femoris
rhabd/o	rod	rhabdoid
rhabdomy/o	striated/skeletal muscle	rhabdomyolysis
rot/o, rotat/o	turn, revolve	rotator
-stasis	standing still, standing	homeostasis
sthen/o, -sthenia	strength	sthenometry
-stroma	supporting tissue of an organ	myostroma
syndesm/o	ligament, connective tissue	syndesmectopia
ten/o	tendon	tenorrhaphy
tenont/o	tendon	tenontography
-therapy, therapeut/o	treatment	kinesiotherapy
ton/o	tone, tension	myatonia

14

Labelling Activity

Label the following diagrams.

1.

2.

3.

14

14

4.

5.

6.

7.

14

8.

14

9.

10.

11.

12.

13. a.

b.

14

14.

15. a.

b.

c.

Answer to Application Question

The following muscles are used for a squat:

- Quadriceps—eccentric contraction during downward movement, extension of the leg at the knee during standing
- Hamstrings—eccentric contraction during downward movement, extension of the hip during standing
- Glutes—eccentric contraction during downward movement, extension of the hip during standing
- Soleus and gastrocnemius—used to stabilize the legs and balance during downward and upward movement
- Abdominals and erector spinae muscles—core muscles used for any multi-muscle movement

involving bending; erector spinae muscles also involved in flexion of the lower back during standing

Exercise with weights is anaerobic exercise that is high intensity and lasts for a short duration; therefore, fast-twitch fibres are primarily used. Fast-twitch fibres primarily use creatine phosphate as the initial fuel source and then use the glucose in the bloodstream from the breakdown of glycogen. Yes, creatine would be a useful supplement to take for this type of exercise. In contract, endurance activities involve slow-twitch fibres that do not use creatine; they primarily break down fat for ATP production.

15 The Nervous System

CHAPTER OUTLINE

15.1 OVERVIEW OF THE NERVOUS SYSTEM

The nervous system is composed of the **central nervous system (CNS)** and the **peripheral nervous system (PNS)** (Figure 15.1). The CNS includes the brain and the spinal cord, and the PNS includes **cranial nerves, spinal nerves,** and all the neurons and sensory receptors outside the CNS. There are billions of neurons and supporting cells in the central and peripheral nervous systems. The brain receives and interprets all sensory information and initiates responses. The spinal cord transmits information to and from the brain and is involved with reflexes. A **nerve** is a bundle of hundreds or thousands of neurons in the peripheral nervous system. The human body has 12 pairs of cranial nerves and 31 pairs of spinal nerves that innervate every tissue in the body. Cranial and spinal nerves receive sensory information and send motor information to specific regions of the body. **Tracts** are bundles of neurons in the central nervous system.

The major divisions of the PNS are the **somatic** nervous system and the **autonomic** nervous system (ANS) (Figure 15.2).

Learn everything you can, anytime you can, from anyone you can, there will always come a time when you will be grateful you did.

Sarah Caldwell

15

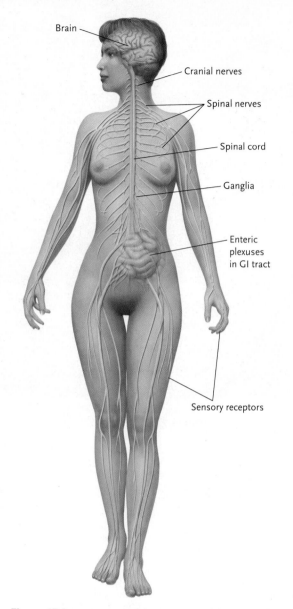

Figure 15.1 Overview of the Nervous System

The nervous system consists of the brain, spinal cord, and peripheral nerves.

The somatic nervous system (SNS) is involved in voluntary control of movements. The somatic nervous system includes receptors, sensory neurons, also called afferent neurons, and the motor neurons, also called efferent neurons, which stimulate skeletal muscle fibres to contract (Chapter 14) (Figure 15.3).

The autonomic nervous system (ANS) is involved in involuntary control of organs, blood vessels, endocrine glands, and exocrine glands. Autonomic afferent neurons bring sensory information from visceral organs to the CNS. The ANS works with the endocrine system to maintain homeostasis, such as body temperature, blood pressure, and blood sugar levels. The efferent neurons in the ANS innervate smooth muscle, cardiac muscle, or glands (Figure 15.3). The ANS is discussed in detail in Section 15.7.

The **enteric nervous system** is a network of neurons that regulates the digestive system (Chapter 20). The enteric nervous system is made up of the **submucosal plexus**, which stimulates the secretion of enzymes, gastric acid, and other substances required for digestion, and the **myenteric plexus**, which stimulates the contraction of the smooth muscles involved in digestion. The enteric nervous system is influenced by the autonomic nervous system. The parasympathetic branch of the ANS stimulates the enteric nervous system. The sympathetic branch inhibits digestive processes.

Plexus is a term that defines a specific network of nerves and neurons. For example, the brachial plexus is the group of nerves in the brachial region that innervates the arm. The submucosal plexus stimulates secretion of digestive enzymes in the GI tract.

In the peripheral nervous system, groups of cell bodies are called **ganglia** (singular, *ganglion*). For example, each dorsal root ganglion in every spinal nerve contains sensory neuron cell bodies (Section 15.4). In the CNS, groups of cells bodies are called **nuclei**

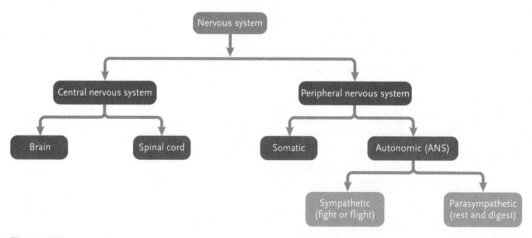

Figure 15.2 Divisions of the Nervous System

The nervous system is divided into the central nervous system and the peripheral nervous system.

A. Somatic nervous system

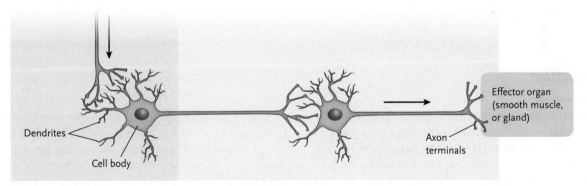

B. Autonomic nervous system

Figure 15.3 Somatic and Autonomic Neurons

Afferent neurons are sensory neurons that bring information to the brain, and efferent neurons are motor neurons that bring information from the brain to muscles, glands, or other neurons.

(singular, nucleus); however, this is not the same as the DNA-containing nucleus within cells. An example of the CNS nuclei is the supraoptic nuclei in the hypothalamus, a group of cells that produce antidiuretic hormone (ADH) (see Chapter 17).

Sensory receptors may be extensions of a sensory neuron or may be specialized receptor cells that signal a sensory neuron. Note that these receptors are not the same as the membrane protein receptors discussed in Chapter 2. Sensory receptors detect specific sensory information. Table 15.1 provides an overview of receptor types and the stimuli they detect.

The following are the functions of the nervous system:

• Detecting sensory stimuli

• Analyzing and integrating sensory information in the CNS

- Responding to environmental and internal stimuli through efferent neurons of the somatic, autonomic, or enteric nervous systems

- Playing a major role in the regulation of various organ systems (homeostasis)
- Reflexes
- Learning, memory, and emotions

TABLE 15.1

Receptor and Stimulus

Receptor	Stimulus
Chemoreceptor	Chemical concentrations, such as hormones, neurotransmitters, and nutrients
Osmoreceptor	Osmolarity, concentration of water and ions
Baroreceptor	Blood pressure; found in certain blood vessels
Photoreceptor	Light; found in the retina
Mechanoreceptor	Stretch and physical movement
Proprioceptor	Body position
Nociceptor	Pain
Thermoreceptor	Temperature; found in the PNS and CNS
Tactile receptors	Touch, pressure, vibration; found in the skin

CONCEPT REVIEW 15.1

1. Describe the major divisions of the nervous system.
2. What tissues are innervated by the ANS?
3. What are the two plexuses of the enteric nervous system, and what are their main functions?

4. Which division of the peripheral nervous system is involved in voluntary movements?
5. Name the stimuli detected by the following types of receptors: chemoreceptor, osmoreceptor, baroreceptor, photoreceptor, mechanoreceptor, proprioceptor, nociceptor, thermoreceptor, tactile receptors.

Application

As you read through the rest of this chapter, consider this scenario. Trevor accidentally steps on a piece of Lego in his bare feet. He feels the pain of the Lego in his foot and has a reflex reaction of pulling his foot up and off the Lego. Follow the entire sensory and motor pathway, and list everything involved from the specific type of receptor to the motor response.

If Trevor feels angry about the Lego being on the floor, what part of his brain is involved in that emotional response? If the anger stimulates his sympathetic nervous system, what kinds of effects occur in his body? Which neurotransmitters play a role in a stress response?

15.2 NERVOUS TISSUE

Neuron Structure

Nervous tissue consists of neurons that transmit electrical signals and the **glial cells**, which support the functioning of the neurons. Neurons and glial cells are found within the CNS and the PNS. Glial cells divide and go through mitosis (cell division), but

neurons do not. We do have some stem cells that can differentiate into neurons if stimulated by injury, but generally, once neuron cells die they are not replaced. A typical motor neuron is shown in Figure 15.4. The cell body contains all the same organelles as any other cell type, including a nucleus with the DNA, where genes for the production of neurotransmitters are transcribed; mitochondria for ATP production; and rough endoplasmic reticulum for protein synthesis.

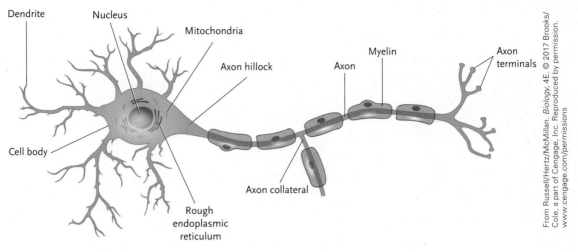

From Russell/Hertz/McMillan. *Biology*, 4E. © 2017 Brooks/ Cole, a part of Cengage, Inc. Reproduced by permission. www.cengage.com/permissions

Figure 15.4   Structure of a Typical Motor Neuron

The **dendrites** are extensions of the cell body that detect stimuli.

The dendrites have membrane receptors (not the same as the sensory receptors discussed in Section 15.1); these are membrane proteins that are linked to ion channels that allow ions to move into or out of the cell, causing a charge change (graded potential, Section 15.3). Action potentials occur in the axon, the main extension of the cell body, and move toward the axon terminals, where neurotransmitters are released and signal another neuron, a muscle, or gland. Axon collaterals are branches of axons that also end with axon terminals. Neurotransmitters are chemicals or proteins that can bind to membrane receptors on other cells and cause muscle contraction, stimulate a gland to secrete hormones, or signal another action potential in another neuron. Some neurotransmitters can also inhibit an action potential in other neurons. The axon hillock is the region where the cell body connects to the axon, and it contains the initial segment, which is the starting point for an action potential. Many neuron axons contain a protective fatty sheath—called the *myelin sheath*—that helps speed up the transmission of action potentials along the axon. The junction of an axon terminal and the membrane of the neuron it innervates is called the *synapse*.

Types of Neurons

There are three main types of neurons: multipolar, bipolar, and unipolar (Figure 15.5). Multipolar neurons have many dendrite extensions from the cell body (Figure 15.4). They are the common structure of the **motor neurons** (efferent) that innervate muscle tissue (skeletal, smooth, or cardiac) or glands, as well as of the much shorter **interneurons** (association neurons) that relay information between sensory and motor neurons in the CNS (Figure 15.3). Motor neurons signal the effector in a negative feedback response (Chapter 1). Motor neuron cell bodies originate in the spinal cord. Most motor neurons that innervate the skeletal muscles are extremely long. The cell bodies are located in the grey matter of the spinal cord, the axons extend the length of the nerve, and the axon terminals synapse onto the target tissue. One motor neuron can be over one metre long.

Unipolar neurons are sensory (afferent) neurons, and they bring information to the spinal cord. The cell body projects out to the side of the axon. Sensory neurons have dendrites in the tissue and receive sensory stimuli either directly or from a separate sensory receptor cell (Section 15.1). The axon extends the length of the nerve and carries signals to the spinal cord, where the cell bodies from each spinal nerve are grouped together in the dorsal root ganglia (Section 15.4), and the axon terminals synapse onto interneurons in the spinal cord.

Bipolar neurons have one main dendrite that extends from the cell body and then branches out. Bipolar neurons, the least common of the three types of neurons, are found mainly in the retina, inner ear, and the olfactory area of the brain.

DID YOU KNOW?

An individual neuron can form connections through its dendrites and axon terminals with 20 000 other neurons. When we learn, exercise, talk, think, experience emotions, and remember things, we reinforce those neural connections and create signalling patterns in our brain that can make it easier to do certain things. Any connections that are not used frequently will break down. We create new connections when we learn, and we maintain these connections as long as they are reinforced.

15

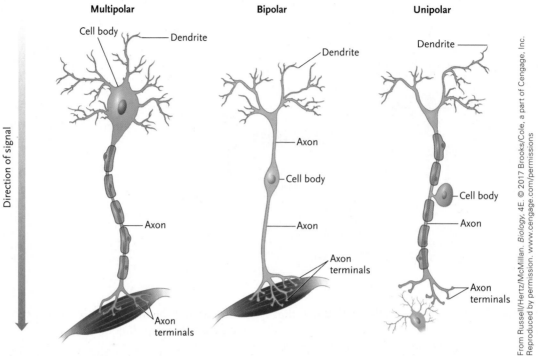

Figure 15.5 Three Types of Neurons

Neurons can be divided into three categories: multipolar, bipolar, and unipolar.

DID YOU KNOW?

Neurons do not undergo cell division. So how do people get brain tumours? Brain tumours are derived from glial cells, most commonly astrocytes, because these cells buffer ions and toxins and take in many metabolites so neurons can function properly. By absorbing toxins, the glial cells are predisposed to DNA mutations that can lead to cancer. Approximately 5% of brain tumours are caused by inherited genetic mutations.

Cells of the CNS

Glial cells are the most abundant cell types in the CNS. In the CNS, there are four types of supporting glial cells: astrocytes, oligodendrocytes, microglia, and ependymal cells (Figure 15.6). Astrocytes are the most abundant glial cell in the CNS and are extremely important for regulating nervous system functions.

- Astrocytes help form the blood-brain barrier by producing extensions—called *end foot processes*—that wrap around the outside of capillary endothelial cells where tight junctions between those endothelial cells separate the circulating blood from the CNS extracellular fluid.

- Astrocytes regulate the external environment of the neurons, including ion concentrations—mainly sodium, potassium, and calcium—during action potentials and when neurotransmitters are released into the synapse.

- Astrocytes can store glycogen and can break it down into glucose (glycogenolysis) and provide it to the neurons when needed, and they remove waste products such as ammonia.

- Astrocytes play a key role in the development of the nervous system by producing nerve growth factors that guide migrating neurons.

- Astrocytes are involved in producing scar tissue during healing of any CNS injuries.

Oligodendrocytes are the type of glial cells that form the myelin sheath that surrounds the axons of many neurons in the CNS. Small interneurons are not myelinated. Oligodendrocytes have branching fatty membrane extensions that can wrap around several axons. Myelin acts like an insulator so that electrical signals cross the neuron cell membrane only in the spaces between the myelin; these spaces are called the *nodes of Ranvier*. In these nodes are the ion channels that propagate the action potential. **Saltatory conduction** is the term that refers to action potentials travelling along myelinated neurons.

Microglia are resident immune cells that are phagocytic and fight infections that occur in the CNS. Microglia also engulf dead cells or cell debris.

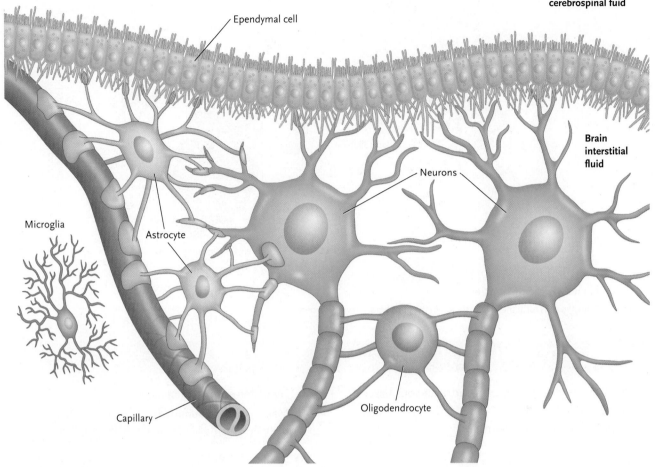

Figure 15.6 Glial Cells

Glial cells in the CNS include oligodendrocytes, astrocytes, microglia, and ependymal cells.

Ependymal cells are found in the choroid plexus of each of the four ventricles of the brain, and they produce cerebral spinal fluid (CSF) from contents in the blood. The composition of CSF is similar to blood plasma, but not exactly the same because some of their ion concentrations differ. The CSF bathes the cells of the CNS with nutrients, ions, and oxygen. CSF is very low in proteins compared to the blood and has a higher chloride concentration compared to blood. CSF is produced continuously, and it circulates through the ventricles and spinal cord and eventually re-enters the circulatory system at the superior sagittal sinus, which is the largest vein in the brain (Chapter 18). There is always approximately 100 to 150 mL of CSF in the CNS at any one time, and we produce an average of 400ml per day.

Glial Cells of the PNS

The peripheral nervous system has two kinds of glial cells: satellite cells and Schwann cells. Satellite cells are similar in function to astrocytes and regulate ions and nutrients in the interstitial fluid surrounding cell bodies in the PNS ganglia, such as the dorsal root ganglia that contain the cell bodies of sensory neurons. Schwann cells are similar in function to oligodendrocytes and produce myelin that surrounds the very long axons of afferent and efferent neurons. Schwann cells wrap around the axons of one neuron only and do not branch and wrap around several axons as do the oligodendrocytes in the CNS (Figure 15.7).

DID YOU KNOW?

Multiple sclerosis is an autoimmune disease that occurs when the immune system targets the myelin sheath as though it was fighting an infection. This process eventually destroys the myelin, and nerve impulses cannot travel at the same velocity nor to the correct locations. It affects sensory and motor neurons and often results in loss of sensations and paralysis.

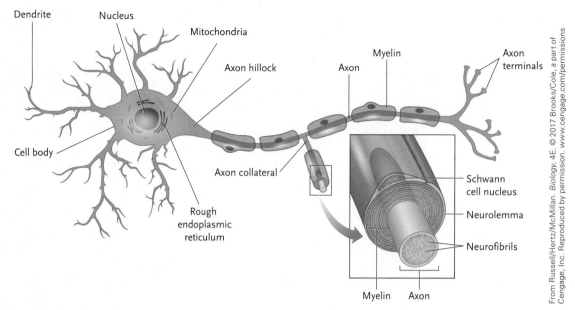

Figure 15.7 Schwann Cell

Schwann cells form the myelin sheath in the peripheral nervous system.

CONCEPT REVIEW 15.2

1. What two cell types are found in nervous tissue?

2. What is the main function of the dendrites? Cell body? Axon? Axon terminals?

3. Name three types of neurons.

4. Make a chart to summarize the main functions of each type of glial cell in the CNS and PNS.

15.3 ACTION POTENTIALS AND GRADED POTENTIALS

Action Potentials

Recall from Chapter 3 that the ion concentrations in the cytoplasm and in the interstitial fluid are not the same. There are more sodium ions and chloride ions in the interstitial fluid, and there are more potassium ions in the cytoplasm. Along with the contribution of charges from proteins and other molecules inside and outside the cell, the charge inside the membrane of the cell is *negative* relative to outside the cell. Also recall that the gradient of sodium and potassium ions is maintained by the **Na$^+$/K$^+$ pump** (Chapter 3). The charge difference between the cytoplasm side and the interstitial fluid side of the cell membrane creates a **concentration gradient** as well as an electrical gradient, and the electrical gradient makes the cell polarized. The charge inside a resting neuron is approximately −70mV, and this is called the *resting membrane potential* (Figure 15.10).

Ions cannot move across a cell membrane without the aid of a membrane protein ion channel. Ion channels in the neuron are regulated so that they are either open or closed depending on stimulation of the neuron. If we poke the back of our hand, tactile sensory receptors in the

skin will stimulate a sensory neuron. The stimulation takes the form of opening sodium ion channels. Because sodium ions (Na$^+$) are positively charged and in higher concentration outside the cell, the opening of Na$^+$ membrane channels causes Na$^+$ to move *into* the cell, down the concentration gradient, *and* down the electrical gradient. When Na$^+$ moves into the cell, the inside of the cell becomes positively charged, which is opposite to the negative charge of the resting membrane potential; this is called **depolarization**, and the cell is no longer polarized. The depolarization and subsequent repolarization of that region of the membrane is called an *action potential*. When the axon depolarizes in one area, it signals Na$^+$ channels in the next segment to open, and, in this way, depolarization occurs along the axon toward the axon terminals. The movement of depolarization along the axon propagates an action potential. The action potential moves along the axon of the presynaptic neuron, causing the release of neurotransmitters that bind to the postsynaptic neuron and stimulate a response (Figure 15.8).

Ion channels open or close in response to a chemical or electrical stimulus. When the stimulus is a chemical, such as a neurotransmitter from another neuron, the channel is called a **chemically gated channel** (also called *ligand-gated*). When an ion channel opens or closes because of a change in the *charge* of the membrane, it is called a **voltage-gated channel**.

Figure 15.8 Presynaptic and Postsynaptic Neurons

Action potentials move down the axon of the presynaptic neuron and cause neurotransmitters to be released. The neurotransmitters bind to the postsynaptic neuron and stimulate a response.

An action potential begins at the initial segment of an axon hillock in response to a change in the charge of the membrane that is caused by a graded potential (described in Section 15.3). The resting membrane potential is approximately −70mV. If the change in charge reaches the threshold level of approximately −55mV, an action potential will occur and voltage-gated Na+ channels will open, allowing Na+ to enter the cell and cause depolarization (Figure 15.9). Action potentials are considered "all-or-none" because they either occur or they do not occur, and this depends on the charge of the axon hillock reaching the threshold level. Action potentials always have the same magnitude, and so a stronger stimulus causes an increase in the *frequency* of the action potentials, not a larger action potential.

Stages of an Action Potential

1. **Depolarization.** Once the cell membrane potential reaches threshold, the voltage-gated sodium channels open, allowing Na+ ions to move into the cell, further increasing the charge to about +30mV (and further depolarizing the membrane).

2. **Repolarization.** The cell membrane reaches +30mV, the Na+ channels become inactivated (Figure 15.10), so no more sodium ion can enter the cell, which stops further depolarization. When the sodium inactivation gate closes, this creates a refractory period so that another action potential cannot be initiated. Then voltage-gated potassium ion channels open. When K+ channels open, K+ moves out of the cell, causing the membrane potential to again become negative; this is repolarization.

3. **Hyperpolarization.** The potassium channels have a slight delay in closing once the cell's charge becomes negative; this means more K+ leaves the cell than is required to repolarize the membrane. So the membrane becomes more negatively charged than it was in the resting state, causing the membrane to be

Figure 15.9 Action Potential

An action potential occurs in three stages: depolarization, repolarization, and hyperpolarization.

KEY

- = Depolarization   = Hyperpolarization
- = Repolarization

Figure 15.10 Depolarization and Repolarization

Depolarization is caused by Na⁺ moving into the cell, and repolarization is caused by K⁺ moving out of the cell.

hyperpolarized. This is important because it means the charge of the cell is further from the threshold level, and this will prevent another immediate action potential, unless the stimulus is strong enough to reach threshold. This refractory period ensures that the action potential moves in one direction down the axon toward the axon terminals.

4. Once the potassium ion channels close, the Na⁺/K⁺ pump helps restore the resting membrane potential by moving K⁺ in and Na⁺ out of the cell.

Note that one action potential in one section of the axon occurs within one to four milliseconds, that is, one to four 1000ths of a second. Also note that in a non-myelinated axon, action potentials continue along the entire length of the axon until they reach the axon terminal; in myelinated axons the action potential only occurs in the nodes of Ranvier.

In Figure 15.11, phase 3 is the repolarization phase. During this phase the sodium inactivation gate prevents any further sodium ions from entering the cell, and therefore, another action potential cannot occur; this is the **absolute refractory period**. Phase 4 is the hyperpolarization phase. During this phase, the membrane is more negative than the normal resting membrane potential. Another action potential could occur at this point but only if the stimulus is strong enough to bring the membrane to the threshold level; this is the **relative refractory period**.

Once an action potential reaches the axon terminal, neurotransmitters are released into the **synaptic cleft**. If the neurotransmitter binds to a muscle fibre, contraction occurs; if it binds to another neuron, it stimulates a **graded potential**.

DID YOU KNOW?

Novocaine, which is a local anesthetic, works by preventing voltage-gated sodium ion channels from opening, and therefore, the sensory neurons can't send sensory information to the brain.

Synapse

Recall from Chapter 14 that an axon terminal that connects with a skeletal muscle fibre is called a *neuromuscular junction*, which is one kind of synapse. The release of neurotransmitters at the synapse is the same whether the axon terminal is synapsing with a smooth muscle, cardiac muscle, a gland, or another neuron. The following is the order of events that occur in the axon terminal after an action potential has been triggered (Figure 15.12):

1. The action potential causes the neuron membrane to depolarize, and the depolarization triggers

Sodium voltage-gated channel

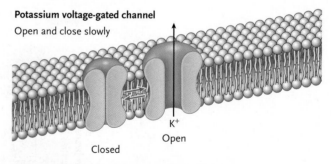

Potassium voltage-gated channel

Figure 15.11 Sodium and Potassium Channels

Sodium voltage-gated channels open and close quickly and have an inactivation gate. Potassium voltage-gated channels open and close slowly and do not have an inactivation gate.

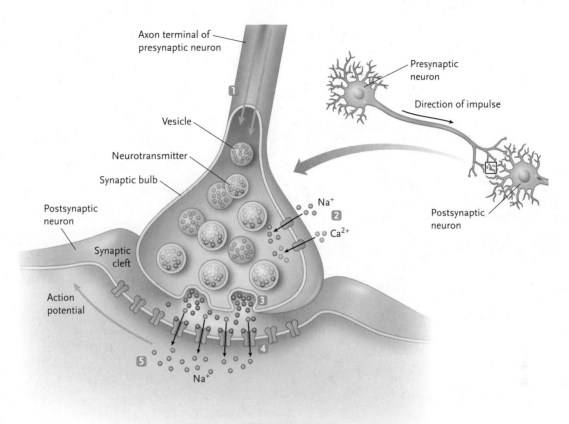

Figure 15.12 The Synapse

The synapse consists of the axon terminal, the synaptic cleft, and another neuron or a muscle cell.

voltage-gated calcium channels in the axon terminal to open and allow Ca2+ to enter the cell.

2. When calcium ions enter the axon terminal, the synaptic vesicles containing neurotransmitters fuse with the cell membrane, and exocytosis of the neurotransmitters occurs.

3. Once the neurotransmitters are released into the synaptic cleft, they bind to *specific* receptors on the next neuron, the **postsynaptic neuron**.

4. The receptors on the postsynaptic neuron are chemically gated ion channels that open when the neurotransmitter binds to them, and this allows ions to enter the next cell, which causes a graded potential.

5. If enough positively charged ions, such as Na^+, enter the postsynaptic neuron, an action potential can be triggered in that neuron (Figure 15.13).

Graded Potentials

A graded potential is a small change in the membrane potential, usually on the dendrites or cell body. Action potentials always cause a depolarization of the axon

membrane when sodium ions move into the cell. A graded potential, however, can be either depolarizing (becoming *positively* charged inside the cell) or hyperpolarizing (becoming more *negative* inside the cell). Also, graded potentials can be caused by positively or negatively charged ions. Graded potentials are given various names based on the location of the graded potential, such as receptor potential, synaptic potential, and pacemaker potential.

When one neuron is signalling another neuron, the one that is releasing the neurotransmitter is the presynaptic neuron, because it is before the synapse. The neurotransmitters released by the presynaptic neuron act on the postsynaptic neuron (Figure 15.13).

For example, if a neurotransmitter from a presynaptic neuron stimulates chemically gated chloride (Cl^-) channels to open in the postsynaptic neuron, Cl^- moves *into* the cell—because of the concentration gradient—and causes it to become more *negative*, hyperpolarized. Therefore, an action potential is *less* likely to occur at the axon hillock because the charge will be further from the threshold level. Neurotransmitters that cause hyperpolarization and decrease the likelihood of an action potential are **inhibitory neurotransmitters**. Inhibitory neurotransmitters cause graded

15

Figure 15.13 Graded Potentials

Graded potentials can be excitatory or inhibitory, and they determine whether or not the postsynaptic cell will have an action potential.

potentials that are called **inhibitory postsynaptic potentials (IPSPs)** (Figure 15.13). If a neurotransmitter stimulates chemically gated sodium channels, Na^+ enters the cell, making the charge more positive (depolarized), which brings the cell membrane closer to the threshold level at the axon hillock and increases the probability of an action potential. Neurotransmitters that cause depolarization and increase the likelihood of an action potential are excitatory, causing **excitatory postsynaptic potentials (EPSPs)**. When multiple presynaptic neurons are affecting a postsynaptic neuron, an action potential will occur only if the threshold level is reached.

Unlike action potentials, graded potentials can be summed, so multiple excitatory or inhibitory input from multiple connecting neurons can influence a single neuron. The single neuron will have an action potential only if the excitatory signals are stronger than the inhibitory signals and cause the cell to reach threshold level. Graded potentials can have **variable amplitude** depending on the amount of stimulation; more stimulation causes a greater depolarization (EPSP) or hyperpolarization (IPSP). Graded potentials do not have a refractory period. Chemically gated ion channels in the dendrites and cell body can be opened whenever neurotransmitters from the presynaptic neuron bind to them.

Stages of Action Potential from Stimulation to Release of Neurotransmitters

1. Resting membrane potential in a neuron is −70mV.
2. EPSPs (graded potential) cause depolarization of the membrane, and if that increase in positive charge inside the cell is enough to reach the threshold level of −55mV, then an action potential can occur.
3. Voltage-gated sodium channels open in the axon, allowing Na^+ to move *into* the cell, resulting in depolarization.
4. When the inside of the cell becomes positively charged, sodium channels close (the sodium inactivation gate causes the absolute refractory period), and voltage-gated potassium channels open, allowing K^+ to move *out* of the cell, causing the cell to become negatively charged inside and resulting in repolarization.
5. The potassium channels are slow to close again once the membrane is repolarized, so the cell becomes a little more negative than the resting membrane potential. This results in hyperpolarization (relative refractory period), which is required to ensure unidirectional movement of action potentials.

6. The Na⁺/K⁺ pump helps keep the distribution of sodium and potassium ions in their correct concentrations; but notice that the Na⁺/K⁺ pump does *not* cause repolarization of the membrane.

7. Steps 3 to 6 repeat all along the axon. As depolarization occurs in one section, the positive charge inside the cell stimulates the next voltage-gated channels to open. This continues along the axon until the axon terminal is reached.

8. When an action potential reaches the axon terminal, it triggers voltage-gated Ca^{2+} channels to open and allow Ca^{2+} to move into the cell.

9. Calcium entering the cell triggers exocytosis of neurotransmitters into the synaptic cleft.

10. Neurotransmitters bind to chemically gated channels on the postsynaptic cell membrane (dendrites or cell body) and cause a graded potential, either an EPSP or an IPSP.

11. The combination of graded potentials from all connecting neurons determines whether the postsynaptic cell reaches threshold level.

Action Potentials in Myelinated Axons

The propagation of action potentials in myelinated axons—saltatory conduction—occurs in exactly the same way as along non-myelinated axons, except that voltage-gated sodium channels are present only in the nodes of Ranvier (Figure 15.14). When the membrane depolarizes at one node of Ranvier, the Na⁺ current flows within the cell near the membrane and alters the charge of the membrane at the next node, triggering voltage-gated channels to open at the next node. This causes the action potentials to occur only in the nodes, which is like taking big steps when walking instead of walking by placing one foot directly in front of the other. Action potentials move faster in myelinated axons and in axons with a large diameter, such as motor neurons.

Neurotransmitters

Over 100 different molecules are neurotransmitters. They are released from neurons and affect other neurons, glands, or muscles. The following is a list of some of the most common. Notice that some of these are also hormones. The only difference, for example, in epinephrine, is that when it is released from a neuron it is a neurotransmitter, and when it is released by the adrenal medulla into the bloodstream, it is a hormone (Chapter 17). Neurotransmitters can be amino acids, derivatives of amino acids, peptides (several amino acids held together by peptide bonds), or proteins (Table 15.2).

Figure 15.14 Saltatory Conduction

Saltatory conduction involves the transmission of an action potential along a myelinated axon.

From Russell/Hertz/McMillan/Fenton/Addy/Maxwell/Hattie/Milson. *Biology*, 2E. © 2013 Nelson Education Ltd. Reproduced by permission. www.cengage.com/permissions

DID YOU KNOW?

Alcohol can decrease the function of the calcium channels in the axon terminals, causing fewer neurotransmitters to be released in certain parts of the brain. When the neurons can't communicate, cognitive abilities are decreased. Alcohol increases the production of serotonin, making people feel more relaxed and sociable. Alcohol also decreases the effects of GABA (inhibitory neurotransmitter), which tends to make people feel more outgoing and spontaneous. Effects of alcohol are dose-dependent, causing symptoms ranging from feeling relaxed to having uncoordinated movements, loss of memory, and engaging in high-risk activities, to unconsciousness. Blood alcohol concentrations of 0.4% can be fatal; when neurons in the brainstem cannot function, breathing cannot be controlled.

DID YOU KNOW?

Some anti-anxiety medications, such as valium (diazepam) or Ativan (lorazepam), act by increasing the effects of GABA. The medications bind to GABA receptors (mimicking GABA) and cause chemically gated chloride ion channels to open. Cl– moves into the cells because of the strong concentration gradient, causing the membrane to become hyperpolarized, thus reducing the probability of an action potential occurring in the postsynaptic neuron.

TABLE 15.2

Neurotransmitters and Their Functions

Neurotransmitter	Function
Acetylcholine (ACh)	Excitatory. Causes muscle contraction; is released by autonomic neurons; is a neurotransmitter in the brain that plays a role in alertness, learning, memory, and motivation
Dopamine	Excitatory. When expressed in the frontal lobe, increases the ability to concentrate; when expressed in the limbic system, plays a role in the "reward" pathway and contributes to addictive behaviours; when expressed in the midbrain, helps control muscle movements Inhibitory. When expressed by the midbrain, it inhibits excessive motor signalling so that we can control motor movements. Parkinson's is caused by a lack of dopamine production in the midbrain.
Inhibitory.	Natural opiates, which decrease pain. Excitatory. Produce feelings of euphoria, such as "runner's high".
Epinephrine and norepinephrine (also hormones)	Excitatory and inhibitory. Involved in autonomic nervous system and the stress response (Section 15.7)
Gamma aminobutyric acid (GABA) Glycine	Inhibitory. Decreases neuron excitability and helps people feel calm Inhibitory. Decreases neuron excitability and helps people feel calm
Glutamate and aspartate	Excitatory. Play important roles in thinking, learning, and memory; excess glutamate associated with feelings of anxiety
Serotonin	Inhibitory. Mostly produced by neurons in the digestive tract during eating; generates feeling of contentment; low levels associated with depression

CONCEPT REVIEW 15.3

1. Where is the higher concentration of sodium, potassium, and chloride in a resting neuron: intracellular or extracellular?

2. What is the difference between a chemically gated and a voltage-gated ion channel?

3. What ion causes depolarization and repolarization of the membrane during an action potential?

4. Why is the hyperpolarization phase of an action potential important?

5. Why is calcium important for proper neuron function?

6. What is saltatory conduction?

7. What two factors increase the speed of transmission of action potentials?

8. Which neurotransmitter is important for (1) addictive behaviour, (2) mood, and (3) muscle contraction?

15.4 SPINAL CORD AND SPINAL NERVES

Sensory information travels to the CNS through afferent (sensory) neurons and from the CNS to effectors (muscles and glands) through efferent (motor) neurons. A sensory neuron carries information from a tissue all the way to the spinal cord before there is a synapse with another neuron. Multiple interneurons synapse with the axon terminals of the sensory neuron, which then communicate with various regions of the brain. Any thoughts or motor movements that originate in the brain must travel through the spinal cord before reaching the target tissue in the body. The spinal cord is a very important connection centre between the brain and the body's tissues (Figure 15.15).

Anatomy of the Spinal Cord

The spinal cord is located within the vertebral foramen of each vertebra (Chapter 13) and runs from the foramen magnum to the L1 and L2 vertebrae. The spinal cord is highly protected by bone, adipose tissue in the epidural region, the CSF produced by the ependymal cells, and layers of connective tissue called **meninges** (Figure 15.16). Three layers of meninges protect the spinal cord and extend to cover the brain. The **dura mater** is the toughest dense, irregular, connective tissue outer layer. Between the dura mater and the vertebral bone is the **epidural space**, which contains blood and lymphatic vessels, the nerve roots, and fat that cushions the spinal cord. Having an "epidural" means that an anesthetic is injected into this space to block action potentials in the neurons and therefore decrease or block sensations in the body below the point of injection;

Figure 15.15 Neural Signalling through the Spinal Cord

Signals travel to and from the brain through the spinal cord.

this is often used during labour and the delivery of babies. The middle layer of the meninges is the **arachnoid mater,** which looks like a spider web and consists of collagen and elastin fibres. The innermost layer, called the **pia mater,** covers the outside of the spinal cord and brain. The pia mater, like the arachnoid mater, consists of collagen and elastin fibres, but it forms a sheet (Figure 15.16). The space

between the pia mater and the arachnoid mater—called the **subarachnoid space**—contains blood vessels and cerebral spinal fluid. The CSF flows through the arachnoid space and surrounds the spinal cord with nutrients. Molecules in the arachnoid space can diffuse through the pia mater to reach the neurons in the spinal cord.

The spinal cord has a central canal that carries CSF. The posterior (dorsal) side of the spinal cord has a groove, called the **posterior median sulcus,** and an anterior (ventral) groove, called the **anterior median fissure** (Figure 15.17).

The spinal cord is composed of grey matter and white matter and is divided into right and left halves. **Grey matter** includes neurons that are short interneurons with no myelin, and also the cell bodies, dendrites, and axon terminals of myelinated neurons. The **white matter** consists of just the myelinated axons, which look white. The grey matter forms a butterfly shape in the centre of the spinal cord. This is where sensory neurons entering the spinal cord synapse with interneurons that eventually signal the brain, and where motor information from the brain synapses with the cell bodies of motor neurons that signal the tissues (Figure 15.17). Spinal nerves are located between each vertebra and leave the CNS through the intervertebral foramen on the right and left side of the vertebral column. Each spinal nerve branches inside the vertebral column into anterior and posterior roots—called the *dorsal* and *ventral root*, respectively (Figure 15.13).

Sensory information is brought to the spinal cord through the **dorsal root** (posterior side of the spinal cord) to the **dorsal horn** (Figure 15.17). The unipolar sensory neuron cell bodies are located in the dorsal root ganglion, and the axon terminals synapse with interneurons in the dorsal horn. Those interneurons then synapse with the dendrites of somatic or autonomic motor neurons, as well as with long myelinated

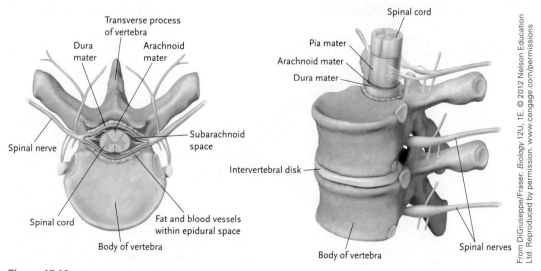

Figure 15.16 The Spinal Cord

The spinal cord contains the same meningeal layers as the brain: the outermost is the dura mater; the middle layer is the arachnoid mater; and the inner layer, the pia mater, covers the spinal cord.

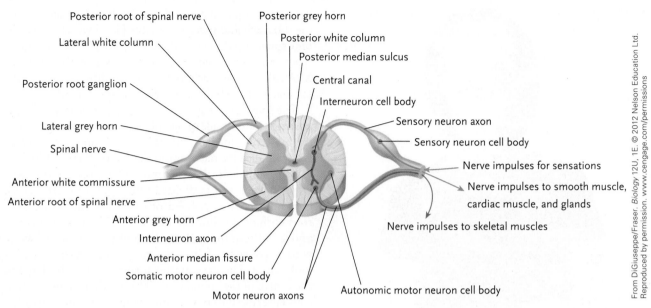

Figure 15.17 Structures in a Transverse Section of the Spinal Cord

From DiGiuseppe/Fraser. *Biology* 12U, 1E. © 2012 Nelson Education Ltd. Reproduced by permission. www.cengage.com/permissions

neurons—the **ascending tracts**—that travel through the white matter to the brain.

Motor information is brought from the brain to the spinal cord through the white matter axons—the descending tracts—that synapse with interneurons and motor neurons in the ventral horn (anterior side of spinal cord). The cell bodies of the motor neurons are in the ventral horn, and the axons extend to the cell bodies through the ventral root to the spinal nerve, where the axon terminals synapse with a target muscle or gland in the tissues. Lateral horns contain cell bodies of autonomic neurons, only in the thoracic region.

Somatic and autonomic sensory and motor information goes to and leaves from specific regions of the spinal cord (Figure 15.18). Sensory information from the right side of the body goes to the right side of the spinal cord, and left side information goes to the left side of the cord. Likewise, motor information going to the right side of the body initiates in the right side of the cord; the left side of the body is controlled by the left side of the cord. The information crosses over in the cord or in the brainstem and is interpreted by the *opposite* side of the brain.

Spinal Nerves

The spinal cord extends from the **medulla oblongata** of the brainstem, at the first cervical vertebra, to the **conus medullaris**, at the beginning of the lumbar vertebrae. There the spinal nerves branch out in a structure that

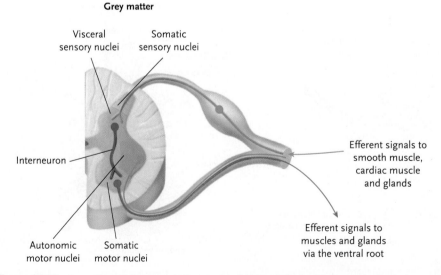

Figure 15.18 Somatic and Autonomic Sensory and Motor Information
Somatic and autonomic sensory and motor information is contained in specific regions of the spinal cord.

looks like a horse's tail and is called the **cauda equina**. The spinal cord has two enlarged regions: the **cervical enlargement** that contains many neurons that innervate the upper limbs, and the **lumbar enlargement** that contains neurons that innervate the lower limbs (Figure 15.19).

Each spinal (and cranial) **nerve** is covered by layers of connective tissue in a way that's similar to how muscle and its bundles of muscle fibres are organized (Figure 15.19). Nerves contain groups of axons from sensory and motor neurons that may or may not be myelinated. The outermost connective tissue cover protects the whole nerve and is called the **epineurium**. Within the nerve are bundles of axons, called *fascicles* (just like bundles of muscle fibres), and these are wrapped in the **perineurium**. Each fascicle contains bundles of individual axons that are covered by the **endometrium**; underneath is either the neuron cell membrane or the protective myelin sheath.

There are 31 pairs of spinal nerves: 8 pairs of cervical nerves (note that there are only 7 cervical vertebrae), 12 of thoracic, 5 of lumbar, 5 of sacral, and 1 pair of coccygeal nerves. Spinal nerves from T1 to L2 contain neurons of the sympathetic component of the autonomic nervous system.

Some of the spinal nerves form a network of nerves called a *plexus* (Figure 15.20 and Figure 15.21). The

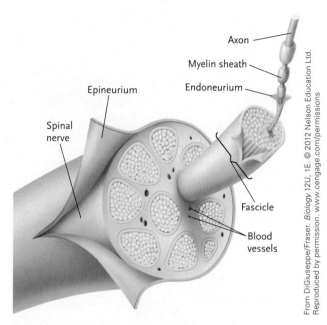

From DiGiuseppe/Fraser. *Biology 12U*, 1E. © 2012 Nelson Education Ltd. Reproduced by permission. www.cengage.com/permissions

Figure 15.19   Structure of a Nerve

Nerves are made up of bundles of neurons and layers of protective connective tissue. The epineurium surrounds the nerve, the perineurium surrounds the fascicles, and the endoneurium surrounds each axon.

Figure 15.20   Spinal Nerves

There are 31 pairs of spinal nerves.

15

Upper limb

C5
C6
C7
C8
T1

Axillary nerve
Radial nerve
Musculocutaneous nerve
Humerus
Median nerve

Superficial branch
of radial nerve

Ulnar nerve

Radius

Median nerve

Ulna

Lower limb

L2
L3
L4

Femoral nerve

Superior gluteal nerve
Inferior gluteal nerve

Pudendal nerve

Sciatic nerve

Common fibular nerve

Superficial fibular nerve

Deep fibular nerve

Tibial nerve

Plantar branches

Figure 15.21 Nerves of the Upper and Lower Limbs

cervical plexus (C1–C5) contains neurons that innervate the head, neck, and shoulders and also the **phrenic nerve** that innervates the diaphragm. The **brachial plexus** contains nerves that innervate the upper limbs, neck, and shoulders. Nerves of the brachial plexus include the **axillary nerve** that innervates the deltoid and rotator cuff muscles, the **musculocutaneous nerve** that innervates the biceps and brachialis, the **median nerve** that innervates the forearm flexor muscles that affect the thumb and first three fingers, the **radial nerve** that innervates the thumb and first finger, and the ulnar nerve that innervates the last two fingers.

The **lumbar plexus** contains nerves that innervate the abdominal wall, genitals, and parts of the lower limbs. The **femoral nerve** innervates the quadriceps and hip flexors, and the **obturator nerve** innervates the hip joint and adductor muscles. The **sacral plexus** contains nerves that innervate the lower limbs, genitals, and gluteal muscles. The **superior and inferior gluteal nerves**

innervate the gluteal muscles. The sciatic nerve innervates the hamstrings and then divides at the knee into the **tibial nerve**, which innervates the calves and feet, and the **common fibular nerve** that innervates the tibialis anterior (Figure 15.21).

DID YOU KNOW?

Approximately 50% of people who are diabetic will suffer from diabetic neuropathy, which is damage to any of the peripheral nerves, including sensory and motor nerves, and also nerves of the autonomic nervous system. Depending on the nerves that are affected, symptoms can include numbness and tingling, burning pain sensations, trouble swallowing, facial drooping, urinary incontinence, and muscle twitches.

CONCEPT REVIEW 15.4

1. List the meninges from innermost layer to the outermost layer.

2. What is found within the epidural space and within the arachnoid space?

3. What neuron structures are found in grey matter and white matter?

4. Explain how sensory information enters the spinal cord and motor information leaves the spinal cord.

5. Describe the structure of a nerve, including the connective tissue coverings.

6. How many pairs of spinal nerves do humans have in the cervical, thoracic, lumbar, sacral, and coccygeal regions of the spinal column?

7. Fill in the following chart:

Nerve	Vertebral Region	Tissue Innervated
Phrenic	Cervical	Diaphragm
Axillary		
Musculocutaneous		
Median		
Radial		
Ulnar		
Obturator		
Pudendal		
Femoral		
Sciatic		
Tibial		
Fibular		
Superior and inferior gluteal		

15.5 REFLEXES

A **reflex** is a rapid and involuntary response to a stimulus, also called a **reflex arc**. Reflexes are protective mechanisms that prevent damage to any part of the body.

Monosynaptic reflexes involve a sensory neuron with *one* synapse onto a motor neuron in the spinal cord, and this causes a muscle to contract. For example, a **stretch reflex** involving the patellar tendon—called the *knee-jerk response* or **patellar reflex**—is monosynaptic. A common test in a doctor's office to determine the integrity of the peripheral nervous system is to tap the patellar ligament to cause the leg to extend. The quadriceps muscles are all connected by the quadriceps tendon, which encloses the patella and inserts on the tibial tuberosity on the proximal, anterior region of the tibia bone. Tapping the patellar ligament pulls on the quadriceps muscles making them *stretch*. The stretch is detected by sensory neurons in the quadriceps muscles; the signal is sent to the spinal cord, where the axon terminals of the sensory neuron synapse directly onto a motor neuron that innervates the quadriceps muscle and causes contraction, which causes the leg to extend (Figure 15.22). In all types of reflexes, some neurons connect to interneurons in the spinal cord that send the sensory information to the brain, and so we become conscious of the movement.

Polysynaptic reflexes involve more than one synapse from sensation to response, such as the **withdrawal reflex**. Again, with any type of reflex, there are always neurons that synapse with neurons of the ascending, white matter tract, and these carry the sensory information to the brain. A withdrawal reflex protects the body from damage, and the sensory neuron involved detects pain. For example, if you touch something hot, your hand will jerk away from the heat source before you are consciously aware of the pain because the sensory neuron synapses in the spinal cord with an interneuron, which immediately sends a motor response through a motor neuron, causing the biceps to contract (Figure 15.23). Often, **reciprocal innervation** occurs

15

Figure 15.22   Monosynaptic Reflex

Monosynaptic reflexes involve the signalling of a sensory neuron directly to a motor neuron in the spinal cord.

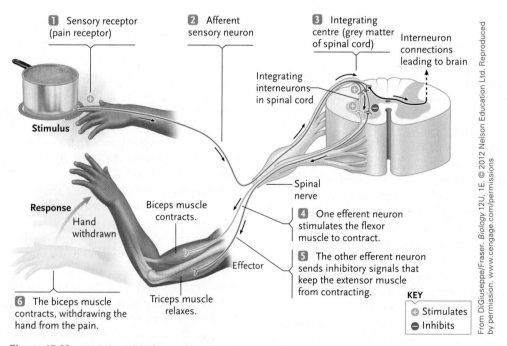

Figure 15.23   Withdrawal Reflex and Reciprocal Innervation

The withdrawal reflex involves the contraction of a muscle and the inhibition of contraction of its antagonist. Reciprocal innervation involves the opposite muscle contracting on the opposite side of the body.

From DiGiuseppe/Fraser. *Biology* 12U, 1E. © 2012 Nelson Education Ltd. Reproduced by permission. www.cengage.com/permissions

during a withdrawal reflex; this means the antagonistic muscle is inhibited to ensure that the agonist can contract. Recall from Chapter 14 that at a neuromuscular junction, acetylcholine is released from the motor neuron, and the neurotransmitter is always stimulatory, causing contraction. So a motor neuron cannot inhibit muscle contraction. The inhibition takes place in the spinal cord by the interneuron.

CONCEPT REVIEW 15.5

1. What is the difference between a monosynaptic and polysynaptic reflex?
2. Why do humans have reflexes?
3. Describe the events that occur in a patellar reflex.
4. Describe the events that occur in a withdrawal reflex.

15.6 BRAIN AND CRANIAL NERVES

The brain is protected by the bones of the skull, three layers of meninges, and cerebral spinal fluid (CSF). The bones of the skull are discussed in Chapter 13. The meninges are the same as those protecting the spinal cord (Figure 15.24). The outermost layer, the dura mater, is the thick, tough layer that lies next to the skull bones and contains blood vessels. The middle layer, the arachnoid mater, has the same spider web appearance as this layer surrounding the spinal cord. The innermost layer, the pia mater, lies directly next to the brain. The pia mater contains blood vessels that branch into and throughout the brain. In between the arachnoid mater and the pia mater is the subarachnoid space, which contains CSF that bathes the brain with nutrients and protects it by absorbing shock when we move around and jump.

Like the spinal cord, the brain has areas of grey matter and white matter. The white matter is made up of myelinated axons, and the grey matter contains cells bodies, dendrites, unmyelinated axons, and axon terminals (Figure 15.25).

Figure 15.24 The Brain
White matter and grey matter

15

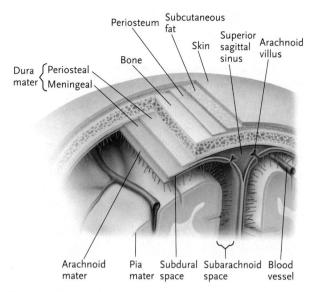

Figure 15.25 Meninges

There are three layers of meninges: the dura mater, arachnoid mater, and the pia mater.

Recall from Section 15.2 that astrocytes are glial cells that have extensions that wrap around the capillaries in the brain and help form the blood-brain barrier (mainly due to the tight junctions between the capillary endothelial cells) (Figure 15.26). The blood-brain barrier prevents the brain from exposure to any toxins that might be in the bloodstream. The CSF is produced by ependymal cells (Section 15.2) in a structure called the **choroid plexus** within each ventricle

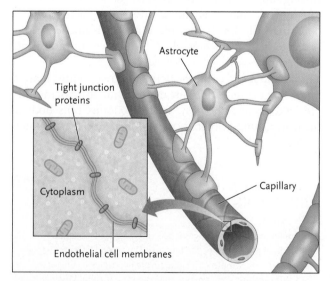

Figure 15.26 Blood-Brain Barrier

The blood-brain barrier is formed by tight junctions between vascular endothelial cells and the end foot processes of astrocytes.

(Figure 15.27). The choroid plexus is a region where the ependymal cells lie close to blood vessels, and they extract water and nutrients from the blood plasma to produce CSF. As the CSF flows in the arachnoid space throughout the CNS, it eventually reaches the **arachnoid villi** where it re-enters the circulatory system through a vein called the **superior sagittal sinus**. The human body produces approximately 400 mL of CSF each day, with approximately 150 mL in the CNS at any one time.

The brain can be divided into four main regions: the diencephalon, brainstem, cerebellum, and cerebrum (cerebral cortex) (Figure 15.28). The structure and function of each region is described in the following sections.

Diencephalon

The **diencephalon** includes the thalamus, hypothalamus, and the pineal gland. The **thalamus** is well known as the "sensory relay station" where all peripheral sensory signals are sent before being distributed to other regions of the brain, such as the cerebral cortex, for interpretation. The thalamus also plays an important role (along with the cerebellum) in controlling the coordination of skeletal muscle contractions during complex movements. This is because sensory information from the muscles, including amount of stretch, force of contraction, and body position, are sent to the thalamus, and that information is used to fine tune body movements.

The hypothalamus is an endocrine gland that produces releasing hormones that stimulate or inhibit the **pituitary gland**. The hormones of the hypothalamus and the pituitary gland are involved in essential homeostatic mechanisms—such as hunger, growth, temperature regulation, water and ion balance, regulation of the reproductive system—and are part of the autonomic nervous system (Section 15.7). The hypothalamus and pituitary gland are explained in Chapter 17.

The **pineal gland**, a small gland located in the posterior region of the diencephalon, produces a hormone called *melatonin*. **Melatonin** is produced in low light, like in the evening when the sun goes down, and it induces sleep. The pineal gland plays an important role in regulating our daily sleep-wake circadian rhythm. Interestingly, the pineal gland also produces a hallucinogenic molecule called *DMT* (dimethyltryptamine) during REM sleep (rapid eye movement) that makes us have dreams.

Figure 15.27 Ventricles of the Brain
Ventricles are hollow cavities within the brain. Ependymal cells within each choroid plexus in each ventricle produce cerebral spinal fluid.

DID YOU KNOW?

The pineal gland, located outside the blood-brain barrier, receives abundant blood flow, so it is exposed to all substances in the bloodstream. This explains why the pineal gland is so reactive to psychoactive substances. Secretions of the pineal gland affect how we perceive things. For example, the pineal gland is highly impacted by drugs that cause hallucination and is responsible for dreaming. Light also has a strong influence on the pineal gland and thereby affects mood, alertness, and perceptions. Some people develop seasonal affective disorder (SAD) during the winter when days are short and dark, and light therapy can be a beneficial treatment for many people.

Brainstem

The **brainstem** includes the midbrain, pons, and medulla oblongata (Figure 15.29). It is the most primitive brain region and is essential for life. All information travelling through neurons in the spinal cord passes through the brainstem before reaching any other part of the brain. Throughout the brainstem, there are neurons that regulate **consciousness:** the *reticular activating system (RAS),* also known as the *reticular formation.* This region includes sensory neurons in ascending tracts to the cortex; the sensory information stimulates the RAS and causes us to wake up and be conscious. The RAS is inhibited with general anesthetic during surgery. The RAS and the thalamus allow us to focus attention on one specific

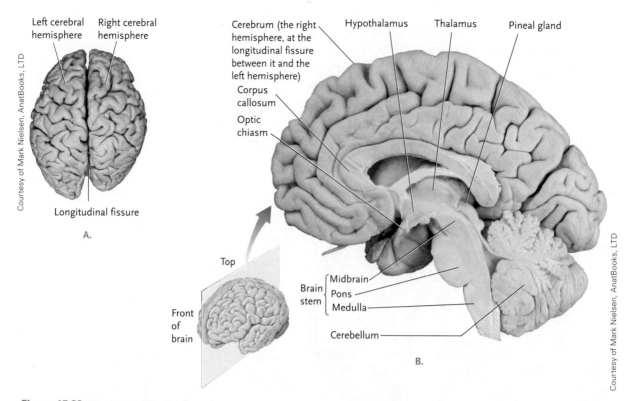

Courtesy of Mark Nielsen, AnatBooks, LTD

Courtesy of Mark Nielsen, AnatBooks, LTD

Figure 15.28 Regions of the Brain

The regions of the brain include the diencephalon, brainstem, cerebellum, and cerebrum.

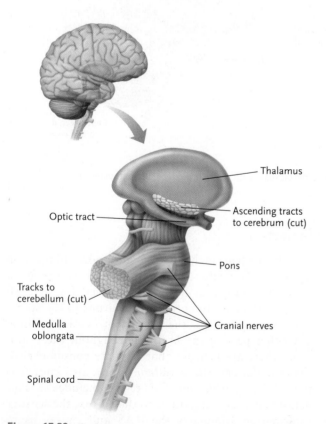

Figure 15.29 Brainstem

The brainstem consists of the midbrain, pons, and medulla oblongata.

stimulus while we are surrounded by multiple stimuli, such as listening to one person talk in a crowded noisy room.

At the top of the brainstem is the **midbrain**, which connects the pons to the diencephalon. The midbrain contains nuclei—groups of cell bodies in the CNS with similar functions—called **substantia nigra**. In conjunction with the cerebellum and the thalamus, these nuclei regulate motor movements. The substantia nigra neurons produce the neurotransmitter **dopamine**, which acts on a region of the cerebrum called the **basal nuclei** (explained below) to control excess motor movements. **Parkinson's disease** is caused by a loss of neurons in the substantia nigra and results in a decreased ability to control motor movements.

The **pons** is located in the brainstem between the midbrain and the medulla oblongata. The pons receives signals from the motor cortex of the cerebrum and directs them to the cerebellum. It also helps in the relay of information that allows the cerebellum to send motor signals to the muscles in a smooth and coordinated way. Other nuclei in the pons play a secondary role in the regulation of breathing. The medulla oblongata contains the main respiratory centre, but the pons helps smooth out respiratory movements and is involved when we voluntarily hold our breath (Chapter 19).

The medulla oblongata is the inferior region of the brainstem and contains nuclei that regulate the rhythm of breathing (Chapter 19) and the rate and force of heart contraction (Chapter 18). The medulla oblongata controls reflexes, such as for vomiting, coughing, and sneezing.

Cerebellum

The **cerebellum** is found at the base of the brain, posterior to the brainstem. The cerebellum plays a key role in coordinating, learning, and remembering complex motor movements, such as walking, running, swimming, and dancing. The cerebellum receives signals from many parts of the brain, including the thalamus, the pons, the motor cortex, the vestibular system, and the visual cortex. The signals received by the cerebellum serve to increase precision and accuracy of movements, maintain posture and balance during complex movements, and allow for fine motor control, such as in writing with a pen, or typing, or controlling facial muscles involved in talking. People with **cerebellar disease** have clumsy and uncoordinated movements and cannot learn new movements.

DID YOU KNOW?

A standard test for cerebellar function is also used to determine if a person is inebriated from alcohol. The test involves touching the nose with the tip of the finger starting with the arm fully extended. A person who has a healthy cerebellum, and is sober, can touch the nose in one fluid motion; whereas a person with cerebellar damage, or someone intoxicated, will reach for the nose slowly and erratically. Alcohol consumption depresses the function of neurons throughout the nervous system, including the cerebellum, causing uncoordinated movements.

Cerebrum

The **cerebrum** consists of the cerebral cortex, which is the outer surface of the brain, and the subcortical region, which lies beneath the cortex. The cerebrum is made up of white matter and grey matter, like the spinal cord. However, the grey matter in the brain is located in the cortex, and the white matter is in the subcortical region; this is opposite to their positions in the spinal cord.

The **subcortical region** of the brain contains several important areas, including the basal nuclei and the amygdala. The basal nuclei are sometimes called the *basal ganglia*. But recall that in the CNS groups of cell bodies are called *nuclei*, and in the PNS groups of cell bodies are called *ganglia*. The basal nuclei consists of several groups of neurons that are primarily involved in regulating skeletal muscle movements, posture, and complex behaviour patterns (Figure 15.30). The basal nuclei control movements in conjunction with the midbrain, frontal lobe and motor cortex, thalamus, and cerebellum. One of the most well-known of the basal nuclei is the **nucleus accumbens**, which has been shown to play a role in motivation, aversion, incentive, compulsions, impulsivity, pleasure, positive reinforcement, reward and addictions. One of the main neurotransmitters released by neurons in this region is dopamine (Section 15.3). It is thought that addictive drugs such as nicotine, cocaine, and amphetamines increase the amount of or response to dopamine. Another well-known basal nucleus is the **caudate nucleus**, which plays an important role in ability to concentrate, learning, memory, and feeling romantic love and attachment. The **amygdala**, a subcortical nucleus, is a component of the limbic system. It is involved in emotions, fear, aggression, phobias, pleasure, and complex social and sexual interactions.

The **limbic system** consists of a group of structures, including regions of the thalamus, hypothalamus, amygdala, hippocampus (part of the cerebral cortex), olfactory bulb (detects sense of smell), fornix (fibre tracts that connect the limbic system structures), and the cingulate gyrus (located above the corpus callosum) (Figure 15.31). In combination, these structures play a major role in our emotions, learning, memory, and social interactions. Because the olfactory bulb is part of the limbic system, the sense of smell is highly linked to memories. You may have noticed that a particular scent causes you to recall past events in great detail. This strong connection between scents and emotions is also the basis of the perfume industry. The limbic system is like a primitive brain. It functions to protect us by reacting rapidly to emotional sensory stimuli: for example, getting angry and becoming aggressive, or feeling sad and wanting to cry. Our limbic system reactions are highly modified by the frontal lobe of the cerebral cortex, which develops as we live in social environments. For example, we may feel angry but still respond in a socially appropriate way. A baby or a two-year-old uses primarily their limbic system with very little input from their frontal lobe.

The **cerebral cortex** differentiates humans from other animals. It has a highly folded surface that increases the surface area of grey matter. An indentation is called a **sulcus**, and the protruding portion is a **gyrus** (Figure 15.32a). A deep fissure—called the **longitudinal fissure**—runs along the centre of the brain and splits the brain into the right and left hemispheres. A groove along the centre of the brain divides the brain into anterior and posterior regions (separates the frontal lobe from the parietal lobes); this groove is called the **central sulcus**. The protruding

15

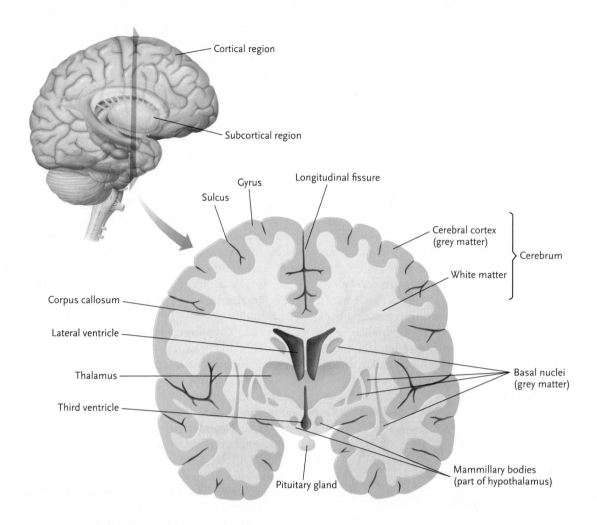

Figure 15.30 Cerebrum

The cerebrum is divided into the cortex and the subcortical regions.

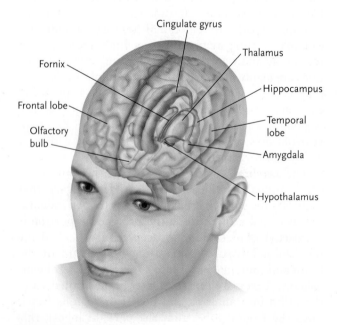

Figure 15.31 Limbic System

The limbic system is a group of structures that are involved in emotions.

region immediately anterior to the central sulcus is the **pre-central gyrus**, also known as the *motor cortex*. The protruding region immediately posterior to the central sulcus is the **post-central gyrus**, also known as the *sensory cortex*. The cerebral cortex is made up of the following lobes, which are named after the skull bones they are associated with: frontal lobe, left and right parietal lobe, left and right temporal lobe, and the occipital lobe and insula (Figure 15.32b).

The right and left hemispheres of the brain are connected by the corpus callosum in the subcortical region. The motor cortex and sensory cortex on one side of the brain control and receive input from the opposite side of the body. The right and left sides of the brain communicate through the corpus callosum, but sensory information crosses to the opposite side either in the spinal cord, the midbrain, or the thalamus.

The **frontal lobe** contains the motor cortex, which is where conscious control of skeletal muscles begins.

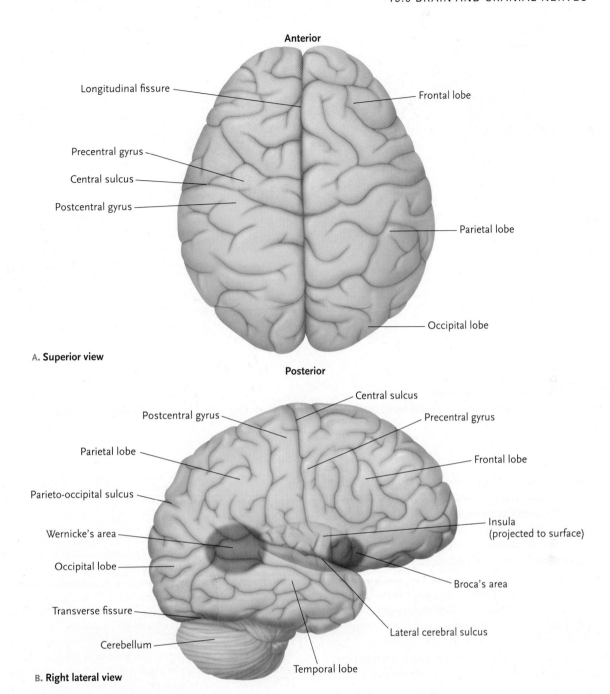

Anterior

Longitudinal fissure

Frontal lobe

Precentral gyrus

Central sulcus

Postcentral gyrus

Parietal lobe

Occipital lobe

A. **Superior view**

Posterior

Central sulcus

Postcentral gyrus

Precentral gyrus

Parietal lobe

Frontal lobe

Parieto-occipital sulcus

Wernicke's area

Insula
(projected to surface)

Occipital lobe

Broca's area

Transverse fissure

Cerebellum

Lateral cerebral sulcus

Temporal lobe

B. **Right lateral view**

Figure 15.32 Cerebral Cortex
(A) Superior view and (B) lateral view

The neurons in specific regions of the motor cortex as well as input from the cerebellum and basal nuclei control voluntary movements. The motor cortex has specific designated regions that associate with specific parts of the body (Figure 15.32). The **pre-frontal cortex**, located just anterior to the motor cortex, is involved in the planning of motor movements; it is the region of the brain where we make the decision to initiate a movement. The frontal lobe has many other important roles, including higher-thinking functions,

such as setting goals, planning, reasoning, making decisions, concentrating on a task, and also short-term memory. The frontal lobe plays an important role in self-control and regulates the expression of emotions that are triggered by the limbic system. The frontal lobe also contains a very important region in the left hemisphere, called **Broca's area**, which is a language centre. The Broca's area controls the motor movements involved in speaking.

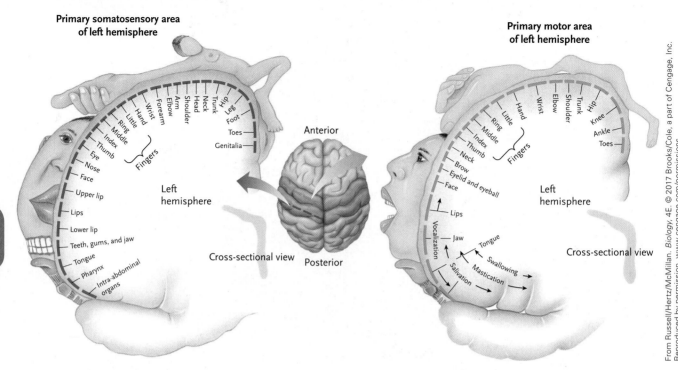

Figure 15.33 Motor Cortex and Somatosensory Cortex

The motor cortex and the somatosensory cortex each have a specific distribution of neurons that is linked with specific regions of the body.

From Russell/Hertz/McMillan. *Biology*, 4E. © 2017 Brooks/Cole, a part of Cengage, Inc. Reproduced by permission. www.cengage.com/permissions

The **parietal lobe** contains the sensory cortex, also called the **somatosensory cortex** because it receives input from sensations coming from the skin, muscles, and joints, but not the eyes or ears or other special senses. The sensory cortex, like the motor cortex, has specific, designated regions that associate with specific regions of the body (Figure 15.33). The parietal lobe has many other important functions, such as *integrating* sensory information from the somatosensory cortex, the **occipital lobe**, which receives visual information, and the **temporal lobe**, which receives auditory information. The parietal lobes use the incoming sensory information for visual-spatial processing: being able to recognize distance and changes in the environment that could impact our motor movements, such as avoiding a pothole when jogging or grasping an object. The right and left hemispheres of the parietal lobes have some differences. The left hemisphere is more involved in recognizing numbers and patterns and language, and the right side is more involved in spatial relationships and interpreting images such as a map. The parietal lobe is also the region of the brain involved in human self-awareness and an awareness of being part of a larger universe.

The temporal lobe contains the auditory cortex, which receives information about sound from the auditory nerve, and *association areas* that interpret sounds, including speech. The temporal lobe has a region called

DID YOU KNOW?

People with attention deficit disorder have less dopamine released in neurons of the frontal lobe and the caudate nucleus that are involved in concentration, learning, memory, and focusing on a task. With decreased dopamine in the frontal lobe and caudate nucleus, a person is easily distracted, has poor short-term memory, and difficulty completing tasks. Ritalin and other similar medications stimulate dopamine secretion in the frontal lobe and caudate nucleus, which helps increase concentration. Children are normally easily distracted; this usually decreases as they grow older and practise concentrating. The more a person thinks, learns, and concentrates, the more the brain adapts to doing this—just like exercising causes muscles to be stronger. For the most part, children grow out of ADD, usually by the time they reach the late teens.

Wernicke's area, which is important for comprehending the meaning of spoken language; this is not to be confused with Broca's area in the frontal lobe that allows us to physically speak. The temporal lobe is the region of the brain that stores memories of names and different sounds. The hippocampus, a structure located within each temporal lobe, plays a key role in long-term memory. The hippocampus is one of the structures of the limbic system involved in emotions, and so long-term memory

TABLE 15.3

Function and Point of Origin of the Cranial Nerves

Nerve	Function	Origin
I Olfactory	Sense of smell	Cerebrum
II Optic	Vision	Cerebrum
III Oculomotor	Movement of eyelid, eyeball, lens, pupil constriction	Midbrain
IV Trochlear	Movement of eyeball	Midbrain
V Trigeminal (ophthalmic, maxillary, and mandibular branch)	Sensations in head, face, and jaw; motor control of chewing	Pons
VI Abducens	Movement of eyeball	Pons
VII Facial	Taste, facial expression, tears, salivation, and facial sensations	Pons
VIII Vestibulocochlear	Equilibrium and hearing	Lateral to facial nerve
IX Glossopharyngeal	Taste, swallowing, speech, and salivation	Medulla oblongata
X Vagus	Taste, sensations from pharynx, swallowing, coughing, voice, GI tract smooth muscle contraction, heart rate reduction, digestive secretion	Medulla oblongata
XI Accessory	Swallowing and movement of head and shoulders	Medulla oblongata
XII Hypoglossal	Speech and swallowing (tongue muscles)	Medulla oblongata

and learning is enhanced when it is linked with an emotional experience.

The occipital lobe contains the primary visual cortex and association areas that interpret light and colours and recognize objects. The occipital lobe also plays a role in interpreting motion. Information about our body position and movement occurs in both the occipital lobe and the parietal lobe, which receives sensory input from joints and muscles as well as the vestibular system. When information about body position appears to the brain as different than usual, such as standing in a rocking boat, the visual information tells the occipital lobe that the body is *stable*, but the information from contracting muscles and moving joints, as well as the vestibular system in the inner ear, tells the parietal lobe the body is *moving*; this incongruence causes motion sickness.

The final lobe of the brain is the **insula**, located in the folds deep to the temporal lobe. The insula is involved in recognizing pain, as is the somatosensory cortex. However, the insula seems to be more involved in how we perceive the pain, compared to the sensory cortex that interprets the location and intensity of the pain. The insula also interprets gastric and bladder distention.

The motor cortex and the somatosensory cortex have very similar brain maps that associate with specific areas of the body (Figure 15.33). The parts of the body that have the most corresponding brain area have the highest amount of motor movement and sensitivity. On the motor cortex, there is a large amount of brain tissue dedicated to the face, mouth and lips, and hands. This large area relates to our need for the very extensive and finely controlled muscle movements in facial expressions, moving the lips and mouth during chewing and talking, and in all the detailed movements of our hands. The same is true for the sensory cortex. There is a lot of brain tissue dedicated to receiving sensory information from highly sensitive areas—lips and hands (Chapter 16.1).

Cranial Nerves

There are 12 pairs of cranial nerves that originate in the brain but are part of the peripheral nervous system. Table 15.3 summarizes the main functions of each cranial nerve. Cranial nerve I, the olfactory nerve that is not really a true nerve, stays within the CNS and carries information from the olfactory epithelial cells (sense of smell) to the olfactory bulb. Similarly, cranial nerve II is the optic nerve that carries information from the eyes to the occipital lobe and is not a true nerve. Cranial nerves III to XII originate from the brainstem and carry both motor and sensory information to specific tissues in the body (Figure 15.34).

15

KEY

— Sensory fibres
— Motor fibres

Olfactory nerve (I)
Mucosa of
nasal cavity

Optic nerve (II)

Retina

Oculomotor nerve (III)
Extrinsic eye muscles;
ciliary muscle;
muscles of iris

Trochlear nerve (IV)
Superior oblique
(extrinsic eye muscle)

Abducens nerve (VI)
Lateral rectus
muscle

Motor–muscles
of mastication

Sensory–face
and head

**Trigeminal
nerve (V)**

Motor–muscles
of face and scalp;
salivary and
tear glands

Sensory–taste buds
on anterior tongue

Facial nerve (VII)

**Vestibulocochlear
nerve (VIII)**

**Glossopharyngeal
nerve (IX)**

Motor–muscles of
pharynx; paratid gland

Cochlea, vestibule,
and semicircular
canals of inner ear

Sensory–taste buds
on posterior tongue;
receptors in pharynx
and carotid sinus

Vagus nerve (X)

**Accessory
nerve (XI)**

**Hypoglossal
nerve (XII)**

Tongue muscles

Muscles of larynx,
pharynx, soft palate,
shoulder, and neck

Motor–muscles of
pharynx and layrnx;
thoracic and
abdominal organs

Sensory–taste buds
on tongue and
pharynx; thoracic
and abdominal
organs

Figure 15.34 Cranial Nerves
There are 12 pairs of cranial nerves.

CONCEPT REVIEW 15.6

1. Describe the features that help protect the brain.

2. What cell type produces CSF, and where in the brain does this occur?

3. Where is CSF reabsorbed back into the bloodstream?

4. Name the four regions of the brain, and list the main structures found in each region.

5. What are the main functions of the thalamus and pineal gland?

6. What is the role of the reticular activating system? Where is it located?

7. What is the role of the substantia nigra in the midbrain?

8. Name two functions of the pons.

9. What two major organ systems are regulated by the medulla oblongata?

10. What is the function of the cerebellum?

11. Describe the main function of the basal nuclei and amygdala.

12. What is the function of the limbic system?

13. Name the lobe of the brain, and any specific region, if applicable, that is primarily involved in each of the following functions:

    a. Recognizing the sound of someone's voice
    b. Being aware of your body position
    c. Having the physical ability to speak
    d. Deciding to walk across a room
    e. Calculating the cost of your school supplies
    f. Recognizing the difference between red and blue paint
    g. Remembering a childhood experience
    h. Concentrating on a difficult task
    i. Understanding language
    j. Interpreting the pain of a bee sting as being more severe because of a fear of bees

14. Make a chart to list the main functions of each cranial nerve.

15.7 AUTONOMIC NERVOUS SYSTEM

The autonomic nervous system is the involuntary, unconsciously controlled component of the nervous system that regulates the smooth muscle, cardiac muscle, and some glands. The ANS has two divisions: the **sympathetic nervous system**, also known as the *fight-or-flight response*; and the **parasympathetic nervous system**, also known as the *rest-and-digest response*. Most tissues receive input from both systems; usually one division is stimulatory and the other is inhibitory. Table 15.4 describes the functions of each system in relation to various tissues.

The ANS neural pathway from the CNS to the tissues involves two neurons. The first neuron leaves the CNS through a cranial or spinal nerve and synapses with a second neuron in an **autonomic ganglion** (Figure 15.35). The first neuron is the **preganglionic**, and the second neuron is the **postganglionic**. One preganglionic neuron may synapse with many postganglionic neurons. Sympathetic preganglionic neurons are shorter than parasympathetic neurons, and the ganglia are often located close to the spinal cord. By contrast, parasympathetic ganglia are located very close to or within the target tissue. All ANS preganglionic neurons release acetylcholine (ACh) at the first synapse.

TABLE 15.4

Comparison of the Somatic and Autonomic Nervous Systems

Somatic	Autonomic
Voluntary	Involuntary
Innervates skeletal muscle	Innervates smooth and cardiac muscle and some glands
Always excitatory	Can be excitatory or inhibitory
Neurotransmitter is acetylcholine (ACh).	Neurotransmitter can be acetylcholine or norepinephrine (NE). Epinephrine is released as a hormone by the adrenal gland.
One motor neuron connects the spinal cord with the muscle fibre.	Two neurons—one preganglionic, the other postganglionic—connect the CNS to the muscle or gland.

15

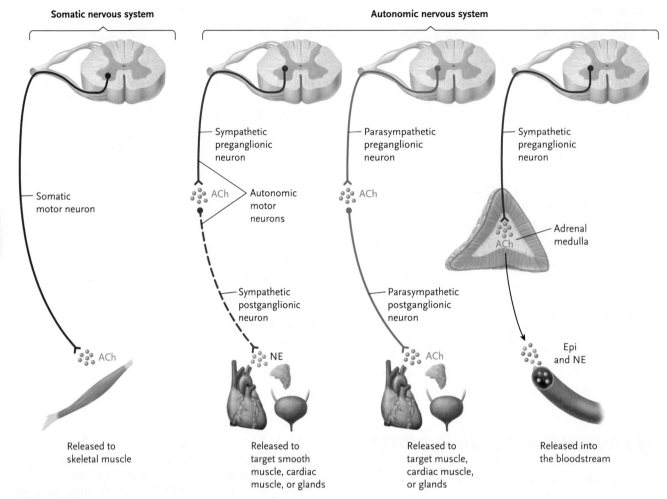

Figure 15.35 Somatic and Autonomic Nervous Systems

Somatic and autonomic nervous systems release specific neurotransmitters onto specific cell types to cause various effects.

Parasympathetic postganglionic neurons release acetylcholine on the target muscle or gland, whereas sympathetic neurons release either **norepinephrine** or epinephrine. The sympathetic nervous system preganglionic neurons can signal the **adrenal medulla** to release epinephrine as a hormone into the bloodstream, which causes the same, but longer-lasting, effects as the postganglionic neuron releasing norepinephrine directly onto target tissues.

Neurons of the sympathetic nervous system leave the CNS from the thoracic and upper lumbar spinal nerves, and the parasympathetic neurons leave the CNS through cranial or sacral spinal nerves (Figures 15.20 and 15.34). The functions of the sympathetic nervous system include effects on the tissues required for a stress response, such as increasing heart rate; dilating bronchial tubes to increase the amount of oxygen that enters the blood for energy production; and increased blood flow to skeletal muscles (Figure 15.36). The stress response is triggered when we need to "fight or flee" a dangerous situation, and through evolution

this has become very beneficial for a rapid physical response. However, in a modern technological world where we no longer need to hunt or gather food, or fight for territory, our stress response tends to be triggered primarily during exercise and emotional stress. It is healthy and beneficial for the sympathetic response to be activated during exercise, when the increased blood sugar is used to contract muscles, but chronically activating the sympathetic response for emotional, non-physical stress can be detrimental in the long term since the increased blood sugar is not used to make ATP.

The functions of the parasympathetic nervous system involve stimulating digestion and relaxation, such as increasing blood flow and smooth muscle contraction in the gastrointestinal tract, slowing heart rate, and promoting the storage of nutrients. The two divisions of the ANS are always functioning, and one system is more or less dominant at any point in time (Figure 15.36). Table 15.5 lists the effects of the sympathetic and parasympathetic nervous systems.

Sympathetic division

Parasympathetic division

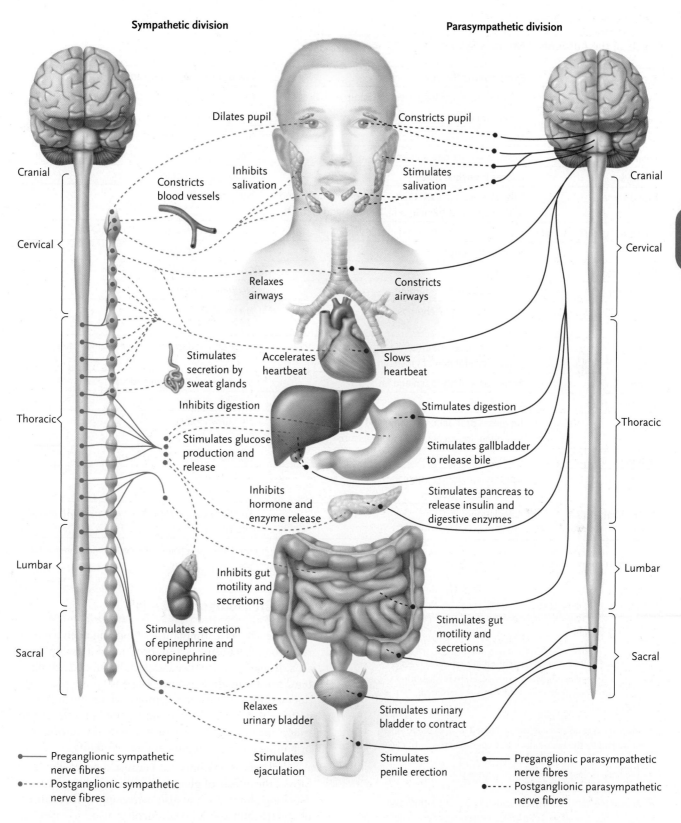

Cranial

Cervical

Thoracic

Lumbar

Sacral

Cranial

Cervical

Thoracic

Lumbar

Sacral

Dilates pupil

Constricts pupil

Inhibits salivation

Constricts blood vessels

Stimulates salivation

Relaxes airways

Constricts airways

Stimulates secretion by sweat glands

Accelerates heartbeat

Slows heartbeat

Inhibits digestion

Stimulates digestion

Stimulates glucose production and release

Stimulates gallbladder to release bile

Inhibits hormone and enzyme release

Stimulates pancreas to release insulin and digestive enzymes

Inhibits gut motility and secretions

Stimulates gut motility and secretions

Stimulates secretion of epinephrine and norepinephrine

Relaxes urinary bladder

Stimulates urinary bladder to contract

Stimulates ejaculation

Stimulates penile erection

● —— Preganglionic sympathetic nerve fibres
● ---- Postganglionic sympathetic nerve fibres

● —— Preganglionic parasympathetic nerve fibres
● ---- Postganglionic parasympathetic nerve fibres

Figure 15.36 Sympathetic and Parasympathetic Nervous Systems

The sympathetic nervous system causes the fight-or-flight response, and the parasympathetic nervous system causes the rest-and-digest response.

TABLE 15.5

Effects of the Autonomic Nervous System

Tissue	Sympathetic Response	Parasympathetic Response
Eyes	Dilates pupils and increases far vision	Constricts pupils and increases near vision
Lacrimal glands	No known effect	Production of tears
Salivary glands	Decreases production of saliva	Increases production of saliva
Heart	Increases heart rate and force of contraction	Decreases heart rate and force of contraction
Airways	Dilates bronchial tubes	Constricts bronchial tubes.
Blood vessels	Dilates vessels in skeletal muscles, heart, lungs, and brain; constricts them in skin, kidneys, and GI tract	No known effect
Smooth muscle in GI tract	Decreases contraction and motility	Increases contraction and motility
GI tract	Inhibits secretions	Stimulates secretions
Gallbladder	Stores bile	Contracts and releases bile
Sweat glands	Increases sweat production	No known effect
Erector pili muscles in skin	Contraction—goosebumps	No known effect
Pancreas	Inhibition of secretion of digestive enzymes and insulin; production of glucagon	Secretion of digestive enzymes and insulin
Liver	Breakdown of glycogen and fat	Synthesis of glycogen and fat
Adipose tissue	Breakdown of fat	No known effect
Kidney	Increases production of renin and therefore water retention and increased blood pressure	No known effect
Bladder	Relaxes, no urination*	Contracts and causes urination
Uterus	Contraction during pregnancy; inhibits contraction in non-pregnant women	No known effect
Reproductive tract	Ejaculation	Erection, vasodilation of blood vessels in penis/clitoris

* Note that during a sympathetic nervous system response, the blood pressure increases, which can cause increased urine production and the urge to urinate; however, the parasympathetic nervous system is what stimulates urination.

DID YOU KNOW?

One of the healthiest ways to deal with emotional stress is to engage in regular, vigorous exercise. Here is a list of some of the benefits of exercise to relieve emotional stress.

- Muscles take up and use the increased level of blood sugar produced by the liver— during stress or exercise—to make ATP.
- The brain releases endorphins that make you feel good— known as *runner's high*.
- Dopamine increases, which helps you to concentrate and focus your attention. Try a 30-second sprint or do 10 push-ups the next time you need to take a study break,
- Exercise increases other molecules involved in elevating mood, such as serotonin and oxytocin.
- Exercise results in a burst of sympathetic nervous system activity and cortisol release that quickly dissipates and leads you to feeling calm but energized.

How Do Drugs Affect the Brain?

- Caffeine acts as a stimulant by blocking adenosine receptors. Adenosine is a by-product of action potentials. It makes us feel tired, especially after thinking hard for a long time. Neurons can adapt by producing more adenosine receptors, which means more caffeine is required for the same stimulant effect over time. Since caffeine can also stimulate the release of dopamine, it has a slightly addictive quality.

- Alcohol acts as a depressant because it generally blocks the effects of glutamate, which is stimulatory. Blocking glutamate and also increasing GABA causes us to feel calm and relaxed. Alcohol also increases serotonin, which makes us feel content and in a good mood, as well as feeling more social. Alcohol can bind to calcium channels in axon terminals and inhibit the release of neurotransmitters in the cerebellum, causing loss of coordination. Alcohol can inhibit neurotransmitter release in many areas of the brain, including the medulla oblongata. Excessive alcohol consumption leading to alcohol poisoning can cause death due to the suppression of the medulla respiratory centre and the inhibition of the reticular activating system, which affects consciousness.

- Nicotine acts as a stimulant by binding to nicotinic acetylcholine receptors. Acetylcholine is excitatory and important for cognitive functions in the brain. Nicotine stimulates dopamine release, which both gives it addictive qualities and brings about a relaxing feeling, even though it is a stimulant. Nicotine also increases epinephrine, serotonin, and endorphins.

- Marijuana contains the psychoactive chemical TCH (delta-9-tetrahydrocannabinol), which binds to cannabinoid receptors in the brain. Most people experience a relaxing effect, but some experience the opposite effect and feel anxiety and paranoia. Regions of the brain with high concentrations of cannabinoid receptors are the hippocampus (can affect memory), and the cerebellum and basal ganglia (can affect coordination of muscle movements). Marijuana can be used for medical purposes and has been effective in relieving pain in people with MS and cancer.

- LSD and other hallucinogens bind to serotonin receptors in various parts of the brain, including the subcortical nuclei that affect perceptions of sensory stimuli. As a result, people often see, hear, and feel things that are not there. The hallucinogenic drugs include ecstasy, PCP, psilocybin (magic mushrooms) and peyote.

- Cocaine, methamphetamine, and other stimulants increase the release of dopamine in the frontal lobe and caudate nucleus, which increases the ability to concentrate. For this reason, stimulants such as Ritalin and Adderall are used to treat ADHD (attention-deficit/hyperactivity disorder). Stimulants also increase dopamine in the limbic system. When consumed in very large amounts, such as the use of meth or cocaine, stimulants produce a euphoric high and can be very addictive.

- Opioids include drugs such as heroine, OxyContin, Vicodin, methadone, and Dilaudid. These drugs bind to opioid receptors in the brain and inhibit the transmission of pain signals. They also decrease the production of norepinephrine and act as a central nervous system depressant. Opioids also trigger the release of dopamine in the limbic system. In high amounts, the excess dopamine can cause euphoria, which is why opioids are highly addictive. Dependence on opioids can occur very quickly, and withdrawal symptoms resemble having a severe flu.

- Ecstasy is a psychoactive chemical called *MDMA* (3,4-methylenedioxymethamphetamine). MDMA slows the reuptake of serotonin, oxytocin, norepinephrine, and dopamine and increases feelings of empathy, love, sociability and bonding (oxytocin), heightened sensations (serotonin), increased energy (norepinephrine), and euphoria (dopamine).

- Benzodiazepines are sedatives or tranquilizers that include drugs such as valium, diazepam, lorazepam, and clonazepam. These drugs are used to treat panic and anxiety disorders and insomnia because they mimic the inhibitory neurotransmitter GABA and cause a relaxing effect. Tolerance (reduced effectiveness) to these drugs can develop with extended use.

CONCEPT REVIEW 15.7

1. What types of tissues are innervated by the somatic and autonomic nervous system?

2. Which neurotransmitter is released by the preganglionic neurons in the ANS?

3. What neurotransmitters are released by the sympathetic and parasympathetic postganglionic neurons?

4. Name any five tissues that are stimulated by the sympathetic nervous system.

5. What happens in the liver during a fight-or-flight response compared to a rest-and-digest response?

6. Where do the preganglionic neurons of each ANS division originate in the CNS?

7. How is the adrenal gland involved with the autonomic nervous system?

THE EFFECTS OF AGING

As we age, our central and peripheral nervous systems go through natural changes. Some of the important changes include the following:

1. Neurons are continuously lost throughout our life. The loss of neurons can decrease reflexes, sensations, senses, and motor control over movements.
2. Brain volume decreases.
3. Neuron action potentials move more slowly along axons.
4. Waste products from metabolism can build up and interfere with action potentials.
5. Fewer neurons, increased metabolites, and slower action potentials decrease the ability to think, learn, and remember.
6. Some people experience dementia, which is not part of normal aging but is very common and involves mild to extreme loss of cognitive function.

These changes are slowed or accelerated depending on our health, diet, blood sugar level, amount of exercise, exposure to

(Continued)

toxins or drugs, genetic predispositions, positive or negative emotions, stress, and learning. The following are some ways to slow these effects of aging:

1. Control blood sugar by minimizing the consumption of refined sugar. Sugar increases inflammation in the whole body, and high blood sugar impairs normal cognitive functions.
2. Engage in exercise to increase blood flow, which helps bring nutrients to the brain and flush away metabolic waste.

3. Get enough sleep, which increases the production of CSF and helps flush away metabolic waste. Lack of sleep increases the production of stress hormones—primarily cortisol—that have a significant negative impact on cognitive functions.
4. For mental exercise, continue to learn and challenge your brain; this ensures that synapses between neurons are not lost.

CHAPTER REVIEW

Medical Terminology

Term	Definition	Example
astr/o	star, star shaped	astrocyte
cerebell/o	cerebellum	cerebellospinal
cerebr/o	cerebrum, brain	cerebrovascular
dipl/o	double	diploid
drom/o,	running	syndrome
encephal/o	brain	encephalitis
esthesi/o,	sensation, feeling	anesthesia
gangli/o,	ganglion	gangliocytoma
gli/o	glue, neuroglia	gliocyte
hydr/o	water, hydrogen	hydrophobic
hygr/o	moisture	hygroblepharic
hypn/o	sleep	hypnogenic
hypothalam/o	hypothalamus	hypothalamo-hypophysical
kerat/o	cornea, horny tissue	keratomalacia
klept/o	theft, stealing	kleptomania
-lemma	confining membrane	epilemmak
-lepsy	seizure	epilepsy
-lexia	speech, word	bradylexia
log/o, log, -logue	word, speech, thought	logorrhea
-mania	madness, obsessive, preoccupation	hypomania
medull/o	medulla, marrow	medulloblast
mening/o	meninges, membranes	meningitis
ment/o	mind	dementia
-mnesia	memory	amnesia
myel/o	bone marrow, spinal cord	myeloblast
neur/o	nerve	neurotransmitter
-noia	mind, will	paranoia
-paresis	partial paralysis	hemiparesis
-phobia, phob/o	fear, aversion	hydrophobic
phren/o	mind, diaphragm	tachyphrenia

-piesis	pressure	piesesthesia
platy-	broad, flat	platycoria
-plegia	paralysis	quadriplegia
poli/o	grey (matter)	poliomyelitis
-pore	opening, passageway	neuropore
-praxia	action, activity	parapraxia
psych/o	mind	psychokinesis
psychr/o	cold	psychrophobia
radicul/o	nerve root	radiculitis
rhiz/o	root	rhizotomy
schiz/o	split, division	schizophrenia
somn/l, -somnia	sleep	insomnia
syring/o	tube, fistula	syringomyelocele
-taxia, tax/o	arrangement, coordination	ataxia
-tropia	to turn	anatropia
uve/o	uvea	uveoplasty
vag/o	vagus nerve	vagolysis

Labelling Activity

Label the following diagrams.

1.

2.

3.   a.   b.

4.

5.

15

6.

7.   a.   b.

c.

8.

Stimulus

1 _____

2 _____

3 _____

4 _____

5 _____

6 _____

KEY
⊕ Stimulates
⊖ Inhibits

From DiGiuseppe/Fraser. *Biology* 12U, 1E. © 2012 Nelson Education Ltd.
Reproduced by permission. www.cengage.com/permissions

9.

10.

Top

Front
of
brain

15

Courtesy of Mark Nielsen, AnatBooks, LTD

Courtesy of Mark Nielsen, AnatBooks, LTD

11.

12.

13.

Answer to Application Question

Pain and pressure receptors in Trevor's foot detect the sensory stimulation caused by his stepping on a piece of Lego. Sensory (afferent) neurons send the information to the spinal cord, where a withdrawal reflex reaction causes motor information to be sent to the leg muscles, which then contract so that the foot is immediately removed from the Lego.

Sensory information is also sent to the brain by sensory pathways (Chapter 16) that connect with Trevor's sensory cortex, located in the parietal lobe. A specific region of the parietal lobe detects information from his foot, so that Trevor is aware that his foot has pain. Various parts of the brain integrate sensory information: the cerebellum, which maintains coordination of muscles as he pulls his foot away from the Lego; the midbrain, which regulates muscle movements; and the frontal lobe, which stimulates voluntary motor movements. If Trevor becomes angry, his autonomic nervous system will also be activated, causing a sympathetic nervous system response (fight-or-flight response). The sympathetic nervous system sends autonomic information to the body via cell bodies of the lateral grey matter, causing effects such as increased heart rate, bronchodilation, inhibited digestion, dilation of pupils, and dilation of blood vessels in the skeletal muscle, heart and lungs. The main neurotransmitters involved in the sympathetic nervous system response include acetylcholine, released by preganglionic neurons; and epinephrine and norepinephrine, released by postganglionic neurons and the adrenal medulla.

16 The Sensory System

CHAPTER OUTLINE

16.1 TOUCH, TEMPERATURE, AND PAIN

The sensory system allows the brain to perceive environmental stimuli such as touch, pressure, vibration, proprioception, temperature, and pain. These stimuli are detected by the peripheral nervous system through specific receptors (Chapter 1).

- Tactile receptors, which detect touch, pressure, and vibration
- Proprioceptors, which detect body position and movement
- Thermoreceptors, which detect temperature
- Nociceptors and free nerve endings, which detect pain (Note that free nerve endings are dendrites, not axon terminals.)

The sensory receptors in the skin include the following (see Figure 16.1):

- **Meissner's corpuscle,** which detects touch and pressure
- **Merkle's corpuscle,** which detects touch and pressure
- **Free nerve endings,** some of which detect pain (nociceptors), others detect temperature
- **Pacinian corpuscles,** which detect vibration and deep pressure
- **Ruffini corpuscles,** which detect skin stretching
- **Hair follicles,** which detect touch and vibration

You learn something every day if you pay attention.

Ray LeBlond

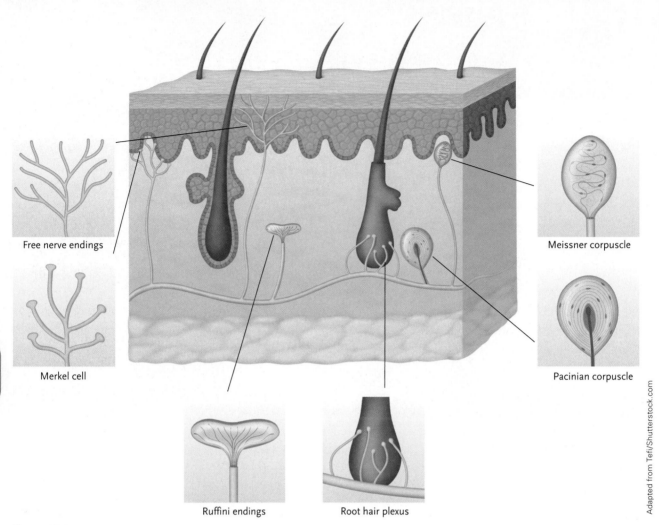

Free nerve endings

Merkel cell

Meissner corpuscle

Pacinian corpuscle

Ruffini endings

Root hair plexus

Adapted from Tefi/Shutterstock.com

Figure 16.1

Tactile receptors detect touch, pressure, and vibration.

The sensory system is made up of receptors, sensory (afferent) neurons, and an integrating centre (brain). Sensory receptors cause the depolarization of afferent neurons (Chapter 15) that send action potentials to the various regions of the brain for interpretation. How we interpret sensory information is called *perception*. Sensory information can be conscious, such as the sensation of touch or pain, or it can be unconscious, such as blood pressure or blood ion concentrations. The pattern of action potentials is a code that provides information about the type, location, and intensity of the sensation.

Thermoreceptors are sensory neurons located throughout the skin, cornea, and bladder. Different thermoreceptors respond to heat and cold. Very hot or very cold temperatures can stimulate free nerve endings and is detected as pain. We also have thermoreceptors in the hypothalamus that respond to core body temperature changes.

Proprioceptors detect body position and movement. We have proprioceptors in the skeletal muscles (spindle fibre, Chapter 14), tendons (Golgi tendon organ), and joint capsules. We also detect body position and movement through the vestibular system (Section 16.3). Proprioception is required for walking in the dark without falling and for learning new movements, such as playing a sport or musical instrument. People with an amputated limb may feel the presence of that limb due to proprioceptive memory, a phenomenon called *phantom limb syndrome*.

DID YOU KNOW?

Some people do not have proprioception and therefore have no awareness of their body position and have to use visual information to move appropriately. In severe cases, these people feel as though they are detached from their body. A milder form of this can occur in people who take excessive amounts of vitamin B6, or those who have had a certain type of chemotherapy.

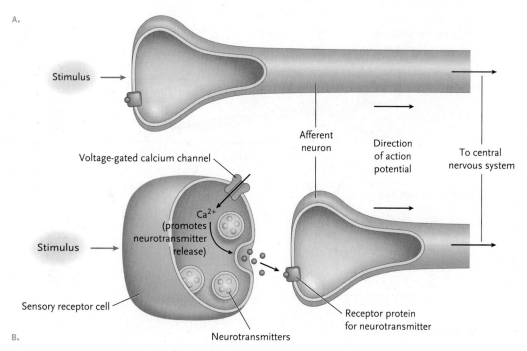

Figure 16.2 Sensory Receptors

Sensory receptors can be (A) a sensory neuron or (B) a sensory cell.

Receptor Potential

Recall from Chapter 15 that graded potentials are caused by neurotransmitters that depolarize or hyperpolarize the dendrites and cell body of a neuron, which leads either to the occurrence of an action potential, in the case of an EPSP, or to the lack of an action potential, in the case of an IPSP. Receptor potentials are graded potentials that occur in a sensory receptor cell or directly on the dendrite of a sensory neuron (Figure 16.2).

Receptor potentials are graded, meaning that an increase in the intensity of the stimulus will cause an increase in the *amplitude* of the receptor potential, which will then cause an increase in the *frequency* of action potentials (Figure 16.3). Receptor potentials are generally depolarizing, but can be hyperpolarizing. During a depolarizing receptor potential, sodium ions enter the cell; during a hyperpolarizing receptor potential, potassium ions leave the cell.

Adaptation

Sensory receptors adapt to continuous stimuli by decreasing the amount of depolarization, even when the stimulus is still present. This is why we are not constantly aware of our clothing or jewelry touching our skin. When there is a change in the stimulus, the receptors will once again generate receptor potentials. Notice in Figure 16.3 that receptor potentials decrease over time, and therefore the frequency of action potentials also decreases.

Receptive Field

A sensory neuron can have a large or small receptive field, depending on the number of dendrites in the sensory area (Figure 16.4). A small receptive field has very few dendrites for detecting stimuli, and this increases the ability to precisely locate a stimulus. On the other hand, a large receptive field has more dendrites covering a larger area, which reduces the ability to precisely locate a stimulus. Areas of the body that require us to determine a specific location of a sensory stimulus are more sensitive to stimuli and have smaller receptive fields: for example, the lips, tongue, and fingertips.

Application

As you read through the rest of this chapter, consider this scenario. Gertrude was riding her road bike on a long country road when a distracted driver hit her. She was thrown from her bike and landed in a ditch at the side of the road. She had severe injuries, including damage to the right dorsal column in her lower back because a disc in a lumbar vertebra had shifted in her spine. What sensory input would be affected by this injury? What parts of her body would be affected and why?

16

Figure 16.3 Stimulus Intensity and Receptor Potentials

As stimulus intensity increases 1. there will be an increase in receptor potential amplitude and 2. an increase in action potential frequency.

Neural Pathways

Sensory information reaches the brain by transmission through three neurons that extend from the sensory receptor to the somatosensory cortex. Different receptors signal different neural pathways, so specific information goes to the correct region of the brain for interpretation. Two main pathways carry different sensory information to the brain: the anterolateral spinothalamic tract (Figure 16.5a), and the dorsal column-medial lemniscal tract (Figure 16.5b). Sensory information from the left side of the body enters the right side of the brain, and information from the right side of the body enters the left side of the brain. Recall from Chapter 15 that the somatosensory cortex is located in the anterior region of the parietal lobes

(Figure 16.6). Also recall that the thalamus is an important sensory relay area, where all peripheral sensory information enters and is then relayed to the appropriate region in the somatosensory cortex.

Anterolateral Tract

The anterolateral tract carries pain and temperature information. Neurons from the periphery enter the spinal cord dorsal root ganglion (recall from Chapter 15 that all sensory information enters through the dorsal root, and all motor information leaves through the ventral root). The first sensory neuron synapses with another neuron in the spinal cord. The second neuron crosses to the opposite side of the spinal cord, ascends

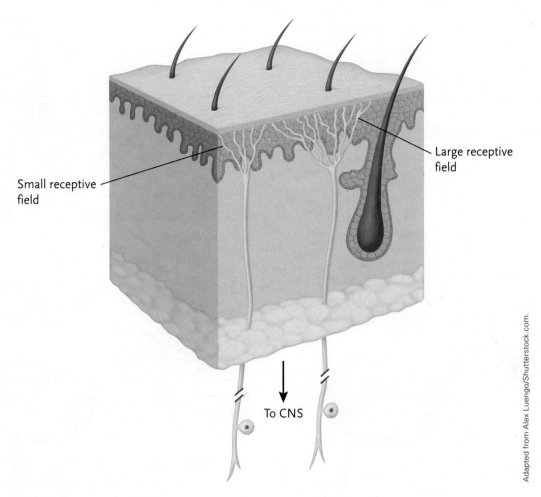

Small receptive field

Large receptive field

To CNS

Adapted from Alex Luengo/Shutterstock.com.

Figure 16.4 Receptive Fields
Areas of the body that have small receptive fields have a greater ability to precisely locate a stimulus, compared to areas of the body with larger receptive fields.

through the anterolateral white matter tract, and synapses with a neuron in the thalamus. The third neuron then transmits information to the correct region in the somatosensory cortex.

Dorsal-Medial Tract

The dorsal-medial tract carries touch, pressure, limb position, and body movement information. Neurons from the periphery enter the spinal cord dorsal root ganglion and carry information through the dorsal-medial white matter tract to the medulla oblongata. The second neuron crosses to the opposite side in the medulla oblongata and synapses with a neuron in the thalamus. The third neuron then transmits information to the correct region in the somatosensory cortex.

The Somatosensory Cortex

The somatosensory cortex is located in the anterior portion of the parietal lobes on the postcentral gyrus,

just posterior to the central sulcus (Figure 16.6a). Recall from Chapter 15 that the motor cortex is located in the posterior region of the frontal lobe on the precentral gyrus. Similar to motor cortex, the somatosensory cortex is arranged in a very specific way so that different portions are dedicated to receiving and interpreting information about a specific region of the body. A particular view of the topographical representation of the body regions on the somatosensory cortex is called the **homunculus** (Figure 16.6b). The most sensitive areas of the body—such as lips, tongue, and hands—require more brain area to process sensory information.

Association areas on each lobe of the cerebral cortex are involved in the interpretation and processing of sensory information. Perceptual processing of various stimuli involves learning, memory, language, and emotions. We can determine the difference between stimuli that are harmless and those that are threatening by combining information from immediate sensations and past experiences. Take, for example, the experience

Figure 16.5 Neural Pathways

(A) The anterolateral tract transmits pain and temperature information. (B) The dorsal-medial tract transmits touch, pressure, limb position, and body movement information.

of having a blood test, which involves a needle being inserted into a vein. That touch and pain sensory information can be interpreted very differently by different people. One person may view the needle as a quick poke of the skin, whereas someone else may experience extreme anxiety. Many factors affect the perception of that stimulus, including visual and auditory information, such as seeing the body language and confidence level of the nurse taking the blood and hearing soft, relaxing music. Each of these can help reduce the perception of pain during the procedure. Past experience also plays a large role in the interpretation of different stimuli. If the experience of a blood test in the past was very painful and difficult, a future blood tests will likely be perceived as painful as well.

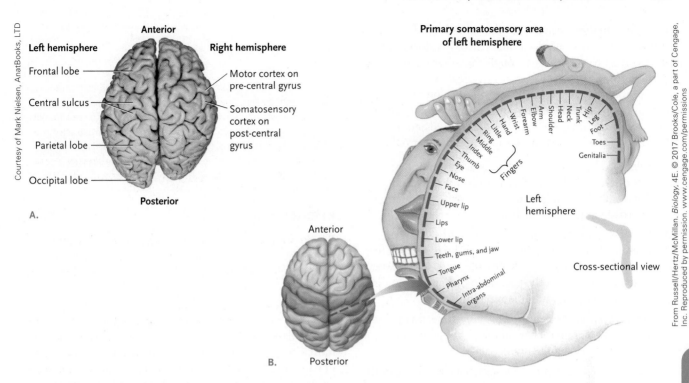

Courtesy of Mark Nielsen, AnatBooks, LTD

From Russell/Hertz/McMillan. *Biology*, 4E. © 2017 Brooks/Cole, a part of Cengage, Inc. Reproduced by permission. www.cengage.com/permissions

Figure 16.6 The Homunculus

(A) The somatosensory cortex is on the postcentral gyrus of the parietal lobes. **(B)** The homunculus depicts the regions of the somatosensory cortex that process information from each region of the body.

Pain

Pain is detected by sensory neurons called *nociceptors*, also called *free nerve endings*. Free nerve endings are dendrites on unipolar neurons (Chapter 15, Figure 15.5). Nociceptors depolarize and send sensory pain information in response to any of the following stimuli:

- Mechanical deformation, such as twisting or damaging the skin (such as a cut), breaking or spraining any body part, and intense pressure
- Extreme cold or hot temperatures, detected as pain

- Chemicals, such as strong acids
- Inflammatory molecules, produced by immune cells or damaged cells
  a. prostaglandins
  b. potassium
  c. histamine

Sensory neurons that transmit pain information release a number of neuropeptides. At least 20 have been identified, but the functions of each are not yet well known. The two main neurotransmitters released by nociceptors are glutamate and **substance P** (Figure 16.7).

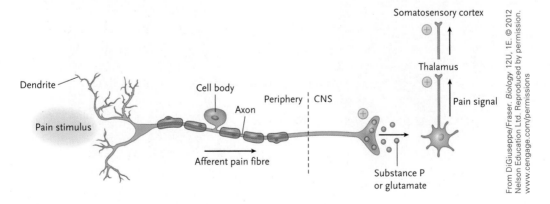

From DiGiuseppe/Fraser. *Biology* 12U, 1E. © 2012 Nelson Education Ltd. Reproduced by permission. www.cengage.com/permissions

Figure 16.7 Transmission of Pain Information

Sensory neurons that transmit pain information release the neurotransmitters glutamate and substance P.

Pain information travels to the spinal cord through the dorsal root—as does all sensory information—then travels to the brain through the anterolateral tract (Figure 16.5).

An **analgesic** is a medication that inhibits the transmission of pain. Our brain produces endogenous **opioids** that decrease pain—such as enkephalins and endorphins. Different types of medications are used to treat pain.

- Tylenol. The active ingredient is acetaminophen. Although how this relieves pain is not fully understood, the thought is that it reduces the production of prostaglandins in the brain.
- NSAIDs (non-steroidal anti-inflammatory drugs). These include aspirin, ibuprofen, and naproxen. They work by inhibiting an enzyme called *cyclooxygenase*, which is involved in the production of prostaglandins that increase pain and inflammation.
- Opioids. These include codeine, morphine, methadone, oxycontin, fentanyl. They inhibit pain by binding to opioid receptors in the brain.
- Anesthetics. Local anesthetics, such as lidocaine, block sodium channels on sensory neurons so that action potentials can't be triggered. General anesthetics act on the reticular activating system (RAS) in the brain; inhibiting the RAS causes unconsciousness.

DID YOU KNOW?

Vigorous exercise can stimulate the release of endorphins: *endo* means endogenous, "made in the body," and *orphin* means "morphine-like substance." Endorphins are a group of three different compounds that bind opioid receptors and inhibit pain and can cause feelings of euphoria. This is the substance released during a "runner's high," and some people have this experience during strenuous workouts. Laughing can also stimulate the release of endorphins.

Referred Pain

We do not usually perceive very specific sensory information from our internal organs. For example, experiences of a "stomach ache" are typically very general, and the pain could be originating in the stomach, small intestine, or large intestine. We also cannot perceive specific pain in our liver or pancreas. Sensory information from these organs does go to the brain, but we are not conscious of most visceral stimuli. Sometimes pain can be experienced in regions of the body where the pain stimulus is not originating. For example, someone having a heart attack can feel pain in the left arm. This is called **referred pain**, which is due to the convergence of sensory neurons from the periphery and from the visceral organs onto some of the same interneurons in the spinal cord (Figure 16.8).

A. **Anterior view**   B. **Posterior view**

Figure 16.8 Referred Pain

Areas of referred pain and where this is felt on the body are shown in (A) anterior view and (B) posterior view.

1. List the tactile receptors found in the skin and what type of sensation they detect.

2. What do proprioceptors detect?

3. How does an increase in intensity of a stimulus affect action potentials?

4. Which areas of the body have smaller receptive fields, and why?

5. What type of sensations are sent to the brain using (a) the anterolateral and (b) the dorsal-medial tracts?

6. Which information transmitted along the spinothalamic tract crosses over in the medulla oblongata?

7. In which lobe of the brain is the somatosensory cortex located?

8. What kind of stimuli is detected by nociceptors?

9. Name two neurotransmitters involved in transmitting pain signals.

10. What is an analgesic?

11. Give an example of an NSAID.

12. Where can pain from the gallbladder and heart be felt?

16.2 VISION

The visual system contains photoreceptors that detect light and convert it into action potentials, which are sent through the optic nerve to the primary visual centre in the occipital lobe for interpretation (Chapter 15). Photoreceptors in the eye are responsive to only a very small portion of the **electromagnetic spectrum** (Figure. 16.9). Radiant energy is described in terms of wavelength and frequency. The longer the wavelength, the slower the frequency and the lower the energy. Figure 16.9 shows that the visible light spectrum ranges from 400 nm–700 nm, and that different wavelengths are perceived as different colours.

Anatomy of the Eye

The eye is a fluid-filled filled organ located in the orbits of the skull. The eye is composed three layers (Figure 16.10).

1. The outermost layer is the **sclera**, which consists of fibrous connective tissue, and the **cornea**, which is the clear covering in front of the eye. Tendons of muscles attach to the sclera to control eye movements.

2. The middle layer is the **choroid**, which is highly vascular and contains a large amount of melanin that absorbs light. Anterior to the choroid is the **ciliary body**, which is composed of smooth muscles that control the shape of the **lens**. The lens changes shape depending on whether we are focusing on an object that is near or far. The **iris** is the coloured portion of the front of our eye and contains the **pupil** in the centre. Smooth muscles in the iris control pupil dilation. Pupils dilate in low light, during an effort to see far distances, and during a fight-or-light response.

3. The innermost layer consists of the **retina**, which contain the photoreceptors called **rods** and **cones**. The rods detect black-and-white images and function primarily in low light. The cones detect colours and

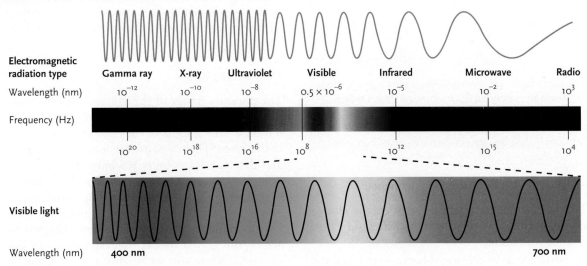

Figure 16.9 The Electromagnetic Spectrum
Visible light ranges between 500 nm and 700 nm.

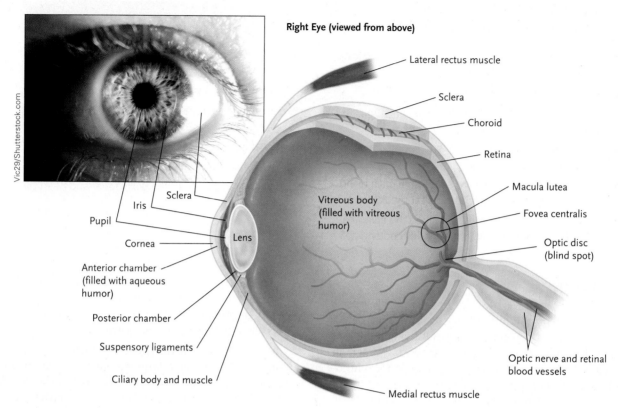

Vic29/Shutterstock.com

Right Eye (viewed from above)

Lateral rectus muscle

Sclera

Choroid

Retina

Macula lutea

Fovea centralis

Optic disc (blind spot)

Vitreous body (filled with vitreous humor)

Optic nerve and retinal blood vessels

Medial rectus muscle

Sclera

Iris

Pupil

Cornea

Lens

Anterior chamber (filled with aqueous humor)

Posterior chamber

Suspensory ligaments

Ciliary body and muscle

Figure 16.10 Anatomy of the Eye

function primarily in brighter light. The **macula lutea**, also called the **fovea**, is a small region at the back of the retina that contains very few blood vessels; it has a high concentration of cones and is responsible for very detailed central vision. The **optic disc** is the region of the retina where the neurons carrying information from the photoreceptors join the optic nerve. There are no rods or cones in the optic disc area and this area forms a small **blind spot**.

DID YOU KNOW?

The retina is attached to the choroid, which contains high amounts of blood vessels that supply the retina with nutrients and oxygen. Sometimes head trauma can cause the retina to detach (called *retinal detachment*) from the choroid, and this causes blindness. If this injury is caught quickly, the retina can be repaired with laser surgery.

The eye is divided into two compartments:

1. The anterior compartment is between the cornea and the lens, and is filled with a clear fluid called the **aqueous humor.**

2. The posterior compartment is between the lens and the retina, and is filled with a jelly-like fluid called the **vitreous humor.**

DID YOU KNOW?

Here's how you can find your blind spot. Draw a small dot on a piece of paper and a small + sign about five inches to the right of the dot. Hold the paper approximately 20 inches from your eyes, close your left eye, and look at the dot with your right eye. You should be able to see the + sign even though you are looking at the dot. Move the paper closer to your eyes; at some point the + sign will disappear. This is the point when the + sign is focused on your blind spot, and it seems to disappear.

Photoreceptors

The photoreceptors of the eye are the rods and cones. The retina contains about 120 million rods and about 6 million cones; most of the rods and cones are located in the fovea. The cones are responsible for visual acuity and detection of colour. We have three types of cone cells that contain proteins that absorb photons of light in either the short, medium, or long wavelengths of the visible spectrum. Rods are stimulated by all wavelengths in the visible spectrum in low light, but with less acuity; this is why we can't see colours in low light. For example, when you walk into a dark room, it takes approximately 7 to 10 seconds for the rods to be activated, and by then your eyes will have "adjusted" to the darker room. Nocturnal animals, such as owls, have double or triple the number of rods as humans.

Photoreceptors are composed of an outer segment called the *disc*, an inner segment that contains a large number of mitochondria, and a synaptic terminal that releases glutamate onto bipolar retinal neurons (Figure 16.11). The bipolar neurons are connected to **ganglion cells**, which play an important role in circadian rhythms. Low light stimulates the brain to produce sleep-inducing melatonin, and sunlight stimulates alertness.

The photopigments in the discs are stimulated by light photons. In rod cells the photopigment is called **rhodopsin**. In cone cells there are three types of photopigments, known as **photopsins**, and each one reacts to different wavelengths. The pigments that react to shorter wavelengths are responsible for us seeing the colour blue, those that react to medium wavelengths give us the colour green, and those that react to the longer wavelengths give us red.

The activation of rods and cones actually causes *hyperpolarization*. When not being stimulated by any light, they depolarize continuously and release glutamate onto bipolar cells. When light stimulates the photoreceptors, the pigment protein changes shape and causes the ion channels to close, thereby preventing the influx of positively charged ions, which then causes the rod or cone to hyperpolarize. Hyperpolarization prevents the release of glutamate onto the bipolar cells; this causes bipolar cells to depolarize and send signals to the optic nerve.

The Lens

The lens of the eye functions a lot like the lens of a camera. It changes shape to allow us to focus on objects that are near or far away. The shape of the lens is controlled by ciliary muscles, and this makes the light focus on the retina. When we look at far-away objects, the pupil dilates, and the lens becomes longer and thinner. When we look at close-up objects, such as when reading, the pupils constrict, and the lens becomes thicker and rounder (Figure 16.12).

Farsighted and Nearsighted

Sometimes the shape of the eyeball or the lens can affect how much a person is able to focus on near and far objects.

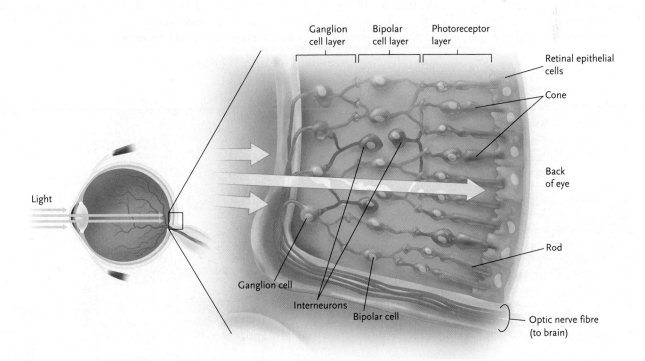

Figure 16.11 Structure of the Retina

The retina contains photoreceptors, bipolar cells, and ganglion cells.

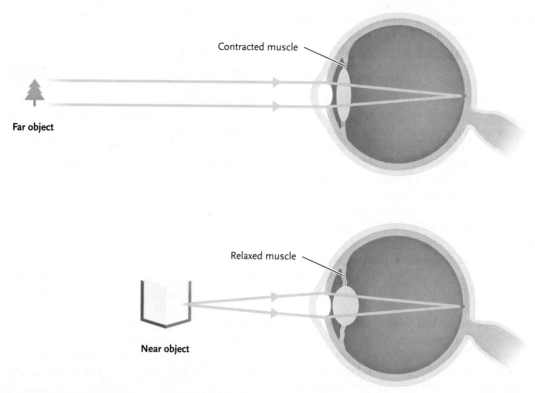

Contracted muscle

Far object

Relaxed muscle

Near object

Figure 16.12 Lens Shape

Changes in the shape of the lens affect our ability to focus on near and far objects.

Nearsightedness (also called *myopia*) occurs if the eyeball is slightly elongated, which causes the focal point of a far object to be in front of the retina (Figure 16.13). Nearsightedness can also occur if the lens is too concave. **Farsightedness** (also called *hyperopia*) occurs if the eyeball is slightly shortened (front to back), which causes the focal point of a near image to be behind the retina. Nearsightedness and farsightedness can be corrected with eye glasses or contact lenses that adjust how the focal point of images will concentrate on the retina.

DID YOU KNOW?

Colour blindness occurs when a person is unable to differentiate red and green or blue and yellow. Red-green colour blindness is the most common type of colour blindness and can be the result of an inherited mutation carried on the X chromosome (Chapter 6). This X-linked characteristic accounts for why colour blindness is much more common in males than females. The Canadian Association of Optometrists says 1 in 10 males are colour blind. As mentioned earlier, there are three types of cones, each with a different photopigment that detects light in either the blue, green, or red wavelengths. A person who lacks cone cells that respond to a certain wavelength will not be able to perceive that colour. Look at this image: if you can see a number in the circle, you do not have red-green colour blindness.

mf to/Getty Images

| Normal | Nearsighted | Farsighted |

Figure 16.13   Farsighted versus Nearsighted

CONCEPT REVIEW 16.2

1. What wavelengths of the electromagnetic spectrum can we detect as colour?

2. Which layer of the eye is highly vascular and contains melanin that absorbs light?

3. Which part of the eye changes shape and causes images to focus on the retina?

4. What do rods and cones detect?

5. What type of fluid is found in the anterior and posterior compartments?

6. What type of photopigment is found in the rods?

7. What is an important role of ganglion cells?

8. What happens to the lens when we focus on an object that is close up versus far away?

9. What are two factors that can cause nearsightedness?

16.3 HEARING

The auditory system contains receptors that convert vibration energy into action potentials that are sent through the auditory nerve to the primary auditory centre in the temporal lobe for interpretation. Sound is described in terms of frequency (Hz), which is perceived as pitch, and decibels (dB), which is loudness (Figure 16.14). The higher the frequency of sound waves, the higher the pitch sounds. Higher decibels equal more volume. Humans can hear sound waves in the range of 20–20 000 Hz; 20 Hz are very low base notes, and 20 000 Hz are very high-pitched notes. Dogs can hear up to 45 000 Hz, and dolphins can hear up to 120 000 Hz. Human hearing is most sensitive at ranges of 1–65 decibels. Sounds more than 85 dB can harm your hearing, and sounds more than 120 dB are perceived as pain and can cause severe damage to the eardrum.

Anatomy of the Ear

The ear includes sensory organs for hearing as well as the vestibular system that detects body position and movement. The ear can be described in terms of three main regions (Figure 16.15).

1. Outer ear. Called the **pinna** or **auricle**, this part of the ear consists of skin-covered folds of cartilage that surround the ear canal. Sound waves hit this surface and help the brain determine the direction of sound. Sound is transmitted through the **auditory canal** to the middle ear. The ear canal contains ceruminous glands (Chapter 12) that produce wax that helps trap debris and helps to prevent infections.

2. Middle ear. This part consists of the **tympanic membrane** (ear drum) and the **ossicles**, which are the three smallest bones in the body: the **malleus**, **incus**, and **stapes**. Sound vibrations hit the tympanic membrane, causing it to vibrate, which causes the ossicles to move, transmitting the pressure energy to the **oval window**, which is connected to the fluid-filled inner ear. The middle ear ossicles amplify the sound vibrations and transmit them to the inner ear. The middle ear also contains the opening to a tube called the **Eustachian tube**, which connects the middle ear to the throat. This tube is necessary for regulating the air pressure within the middle ear so that it is equal to that of the atmosphere. If you have ever felt your ears "pop," such as in an airplane or diving deep into a pool, you have felt a pressure difference between your middle ear and the environment. Yawning and swallowing help this tube open and allow air to move in or out of the middle ear to equalize the pressure.

16

A.

0	
10 dB	Normal breathing
20 dB	Wind-up clock ticking
40 dB	Humming refrigerator
50–65 dB	Normal conversation
70 dB	Hair dryer
75 dB	Dishwasher
80–90 dB	City traffic
85 dB	Lawnmower
100 dB	Train
110 dB	Car horn
110 dB	Power saw
115–120 dB	Rock concert
125 dB	Ambulance siren
140 dB	Fireworks
165 dB	Shotgun
180 dB	Rocket take-off
194 dB	Shatters bones

B.

Figure 16.14 Sound Described in Terms of Frequency and Volume

(A) Slow-frequency waves produce sounds of lower pitch, and fast-frequency waves produce higher pitch. (B) The volume or loudness of various sounds are shown in decibels.

3. Inner ear. This part consists of the **cochlea**, which responds to sound vibrations, and the **vestibular apparatus**, which detects body position and movement. The cochlea contains the **organ of Corti**, which is the sensory epithelium that transforms vibration energy into action potentials that are sent to the primary auditory cortex through the **auditory nerve**. The vestibular apparatus contains the vestibule and the fluid-filled semicircular canals that contain hair cells; these transmit movement into action potentials that are sent through the **vestibular nerve** to several parts of the brain, including the brainstem, thalamus, and cerebellum.

DID YOU KNOW?

When scuba diving, as you descend, the pressure surrounding your body increases because of the pressure of the water. At a depth of 10.3 metres (33 ft) the pressure of the water is equal to 1 atmosphere of air pressure (at 20 km above sea level). During your descent, the air in the middle ear compresses and causes a negative pressure. If this pressure is not equalized, the tympanic membrane can break. Air needs to be pushed into the middle ear to compensate for the drop in pressure. The easiest way to push air into the middle ear through the Eustachian tube is to plug your nose and try to push air out your plugged nose; this causes air to move from your throat into the middle ear. When you ascend to the surface after the dive, the external pressure decreases, which causes the air in the middle ear to expand. During your ascent, the positive air pressure in the middle ear is naturally released through the Eustachian tube to your throat.

Hearing Pitch and Volume

Mechanical vibrations are transmitted from the middle ear to the oval window by the stapes. The pressure exerted by the stapes is transmitted to the fluid in the cochlea (Figure 16.16). Vibrations transmitted to the cochlear fluid cause the basilar membrane to vibrate. Different frequencies of vibrations are detected as high or low pitch at different locations along the basilar membrane of the cochlea (Figure 16.17). High frequencies are detected as higher pitches at the beginning, or base, region of the cochlea, and lower frequencies are detected as progressively lower notes along the basilar membrane toward the apex. Neurons in each region of the cochlea are connected to the hair cells that convert the vibration energy into action potentials and send it to the brain so the brain can perceive different sounds. Volume is detected based on the amplitude of the sound waves. Frequency determines pitch, and amplitude determines volume.

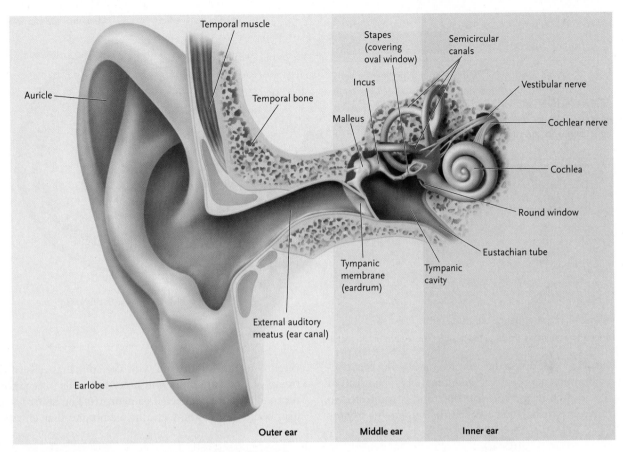

Figure 16.15 Anatomy of the Outer, Middle, and Inner Ear

The ear contains sensory organs for hearing and the vestibular system to detect body position and movement.

Adapted from Blausen.com staff (2014). "Medical gallery of Blausen Medical 2014". *WikiJournal of Medicine* 1 (2). DOI:10.15347/wjm/2014.010. ISSN 2002-4436. Wikipedia.

KEY
- Bony labyrinth
- Membranous labyrinth

Figure 16.16 The Cochlea

The stapes transmits vibration energy to the cochlea through the oval window.

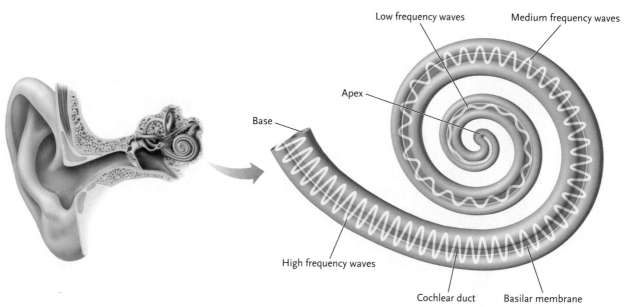

Figure 16.17 Frequency Vibration in the Cochlea
Different frequencies are detected by different regions of the cochlea.

The organ of Corti is the receptor for hearing that converts vibration energy into action potentials. The organ of Corti is located in the cochlear duct and is connected to specialized epithelial tissue called the *basilar membrane* (Figure 16.18). Intermixed with the epithelial cells of the organ of Corti are inner and outer hair cells (mechanoreceptors) that are connected to neurons. The hair cells are also connected to the tectorial membrane that is within

Figure 16.18 Structure of the Organ of Corti
The organ of Corti converts vibrations into action potentials.

the cochlear duct fluid. Vibrations move through the fluid, move the tectorial membrane, and move the hair cells. Activation of the hair cells causes them to open voltage-gated channels that let positively charged potassium and calcium ions enter the hair cells and depolarize them. Depolarization causes voltage-gated calcium channels to open and the hair cells release glutamate. Glutamate is an excitatory neurotransmitter that depolarizes neurons, which then send action potentials through the cochlear nerve to the auditory cortex in the temporal lobe.

The Vestibular System

The vestibular system consists of the semicircular canals and the vestibule. There are three endolymph-filled semicircular canals: the anterior (superior), (lateral (horizontal), and the posterior canals. Each canal connects to the vestibule at regions called **ampullae** that contain the **crista**. The vestibule also contains **otolithic organs**, which are two saclike regions, called the **utricle** and the **saccule** (Figure 16.19a). Each crista contains receptor hair cells that have **stereocilia** projecting into a gelatinous region called the **cupula** (Figure 16.19b). The hair cells are connected to neurons that send action potentials through the vestibular nerve.

Whenever we move our head, fluid in the vestibular apparatus moves, and this movement of the fluid causes the stereocilia to move (Figure 16.19c). The direction of the movement of the fluid affects the direction the stereocilia move. The three semicircular canals are positioned in all three planes so that any type of movement can be detected by the brain. The speed and magnitude of the movement affects the frequency of action potentials. When stereocilia

bend in one direction, hyperpolarization occurs in the hair cell receptors, and when they bend in the opposite direction, depolarization occurs. Depending on the plane of movement and, therefore, which semicircular canal is involved, the brain perceives the direction of movement based on the conversion of signals from the stimulated hair cell receptors. Neurotransmitters are released from the hair cells at *rest*, but movement causes action potentials to increase (depolarization) or decrease (hyperpolarization).

B.

C.

Figure 16.19 Vestibular Apparatus

The vestibular apparatus consists of the semicircular canals that contain crista in the ampullae (A), and these contain specialized hair cell receptors that bend based on body movements (B and C).

The utricle and saccule also contain hair cells and play a role in the detection of movement. When the body is at rest, the utricle sends information to the brainstem about head position and orientation. During acceleration, the saccule sends information to the brainstem. The otolithic organs combined give the brain information about how fast the body is moving and in which direction—backward, forward, up, down, left, or right. The utricle and saccule initiate signals for eye movements so we can focus while moving, and they also signal the muscles that affect posture so we can maintain balance. We detect only acceleration or deceleration. When the body is moving at a constant velocity with no visual input, we do not feel like we are moving. For example, in an elevator that is moving at a constant speed, we cannot feel our body moving, but when it is accelerating or decelerating we sense the movement.

The brainstem receives information about body position and movement from the vestibular system as well as from receptors in the eyes, joints, muscles, and skin.

DID YOU KNOW?

Motion sickness occurs when information from the visual system and the vestibular system do not tell the brain the same information. For example, if you are on a boat and the water is rough, the movement of your body moving with the boat causes the fluid in the vestibular system to send information about your movement and body position to the brain. However, if you are looking at an object in the boat, your eyes do not detect the movement of the boat because you and the boat are moving together. The brain perceives this mismatch of information as a toxin in your body, as if you were poisoned. This is why the brainstem stimulates vomiting. Your brain is trying to help you get rid of the poison that is disturbing your nervous system. When you focus your vision on objects outside of the boat, your brain will receive more appropriate movement signals. Seeing the moving water or other objects outside the boat provides visual information that is similar to the information the vestibular system is sending to the brain, which makes vomiting less likely.

CONCEPT REVIEW 16.3

1. What frequency of sound waves are audible to humans?

2. What level of loudness in decibels can cause damage to the ear?

3. What is another name for the eardrum?

4. What are the three bones in the middle ear that amplify sound vibrations?

5. What is the function of the Eustachian tube?

6. What is the structure of the inner ear that detects body position and movement?

7. In the cochlea, which area detects lower-pitched sounds, the base or the apex?

8. What type of receptors are the hair cells in the organ of Corti?

9. What neurotransmitter stimulates neurons that send information through the auditory nerve?

10. What region of the brain contains the primary auditory cortex?

11. What is the role of the utricle and saccule?

16.4 TASTE

The sense of **taste**—the gustatory system—enables the brain to determine beneficial versus harmful substances. The taste buds, as well as smell, texture, and temperature are all used to determine taste. There are thousands of visible bumps on the surface of the tongue, called **papillae**. Each papilla contains hundreds of taste buds, and each taste bud contains approximately 100 taste receptors. Taste buds are located on the tongue as well as on the roof and sides of the mouth, and the throat.

Humans differentiate five basic tastes, and each taste is detected by specific taste receptors:

1. **Sweet** is produced by the presence of sugars, aldehydes and ketones, such as sucrose, which can include artificial sweeteners such as sucralose or saccharin. Sweet taste is perceived as pleasurable because serotonin—which makes us feel good and content—is produced when we eat sweet foods. There was an evolutionary benefit to consuming sweet foods because they are high-energy foods.

2. **Sour** is produced by substances that are acidic and contain hydrogen ions, such as citric acid (in fruit), acetic acid (vinegar), or lactic acid (in yogurt).

3. **Salty** is produced by the presence of sodium ions (salt is sodium chloride, $NaCl$), but potassium ions can also produce a weaker salty taste. Potassium-chloride is sometimes used as a salt substitute.

4. **Bitter** is produced by substances such as coffee, cocoa, and alcoholic beverages. Bitter taste is generally perceived by the brain as unpleasant, and

there was an evolutionary benefit to avoid bitter foods because most bitter substances are toxic.

5. **Savory (umami)** is produced by substances containing amino acids, particularly glutamine, found in foods such as cheese and meat. Savory is perceived by the brain as pleasant because it is beneficial to consume proteins. Monosodium glutamate (MSG) is a chemical added to many processed foods to increase the pleasurable savory taste.

Taste Receptors

Taste receptors are chemoreceptors located within taste buds located inside the papillae on the tongue. Taste receptors can only bind substances dissolved in liquid, which is one reason why we produce saliva. by way of the taste pore, different molecules bind to different types of taste receptors, which trigger a depolarization of the **gustatory cell** (Figure 16.20). **Basal cells** continually divide and replace the gustatory cells. Gustatory cells trigger action potentials in the associated afferent (sensory) neuron. Taste information is sent to the somatosensory cortex via four cranial nerves:

1. Facial nerve (VII) carries taste information from the anterior two-thirds of the tongue.

2. Glossopharyngeal nerve (IX) carries information from the posterior third of the tongue.

3. Vagus nerve (X) carries information from the back of the oral cavity.

4. Trigeminal nerve (V) carries information regarding texture, spiciness (hot), and food temperature, which contribute to the overall sensation of taste. Substances that are perceived as hot include hot peppers (capsaicin), ginger, garlic, and horseradish. Substances such as peppermint and menthol are perceived as cold.

DID YOU KNOW?

Some people have a heightened sense of taste—hypergeusia. They tend to have more papillae, therefore more taste receptors, and are satisfied with less fat and sugar consumption. They don't like bitter food and eat less of it, but the food they do eat tastes more bitter to them than to other people, so they tend to eat more salt because that helps decrease the bitterness. Other people have a temporary or permanent loss of taste. The most common reasons for this include respiratory infections, allergies, smoking, deficiency in vitamin B12 or zinc, or an autoimmune disease called *Sjogren's syndrome*, where saliva production is dramatically reduced.

16

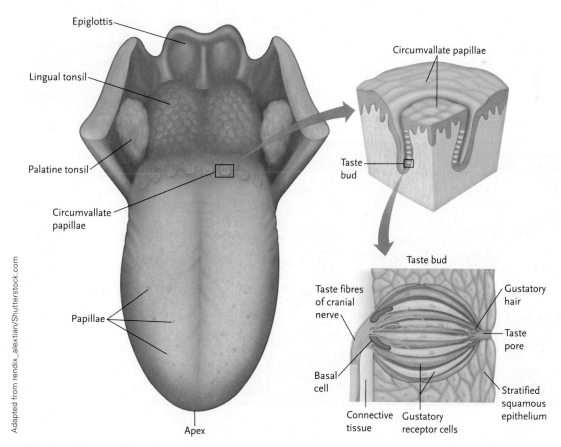

Adapted from rendix_alextian/Shutterstock.com

Figure 16.20 Taste Buds

Taste receptors are located in taste buds.

16.5 SMELL

The sense of smell—the **olfactory system**—enables the brain to detect odours. Odours are detected by chemoreceptors located in the roof of each nasal cavity—the olfactory epithelium (Figure 16.21). When we smell something, we are inhaling molecules that have been released into the air from a particular substance. In addition, the sense of smell and taste are highly interconnected. Consider smelling a loaf of freshly baked bread: some of the molecules we inhale also land in our mouth on our taste receptors, so we can smell and taste the fresh bread at the same time.

Humans have an estimated 5 million chemoreceptors in the olfactory epithelium, whereas dogs have over 200 million and therefore a much keener sense of smell than humans. We can detect thousands of different odours because we likely have thousands of different types of olfactory chemoreceptors. These chemoreceptors adapt very rapidly, and after only approximately one minute, most odours are no longer perceived.

Molecules in the air are inhaled through the nasal passages and travel into the superior nasal concha, where they dissolve in mucus and then bind to the **olfactory cilia** chemoreceptors (Figure 16.21). Olfactory receptors are replaced every few months by stem cells. The binding of molecules to the cilia causes an action potential in the olfactory sensory neuron. The axons of sensory neurons travel through

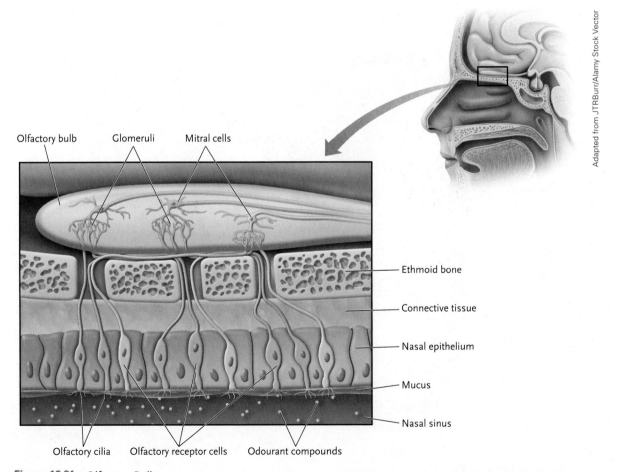

Adapted from JTRBurr/Alamy Stock Vector

Olfactory bulb   Glomeruli   Mitral cells

Ethmoid bone

Connective tissue

Nasal epithelium

Mucus

Nasal sinus

Olfactory cilia   Olfactory receptor cells   Odourant compounds

Figure 16.21 Olfactory Bulb

Olfactory chemoreceptors in the roof of the nasal cavities detect odours and send sensory information to the olfactory bulb.

openings in the ethmoid bone and synapse onto neurons in the **olfactory bulbs,** which lie on the inferior surface of the frontal lobe. The neurons in the olfactory bulb travel through the olfactory nerve (cranial nerve I) to the olfactory cortex of the brain, which is part of the limbic system. Olfactory sensory information is the only sensory input that does not travel to the thalamus.

Many brain regions receive olfactory information, such as the following:

- Limbic system—is involved with emotions and long-term memory. Certain odours are often associated with memories because the olfactory cortex is part of the limbic system.
- Reticular formation—is involved with consciousness and receives information from all the senses. It is important for any changes in light, sound, touch, body position, or odours to alert the brain in case we need to avoid certain stimuli.

- Hypothalamic nuclei—trigger hunger, so smelling foods can stimulate the hypothalamus to initiate the cephalic response (beginning digestion before consumption of food). The perception of food odours is more acute when we are hungry.
- Amygdala (also part of the limbic system)—is involved in autonomic responses to odours, which can stimulate aggressive or sexual behaviours.

DID YOU KNOW?

The senses of taste and smell are highly interconnected. If you have a cold or an allergy that causes a buildup of mucus in the nasal sinuses, preventing inhaled molecules from binding to the olfactory cilia, you will not only lose your sense of smell but also have a difficult time tasting your food.

CONCEPT REVIEW 16.5

1. What type of receptors detect odours?
2. Where are the olfactory cilia located?
3. Does olfactory information travel to the thalamus as does the information from all the other senses?

4. Name three regions of the brain that receive olfactory information.

EFFECTS OF AGING

As we age, the general and special senses tend to decline. Below are the most common changes that occur for each of the special senses.

Vision
- The number of photoreceptors decreases: a decrease in the number of rods leads to poor night vision, and a reduction in the number of cones decreases the ability to acutely detect colour.
- A loss of elasticity of the lens causes a loss of near vision, and it is very common for people over the age of 40 to begin to require the use of reading glasses.
- Cataracts can form, which is a clouding of the lens that causes objects to appear faded or blurry.
- Elderly people with hypertension can have increased intraocular pressure, which can lead to glaucoma.

Hearing
- A loss of elasticity of the tympanic membrane causes difficulty in hearing higher pitched sounds.

- Age-related hearing loss is called *presbycusis* and results from degeneration of the cochlea or the auditory nerves.
- Almost 50% of adults over the age of 65 have some amount of hearing loss.

Taste
- The sense of taste declines with age because the number of papillae decreases, and, therefore, the number of taste buds and taste receptors decreases.
- Older people tend to say food tastes bland or they can tolerate spicier foods, whereas children tend to say food tastes too spicy or too strong.

Smell
- The ability to detect odours decreases with age, and this contributes to a decline in the sense of taste.
- Olfactory receptors die and are not replaced as rapidly, and the existing receptors become less sensitive.

CHAPTER REVIEW

Medical Terminology

Term	Definition	Example
acoust/o	hearing, sound	acoustics
-algesia, alges/o	pain sensitivity	analgesia
audi/o, audit/o	hearing	audiometer
aur/o, auricul/o	ear	auricular
blephar/o	eyelid	blepharorrhaphy
cochle/o	cochlea	cochleitis
conjunctiv/o	conjunctiva	conjunctivitis
cor/o	pupil	corectasia
corne/o	cornea	corneosclera
-cusis	hearing	presbycusis
cycl/o	ciliary body, circular	cyclodialysis
dacry/o	tear	dacryoadenectomy
ir/o, irid/o	iris	iridemia
lacrim/o	tear, lacrimal duct	lacrimotomy
-lucent	light-admitting	radiolucent
lumin/o	light	luminescence
myring/o	eardrum	myringomycosis
narc/o	numbness	narcolepsy
ocul/o	eye	oculonasal
ophry/o	eyebrow	ophryitis
ophthalm/o	eye	ophthalmodynia
-opia, -opsia	vision	heteropsia
opt/o	eye, vision	optometer
ot/o	ear	otology
palpebr/o	eyelid	palpebritis
phac/o	lens	phacocele
phak/o	lens	phakoma
phot/o	light	photophobia
pupill/o	pupil	pupillatonia
retin/o	retina	retinotoxic
son/o	sound	sonometer
staped/o	stapes	stapedectomy
tympan/o	eardrum	tympanic membrane
vitre/o	glassy, vitreous body	vitreocapsulitis

Labelling Activity

Label the following diagrams.

1. a.

b.

2.

Courtesy of Mark Nielsen, AnatBooks, LTD

16

3.

4.

Answer to Application Question

Sensory information enters the dorsal roots, and motor information leaves through the ventral roots. Since Gertrude damaged her right dorsal column, no tactile sensory information would go to her brain. However, her motor abilities would be intact. The type of sensory information that travels through the dorsal spinothalamic tract carries fine touch and proprioceptive information. The information entering the spinal cord does not cross to the opposite side until it reaches the medulla oblongata; therefore, damage to the right dorsal tract would affect sensory information from the right side of the body below the spinal injury.

16

17 The Endocrine System

CHAPTER OUTLINE

I am always ready to learn although I do not always like being taught.
Winston Churchill

17.1 FUNCTIONS OF THE ENDOCRINE SYSTEM

The endocrine system is one of the two major communication systems of the body. The nervous system communicates via neurotransmitters released by neurons, and the endocrine system communicates via hormones released by particular cells into the bloodstream. Both the nervous system and the endocrine system are key components of homeostasis. The nervous system responds to stimuli in the environment through various receptors (such as those for pain, temperature, pH, pressure), and neurotransmitters target other neurons, muscle cells (skeletal, smooth, or cardiac), or glands (including adrenal glands, salivary glands). The nervous system response time from stimulus to target cells is extremely rapid and only lasts for a brief period of time. Think of a moment when something startled you and you experienced an instant electrical feeling flood your body; that is a nervous system response. The lingering effects are due the epinephrine released from the adrenal gland into the bloodstream as a hormone.

The endocrine system involves many glands that produce hormones that can target any cell type in the body. Generally, the hormone response is initiated more slowly than the nervous system response, but the effects last much longer. It is essential that the nervous system and the endocrine system cooperate to ensure proper reactions to changes in the external environment. The major endocrine glands of the body include the hypothalamus, pituitary gland, pineal gland, thyroid and parathyroid glands, thymus, pancreas, adrenal glands, testes and ovaries (Figure 17.1). Many cells that are not considered endocrine glands also produce hormones; these are discussed in Section 17.7.

As a point of comparison, exocrine glands are part of the epithelial tissues that secrete substances (enzymes, sweat, oil, mucus) through a duct into a hollow cavity or to the surface of the body (Chapter 11). For example, digestive enzymes are released into the small intestine, and sweat glands secrete sweat onto the surface of the skin.

Functions of Hormones

Hormones have many functions in the body, depending on the cell type they bind to and what signalling processes occur within the particular cell. The specific functions of particular hormones are covered in Sections 17.3 to 17.7, but the general functions of hormones include the following:

- Homeostatic regulation of blood sugar levels, water balance, blood calcium levels, blood pressure, thirst,

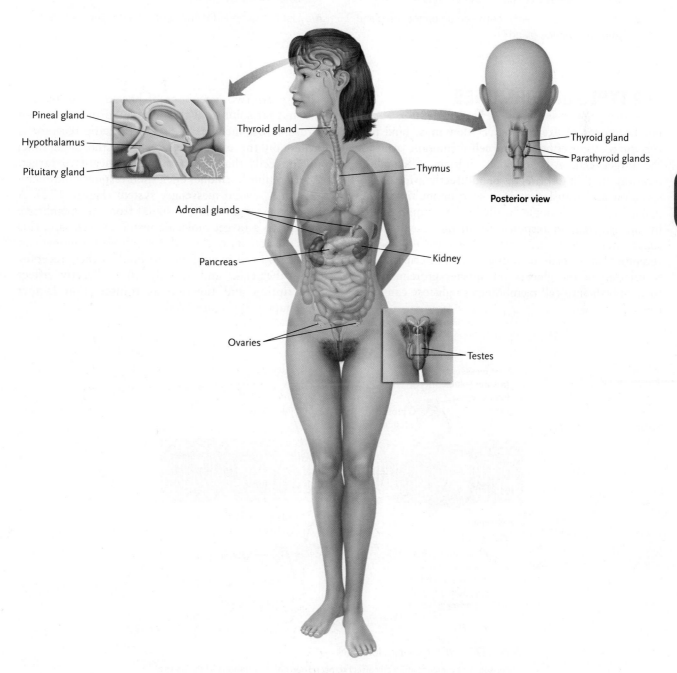

Figure 17.1   Organs of the Endocrine System

hunger, blood cell production, as well as the regulation of other hormones in the blood

- Growth, metabolism, and energy production
- Reproductive functions such as secondary sex characteristics, lactation, childbirth, and development of sperm or eggs

- Stress response
- Regulation of digestion
- Regulation of circadian rhythms such as sleep-wake cycles

CONCEPT REVIEW 17.1

1. What are the similarities and differences between the nervous system and the endocrine system?

2. What is the difference between an endocrine gland and an exocrine gland?

3. List the major organs of the endocrine system and where they are located.

4. List the general functions of the endocrine system.

17.2 TYPES OF HORMONES

For hormones to exert an effect they must bind to a *specific* receptor either on the cell membrane or inside the target cell. All hormones initiate a signalling pathway that either directly or indirectly affects transcription and translation of a gene or multiple genes (Chapter 9). For example, **insulin** is a hormone secreted by the pancreas in response to an increase in blood sugar after we eat food. Insulin binds to very specific receptors on all cells to initiate the transcription and translation of the glucose-transporter protein, which then embeds into cell membranes so glucose can enter the cells.

There are two categories of hormones, water soluble and fat soluble. **Water-soluble hormones** cannot diffuse across membranes, so the specific receptor is located on the extracellular side of the membrane: only on cells that the hormone is meant to target. Water-soluble hormones elicit their response indirectly through a second messenger system (Figure 17.2). A second messenger relays a signal from the membrane receptor to a target molecule inside the cell, and this alters the activity of the cell. **Fat-soluble hormones** can easily cross cell membranes to bind to their receptor inside the cell, and usually they directly affect transcription and function as **transcription factors** (see Chapter 9) (Figure 17.3).

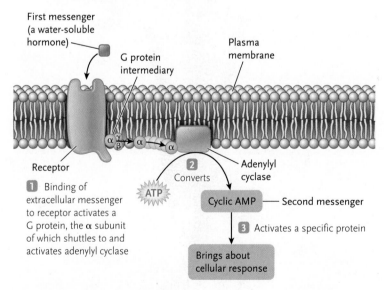

Figure 17.2 Water-Soluble Hormones

Water-soluble hormones indirectly affect transcription and translation by binding to a membrane surface receptor that triggers a second messenger signalling pathway.

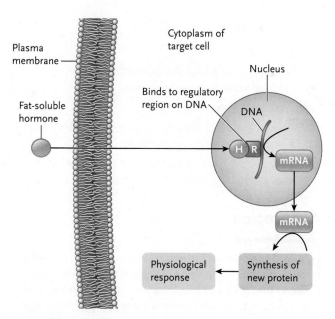

Figure 17.3 Fat-Soluble Hormones

Fat-soluble hormones directly affect transcription and translation by binding to receptors within the cell nucleus.

Water-soluble hormones are modified amino acids, small peptides, or proteins. The four most common categories of water-soluble hormones are the following:

1. Catecholamines—dopamine, epinephrine, and norepinephrine
2. Pancreatic hormones—insulin and glucagon
3. Pituitary hormones
4. Hypothalamic hormones

Fat-soluble hormones are steroids produced from cholesterol (Figure 17.4), except thyroid hormones, which are fat-soluble but are made from tyrosine. Recall from Chapter 2 the many important functions of cholesterol. The five most common fat-soluble hormones are the following:

1. Thyroid hormones—T_3 and T_4
2. Cortisol
3. Aldosterone
4. Androgens—testosterone, estrogen, and progesterone
5. Calcitriol (active Vitamin D_3)

Figure 17.4 Steroid Hormones Derived from Cholesterol

The following hormones are produced from cholesterol: testosterone, estradiol, progesterone, aldosterone, cortisol, and vitamin D.

CONCEPT REVIEW 17.2

1. How does a hormone target a specific cell type?
2. What is the main difference in how water-soluble and fat-soluble hormones affect their target cell?
3. What is a second messenger?
4. What is the general effect of all hormones on their target cell?
5. What types of molecules do our cells use to produce water-soluble hormones?
6. Why is cholesterol important?
7. Give examples of important water-soluble and fat-soluble hormones in our body.
8. How does the structure and solubility of thyroid hormones differ from the other hormones?

DID YOU KNOW?

We make some very important molecules from amino acids other than proteins. Tryptophan and tyrosine are the amino acids used to produce hormones. Tryptophan is an essential amino acid, so it has to be acquired through the diet (mostly poultry, eggs, and dairy, but also fish and beef), and it is used to produce serotonin, which is both a hormone and a neurotransmitter and involved in feeling happy and content. Tyrosine is the amino acid used to produce dopamine, thyroid hormones, and the catecholamines. Tyrosine is not an essential amino acid, so our cells can produce it from other amino acids.

Application

As you read through the rest of this chapter, consider this scenario. Helen is a college student that would like to lose 9 kg (20 lb). As you read through the next sections, note which hormones would help Helen lose weight. How can the production of those hormones be increased? Which hormones would decrease her ability to lose weight? How can the production of those hormones be reduced? Any hormones that are involved in metabolism, blood sugar regulation, or stress can affect weight gain or weight loss.

17.3 HYPOTHALAMUS AND PITUITARY

The hypothalamus is the control centre of the brain and is involved in many aspects of homeostasis by being part of both the nervous system (composed of neurons in the brain) and the endocrine system (secretes hormones into the bloodstream). The hypothalamus is directly connected to the posterior pituitary gland via neurons and to the anterior pituitary via the *hypophyseal portal blood system* (Figure 17.5). The hypothalamus produces releasing hormones or inhibiting hormones that travel through the hypophyseal portal system to the anterior pituitary gland, and these hormones then signal the pituitary to produce specific hormones (Table 17.1). The hypothalamus also produces two hormones, oxytocin and antidiuretic hormone (ADH), also called *vasopressin*, which it transports to the posterior pituitary gland. The posterior pituitary does not produce any hormones itself. The neuronal cell bodies are located in the hypothalamus where the hormones are produced. These hormones are then transported along the axons and released and stored in the axon terminals in the posterior pituitary until they need to be released into the bloodstream. Behind the hypothalamus is the pineal gland that produces the hormone melatonin, which is involved in inducing sleep (Chapter 15).

Posterior Pituitary Hormones

Oxytocin Oxytocin is produced in the hypothalamus, stored in the posterior pituitary gland, and released during childbirth. The main functions of oxytocin are to cause uterine contractions and the ejection of breast milk from the mammary glands during breast-feeding. Uterine contractions and oxytocin release is an example of a positive feedback mechanism since the contracting uterus further stimulates the release of more oxytocin, which causes the uterus to contract, and so on. Oxytocin also acts as neuromodulator in the brain. It plays a role in maternal bonding with the infant, bonding and feelings of love for another person, and feelings for another person such as empathy. It also plays a role in orgasm, the regulation of the immune response, and decreases anxiety. Oxytocin's role in decreasing inflammation and promoting wound healing supports the idea that positive social interactions are good for our health.

Antidiuretic Hormone (ADH) Antidiuretic hormone (ADH), also called *vasopressin*, is produced in the hypothalamus and stored in the posterior pituitary gland. ADH is released from the posterior pituitary gland when blood volume/blood pressure is low. An antidiuretic has the opposite effect of a diuretic. Diuretics are substances that cause water to be excreted by the kidneys. ADH acts on the distal tubule and collecting ducts of the kidneys to reabsorb water, which restores blood volume and therefore blood pressure (see Chapter 21, Section 5). ADH causes blood vessels to constrict to maintain blood pressure during periods of lower blood volume.

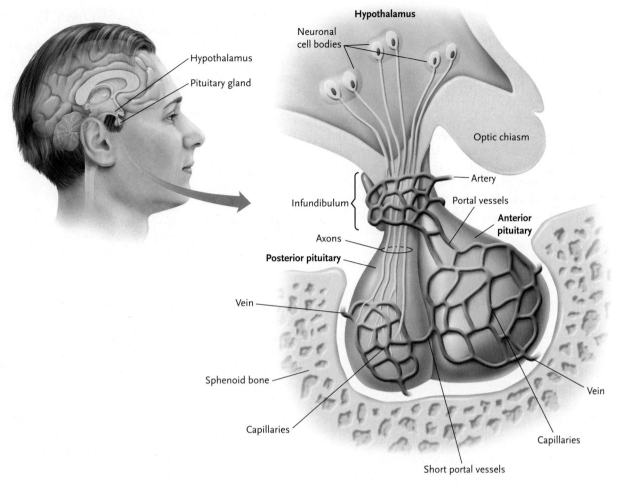

Figure 17.5   Anatomy of the Hypothalamus–Pituitary Gland Interactions

TABLE 17.1

Hormone Action on the Pituitary Gland

Hypothalamus Hormone	Action on the Anterior Pituitary Gland
Growth hormone-releasing hormone (GHRH)	Secretion of human growth hormone (GH)
Growth hormone-inhibiting hormone (GHIH)	Inhibition of GH
Thyrotropin-releasing hormone (TRH)	Secretion of thyroid-stimulating hormone (TSH)
Corticotropin-releasing hormone (CRH)	Secretion of adrenocorticotropic hormone (ATCH)
Gonadotropin-releasing hormone (GnRH)	Secretion of follicle-stimulating hormone (FSH) Secretion of luteinizing hormone (LH)
Prolactin-releasing hormone (PRH)	Secretion of prolactin (PRL)
Prolactin-inhibiting hormone (PIH)	Inhibition of PRL

DID YOU KNOW?

The drug ecstasy is referred to as the "love drug" because it causes the hypothalamus to release large amounts of oxytocin into the blood.

DID YOU KNOW?

Oxytocin is used to induce labour in pregnant women that are passed full term. It can also be given to control bleeding after delivery because it causes the uterus to contract.

DID YOU KNOW?

Drinking alcohol and caffeine decreases the amount of ADH produced (dose-dependent response) and can increase the excretion of water in the urine. One of the main causes of a hangover is the dehydration effect of alcohol due to the inhibition of ADH.

Anterior Pituitary Hormones

Growth Hormone

Human growth hormone (GH) (also called *somatotropin*) is a protein hormone that is released from the anterior pituitary gland in response to GHRH produced by the hypothalamus. The main stimuli for the production of GHRH is deep sleep, high-intensity exercise, periods of increased androgen production (growth periods during childhood and puberty), and fasting (period between meals). Adolescents produce twice as much growth hormone as an adult. Even during adulthood, GH is very important for maintaining healthy cell division to replace old and damaged cells; a lack of GH results in increased signs of aging. As people age, less GH is secreted, and so it is very important to include exercise in a daily routine.

Growth hormone performs the following functions in the body:

- Promoting bone growth (height) in children and adolescents. GH causes calcium to be incorporated into bones, and decreases osteoporosis in adults.
- Stimulating protein synthesis. This occurs particularly in skeletal muscle where protein is required to increase muscle strength.
- Aiding metabolism. GH causes the liver to break down glycogen and fat and thereby increases the level of blood sugar so fuel is available for cells.
- Stimulating the immune system. Stress hormones inhibit growth hormone and inhibit the immune system.

Excessive GH production during childhood and puberty can cause gigantism, too much GH produced after puberty causes acromegaly, and too little GH production causes dwarfism. Where early diagnosis is possible, these conditions are much less common. Dwarfism can now be prevented in children with exogenous growth hormone produced through genetic engineering.

DID YOU KNOW?

If you have a mild cold without a fever, moderate- to high-intensity exercise can help you get rid of the cold faster because you will stimulate the production of growth hormone and give your immune system a boost. Exercising while you have an infection that causes a fever will divert too much energy from your immune system and can make you feel worse.

Thyroid-Stimulating Hormone

The anterior pituitary gland produces thyroid-stimulating hormone (TSH) in response to the secretion of thyrotropin-releasing hormone (TRH) from the hypothalamus. The hypothalamus releases TRH when the blood levels of thyroid hormones T_3 and T_4 are low. TSH acts on the thyroid gland and is regulated, through negative feedback, by the concentration of thyroid hormones in the blood; a high level of circulating thyroid hormones inhibits the release of TSH from the anterior pituitary gland.

DID YOU KNOW?

Having low blood levels of thyroid hormones can contribute to depression because T_3 and T_4 play a role in the release of serotonin, also referred to as "the good mood molecule."

Adrenocorticotropic Hormone

The anterior pituitary gland releases adrenocorticotropic hormone (ACTH) in response to the secretion of corticotrophin-releasing hormone (CRH) from the hypothalamus. ACTH acts on the adrenal cortex and regulates the production of glucocorticoids, mainly cortisol, in response to stress. ACTH causes cortisol to be produced by the zona fasciculata region of the adrenal cortex from cholesterol, and it stimulates adrenal cells to take up cholesterol from low-density lipoproteins (LDLs) in the blood.

Follicle-Stimulating Hormone (FSH) and Luteinizing Hormone (LH)

FSH and LH have important roles in the stimulation of the gonads during puberty and reproduction. FSH is the primary hormone that stimulates development of the sperm in the testes and oocytes in the ovaries. LH stimulates the gonads to produce the hormones testosterone, estrogen, and progesterone. Males and females secrete all three hormones; however, females have much more of the aromatase enzyme that converts testosterone into estrogen.

Prolactin

Prolactin (PRL) is produced in the anterior pituitary gland and released in response to the secretion of prolactin-releasing hormone (PRH), which is produced in the hypothalamus. Prolactin is blocked by prolactin-inhibiting hormone (PIH), also produced in the hypothalamus. The primary role of prolactin is in lactation. During pregnancy it stimulates the growth of the female mammary glands and their production of milk after delivery of the baby. Recall that the posterior pituitary hormone oxytocin causes the ejection of the milk. As progesterone levels decrease at the end of pregnancy, the hypothalamus begins to produce PRH. Prolactin also plays a role in surfactant production in the fetus at the end of the pregnancy, and this helps the mother's immune system not reject the infant.

Prolactin is produced in both men and women after orgasm, causing a satisfied, relaxed feeling. Inappropriately high prolactin levels can cause impotence in males. In females, prolactin inhibits the production of FSH and LH during lactation after a baby is born. This biological mechanism is meant to prevent another pregnancy before the newborn has developed enough to no longer need the mother's breast milk for nutrition. However, this inhibition is weak, so some FSH and LH is still produced, and many pregnancies occur while mothers are still nursing their newborns.

CONCEPT REVIEW 17.3

1. Which hormones are made in the hypothalamus and stored in the posterior pituitary?

2. List the hypothalamus hormones and their effect on the anterior pituitary gland.

3. List the main functions of oxytocin and ADH.

4. What are the functions of growth hormone? Does GH have any function in adults?

5. What are the effects of too much or too little growth hormone?

6. What endocrine glands are stimulated by the anterior pituitary hormones TSH, ACTH, FSH, and LH?

7. What is the function of prolactin?

17.4 THYROID AND PARATHYROID

The thyroid gland is located below the larynx. It consists of thyroid follicles composed of follicular cells that secrete the hormones thyroxine (T_4) and triiodothyronine (T_3), and parafollicular cells that secrete calcitonin (Figure 17.6). Follicular cells surround thyroid follicles, and the smaller parafollicular cells are located between the follicles.

Thyroid Hormones

Thyroid hormones are produced by the thyroid gland in response to stimulation by the anterior pituitary hormone TSH, and they are extremely important regulators of metabolism. TSH production is stimulated by exercise and stress so that energy is available to ATP production. They act on every cell in the body because all cells must make ATP. Triiodothyronine (T_3) and thyronine (T_4) are produced from the amino acid tyrosine and the trace mineral iodine (Figure 17.7). T_3 contains three iodine atoms and T_4 contains four iodine atoms. The thyroid gland is the only tissue in the body that requires iodine. Selenium is also a very important micronutrient required for thyroid hormone synthesis, and selenium deficiency can cause severe thyroid malfunction. This micronutrient is found in foods such as nuts (especially Brazil nuts), fish, wheat bran, pork, shellfish, eggs, and mushrooms. The follicular cells of the thyroid gland produce both T_3 and T_4. However, once these hormones circulate in the blood and are taken up by cells throughout the body, most T_4 is converted into T_3, which is the more active form. Although thyroid hormones are produced from

Anterior view

Figure 17.6 Thyroid Gland

The thyroid gland is made up of follicular cells that secrete the thyroid hormones T_3 and T_4, and parafollicular cells that secrete calcitonin.

Figure 17.7 Thyroid Hormones

Thyroxine (T$_4$) and triiodothyronine (T$_3$) are thyroid hormones produced from the amino acid tyrosine.

the amino acid tyrosine, they are not water soluble and must be transported through the blood on plasma proteins. Once they reach a target cell, they enter the cell through a membrane transporter protein and act on the receptors inside the cell that affect gene transcription.

DID YOU KNOW?

Table salt is "iodized" because it contains iodine in order to prevent deficiencies that can lead to hypothyroidism. Before iodization of salt, millions of people in the world had severe iodine deficiency that caused goiter. Iodine is common in soils located near oceans, but many inland regions lack iodine. In the early 1920s, iodine was routinely added to table salt in North America, and thyroid disease dramatically decreased.

The thyroid hormones have the following functions in the body:

- Regulating oxygen use by all cells during cellular respiration
- Increasing basal metabolic rate (BMR) by increasing the production of ATP from glycogen and fats
- Increasing the synergistic effects of epinephrine and norepinephrine in the breakdown of glycogen and fat for energy production
- Stimulating the production of growth hormone, especially important during fetal development
- Increasing heart rate, breathing rate, and cardiac output so that more oxygen can be delivered to cells

Abnormal Functioning of the Thyroid Gland

Hypothyroidism is a condition caused by insufficient production of thyroid hormones. People with hypothyroidism are not able to utilize nutrients to produce energy, and so they feel very tired, have low energy, often gain weight, feel cold (since body heat is produced when ATP is produced), often have dry skin and brittle nails, and sometimes experience depression. Thyroid hormones play a role in the production of serotonin, and a lack of serotonin leads to depression. The most common causes of hypothyroidism are iodine deficiency, selenium deficiency, any malfunction of the hypothalamus or anterior pituitary gland, extreme stress, and anorexia. Note that very low-calorie diets cause the body to stop producing thyroid hormones; if nutrients aren't available for energy production, the body won't produce the necessary hormones to break down nutrients to produce energy. Eating properly and exercising regularly helps ensure sufficient thyroid hormone production.

Hashimoto's disease is an autoimmune disease in which the immune cells destroy the thyroid cells. The lack of thyroid hormone stimulates an increase in TSH secretion from the anterior pituitary gland and this can lead to goiter. Goiter can occur in cases of both hypothyroidism and hyperthyroidism.

Hyperthyroidism is a condition caused by excessive production of thyroid hormones. People with hyperthyroidism feel nervous, irritable, hot; have increased heart rate and difficulty sleeping; usually have weight loss; and sometimes have protruding eyes (exophthalmos) from the swelling of the fat and muscle tissue behind the eyes. The most common cause of hyperthyroidism is Graves' disease, which is an autoimmune disease (Figure 17.8). Antibodies that act against the thyroid-stimulating hormone (TSH) bind to the TSH receptors instead of TSH,

Figure 17.8 Graves' Disease

Typical appearance of an individual with Graves' disease experiencing exophthalmos.

and they continually stimulate the thyroid to produce thyroid hormones, which causes the thyroid gland to enlarge (goiter), and further increases thyroid hormone production.

DID YOU KNOW?

Radioactive iodine can be used to specifically destroy cancerous or overactive thyroid cells because no other cells in the body take up iodine.

Calcitonin

Calcitonin (CT) is the hormone produced by the parafollicular cells of the thyroid gland. The main function of calcitonin is to prevent calcium (Ca^{2+}) blood levels from being too high; it balances the effects of parathyroid hormone, which increases blood calcium levels. Calcitonin secretion from the parafollicular cells is regulated by calcium blood levels. When levels increase, calcitonin is produced and inhibits osteoclast activity. Osteoclasts are the bone cells that break down the mineral matrix and release stored calcium and phosphate into the blood (Chapter 13).

The functions of calcitonin include the following: inhibiting osteoclast activity, inhibiting calcium absorption in the small intestine, and inhibiting reabsorption of calcium in the kidney. Because it prevents bones from losing calcium, calcitonin can be used to treat **osteoporosis**.

Parathyroid Hormone

Parathyroid hormone (PTH) is produced by the chief cells in the parathyroid glands, which are located on the posterior side of the thyroid gland, and is the major regulator of blood calcium (Ca^{2+}), magnesium (Mg^{2+}), and phosphate ions (PO_4^{3-}) in the blood (Figures 17.9 and 17.10).

The functions of PTH include the following:

1. Stimulating osteoclasts to release calcium and phosphate into the bloodstream
2. Stimulating kidney cells to increase reabsorption of Ca^{2+} and Mg^{2+}
3. Increasing the excretion of PO_4^{3-} by the kidneys
4. Increasing the production of active vitamin D (calcitriol) by the kidneys, which then increases the absorption of calcium in the small intestines

PTH is stimulated by low blood calcium levels. Very low calcium levels caused by calcium deficiency can cause hypersecretion of PTH, which causes excessive bone demineralization and osteoporosis. Calcium deficiency can be caused by insufficient dietary intake of calcium, vitamin D deficiency, or renal failure. Foods high in calcium include dairy products, such as milk, cheeses and yogurt; leafy green vegetables; fish; beans; and nuts.

Parathyroid glands (behind thyroid gland)

Right superior parathyroid gland

Thyroid gland

Left superior parathyroid gland

Right inferior parathyroid gland

Left inferior parathyroid gland

Esophagus

Trachea

Posterior view

Figure 17.9 Parathyroid Glands

Four small parathyroid glands are located on the posterior side of the thyroid gland. PTH is the major hormone that regulates blood calcium, phosphate, and magnesium ion levels.

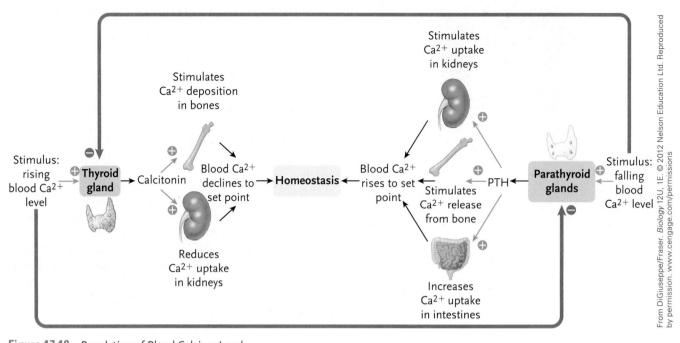

Figure 17.10 Regulation of Blood Calcium Levels

Blood calcium levels are regulated by calcitonin and parathyroid hormone.

From DiGiuseppe/Fraser. *Biology* 12U, 1E. © 2012 Nelson Education Ltd. Reproduced by permission. www.cengage.com/permissions

CONCEPT REVIEW 17.4

1. Which hormones are produced by the follicular and parafollicular cells of the thyroid gland?

2. What is the difference between T_3 and T_4? Which is the more active hormone?

3. What amino acid is used by the thyroid to produce T_3 and T_4?

4. What two dietary minerals are required for the production of thyroid hormones? What foods contain them?

5. Are thyroid hormones fat soluble or water soluble? How are they transported through the blood?

6. List the functions of thyroid hormones.

7. What are the symptoms of hypothyroidism and hyperthyroidism?

8. What are the most common causes of hypothyroidism and hyperthyroidism?

9. What is the function of calcitonin? What cell types does it act on?

10. What are the functions of parathyroid hormone? What cell types does it act on?

11. How can low calcium cause osteoporosis?

12. What foods are high in calcium?

17.5 ADRENAL GLANDS AND THE STRESS RESPONSE

The adrenal glands are located above the kidneys. They consist of the outer adrenal cortex and the inner adrenal medulla (Figure 17.11). The adrenal cortex is divided into three zones. The outer zona glomerulosa secretes mineralocorticoids, mainly aldosterone; the middle zona fasciculata secretes glucocorticoids, mainly cortisol; and the inner zona reticularis secretes androgens, mainly dehydroepiandrosterone (DHEA) (Figure 17.12). The adrenal medulla is controlled by the autonomic nervous system, and stimulation from sympathetic

preganglionic neurons causes the medulla to secrete epinephrine and a small amount of norepinephrine.

Aldosterone

Aldosterone is the main mineralocorticoid produced by the adrenal cortex. A steroid hormone made from cholesterol (Figure 17.4), aldosterone is produced when blood pressure is low and water needs to be reabsorbed. Its production is under the control of the renin-angiotensin system (see Chapter 21, Section 5). It causes the distal tubule and collecting duct of nephrons in the kidney to produce sodium ion channels; this causes more sodium to be reabsorbed into the bloodstream. When more

Figure 17.11 Adrenal Glands

The adrenal glands are located above each kidney.

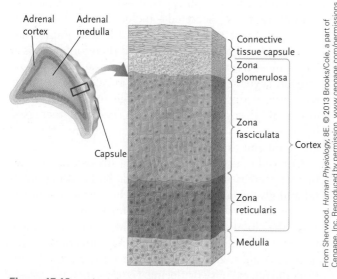

Figure 17.12 Adrenal Cortex

The adrenal cortex consists of three layers that produce different hormones. The zona glomerulosa produces mineralocorticoids, the zona fasciculate produces glucocorticoids, and the zona reticularis produces androgens.

sodium is reabsorbed, more water is also absorbed because of osmotic pressure gradients. And increase in water reabsorption brings blood volume, and therefore blood pressure, back to normal. Aldosterone also causes the nephrons to secrete potassium ions.

Cortisol

Cortisol is the main glucocorticoid produced in the adrenal cortex in response to the anterior pituitary hormone adrenocorticotropic hormone (ACTH). Cortisol is a steroid hormone produced from cholesterol (Figure 17.4). It is released during a 24-hour daily cycle,

as well as during stress. The functions of cortisol include helping the body access stored nutrients so that cells can make energy (ATP). By stimulating glycogen and fat breakdown in the liver and in fat cells, cortisol increases blood sugar. It also inhibits the immune response in relation to the amount of cortisol secreted; that is, immune suppression is dose dependent. Normal cortisol secretions do not prevent the immune system from functioning, but they do act as a brake so the immune system doesn't overreact.

In people who have a regular sleep schedule, their daily cortisol level begins to rise a couple hours before they wake up in the morning. This has a beneficial effect because it helps increase their blood sugar level—which lowered during the period of sleeping and not eating—so they have energy in the morning until they eat breakfast (Figure 17.13). Cortisol secretion increases during a stress response, along with epinephrine and norepinephrine.

Cushing's Syndrome

Cushing's syndrome results from an excess of cortisol secretion. An excess secretion can be caused by conditions that increase the production of corticotropin-releasing hormone (CRH) by the hypothalamus or that increase the release of adrenocorticotropic hormone (ACTH) by the anterior pituitary gland. It can also be caused by long-term glucocorticoid drug treatment, which is used to treat chronic inflammatory diseases because of its immune suppression abilities. People with Cushing's syndrome have symptoms that result from excessive breakdown of fats, glycogen, and bone minerals; redistribution of stored body fat to the abdomen; and decreased production of new proteins. This condition leads to hyperglycemia, muscle weakness, osteoporosis, type-2 diabetes, immune suppression, poor wound healing, thin skin, bruising, depression, moodiness, fat redistribution from limbs to trunk and face, "moon face," "buffalo hump," hair loss, decreased fertility, fatigue, insomnia, menstrual irregularities, water retention, and high blood pressure. People who have

Figure 17.13 Daily Cortisol Secretions

Daily cortisol secretions are highly linked to circadian rhythm.

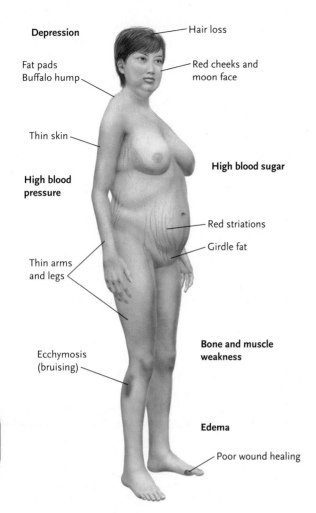

Depression

Hair loss

Fat pads
Buffalo hump

Red cheeks and
moon face

Thin skin

High blood sugar

High blood
pressure

Red striations

Girdle fat

Thin arms
and legs

Ecchymosis
(bruising)

Bone and muscle
weakness

Edema

Poor wound healing

Figure 17.14 Cushing's Syndrome
Oversecretion of cortisol or long-term treatment with glucocorticoid medication can cause Cushing's syndrome.

chronic stress can experience these symptoms as well, although less extremely than Cushing's Syndrome; this helps explain why chronic stress is a major risk factor for chronic diseases. (Figure 17.14).

Androgens

Androgens are secreted by the adrenal cortex in both males and females. In males, the testosterone produced by the testes during puberty plays a much more significant role in the development of secondary sex characteristics and sperm development than dehydroepiandrosterone (DHEA). In females, the adrenal cortex androgens contribute to libido and are converted into estrogens; after menopause, this is the only source of estrogen. Overstimulation of the adrenal glands can cause increased androgen production in some women, and, although rare, can lead to masculinization.

Adrenal Medulla and the Stress Response

The adrenal medulla secretes epinephrine and norepinephrine (NE)—also called *adrenaline* and *noradrenaline*,

respectively—that are released due to stimulation of the preganglionic neurons of the sympathetic nervous system during a stress response (Chapter 15). Each adrenal medulla secretes primarily epinephrine, but also small amounts of norepinephrine; the neurons of the sympathetic system release primarily norepinephrine. Note that epinephrine and norepinephrine are hormones when released into the bloodstream by the adrenal medulla, and they are neurotransmitters when released by neurons of the autonomic nervous system. Recall from Chapter 15 that the sympathetic nervous system controls the fight-or-flight response that's activated during physical or emotional stress. When stressed, epinephrine and norepinephrine help to break down stored fat to increase blood sugar that can be used to make ATP. These hormones also help cells utilize oxygen for the production of ATP during a fight-or-flight situation. Stressors that stimulate this response include fear, intense exercise, surgery, infection, illness, injury, pain, sleep deprivation, low blood sugar, cold temperatures, and any strong emotional response.

The effects of epinephrine—the main catecholamine that is secreted—include the following: increased heart rate; bronchiole dilation; pupil dilation; increased water retention and therefore increased blood pressure; vasodilation and increased blood flow in liver, skeletal muscles, cardiac muscle, and adipose tissue; vasoconstriction in the digestive tract; breakdown of glycogen and fat and therefore increased blood sugar, since epinephrine increases the effects of glucagon and thyroid hormones.

Cortisol is also secreted during a stress response and further increases blood sugar levels by breaking down fat and glycogen. This response is beneficial and adaptive for short-term physical stress: for example, at times when the blood sugar would be used, when the immune response wouldn't overreact during infection, when healing could take place, or when muscles could contract during exercise or fleeing.

Long-term emotional stress, however, is very damaging to the body because the level of blood sugar remains high since it is not used by cells; protein synthesis cannot occur properly, so cells are not replaced; the immune system becomes overly suppressed, which leads to increased infections; and muscle-wasting and weakness can occur.

DID YOU KNOW?

Chronic cortisol secretion from emotional stress inhibits the production of proteins, including collagen and elastin in the connective tissue layers of the skin. This leads to premature aging of the skin, wrinkles, and loss of elasticity. Cortisol secretion also inhibits the immune response, so chronic emotional stress results in an increased risk of infection such as the common cold or influenza.

CONCEPT REVIEW 17.5

1. What hormones are secreted from the adrenal cortex and the adrenal medulla?

2. What are the three zones of the cortex, and what types of hormones are secreted from each?

3. What is the most common mineralocorticoid, glucocorticoid, and androgen?

4. How does the autonomic nervous system affect the adrenal gland?

5. What are the main functions of aldosterone?

6. What are the main functions of cortisol, and when does your body secrete it?

7. What condition can occur if too much cortisol is secreted? Describe the symptoms.

8. What is the function of cortical androgens in males and females?

9. What hormones are released during a stress response?

10. Describe how epinephrine acts on various parts of the body during a stress response.

17.6 PANCREAS

The pancreas is both an endocrine and an exocrine gland. The exocrine functions of the pancreas include secreting enzymes and bicarbonate through the pancreatic duct into the small intestine during digestion. The endocrine functions include secreting the hormones insulin and glucagon into the blood to regulate blood glucose levels. The pancreas is located in the abdominopelvic cavity, inferior and posterior to the stomach, within the curve of the first portion of the small intestine, the duodenum (Figure 17.15). The endocrine portion consists of

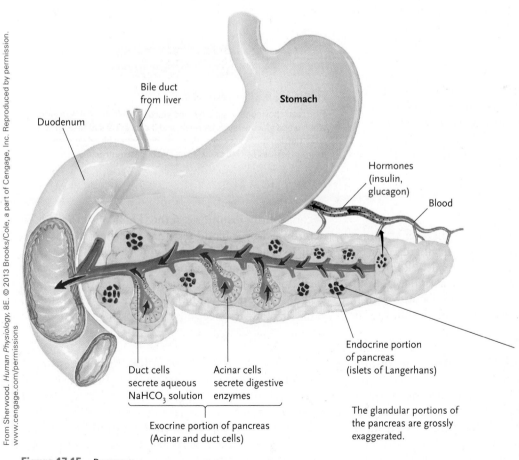

Bile duct from liver

Stomach

Duodenum

Hormones (insulin, glucagon)

Blood

Duct cells secrete aqueous NaHCO₃ solution

Acinar cells secrete digestive enzymes

Endocrine portion of pancreas (islets of Langerhans)

Exocrine portion of pancreas (Acinar and duct cells)

The glandular portions of the pancreas are grossly exaggerated.

Figure 17.15 Pancreas

The pancreas consists of exocrine cells that produce digestive enzymes and endocrine cells that produce hormones. The alpha islet cells produce glucagon, and the beta islet cells produce insulin.

Robert Markus/Science Source

pancreatic islets called the *islets of Langerhans*, which are made up of alpha cells that secrete glucagon, and beta cells that secrete insulin.

Insulin is produced after we eat food, Figure 17.16 during what's called the *absorptive phase*. Different foods stimulate different amounts of insulin secretion; this variability between foods is called the *glycemic index* (Chapter 2). The foods that increase insulin secretion the most are sugars and starch, but eating protein and fat also cause some release of insulin. Insulin signals the body's cells to take up available nutrients from the bloodstream and store anything that isn't needed immediately. All cells respond to insulin by taking in glucose from the blood to use for ATP production. Excess nutrients are taken up by the liver, muscle, and fat cells. The liver and muscle cells store glucose as glycogen, the liver and fat cells store excess glucose, amino acids, and fatty acids as triglycerides.

Glucagon is produced between meals when blood sugar levels decrease: known as the *fasting phase*. Glucagon causes cells to break down the stored glycogen and triglycerides in the liver, adipose tissue, and muscle cells so that glucose becomes available for ATP production. Normal blood sugar levels remain quite stable after eating, as well as during exercise or fasting: that is, between 4.5 and 5.5 mmol/L. When blood sugar levels are low, some epinephrine is secreted, and this also causes cells to break down stored nutrients to increase blood sugar. In fact, epinephrine can cause us to feel jittery when we are hungry and blood sugar is decreased. Notice that several hormones are involved in breaking down stored nutrients to increase blood sugar so our cells have a constant supply of glucose for ATP production: thyroid hormone, epinephrine, cortisol, growth hormone, and glucagon.

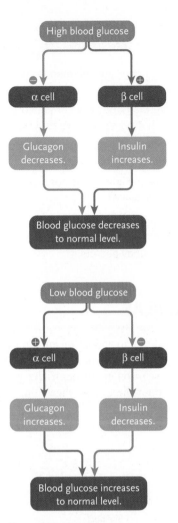

Figure 17.16 Pancreatic Hormones

Pancreatic hormones regulate blood glucose levels. Insulin is released when blood glucose levels increase, and it causes cells to take up and store nutrients. The glucagon that's released when blood glucose levels decrease causes cells to break down and release nutrients.

Diabetes

Type-1 diabetes is an autoimmune disease; immune cells destroy the beta islet cells, and insulin cannot be produced by the pancreas. This is also known as *insulin-dependent diabetes mellitus* (IDDM), and careful monitoring of blood sugar levels and insulin injections are required to maintain blood glucose at optimum levels.

Type-2 diabetes is also known as *non-insulin-dependent diabetes mellitus* (NIDDM) or *insulin resistance*. Eating a lot of sugar and starch, and therefore having constant high insulin levels, leads to down-regulation of insulin receptors on target cells. This causes glucose to stay in the bloodstream instead of being taken up and stored by cells. Type-2 diabetes progresses gradually and worsens as insulin receptors down-regulate further, until eventually the person becomes fully insulin resistant. This condition can be determined by a blood test that measures glucose tolerance. People with impaired glucose tolerance excrete glucose in the urine, which leads to increased water excretion and therefore excessive thirst; weakness and fatigue; impaired wound healing; increased tooth, skin, or bladder infections; and possibly tingling and numbness of extremities.

Prevention of Type-2 Diabetes Insulin resistance will not occur if insulin secretions remain in the normal physiological range. This requires a moderate intake of sugar and starch (that breaks down into glucose); balanced meals that include healthy fats and fibre, which slow digestion and absorption and therefore prevent large insulin secretions; and exercise to use up blood glucose for ATP production.

CONCEPT REVIEW 17.6

1. What are the exocrine functions of the pancreas?

2. What hormones are secreted by the alpha and beta islet cells of the pancreas?

3. When is insulin secreted, and what are the effects on cells in the body? Which cells can insulin act on?

4. What is the glycemic index?

5. When is glucagon secreted, and what are the effects on cells in the body?

6. How and where are excess glucose, fat, and protein stored in the body?

7. Explain the difference between type-1 and type-2 diabetes.

8. Can type-2 diabetes be prevented?

17.7 OVARIES AND TESTES

Hormones produced in the gonads are stimulated by the anterior pituitary hormone known as *luteinizing hormone* (LH) (Figure 17.17). LH causes the testes to produce testosterone and the ovaries to produce estrogen and progesterone. Sperm and egg development—spermatogenesis and oogenesis, respectively—are stimulated by the anterior pituitary hormone known as *follicle-stimulating hormone* (FSH).

Testosterone, the primary androgen in males, is produced by the Leydig cells in the testes in response to LH (Figure 17.18). Testosterone is considered an anabolic steroid hormone because it causes growth and is

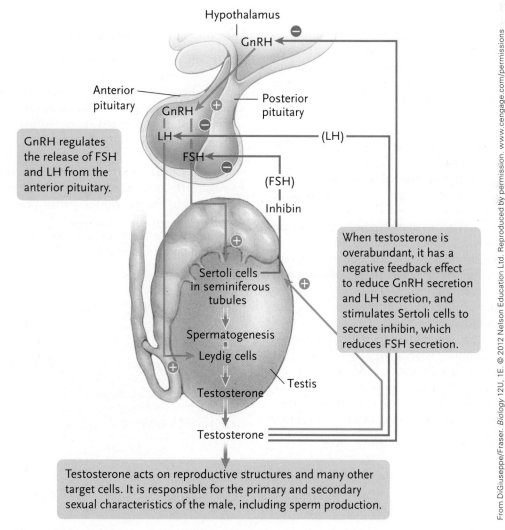

GnRH regulates the release of FSH and LH from the anterior pituitary.

When testosterone is overabundant, it has a negative feedback effect to reduce GnRH secretion and LH secretion, and stimulates Sertoli cells to secrete inhibin, which reduces FSH secretion.

Testosterone acts on reproductive structures and many other target cells. It is responsible for the primary and secondary sexual characteristics of the male, including sperm production.

Figure 17.17   Male Hormone Production

Testosterone is produced by the Leydig cells in response to FSH, which is produced in response to GnRH.

Figure 17.18 shows labels: Leydig cell, Sertoli cell, Germ cell, Spermatid, Seminiferous tubule

Courtesy of Wendi Roscoe

Figure 17.18 Anatomy of the Testis

Sertoli cells and germ cells are located within the seminiferous tubules, and Leydig cells are located between the seminiferous tubules.

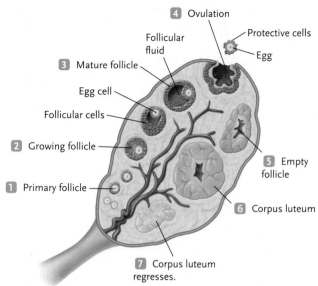

Figure 17.19 shows labels: 4 Ovulation, Follicular fluid, Protective cells, Egg, 3 Mature follicle, Egg cell, Follicular cells, 2 Growing follicle, 5 Empty follicle, 1 Primary follicle, 6 Corpus luteum, 7 Corpus luteum regresses.

Figure 17.19 Anatomy of the Ovary

In the ovary, the oocyte develops in stages from a primary follicle to a mature follicle that is ready for ovulation.

involved in building up tissues. The main functions of testosterone are increasing protein synthesis; increasing muscle mass; increasing bone density and bone maturation; and development of sex organs and secondary sex characteristics, such as facial and axillary hair growth and deepening voice.

Estrogen is produced by the granulose cells that surround the developing oocyte in the ovary because of stimulation by LH (Figure 17.19). Each month as the follicle grows, the surrounding granulosa cells produce more and more estrogen, causing a surge of LH at approximately day 14 of the cycle, and this prompts ovulation. After ovulation, the corpus luteum secretes progesterone, which increases the growth of the uterine lining in preparation for a potential pregnancy. If pregnancy occurs, the corpus luteum and eventually the placenta continues to produce progesterone until labour begins. If pregnancy does not occur, the corpus luteum breaks down and stops producing progesterone, which then stimulates menses—when the uterine lining is shed (Figure 17.20). This cycle occurs approximately every 28 days in females starting from puberty until menopause. Note that males and females both produce testosterone, estrogen, and progesterone, but in different amounts. See Figure 17.20.

Functions of Estrogen and Progesterone

The main functions of estrogen include the development of secondary sex characteristics during puberty; the LH surge when estrogen increases during the menstrual cycle, which leads to ovulation and growth of uterine lining; increased bone formation—explains why osteoporosis is more common in menopausal women; and increased blood clotting—explains why women on birth control have higher risk of stroke. Estrogen affects the transcription of more than 100 different genes. The main function of progesterone is to help increase the uterine lining before fertilization, but its primary role is to support pregnancy and support fetal development in many ways.

DID YOU KNOW?

Birth control pills prevent pregnancy because the estrogen and progesterone in the birth control pills cause the blood level of those hormones to be similar to that during pregnancy; so ovulation is inhibited. Different types of birth control pills have varying levels of estrogen and progesterone.

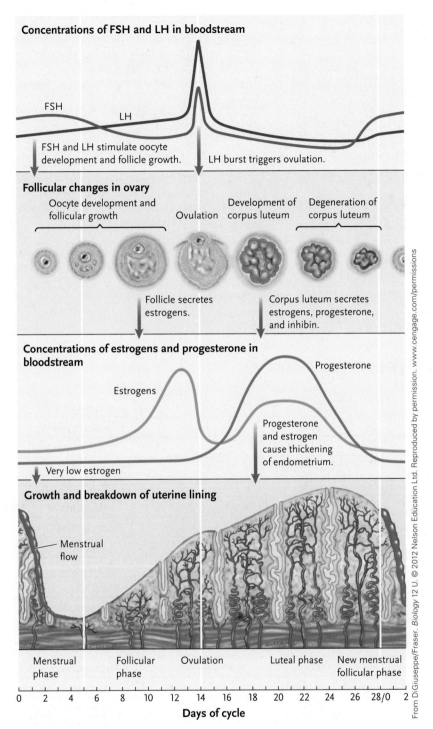

Concentrations of FSH and LH in bloodstream

FSH

LH

FSH and LH stimulate oocyte development and follicle growth.

LH burst triggers ovulation.

Follicular changes in ovary

Oocyte development and follicular growth

Ovulation

Development of corpus luteum

Degeneration of corpus luteum

Follicle secretes estrogens.

Corpus luteum secretes estrogens, progesterone, and inhibin.

Concentrations of estrogens and progesterone in bloodstream

Progesterone

Estrogens

Progesterone and estrogen cause thickening of endometrium.

Very low estrogen

Growth and breakdown of uterine lining

Menstrual flow

Menstrual phase

Follicular phase

Ovulation

Luteal phase

New menstrual follicular phase

0 2 4 6 8 10 12 14 16 18 20 22 24 26 28/0 2

Days of cycle

From DiGiuseppe/Fraser. *Biology 12 U.* © 2012 Nelson Education Ltd. Reproduced by permission. www.cengage.com/permissions

Figure 17.20 Female Hormone Cycle

Female hormones are secreted in differing amounts in the follicular phase and the luteal phase.

CONCEPT REVIEW 17.7

1. What stimulates the production of testosterone, estrogen, and progesterone?

2. What enzyme, found in higher abundance in females than in males, causes females to have more estrogen than males?

3. What are the functions of testosterone, estrogen, and progesterone?

4. What is the role of FSH?

5. What causes ovulation?

EFFECTS OF AGING

As we age there is a general decrease in the production of many hormones and hormone receptors, so endocrine function tends to decrease.

The following hormones tend to decrease as we age:

- Estrogen in women is not produced in the ovaries once menopause occurs.
- Testosterone in men decreases with age, but exercise can stimulate its production.
- Melatonin, which induces sleep, can decrease, so people tend to sleep fewer hours or wake up more frequently during the night.

- Growth hormone production decreases, leading to loss of muscle mass and bone density, but exercise and intermittent fasting can slow the decline.
- Thyroid hormones tend to decrease, leading to a decrease in basal metabolic rate and a tendency to gain weight and to produce less body heat, which is a by-product of ATP production.

The following hormone tends to increase as we age: Parathyroid hormone tends to increase to help maintain blood calcium levels when calcium absorption in the intestines becomes reduced. PTH causes the release of minerals from the bones, and without regular exercise, bones can become less dense, and the risk of osteoporosis increases.

CHAPTER REVIEW

Medical Terminology

Term	Definition	Example
aden/o	gland	adenohypophysis
adren/o	adrenal glands	adrenomegaly
-comat/o	deep sleep	comatose
cortic/o	cortex	corticoadrenal
crin/o	secrete, separate	crinogenic
duct/o	to lead	conduction
ex-	out, away from	exophthalmos
ger/o, geront/o	aged, old age	geriatrics
hirsut/o	hairy	hirsutism
hormon/o	hormone	hormonopoiesis
-ism	condition or disease	hypothyroidism
medull/o	medulla, marrow	medulloadrenal
myx/o	mucus	myxedema
pancreat/o	pancreas	pancreatolithectomy
phe/o	dusky	pheochromocytoma
pineal/o	pineal gland	pinealopathy
pituitar/o	pituitary gland	pituitarism
thym/o	thymus gland	thymotoxin
thyr/o	thyroid gland	thyrocele
tox/o, toxic/o	poison	toxin
-trophy, troph/o	nourishment, growth	hypertrophy

Labelling Activity

Label the following diagrams.

1.

2.

3.

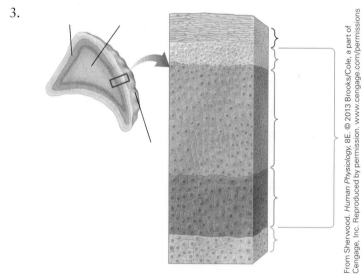

17

Answer to Application Question

The following hormones promote weight loss:

- Growth hormone—stimulates the liver to break down glycogen and fat; stimulated by exercise, deep sleep, and periods of intermittent fasting

- Thyroid hormones—stimulates cells to break down glycogen and fat to produce ATP; stimulated by exercise or a stress response; need to ensure no deficiency in tyrosine, iodine, or selenium

- Epinephrine and norepinephrine—increase heart rate and oxygen utilization for producing ATP; produced during stress response and exercise

- Glucagon—stimulates the liver to break down glycogen and fat; stimulated by fasting

- Androgens—stimulate cells to utilize ATP to synthesize proteins; stimulated by exercise

The following hormones inhibit weight loss:

- Insulin—stimulates the conversion of excess glucose into glycogen and fat; production stimulated by eating foods with sugar and starchy carbohydrates, which cause the greatest insulin production and fat storage

- Cortisol—in normal amounts, stimulates fat and glycogen breakdown, such as in the early morning; important for keeping blood sugar stable during periods of fasting; in excess, such as during chronic stress, leads to insulin resistance and increased storage of visceral fat in the abdomen

18 The Cardiovascular System

CHAPTER OUTLINE

I had discovered that learning something, no matter how complex, wasn't hard when I had a reason to want to know it.

Homer Hickam

18.1 OVERVIEW OF THE CARDIOVASCULAR SYSTEM

The cardiovascular system consists of the heart that pumps the blood; the blood vessels that transport the blood; and the blood, which contains red and white blood cells, platelets, and plasma. This **circulatory system** brings the blood very close to each and every cell in the body and this allows an optimal exchange of oxygen, carbon dioxide, nutrients, and waste products across cell membranes. The circulatory system is arranged into two circuits: the **pulmonary circulation** brings blood to the lungs to pick up oxygen and remove carbon dioxide, and the **systemic circulation** brings oxygenated blood to all the body cells that take up the oxygen and get rid of carbon dioxide (Figure 18.1).

The cardiovascular system (CVS) performs many functions in the body, including the following:

- Transporting oxygen from the lungs to the tissues
- Transporting carbon dioxide from the tissues to the lungs

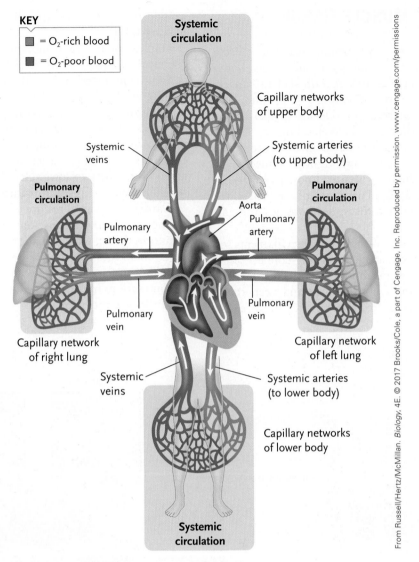

KEY

☐ = O₂-rich blood

☐ = O₂-poor blood

Systemic circulation

Capillary networks of upper body

Systemic veins

Systemic arteries (to upper body)

Pulmonary circulation

Pulmonary circulation

Pulmonary artery

Aorta
Pulmonary artery

Pulmonary vein

Pulmonary vein

Capillary network of right lung

Capillary network of left lung

Systemic veins

Systemic arteries (to lower body)

Capillary networks of lower body

Systemic circulation

Figure 18.1   Organization of the Circulatory System

- Distributing nutrients from the digestive system to the tissues
- Distributing waste products from the tissues to the urinary system
- Delivering immune cells and signalling molecules to the tissues
- Regulating **blood pressure** through hormonal and local vascular control mechanisms
- Regulating body temperature through the heat-absorbing properties of plasma and vascular shunting

CONCEPT REVIEW 18.1

1. What are the functions of the cardiovascular system?

2. What are the main components of the cardiovascular system?

3. What is the difference between the pulmonary circulation and systemic circulation?

18.2 CARDIAC MUSCLE TISSUE

The heart muscle, or myocardium (cardiac muscle tissue), is one of the most unique tissues in the body. This is because it has features of both skeletal muscle and smooth muscle tissue (Chapters 11 and 14). **Cardiac myocytes** (cardiac muscle cells, also called *muscle fibres*) are shorter than skeletal muscle fibres. They contain many mitochondria and one to three nuclei. They are branched and tightly connected at specialized intercellular structures known as *intercalated discs* (Figure 18.2). Two types of structures are found within intercalated discs: desmosomes and gap junctions (Figure 18.3).

Desmosomes consist of interlocking filaments of extracellular proteins that mechanically connect adjacent cardiac myocytes: much like velcro. This intercellular anchoring arrangement stabilizes and strengthens the muscle cells; this is an important feature because

muscle fibres must continually contract in unison to develop force.

Gap junctions are membrane proteins that can form **chemical synapses**, which are channels that chemically connect cells together and allow the *direct* transfer of ions between cells. This direct transfer of ions makes it possible for action potentials to rapidly spread throughout the mass of myocardial cells so that large groups of muscle cells contract simultaneously. The two upper chambers, the right and left atria, beat as one coordinated unit. The two lower chambers, the right and left ventricles, beat as another distinct coordinated unit.

Cardiac muscle fibres are involuntary as their contraction is not under conscious control. Cardiac muscle is autorhythmic, which means that specialized cells in a particular region of the right atrium—the **sinoatrial node (SA node)**—act as a pacemaker to control the rate of contraction; however, the autonomic nervous system can influence the heart rate.

Biophoto Associates/ Science Source

From Russell/Hertz/McMillan. *Biology*, 4E. © 2017 Brooks/Cole, a part of Cengage, Inc. Reproduced by permission. www .cengage.com/permissions

Figure 18.2 Cardiac Muscle Cells

Cardiac muscle cells are branched, striated, and contain many mitochondria.

Figure 18.3 Intercalated Discs

Intercalated discs contain desmosomes and gap junctions that are important for holding cardiac muscle cells together and for transmitting electrical signals during contraction.

CONCEPT REVIEW 18.2

1. What are the features of cardiac muscle tissue?
2. How do desmosomes and gap junctions contribute to the function of the intercalated discs?
3. Explain how the heart is autorhythmic.

18.3 ANATOMY OF THE HEART

The heart is located superior to the diaphragm and near the midline of the thoracic cavity. The heart is enclosed in the mediastinum, which is the area between the lungs that extends from the sternum to the vertebral column and between the first rib and the diaphragm (Figure 18.4). Approximately two-thirds of the heart is left of the midline.

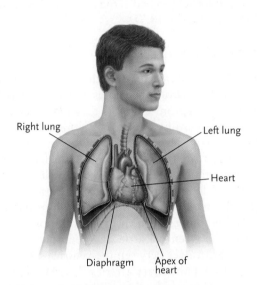

Figure 18.4 Position of the Heart in the Body

The heart is located in the mediastinum of the thoracic cavity.

The heart has four chambers. The thin-walled right and left atria are separated from the thick-walled right and left ventricles by a fat-filled depression called the **coronary sulcus**: also known as the **atrioventricular groove** (Figure 18.5). This deep groove contains several blood vessels that supply the heart; these are the **coronary arteries**. The superior part of the heart is known as the *base*. The narrower, inferior part of the heart is the apex.

Blood vessels are named according to the direction of blood flow. **Veins** always transport blood *toward* the heart, and **arteries** always transport blood *away* from the heart. The *great vessels* include the inferior and superior vena cavae, the pulmonary arteries and veins, and the aorta. The **superior vena cava** returns blood to the right atrium from regions superior to the heart, such as the head, neck, face, and upper limbs. The **inferior vena cava** returns blood to the right atrium from the trunk and lower limbs. Blood returning from the coronary veins is collected in a thin-walled sac called the **coronary sinus**, located on the posterior wall of the right atrium. The coronary sinus has an opening that returns most of the coronary blood directly into the right atrium. The right and left **pulmonary arteries** receive blood from the right ventricle to bring deoxygenated blood to the right and left lungs, respectively. Two **pulmonary veins** from each lung return oxygenated blood to the left atrium. The **aorta** is the largest artery in the body, and it leaves the left ventricle as a short segment called the **aortic arch**. It then descends through the posterior trunk

Figure 18.5 External Features of the Heart

of the body as the **descending aorta** and supplies blood to all parts of the body, except the lungs.

Layers of the Heart

The heart is contained within a double-walled membrane structure called the **pericardium**. The pericardium attaches to the diaphragm and the great vessels, and these attachments secure the heart in the mediastinum. The **parietal pericardium** forms the outer wall surrounding the actual cardiac muscle, and the **visceral pericardium** form the inner wall (Figure 18.6). Because the visceral part of the membrane forms the outer layer of the wall of the heart, it is also known as the **epicardium**. The space between the two layers is called the *pericardial cavity*, which is filled with a small volume of lubricating **serous fluid** that prevents friction between the two membranes when the heart is pumping. Next to the epicardium is the heart muscle: the **myocardium**. The myocardium is the thickest where it surrounds the left ventricle; this is because that region must generate the most force to move blood out of the heart to the systemic circulation. The right ventricle moves blood to the pulmonary circulation, and so it is under significantly less pressure than the systemic circulation. The atria have a very thin myocardial layer because blood moves from the atria into the ventricles mostly by the force of gravity. Atrial contraction contributes only a small amount of blood

volume. Inside each of the chambers—the two atria and the two ventricles—is the final layer of the heart: the **endocardium**. The endocardium is composed of simple, squamous epithelial tissue since it is covering and lining a hollow organ. The myocardium is muscle tissue, and the pericardium is a combination of epithelial tissue and connective tissue. The endocardium is continuous with the inside of the blood vessels that are connected with the heart: the vena cavae, aorta, and the pulmonary arteries and veins. This inner layer is called the **endothelium** and is the innermost layer of all blood vessels (Figure 18.7).

Internal Features of the Heart

The **interatrial septum** separates the two atria, and the **interventricular septum** separates the two ventricles. The interatrial septum contains a shallow depression called the **fossa ovalis**, which is a remnant of a fetal structure known as the **foramen ovale** that allows fetal blood to flow from the right atrium to the left atrium.

Valves maintain unidirectional blood flow in veins so that blood does not flow backward. These are found in large- and medium-diameter veins and in the heart. There are four valves in the heart: two atrioventricular (AV valves) between the atria and ventricles, and two semilunar (SL) valves between the ventricles and either the aorta or the pulmonary arteries. The AV valves prevent the backflow of blood from the ventricles into the

Endocardium   Myocardium   Epicardium (visceral layer)

Pericardial cavity   Pericardium

Endocardium
Myocardium
Epicardium (visceral pericardium)

Articulate Graphics/IllustrationOnLine.com

Figure 18.6 Layers of the Heart

The layers of the heart include the endocardium, which covers the inside of the atria and ventricles; the myocardium, which is the heart muscle tissue; and the epicardium, which is the outer covering of the heart.

atria when the ventricles contract. The right AV valve is the **tricuspid** (consists of three cusps) (Figure 18.8). The left AV valve is the bicuspid (consists of two cusps): also called the **mitral valve**. The cusps of the

atrioventricular valves are connected to the chordae tendineae, which are connected to papillary muscles in the ventricles. When the papillary muscles contract, they pull on the chordae tendineae to close the valve

Figure 18.7   Internal Structures of the Heart

The heart contains four chambers and four valves that blood flows through. Arteries carry blood away from the heart and veins carry blood to the heart.

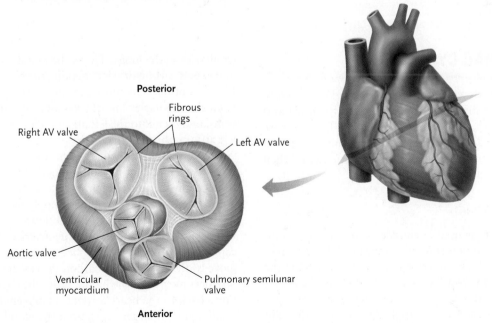

Figure 18.8   Heart Valves

The heart valves include the atrioventricular valves (tricuspid and bicuspid) and the semilunar valves (aortic and pulmonary).

and prevent the backflow of blood. The contraction of the ventricles produces a lot of force to move the blood into the aorta or pulmonary arteries, and the chordae tendineae prevent the valves from flipping inside the atria. The **aortic semilunar (SL) valve** is located between the left ventricle and the aortic arch, and the **pulmonary semilunar valve** is located between the right ventricle and the pulmonary arteries. The semilunar valves do not have chordae tendineae. Instead, they each have three concave, half-moon-shaped valves that are attached to the wall of the aorta and to the pulmonary trunk just before it branches into the pulmonary arteries.

Heart valves open and close based on the pressure gradient across the valve. Blood always moves toward the area that has less pressure. When the atria contract, the pressure in the atria is higher than the pressure in the ventricles, and blood moves from atria to ventricles. The atrioventricular valves open during atrial contraction and then close once the blood has moved into the ventricles. When the ventricles contract, the semilunar

valves open and allow blood to move into the aorta and pulmonary arteries.

Heartbeat sounds are caused by the closing of the valves. The first heart sound—"lub"—is from the closing of the atrioventricular valves, and the second heart sound—"dub"—is from the closing of the semilunar valves. If a valve does not fully close, blood can leak back into the atria or ventricles, depending on which valve is damaged, causing a sloshing sound called a **heart murmur**.

DID YOU KNOW?

Leaking heart valves can be caused by high blood pressure (hypertension) because of the excess stress put on the heart and valves. People with heart murmurs may have no symptoms, or may feel tired (due to inefficient circulation enof oxygen), or may have a fluttering feeling in the chest.

CONCEPT REVIEW 18.3

1. Describe the external structures of the heart.
2. Where are the coronary arteries located?
3. Briefly describe the heart chambers and large vessels.
4. What is the role of the pericardium?
5. Where are the myocardium and endocardium located?

6. Name the four heart valves.
7. Which valves have chordae tendineae? Why are these important?
8. Explain how the heart valves create the heartbeat sounds?
9. What causes the valves to open and close?

18.4 CARDIAC CYCLE

Blood always flows from areas of higher pressure toward areas of lower pressure—through the heart and the blood vessels and back to the heart. The pressure differences are due to the heart muscle contracting and pushing blood from the atria to the ventricles and then from the ventricles to the blood vessels.

Blood Flow Circuit

The right atrium is usually considered the starting point for studying blood flow throughout the body. Blood comes from the superior and inferior vena cavae, which are the largest veins in the blood flow circuit and have the lowest pressure of all the vessels. Blood flows into the right atrium, through the tricuspid (right AV valves) to the right ventricle. When the ventricles contract, the force of the contraction causes significant pressure. Blood is pumped from the right ventricle through the pulmonary semilunar valve to the pulmonary arteries, to the **arterioles**, to the

capillaries in the lungs. There, the blood picks up oxygen and expels carbon dioxide, making the blood oxygenated. Blood continues to move toward lower pressure areas: from the capillaries into the **venules**, into the pulmonary veins, and then into the left atrium. Blood moves from the left atrium, through the bicuspid (left AV valve) to the left ventricle. Contraction of the left ventricle generates the most force, causing the highest blood pressure in the body to occur in the aortic arch. As blood flows through the aortic arch, two small openings supply blood to the coronary arteries (Figure 18.8). Blood flows from the aorta to vessels of the systemic circulation: arteries, arterioles, and capillaries. In the capillary beds, oxygen and nutrients move to the tissues, and carbon dioxide and other waste products move into the blood. From the capillaries, blood flows back to the heart through small venules, leading to small and then larger veins, until the blood again reaches the superior and inferior vena cavae (Figure 18.9).

The mechanical events associated with one heartbeat make up the **cardiac cycle**. These events include the contraction and relaxation of the different chambers of the

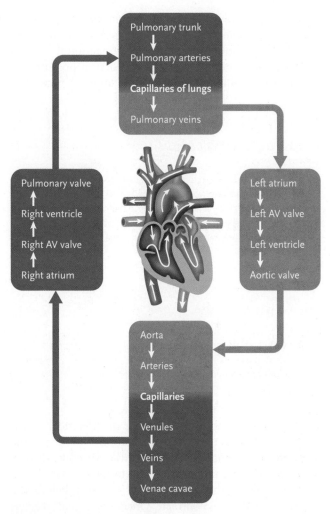

Figure 18.9 Blood Flow Circuit

Blood flows from the right side of the heart to the lungs and from the left side of the heart to the rest of the body through arteries, arterioles, and capillaries, where gas exchange occurs. Blood flows from the tissues back to the heart through venules, and then through veins to the inferior and superior vena cavae.

heart, the opening and closing of heart valves, and the entry and exit of blood. When the heart contracts, the atria contract together, and the ventricles contract together; so blood moves to the pulmonary and systemic circulation simultaneously. The cardiac cycle consists of the following distinct stages:

1. When the atria contract, the AV valves are open, and blood moves into the ventricles. This stage is called **atrial systole**. Note that at this point the ventricles are almost full from blood flow due to gravity; atrial contraction adds the last 10 to 20 mL.

2. Atrioventricular valves close due to the rising pressure in the ventricles as they begin to contract. The closing of the AV valves prevents backflow of blood into the atria and makes the first heart sound: lub.

3. When the atria relax, blood flows into the right atrium from the superior and inferior vena cavae, and into the left atrium from the pulmonary veins. This stage is called **atrial diastole**.

4. At the same time as the atria relax, the ventricles contract. The AV valves are closed so blood cannot move back into atria, and the semilunar valves open when the pressure in the contracting ventricles increases to a threshold level that forces the high-pressure blood flow into the aorta (from the left ventricle) and the pulmonary arteries (from the right ventricle). This stage is called **ventricular systole**.

5. Semilunar valves close. After ventricular contraction, a small amount of blood in the aorta and pulmonary arteries moves back toward the heart because of gravity, and this causes the SL valves to close. This closing prevents backflow of blood into the ventricles and makes the second heart sound: dub.

6. When the ventricles relax, the pressure in the ventricles decreases, which causes the AV valves to open. This stage is called **ventricular diastole**. Blood can begin flowing from the atria into the ventricles *before* the atria contract (back to step 1).

CONCEPT REVIEW 18.4

1. Trace the path of a red blood cell as it travels from the right atrium throughout the body and back to the right atrium.

2. How much of the final ventricular blood volume is due to atrial systole?

3. What causes the AV valves to open and close, and when does this occur?

4. What causes the SL valves to open and close, and when does this occur?

5. What do *systole* and *diastole* refer to?

DID YOU KNOW?

The human body has about 100 000 km of blood vessels, enough to wrap around Earth two-and-a-half times!

18.5 ELECTRICAL CONDUCTION

The **conducting system** of the heart begins at the SA node located in the posterior region of the right atrium, inferior to the opening for the superior vena cava. The SA node can spontaneously depolarize, a trait called

autorhythmicity. These specialized cells do not have a steady resting membrane potential as do other cells in the body; however, they slowly depolarize, and this is called the **pacemaker potential**, which leads to an action potential, which leads to cardiac muscle contraction.

Recall from Chapter 15 that in a neuron, an action potential occurs if there is a stimulus that causes the inside of the cell to depolarize; this usually occurs when sodium ions enter the cell. Once a threshold level is reached, an action potential occurs as sodium ions move into the cell, then repolarization occurs when potassium leaves the cell. Action potentials in the SA node are slightly different. In the SA node, sodium ions begin to leak into the cell as soon as the cell is repolarized to approximately −60 mV. As sodium ions slowly move *in*, the cell becomes more positively charged (pacemaker potential), and once it reaches approximately −40 mV, voltage-gated calcium channels open, allowing calcium ions to move *into* the cell, further depolarizing the cell (action potential) to approximately +10 mV. Once the inside of the cell is positively charged, voltage-gated potassium channels open, and potassium ions move *out* of the cell, causing repolarization. Then the cycle continues without the need for any external stimulation (Figure 18.10).

The autonomic nervous system can influence heart rate, but there is no nervous tissue within the heart itself. Without any external stimulation, the SA node has a firing rate of approximately 100 beats per minute (bpm), but our regular heart rate is usually lower than that due to nervous system and endocrine interaction. For example, the sympathetic nervous system will increase heart rate by releasing norepinephrine, which increases the opening of sodium channels in the SA node, causing the pacemaker depolarization phase to occur faster. The parasympathetic nervous system will release acetylcholine that slows the opening of sodium channels, so it takes longer for the action potential to occur. A typical resting heart rate is 60 to 75 bpm.

> **DID YOU KNOW?**
>
> If the SA node is damaged, the AV node can become the pacemaker. However, the heart rate would drop to 40 to 60 bpm, which may not be adequate to oxygenate the brain. In that case, an artificial pacemaker can be surgically implanted into the heart to deliver the electrical stimulation required to cause the heart to contract.

> **DID YOU KNOW?**
>
> The heart can beat for hours after being removed from the body if it is kept in oxygenated saline solution. This factor is very beneficial for people needing a heart transplant.

Once the SA node depolarizes, the electrical signal travels throughout the atrial muscle fibres and then to the **atrioventricular node (AV node)**. At the AV node, there is a 100 ms delay in the transmission to the bundle of His, and this allows the atria to fully contract before the ventricles begin to contract. Once the signal reaches the **bundle of His** (pronounced "hiss")—also called **atrioventricular bundle**—it travels along the right and left **bundle branches** to the **Purkinje fibres** at the apex of the heart. Action potentials travel quickly in the large-diameter Purkinje fibres and cause the ventricular muscle cells to contract from the bottom up, which pushes blood from the ventricles out to the aorta and pulmonary arteries.

ECG

Action potentials in the conducting cells of the heart stimulate the contraction of the heart muscle fibres. An **electrocardiogram (ECG)** is a method of measuring the electrical currents caused by the movement of ions throughout the heart during the cardiac cycle (Figure 18.11). There are three distinct waves associated with the heart contractions.

1. The **P wave** represents the atrial depolarization that causes contraction of the atria.

2. The **QRS** represents ventricular depolarization that causes contraction of the ventricles. Although not visible on the ECG, at this time the atria are repolarizing and relaxing.

3. The **T wave** represents ventricular repolarization and triggers relaxation.

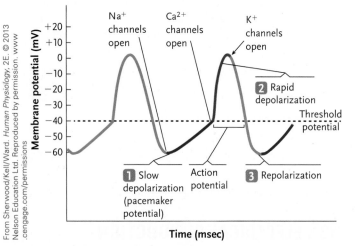

From Sherwood/Kell/Ward. *Human Physiology*, 2E. © 2013 Nelson Education Ltd. Reproduced by permission. www .cengage.com/permissions

Figure 18.10 Pacemaker Potential

The pacemaker potential involves (1) slow depolarization caused by sodium ions moving into the cell; (2) rapid depolarization, caused by calcium ions moving into the cell; (3) repolarization, caused by potassium ions moving out of the cell.

Figure 18.11 Conduction System of the Heart

(A) Anatomical structures and (B) ECG patterns

Variations in ECG patterns can be used to diagnose malfunctions of the cardiac conduction system such as **arrhythmias**, which are irregular heart contractions. Premature ventricular contractions (PVCs) are relatively common, harmless, irregular heartbeats that can feel like a flutter in your chest. During a PVC the heartbeat is initiated by the Purkinje fibres rather than the SA node, so the ventricles contract before they are filled with blood from the atria; this is perceived as a skipped heartbeat. A PVC can occur during exercise or stress and may be due to a temporary drop in oxygen, much like a twitch or cramp can occur in a skeletal muscle. Figure 18.12 shows an ECG with a PVC occurring.

An overview of the major events occurring during one cardiac cycle is shown in Figure 18.13 and summarized as follows:

1. P wave occurs when depolarization of the atria leads to contraction of the atria.

2. Slight pressure increases in ventricle due to increased blood volume.

3. Approximately 10 to 20 mL of blood volume is added to the ventricles due to atrial contraction.

4. **End diastolic volume (EDV)** occurs when the ventricles are full; this is the volume of blood in the ventricles after ventricular diastole (relaxation).

5. The first heart sound occurs from the closing of the AV valves after the atria have contracted. Closing the AV valves prevents the backflow of blood from the ventricles to the atria once the ventricles begin to contract and ventricular pressure is higher than in the atria.

6. The first isovolumetric phase occurs when the AV valves are closed and the semilunar valves have not yet opened, and so the volume of blood in the ventricles does not change.

Figure 18.12 Premature Ventricular Contraction

ECG showing a PVC

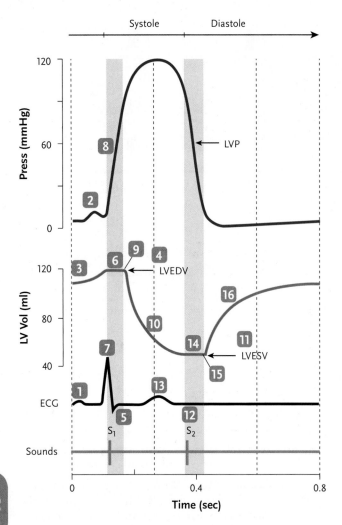

Figure 18.13 Overview of Cardiac Cycle

The cardiac cycle is shown in terms of ECG, heart sounds, and blood volume and pressure in the left ventricle.

7. QRS phase occurs when the ventricles depolarize, which leads to contraction of the ventricles.

8. Contraction of the ventricles causes the pressure in the ventricles to become higher than the pressure in the aorta.

9. Once ventricular pressure exceeds aortic pressure, the semilunar valves open.

10. Blood is pumped out of the left ventricle into the aorta and out of the right ventricle into the pulmonary arteries, and the volume of blood in the ventricles decreases.

11. End systolic volume (ESV) occurs when the ventricles are almost empty; this is the volume of blood left in the ventricles after systole (contraction). Note, as shown in Figure 18.13, approximately 50 mL of blood volume remains in the left ventricle after contraction.

12. Semilunar valves close once the pressure in the ventricles decreases, causing the second heart sound.

13. The second isovolumetric phase occurs when the volume in the ventricles is not changing because the semilunar valves have just closed and the AV valves have not yet opened.

14. T wave occurs when the ventricles repolarize, the muscle relaxes, and the pressure inside the ventricle decreases.

15. Atrioventricular valves open because the pressure is now higher in the atria than in the ventricles.

16. Ventricles begin filling with blood as soon as the AV valves open, and before the atria contract at the next P wave.

CONCEPT REVIEW 18.5

1. What structures make up the conducting system of the heart?

2. What is autorhythmicity?

3. What three ions play an important role in the pacemaker potential?

4. How is the pacemaker potential different from an action potential in a neuron?

5. What is happening in an ECG during the P wave? QRS? T wave?

6. What causes the first and second heart sounds?

7. What is an arrhythmia?

8. What is happening during the first and second isovolumetric phases of the cardiac cycle?

9. When does most of the blood enter the ventricle?

10. What is end diastolic volume and end systolic volume?

18.6 CARDIAC OUTPUT

Cardiac output refers to the total amount of blood ejected from the left ventricle per minute. It is determined by the number of beats per minute (heart rate) and the amount of blood ejected with each contraction (*stroke volume*).

cardiac output (CO) = heart rate (HR) × stroke volume (SV)

For an adult the average resting heart rate is approximately 70 bpm. An adult's average stroke volume is approximately 70 mL. Stroke volume is determined by the amount of blood in the ventricle after diastole (relaxation): that is, the end diastolic volume (EDV) minus the end systolic volume (ESV), which is the amount of blood left in the ventricle after systole (contraction).

stroke volume = EDV − ESV

Given an average EDV of 120 mL and an average ESV of 50 mL,

stroke volume = 120 mL − 50 mL

stroke volume = 70 mL

If we assume a resting heart rate of 70 beats per minute and a stroke volume of 70 mL per beat, the cardiac output at rest is calculated as follows:

$$\text{cardiac output} = \frac{70\ beats}{1\ min} \times \frac{70\ mL}{beat} \times \frac{1L}{1000\ mL}$$

cardiac output = 4900 mL/min

cardiac output = 4.9 L/min

There are approximately 5 to 6 L of blood in the body (depending on body weight). Therefore, at rest the heart is pumping nearly the entire blood volume each and every minute. It is possible to estimate total blood volume. There is approximately 75 mL of blood for each kilogram of body weight. Blood volume can be up to 15% less in obese people.

Example: 80 kg male
Blood volume = 80 kg × 75 mL
Blood volume = 6000 mL = 6 L

Cardiac Output during Exercise

With moderate exercise, we might have an increase in heart rate from 70 bmp to 100 bpm, and an increase in stroke volume from 70 mL to 100 mL. That would increase cardiac output to 10 L/min. This is known as **cardiac reserve**. Most people have a cardiac reserve of two to three times their resting value (approximately 10–15 L/min). Endurance-trained elite athletes may have a cardiac reserve of more than four times their resting value (20+ L/min).

Factors That Affect Cardiac Output

Any factor that increases heart rate or stroke volume increases cardiac output up to a maximum (Figure 18.14).

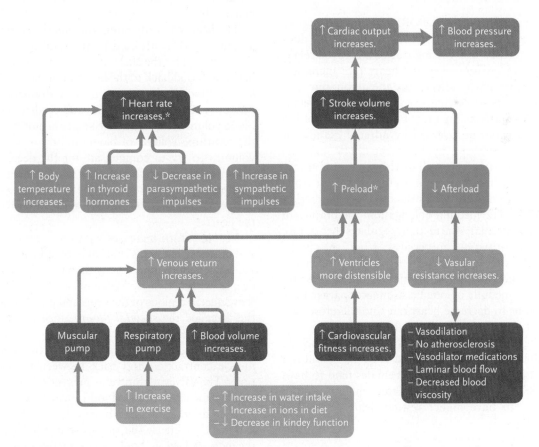

*Once heart rate increases past 200 to 220 bpm, then stroke volume decreases and contractions are inefficient, causing cardiac output to decrease.

Figure 18.14 Factors Affecting Cardiac Output
Factors that increase heart rate or stroke volume will increase cardiac output.

Once the heart rate gets very high (190–200 bpm) the heart becomes inefficient, and the stroke volume decreases significantly. Heart rate varies on average between 60 and 180 bpm. Athletic individuals tend to have a lower resting heart rate because their heart is stronger and, therefore, more blood volume is pumped with each contraction. Cardiac output can be maintained with a lower heart rate if the stroke volume is higher. Stroke volume varies between 70 and 120 mL.

DID YOU KNOW?

We can estimate our maximum heart rate by subtracting our age from 220. Someone who is 20 years old could have a maximum heart rate of 200 bpm. This is just an estimation, more athletic people can have a higher maximum heart rate compared to non-athletic people. When a person's maximum heart rate is exceeded, stroke volume becomes dramatically reduced, not enough oxygen circulates to fuel the high-energy demands of the heart, and the heart can fail.

Factors That Affect Heart Rate

When heart rate increases, the QRS depolarization time and the time required for the ventricles to contract does not change relative to a resting state; rather, the relaxation period decreases when the heart rate increases. Heart rate is primarily affected by the autonomic nervous system (ANS). During exercise, stress, or relaxation, the ANS affects the rate of depolarization of the SA node. Epinephrine released by adrenal glands, or norepinephrine released by the sympathetic neurons increase the heart rate as well as the force of contraction. Acetylcholine from parasympathetic neurons slows heart rate and decreases the force of contraction (Section 17.5). The medulla oblongata in the brainstem is the cardiovascular centre that regulates heart rate based on information it receives from the body. Peripheral baroreceptors detect blood pressure in the aortic arch and the carotid arteries. If blood pressure increases, the medulla oblongata stimulates parasympathetic neurons to decrease heart rate and therefore cardiac output, which then decreases blood pressure.

Other factors that can increase heart rate include thyroid hormones; increased blood Na^+, Ca^{2+}, and K^+; and increased body temperature. Females tend to have slightly higher resting heart rates than males, and the elderly tend to have higher resting heart rates than younger adults.

Factors That Affect Stroke Volume

The strength of contraction is directly proportional to the amount of stretch in the myocardial fibres: a relationship known as the **Frank–Starling law**. A larger blood volume increases the amount the muscle fibres stretch, which increases the strength of the contraction. Any increase in the end diastolic volume will increase ventricular contraction, which leads to an increase in stroke volume. The sympathetic nervous system can increase the force of contraction, which also increases the stroke volume.

The volume of blood returning to the atria from the body and from the pulmonary circulation is called the **preload**. An increase in preload increases EDV and, therefore, increases stroke volume. The stroke volume is also dependent on the systemic vascular resistance, which is called **afterload**. People who have constricted blood vessels due to atherosclerosis (Section 17.11) have greater vascular resistance and therefore reduced stroke volume and cardiac output.

Preload is affected primarily by **distensibility** of the ventricles and by **venous return**, which is the amount of blood returning to the heart. *Distensibility* refers to how readily the heart wall expands to accommodate a returning volume of blood. A heart that is rigid, or one that has extensive scarring as a result of a heart attack, does not easily expand. As a result, a limited amount of blood can enter the heart, and this reduces stroke volume and cardiac output (at a given heart rate).

The heart can only pump the blood that returns to it. Venous return is affected by the total blood volume in the body and by the skeletal and respiratory pumps that help pull blood back to the heart from the systemic circulation. **Blood volume** can have a dramatic effect on venous return, stroke volume, and cardiac output. Low blood volume, as occurs with severe hemorrhage, reduces the returning volume of blood and decreases the stroke volume. To try to compensate for the decreased blood volume, the heart rate increases in an effort to maintain an optimal cardiac output. Raising the legs can help overcome the force of gravity and help return blood to the heart.

Venous return increases with the contraction of skeletal muscles. The veins that carry blood back to the heart are located more superficially than the arteries, which are deeper within tissues. When skeletal muscles contract, they squeeze regions of the veins and push blood through; this is called the *skeletal pump*. Since the blood in the veins is under the lowest pressure of any of the other blood vessels, blood moves forward in the veins during a muscle contraction and back between muscle contractions. The valves within veins prevent the backflow from moving too far, so the blood moves forward from one valve to another in small bursts (Figure 18.15). Exercise can greatly increase the venous return (preload) because the increased skeletal muscle contractions force the blood through the veins and back to the heart. This is one reason why cardiac output increases during exercise.

Venous return also increases with increased respiration. This is due to the pressure gradient created

Figure 18.15 Skeletal Pump

The *skeletal pump* refers to how the contraction of the lower leg muscles helps push blood through the veins and back toward the heart.

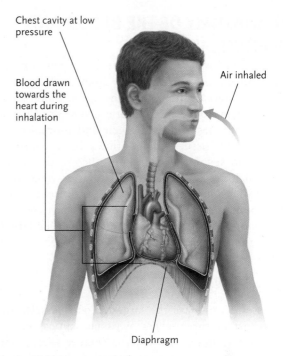

Figure 18.16 Respiratory Pump

The respiratory pump helps move blood through veins back toward the heart, with each inhalation when the pressure in the thoracic cavity decreases.

between the thoracic and abdominal cavities with each breath, when the diaphragm contracts and relaxes (Figure 18.16). When we inhale, the diaphragm contracts and moves downward, increasing the volume of the thoracic cavity and *decreasing* intrathoracic pressure; this is called the *respiratory pump*. Recall that blood flows through vessels from areas of higher to lower pressure. When the pressure in the thoracic cavity decreases with each inhalation, blood moves from the area of higher pressure in the large abdominal veins toward the inferior vena cava in the thoracic cavity.

Factors That Increase Cardiac Output Cardiac output increases when heart rate increases, which is affected by the following:

- An increase in sympathetic stimulation by epinephrine and norepinephrine
- A decrease in parasympathetic stimulation
- Exercise

- An increase in sodium, potassium, and calcium levels
- An increase in body temperature

Cardiac output increases when stroke volume increases, which is affected by the following:

- An increase of venous return (preload)
- Exercise, by means of skeletal and respiratory pumps
- An increase in blood volume, for example, due to water retention when sodium level increases
- An increase in ventricular contractility and distensibility
- A decrease in vascular resistance (afterload)

CONCEPT REVIEW 18.6

1. What is cardiac reserve?
2. Bill has an EDV of 130 ml, an ESV of 50 mL, and a heart rate of 75 bpm. What is his cardiac output?
3. How does the sympathetic and parasympathetic nervous system affect heart rate?
4. How do epinephrine and thyroid hormones affect cardiac output?
5. How would an increase in vascular resistance affect stroke volume?

6. What is preload and afterload?
7. Explain the Frank–Starling law.
8. What factors affect venous return?
9. How do the skeletal and respiratory pumps increase venous return?
10. What factors increase cardiac output?

18.7 ANATOMY OF THE BLOOD AND LYMPHATIC VESSELS

Arteries always move blood *away* from the heart, and the largest artery in the body is the aorta. Blood flows from the aorta, to the main arteries, to arterioles, and then to capillaries—where oxygen and nutrients can move into the tissues. Blood flows back from the capillaries to venules, to veins, and then to the largest veins, which are the superior and inferior vena cavae (Figure 18.17). Our total blood volume at any one time is mostly found in the veins. The capillaries have the least total amount of blood, although they have the largest cross-sectional area compared to any other type of blood vessel (Figure 18.18).

Arteries and Veins

Arteries and veins are the largest blood vessels, and they consist of three layers of tissue: tunica intima, tunica media, and tunica externa (Figure 18.19). The innermost layer, the **tunica intima**, is composed of simple squamous epithelial tissue. This tissue, the endothelium, is a basement membrane that surrounds all epithelial tissue; it is made of proteins that connect the epithelial layer to connective tissue. Next to the basement membrane in arteries, but not in veins, is a connective tissue layer that is high in elastic proteins. This layer is necessary in arteries, because the arteries must stretch with each heartbeat, and the elastic layer helps the artery move back to its original shape. The middle layer, the **tunica media**, consists of smooth muscle tissue, and arteries also have another layer of elastic connective tissue. The smooth muscle layer is larger in arteries because they must contract to help move the blood through the vessels with each heartbeat. The outermost layer, the **tunica externa**, consists of a thick layer of loose connective tissue (areolar and reticular).

Of all the blood vessels, arteries must withstand the highest pressure. Veins withstand relatively low blood pressure but have much more blood volume. The elastic layer and larger smooth muscle layers are needed for arteries to contract and move blood to the arterioles. As the bolus of blood moves through the blood vessels with each heartbeat, we can feel a pulse in the large arteries

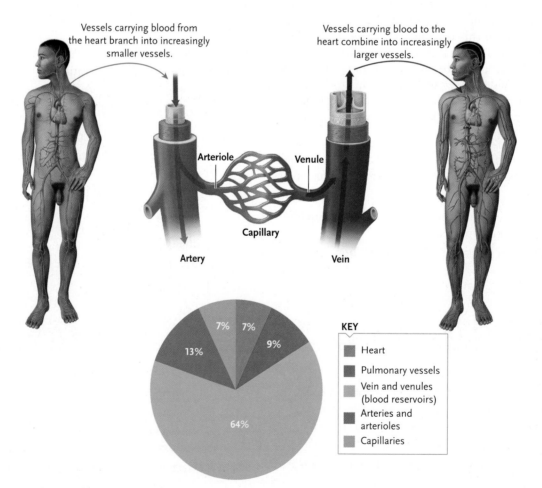

Figure 18.17 Overview of Blood Vessels

Most of the blood volume is in the veins at any given time.

From Russell/Hertz/McMillan. *Biology*, 4E. © 2017 Brooks/Cole, a part of Cengage, Inc. Reproduced by permission. www.cengage.com/permissions

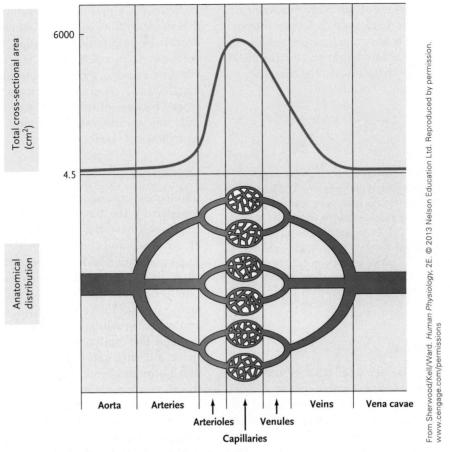

From Sherwood/Kell/Ward. *Human Physiology*, 2E. © 2013 Nelson Education Ltd. Reproduced by permission. www.cengage.com/permissions

Figure 18.18 Cross-Section of Blood Vessels

Capillaries have the greatest cross-sectional area compared to all other blood vessels.

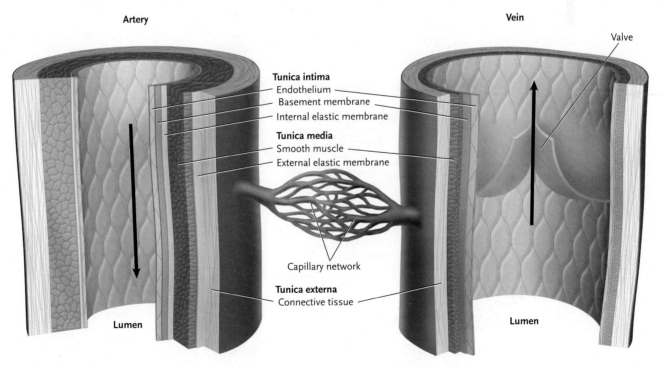

Figure 18.19 Arteries and Veins

Arteries and veins are made up of three layers: the tunica intima, tunica media, and tunica externa.

that are close to the surface of the body, but veins do not have a pulse. Veins have a much larger internal diameter because they hold a much larger volume of blood, and the veins have valves that prevent the backflow of blood as it moves back toward the heart.

Arterioles and Venules

Arterioles and venules are like small arteries and veins. The larger arterioles and venules contain the same layers as arteries and veins, but each of the layers is smaller. The smallest arterioles consist only of inner endothelium and basement membrane, a very small smooth muscle layer, and an outer loose connective tissue layer. The smallest venules consist only of inner endothelial cells and basement membrane and outer connective tissue (Figure 18.20).

Arterioles play a very important role in the regulation of blood flow to specific regions of the body. They do this by responding to signals from the autonomic nervous system and the endocrine system. For example, when we exercise, more blood is required in the skeletal muscles and is diverted away from the digestive system, mostly because of **vasodilation** of arterioles in the skeletal muscles and **vasoconstriction** of arterioles in the gastrointestinal tract.

Capillaries

Capillaries consist of inner endothelium cells surrounded by the basement membrane (Figure 18.21). Capillaries have the most cross-sectional area of any blood vessel, and come within 100 μm of almost every single cell in our body (except the avascular cells of the outer layers of the skin, cartilage, and the cornea). Capillaries have the important function of exchanging materials with the cells. Oxygen, nutrients, hormones, and other signalling molecules leave the bloodstream through the capillaries, diffuse across the interstitial space, and then enter the cells. Recall from Chapter 17 that hormones bind only to cell types that contain a specific receptor. Carbon dioxide and waste products leave the cells, enter the capillaries, and are expelled or broken down once they circulate to the lungs, liver, or kidneys. The movement of oxygen from the capillaries to the cells and carbon dioxide from the cells to the capillaries is called **gas exchange**. Blood flow regulation can occur at the level of the capillaries as well as the arterioles. Not every capillary bed in the body

Figure 18.20 Arterioles and Venules

Arterioles and venules contain endothelial cells, a basement membrane, a layer of smooth muscle, and connective tissue.

Figure 18.21 Capillaries

Capillaries are the site of exchange of gases and nutrients.

contains the same amount of blood all the time. Tissues that need more blood oxygen and nutrients will receive more blood flow. Each capillary bed has precapillary sphincters, which are small circular smooth muscles that block blood flow when they contract (Figure 18.22). Precapillary sphincters contract or relax depending on the amount of blood flow required. For example, when the body temperature decreases, the precapillary sphincters in the deeper layers of the skin contract and decrease blood flow to prevent excessive heat loss.

In capillary beds, fluid leaves the blood vessels at the arteriole end, and some of it returns to circulation at the venule end. This loss of fluid is a normal process that helps move nutrients from the bloodstream to the cells. Depending on where the fluid is located, it has different names: (1) When the fluid is inside the capillaries, it is called **plasma**, which is the liquid portion of the blood that contains dissolved nutrients, ions, vitamins, and minerals; (2) when some of this fluid moves out of the capillaries and into the interstitial (intercellular) space, the fluid is called *interstitial fluid*, which continually bathes the cells in nutrients and dissolved oxygen so that each cell can readily take up whatever it needs; (3) when excess fluid either returns to the venule end of the capillary bed or is taken up by lymphatic vessels, the fluid is called **lymph**.

Why Fluid Leaks from the Circulatory System

The movement of fluid out of the capillaries into the interstitial space increases as blood pressure increases (Figure 18.23a). This is called **hydrostatic pressure**, which is the physical pressure of the fluid moving through the blood vessels (see Figure 18.23b, large arrow at arteriole end). During exercise, when blood pressure increases, more fluid is pushed from the capillaries into the interstitial space, increasing hydrostatic pressure in the interstitial space. All the fluid from the blood does not leave the vessels because certain proteins (mostly **albumin**) cause an **osmotic pressure** that keeps fluid inside the capillaries (see Figure 18.23b, small arrow at arteriole end). As blood moves through the capillary network toward the venule end, the osmotic pressure in the capillaries increases due to the loss of fluid along the way, but

Figure 18.22 Precapillary Sphincters

Precapillary sphincters regulate blood flow into various capillary beds.

the proteins stay inside the blood vessels. The increased osmotic pressure at the venule end causes some interstitial fluid to re-enter the capillaries (see Figure 18.23b, larger arrow at venule end). Also at the venule end, fluid re-enters the capillary because of the increased hydrostatic pressure of fluid in the interstitial space (see Figure 18.23b, smaller arrow on venule end). In Chapter 21, the movement of fluid out of capillaries is discussed in relation to the urinary system.

About 20 L of fluid moves from the cardiovascular system into the interstitial spaces every day. All but about three or four litres of it re-enters the capillaries, and the rest must be taken up by the lymphatic vessels. If fluid builds up in the interstitial space, tissues swell; this is called **edema**. Excessive edema causes blood pressure to drop because the fluid is moving from the blood vessels into the tissues; in extreme cases this can be fatal.

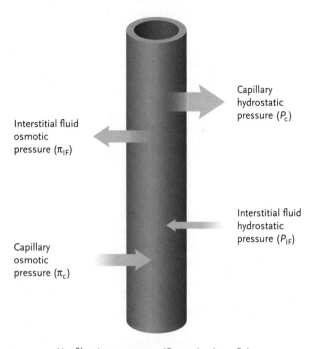

Capillary hydrostatic pressure (P_c)

Interstitial fluid osmotic pressure (π_{IF})

Interstitial fluid hydrostatic pressure (P_{IF})

Capillary osmotic pressure (π_c)

A. Net filtration pressure = $(P_c + \pi_{IF}) - (\pi_c + P_{IF})$

DID YOU KNOW?

Eating a lot of salt can cause edema because sodium ions are in higher concentrations just outside of cells, in the interstitial space. This causes an osmotic gradient that pulls water into the interstitial space. In a healthy person, the kidneys filter and excrete excess sodium, so persistent edema could be an indication that the kidneys are not functioning optimally.

Arteriole

Smooth muscle

Precapillary sphincter

Tissue cell

Lymph capillary

Interstitial fluid

P_c

π_c

π_{IF}

P_{IF}

Venule

B.

Figure 18.23 Movement of Fluid

Movement of fluid into and out of capillaries is due to hydrostatic and osmotic pressures.

Lymphatic Vessels

The lymphatic system plays three important roles: (1) returning fluid from the interstitial tissue back to the circulatory system, (2) transporting the fats absorbed in the digestive tract to the circulatory system, and (3) making up the majority of the immune system (Chapter 23). Lymphatic vessels are composed of endothelium and a basement membrane (as in capillaries), but they also contain some smooth muscle cells (as in small arterioles) and valves (as in veins) to prevent backflow of lymph fluid as it moves through the vessels (Figure 18.24). Lymph moves through the lymphatic system by contraction of the skeletal muscles and also by contraction of smooth muscle cells that surround the lymph vessels. Fluid moves from regions of higher pressure to lower pressure—just like blood flow—through the vessels until it reaches the lymphatic ducts that drain into the venous system. The lymphatic system of the right, upper side of the body drains into the **right lymphatic duct**, then into the **right subclavian vein**. The lymphatic system of the rest of the body drains into the **thoracic lymphatic duct**, then into the **left subclavian vein** (Figure 18.25).

Summary of Lymph Flow plasma → interstitial space → lymphatic vessel → lymph ducts → subclavian veins of the circulatory system

Figure 18.24 Anatomy of a Lymph Vessel

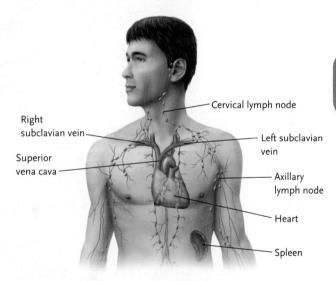

Figure 18.25 Drainage Paths for the Lymphatic System

Lymph drains back into the circulatory system via the right and left subclavian veins.

DID YOU KNOW?

Massage therapists can help with lymphatic drainage by massaging and pushing lymph through the vessels toward the subclavian veins; this process is called *manual lymphatic drainage*. Increasing lymphatic drainage increases blood circulation because blood volume and venous return increases. It is also believed that manual lymphatic drainage can help fight colds and flus. This is because lymph nodes filter the fluid that moves through lymph vessels, and this is where immune cells can fight infections.

CONCEPT REVIEW 18.7

1. Describe each of the three layers of an artery.
2. How are arteries and veins different?
3. Which blood vessels withstand the most blood pressure?
4. Which blood vessels have valves?
5. Where does gas exchange occur?
6. What is the importance of precapillary sphincters?
7. What is the difference between plasma, interstitial fluid, and lymph?
8. What is edema?
9. How is lymph moved through the lymphatic vessels back to the circulatory system?

18.8 CIRCULATORY ROUTES

Blood vessels are organized in terms of their specific routes to all the tissues of the body (Figure 18.26). The right ventricle moves *deoxygenated* blood through the pulmonary semilunar valve into the pulmonary trunk, which divides into left and right pulmonary arteries, and then to the arterioles and to the capillary beds in the lungs (Figure 18.27). *Oxygenated* blood flows back through the pulmonary venules into the right and left pulmonary veins.

Blood is then pumped from the left ventricle to the aorta, where all other systemic arteries branch off (Figure 18.28). Blood flows from the aorta, which is the largest artery in the body. The main regions of the aorta are the ascending aorta, the aortic arch, the thoracic aorta, and the abdominal aorta—also called the *descending aorta*. From the aorta, major arteries branch to the upper and lower body. The right and left coronary arteries branch off from the ascending aorta and supply the heart muscle.

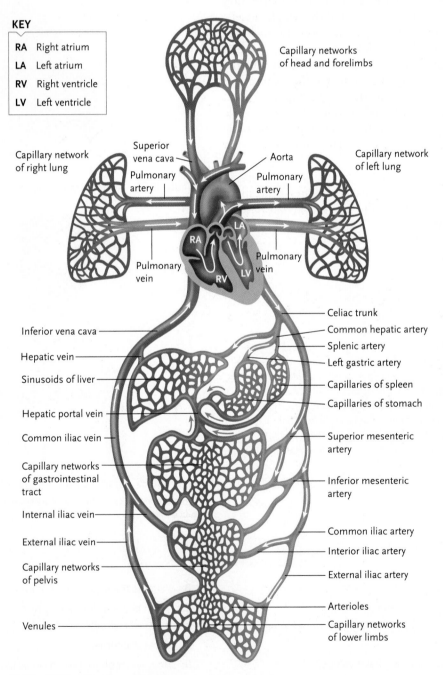

KEY

RA	Right atrium
LA	Left atrium
RV	Right ventricle
LV	Left ventricle

Capillary networks of head and forelimbs

Capillary network of right lung

Superior vena cava

Aorta

Pulmonary artery

Pulmonary artery

Capillary network of left lung

Pulmonary vein

Pulmonary vein

Inferior vena cava

Hepatic vein

Sinusoids of liver

Hepatic portal vein

Common iliac vein

Capillary networks of gastrointestinal tract

Internal iliac vein

External iliac vein

Capillary networks of pelvis

Venules

Celiac trunk

Common hepatic artery

Splenic artery

Left gastric artery

Capillaries of spleen

Capillaries of stomach

Superior mesenteric artery

Inferior mesenteric artery

Common iliac artery

Interior iliac artery

External iliac artery

Arterioles

Capillary networks of lower limbs

Figure 18.26 Circulatory Routes

Blood vessels are organized into various circulatory routes throughout the body.

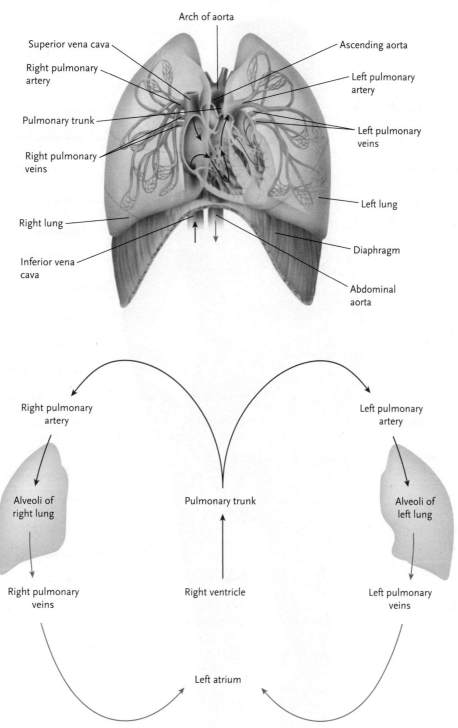

Figure 18.27 Pulmonary Circulation
Pulmonary circulation allows the blood to pick up oxygen and remove carbon dioxide in the lungs.

From the aortic arch, the **brachiocephalic trunk** branches to the **right subclavian artery** and the **right carotid artery**. The **left subclavian artery** and the **left carotid artery** branch directly from the aortic arch. The subclavian arteries supply the arms, and the carotid arteries supply the head and neck.

From the thoracic aorta, the bronchial arteries supply the bronchi in the lungs. The **esophageal arteries** supply the esophagus. The **intercostal arteries** supply the chest and intercostal muscles. The **superior phrenic arteries** supply the superior region of the diaphragm.

From the **abdominal aorta**, the inferior phrenic arteries supply the inferior region of the diaphragm. Blood flows to the digestive organs through the **celiac trunk** and the **superior and inferior mesenteric arteries**. The celiac trunk branches into the **common hepatic**

Left internal carotid
Left external carotid
Left vertebral
Right common carotid
Left common carotid
Right subclavian
Left subclavian
Brachiocephalic trunk
Arch of aorta
Ascending aorta
Left axillary
Right brachial
Thoracic aorta
Abdominal aorta
Left gastric
Celiac trunk
Splenic
Common hepatic
Left renal
Right renal
Superior mesenteric
Right ulnar
Left gonadal (testicular or ovarian)
Right radial
Inferior mesenteric
Right deep palmar arch
Left common iliac
Right superficial palmar arch
Left common palmar digital
Left proper palmar digital
Left external iliac
Right deep femoral
Left internal iliac
Left femoral
Left deep femoral
Left popliteal
Left anterior tibial
Left posterior tibial
Left fibular (peroneal)
Left dorsal (dorsalis pedis)
Left arcuate
Left dorsal metatarsal
Left dorsal digital

Figure 18.28 Major Arteries of the Body

Figure 18.29 Abdominal Arteries

artery to supply the liver, stomach, duodenum, and pancreas; the **left gastric artery** to supply the stomach and esophagus; and the splenic artery to supply the **spleen**, stomach, and pancreas (Figure 18.29). The superior mesenteric arteries supply the small intestine, pancreas, and the ascending and transverse colon. The inferior mesenteric arteries supply the transverse and descending colon, as well as the rectum. All blood that flows through the gastrointestinal tract circulates to the liver through the **hepatic portal vein** before it flows to the inferior vena cava.

Also branching from the abdominal aorta are the **renal arteries** that supply the kidneys, the **gonadal arteries** that supply the ovaries or testes, and the **common iliac arteries**. The common iliac arteries branch into the internal and external arteries. The **internal iliac artery** supplies the uterus or prostate, gluteal muscles, and the bladder. The **external iliac artery** supplies the lower limbs.

The right common carotid artery branches from the brachiocephalic trunk, and the left common carotid branches directly from the aortic arch. The common carotid arteries branch into the internal and external carotid arteries that supply the brain, and these branch off to arteries that supply the face (Figure 18.29).

The internal carotid arteries supply structures inside the skull, brain, eyes, and ears. The inferior surface of the brain has an arrangement of blood vessels called the **circle of Willis** (cerebral arterial circle). These vessels include the internal carotid arteries, the anterior and posterior cerebral arteries, and the anterior communicating artery (Figure 18.30). The circle of Willis

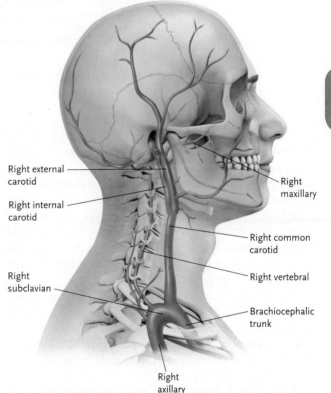

Figure 18.30 Arteries of the Head and Neck

regulates the blood pressure inside the brain, and it also provides redundant or excess circulatory routes so that the brain can still receive blood flow if some vessels are damaged (Figure 18.31).

Vertebral

Basilar

Superior
cerebellar

Pituitary
gland

Middle
cerebral

Internal
carotid

Anterior
cerebral

Posterior
inferior
cerebellar

Anterior
inferior
cerebellar

Posterior
cerebral

Posterior
communicating

Anterior
cerebral

Anterior
communicating

Cerebral arterial circle

Figure 18.31 Arteries of the Brain

The right brachiocephalic trunk branches to the right subclavian artery, and the left subclavian artery branches directly from the aortic arch. The subclavian arteries pass through the **axillary area,** and at this point the name changes to **axillary artery.** The axillary artery continues into the upper arm and becomes the **brachial artery.** This is the artery that is used to determine blood pressure using a sphygmomanometer. At the elbow, the brachial artery branches into the **radial artery** and the **ulnar artery** (Figure 18.32).

The external iliac artery becomes the femoral artery in the thigh; this becomes the **popliteal artery** once it reaches the knee. The popliteal artery branches into the **anterior tibial artery** and the **posterior tibial artery,** and the fibular artery branches from the posterior tibial artery (Figure 18.33).

After blood flows from the capillaries where it exchanged gases within the cells, the deoxygenated blood travels from the capillary beds of each tissue to the small venules that lead to larger venules and eventually to the major veins (Figure 18.34). The major veins carry the blood to the superior or inferior vena cava and back to the right atrium. Blood from the cardiac muscle—which was initially supplied by the coronary

arteries—flows to the coronary sinus and then directly back to the right atrium.

The cerebral spinal fluid (CSF), produced in each choroid plexus of the four fluid-filled ventricles of the brain, circulates through the central nervous system in the subarachnoid space and the central canal of the spinal cord. It bathes the central nervous tissue with specific nutrients and oxygen and removes waste. The CSF is continually produced, circulates through the CNS, and drains into the superior sagittal sinus and the **transverse sinus:** the largest veins in the brain (Figure 18.35). Blood drains from the venous sinuses in the brain to the **internal jugular veins,** which then drains into the brachiocephalic vein. Blood from the scalp and face drain into the **external jugular veins,** which then drains into the sunclavian vein, which then drains into the brachiocephalic vein. The brachiocephalic vein drains into the superior vena cava. (Figure 18.36).

Blood moves from the upper limb capillaries to venules, and then to veins. The veins of the upper arm are named just like the arteries. Blood moves from the deeper tissues to the **radial vein** and the **ulnar vein,** and then into the **brachial vein,** which merges with the axillary vein, which becomes the brachiocephalic. The

Figure 18.32 Arteries of the Upper Limb

Figure 18.33 Arteries of the Lower Limb

brachiocephalic vein drains into the subclavian vein, then to the superior vena cava and to the right atrium (Figure 18.36). Blood moves from the superficial regions through the **median antebrachial vein**, the **cephalic vein**, and the **basilica vein** in the forearm, to the **median cubital vein** in the elbow region (Figure 18.37). This is the most common vein used for inserting an IV or drawing blood, but any of the superficial veins can be used. The cephalic vein continues through the upper arm and, along with the median cubital vein, drains into the axillary vein.

The lower limb also carries blood through deep and superficial veins. The deep veins include the anterior and posterior tibial veins, and the popliteal and femoral veins—continuous with popliteal veins but located in the thigh. The femoral veins drain into the external iliac veins, then the common iliac, and then the inferior vena cava. The superficial veins include the **small saphenous vein** and the **great saphenous vein**. The small saphenous vein drains into the popliteal vein. The great saphenous vein is one of the longest in the body, extending the length of the lower limb and draining into the femoral veins (Figure 18.38).

Blood that flows through the organs of the digestive tract pick up water and nutrients that have been absorbed mainly by the small and large intestines. This nutrient-rich blood first circulates to the liver before continuing to the inferior vena cava and back to the heart. It is very important that the blood flows from the digestive organs through the hepatic portal vein to the liver so that the liver can process and store certain nutrients (Chapter 20). Blood flows from the tissue capillaries to the splenic and mesenteric veins, to the hepatic portal vein, and then to the liver. Once nutrients and molecules have been processed by the liver, blood flows through the hepatic vein to the inferior vena cava (Figure 18.39).

Superior sagittal sinus
Inferior sagittal sinus
Straight sinus
Transverse sinus
Sigmoid sinus

Left internal jugular
Left external jugular
Left subclavian
Left brachiocephalic

Superior vena cava
Right axillary
Right cephalic
Right brachial
Right hepatic
Right basilic
Right median cubital
Right radial
Right ulnar
Right median antebrachial

Coronary sinus
Great cardiac

Hepatic portal
Splenic
Superior mesenteric
Inferior mesenteric

Inferior vena cava
Left common iliac
Left internal iliac
Left external iliac

Right palmar venous plexus
Right palmar digital
Right proper palmar digital

Left femoral

Left great saphenous

Left popliteal

Left small saphenous

Left anterior tibial

Left posterior tibial

Left dorsal venous arch
Left dorsal digital

Figure 18.34   Major Veins of the Body

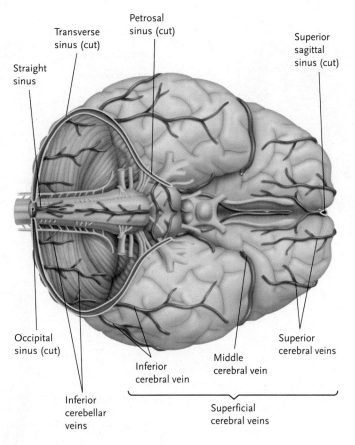

Figure 18.35 Veins of the Brain

Figure 18.36 Veins of the Head and Neck

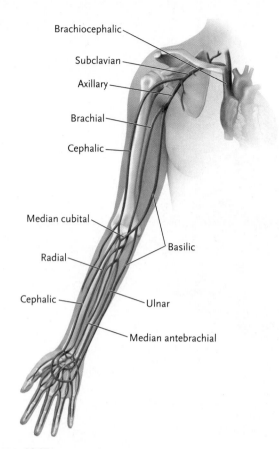

Figure 18.37 Veins of the Upper Limb

18

DID YOU KNOW?

The great saphenous vein can be removed and used to replace coronary arteries during coronary bypass surgery.

Figure 18.38 Veins of the Lower Limb

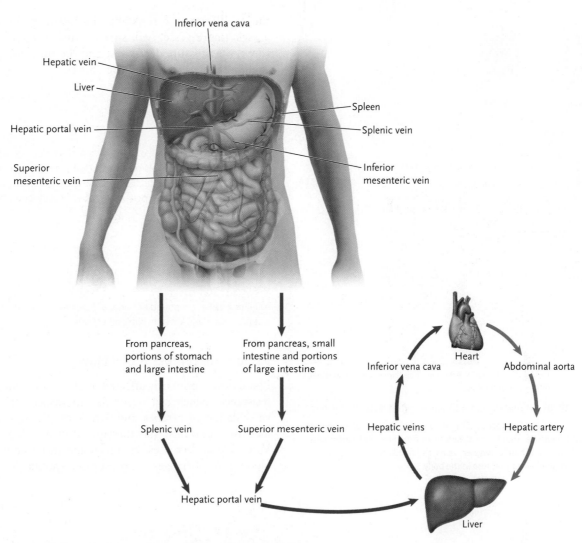

Figure 18.39 Abdominal Veins and the Hepatic Portal System

CONCEPT REVIEW 18.8

1. Why is the circle of Willis important?

2. How does the CSF drain from the ventricles of the brain into the circulatory system?

3. Why is the hepatic portal system important?

18.9 VASCULAR RESISTANCE AND BLOOD PRESSURE

Three factors relate to the movement of blood through the blood vessels: flow, pressure, and resistance. **Flow** is the amount of blood volume; **pressure** is the force of that blood volume on the walls of the blood vessels; and resistance is the force that opposes blood flow. The following equation is used to determine how these three factors relate:

flow = change in pressure / resistance

Flow increases if the pressure increases or if the resistance decreases. When the heart contracts strongly,

pressure and flow increase. When the blood vessels in a tissue dilate, more blood flows into that tissue.

Blood always flows from areas of higher pressure to areas of lower pressure. The highest pressure is in the aorta because of the contraction of the heart. As blood flows from arteries to arterioles to capillaries, the pressure and flow continue to decrease (Figure 18.40).

Resistance is largely determined by the *diameter* of the blood vessels. As blood moves from arteries to arterioles, the vessel diameter decreases; therefore, resistance increases, causing flow to decrease. Recall that arteries and arterioles branch into many vessels that are progressively smaller, and the capillaries have the largest cross-sectional area. As blood flows from the aorta to

Figure 18.40   Blood Pressure Changes in the Vascular System

Blood always flows from areas of higher pressure to areas of lower pressure, which ensures that blood always flows from arteries to arterioles to capillaries, and then to venules, veins, and finally the vena cavae, which have the lowest pressure in the body.

the capillaries, the resistance increases in the smaller vessels (more vessel wall surface area to oppose blood flow), the pressure decreases, and the flow decreases in each individual vessel (Figure 18.41).

DID YOU KNOW?

When a person gains a lot of body weight, their blood vessel growth increases to supply oxygen and nutrients to the new tissue. This increase in total blood vessel length increases vascular resistance, and this is one reason why obesity contributes to cardiovascular disease.

DID YOU KNOW?

Gaining one extra pound of body fat leads to the growth of approximately 15 km of new blood vessels.

Factors That Affect Blood Flow

Blood flow is primarily affected by changes in blood vessel resistance, which can occur due to vasoconstriction or vasodilation of arteries and arterioles. Vasoconstriction increases resistance and therefore decreases blood flow. Vasodilation decreases resistance and therefore increases blood flow. Changes in resistance, specifically due to

Figure 18.41   Blood Vessels Branching from Large to Small Vessels

Resistance increases if a vessel becomes constricted, but resistance does not increase when blood moves into smaller vessels because the vessels branch and the overall cross-sectional area is increased.

constriction or dilation of blood vessels, are the primary means of regulating blood flow to specific tissues.

Other factors that influence blood flow include (1) metabolic requirements of tissues, (2) body temperature, (3) blood volume, (4) **blood viscosity**, (5) cardiovascular disease, and (6) medications.

1. *Metabolic requirements.* Tissues that need an increase in oxygen and nutrients require greater blood flow. For example, during exercise the skeletal muscles require more blood flow than the digestive system. Blood vessels carrying blood to the skeletal muscle dilate (decreasing resistance), and blood vessels leading to the digestive tract constrict (increasing resistance).

2. *Body temperature.* When our body temperature increases above 37°C, a negative feedback mechanism takes effect (Chapter 1). Thermoreceptors detect an increase in temperature; this information is carried to the hypothalamus (integrating centre); and the hypothalamus signals effectors, such as sweat glands and blood vessels. When the body needs to lose heat, the blood vessels in the skin dilate so more blood flows to the surface of the body and heat can be removed. When body temperature decreases, surface blood vessels constrict, and heat is retained within the body.

3. *Blood volume.* An increase in blood volume increases blood flow to all areas of the body because the increased blood volume increases cardiac output. Anything that increases cardiac output increases blood flow (Section 18.6). Blood volume fluctuates normally within a small range due to the dietary intake of fluids and ions, mainly salt. In healthy people the kidneys quickly regulate blood volume by excreting or retaining excess water and ions. People with high blood pressure need to closely monitor the amount of salt in their diet (see Section 18.10 regarding hypertension).

4. *Blood viscosity. Viscosity* refers to the thickness of the blood. Increased viscosity leads to increased resistance; thicker blood requires more pressure to move through the blood vessels. Our blood is composed of cells and plasma—the water portion that contains dissolved ions and nutrients. The viscosity is determined primarily by the amount of red blood cells (hematocrit level). Under normal circumstances the hematocrit level does not change very much, and this is why blood vessel diameter is the primary factor of resistance. Anemia (decreased number of red blood cells) can also decrease resistance.

5. *Cardiovascular disease.* Atherosclerosis is the primary cause of cardiovascular disease, which is the buildup of plaque (mostly fat and immune cells)

A. Laminar flow

B. Turbulent flow

Figure 18.42 Laminar Flow and Turbulent Flow

(A) Laminar flow is the smooth, healthy flow of blood through vessels. **(B)** Turbulent flow occurs when vessel abnormalities, such as accumulation of plaque, cause blood to flow in a chaotic manner.

within the walls of the blood vessels. When this happens, the inside of the blood vessel walls are no longer smooth and uniform, and this affects how blood flows through a vessel. When the vessel is healthy and smooth inside the lumen, blood flows evenly—laminar flow (Figure 18.42). With atherosclerosis, the uneven surface of the lumen of a blood vessel causes the blood to flow in a random pattern—turbulent flow—that increases resistance and decreases flow to that tissue. Atherosclerosis can also cause clot formation, leading to heart attack or stroke because turbulent blood flow activates platelets.

6. *Medications* that affect blood flow in turn affect blood volume, vascular resistance, or blood pressure. Diuretics are a class of medications that decrease blood volume and therefore decrease blood pressure by increasing water excretion by the kidneys. Vasodilators decrease vascular resistance by causing vasodilation and, therefore, they decrease blood pressure.

DID YOU KNOW?

Athletes have been known to engage in "blood doping," which means they take the hormone erythropoietin to stimulate the bone marrow to increase production of red blood cells (Chapter 13). This lets them carry more oxygen and, therefore, increase their endurance ability. The problem with blood doping is that excessive production of red blood cells increases the viscosity of the blood, which increases resistance and puts a lot of strain on the cardiovascular system. Many athletes have died from heart attack or stroke because the blood is simply too thick to circulate properly.

18

CONCEPT REVIEW 18.9

1. How is blood flow related to blood pressure and blood vessel resistance?

2. What are the main factors that increase resistance in blood vessels?

3. List six important factors that affect blood flow?

4. Discuss the effects of temperature on blood flow?

5. How is resistance affected by an increased hematocrit level?

6. Explain laminar and turbulent blood flow and how cardiovascular disease can cause turbulent flow.

7. Discuss the effects of diuretics and vasodilators on blood pressure?

18.10 BLOOD PRESSURE REGULATION

Blood pressure in the arteries is higher during **systole** (contraction) and lower during **diastole** (relaxation). A typical, healthy resting blood pressure is 120/80 mmHg, where 120 is the systolic pressure and 80 is the diastolic pressure. Blood pressure and blood flow in the body is regulated by changes in heart rate, stroke volume, blood volume, and vascular resistance (vasoconstriction and vasodilation). Blood flow and pressure in various tissues depends on the metabolic needs of that tissue; during exercise, more blood is diverted to the skeletal muscles; after eating a meal, more blood is diverted to the gastro-intestinal system. All the factors that increase cardiac output also increase blood pressure (Figure 18.14).

The body uses negative feedback mechanisms to regulate blood pressure. The cardiovascular centre is in the medulla oblongata, the inferior region of the brainstem, and it regulates heart rate and stroke volume through the autonomic nervous system (ANS). Recall from Chapter 1 that a negative feedback loop involves sensory receptors, sensory (afferent) neurons, integrating centre (in this case, the medulla oblongata), motor (efferent) neurons, and effectors.

Receptors that detect blood pressure are called *baroreceptors*, and peripheral baroreceptors are located in the aortic arch and the carotid arteries. These are key locations, since the coronary arteries branch from the aorta, and the carotid arteries supply the delicate brain tissue, so monitoring blood pressure in these locations is very important. Baroreceptors are a type of mechanoreceptor that detects how much the arteries stretch when blood pressure increases. When the stretch of these receptors increases, the rate of action potentials sent to the medulla oblongata increases, indicating that blood pressure is increased. Chemoreceptors in the aortic arch and carotid arteries detect blood levels of oxygen, carbon dioxide, and hydrogen ions. When oxygen is low, carbon dioxide is high or H^+ concentration is high, and more action potentials reach the medulla oblongata (Chapter 19). Proprioceptors located in the muscles and joints also play a role in blood pressure regulation by detecting and providing information about body position. When we stand up from a laying down position, the blood vessels in our lower body constrict to prevent blood from being pulled into the lower body and away from the brain. When we begin exercising, the movement of the body signals the medulla oblongata to increase our heart rate *before* it is actually needed; this is known as an **anticipatory response** that prepares the body for exercise.

The medulla oblongata sends signals through sympathetic neurons when the heart rate or force of heart contraction needs to increase and signals through the parasympathetic neurons when the heart rate needs to decrease (Figure 18.43).

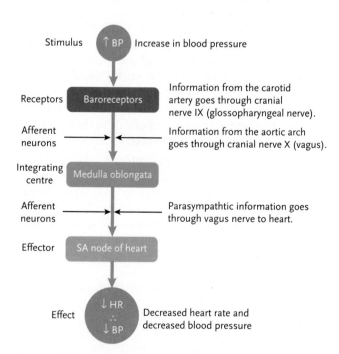

Figure 18.43 Blood Pressure Regulation

Baroreceptors detect blood pressure in the aortic arch and carotid arties. Changes in pressure are regulated by the cardiovascular centre: the medulla oblongata in the brainstem.

Information from the baroreceptors and chemoreceptors in the carotid artery travels to the medulla oblongata through afferent neurons in the **glossopharyngeal nerve**, and information from the aortic arch travels through afferent neurons in the **vagus nerve**. Information from proprioceptors travels through the peripheral sensory nervous system to the spinal cord, at which point there are synapses to neurons that signal the medulla oblongata. The cardiovascular centre in the medulla oblongata interprets the information (action potential frequency) from the sensory input. The cardiovascular centre then sends signals through autonomic efferent neurons to the effectors.

Efferent information can travel from the medulla oblongata through sympathetic or parasympathetic neurons to the target tissues. Parasympathetic neural signals travel to the SA node of the heart through the vagus nerve to decrease heart rate. Sympathetic system information travels through the sympathetic spinal nerves to the SA node and ventricles of the heart to increase heart rate and increase the force of contraction of the ventricles. Recall from Chapters 15 and 17 that preganglionic neurons of the sympathetic nervous system can target the adrenal medulla to produce epinephrine hormones that enter the bloodstream. The hormonal response is slightly slower than the response of the nervous system, but it targets many different cell types. Efferent signals also target many different blood vessels, primarily arterioles, to constrict or dilate.

The primary effectors involved in blood pressure regulation include the SA node and ventricle myocardium of the heart, blood vessels, and the kidneys (targeted by hormones). The following factors are involved in blood pressure regulation.

When blood pressure increases,

- parasympathetic stimulation decreases heart rate;
- a lack of sympathetic stimulation decreases heart rate and the force of ventricular contraction;
- parasympathetic stimulation causes vasodilation, which causes the kidneys to excrete more water.

When blood pressure decreases,

- sympathetic stimulation increases heart rate and the force of ventricular contraction;
- sympathetic stimulation causes veins to constrict, which increases venous return, cardiac output, and blood pressure;
- sympathetic stimulation causes vasoconstriction; which causes the kidneys to reabsorb more water.

Hormones Involved in Regulating Blood Pressure

Antidiuretic hormone, released by the hypothalamus, increases blood volume and therefore blood pressure by signalling the distal tubule and collecting duct of the kidney to increase water reabsorption.

Aldosterone, released by the adrenal cortex, increases blood volume through the reabsorption of sodium, and therefore water by osmosis, in the distal tubule.

Epinephrine and **norepinephrine**, released by adrenal medulla, increases blood pressure by increasing vascular resistance through vasoconstriction.

Angiotensin II, formed by the renin-angiotensin system (RAS), increases blood pressure by stimulating the production of aldosterone and by causing vasoconstriction.

Atrial natriuretic peptide (ANP), released by the atria of the heart when blood pressure is too high, causes the kidneys to excrete sodium and, therefore, lowers blood volume and blood pressure; it also lowers blood pressure by causing vasodilation.

18

CONCEPT REVIEW 18.10

1. What is a normal, healthy, blood pressure?
2. Name the major components of a negative feedback loop.
3. What is the function of the three main types of receptors involved in detecting changes in blood pressure?
4. Explain how sensory information travels from the carotid arteries and aortic arch to the brain.

5. Where is the cardiovascular centre?
6. How is the sympathetic nervous system involved in increasing blood pressure?
7. Name each important hormone that is involved in regulating blood pressure. Make a chart to show where each hormone is produced and how it affects blood pressure.

Application

Heart disease is the second leading cause of death in Canada. It affects over 2 million people and kills approximately 50 000 people per year. Heart disease is caused by atherosclerosis, which is a buildup of plaque in blood vessel walls. Atherosclerosis, which can begin as early as age 20 and show no symptoms for decades, is one the most preventable diseases. As you read Section 18.11, make a list of lifestyle factors that you could incorporate into your life to prevent atherosclerosis. For each factor, explain why it is beneficial.

18.11 CARDIOVASCULAR DISEASE

Cardiovascular disease is the leading cause of death in North America, and it is almost completely preventable. Cardiovascular disease is caused by a buildup of fats, minerals, and immune cells (**plaque**) inside the walls of arteries. This buildup is called **atherosclerosis**. Because the immune cells are involved, cardiovascular disease is also considered a **chronic inflammatory disease**. Recall the layers of an artery wall: plaque builds up underneath the inner endothelial cell layer, not inside the lumen of the blood vessel (Figure 18.44). Plaque is primarily made up of oxidized or damaged **low-density lipoproteins (LDLs)**. LDLs are normal cholesterol-protein molecules produced by the liver to transport dietary triglycerides from the liver to tissues. LDLs are often referred to as the "bad" cholesterol because high blood LDLs indicate that more fat is being stored than used for fuel and is considered a risk for heart disease. If we eat more of any macronutrient (protein, carbohydrates, or fat) than we need, the excess will be converted into LDLs by the liver and transported to adipose tissue to be stored. Furthermore, trans fats are not easily taken up by our cells and can stay circulating on LDLs, causing high LDL levels. HDLs also transport fats through the bloodstream, but they carry fat from the adipose tissue back to the liver to be broken down and used for energy. HDLs can remove fat from blood vessel walls when the fat is needed to produce energy; therefore, HDLs are referred to as the "good" cholesterol and correlate with a decreased risk of heart disease. HDL levels can be increased by exercising or intermittent fasting.

Recall from Chapter 3 that molecules that have both a hydrophilic and a hydrophobic component are called *amphipathic*. HDLs and LDLs are **amphipathic**, so they can bind to the triglycerides and cholesterol as well as circulate through the water-based blood. Damaged LDLs (such as LDLs containing trans fats) migrate into the artery wall and begin to oxidize. This causes tissue damage, which stimulates an inflammatory response. The immune cells, mostly phagocytic macrophages, then migrate into the wall of the artery to try to remove the damaged fats. Macrophages that engulf a large amount of oxidized fats become known as **foam** cells. As plaque increases, the diameter of the artery lumen decreases, which increases resistance for two reasons: (1) the lumen becomes narrowed, and (2) an uneven endothelial surface causes turbulent blood flow (Figure 17.41). As resistance increases, overall blood pressure increases, resulting in high blood pressure, known as **hypertension** (Table 18.1). Atherosclerosis is the underlying cause of cardiovascular disease, including heart attack

Figure 18.44 Atherosclerosis

Atherosclerosis is caused by the buildup of plaque within the wall of the large vessels.

TABLE 18.1

Blood Pressure and Hypertension		
Category	Systolic (mmHg)	Diastolic (mmHg)
Healthy	120 or less	80 or less
Prehypertension	121–139	81–89
Hypertension	140–159	90–99
Severe hypertension	160 or higher	100 or higher

(myocardial infarction), stroke, congestive heart failure, pulmonary embolism, deep vein thrombosis, and kidney failure. Atherosclerosis can progress for 10 to 20 years without symptoms.

DID YOU KNOW?

Did you know that an ophthalmologist or optometrist can use an ophthalmoscope to see the blood vessels in your retina and can tell if atherosclerosis is occurring? If atherosclerosis is visible in the retinal blood vessels, it is also happening in all the other blood vessels, including coronary arteries, which can lead to a heart attack, and vessels in the brain, which can lead to a stroke. Retinal vasculature is being used more frequently as a biomarker for assessing a person's risk of cardiovascular disease.

Risk Factors for Cardiovascular Disease

The exact causes of cardiovascular disease are not known, but the following risk factors are highly linked to the condition:

- High blood LDL and triglyceride levels
- Low blood HDL levels
- Obesity, particularly abdominal fat, which is significantly correlated with atherosclerosis and high blood pressure
- Diet, particularly high sugar intake (sugar is converted into fat by the liver) and trans fats in many processed foods, which increase blood LDLs more than any other type of fat
- Lack of physical activity, which contributes to obesity and low HDLs
- Type-1 or type-2 diabetes, which doubles the risk of heart attack or stroke
- High blood levels of **c-reactive protein (CRP)**, which is a strong indicator of chronic inflammation
- High blood homocysteine level, which affects the lining of blood vessels, thereby contributing to atherosclerosis. Elevated homocysteine can be due to a lackB vitamins, common in alcoholics.
- Smoking, which increases blood pressure and, due to the toxins in cigarettes, damages blood vessel walls by circulating fats and increasing plaque deposits
- Stress, which increases blood sugar and blood pressure; possibly linked to behaviours such as smoking, drinking, and overeating, and also to stress hormones such as cortisol and epinephrine
- Menopause, related to decreased estrogen in women, but no higher risk for women with a healthy lifestyle. Estrogen activates endothelial cells to release nitric oxide, which causes vasodilation and decreases blood pressure. Estrogen also affects

gene and protein expression that indirectly influences blood lipid levels.

- Excessive alcohol consumption, that is, more than two drinks per day for women, more than three drinks per day for men. Excessive alcohol consumption is correlated with an increases in triglyceride-rich lipoproteins in the blood. Excessive alcohol intake damages the liver and correlates with a decrease in beneficial HDL levels.
- Genetic predisposition
- Ethnicity, with greater risk of high blood pressure and diabetes among First Nations people and those of African or South Asian descent

Metabolic Syndrome

According to the International Diabetes Federation, people with **metabolic syndrome** have significantly increased risk of death from heart attack or stroke. A person is considered to have metabolic syndrome if they have central obesity (abdominal fat)—in females, more than 35 inch waist; in males, more than 40 inch waist—as well as two of the following:

- high blood triglycerides
- low HDLs
- high resting blood pressure—systolic > 130 mmHg, or diastolic > 85 mmHg
- high fasting blood glucose—greater than 5.6 mmol/L, or diagnosed diabetes

Complications of Atherosclerosis

Myocardial Infarction Atherosclerosis or a blood clot in the coronary arteries that supply the heart muscle leads to **ischemia**, which is decreased blood flow to the myocardium. If one or more coronary arteries become fully blocked, areas of the heart muscle do not receive oxygen (**hypoxia**), can't make ATP, and so cannot contract. Typical symptoms of a heart attack include severe squeezing chest pain that may radiate down the left arm, nausea, weakness, and difficulty breathing. Often people with advanced atherosclerosis have **angina** for a period of time before they have a heart attack. Angina involves bouts of chest pain that diminish after several minutes, and it is a sign that significant artery blockage exists.

Stroke There are two types of strokes: (1) **ischemic**, where the blood supply to areas of the brain has been decreased by a blockage or a clot that has travelled from another area of the body (Figure 18.45a); (2) **hemorrhagic**, where capillaries have broken and blood flows out of the circulatory system into the brain, putting pressure on the brain tissue, and preventing normal blood circulation from reaching other brain areas (Figure 18.45b). Symptoms of stroke are often noticed on one side of the body, since the left side of the brain communicates

Figure 18.45 Types of Stroke

(A) An ischemic stroke, usually caused by a clot, occurs because there is a lack of blood flow and, therefore, oxygen to a part of the brain.
(B) A hemorrhagic stroke is caused by damaged blood vessels.

where the damage has occurred, but common symptoms include one-sided weakness, numbness, or paralysis, slurred speech, or confusion. Sometimes people experience **transient ischemic attacks (TIAs)**, sometimes referred to as "mini strokes," which are similar to angina in that the symptoms resolve on their own; this condition indicates that significant blood vessel damage exists.

Kidney Failure The most common cause of kidney failure is chronic hypertension, also called **end stage renal disease**. Constant elevated blood pressure causes damage to the glomerular capillaries that filter blood, until eventually the kidneys cannot function to excrete waste, which builds up in the bloodstream and can quickly cause death.

Medication for Treating Hypertension

Atherosclerosis can be significantly decreased with lifestyle changes, such as eliminating trans fats from the diet and strictly limiting processed foods that contain trans fats, daily exercise, monitoring blood sugar, quitting smoking, losing weight, and managing stress. When the disease is advanced, the following types of medication may be used to decrease blood pressure and stress on the heart:

- **Diuretics** that increase water excretion and decrease blood volume
- **Beta blockers** that decrease the effects of epinephrine and therefore decrease heart rate and force of contraction
- **Alpha blockers** that decrease the effects of norepinephrine, which affects vascular tone and therefore decreases preload
- **Calcium channel blockers** that decrease the force of heart contraction, because calcium causes depolarization of the SA node as well as contraction in muscle fibres
- **Vasodilators** that dilate blood vessels, such as nitroglycerin that's taken to provide rapid relief from angina
- **ACE inhibitors**, which block the effects of angiotensin converting enzyme that converts angiotensin I into angiotensin II, and which leads to the production of aldosterone

with the right side of the body and the right side of the brain communicates with the left side of the body. A person's symptoms depend on the location in the brain

CONCEPT REVIEW 18.11

1. What is atherosclerosis?
2. Explain the difference between LDLs and HDLs, and how HDLs can be increased and LDLs decreased.
3. Make a chart of risk factors and how each contributes to cardiovascular disease.
4. What are some of the major complications of long-term atherosclerosis?
5. What is metabolic syndrome?
6. What is the difference between angina and a heart attack?
7. What are TIAs?
8. What can you do to prevent cardiovascular disease?
9. Make a chart that lists the main function of the major classes of medications used to treat cardiovascular disease.

18.12 COMPONENTS OF THE BLOOD

The blood is composed of cells and plasma. The average total blood volume is 5 to 6 L, depending on body size. The main functions of the blood are to transport nutrients, oxygen, ions, and hormones to the body's cells and to transport waste to the liver and kidneys for excretion. The blood is also involved in regulating body temperature. As more blood flows to the surface of the body, more heat can be removed. Blood is also involved in protecting the body from infections because it contains white blood cells and antibodies that fight bacterial, viral, parasitic, and fungal infections (Chapter 24). Blood contains platelets that form clots when there is tissue injury.

Plasma

The plasma consists of water, ions, vitamins, nutrients (glucose, fatty acids, amino acids), hormones, plasma proteins, some dissolved oxygen and carbon dioxide, and waste products. Plasma makes up approximately 55% of the total blood volume (Figure 18.46). Plasma is mostly water—approximately 92% of the plasma. The most important ions dissolved in the blood plasma include Na^+, K^+, Ca^{2+}, HCO_3^-, H^+, Mg^{2+}, and Cl^-. These ions are collectively called **electrolytes**, and they must be in balance to maintain the functioning of all cells, specifically the nervous system.

Some important plasma proteins include albumin, which is the main protein that creates the osmotic gradient that keeps fluid in the circulatory system; it makes up approximately 54% of all plasma proteins. **Antibodies**, also called *immunoglobulins*, are produced by specific white blood cells—called *B cells*—that can bind to

specific pathogens during an infection. Complement is a plasma protein produced by the liver and also an important component of the immune response. **Fibrinogen** is essential for the formation of blood clots. Blood **serum** is a term that describes blood plasma without fibrinogen, and it is removed by allowing clotting. **Lipoproteins** are produced by the liver and consist of proteins and cholesterol; they function to transport most fats through the bloodstream. Hormones make up approximately 1% of the plasma proteins (some hormones are lipid soluble). They are transported by means of the circulatory system from the organ where it is produced to its target cell type (Chapter 17).

Cells

Platelets, also called *thrombocytes*, are *fragments* of cells, and they are produced in the bone marrow. **Red blood cells**, also called **erythrocytes**, make up most of the remaining 45% of the total blood volume, and this is called the *hematocrit level*. A hematocrit level below 45% indicates anemia (low red blood cells). There are several types of anemia, including the following:

- Pernicious anemia—caused by a lack of vitamin B12
- Iron-deficiency anemia
- Hypovolemic anemia—caused by excessive blood loss
- Aplastic anemia—associated with a lack of production of red blood cells in the bone marrow
- Hemolytic anemia—caused by increased destruction of red blood cells
- Sickle cell anemia—inherited disease that causes malformed hemoglobin

Red blood cells contain a protein called *hemoglobin* that consists of four proteins plus iron and carries oxygen from the lungs to the tissues. Red blood cells are produced in the bone marrow by a process called *hematopoiesis*. During **hypoxia**, when oxygenation of cells decreases, the kidneys secrete the hormone **erythropoietin**, which stimulates hematopoiesis.

As red blood cells mature they lose their nucleus, so there is more room for hemoglobin-iron complexes to carry oxygen. A red blood cell lives for approximately 120 days and then is destroyed by the spleen or the liver. When the red blood cells are broken down, the amino acids are recycled and used to make other proteins, and much of the iron is recycled back to the bone marrow to make new red blood cells. The **heme** portion of the red blood cells is broken down, and **bilirubin** is formed. Bilirubin circulates to the liver and is added to the **bile** and then excreted into the small intestine when the gallbladder releases bile to be released as waste into the GI tract. Because of the bright yellow colour of bilirubin, if the liver is not functioning properly, the bilirubin stays in the body, giving the skin a yellow appearance: called **jaundice**.

Plasma (55–58%)

Platelets and white blood cells

Red blood cells (42–45%)

Figure 18.46 Components of Blood

Blood is composed of cells and plasma, and red blood cells make up 45% of the total blood volume.

Common Blood Values Blood testing is an important way for doctors to determine our state of health. The following are some of the most common blood values:

Hematocrit level (males, 40–54%; females, 37–47%). Low hematocrit levels indicate low red blood cells or anemia.

Creatinine (males, 8.8–17.6 mmol/d; females, 7.0–15.8 mmol/d). Creatinine is a waste product of muscle metabolism, and elevated levels indicate impaired kidney function.

Arterial blood pH (7.35–7.45). Venous blood is slightly more acidic than this because of higher carbon dioxide levels. Blood pH is extremely tightly regulated by the respiratory system and kidneys.

Blood glucose (3.9–6.1 mmol/L). Blood sugar fluctuates during the day depending on when and what we eat and how active we are.

HbA1C (Normal levels below 5.7%). HbA1C is an indication of the amount of glycation occurring on the hemoglobin of red blood cells. Levels higher than 6.5% indicate diabetes, and levels between 5.7% and 6.4% indicate prediabetes.

Blood cholesterol (healthy levels)
- LDLs (low density lipoproteins, "bad" cholesterol) < 3.4 mmol/L
- HDLs (high density lipoproteins, "good" cholesterol) > 0.91 mmol/L
- triglycerides > 2.2 mmol/L

Homocysteine (lower than 7.2 μmol/L). High levels of this amino acid are associated with increased risk of heart disease.

Ferritin (20–200 mg/L). Ferritin is an intracellular protein that stores iron, but small amounts are found in the blood, and it is used as an indicator of the total amount of iron in the body.

Leukocytes ($3.5–12.0 \times 10^9$/L). High leukocyte counts can indicate an infection or possibly leukemia.

Platelets ($150–400 \times 10^9$/L). Low platelet count is a condition called *thrombocytopenia*.

Blood urea nitrogen (BUN) (2.9–8.2 mmol/L). Urea is produced in the liver from the ammonia produced during the breakdown of proteins. BUN levels can give an indication of liver and kidney function.

Blood Types

Red blood cells contain antigens on the surface of the membranes. These antigens give us a blood type, which can be type A, B, AB, or O. Type A blood has type A antigens and B antibodies, type B has B antigens and A antibodies, type AB has both antigens but no antibodies, and type O has neither antigen but makes both antibodies (Table 18.2). If someone with A antigens receives blood with B antigens, their B antibodies will bind to the B antigens and cause coagulation of the blood that can be fatal. It is essential for

TABLE 18.2

Blood Type Antigens and Antibodies

Blood Type	Antigens	Antibodies
Type A	A	B
Type B	B	A
Type AB	Both	Neither
Type O	Neither	Both

TABLE 18.3

Blood Type Donations

Blood Type	Can donate to	Can receive from
Type A	A and AB	A and O
Type B	B and AB	B and O
Type AB	AB	AB, A, B, and O (universal receiver)
Type O	A, B, AB, and O (universal donor)	O

people receiving blood to receive the correct type (Table 18.3)

Red blood cells also have an Rh factor; it is called *Rh* because it is an antigen that was first discovered in rhesus monkeys; it is also called *D antigen*. Someone who has the Rh antigen is said to be positive, and someone without the Rh factor is negative. Approximately 85 to 95% of the population is Rh+, depending on ethnicity. Rh factor incompatibility can occur if the mother is Rh– and the fetus is Rh+. In this case, the mother will produce Rh antibodies that can cross the placenta and cause hemolytic disease in the fetus. This is preventable with anti-D antigen antibodies during pregnancy.

White blood cells, also called **leukocytes**, are cells involved in the immune response. Like red blood cells, white blood cells are produced in the bone marrow through the process of hematopoiesis. White blood cells can be divided into two categories: **granular leukocytes** and **agranular leukocytes**. When stained and viewed under a microscope, granular leukocytes have visible cytoplasmic vesicles that contain enzymes that digest microorganisms.

Granular Leukocytes

Neutrophils, the most abundant type of white blood cells, are most often the first cell type at an area of infection. They secrete chemicals that neutralize the infected area, primarily killing bacteria with enzymes and oxidants, such as hydrogen peroxide.

Basophils, the least common of the white blood cells, release histamine during allergic reactions.

Eosinophils, are involved in allergic reactions, as well as fighting multicellular parasites, such as worms.

Natural killer cells are large granular lymphocytes, that directly kill virus-infected body cells and cancer cells, and they play important roles in both the innate and adaptive immune response.

Agranular Leukocytes

Monocytes, the largest of the white blood cells, migrate from the bloodstream to tissues and then differentiate into mature **macrophages** and **dendritic cells**, which are very important antigen-presenting cells.

B cells are lymphocytes that are involved in the adaptive immune response, respond to specific antigens, and differentiate into plasma cells that secrete antibodies.

T cells are lymphocytes that are involved in the adaptive immune response, respond to specific antigens, and differentiate into helper T cells and cytotoxic T cells.

CONCEPT REVIEW 18.12

1. What are the four main functions of the blood?
2. What are electrolytes?
3. What are the main functions of plasma proteins?
4. What is the role of platelets?
5. What is a healthy hematocrit level? How does a person feel if they have a low hematocrit level?
6. What is hematopoiesis, and what hormone stimulates this process?
7. What mineral is important for the production of red blood cells?
8. What happens to the cell components in old, worn-out erythrocytes?
9. List the four ABO blood types and who they can donate to or receive from.
10. Describe the main functions of each type of leukocyte.

EFFECTS OF AGING

Most of the effects of declining cardiovascular function that we associate with aging are actually not due to aging specifically but are caused by an accumulation of years of poor lifestyle factors, such as consuming inflammatory sugar, trans fats, and processed foods, smoking, being overweight, and not exercising. Very few noticeable changes to the cardiovascular system occur with age in healthy, active people.

The following are some of the age-related changes that do occur:

1. The SA node (pacemaker) loses some cells, and this may result in a slightly lower resting heart rate and an increase in arrhythmias.
2. Heart valves tend to become stiffer and thicker, and this results in heart murmurs, which are common in elderly people.
3. Lipofuscin, a brownish pigment that builds up in cardiac and smooth muscle cells, is produced from the break down of red blood cells.
4. Baroreceptors that detect blood pressure may become less sensitive with age, making orthostatic hypotension more common in elderly people.
5. Capillary cells can become thicker, which decreases the rate of nutrients moving from the bloodstream to the tissues and can result in slower wound healing.
6. Blood vessels can become less flexible, due to decreased extracellular matrix proteins such as collagen and elastin (see Chapter 3, Effects of Aging), and this can result in a slightly higher resting blood pressure.
7. Total blood volume decreases.
8. Production of new white blood cells in the bone marrow decreases, and this can result in lower resistance to infections.

CHAPTER REVIEW

Medical Terminology

Term	Definition	Example
heme-	blood	hemoglobin
-poiesis	formation	hematopoiesis
agglutin/o	clumping	agglutinin
angi/o	vessel	angiogenesis
aort/o	aorta	aorta
arteri/o	artery	arteriosclerosis
arteriol/o	small artery	arteriole
ather/o	fatty substance, plaque	atheroma
bar/o	weight, pressure	baroreceptor
bas/o, basi/o	base, foundation	basophil
brady-	slow	bradycardia
capill-	resembling hair	capillary
cardi/o	heart	cardioptosis
coagul/o	clotting	coagulopathy
coron/o	heart	coronary
-crit	separate	thrombocytocrit
-emia	blood condition	anemia
eosin/o	red	eosinophil
erythr/o	red	erythrocyte

TERM	DEFINITION	EXAMPLE
ferr/i, ferr/o	iron	ferrometer
fil/I, fil/o	thread	filopressure
-gram	to record, picture	angiogram
hem/o, hemat/o	blood	hematopoeisis
isch/o	suppress, restrain	ischemia
kal/i	potassium	hyperkalemia
leuk/o	white	leukocytotoxin
-megaly	enlargement	atriomegaly
mi/o	less, smaller	miocardia
-motor	movement, motion	venomotor
myel/o	bone marrow, spinal cord	myelocytosis
natr/o	sodium	hypernatremia
palpit/o, palpitat/o	flutter, throbbing	palpitation
-pheresis	removal	plateletpheresis
-phil, -philia	affinity for, tendency toward	hemophilia
phleb/o	vein	phleborrhexis
-phoresis	bearing, transmission	electrophoresis
-plasty	surgical correction/repair	angioplasty
rhe/o	flow, current, stream	rheocardiography
schist/o,	split, cleft	schistocyte
scler/o	hard	scleroderma
-sclerosis	hardening	atherosclerosis
ser/o	serum, serous	serosanguineous
sider/o	iron	sideropenia
-spasm	involuntary contraction	vasospasm
spher/o	round, sphere	spherocytosis
sphygm/o	pulse	sphygmascope
-sphyxia	pulse	asphyxia
-stenosis	narrowed, constricted	aortostenosis
strept/o	twisted, curved	streptococcus
tachy-	fast	tachycardia
tel/e	end, distant	telecardiography
-tension	stretched, strained	hypertension
thromb/o	clot, thrombus	thromboelastogram
varic/o	varicose veins	varicophlebitis
vas/o	vessel, vas deferens	vasophypotonic
vascul/o	vessel	vasculitis
ven/o	vein	venography
ventricul/o	ventricle of the heart or brain	ventriculogram
venul/o	venule	venular
-volemia	blood volume	normovolemia

18

Labelling Activity

Label the following diagrams.

1.

2.

Articulate Graphics/IllustrationOnLine.com

3.

4.

5.

6.

7.

8.

9.

10.

11.

12.

18

Answer to Application Question

- Avoid eating trans fats; they increase LDLs and inflammation in blood vessels.
- Avoid processed foods; they are high in salt, sugar, starch, and trans fats and low on micronutrients.
- Eat less sugar; excess sugar is converted by the liver into fat, which is then transported on LDLs.
- Eat more plant foods; they contain antioxidants that help get rid of free radicals that contribute to inflammation in blood vessels.
- Maintain healthy body weight; belly fat/visceral fat contributes to increased blood pressure

and is a significant risk factor for atherosclerosis. and

- Exercise at least three to four times per week; this increases HDLs, burns calories, helps control body weight, and helps to reduce stress.
- Don't overconsume alcohol; this causes depletion of B vitamins, which can cause homocysteine levels to increase.
- Manage stress; this causes release of cortisol, which increases belly fat and blood sugar, and inhibits the immune system.

19 The Respiratory System

19.1 OVERVIEW OF THE RESPIRATORY SYSTEM

The respiratory system is closely connected with the cardiovascular system, and, in combination, these systems take in and transport oxygen to all tissues in the body and remove carbon dioxide. Structures in the respiratory system include the nose, mouth, pharynx, larynx, trachea, bronchi, bronchioles, and alveoli (air sacs) (Figure 19.1). The movement of air into and out of the lungs is called **ventilation**. Exchange of gases in the alveoli is **respiration**. Breathing in is called **inhalation** or *inspiration*, and breathing out is called **exhalation** or *expiration*. Once oxygen has circulated to the tissues and is used by cells to make ATP, that process is called *cellular respiration* (see Chapter 4).

The respiratory system has the following functions in the body:

- Taking in oxygen and removing carbon dioxide
- Regulating blood pH by removing carbon dioxide
- Warming and moistening inhaled air
- Filtering particles from inhaled air
- Providing a sense of smell
- Producing sound by moving air past the vocal cords

I have not failed. I've just found 10,000 ways that won't work.

Thomas A. Edison

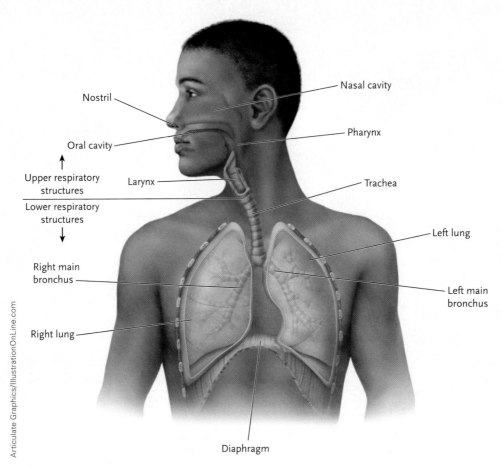

Articulate Graphics/IllustrationOnLine.com

Figure 19.1 Structures of the Respiratory System

Structures in the respiratory system include the nose, mouth, pharynx, larynx, trachea, bronchi, bronchioles, and alveoli (air sacs).

CONCEPT REVIEW 19.1

1. What are the functions of the respiratory system?

19.2 ANATOMY OF THE RESPIRATORY SYSTEM

The respiratory system can be divided *structurally* into the upper and lower respiratory system. The upper structures include the nose, mouth, pharynx, and larynx, and the lower structures include the trachea, bronchi, bronchioles, and alveoli. The respiratory system can be divided *functionally* into the **conducting zone** and the **respiratory zone** (Figure 19.2). The nose, mouth, pharynx, larynx, trachea, bronchi, bronchioles, and terminal bronchioles make up the conducting zone; these structures are involved in the movement of air into and out of the lungs. The respiratory bronchioles, alveolar ducts, and alveoli make up the respiratory zone, which is where gas exchange occurs.

The nose and mouth are the entrances to the respiratory system (Figure 19.3). The external nose consists of the nasal bones, hyaline cartilage, and skin, and its two openings—the external nares (nostrils)—lead into the nasal cavity. The nasal cavity is separated by a septum, which is a perpendicular plate of the ethmoid bone. Air flows through the nasal cavity, which consists of the superior, middle, and inferior **nasal conchae**—also called **turbinates**—and through the frontal, ethmoid, maxillary, and sphenoid sinuses. As air flows through these passageways, which are lined with mucus membranes, it is warmed by heat transferred from the blood vessels and filtered by the hairs and cilia that line the upper airways. The mucus membrane of the superior conchae contains receptors for detecting scents. Therefore, this membrane is also called the *olfactory epithelium*.

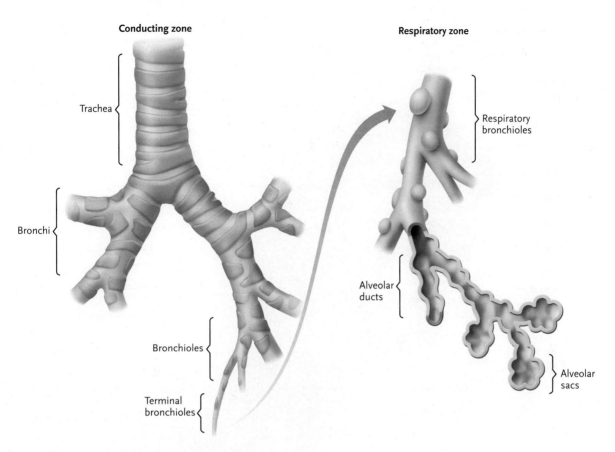

Conducting zone

Trachea

Bronchi

Bronchioles

Terminal bronchioles

Respiratory zone

Respiratory bronchioles

Alveolar ducts

Alveolar sacs

Figure 19.2 Functional Divisions of the Respiratory System
The conducting zone includes structures that transport air to the respiratory zone, which is where gas exchange occurs.

Air travels from the nasal cavity to the **pharynx** (throat). The pharynx is divided into an upper portion, called the **nasopharynx**; the middle portion where food also travels, called the **oropharynx**; and the lower portion called the **laryngopharynx**. Within the nasopharynx is the opening of the auditory tube, also called the **Eustachian tube**. This tube leads to the middle ear and is important for equalizing air pressure within the inner ear. When swallowing food, the **epiglottis** covers the opening to the larynx so that food does not enter the respiratory structures (Figure 19.3).

The **larynx**, also called the *voice box*, contains the **vocal cords** (Figure 19.4) and connects the pharynx to the trachea (Figure 19.5). The mucus membrane of the larynx forms the vocal cords, and it contains elastic connective tissue that's connected to cartilage and also muscles that are under voluntary control. When these muscles contract, the vocal cords become pulled to a certain degree and produce sounds. As air flows over the vocal cords during exhalation, it causes them to vibrate, in a way similar to plucking strings on a guitar. When males produce an increased amount of testosterone during puberty, their vocal cords become thicker and longer and produce a lower vocal sound. Everyone's voice is slightly different because of tiny variations in the thickness and length of the vocal cords, as well as

the shape and size of the nasal cavity, pharynx, and larynx. The larynx is protected by hyaline cartilage (also called *thyroid cartilage*); known as the *Adam's apple*, it can be palpated in the neck.

The **trachea**—or *windpipe*—is located anterior to the esophagus between the C6 and T5 vertebrae and is protected by 16 to 20 incomplete rings of *hyaline cartilage* (Figure 19.5). The cartilage keeps the airway open and prevents the soft tissues from collapsing inward so that breathing is not obstructed. The trachea is lined with a mucus membrane composed of pseudostratified, ciliated, columnar, epithelial tissue (Chapter 11). The ciliated epithelial tissue contains **goblet cells** that produce **mucus** that protects the lungs by trapping inhaled particles and bacteria, and the cilia continually move the mucus up toward the pharynx so it can be swallowed or coughed out of the lower airways. The lower airways also contain immune cells that fight infectious organisms that enter the lungs.

The trachea branches at T5 into the left and right **bronchi**, which enter the lungs (Figure 19.6). The trachea and bronchi have supportive cartilage rings, whereas the smaller conducting zone and respiratory zone structures do not. The bronchi branch into many smaller and smaller **bronchioles**. The smallest conducting-zone bronchioles are called **terminal**

19

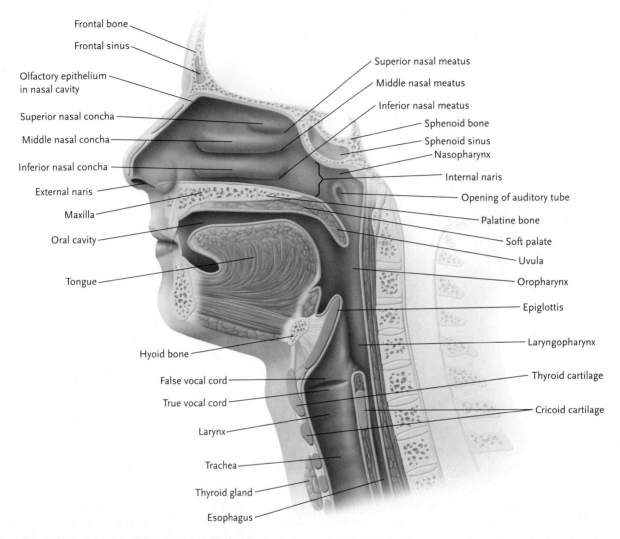

Figure 19.3 Upper Respiratory Structures

The upper respiratory structures include the mouth, nasal passages, larynx, and pharynx.

CNRI/Science Photo Library/Getty Images

Figure 19.4 Vocal Cords

The vocal cords are composed of elastic connective tissue.

Anterior   Posterior

normaals/Getty Images

Figure 19.5 Cartilage of the Larynx

The larynx is made up of thyroid cartilage, cricoid cartilage, and tracheal cartilage.

bronchioles, which lead to the first of the respiratory structures. Gas exchange occurs in the respiratory structures, which include the **respiratory bronchioles,** the **alveolar ducts,** and the **alveoli.**

The lungs are divided into the right and left side, and each side is divided into **lobes.** The right lung contains three lobes (Figure 19.6). The left lung has two lobes and

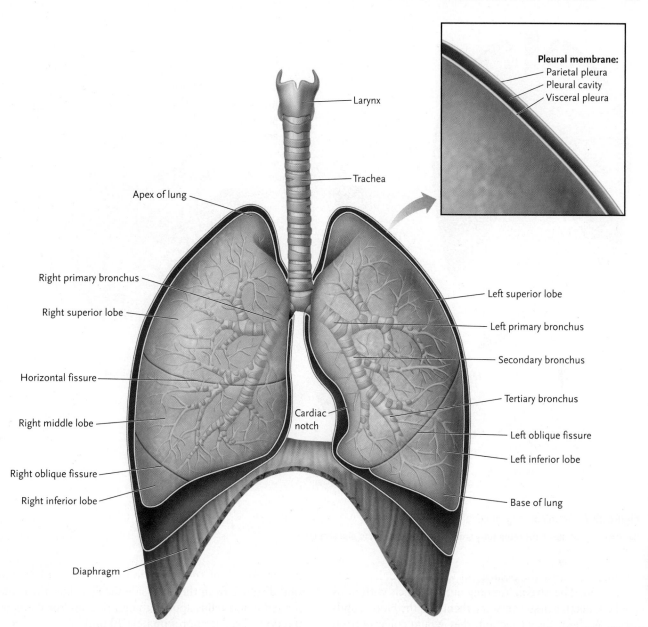

Figure 19.6 Lower Respiratory Structures

The lower respiratory structures include the trachea, bronchi, bronchioles, and alveoli.

the **cardiac notch**, where the heart is located. On the right side, the superior and middle lobes are separated by the **horizontal fissure**, and the middle and inferior lobes are separated by the **oblique fissure**. On the left side, the left oblique fissure separates the superior and inferior lobes.

The lungs are surrounded by mucus membranes called **pleura**. The membrane that directly surrounds the lungs is the **visceral pleura**. The **parietal pleura** covers the inside of the thoracic cavity (Figure 19.6). The space between the two pleural membranes is very small and is called the *pleural cavity*, which is filled with fluid that prevents friction during breathing when the lungs expand and contract.

The bronchioles branch into smaller and narrower bronchioles that lead to the smaller terminal and respiratory bronchioles, which lead to the alveoli. Each lobe of the lungs has **lobules** that contain respiratory structures, arterioles, venules, and lymphatic vessels. The respiratory structures are surrounded by capillaries so that gas exchange can occur efficiently (Figure 19.7). Bronchioles are surrounded by smooth muscle cells that can contract to cause constriction of the bronchiole tubes, or relax to cause dilation. The smooth muscles cells are controlled by the autonomic nervous system.

The alveoli consist of two types of epithelial cells. **Type I alveolar cells** are simple squamous epithelial cells that make up the majority of the alveoli and are the site of gas exchange. **Type II alveolar cells** are specialized epithelial cells that produce and release **surfactant** into the alveoli. Surfactant is an amphipathic fluid composed of phospholipids and lipoproteins that have both polar and nonpolar regions. The surfactant relieves the surface

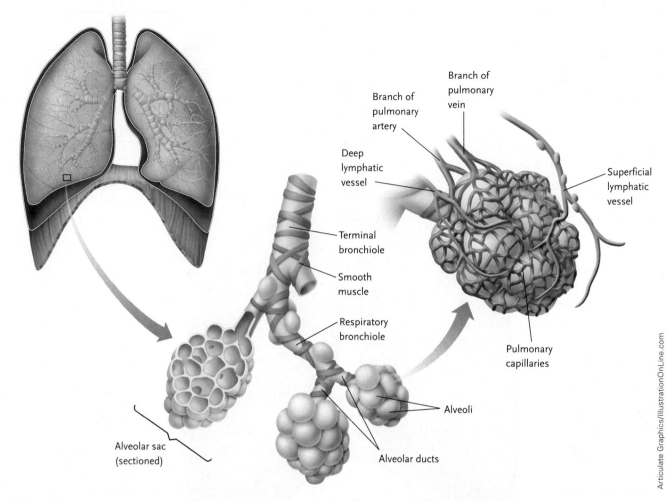

Figure 19.7 Terminal Bronchioles and Alveoli

Gas exchange occurs in the respiratory bronchioles, alveolar ducts, and alveoli.

Articulate Graphics/IllustrationOnLine.com

tension caused by the polarity of water molecules inside the alveoli. The alveoli are only one cell thick with minimal connective tissue between them and the blood capillaries. Without the surfactant, they would collapse from the pull of the hydrogen bonds in the water molecules, even though there is just a very small film of water in the lungs (see Chapter 2, Figure 2.1). Inhaled oxygen must dissolve in that water film before it can diffuse across cell membranes into the red blood cells. A molecule of oxygen must cross several alveolar epithelial cell membranes, pass through a small layer of connective tissue, cross the capillary endothelial cell membranes, and finally cross the membrane of the red blood cell (Figure 19.8). The total distance from the air inside the alveoli to the inside of a red blood cell is approximately 0.2 μm (the diameter of a typical cell is approximately 10 μm).

DID YOU KNOW?

Taking a deep breath causes the lungs to stretch, and this stimulates the type II alveolar cells to produce surfactant, which decreases surface tension and makes breathing easier.

Application

As you read through the rest of the chapter, consider this scenario. Greg is going to train for a marathon at the Swiss Olympic Training Base in the Alps. The altitude is 1856 m (6089 ft). The percent of oxygen is approximately 16.5%, and the atmospheric pressure is approximately 600 mmHg. Calculate the partial pressure of oxygen in the atmosphere at the training camp. Until Greg has adapted to the higher altitude, do you think his tidal volume will change? What will his alveolar pressure be at rest? With decreased oxygen in the air, how will his hemoglobin saturation be affected? What kinds of regulatory changes will occur in Greg's respiratory system?

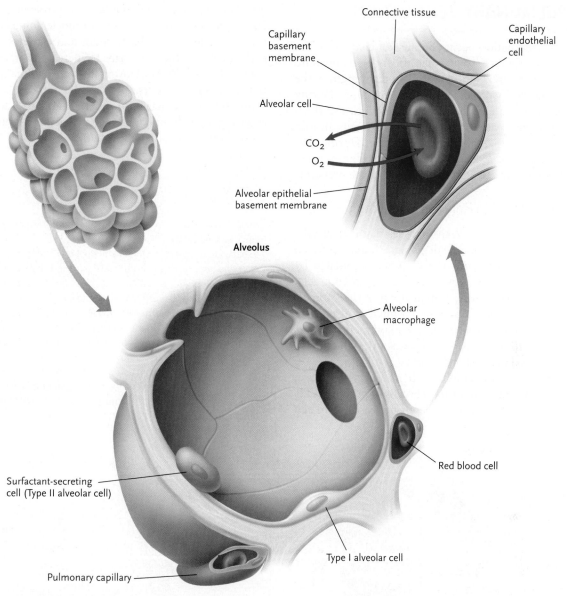

Figure 19.8 Gas Exchange

Gas exchange occurs across the type I alveolar cells of the alveoli.

CONCEPT REVIEW 19.2

1. What is the difference in function between the conducting zone and the respiratory zone?

2. Why is it important for air to flow through the nasal conchae?

3. What are the three regions of the pharynx?

4. Why are the Eustachian tubes important?

5. How do the vocal cords produce sounds?

6. Name two things that prevent particulate matter such as dust from entering the lungs.

7. What is the difference between the visceral pleura and the parietal pleura?

8. Why is the pleural cavity filled with fluid?

9. Which alveolar cell type produces surfactant?

10. Why is surfactant necessary?

19.3 PULMONARY VENTILATION

Pulmonary ventilation occurs because of changes in the volume of the thoracic cavity that affect the air pressure and cause air to move in or out. The diaphragm is the primary muscle involved in relaxed breathing. During diaphragm contraction, thoracic volume increases, which causes a brief decrease in pressure, and this pulls air into the lungs, quickly equalizing the pressure. During diaphragm relaxation, a decrease in thoracic volume causes the air to move out, also quickly equalizing the pressure.

Muscles Involved in Relaxed Breathing

During *relaxed* inspiration, the muscles that increase the thoracic cavity volume are the diaphragm and the **external intercostal muscles** (Figure 19.9a). When stimulated by neurons in the phrenic nerve, the diaphragm contracts, causing it to move down, thus increasing the thoracic volume (Figure 19.9b). The external intercostal muscles are located between the ribs, and their contraction causes the ribs to move upward and outward, which also increases the thoracic volume.

During *forced* inhalation, such as during exercise, more oxygen is required for muscle cells to produce ATP. Therefore, breathing rate and depth increases to match demand, and additional muscles play a role in breathing. During forced inspiration, the **sternocleidomastoid** contracts to help elevate the sternum and clavicles, and the **scalenes** elevate the upper ribs to increase thoracic cavity volume (Figure 19.9a). During a severe asthma attack, it is also possible to contract pectoralis muscles, which are connected to the sternum and the humerus. A person enduring an asthma attack also raises their shoulders to try to further increase lung volume.

DID YOU KNOW?

An inability to catch your breath for a long time after exercise can be an indication of early heart disease. The heart contracts to circulate oxygenated blood to all the body's cells so they can have enough oxygen to produce ATP. When the cells do not receive the oxygen, your breathing centre (medulla oblongata and pons) signals the respiratory muscles to increase your rate and depth of breathing, making you feel short of breath. Often the only symptom of a heart attack, particularly in females, is not being able to catch one's breath for a long time after exercise. A good measure of cardiovascular fitness is how easily you can regain normal breathing after intense exercise. Although shortness of breath is the most common symptom of a heart attack, many pulmonary problems can also cause shortness of breath; in either case, medical attention is required.

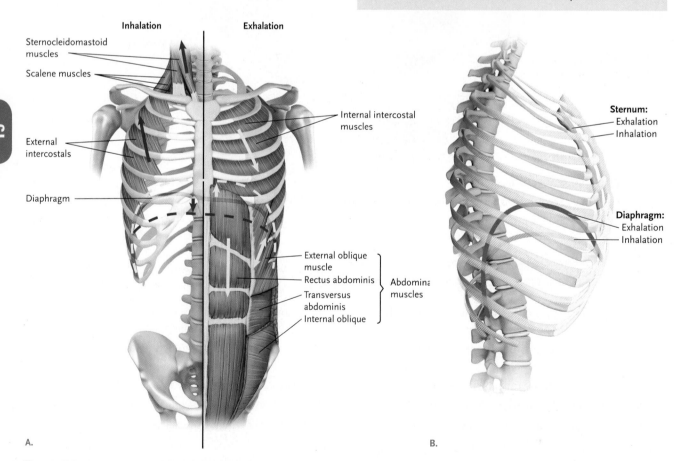

Figure 19.9 Structures Involved in Pulmonary Ventilation

(A) Muscles involved in pulmonary ventilation and **(B)** changes in the thoracic cavity and diaphragm during inhalation and exhalation

During *relaxed* exhalation, there is no muscle contraction; the diaphragm and the external intercostals simply relax, which causes the thoracic volume to decrease, which causes air to leave the lungs. The lung tissues have a high amount of elastic proteins that recoil and aid the movement of air out of the lungs.

During *forced* exhalation, such as exercise, sneezing, or coughing, other muscles contract to increase the rate of exhalation by causing the thoracic volume to decrease faster. The muscles involved are the internal intercostals and abdominal muscles (Figure 19.7). When contracted, the internal intercostal muscles—which lie in an oblique angle opposite to the external intercostals—cause the ribs to move down and inward, which decreases thoracic volume. The **external obliques** also move the ribs down and inward. When the **rectus abdominis**, **transverse abdominis**, and **internal obliques** contract, abdominal pressure increases, which forces the diaphragm to move upward and thereby decreases thoracic volume.

Changes in Thoracic Pressure during Breathing

Boyle's law is a gas law that states that in a closed container with a constant number of molecules at a constant temperature, the pressure is inversely proportional to the volume. If the volume increases, the pressure decreases; if the volume decreases, the pressure increases (Figure 19.10). The amount of pressure is determined by the amount of force of the molecules on the container. In our lungs, the molecules are the components of air (see Table 19.1), and the container is the lungs, which are in close contact with the pleural membranes of the thoracic cavity. When we breathe, a pressure change of 2 mmHg (millimetres of mercury) is equivalent to approximately 500 mL of air.

As we breathe, the pressure in the lungs is very close to constant because as soon as the pressure drops or increases, air moves into or out of the lungs and, therefore, quickly equalizes the pressure (Figure 19.11). The air pressure in the environment is the **atmospheric pressure**. At sea level atmospheric pressure is approximately 760 mmHg; this measurement changes with altitude and various weather patterns. The pressure inside the lungs is the **alveolar pressure**, and the pressure between the visceral and parietal pleura is the **intrapleural pressure**. The changes in volume and pressure during breathing are described in three stages:

1. No change in volume. Any time you are not changing the volume of air in your lungs, the alveolar pressure is equal to the atmospheric pressure, whether you are holding a lung full of air or if your lungs are empty. If the atmospheric pressure is 760 mmHg, and the thoracic volume is constant, the alveolar pressure will also be 760 mmHg. The intrapleural pressure must always be *lower* than the alveolar pressure—also called *negative pressure*—so the lungs do not collapse. At rest, the intrapleural pressure is −4 mmHg compared to the alveolar and atmospheric pressure.

2. Inhalation. During inhalation, muscles contract, causing the thoracic volume to *increase*, which causes the alveolar pressure to *decrease* relative to the atmospheric pressure (Figure 19.11). Since the lungs are not a closed container, the decrease in pressure causes more air molecules to move into the alveoli and equalize the pressure in the larger volume. During inhalation there is a corresponding change in the intrapleural pressure, so a difference of 4 mmHg ensures that the lungs remain open as the pressure decreases inside the alveoli (very briefly since the incoming air immediately equalizes the pressure).

3. Exhalation. During exhalation, muscles relax, and this causes the thoracic volume to *decrease*, which makes the alveolar pressure *increase* relative to the atmospheric pressure. The increase in pressure causes air molecules to move out of the lungs, equalizing the pressure. During exhalation, the intrapleural pressure remains negative compared to the alveolar pressure to ensure that the lungs do not collapse.

The ability of the lungs to expand with inhalation and retract with exhalation is dependent on two important factors, **compliance** and **recoil**, which are determined by the health of the extracellular matrix proteins (connective tissue), such as collagen and elastin, as well as the amount of surfactant required to decrease surface tension inside the alveoli. In healthy lungs, compliance is high, and the lungs can stretch and expand to allow a large volume of air to enter the lungs if needed, such as during exercise. A person with decreased compliance has decreased lung capacity and cannot take in as much air. Healthy recoil allows air to easily move out of the lungs during exhalation, without the need for internal intercostal or abdominal muscles to contract to force the air out of the lungs.

Figure 19.10 Boyle's Law

Boyle's law states that volume is inversely proportional to pressure.

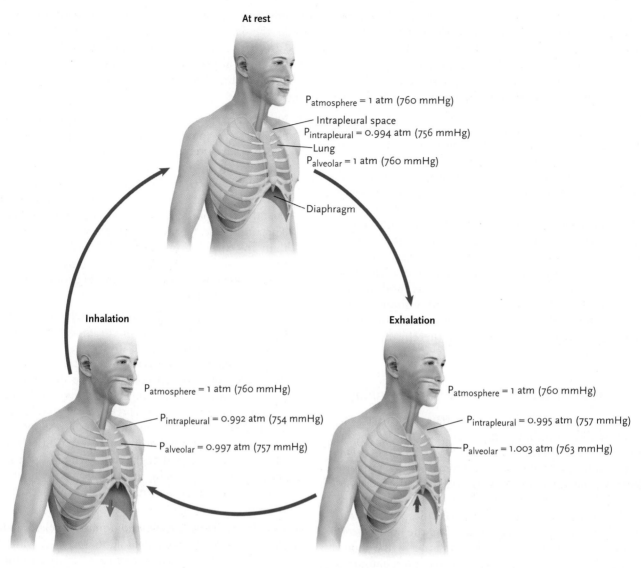

Figure 19.11 Pulmonary Ventilation

Inhalation occurs when the volume in the thoracic cavity increases, causing the pressure inside the lungs to decrease.

Lung Volumes

When we are sitting at rest, we breathe a certain volume of air in and out. However, we can modify that amount by taking an extra deep breath in or by forcing extra air out after a regular exhalation. Each of these volumes of air can be labelled (Figure 19.12).

1. The volume of air in one inhaled breath or exhaled breath is the tidal volume, which is equal to approximately 8 to 10 mL/kg of body weight. Therefore, lung volume is directly proportional to body size. A person weighing 65 kg (143 lb) will have a tidal volume of approximately 520 mL.

2. After taking a normal breath in and continuing to breathe in as much as possible, the lung volume is approximately 3 L; this is called the **inspiratory reserve volume**. The inspiratory reserve volume is

about six times greater than the resting tidal volume.

3. The sum of the tidal volume and the inspiratory reserve volume is called the **inspiratory capacity**.

4. After taking a normal breath out and continuing to breathe out as much as possible— which requires contraction of the intercostal and abdominal muscles—the lung volume is approximately 1.2 L; this is called the **expiratory reserve volume**.

5. The amount of air that we can move into and out of our lungs by breathing in and out as much as possible is approximately 4.7 L, but this varies with age, body size, and level of health; this amount is called the **vital capacity**. Vital capacity is equal to the sum of tidal volume, inspiratory reserve, and expiratory reserve.

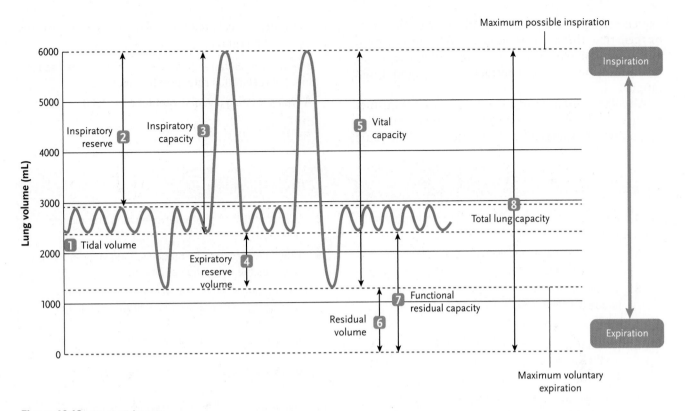

Figure 19.12 Lung Volumes

Lung volumes can vary based on how much we consciously breathe in or out. Lung volumes can be altered by disease.

6. Some air is always left in the lungs, even after we have exhaled as much as possible. Approximately 1.2 L remains within the lungs that we cannot exhale; this amount is called the **residual volume**.

7. The sum of the expiratory reserve and the residual volume is called the **residual capacity**, which is approximately 2.4 L; this amount is significantly higher than the fresh air coming into the lungs with each 500 mL tidal volume.

8. The total maximum amount of air that the lungs can hold is approximately 5.8 to 6 L. This is called the **total lung capacity**: the sum of the total amount of air we can breathe in (vital capacity) plus the residual volume.

DID YOU KNOW?

One test of pulmonary function is called the *forced expiratory volume* (FEV). It involves breathing in the inspiratory reserve volume and then breathing out as fast and as much as possible into a bronchospirometer that measures lung volumes. This method is used as a diagnostic tool to determine if someone has an obstructive pulmonary disease that involves bronchoconstriction, such as asthma. A healthy FEV is 80% of vital capacity in one second. A person with an obstructive pulmonary disease will have a lower FEV percentage due to an inability to breathe out as much air and as fast.

Certain categories of individuals have larger lung volumes than others. People with larger lung volumes include those who are tall, those living at high altitudes, those with normal weight compared to those who are obese, men more than women, athletes more than non-athletes, non-pregnant women compared to pregnant women, and those who are healthy compared to those with restrictive lung diseases.

Dead Space

The body's respiratory structures are within either the conducting zone or the respiratory zone. The conducting zone only transports air; gas exchange occurs only in the respiratory zone, which includes the terminal bronchioles, alveolar ducts, and alveoli. The air in the conducting zone does not contribute to gas exchange and is therefore called **anatomical dead space**. Approximately 30% of each tidal volume stays within the conducting zone. After a normal inhalation of 500 mL of air, only 350 mL is new air from outside; 150 mL is already in the conducting zone from the previous breath. This is why we can't swim underwater by breathing through a garden hose, which would simply create a very long dead space, where eventually there would be no oxygen available for gas exchange. This also explains why it is important to take a deep breath every once in a while: to exchange the air in the dead space to ensure there is always enough oxygen for gas exchange. During

19

exercise, deeper and longer breaths deliver the extra oxygen that the muscles need for ATP production, and this can make a very significant difference in our ability to continue to contract muscles and resist fatigue.

Alveolar ventilation refers to the amount of air that reaches the alveoli in one minute. It is similar to cardiac output, which is amount of blood ejected by the heart per minute. To calculate alveolar ventilation, use the following equation:

alveolar ventilation = [tidal volume (mL) − dead space (mL)] × respiration rate (breaths/min)

A typical resting alveolar ventilation = (500 − 150) × 12 = 4.2 L/ min)

CONCEPT REVIEW 19.3

1. Which muscles are involved in relaxed inhalation and relaxed exhalation?

2. What muscles are involved in inhalation during exercise or an asthma attack?

3. Which muscles would contract when you have forced exhalation such as coughing?

4. What is Boyle's law?

5. How are lung volume and pressure related, and what changes in volume and pressure occur during inhalation?

6. Suppose you take a very deep breath in and hold it. Assuming you are at sea level, what is the atmospheric pressure, the alveolar pressure, and the intrapleural pressure?

7. Why does the intrapleural pressure have to be negative in relation to the alveolar pressure?

8. What is the difference between lung compliance and recoil?

9. Provide the name of a particular lung volume that matches the following definitions:

   a. amount of air inhaled in one normal breath
   b. total amount of air in lungs after a full deep breath in
   c. amount of air further exhaled after a normal exhalation
   d. amount of air in the lungs after exhaling as much air as possible

10. Name three factors that affect lung volume.

11. What is dead space?

12. What is the alveolar ventilation if the tidal volume is 600 ml, dead space is 190 ml, and respiratory rate is 14 breaths per minute?

19.4 GAS EXCHANGE AND TRANSPORT OF OXYGEN AND CARBON DIOXIDE

Gas Exchange

When the oxygen in the inhaled air reaches the respiratory zone, oxygen dissolves in the small amount of water that's in the alveoli, and then moves by diffusion through the alveolar epithelial cell membranes into the bloodstream. Inhaled air is a mixture of gasses of which oxygen makes up approximately 21% (Table 19.1).

TABLE 19.1

Composition of Air	
Molecule	Percentage
Nitrogen	78.08
Oxygen	20.9
Argon	0.9
Carbon dioxide	0.03
Ozone	0.01
Other trace elements	~0.5

Dalton's law states that the pressure of each gas in a mixture is independent of the pressure exerted by the other gases. If atmospheric pressure at sea level is 760 mmHg, oxygen will exert 21% of that pressure: this percentage is called a **partial pressure**. The partial pressures of all the gases in the mixture, add up to the total pressure.

Partial pressure of oxygen in the atmosphere can be calculated as follows:

$P_{02} = 760 (0.209) = 158.8$ mmHg

During respiration, oxygen and carbon dioxide move down their partial pressure gradients from an area of higher concentration to an area of lower concentration (Figure 19.13). As we breathe in, the concentration of oxygen is higher in the inhaled air compared to that in the bloodstream, so oxygen moves into the blood in the alveoli. The partial pressure of oxygen in the air is approximately 159 mmHg, but as it enters the lungs and mixes with the air in the dead space, the concentration of oxygen decreases, and the partial pressure of oxygen in the alveoli is approximately 100 mmHg (see Figure 19.11).

The blood flowing through the capillaries is *deoxygenated* until oxygen begins to diffuse across the alveolar membranes into the capillaries. The deoxygenated blood in the alveolar capillaries has a partial

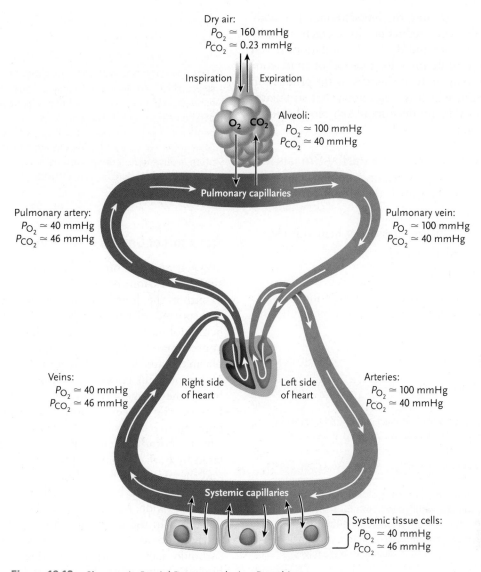

Figure 19.13 Changes in Partial Pressures during Breathing

pressure of approximately 40 mmHg, which is lower than the alveolar partial pressure of 100 mmHg. As oxygen moves into the blood vessels, the partial pressure of the blood increases to approximately 100 mmHg, and at this point the blood is *oxygenated*. Oxygenated blood flows from the lungs back to the left atrium of the heart and is pumped from the left ventricle to the rest of the body. Once the oxygenated blood reaches any cells with low oxygen, the oxygen in the blood again moves toward the area of lower concentration, and into the cells. As the oxygen moves from the capillaries into the body cells, the blood becomes deoxygenated, and the partial pressure gradually lowers to 40 mmHg. The deoxygenated blood circulates back to the lungs, and the cycle continues such that oxygen from inhaled air continuously moves from the alveoli into the blood and then into the body's cells to be used for the production of ATP.

Carbon dioxide gas also moves from areas of higher concentration to areas of lower concentration. As cells take up and consume oxygen to produce ATP, carbon dioxide is produced as a waste product that diffuses into the capillaries within the tissues. As carbon dioxide moves into the capillaries, the partial pressure of carbon dioxide becomes approximately 46 mmHg in the deoxygenated blood, which is then circulated to the right side of the heart. Blood then circulates to the lungs where the CO_2 moves down the concentration gradient into the alveoli. The air in the alveoli contains a mixture of inhaled air and air recycled within the dead space, and so the partial pressure of the carbon dioxide is approximately 40 mmHg in the alveoli.

If you were to breathe air that has less than 21% oxygen, such as at high altitudes, the partial pressure of oxygen would be lower, and therefore the partial pressure of oxygen in the alveoli would be lower, and less oxygen would diffuse into the bloodstream. The body compensates for this reduction in oxygen initially by increasing the rate and depth of ventilation. If you stay at a high altitude for a long period of time, your body will also compensate by increasing the production of erythropoietin, which is a hormone that stimulates the bone marrow to produce more red blood cells to carry the oxygen in the blood.

Conditions That Decrease the Partial Pressure of Oxygen in the Alveoli

Certain conditions cause the partial pressure of oxygen in the alveoli to decrease, including

1. decreased atmospheric pressure at high altitude,
2. decreased ventilation, and
3. increased metabolism.

When metabolism increases ventilation has to increase to ensure that enough oxygen molecules diffuse into the blood.

DID YOU KNOW?

Yawning is stimulated by an increase in blood CO_2 levels, which occurs more often when you are tired and do not breathe deeply. The carbon dioxide levels increase and trigger the yawning reflex, causing you to take a deep breath in (much deeper than the amount of dead space volume), which decreases the alveolar carbon dioxide partial pressure. While yawning, you also hold your breath for a second after the deep inhalation, which allows more time for CO_2 to diffuse from the blood into the alveoli before you exhale. Yawning allows for more carbon dioxide to be removed from your blood.

Hyperventilation occurs when the ventilation rate increases beyond what is necessary for removing CO_2 and acquiring O_2. Hyperventilation can cause the blood CO_2 level to drop below the normal concentration, causing hypocapnia (low CO_2), which can lead to an abnormal increase in blood pH; this is called *respiratory alkalosis.* Symptoms include dizziness; tingling in fingers, toes, and face; weakness; and possibly fainting.

Hypoventilation occurs when the ventilation rate decreases below what is necessary for acquiring O_2 and removing CO_2. Hypoventilation can cause the blood CO_2 levels to be too high, a condition called *hyper-capnia,* which can lead to respiratory acidosis. Hypoventilation can result from different conditions, such as holding your breath; drugs such as morphine, alcohol, stroke; a tumour affecting the brainstem; or altitude sickness. You cannot die from holding your breath because your respiratory centre (medulla oblongata) will stimulate breathing as soon as CO_2 levels reach a certain point.

DID YOU KNOW?

Sleep apnea refers to a periodic absence of breathing during sleep, which can last several seconds or up to one or two minutes, and may occur many times a night without the person being aware. This condition is more common in people who are overweight, smokers, or borderline diabetics, compared to people without these conditions. Symptoms often include loud snoring, gasping or choking sensations during sleep, daytime fatigue, difficulty concentrating, and impaired memory.

Transport of Gasses in the Blood

Oxygen and carbon dioxide cross all membranes by simple diffusion; no membrane proteins are required in either the lungs or the tissues. Most oxygen is transported in the bloodstream by being bound to hemoglobin proteins in red blood cells. The rest, approximately 1.5%, is carried dissolved in the plasma. Each hemoglobin molecule consists of four globin proteins containing heme and an iron ion that binds the oxygen (Figure 19.14). Heme is a large heterocyclic molecule containing iron, and it is a red pigment that turns bright red when oxygen is bound. The function of heme and oxygen-binding is regulated by both the partial pressure gradient of oxygen and the pH of the blood (Figure 19.15).

The pH of the blood is largely due to the amount of carbon dioxide. As we exhale carbon dioxide, the pH of the blood in the lungs becomes slightly more alkaline (carbon dioxide is acidic because of the production of carbonic acid). As we inhale (Figure 19.16), oxygen binds to hemoglobin, forming oxyhemoglobin but has no impact on blood pH. As the blood circulates to the tissues, the blood pH decreases because the cells

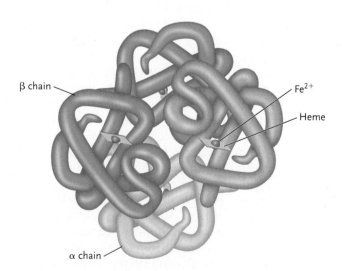

β chain Fe^{2+} Heme   α chain

Figure 19.14 Hemoglobin

Hemoglobin is the protein in red blood cells that binds to oxygen.

Figure 19.15 Hemoglobin Saturation Curve

The hemoglobin saturation curve is an important tool for understanding how the blood carries and releases oxygen.

produce CO_2 that enters the blood. Cells continually use oxygen to make ATP, so the partial pressure of oxygen in the cells always becomes lower than in the blood. Consequently, oxygen leaves the hemoglobin and enters the cells.

Body temperature also affects the binding of oxygen to hemoglobin. During exercise, when body temperature increases, more oxygen is released from hemoglobin in the tissues. During exercise, blood pH decreases because more CO_2 is produced, and this causes an

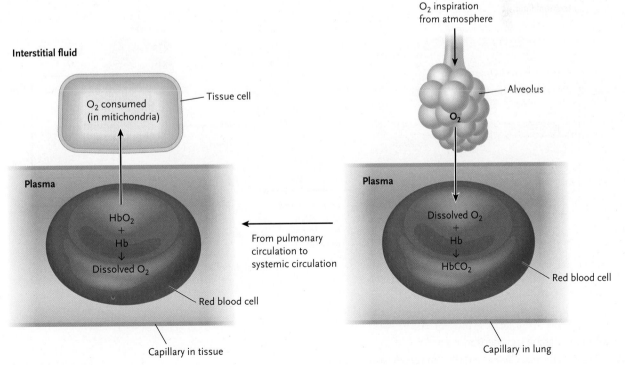

Figure 19.16 Oxygen Transport

Oxygen is transported by being bound to hemoglobin or dissolved in the blood plasma.

increase in the amount of oxygen that dissociates from the hemoglobin in the muscles. Exercise also increases the amount of oxygen consumed, causing more oxygen to be released from hemoglobin in the tissues. Note that when oxygen dissociates from hemoglobin in the tissues and enters the cells, the hemoglobin in the red blood cells is still approximately 78% saturated.

Unlike oxygen, carbon dioxide is transported in the following three ways (Figure 19.17):

1. 7 to 10% transported dissolved in plasma
2. 25 to 30% transported bound to hemoglobin
3. 60 to 65% transported in the form of bicarbonate ions

Carbon dioxide dissolves in water much easier than oxygen does, and therefore some carbon dioxide can be transported dissolved in the plasma. About 25% of the carbon dioxide can be transported on hemoglobin. In this case, CO_2 binds to amino groups that make up the hemoglobin molecules—not to the iron ion, like oxygen—and forms carbaminohemoglobin ($HbCO_2$). The majority of CO_2 is carried within the red blood cell in the form of bicarbonate ions (HCO_3^-), which are formed according to the following chemical equation:

$$CO_2 + H_2O \rightarrow H_2CO_3 \rightarrow HCO_3^- + H^+$$

In the tissues, carbon dioxide and water form carbonic acid (H_2CO_3) in the presence of an enzyme, called **carbonic anhydrase**, that's found inside red blood cells. Carbonic acid dissociates into bicarbonate (HCO_3^-) and hydrogen ions (H^+). This is why an increase in CO_2 in the blood causes a decrease in blood pH, and this plays an important role in the regulation of breathing; when blood pH becomes more acidic, breathing rate and depth increases. Bicarbonate ions diffuse out of the red blood cells and are transported in the plasma. Because bicarbonate ions are negatively charged and leave the red blood cell, the charge of the red blood cell (RBC) has to be balanced. This is accomplished by Cl^- ions moving into the red blood cells; this is called the **chloride shift** (Figure 19.17).

When blood circulates to the lungs, the CO_2 dissolved in the plasma diffuses across the RBC and alveolar membranes to reach the inside of the alveoli, and is then exhaled. Any CO_2 bound to the hemoglobin amino acids is released and also exhaled. Lastly, the reverse bicarbonate reaction occurs. The bicarbonate ions move back into the red blood cells, chloride ions move back out and into the plasma, and the bicarbonate and H^+ ions combine to form carbonic acid. Carbonic anhydrase then converts carbonic acid back into CO_2 and H_2O, which is exhaled.

Figure 19.17 Carbon Dioxide Transport

Carbon dioxide is primarily transported in the form of carbonate.

1. How is Dalton's law different from Boyle's law?

2. Using Table 19.1, calculate the partial pressure of each gas in air at an atmospheric pressure of 760 mmHg.

3. Describe the changes in the partial pressure of oxygen in the inhaled air, the alveoli, and the blood within the lungs compared to the blood within body tissues.

4. Name three factors that can affect the partial pressure of oxygen in the alveoli and therefore affect the amount of oxygen that diffuses into the blood.

5. Describe the changes in the partial pressure of carbon dioxide in the blood within the tissues compared to the blood within the lungs, the alveolar air, and the atmospheric air.

6. Which of the following causes an increase in the amount of oxygen that dissociates from hemoglobin?

    a. An increase or decrease in pH

    b. An increase or decrease in partial pressure gradient

    c. An increase or decrease in body temperature

7. Describe how carbon dioxide is transported from the tissues to the lungs.

8. Why is the chloride shift important?

19.5 REGULATION OF BREATHING

The medulla oblongata and the pons, located in the mid-to-lower part of the brainstem, are both involved in the regulation of breathing. The medulla oblongata is the primary respiratory control centre. The medulla oblongata controls inspiration by signalling the diaphragm and external intercostal muscles to contract, causing inhalation. Motor neurons that lead from the medulla oblongata to the diaphragm are located in the phrenic nerve. The medulla oblongata consists of pacemaker cells that regulate the *rhythm* of breathing. It can also signal the muscles involved in forced expiration, such as the abdominals and the internal intercostals.

The pons, located above the medulla oblongata, influences the medulla oblongata by helping to smooth the transitions between inhalation and exhalation. It can affect the rate of breathing by sending inhibitory signals to the medulla oblongata or by stimulating the medulla to cause deeper and longer breaths.

Stretch receptors, located within the smooth muscles of the airways, stretch as the lungs expand. When stretch increases, sensory signals are sent to the medulla oblongata to inhibit the dorsal respiratory group (DRG), which normally stimulates inhalation. Then further depolarization of the motor neurons stops, which stops contraction of the inspiratory muscles. When you inhale as much as you possibly can, and once your lungs reach their inspiratory capacity, the stretch receptors inhibit the dorsal respiratory neurons in the medulla oblongata to prevent damage to the air sacs.

Summary of Events during Inhalation and Exhalation

During inhalation the following processes occur:

1. The medulla signals *inspiratory* muscles to contract; this may include the scalene muscles and the sternocleidomastoid if inhalation is forceful.

2. The pons helps smooth the transition to exhalation.

3. The pons can stimulate an increase in the rate and depth of inhalation.

4. Stretch receptors prevent overinflation of the lungs by inhibiting the DRG when the lungs are full.

During exhalation the following processes occur:

1. The medulla can stimulate the muscles involved in forceful exhalation, such as the internal intercostals and the abdominals.

2. The pons can inhibit the inspiratory muscles and prevent inhalation.

3. Inspiratory muscles relax and lung recoil pushes air out of the lungs during relaxed exhalation.

Factors That Influence the Medulla Oblongata and the Pons

- Signals from the cerebral cortex can stimulate or inhibit the neurons in the medulla oblongata. As a result, we can voluntarily hold our breath, take a deep breath, or change our rate of breathing.

- The limbic system, also referred to as the "emotional" part of the brain, is involved in emotions that impact breathing: for example, laughing, crying, or expressions of pain.

19

- The autonomic nervous system is involved because a sympathetic nervous system response increases the rate and depth of breathing, and a parasympathetic nervous system response decreases the rate and depth of breathing.

- Irritation to airway structures, such as chemical irritants or blowing air, can signal a reflex that inhibits the medulla oblongata and prevents contraction of inspiratory neurons, which prevents the inhaling of damaging chemicals into the delicate alveoli. Have you ever walked outside into a large gust of wind and lost your breath for a second? Your medulla oblongata was inhibited to prevent the possibility of overinflation of the lungs in very high wind.

- Temperature is a factor because increased body temperature, such as during a fever, increases the rate of breathing.

- Medications such as general anaesthetic, morphine, and alcohol inhibit the medulla oblongata and depress respiration, but amphetamines stimulate respiration.

- Blood levels of O_2, CO_2, and H^+ are involved because breathing rate and depth increases if oxygen decreases, or if carbon dioxide or hydrogen ions increase.

Peripheral and Central Chemoreceptors

Peripheral chemoreceptors detect blood levels of O_2, CO_2, and H^+, which fluctuate based on metabolic rate. We need oxygen to produce ATP, and carbon dioxide is produced as a waste product. Recall that as carbon dioxide levels increase, blood pH decreases; it becomes more acidic as CO_2 is converted into carbonic acid. Whenever metabolism increases, ATP production increases, and so does the amount of CO_2 and H^+ produced, which then enter the blood. Exercise increases the need for ATP production, and, therefore, our rate and depth of breathing increases during exercise. High-intensity exercise requires more ATP and increases breathing more than low-intensity exercise. Peripheral chemoreceptors are located in key areas: the aortic arch and the carotid arteries (Figure 19.18). They detect arterial blood levels in the blood that is leaving the heart and circulating to the body and the brain. They are located very close to the peripheral baroreceptors that detect blood pressure (Chapter 18). The medulla oblongata contains the **central chemoreceptors** and also detects blood levels of O_2, CO_2, and H^+.

Central chemoreceptors in the medulla oblongata primarily respond to increased H^+ in the extracellular fluid of the brain, which is caused by an increase in CO_2 due to increased energy requirements. The peripheral receptors respond to increased H^+, as well as directly detecting increased CO_2 and reduced O_2. The receptors that detect these molecules are directly connected to sensory neurons that send information

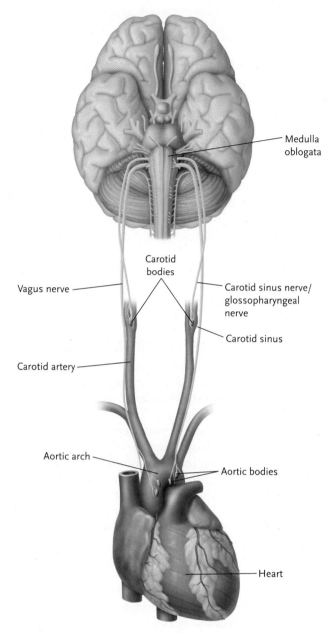

Figure 19.18 Peripheral Chemoreceptors

Peripheral chemoreceptors detect blood oxygen and carbon dioxide levels.

to the brainstem. The overall strongest stimulus for increasing breathing is an increase in carbon dioxide. If you hold your breath as long as possible, the buildup of CO_2 stimulates the medulla oblongata to signal inspiratory muscle contraction and eventually forces you to breathe.

Regulation of Breathing during Exercise

Stimulus → receptors → integrating centre → effectors → effect

1. Exercise requires an increase in ATP for skeletal muscle contraction (Figure 19.19).

Figure 19.19 Negative Feedback Control of Breathing
The medulla oblongata is the primary respiratory centre.

2. ATP production requires oxygen from the blood to be taken up by muscle cells and produces carbon dioxide.

3. Peripheral receptors detect a decrease in blood O_2 (hypoxia), an increase in CO_2, and a decrease in pH (increased formation of H^+). Central receptors detect increased CO_2 in brain extracellular fluid.

4. Sensory information from the receptors is sent through afferent neurons to the medulla oblongata.

5. The medulla oblongata signals motor neurons to cause the alternating contraction of inspiratory muscles and expiratory muscles (depending on the intensity of the exercise).

6. Increased contraction of inspiratory muscles (and possibly expiratory muscles during high-intensity exercise) increases the rate and depth of breathing.

7. An increase in breathing causes blood oxygen levels to increase, and blood carbon dioxide levels to decrease. Decreased carbon dioxide causes a decrease in free hydrogen ions and therefore increases blood pH.

CONCEPT REVIEW 19.5

1. Describe the function of the dorsal and ventral neurons of the medulla oblongata.

2. How does the pons influence breathing?

3. How is overinflation of the lungs prevented?

4. Make a list of some factors that can affect the rate or depth of breathing.

5. Where are the central and peripheral chemoreceptors located?

6. What is detected by the central and peripheral chemoreceptors?

7. Identify the stimulus, receptor, integrating centre, and effectors involved when breathing increases during exercise.

19.6 COPD AND ASTHMA

Chronic obstructive pulmonary disease (COPD) is a chronic, progressive, lung disease caused primarily by smoking or long-term breathing of pollutants or chemical irritants (Figure 19.20). Cigarette smoking is the leading cause of COPD, and the two main types of COPD are **emphysema** and **chronic bronchitis**. People with COPD have damage in both the conducting and respiratory zones of the respiratory system, and this is

Healthy alveoli

Bronchiole

Alveoli

Alveoli with COPD

1 Mucus accumulates in the bronchiole tubes.

2 Alveoli become damaged, resulting in larger and fewer alveoli.

Articulate Graphics/IllustrationOnLine.com

Figure 19.20 Chronic Obstructive Pulmonary Diseases (COPD)
COPD includes emphysema and bronchitis.

characterized by progressive difficulty in breathing. In emphysema, the elastic recoil of the air sacs is lost, and exhaling becomes very difficult. Damage to the air sacs leads to fewer and larger alveoli, and this eventually causes a dramatic decrease in gas exchange. In chronic bronchitis, inflammation causes swollen, irritated airways with significantly more mucus production, often causing a chronic cough. Most people with COPD have both emphysema and chronic bronchitis. There is no cure for COPD, but medications may help decrease symptoms.

Asthma is a chronic disease caused by inflammation and narrowing of the airway's conducting zones. It is also known as **bronchoconstriction** and results in difficulty breathing (Figure19.21). In people who have asthma, the immune system responds to stimuli

Normal bronchiole

Smooth muscle

Lining

Bronchiole with asthma

Inflammation

Muscle contraction

Lining

Mucus

Articulate Graphics/IllustrationOnLine.com

Figure 19.21 Asthma
Asthma is an inflammatory disease of the airways.

that normally it should not react to. In a way similar to an allergic reaction, common particles stimulate an inappropriate immune reaction. The exact cause is not known, but some of the triggers are stress, change in air temperature, and exercise. The inflammatory response causes swelling of the lining of the airways and mucus production, which can obstruct airflow. Also, smooth muscle contractions in the bronchi and bronchioles narrow the diameter of respiratory structures in the conducting zone. Symptoms include difficulty breathing, which requires the use of additional inspiratory muscles—such as the scalenes and sternocleidomastoid muscles—to increase lung volume; coughing, wheezing, and chest tightness due to increased mucus and narrow airways; tachycardia, which means the heart rate increases to compensate for the decreased oxygen in the blood and the effort to increase cardiac output to meet the demands of the cells; and fatigue due to less oxygen reaching the cells and less ATP production. A severe asthma attack can cause anxiety because of the suffocating feeling of the lack of oxygen. Asthma can be treated with **bronchodilators** and anti-inflammatory medications. People with asthma must learn what their specific triggers are and then try to avoid or minimize exposure.

There are four types of hypoxia, and each is characterized by inadequate oxygen delivery to the tissues.

1. Anemic hypoxia involves poor delivery of oxygen because of either too few red blood cells or abnormal hemoglobin.
2. Ischemic hypoxia involves impaired blood circulation.
3. Histotoxic hypoxia involves the inability of the body's cells to use O_2 (cyanide poisoning causes this).
4. Hypoxemic hypoxia involves reduced arterial O_2, which can result from a lack of ventilation due to diseases such as COPD or asthma.

CONCEPT REVIEW 19.6

1. What two main diseases are considered chronic obstructive pulmonary diseases?
2. What is different between chronic bronchitis and emphysema?
3. What is the primary cause of COPD?
4. What are the main symptoms of asthma?
5. Name two types of medications that can be used to treat the symptoms of asthma.
6. What are the main causes of hypoxia?

EFFECTS OF AGING

The respiratory system undergoes some normal changes as we age. The chest wall and thoracic vertebrae become less dense and can deform slightly, which impairs respiratory compliance, making inhalation slightly more difficult. Also, the ribcage can change shape and thereby lose its ability to expand and contract. These changes are not usually noticeable, unless they are accompanied by a respiratory disease such as COPD.

A decrease in connective tissue proteins, such as collagen and elastin, can cause the alveoli to become larger and have less recoil, making exhalation slightly more difficult. Again, these changes are not usually noticeable.

As we age, muscle strength tends to decrease, and this can make it more difficult to cough or have strong contraction of expiratory muscles. If ability to cough decreases, mucus can build up in the airways.

The alveolar dead space increases with age; therefore, deeper inhalations may be required to have sufficient oxygen available for diffusion into the blood.

Respiratory receptors become less responsive and can affect the ability to increase or decrease breathing as quickly as necessary. A sensation of dyspnea (shortness of breath) tends to increase in older people, especially when they are physically active.

Vital capacity decreases with age because collage and elastin protein production decreases, which affects compliance (needed for inhalation) and recoil (needed for exhalation).

Lastly, as we age, immune function tends to decrease, and this can cause an increase in susceptibility to respiratory infections. Older people have a higher risk of developing viral or bacterial pneumonia as a complication of other respiratory infections such as common colds or influenza.

The effects of aging on our respiratory system can be reduced by decreasing or avoiding the inhalation of toxins (e.g., by not smoking), and by staying physically active. Physical activity increases lung function and lung capacity; for example, moderate-to-high-intensity exercise can increase lung capacity by 5 to 15%.

CHAPTER REVIEW

Medical Terminology

Term	Definition	Example
aspir/o, aspirat/o	inhaling, removal	aspiration
blenn/o	mucus	blennothorax
brachy-	short	brachypnea
bronch/o	bronchus	bronchodilation
bronchiol/o	bronchiole	bronchiolitis
-capnia, capn/o	carbon dioxide	hypercapnia
epiglott/o	epiglottis	epiglottitis
lal/o, -lalia	speech, babble	laliatry
laryng/o	pertaining to the lower throat	larynx
nas/o	nose	nasolabia
osm/o, -osmia	sense of smell, odour, impulse	anosmia
osphresi/o, -osphresia	sense of smell, odour	osphresiometer
ox/o, -oxia	oxygen	hypoxia
pector/o	chest	pectoralgia
phas/o, -phasia	speech	dysphasia
phon/o, -phonia	voice, sound	rhinophonia
phren/o	mind, diaphragm	phrenalgia
-pnea	breathe	hyperpnea
pneum/o	lung, air	pneumopexy
pneumon/o	lung, air	pneumonomycosis
-ptosis	prolapse, drooping	laryngoptosis
-ptysis	spitting	hemoptysis
pulmon/o	lung	pulmonologist
respir/o, respirat/o	breathe, breathing	respirator
rhin/o	nose	rhinolithiasis
sept/o	partition	septorhinoplasty
sinus/o	cavity, sinus	sinusotomy
spir/o	breathe	bronchospirometer
steth/o	chest	stethoscope
therm/o	heat	thermopolypnea
trache/o	trachea	tracheotomy
traumat/o	trauma, injury, wound	traumatopnea
xen/o	strange, foreign	xenophonia

19

Labelling Activity

Label the following diagrams.

1.

2.

Articulate Graphics/IllustrationOnLine.com

19

3.

4.

Answer to Application Question

If the percent of atmospheric oxygen is 16.5% and the atmospheric pressure is 600 mmHg, the partial pressure of oxygen is $(600 \times 0.165) = 99$ mmHg. Yes, his tidal volume will increase, because he needs to inhale more to have enough oxygen to diffuse into his blood. If the atmospheric pressure is 600 mmHg, his alveolar pressure will be the same. With a decrease in oxygen, more oxygen dissociates from the hemoglobin at the tissues. With lower oxygen, the medulla oblongata is stimulated to increase the respiration rate, and erythropoietin is produced to stimulate the bone marrow to produce more red blood cells.

20 The Digestive System

Whenever you find yourself on the side of the majority, it is time to pause and reflect.

Mark Twain

20.1 OVERVIEW OF THE DIGESTIVE SYSTEM

The digestive system consists of the gastrointestinal tract and the accessory organs. The **gastrointestinal tract** (GI tract) is the tube that food travels through: the mouth, esophagus, stomach, **small intestine**, **large intestine**, and the anus (Figure 20.1). The **accessory organs** help the digestive process by producing enzymes, bile, or bicarbonate ions; they include the salivary glands, liver, gallbladder, and pancreas.

Six basic processes occur in the GI tract: (Figure 20.2).

1. **Ingestion** occurs when we take in and swallow food.

2. **Mechanical digestion** refers to the physical mixing and breaking down of food into smaller pieces. This occurs when we chew, as well as during muscular contractions in the stomach and small intestine. Muscular contractions also propel food through the digestive tract.

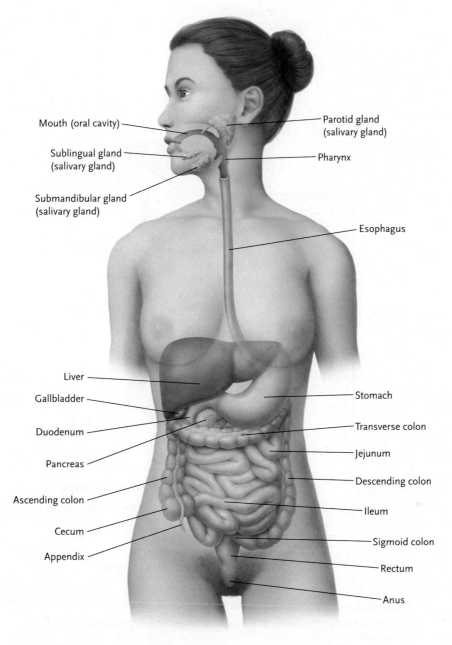

Mouth (oral cavity)

Sublingual gland
(salivary gland)

Submandibular gland
(salivary gland)

Parotid gland
(salivary gland)

Pharynx

Esophagus

Liver

Gallbladder

Duodenum

Pancreas

Ascending colon

Cecum

Appendix

Stomach

Transverse colon

Jejunum

Descending colon

Ileum

Sigmoid colon

Rectum

Anus

Figure 20.1 Overview of the Digestive System

The structures of the GI tract include the mouth, pharynx, esophagus, stomach, small intestine, large intestine, and the anus. The accessory organs include the salivary glands, liver, gallbladder, and pancreas.

3. **Chemical digestion** involves digestive enzymes that chemically break bonds between molecules—such as the peptide bonds in proteins. In this way macromolecules are broken down into monomers (amino acids, glucose, fatty acids, and nucleotides).

4. **Secretion** involves the release of water, ions (mostly sodium), gastric acid, enzymes, bile, and bicarbonate into various regions of the gastrointestinal tract to aid the mechanical and chemical digestion process. Most of the water and ions become reabsorbed during the absorption process.

5. **Absorption** involves the process of moving digested nutrients and water and ions from the lumen of the digestive tract into the bloodstream.

6. **Defecation** is the removal of undigested matter—called *feces*—from the body through the anus.

The overall functions of the digestive system include the following:

• Physically and chemically breaking down ingested food

• Absorbing digested nutrients into the bloodstream

Figure 20.2 Digestive System Processes

Processes of the digestive system include ingestion, digestion, secretion, absorption, and defecation.

- Removing undigested waste
- Producing intrinsic factor, which is required for the absorption of vitamin B12
- Housing symbiotic bacteria—the normal flora—mostly in the large intestine

DID YOU KNOW?

The human digestive tract is approximately 5 m long. If removed after death when it has no muscle tone, it can stretch to 9 m (30 ft)

CONCEPT REVIEW 20.1

1. What are the organs of the gastrointestinal tract?

2. What are the accessory organs?

3. Describe the six main processes of the digestive system.

4. List the main functions of the digestive system.

20.2 TISSUES OF THE DIGESTIVE SYSTEM

The digestive tract consists of all four types of tissue: epithelial, connective, muscle, and nervous. From its innermost layer to the outside layer, the GI tract includes the mucosa, submucosa, muscularis, and serosa (Figure 20.3).

The **mucosa**, the inner lining of the digestive tract, is a mucus membrane composed of epithelial cells—mostly simple columnar epithelial cells in the small intestine—required for absorption of nutrients. The mucosa also contains many capillaries and small lymph vessels that extend into every villus (Figure 20.3). The mucosa is connected to a layer of areolar connective tissue, called the **lamina propria**. This layer is surrounded by a thin layer of smooth muscle tissue, called the **muscularis mucosa**, that is innervated by the myenteric plexus.

The **submucosa** is the layer next to the mucosa, and it contains areolar connective tissue, larger blood and lymphatic vessels, and a network of neurons called the *submucosal plexus*. The neurons that innervate the GI tract are collectively called the *enteric nervous system*, which is part of the autonomic nervous system. A *plexus* is a branching network of neurons. The digestive tract has two: the submucosal plexus and the myenteric plexus in the muscularis layer. The submucosal plexus detects stimuli, such as stretch in the GI tract, and it controls the secretion of substances into the digestive tract.

DID YOU KNOW?

Approximately 70% of the immune system lymph nodes are located in the region of the digestive tract because of the amount of exposure to infectious organisms in food. This part of the immune system is called the *gut-associated lymphatic tissue* (GALT).

The **muscularis** layer contains two thick layers of primarily smooth muscle: an inner circular layer that

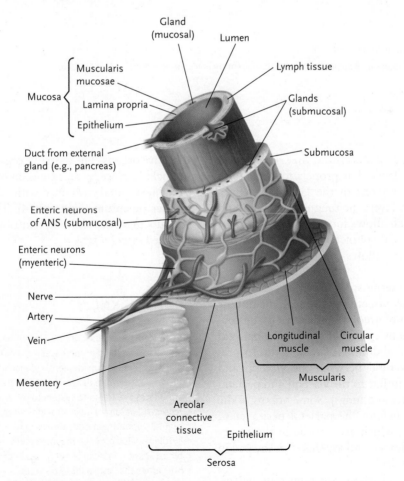

Figure 20.3 Layers of the Digestive Tract

The layers of the digestive tract include connective tissue, smooth muscle, nerves, and the epithelial cell lining.

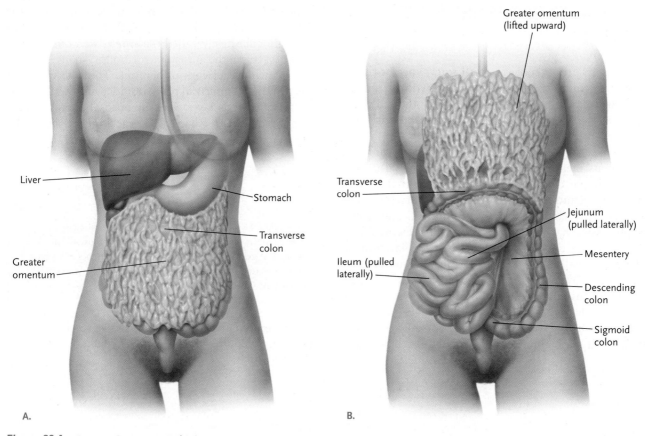

Figure 20.4 Greater Omentum and Mesentery

(A) The greater omentum is composed of adipose tissue. (B) The mesentery holds the abdominal organs in place.

contracts to mix food contents within the GI tract, and an outer longitudinal layer that propels food through the GI tract. Certain regions of the GI tract contain skeletal muscle in this layer: the tongue, pharynx, and upper esophagus, which allows for voluntary control of swallowing. The external sphincter of the anus is also skeletal muscle, which allows voluntary control of defecation.

The **serosa** is the outermost layer of the gastrointestinal tract and is the **visceral peritoneal membrane** (peritoneum) made up of areolar connective tissue covered by simple squamous epithelial tissue. The **parietal peritoneum** covers the abdominal cavity wall. To prevent friction, there is fluid between the visceral and parietal membranes (similar to the visceral and parietal membranes of the heart and lungs). Some organs in the abdominal cavity lie *behind* the parietal peritoneum, including the kidneys, which are located in the posterior portion of the abdomen and are therefore located in the *retroperitoneal cavity*.

Within the abdominal cavity are numerous nerves, blood vessels, and lymphatic vessels that supply the digestive organs. The digestive organs, nerves, and vessels are held in place by an extension of the peritoneum—the **mesentery**—and cushioned with adipose tissue called the **greater omentum** (Figure 20.4). This fat, in addition to the subcutaneous skin fat, contributes to "belly fat" and is called *visceral fat*.

DID YOU KNOW?

Increased abdominal (visceral) fat within the abdominal cavity is linked to a significant increase in the risk of heart disease, type 2 diabetes, depression, dementia, cancer, and sleep disorders. Visceral fat is more dangerous than subcutaneous fat because it stimulates the immune system to provoke an inflammatory response. Inflammatory cytokines interfere with hormones that regulate appetite, mood, and brain functions. Inflammatory cytokines also provoke the movement of oxidized fats into the blood vessels, leading to cardiovascular disease. The two main hormones that cause fat to be stored in the abdomen are insulin (from eating high sugar foods) and cortisol (stress hormone).

1. What are the four major layers of the gastrointestinal tract?
2. Name the three components of the mucosa.
3. What are the functions of the submucosal plexus and the myenteric plexus?
4. What are the functions of the circular and longitudinal muscle layers of the muscularis?

5. What makes up the peritoneum?
6. What is the difference between the visceral peritoneum and parietal peritoneum?
7. What are the functions of the greater omentum and the mesentery?

Application

As you read through the rest of this chapter, consider this scenario. Martha has made a stew for dinner that contains beef, potatoes, peas, carrots, onions, and celery in a bone broth. As Martha eats her meal, where in the digestive tract is mechanical and chemical digestion taking place? Which foods contain the most protein, fat, starch, and fibre? Which enzymes break down each of the foods in her meal? Which hormone is produced that helps her feel full?

20.3 ORGANS OF THE DIGESTIVE SYSTEM

Mouth, Pharynx, and Esophagus

The **mouth** is formed by the lower jawbone (mandible) and upper jawbone (maxilla), the hard palate (anterior maxilla and posterior palatine bones), the muscular tongue, cheeks, gums, and lips. The gums, also called the **gingiva**, cover the joint between the tooth sockets and the teeth; when inflamed, the condition is known as **gingivitis**. The front incisors and the canines tear and cut food; the premolars and molars grind food, giving us the ability to eat a wide range of foods. Adults have a total of 36 teeth: 8 incisors, 4 canines, 8 premolars, 12 molars, and 4 wisdom teeth (Figure 20.5).

The **tongue** is a strong muscle that aids in chewing and swallowing by moving food between the teeth and then rolling it into a ball—called a **bolus**—that can be easily passed through the pharynx and into the esophagus. The tongue contains many taste receptors that detect food flavours such as salty, sweet, bitter, sour, and savoury. *Savoury* is also called *umami* since it detects the amino acid glutamate, for example, in savoury meats. Sensory taste information is sent to the brain via cranial nerves VII, IX, and X. When we swallow, the uvula moves up and covers the opening to the nasal passages, directing food into the pharynx.

The pharynx is the space posterior to the mouth and superior to the opening of the esophagus and trachea. When food moves through the pharynx, the epiglottis covers the opening to the trachea, and so food is directed into the esophagus, located behind the trachea (Figure 20.6). Once the food is in the

esophagus, the muscular contractions are no longer voluntary; from this point on they are controlled by the autonomic nervous system. As the circular and longitudinal muscle layers of the esophagus contract, the bolus of food is propelled downward to the stomach. The wavelike muscle contractions that move food through the entire digestive tract is known as

Figure 20.5 Mouth

Figure 20.6 Swallowing

Swallowing involves wavelike contractions, peristalsis, that move food through the esophagus.

peristalsis. Before food enters the stomach, it moves through the lower esophageal sphincter—also called the *gastroesophageal sphincter* or the *cardiac sphincter.* This is a ring of muscle that relaxes to let food into the stomach and then contracts to prevent the movement of food back into the esophagus.

> **DID YOU KNOW?**
>
> Many animals lick their wounds to promote healing because their saliva contains a molecule called *NGF* (nerve growth factor) that helps wounds heal twice as fast. Humans do not have NGF in their saliva.

Stomach

The stomach has four main parts (Figure 20.7):

1. **Cardia**—the opening of the stomach that's connected to the lower esophageal sphincter

2. **Fundus**—the top portion
3. **Body**—the middle portion
4. **Pylorus**—the inferior region

The stomach is a strong muscular organ with three layers of muscle that contract in different directions to mix and mechanically break down food: the inner oblique layer, the middle circular layer, and the outer longitudinal layer. The muscle layers are smooth muscle controlled by the enteric nervous system. Two sphincters contract to close off each opening to the stomach: the lower esophageal sphincter, and the **pyloric sphincter**. The inner mucosal layer of the stomach—the **rugae**—is highly folded to allow for the stomach to expand. The volume of an empty stomach is approximately 100 to 500 ml, but it can easily expand to hold 1 to 2 L of food.

Once the bolus of food enters the stomach, it is called **chyme**. Food is mainly broken down by mechanical digestion in the stomach from the contractions of the muscular layers, but proteins undergo chemical digestion in the stomach's acid environment. Different cell

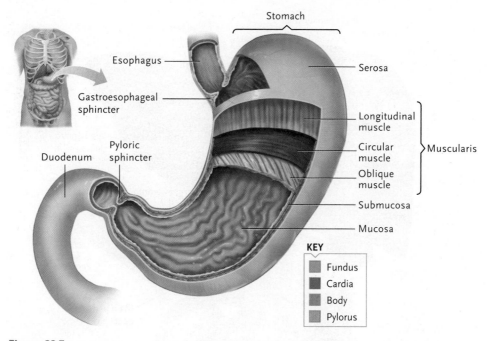

Figure 20.7 Stomach

The stomach can be divided into four regions: the fundus, cardia, body, and pylorus.

types in the mucosal lining of the stomach produce different secretions that aid digestion.

1. Goblet cells produce mucus to protect the lining of the stomach from being digested and from the low pH.

2. Chief cells produce **pepsinogen**, which is activated to form the enzyme **pepsin**, which digests proteins.

3. Parietal cells secrete **hydrochloric acid (HCl)** that kills any pathogens that have been swallowed and converts pepsinogen into pepsin and a substance called **intrinsic factor**, which is needed to absorb vitamin B12 in the small intestine.

Note that mucus, pepsinogen, intrinsic factor, and hydrochloric acid are *exocrine* secretions. *Endocrine* secretions (hormones) are discussed in Sections 20.5 and 20.6. During digestion in the stomach, small amounts of chyme are released at intervals into the small intestine through the pyloric sphincter so that only a small amount enters the small intestine at a time. Only water, ions, and some drugs and alcohol can be absorbed into the bloodstream from the stomach (Section 20.5).

Accessory Digestive Organs

Before considering what happens to the food once it reaches the intestines, it is important to understand the functions of the digestive system's accessory organs: the salivary glands, liver, gallbladder, and pancreas.

Salivary Glands Three pairs of salivary glands—the **parotid**, **submandibular**, and **sublingual** (Figure 20.6)—produce mucus, antibacterial lysozyme, enzymes, water,

and salt. In the mouth, salivary secretions moisten and soften food, which aids both mechanical and chemical digestion. Chewing breaks food into smaller pieces physically. Then two enzymes secreted by the salivary glands begin chemical digestion: **amylase** breaks down complex carbohydrates, and **lipase** begins chemically breaking down fats. Lipase also breaks down bacterial cells walls and plays a role in decreasing tooth decay by controlling the growth of oral microorganisms. **Salivation** is stimulated by the parasympathetic (rest and digest) division of the autonomic nervous system.

Liver and Gallbladder The **liver** is the second largest organ in the body after the skin. It is located in the upper right abdominopelvic quadrant. The liver is connected to the **gallbladder** through the common hepatic duct and the cystic duct (Figure 20.8). The liver produces bile that is stored in the gallbladder. The cells of the liver, called *hepatocytes*, perform all the major functions of the liver. The liver is highly vascularized, containing permeable capillaries called **hepatic sinusoids** between rows of hepatocytes. The nutrients absorbed into the bloodstream throughout the digestive tract are carried to the liver through the hepatic portal vein before they enter the main circulation to the rest of the body. The liver also contains resident immune cells—**Kupffer cells**—that are important for breaking down old or damaged cells, including red blood cells, and any bacteria that may have entered the bloodstream from the GI tract.

During digestion, the bile is released from the gallbladder into the duodenum of the small intestine via the **common bile duct**, which merges with the pancreatic duct, forming the **hepatopancreatic duct**. The duodenum

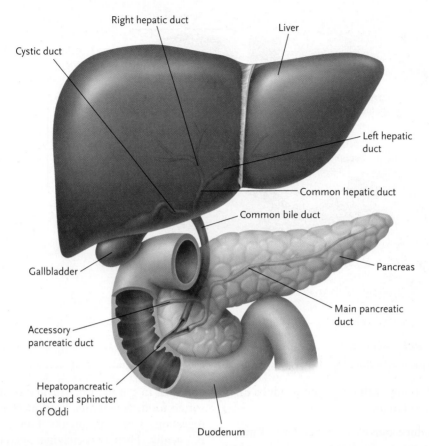

Figure 20.8 Accessory Digestive Organs
Accessory organs include the liver, gallbladder, and pancreas.

controls entry of bile and pancreatic secretions by the **sphincter of Oddi** (Figure 20.8).

The liver performs the following functions:

- Synthesizing bile that is required for fat digestion
- **Glycogenesis**—converting glucose into glycogen
- **Lipogenesis**—converting glucose and amino acids into triglycerides
- **Lipolysis**—breaking down fat and converting fatty acids into **ketones,** which can be converted into acetyl-CoA and then used to produce ATP (Chapter 4)
- **Glycogenolysis**—breaking down glycogen to produce glucose
- **Gluconeogenesis**—breaking down fat or protein to produce glucose
- Producing lipoproteins to transport fats from liver to adipose tissue
- Storing excess fat-soluble vitamins—vitamins A, D, E, and K
- Storing minerals and trace metals
- Producing urea from the breakdown products of proteins (Chapter 21)

- Breaking down old red blood cells (the spleen also does this); the heme component of hemoglobin becoming bilirubin, which is added to the bile and then excreted with other waste from digestion through defecation
- Breaking down many organic and inorganic toxins—such as drugs, alcohol, hormones, and foreign toxins from environmental exposure—and adding the breakdown products to the bile for excretion
- Excreting excess trace metals
- Synthesizing plasma proteins such as albumin, which is important for osmotic regulation and also binding proteins for the transport of fat-soluble hormones in the bloodstream
- Synthesizing fibrinogen, the protein required for blood clotting (Chapter 18)
- Synthesizing **angiotensinogen,** involved in the renin-angiotensin system, to regulate blood pressure (Chapter 21)
- Activating vitamin D—the precursor produced in the skin during sun exposure— that's required for calcium homeostasis and immune regulation (also activated by kidneys)

- Producing insulin-like growth factors (IGFs) in response to growth hormone (Chapter 17)
- Secreting **complement proteins** involved in the immune response (Chapter 23)

DID YOU KNOW?

Bilirubin is a bright yellow pigment that contributes to the greenish colour of bile. If the liver is not functioning properly, such as during a hepatitis infection or cirrhosis of the liver, the bilirubin is not excreted, and it builds up in the tissues as red blood cells are broken down. This causes the yellow pigment to become visible in the skin: a condition called *jaundice*. Bilirubin also causes the yellowish colour of bruises as clotted red blood cells are gradually broken down.

The gallbladder stores bile that is an amphipathic molecule containing polar and nonpolar regions that act like soap to help *mechanically* break down ingested fats (Figure 20.9). Bile is produced by hepatocytes from cholesterol. Bile consists of the bile salts, as well as many breakdown products produced by the liver, water, mucus, and other inorganic salts.

Pancreas The pancreas is located behind and slightly below the stomach in the abdominopelvic cavity. The pancreas consists of glandular epithelial cells; these include a small number of **islet cells** and acinar cells arranged in clusters called **acini**. The pancreas has endocrine and **exocrine** functions. Endocrine functions include the production of the hormone insulin (beta islet cells)

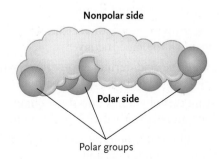

Bile salt (glycocholic acid)

Nonpolar side

Polar side

Polar groups

Figure 20.9 Bile

Bile is required for the digestion of fat.

during the absorptive phase (after eating) and the production of **glucagon** (alpha islet cells) during the fasting phase. The exocrine functions aid digestion. The pancreatic acini cells secrete **pancreatic juice** that contains enzymes, water, salt, and bicarbonate ions. Passing through the sphincter of Oddi, pancreatic juice enters the duodenum via the pancreatic duct, which merges with the hepatopancreatic duct. The enzymes break down proteins, fats, and complex carbohydrates. The bicarbonate ions neutralize the acidic chyme coming from the stomach.

Small Intestine

The small intestine has three parts: the **duodenum**, the first and shortest section; the **jejunum**, the middle region; and the **ileum**, the final and longest portion. The small intestine is approximately three metres long from the pyloric sphincter to the **ileocecal sphincter**, which is the first part of the large intestine. The duodenum receives bile and pancreatic enzymes and also produces some of its own digestive enzymes. The intestinal mucosal epithelium contains goblet cells that produce alkaline mucus. This mucus protects the intestinal lining, helps neutralize the pH, and helps food move through the intestines. The circular and longitudinal muscles of the small intestines contract to mechanically digest food as well as propel it through the GI tract. The intestinal mucosa is highly folded to allow for an extensive surface area for the absorption of nutrients. In addition, the mucosal layer contains **villi** that increase the surface area and microvilli—also called the **brush border**—that further increase the surface area (Figure 20.3). Within each villus are capillaries and a lymphatic vessel called a **lacteal** that take up the absorbed nutrients (Figure 20.10). The ileum also has the important role of absorbing vitamin B12 from the diet; intrinsic factor that is produced by the parietal cells in the stomach is required for B12 absorption.

Large Intestine

The large intestine is larger in diameter than the small intestine, but shorter in length; it is approximately 1.5 m long. It begins with the ileocecal sphincter that joins the **cecum** in the lower right quadrant of the abdominopelvic cavity, which is also the location of the **appendix**. The **ascending colon** travels up the right side of the lower abdominopelvic cavity. The **transverse colon** sits approximately 5 cm above the umbilicus and across the lower part of the **epigastric region** (Figure 1.24). The **descending colon** travels down to the **sigmoid colon**, which connects to the **rectum**, **anal canal**, and the **anus** (Figure 20.11). The internal anal sphincter is regulated by the autonomic nervous system, but the external sphincter consists of skeletal muscle and can be voluntarily controlled.

The large intestine consists of the same tissue layers as described in Section 20.2. As in the small intestine,

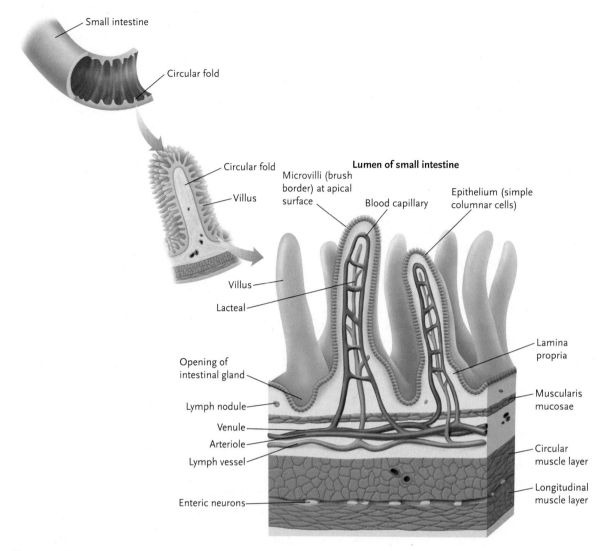

Small intestine

Circular fold

Circular fold

Villus

Lumen of small intestine

Microvilli (brush border) at apical surface

Blood capillary

Epithelium (simple columnar cells)

Villus

Lacteal

Lamina propria

Opening of intestinal gland

Lymph nodule

Venule

Arteriole

Lymph vessel

Muscularis mucosae

Circular muscle layer

Longitudinal muscle layer

Enteric neurons

Figure 20.10 Small Intestine

The small intestine has folds, villi, and microvilli to increase the surface area for the absorption of nutrients.

the epithelial cells are primarily simple columnar cells that absorb water and ions. There are also goblet cells that secrete mucus to lubricate the large intestine and assist the movement of undigested and increasingly solid matter through the final portion of the digestive tract. As the circular muscles in the large intestine contract, it takes on a pouched appearance (Figure 20.11). The large intestine carries out the following functions:

- Absorbing water and ions, although the most water and ions are absorbed in the small intestine
- Storing waste material and excreting it through the process of defecation
- Absorbing vitamins created by the beneficial bacteria—the normal flora—that reside in the large bowel, specifically, B12, thiamine, riboflavin, and vitamin K

Digestive waste—called *feces*—consists mostly of undigested fibre, water, dead intestinal cells, and bacteria.

The bacteria (normal flora) in the large intestines play several very important roles in our overall health:

- Fermenting undigested substrates, mostly soluble fibre
- Regulating the immune system to tolerate food molecules
- Preventing growth of harmful bacteria
- Producing vitamin K and biotin
- Increasing growth of intestinal cells and helping maintain a healthy mucosal layer
- Producing lactic acid during fermentation, which lowers pH and prevents growth of pathogenic species
- Converting carbohydrates into short-chain fatty acids that can be absorbed and used for energy
- Releasing important minerals from food, such as magnesium, calcium, zinc, and iron
- Metabolizing carcinogens, thereby having a significant role in preventing colon, breast, and prostate cancers

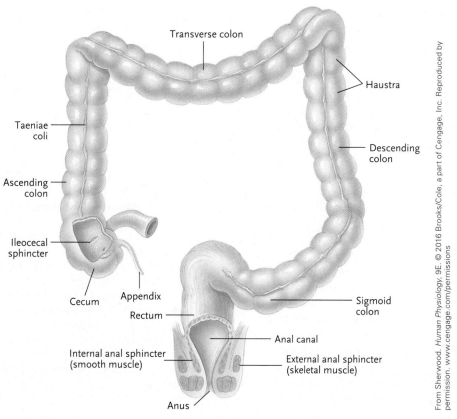

Figure 20.11 Large Intestine

The large intestine stores digestive waste, reabsorbs water, and houses beneficial bacteria.

From Sherwood. *Human Physiology*, 9E. © 2016 Brooks/Cole, a part of Cengage. Inc. Reproduced by permission. www.cengage.com/permissions

DID YOU KNOW?

Mice grown without gut bacteria in a sterile environment need to eat 30% more food to weigh the same as a mouse with normal gut bacteria.

DID YOU KNOW?

Eating fermented foods that contain live bacteria such as sauerkraut or yogurt is very beneficial for the overall health of the digestive system, immune system, and nervous system, and has been shown to improve conditions such as anxiety and depression.

CONCEPT REVIEW 20.3

1. Which bones make up the mouth?
2. What is the gingiva?
3. How many teeth do adults have if they have all of their wisdom teeth?
4. Name the functions of the tongue.
5. What is peristalsis?
6. What is the role of the uvula? Epiglottis?
7. What are the muscle layers of the stomach?
8. What is the role of the pyloric sphincter?
9. What exocrine secretions are produced by the parietal cells, chief cells, and goblet cells?
10. List the functions of the liver.
11. What is bile made from?
12. Describe the endocrine and exocrine functions of the pancreas.
13. Where does the most absorption of nutrients take place?
14. What special features are found in the small intestine to increase surface area for absorption?
15. Name the regions of the large intestine from the ileocecal sphincter to the anus.
16. What are the functions of the large intestine?
17. What are some important functions of gut bacteria?

20.4 DIGESTIVE ENZYMES

Recall from Chapter 2 that enzymes are a group of proteins that catalyze chemical reactions in every cell of the body. The enzymes involved in digestion bind to *specific* substrates and allow hydrolysis reactions to break the chemical bonds that hold together the monomers within the macromolecules (chemical digestion). Table 20.1 lists the macromolecules and their monomers.

Enzymes are secreted in the saliva, stomach, small intestine, and pancreas. See Table 20.2 for a summary of digestive enzymes produced in each organ and the substrates they break down. Chemical digestion begins in the mouth, with amylase breaking starch into smaller chains of glucose molecules and some maltose (disaccharides), and lipase breaking triglycerides into glycerol and fatty acids. Because the food is in the mouth for only a short time, this process of breaking down macromolecules is incomplete at this stage.

In the stomach, the low pH denatures proteins, which become unfolded from their tertiary structure. Pepsin begins breaking the peptide bonds in proteins into peptides, which are short chains of amino acids. No other nutrients are chemically digested in the stomach.

The pancreas secretes several enzymes into the small intestine. In addition, the pancreas secretes bicarbonate ions into the small intestine to neutralize the acidic stomach contents. Enzymes can function in the small intestine only in a neutral pH. Pancreatic **amylase** breaks starch into short chains of glucose and maltose. After bile has emulsified the large fat droplets into small droplets, pancreatic **lipase** breaks down triglycerides into two individual fatty acids and monoglycerides (glycerol with one fatty acid attached). Lipase can act only on the triglycerides on the outside of the droplet, and so it breaks down all the fats gradually. Fats take the longest of all macromolecules to

chemically digest. Therefore, the rate of this step is limited: the more fat in the meal the longer it takes for the meal to be digested. For this reason, it is beneficial to have some healthy fats in a meal so the feeling of fullness lasts longer, but not so much fat that the food stays in the stomach for a long period of time.

DID YOU KNOW?

The most common cause of heartburn is a large meal with a high fat content. Food leaves the stomach and enters the small intestine in small amounts, which must be almost completely digested before chyme from the stomach can enter the intestine. Meals with a high fat content take a very long time to digest. So the stomach continues to mix food for a long time, causing the acidic content to be pushed up into the esophagus through the lower esophageal sphincter, especially after a large meal, and causing damage to the lower esophagus. Heartburn is also more common in people with a large amount of abdominal fat, which increases pressure in the abdominal cavity, forcing stomach contents into the lower esophagus.

The pancreas secretes two enzymes that break down the nucleic acids DNA and RNA that are found in all plant and animal cells. **Ribonuclease** breaks RNA into nucleotides, and **deoxyribonuclease** breaks DNA into nucleotides. Nucleotides can also be broken down into nucleosides, which are just the sugar and base of the nucleotide (Chapter 8). The pancreas produces several enzymes that break down proteins. Pepsin from the stomach does not function once it enters the neutralized pH of small intestine. Pancreatic **trypsin** and **chymotrypsin** break proteins into peptides. Then **carboxypeptidase** breaks the peptides into individual amino acids. It is possible to absorb individual amino acids as well as small peptides, whereas carbohydrates have to be broken down completely into individual monosaccharides to be absorbed.

The small intestine produces enzymes that break down disaccharides: **maltase** breaks down maltose, **sucrase** breaks down sucrose, and lactase breaks down lactose. People who are lactose intolerant do not produce lactase, so this sugar is not absorbed; eventually it reaches the large intestine where the residing bacteria population break it down and produce gas as a byproduct. The small intestine also produces **peptidases** that work with pancreatic carboxypeptidase to break peptides into amino acids.

TABLE 20.1

Macromolecules and Their Monomers	
Macromolecules	**Monomers**
Proteins	Amino acids
Triglycerides	Glycerol and fatty acids
Nucleic acids (DNA and RNA)	Nucleotides
Complex carbohydrates (starch)	Glucose only
Disaccharides	
Sucrose	Glucose and fructose
Lactose	Glucose and galactose
Maltose	Glucose and glucose

TABLE 20.2

Digestive Tract Enzymes and Substrates

Location Where Enzyme is Produced	Enzyme	Substrate the Enzyme Breaks Down
Saliva	Amylase	Starch
Saliva	Lipase	Fats
Stomach	Pepsin	Protein
Pancreas	Amylase	Starch
Pancreas	Lipase	Fats
Pancreas	Ribonuclease	RNA
Pancreas	Deoxyribonuclease	DNA
Pancreas	Trypsin and Chymotrypsin	Proteins
Pancreas	Carboxypeptidase	Peptides
Small intestine	Sucrase	Sucrose
Small intestine	Lactase	Lactose
Small intestine	Maltase	Maltose
Small intestine	Peptidases	Peptides

CONCEPT REVIEW 20.4

1. What are the breakdown products of proteins, carbohydrates, triglycerides, and nucleic acids?
2. Which enzymes break down proteins and peptides?
3. Where are the enzymes produced that break down disaccharides?
4. Where is amylase produced?
5. What enzyme breaks down fat?
6. How does the acidic content of the stomach become neutralized once it reaches the small intestine?

20.5 ABSORPTION

Absorption refers to the movement of nutrients across the simple columnar epithelial cells of the intestine from the lumen into the bloodstream. Nutrients must move across the apical and basolateral membranes of the epithelial cells into the loose areolar connective tissue called the *lamina propria* and then into either a blood capillary or a lacteal (Figure 20.12). Different nutrients are absorbed by different mechanisms, including simple diffusion, osmosis, facilitated diffusion, coupled transport, and active transport (Chapter 3). Anything that is not absorbed continues through the gastrointestinal tract and is eliminated as feces.

Water, Ions, and Water-Soluble Vitamins

Water can be absorbed into the bloodstream by moving through the epithelial cell membranes (osmosis), or between the epithelial cells. Just like the reabsorption of water from the kidney tubule into the bloodstream, as more ions are absorbed more water is absorbed. During

digestion in the small intestine, water and ions are secreted into the lumen to aid digestion. This secreted water and ions as well as the ingested water and ions are absorbed mostly in the duodenum. However, water is also absorbed in the stomach, all sections of the small intestine, and the large intestine. All but approximately 200 ml of water is reabsorbed at some point along the digestive tract.

DID YOU KNOW?

Chicken soup broth helps you to feel better when you are sick simply because it is high in sodium. As we absorb the sodium in the intestine, we absorb more water, which minimizes water loss through diarrhea.

Ions and minerals can move across cell membranes only through protein ion channels or by cotransport with another molecule (Figure 20.13). Each ion moves

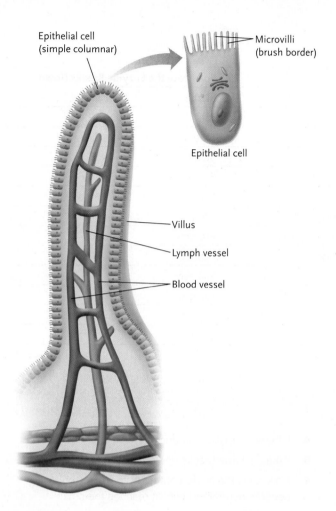

Epithelial cell
(simple columnar)

Microvilli
(brush border)

Epithelial cell

Villus

Lymph vessel

Blood vessel

Figure 20.12 Absorption in the Small Intestine

cell through cotransport with glucose or galactose is pumped out of the cell by the sodium-potassium pump (Chapter 3). Once monosaccharides pass through the basolateral membrane, they enter the capillary in the villus. All blood from the digestive tract flows via the hepatic portal vein to the liver, where the nutrients are processed and stored. Once sugars enter the blood and circulate to the pancreas, the increased blood sugar stimulates the pancreas to produce insulin. Insulin circulates to the cells and signals the production of more GLUT membrane carrier proteins so that more sugar can be absorbed in the intestine as well as taken up by other body cells for ATP production.

Peptides and Amino Acids

Amino acids are absorbed through the apical membrane by cotransport with sodium ions (Figure 20.13), and they move through the basolateral membrane via amino acid carrier proteins. Small peptides of two or three amino acids are also absorbed by cotransport with hydrogen ions. Epithelial cells have peptidases that break down the small peptides into amino acids inside the cell before they are transported through the basolateral membrane. It is important to note that we also break down and absorb amino acids from dead intestinal cells and enzymes used during digestion; this means we do not lose the amino acids that were used to make digestive enzymes. Amino acids move into the capillaries and circulate to the liver.

Nucleotides

Nucleotides are absorbed by cotransport with sodium ions on the apical membrane, and then transported through nucleotide carrier proteins on the basolateral membrane. Like monosaccharides and amino acids, nucleotides are absorbed into the bloodstream, unlike fats that are absorbed into the lacteal.

Fat and Fat-Soluble Vitamins

Fats are mechanically broken down by bile into small fat droplets by bile, and then chemically broken down by lipase into monoglycerides and free fatty acids. Once the triglycerides are enzymatically broken down, they form very small droplets called **micelles**, and then individual fatty acids and monoglycerides can move into the epithelial cells by simple diffusion (Figure 20.14).

Once inside the epithelial cell, fatty acids and monoglycerides re-form into triglycerides. This occurs on the smooth endoplasmic reticulum in the epithelial cell. The triglycerides, along with fat-soluble vitamins,

through a specific ion channel or transporter. Calcium can be absorbed only if vitamin D is present to stimulate the production of calcium channels. Other minerals, such as iron, zinc, potassium, chloride, magnesium, are all absorbed as ions by facilitated transport through their own specific membrane protein.

Water-soluble vitamins, the B vitamins, and vitamin C are cotransported with sodium ions across the apical membrane; the exception is vitamin B12, which requires intrinsic factor for it to be absorbed. Vitamin B12 is absorbed in the last part of the small intestine, the ilium. Some people cannot produce intrinsic factor and will develop pernicious anemia, which can be treated with injections of B12.

Monosaccharides

Glucose and galactose are absorbed on the apical membrane by cotransport with sodium ions (Figure 20.13) and through GLUT (glucose transporter) membrane carrier proteins on the basolateral side. Fructose is transported through both membranes by GLUT carrier proteins. The sodium that enters the

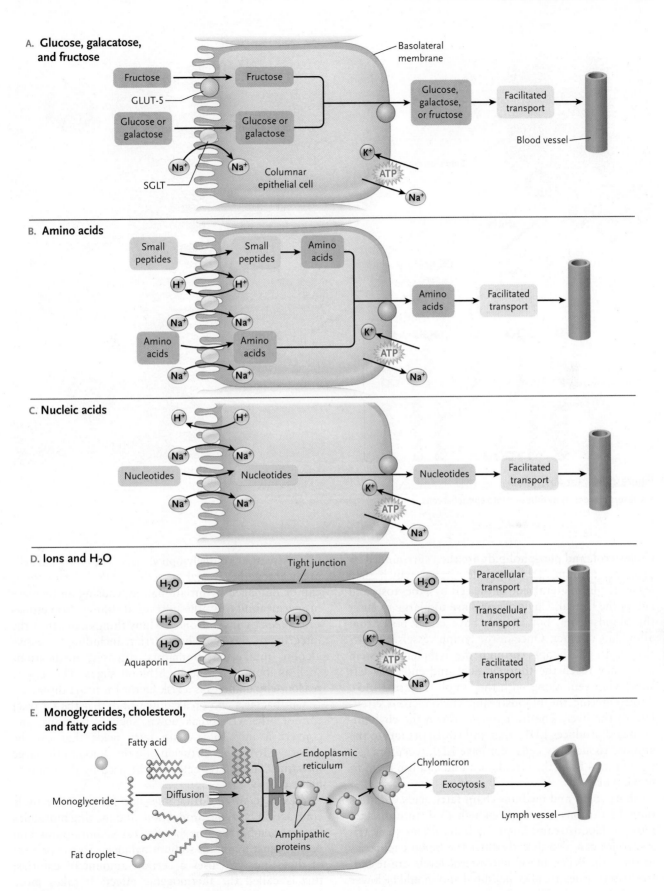

Figure 20.13 Absorption of Nutrients

Absorption of nutrients involves the movement of amino acids, monosaccharides, and fats across the epithelial cell membrane and then into the bloodstream.

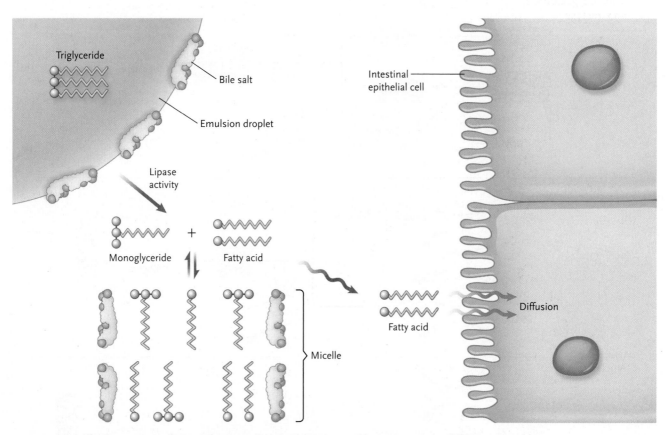

Figure 20.14 Fat Absorption

Fat absorption occurs by diffusion into epithelial cells.

cholesterol, and phospholipids are then surrounded by proteins, forming a large **chylomicron**, which then leaves the basolateral side of the cell by exocytosis and enters the lacteal. Chylomicrons are too large to enter the bloodstream, but lacteals have larger pores that allow fats to enter. Once in the lymph vessels, the fats flow with the lymph through the lymphatic system until they reach the thoracic duct and enter the left subclavian vein along with the lymph (Figure 20.15). After entering the bloodstream, chylomicrons circulate to the liver. The liver breaks down the chylomicron and produces LDLs that will transport fats to the adipose tissue. Any cells that have LDL receptors can take up fats from the circulation and can use them to make ATP.

Short-chain and medium-chain fatty acids with less than 12 carbons are small enough to diffuse directly into the bloodstream. Some cholesterol and some triglycerides can also directly enter the blood capillaries in the villi. When blood cholesterol levels are tested, free triglycerides are also measured and should be lower than 2.2 mmol/L. When the epithelial mucosa becomes damaged, triglycerides are more likely to enter the

bloodstream than the lymph vessels. The epithelial cells are held together by tight junctions, but these junctions can be damaged by inflammation, causing an increase in permeability of the intestinal lining (sometimes called *leaky gut syndrome*). Many things can affect the integrity of the epithelial barrier, including excessive alcohol intake, inflammatory diseases, medications such as ibuprofen, and high blood sugar. This is one reason why diabetes is a risk factor for heart disease.

Gut bacteria can break down fibre that the GI tract cannot otherwise break down, such as fructans, and convert it into short-chain fatty acids that can be absorbed by simple diffusion into the blood in the large intestine. These fatty acids can be used by the brain to make energy.

ATP is required during digestion for smooth muscle contraction, enzyme production, and moving molecules across membranes (the basolateral sodium-potassium pumps must maintain sodium balance). So the process of digesting food uses a certain amount of calories; this is called the **thermogenic effect**. It takes more energy to digest protein than it does to digest fats or carbohydrates.

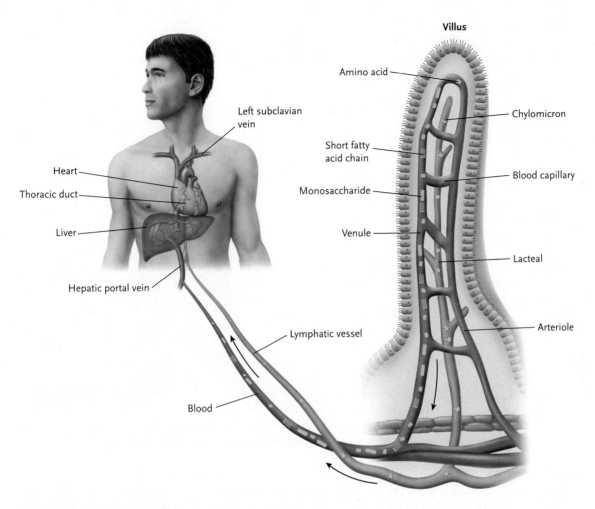

Figure 20.15 Nutrient Transport

The transport of nutrients occurs through the lymphatic and circulatory systems.

CONCEPT REVIEW 20.5

1. How are ions and water absorbed in the small intestine?

2. What vitamin is required for calcium absorption?

3. Which vitamin requires intrinsic factor for absorption?

4. Which nutrients enter the epithelial cells by co-transport with sodium ions?

5. Why are sodium-potassium pumps required in the basolateral membrane?

6. Explain how fats are absorbed into the lymphatic vessels.

7. How are water-soluble and fat-soluble vitamins absorbed?

8. Explain how the intestinal mucosa can be damaged and how this impacts the absorption of nutrients.

9. What is the thermogenic effect?

20.6 REGULATION OF DIGESTION

Digestion occurs in three phases: cephalic phase, gastric phase, and the intestinal phase. The **cephalic phase** is initiated by the sight, smell, or thought of food. It triggers parasympathetic signalling to the submucosal and myenteric plexus to begin secretion of gastric juices and to start muscle contractions in anticipation of ingesting food—causing the stomach to "growl." Salivation also begins during the cephalic stage, and, once food is ingested, the taste of food further increases secretion and motility of the GI tract. *Motility* refers to the contractions of the smooth muscle in the GI tract.

20

The **gastric phase** begins when food enters the stomach. The physical presence of food contacting the walls of the stomach lining signals the gastric secretions and muscle contractions.

The **intestinal phase** begins once the first bolus of chyme enters the duodenum and triggers intestinal motility and the release of pancreatic secretions, bile, and intestinal enzyme secretion. Once the intestinal phase begins, gastric motility begins to decrease to ensure that the intestine is not overloaded.

Foods that contain a large amount of fat require more time to completely digest, whereas simple carbohydrates, such as fruit, are digested very quickly. Certain hormones play a role in the regulation of the rate of digestion. See Table 20.3 for a summary of the main hormones. In the stomach, food enters and stretches the stomach lining, causing the release of the hormone **gastrin**—secreted by *enteroendocrine* cells in the gastric glands in the pyloric region of the stomach (Figure 20.16). Gastrin acts on the parietal cells to produce hydrochloric acid and intrinsic factor. It also stimulates muscle contractions and stimulates the chief cells and mucus cells to produce pepsinogen and mucus, respectively.

As food enters the small intestine, the acidic chyme stimulates the intestinal mucosa to produce the hormone **secretin**. Secretin stimulates the pancreas to release enzymes and also bicarbonate ions, which are responsible for neutralizing the acid from the stomach. As food is chemically broken down, and proteins are broken down into amino acids, and triglycerides are broken down into fatty acids and monoglycerides, the presence of these building blocks stimulates the small intestine to produce the hormone **cholecystokinin (CCK)**. CCK acts on the stomach to slow gastric emptying, causes the gallbladder to contract and the sphincter of Oddi to relax, allowing the release of bile into the small intestine. CCK is the hormone that signals the satiety centre in the hypothalamus and causes

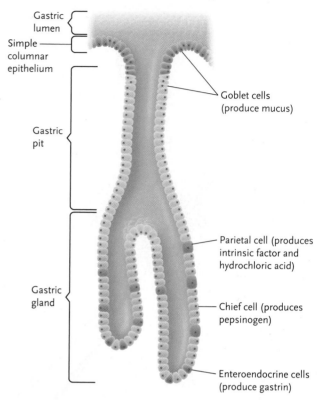

Figure 20.16 Gastric glands

Gastric glands contain cells that secrete hormones.

the feeling of fullness. CCK is not produced until food starts entering the small intestine: that is, during the intestinal phase. Eating quickly can lead to eating much more food than needed, so it is important to eat slowly to avoid overeating.

When glucose—derived from the digestion of starch or simple carbohydrates—is present in the small intestine, glucose-dependent insulinotropic peptide (GIP) is released. GIP stimulates the pancreas to release insulin.

TABLE 20.3

Summary of the Main Hormones Involved in Regulation of Digestion

	Gastrin	CCK (cholecystokinin)	Secretin	GIP (glucose-dependent insulinotropic peptide)
Site produced	Stomach	Small intestine	Small intestine	Small intestine
Stimulus for production	Distension of the stomach, peptides in the stomach	Fat in the small intestine	Acid in the small intestine	Glucose in the small intestine
Functions	Stimulates release of HCl and pepsinogen, and promotes stomach muscle contractions	Stimulates gallbladder contraction to release bile; also causes the feeling of fullness	Stimulates pancreas to release bicarbonate and enzymes	Stimulates pancreas to release insulin

CONCEPT REVIEW 20.6

1. Describe the three phases of digestion.
2. Create a chart to summarize the hormones involved in digestion. Include where the hormone is produced, the stimulus for hormone production, and its function.
3. Which hormone causes us to feel full after eating?

20.7 DISORDERS OF THE DIGESTIVE TRACT

Many disorders can affect the digestive system, including certain infections, autoimmune diseases, cancer, allergies, and inflammation. The following list highlights some of the most common.

- Gastroesophageal reflux disease (GERD). This disease occurs when acidic contents of the stomach are forced up into the lower esophagus, causing irritation and a burning sensation. Acid in the lower esophagus can damage the sensitive tissue and cause ulcerations, inflammation, and possibly cancer. The most common causes are obesity, hiatal hernia, and a diet high in fat. Eating foods very high in fat slows digestion and keeps food in the stomach longer.

- Hiatal hernia. The lower esophageal sphincter is a ring of muscle located at the lower end of the esophagus and just above the stomach. It is meant to contract and keep the stomach contents in the stomach during digestion. When a hernia occurs, the sphincter and a portion of the upper stomach protrude through the diaphragm into the chest cavity. Hernias may not always cause symptoms, but symptoms can include a dull ache in the chest; difficulty swallowing, as though food is staying in the esophagus; and possibly heart palpitations. The most common cause is excessive force in the abdominal cavity, such as from violent vomiting, coughing, lifting heavy weight, or straining with constipation. (See Figure 20.17a.)

- Ulcers. Located along the digestive tract, these sores occur most commonly in the duodenum, but they can occur within the stomach, esophagus, or any region of the small or large intestine. The most common cause is an overgrowth of the bacterium *Helicobacter pylori* (*H. pylori*), and therefore the most common treatment is a course of antibiotics. Other factors can contribute, such as overuse of pain killers, excessive alcohol consumption, and smoking. Smokers are more prone to duodenal ulcers, and alcoholics are more prone to esophageal ulcers. (See Figure 20.17b.)

- Crohn's disease. This inflammatory disease primarily affects the small intestine but can occur in any region of the digestive tract. Crohn's is a chronic disease with no known cure. The most common symptoms include abdominal pain, cramping, diarrhea, nausea and vomiting, fatigue, and weight loss. Chronic inflammation in the bowel is associated with an increased risk of bowel cancer.

A.
David Musher/Science Source

B.
David M. Martin, M.D./Science Source

C.
Gastrolab/Science Source

Figure 20.17 Disorders of the Digestive Tract
(A) Hiatal hernia, (B) ulcer, and (C) diverticulitis

- Ulcerative colitis. This chronic autoimmune disease causes chronic inflammation and ulcerations in the large intestine and often occurs in a relapsing-remitting cycle. There is no known cause or cure for any autoimmune disease (Chapter 23). Symptoms often include cramping, bloody diarrhea, and fatigue. As with Crohn's disease, people with ulcerative colitis have an increased risk of colon cancer.

- Gastritis. This is a general term related to inflammation of the stomach. However, *gastritis* most often refers to an acute infection causing nausea, vomiting, and diarrhea. The most common cause is an infection by a norovirus (Norwalk virus)—often called the *stomach flu*. But gastritis is not caused by influenza. In healthy people, the immune system controls these infections within one to three days.

- Diverticulitis. Small pouches called *diverticula* can form in regions of the bowel where the wall is weakened and inflamed. Symptoms include abdominal pain, nausea, diarrhea, and fever. The cause is unknown, but it is possible that increased pressure from straining with constipation is a risk factor and that a diet high in fibre can help prevent the disease. It was previously thought that people with diverticulitis should avoid nuts and seeds, but research has shown there is no correlation, and, in fact, a higher intake of nuts may help to avoid the development of this disease. (See Figure 20.17c.)

- Irritable bowel syndrome. Symptoms include cramping, abdominal pain, gas and bloating, and possibly constipation or diarrhea. Unlike Crohn's disease and ulcerative colitis, irritable bowel syndrome is not an inflammatory disease. The exact cause is not known, but it may be related to food allergies or sensitivities, infections, malabsorption of nutrients, or abnormalities in the gut flora.

- Hemorrhoids. This condition involves clusters of swollen veins in the anus and rectum. It is caused by increased pressure, most commonly by straining with constipation. A high-fibre diet is very important for preventing constipation.

DID YOU KNOW?

The most common food allergies are peanuts, eggs, and shellfish. Although there has been a dramatic increase in the number of people with food allergies in the last 20 years, no specific reason for this is known.

CONCEPT REVIEW 20.7

1. Which of the disorders of the digestive tract explained in this section are chronic inflammatory diseases?

2. Which digestive disorders can be prevented with a high-fibre diet?

3. Which digestive disorders are most often caused by infectious organisms?

EFFECTS OF AGING

As people age the digestive system can undergo several changes. The rate at which these changes occur, if at all, depends largely on lifestyle factors such as diet, stress, and inflammation.

1. Constipation can occur in elderly people if they consume too little fluids and fibre, are not sufficiently mobile, or take medications that can cause constipation. Constipation is the most common digestive system condition in elderly people.

2. The ability to absorb nutrients can decrease as we age. The production of secretions such as hydrochloric acid, bile, and digestive enzymes decreases. Lifestyle factors, such as high consumption of alcohol, can have a significant impact on the liver and pancreas. A decrease in the ability to digest food can cause symptoms such as bloating, cramps, and gas.

3. Damaged intestinal epithelium due to lifestyle factors such as medications or alcohol can increase food sensitivities.

Elderly people with arthritis or other inflammatory conditions often take anti-inflammatory medications that can be very damaging to the small intestine. When undigested food molecules enter the bloodstream, the immune system will react to certain types of food, such as gluten (wheat and other grains) or lectins (in various vegetables).

4. Polyps are small growths in the colon that can become more common in elderly people. There are no visible symptoms, but with screening, such as colonoscopy, polyps can be removed before they become cancerous.

5. GERD (gastroesophageal reflux disease), otherwise known as *heartburn*, can occur more frequently in older people due to low stomach acid and digestive enzymes, which causes food to stay in the digestive system longer and acid to be pushed into the esophagus by stomach contractions. Other factors include high blood pressure, obesity, eating large meals late at night, eating meals very high in fat, or taking certain medications.

CHAPTER REVIEW

Medical Terminology

Term	Definition	Example
-agra	severe pain	cardiagra
amyl/o	starch	amyloid
antr/o	cavity	antrum
-ase	enzyme	lipase
atel/o	incomplete, imperfect	ateloglossia
bucc/o	cheek	buccal
cheil/o, chil/o	lip	cheiloschisis
chol/e	gall, bile	cholecystokinin
choledoch/o	common bile duct	choledochlithiasis
col/o	colon	colonoscopy
dent/i	tooth	dentist
dips/o	thirst	adipsia
-emesis	vomiting	hematemesis
enter/o	intestines (small intestines)	enteroclysis
gastr/o	stomach	gastric bypass
gingiv/o	gums	gingivoplasty
gloss/o	tongue	glossolalia
gluc/o	glucose, sugar	glucocorticoid
glyc/o	glucose, sugar	glycolysis
hepat/o	liver	hepatosplenomegaly
lapar/o	abdomen, abdominal wall	laparoscope
lingu/o	tongue	retrolingual
odont/o	tooth	anodontia
or/o	mouth	intraoral
-orexia	appetite	anorexia
orth/o	straight, normal, correct	orthodontist
-pepsia	digestion	dyspepsia
phag/o, phagia	eating, ingestion	phagocytosis
-posia	drinking	polyposia
-prandial	meal	postprandial
proct/o	rectum, anus	proctopexy
ptyal/o	saliva	ptyalogenic
pyr/e	portal vein	pyrogen
rect/o	rectum	rectocele
sial/o	saliva	sialolith
-stalsis	contraction	peristalsis
staphyl/o	uvula, grapelike clusters	staphylorrhaphy
stomat/o	mouth	stomatomalacia

(Continued)

Term	Definition	Example
-tresia	opening, perforation	proctotresia
typhl/o	cecum, blindness	typhlectasis
uran/o	palate	uranoschisis
zym/o	enzyme, ferment	lysozyme

Labelling Activity

Label the following diagrams.

1.

2.

3.

4.

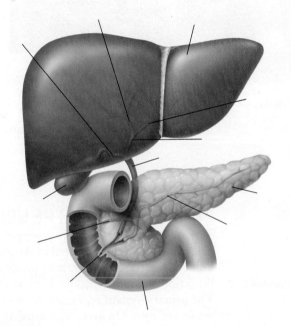

Answer to Application Question

- Mechanical digestion occurs in the mouth, stomach, and small intestine.
- Chemical digestion occurs in the mouth, stomach, and small intestine.
- The most protein is found in the beef. The most fat is in the broth and the beef. The most starch is in the potatoes and carrots. The most fibre is found in the vegetables: peas, carrots, onions, and celery.
- Enzymes that break down protein include pepsin, trypsin, and chymotrypsin. Enzymes that

break down peptides include carboxypeptidase and the small intestine peptidases. The enzyme that breaks down starch is amylase. Maltase will break down maltose that forms when starch is broken down. The enzyme that breaks down fat is lipase. There are no enzymes that break down cellulose; however, healthy intestinal bacteria can break down fibre and produce short-chain fatty acids.

- CCK that is produced during fat digestion will help Martha feel full.

21 The Urinary System

Live as if you were to die tomorrow. Learn as if you were to live forever.

Mahatma Gandhi

21.1 OVERVIEW OF THE URINARY SYSTEM

The urinary system is located in the abdominal cavity and consists of the kidneys, ureters, bladder, and urethra. The kidneys are located in the posterior abdominal cavity behind the digestive organs (Figure 21.1). The urinary system regulates the body's **osmotic composition**, which is water and ions. On average, a normal adult blood volume is 5 L, and the kidneys filter approximately 180 L of blood each day. Therefore, our entire blood volume is filtered more than 30 times a day, and extra water, ions, and waste products are excreted as **urine**.

The urinary system performs the following functions:

- Regulating the blood's water and ion composition
- Excreting the waste products of metabolism
- Regulating blood pressure—the most important organ to do so
- Regulating blood pH by excreting excess H^+ ions
- Releasing two hormones—**calcitriol**, the active form of vitamin D, and **erythropoietin**, which stimulates the bone marrow to produce red blood cells

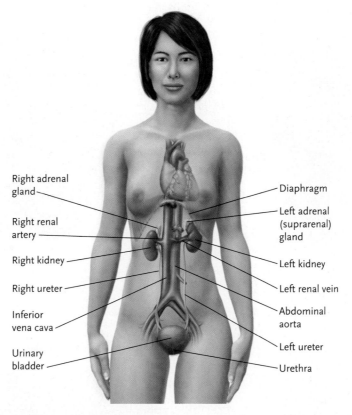

Figure 21.1   Location of the Kidneys
The kidneys are located in the posterior abdominal cavity.

CONCEPT REVIEW 21.1

1. What are the functions of the urinary system?

2. What structures make up the urinary system?

21.2 ANATOMY OF THE URINARY SYSTEM

Each kidney is shaped like a kidney bean, approximately three inches long. The adrenal glands are attached to the superior region of each kidney. The kidneys receive their blood supply from the **renal arteries** that branch from the abdominal aorta. The renal arteries enter each kidney from the medial side at a region called the **renal hilum;** this is the same region where the renal vein, lymphatic vessels, nerves, and the ureter leave the kidney (Figure 21.2a). The kidney is surrounded by a **renal capsule,** a protective covering for the kidneys that consists of connective tissue and is connected to adipose tissue that also helps to protect and cushion the kidneys.

The kidney consists of an outer cortex and an inner medulla. The functional unit of the kidney is the **nephron,** and each kidney contains about 1 million nephrons. Each nephron has a glomerulus located in the cortex region, which is where the initial filtering of the blood occurs. As the filtered fluid moves through the nephron, water and ions are reabsorbed back into the bloodstream,

specifically, into the peritubular capillaries that surround the nephron in both the cortex and medulla regions. The fluid left in the nephron will be excreted as urine.

The medulla region of the kidney has a higher **osmotic gradient** compared to the cortex and is important for increased water reabsorption back into the bloodstream. The renal medulla contains areas called **renal pyramids,** which is where many nephrons converge onto **collecting ducts** and where the final reabsorption of water and ions occurs before the urine is fully formed. The regions of cortex that extend toward the middle of the kidney, in between the pyramids, are called the **renal columns.** Each section of kidney from the capsule to the renal hilum that encompasses one renal pyramid is called a **renal lobe** (Figure 21.2b).

Urine drains from the collecting ducts in the pyramids, through the **minor calyx** and then the **major calyx** to the large **renal pelvis** in the centre of the kidney. The renal pelvis is connected to the **ureters** that carry the urine to the bladder for storage until the bladder is full. The urinary bladder contains an internal layer of **transitional epithelial cells** (Chapter 11). It is surrounded by connective tissue and a muscle layer called the **detrusor muscle** (Figure 21.3). The

21

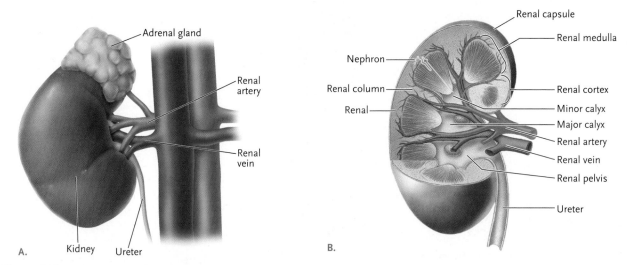

Figure 21.2   Anatomy of the Kidney

(A) External anatomy and (B) internal anatomy

(B) From DiGiuseppe/Fraser. *Biology* 12U, 1E. © 2012 Nelson Education Ltd. Reproduced by permission. www.cengage.com/permissions

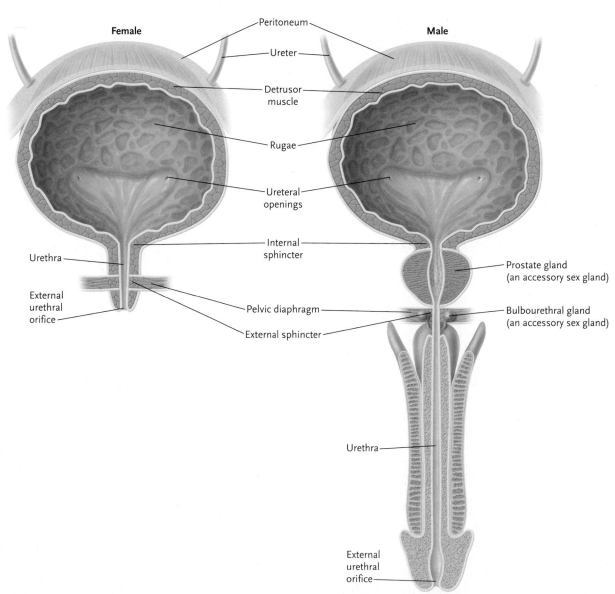

Figure 21.3   Anatomy of the Bladder

Comparison of male and female anatomy of the bladder and accessory structures

Figure 21.4 Blood Flow to the Kidneys

Blood flows into each kidney through the renal artery and leaves through the renal veins.

external urethral sphincter is skeletal muscle and is voluntarily controlled. The **internal urethral sphincter** is controlled by the autonomic nervous system. Urine is excreted from the body through the **urethra.**

When the bladder is full, stretch receptors (mechanoreceptors) trigger a spinal cord reflex, and information travelling through parasympathetic nerves causes the detrusor muscle to contract and the internal sphincter to relax, leading to urination: also called **micturition.**

Blood enters the renal hilum through the renal artery, which branches inside the kidney to the **segmental arteries** and then to the **interlobar arteries** that surround the pyramids. The blood flows toward the cortex of the kidney, where the glomeruli are located (Figure 21.4). From the interlobar arteries, blood flows to the cortex through the **cortical radiate arteries** and finally to the **afferent arteriole,** which branches into the capillaries of each **glomerulus.** This is where filtration occurs. From the glomerulus, the portion of blood that was not filtered continues to the efferent arteriole, which branches to the **peritubular capillaries.** In the peritubular capillaries water and nutrients are reabsorbed from the nephron. The blood flows from the capillaries to the **cortical radiate veins,** to the **arcuate veins** that flow back from the renal pyramids to the **interlobar veins,** and finally back to the large **renal vein** that connects to the inferior vena cava.

CONCEPT REVIEW 21.2

1. What is the difference between the cortex and the medulla of the kidney?

2. List the structures that fully formed urine flows through, from the collecting ducts to the urethra.

3. List the path of blood flow through the kidney from the renal artery to the renal vein.

Application

As you read through the rest of this chapter, consider this scenario. Bert is planning to run a half marathon, which is 21 km. He is hoping to finish the race in two hours. Let's suppose Bert's cardiac output will double during the two hours of running. A normal cardiac output is approximately 5 L/min. His blood pressure will be increased from 120/80 at rest to 160/82 during the exercise. How will the increase in cardiac output and blood pressure affect filtration? Will filtration increase or decrease? How will exercise affect GFR? Will it increase or decrease? Do you think there is anything that Bert could do before the race to try to decrease fluid loss?

21.3 FUNCTIONS OF THE NEPHRON

The nephron is the smallest functional unit of the kidney (Figure 21.5). A nephron consists of several specific regions that have specific functions. Fluid moves through the following kidney structures from the time it leaves the bloodstream until urine is produced: renal corpuscle, which is made up of the glomerulus and the Bowman's capsule; proximal convoluted tubule; descending loop of Henle; ascending loop of Henle; distal tubule; and collecting ducts.

The kidney performs three important tasks (Figure 21.5):

1. Filtration—the movement of fluid from the bloodstream into the Bowman's capsule

2. Reabsorption—the movement of fluid from the tubule back into the bloodstream

3. Secretion—the movement of fluid from the bloodstream (peritubular capillaries) into the tubule somewhere other than the Bowman's capsule.

Excretion is the physical removal of the urine from the bladder through the urethra.

amount of substance excreted = amount filtered − amount reabsorbed + amount secreted

Some substances are completely reabsorbed after filtration, such as glucose and amino acids. Some substances are filtered and then a proportion is reabsorbed, such as sodium, and the rest is excreted. Some substances are completely excreted by filtration and secretion with no reabsorption, such as the breakdown products of medications and other toxins.

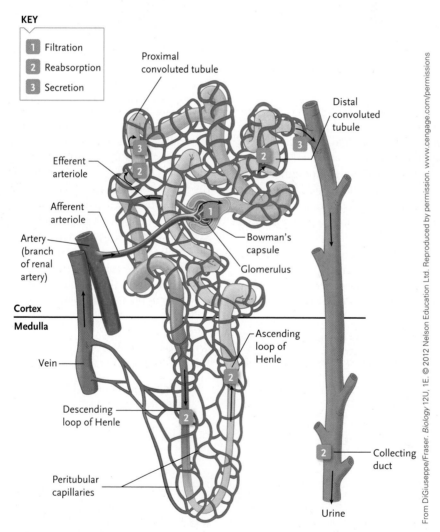

Figure 21.5 Nephron

The nephron is the smallest functional unit of the kidney. The processes of filtration (1), reabsorption (2), and secretion (3) occur in specific regions of the nephron.

Filtration

Filtration occurs in the renal corpuscles. **Podocytes** surround the glomerular endothelium and form a leaky filtration membrane that permits plasma containing water, ions, nutrients, and waste products to move into the **Bowman's space** (Figure 21.6). Blood cells and most plasma proteins remain in the blood because they are too large to pass through the membrane. Once the fluid has moved from the blood to the Bowman's capsule, it is called **filtrate**. Recall from Chapter 18 that when plasma moves from the tissue capillaries to the interstitial space, it becomes interstitial fluid. The same thing occurs in the kidneys, except that the capillaries allow much more fluid to leave the circulatory system and enter the kidney tubules.

The **macula densa** and **juxtaglomerular cells** (Figure 21.6) play a role in the regulation of filtration and excretion. Autonomic neurons innervate the blood vessels in the kidney to regulate blood pressure via vasodilation and vasoconstriction of the arterioles, which have a smooth muscle layer that can contract and relax.

The pressure that causes filtration is the blood pressure in the glomerular capillaries, otherwise known as the **hydrostatic pressure** (P_{GC}) in the glomerular capillaries (Chapter 18). Although blood pressure accounts for some fluid leaving the capillaries in every tissue, there are also forces that oppose filtration (Figure 21.7). Because the plasma proteins are not filtered and they remain in the blood vessels, they exert an **osmotic pressure** (π_{GC}) that prevents too much fluid from leaving the blood circulation. Once the fluid moves into the Bowman's capsule, it exerts a hydrostatic pressure (P_{BC}), which prevents more fluid from entering the capsule. Once the capsule is filled with fluid, no more fluid

KEY

P_{GC} Hydrostatic pressure in glomerular capillaries
P_{BC} Hydrostatic pressure in Bowman's capsule
π_{GC} Osmotic pressure in glomerular capillaries

Figure 21.7 Pressures That Affect Filtration
Hydrostatic and osmotic pressures affect filtration.

can enter until the filtrate has moved into the proximal tubule. The filtrate is isotonic (Chapter 3) to the blood plasma since it is the plasma that enters the Bowman's capsule. So there is no opposing osmotic pressure in the Bowman's capsule. If blood pressure increases, filtration will increase. The total amount of fluid filtered per day is called the **glomerular filtration rate (GFR)**, which amounts to approximately 180 L/day.

Filtration is not selective. Approximately 2.5% of all blood plasma that circulates through the kidneys is filtered into the nephron, including molecules that we do not necessarily want to excrete, such as glucose and amino acids, and most of the water and ions. Therefore, most substances need to be reabsorbed back into the bloodstream.

Reabsorption

Reabsorption involves the movement of water, ions, or other nutrients back into the bloodstream (Table 21.1). The efferent arteriole leaves the glomerulus and then branches into many peritubular capillaries that surround the rest of the nephron and reabsorb molecules. The proximal convoluted tubule is the first portion of the nephron where reabsorption takes place. Most reabsorption occurs in the proximal tubule. All glucose and amino acids are reabsorbed back into the bloodstream in the proximal tubule. There should not be any glucose or amino acids in the urine of healthy people. When

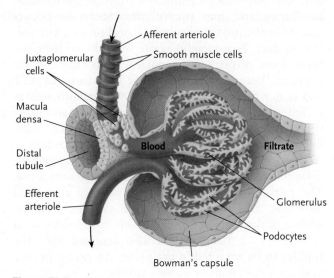

Figure 21.6 Filtration
Filtration occurs in the renal corpuscle.

TABLE 21.1

Substances Reabsorbed in the Proximal Tubule

Substance	Percentage
Water	65
Glucose and amino acids	100
Na^+ and K^+	65
HCO_3^-	80–90
Cl^-	50
Urea	50
Ca^{2+} and Mg^{2+}	variable*
Vitamins	100
Lactate	100

*Calcium and magnesium are reabsorbed when blood levels of those minerals are low and parathyroid hormone is released.

TABLE 21.2

Substances Reabsorbed in Other Regions of the Nephron

Region	Substance
Descending loop of Henle	Water
Ascending loop of Henle	Ions
Distal tubule	Na^+ Cl^- HCO_3^- water
Collecting duct	Water and urea

someone is diabetic, the very high blood sugar levels that enter the filtrate cannot be 100% reabsorbed; therefore, glucose can be detected in the urine. Recall from Chapter 3 that membrane carrier proteins can become saturated, and the rate of movement across the membrane cannot increase with a further increase in solute concentration, which is what happens in the proximal tubule of a diabetic.

The descending loop of Henle is permeable to the passage of water, but impermeable to ions and urea. Approximately 15% of the water that was filtered in the glomerulus is reabsorbed along the descending loop of Henle. As the filtrate moves down the descending loop and more water is reabsorbed, the filtrate becomes more concentrated: hypertonic compared to initial filtrate. The loops of Henle vary in length throughout the kidney; some are short and stay mostly within the cortical region, and others are much longer, reaching into the medulla region. The longer loops reabsorb more water than the cortical nephrons because of the increased osmolarity of the medulla. Due to the osmotic gradient, a higher concentration of ions in the medulla causes more water to be reabsorbed there.

As the filtrate moves up along the ascending loop of Henle, the epithelial cells of the nephron become *impermeable* to water and permeable to ions because of the expression of ion channel membrane proteins. Approximately 20 to 30% of sodium, potassium, and chloride ions are reabsorbed along the ascending loop.

In the distal tubule and collecting duct, hormonal regulation affects the remaining reabsorption of ions and water. The release of parathyroid hormone increases calcium reabsorption, ADH increases water reabsorption, and aldosterone increases sodium reabsorption, which then causes more water to be reabsorbed by osmosis. Aldosterone also increases potassium secretion. The last 9% of the water that was initially filtered

is reabsorbed in the collecting ducts. This leaves the remaining 1% of water in the urine that is excreted, which amounts to approximately 1 to 2 L per day. The distal tubules and the collecting ducts also reabsorb a small amount of urea; although it is a waste product, we excrete only about 40% of the amount filtered. See Table 21.2 for a summary of substances absorbed in the tubule.

DID YOU KNOW?

Animals living in very dry desert regions need to highly conserve their body water. Compared to other animals, they have the longest loops of Henle for maximal water reabsorption. A camel can last about 15 days in the desert without water, and a kangaroo rat can live on just the water in its food without drinking any extra water for its whole life.

Secretion

Secretion involves active transport of **nitrogenous wastes, creatinine,** excess hydrogen and potassium ions, and some medications out from the peritubular capillaries and into the tubules. Secretion occurs mainly in the proximal tubule, distal tubule, and collecting duct. Nitrogenous waste is produced from the breakdown of proteins or nucleic acids that contain nitrogen, because nitrogen is not used to make ATP and is not recycled by our cells. Nitrogenous wastes include **ammonia, urea,** and **uric acid.** Most ammonia is converted into urea. Ammonia is extremely alkaline and toxic, so only very small amounts remain in the blood. Most pyrimidines—the single-ring nucleotides of nucleic acids—are also converted into urea. The purines—adenine and guanine—are converted into uric acid. Two ammonia molecules combine with a molecule of carbon dioxide to produce urea (Figure 21.8). Uric acid is also secreted into the tubules to be excreted in the urine. An excess of uric acid can cause gout or kidney stones.

Creatinine is a waste product produced when muscle cells use creatine phosphate as an energy source (see Chapter 14). H^+ ions are secreted depending on

A. Amino acid (leucine)   B. Ammonia   C. Urea   D. Purine (adenine)   E. Pyrimidine (thymine)   F. Uric acid

Figure 21.8 Waste Products from the Breakdown of Protein and Nucleic Acids

(A) Amino acid—leucine, (B) ammonia, (C) urea, (D) purine—adenine, (E) pyrimidine—thymine, and (F) uric acid

the pH of the blood. The respiratory system and the kidneys are the regulators of blood pH, which must be maintained at a very stable 7.41 to 7.45 in the arteriole blood entering the renal corpuscles. When the blood becomes too acidic, more H^+ ions are secreted into the tubule. Excess K^+ ions are also secreted, mainly in the distal tubules; this is increased when aldosterone is present. Excretion of potassium depends on the blood levels, which are determined from dietary intake. Many medications, such as penicillin, digoxin, and tetracycline, are secreted into the tubule for excretion.

DID YOU KNOW?

Gout is an inflammation of the joints and is considered a type of arthritis; it is caused from an accumulation of uric acid in the joints. This very painful inflammatory disorder can be exacerbated by eating foods that tend to cause high uric acid production, such as herring, tuna, and sardines. So people with gout need to watch out for omega 3 fatty acid supplements that are extracted from small fish. Instead, they can increase omega 3 fatty acids by eating flaxseeds. Omega 3 fatty acids are key to preventing chronic inflammation.

CONCEPT REVIEW 21.3

1. List the order of the structures in the nephron that filtrate flows through.

2. Where do most water and nutrients get reabsorbed?

3. What is significantly different between the ascending and descending loop of Henle?

4. Which nutrients are 100% reabsorbed?

5. What is reabsorbed when the following hormones are present: PTH, ADH, or aldosterone?

6. What are the three types of nitrogenous waste products? Why are they formed?

7. Where does secretion occur?

8. Which molecules are secreted?

21.4 TRANSPORT ACROSS MEMBRANES

Molecules move into or out of the nephron across cell membranes by three different processes in the kidney. Filtration is the movement into the Bowman's capsule; reabsorption is the movement from the nephron tubules into the peritubular capillaries and back into the circulatory system; and secretion is the movement from the peritubular capillaries into the tubules to be excreted (Figure 21.9a).

Any movement into or out of the nephron tubules requires passage through the simple cuboidal epithelial cells that line the tubule. The focus of this section is the reabsorption of important nutrients from the lumen back into circulation (Figure 21.9b). The side of the epithelial membrane that is next to the lumen of the tubule is the **apical**, or luminal, membrane. The opposite side,

close to the peritubular capillaries, is the basolateral membrane. In between, the epithelial cells are tight junctions that hold the cells together tightly so that molecules cannot pass between the cells but must be regulated by membrane proteins. In this way the nephron regulates which molecules and how many of them move back into the bloodstream. The water and molecules left in the tubules are the filtrate that is excreted as urine.

Different nutrients are absorbed by different mechanisms, including simple diffusion, osmosis, facilitated diffusion, coupled transport, and active transport. Water can be reabsorbed into the bloodstream by moving (1) through the epithelial cell membranes (osmosis); (2) through **aquaporin** membrane proteins; or (3) between the epithelial cells by a process called **paracellular transport**. Just like the absorption of water in the small intestine, more water is absorbed in the kidneys as more ions are absorbed (Figure 21.9c).

21

Figure 21.9 Movement across the Tubule Epithelial Cell Membranes

(A) Filtration, reabsorption, and secretion. (B) Substances that are reabsorbed must move from the tubule lumen, through the epithelial cell, and into the bloodstream. (C) Substances cross the tubule epithelial cell membrane through various membrane transport proteins, carriers, or via diffusion.

In healthy people glucose and amino acids are 100% reabsorbed in the proximal tubule by cotransport with sodium on the apical membrane, and then through membrane carrier proteins (GLUT, glucose transporters) on the basolateral membrane (Figure 21.9c). The sodium brought into the cell through cotransport is removed from the cell on the basolateral membrane by the sodium-potassium pump. Potassium ions are moved into the epithelial cells by the Na$^+$/K$^+$ pump and must be *secreted* into the tubule lumen. Aldosterone causes an increase in sodium reabsorption, while at the same time increasing the secretion of potassium. Hormones regulate reabsorption and secretion by affecting the transcription of membrane proteins.

CONCEPT REVIEW 21.4

1. What direction are molecules moving—into or out of the tubule—during filtration, reabsorption, and secretion?

2. Name three ways that water can be reabsorbed.

3. How are glucose and amino acids transported across the apical membrane and the basolateral membrane?

4. Where is the sodium-potassium pump located on tubule epithelial cells?

21.5 REGULATION OF THE URINARY SYSTEM

Hormones that regulate the kidneys affect reabsorption or secretion in the distal tubules and the collecting ducts. Regulation of kidney function is related to the regulation of blood pressure and blood mineral concentrations, such as that of calcium, sodium, potassium, and magnesium. See Table 21.3 for a summary of the hormones involved in regulation blood volume, ions, and minerals.

When sodium or water in the blood is low, blood pressure decreases. A decrease in sodium level is detected by the macula densa cells located in the distal convoluted tubule, which—through the folding of the tubules—is located beside the renal corpuscle where the afferent arteriole enters the glomerulus (Figure 21.6). Macula densa cells send signals to the smooth muscles of the afferent arteriole to decrease glomerular filtration, which decreases urinary output (Figure 21.10).

Figure 21.10 Effects of Decreased Blood Sodium Level
Blood sodium levels are detected by the macula densa cells.

TABLE 21.3

Hormones that Act on the Kidney

Hormone	Stimulus for Production	Production Site	Target Tissue	Function
ADH	Low blood pressure	Hypothalamus, released from posterior pituitary gland	Distal tubules and collecting ducts	Producing aquaporin membrane proteins to increase water reabsorption
Aldosterone	Increased angiotensin II or high blood potassium	Adrenal cortex (zona glomerulosa)	Distal tubules and collecting ducts	Increasing sodium reabsorption, therefore water reabsorption; increasing potassium secretion
PTH	Low blood calcium or magnesium	Parathyroid gland	Distal tubules	Increasing calcium and magnesium reabsorption
ANP	High blood pressure	Atria	Distal tubules and collecting ducts	Decreasing sodium reabsorption, resulting in more water being excreted

21

Renin-Angiotensin System (RAS)

When blood pressure decreases, there is decreased stretch in the baroreceptors (stretch receptors) in the afferent arterioles. This stimulates the secretion of the enzyme **renin** from the juxtaglomerular cells (Figure 21.6). Renin moves into the bloodstream where it cleaves angiotensinogen, produced by the liver, into angiotensin I. Angiotensin I is then cleaved into angiotensin II (Figure 21.11).

Angiotensin II has several functions that help increase blood pressure:

1. Stimulating the adrenal cortex to produce aldosterone
2. Causing vasoconstriction
3. Stimulating the posterior pituitary to release ADH
4. Stimulating the sympathetic nervous system to increase blood pressure
5. Stimulating the reabsorption of sodium and chloride and therefore water
6. Stimulating thirst and cravings for salt

DID YOU KNOW?

There is a class of diuretics called *ACE inhibitors* that decrease blood pressure by preventing the formation of angiotensin II. Without angiotensin II, aldosterone production is not stimulated and therefore more water is excreted, causing blood pressure to decrease.

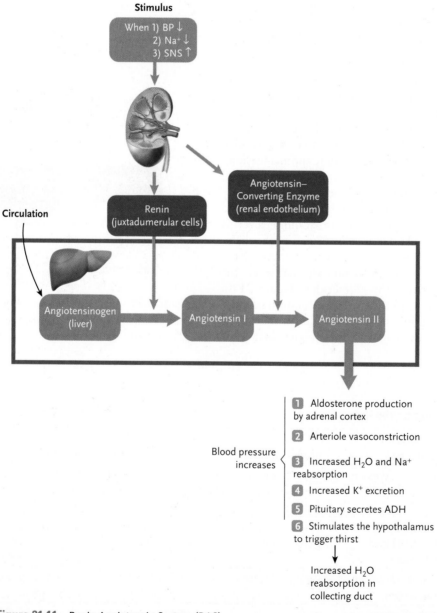

Figure 21.11 Renin Angiotensin System (RAS)

The renin-angiotensin system (RAS) involves the production of hormones that regulate water and ion concentrations.

CONCEPT REVIEW 21.5

1. List each hormone that plays a role in regulating blood pressure, where it is made, what stimulates its production, what tissue it acts on, and its main functions.

2. What is the function of renin?

3. What is the function of angiotensin-converting enzyme?

4. List all the functions of angiotensin II.

21.6 KIDNEY DISEASES

Chronic Kidney Disease

Chronic kidney disease (CKD), the most common kidney disease, is characterized by a gradual loss of kidney function and leads to kidney failure. Approximately 26 million people in North America have CKD. The primary causes are hypertension and diabetes. Kidney function can be determined by the glomerular filtration rate (GFR). As chronic kidney disease progresses, GFR decreases until eventually **renal failure** occurs, and there is very little or no urine output. Kidney function can also be determined by testing for **urine albumin** content. Recall from Chapter 18 that albumin is the main plasma protein involved in maintaining osmotic pressure in the blood vessels and preventing excess fluid from leaking into the interstitial tissues. Proteins are too large to be filtered at the renal corpuscle, and all proteins, cells, and large molecules should not be found in the urine of healthy people. If urine contains protein, kidney damage is present. Chronic hypertension can damage the delicate capillaries of the glomeruli, and high blood sugar from diabetes also damages blood vessels.

When the kidneys cannot properly remove waste products from the blood, regulate blood pH and ion levels, and properly regulate blood pressure, symptoms can become very severe very quickly. Symptoms of advanced kidney disease include nausea, weakness, fatigue, inability to concentrate, decreased appetite, muscle cramping (from lack of ion regulation), increased blood pressure, itching of the skin (from lack of histamine excretion), and edema in the tissues—commonly in the feet and around the eyes—from decreased control of blood pressure. Chronic kidney disease is often treated with blood pressure medication and by controlling diabetes. Once kidney failure occurs, the only treatment options are **dialysis** or a kidney transplant.

DID YOU KNOW?

Without dialysis, renal failure, with no urine output at all, would cause death within one week. People can live with kidney failure and dialysis treatments for many years.

Kidney Stones

Kidney stones affect approximately 10% of the population and are most commonly caused by excess uric acid or calcium-oxalate precipitation. Recall that uric acid is a waste product produced from the breakdown of the purine nucleotides of DNA and RNA molecules. We eat nucleic acids any time we eat plants (fruit and vegetables) or meat, because all cells contain these molecules. Uric acid buildup in the joints causes gout, which is becoming an increasingly common disorder. The buildup of uric acid as either kidney stones or gout is highly linked to metabolic syndrome and also has a strong genetic component.

Kidney stones are said to be one of the most painful health problems to endure. The buildup of uric acid or calcium crystals often form in the renal pelvis and then travel through the very small ureters to the bladder (Figure 21.12). The symptoms depend on the size of the stone. As the stone moves through the ureters, the smooth muscles in the ureters contract to try to help move the stone through; in addition to the inflammation caused from the stone scraping the delicate epithelial cells, this causes severe pain. Pain can radiate from the lower back toward the front lower abdominal groin area. Other symptoms include nausea, blood in urine, fever (from inflammatory response), and painful urination. Small stones often pass without treatment other than pain medication and drinking large amounts of water. Large kidney stones may need to be removed with surgery. Drinking enough water on a regular basis can help prevent the development of large kidney stones. A person with a family history of kidney stones needs to take dietary precautions.

DID YOU KNOW?

People who are prone to calcium-oxalate stones can decrease the amount of oxalate in their diet by rinsing and cooking such green vegetables as spinach, Swiss chard, and broccoli. The vegetables should be cooked in boiling water for at least two minutes to allow most of the oxalates to leave the food and diffuse into the water. Also, by eating enough calcium, any precipitation with oxalates will occur in the digestive tract and be removed with the feces; this will prevent the oxalates from being absorbed into the bloodstream and precipitating in the kidney.

21

A.

Science Photo Library

B.

Steve Sant/Alamy Stock Photo

Figure 21.12 Kidney Stones

Kidney stones are caused by excess precipitation of (A) uric acid and (B) calcium-oxalate.

CONCEPT REVIEW 21.6

1. What are the primary causes of chronic kidney disease?

2. How can chronic kidney disease be prevented?

3. Why would someone with hypertension need to decrease salt intake?

4. What are the symptoms of advanced chronic kidney disease?

5. What are the treatment options?

6. What are the two most common types of kidney stones?

7. What other disease can be caused from a buildup of uric acid?

EFFECTS OF AGING

The urinary system includes the kidneys, ureters, and bladder. Muscle changes can affect the bladder. Specifically, the muscles that control urination tend to become weaker as we age. Therefore, stress incontinence—the leaking of urine from the bladder during coughing or sneezing—is quite common, particularly among women who have given birth. Other changes also occur in the urinary system as we age:

- The amount of kidney tissue decreases.
- The number of nephrons decreases.

- Nephrons become less able to filter, reabsorb and secrete substances, so waste products are not eliminated effectively.
- In people with atherosclerosis, blood flow to the kidneys is decreased.
- In people with high blood pressure, the glomerular capillaries can be damaged and large molecules such as proteins can enter the filtrate.
- The bladder tends to lose elasticity and therefore cannot hold a large amount of urine, which explains why older people tend to have to go to the bathroom frequently.
- In healthy elderly people, kidney function tends to remain fairly normal.

CHAPTER REVIEW

Medical Terminology

Term	Definition	Example
a-, an-	without, not	anuria
-atresia	closure, occlusion	urethratresia
astreto-	closed, lacking an opening	astretocystia
calci/o	calcium	hypocalciuria

cali/o	calyx	pyelocaliectasis
cupr/o	copper	cupruresis
cyan/o	blue	urocyanosis
cyst/o	bladder, cyst	cytogram
-ectasis, -ectasia	dilation, expansion	nephrectasia
fusc/o	dark brown	urofuscohematin
glomerul/o	glomerulus	glomerulopathy
hydr/o	water, hydrogen	hydrolysis
iso-	being equal	isotonic
lith/o	stone, calculus	lithotripsy
ly/o	dissolve, loosen	lyophilic
nephr/o	kidney	nephrotoxic
noct/i	night	noctalbuminuria
py/o	pus	pyocalix
pyel/o	renal pelvis	pyelophlebitis
ren/o	kidney	renogastric
-tripsy	to crush, break	lithotripsy
ur/o	urine	urology
-uresis	urination	diuresis
ureter/o	ureter	ureterocolostomy
urethr/o	urethra	urethrorrhagia
-uria	urine condition	ketonuria
uric/o	uric acid	uric acid stones
urin/o	urine	urinalysis
vesic/o	urinary bladder	vesicle

Labelling Activity

Label the following diagrams.

1.

2.

From DiGiuseppe/Fraser. *Biology 12U, 1E.* © 2012 Nelson Education Ltd. Reproduced by permission. www.cengage.com/permissions

21

3.

4.

KEY

1	Filtration
2	Reabsorption
3	Secretion

From DiGiuseppe/Fraser. *Biology 12U, 1E*. © 2012 Nelson Education Ltd. Reproduced by permission. www.cengage.com/permissions

21

Answer to Application Question

When Bert is running the half marathon his blood pressure will increase. The increased blood pressure will cause an increase in the glomerular filtration rate (GFR). The increased cardiac output will cause more blood volume to circulate through the kidneys and will also increase GFR. The increase in filtration will cause an increase in urine output. However, Bert will also lose a lot of water through sweating. As his blood volume decreases due to fluid loss, his kidneys will regulate blood volume accordingly. If Bert wants to try to retain fluid to prevent excessive water loss and therefore decrease the amount he would need to consume during the run, he could take a salt pill or increase his sodium intake before the marathon. Sodium causes the body to retain water to keep blood osmolarity stable.

22 The Reproductive System

The difference between a successful person and others is not a lack of strength, not a lack of knowledge, but rather a lack of will.

Vince Lombardi

22.1 OVERVIEW OF THE REPRODUCTIVE SYSTEM

The primary role of the reproductive system in both males and females is to produce **haploid gametes** in order to produce offspring. The production of gametes is called **gametogenesis**, which involves meiosis, where **diploid** germ cells go through two cell divisions to create the haploid gametes with 23 chromosomes: **spermatozoa** (**sperm**) in males and **ova** (**eggs**) in females. Gametogenesis occurs within the **gonads**: the **testes** in males and the **ovaries** in females. Gametogenesis in males is called *spermatogenesis* and in females is called *oogenesis*. Both males and females produce steroid hormones that regulate the reproductive process; these are testosterone, estrogen, and progesterone.

The functions of the male reproductive system include spermatogenesis; producing steroid hormones, mainly testosterone; storing and transporting sperm; and adding secretions from accessory glands to the sperm to produce semen.

The functions of the female reproductive system include oogenesis; producing steroid hormones, mainly estrogen and progesterone; and storing and ovulation of ova. The female reproductive system also includes the site of fertilization—the fallopian tubes—and the site of implantation and growth of the fetus—the uterus. The mammary glands produce milk to nourish the newborn.

CONCEPT REVIEW 22.1

1. What are the functions of the male and female reproductive systems?

22.2 MALE REPRODUCTIVE SYSTEM

The testes are the male gonads. They are located within the **scrotum,** which holds the testes away from the body and regulates their temperature to be 35 to 36°C; this is one to two degrees lower than body temperature, which is essential for normal sperm development. The testes consist of **seminiferous tubules,** and this is where spermatogenesis occurs beginning in puberty. The testes are connected to a duct system that transports the mature sperm from the seminiferous tubules (Figure 22.1).

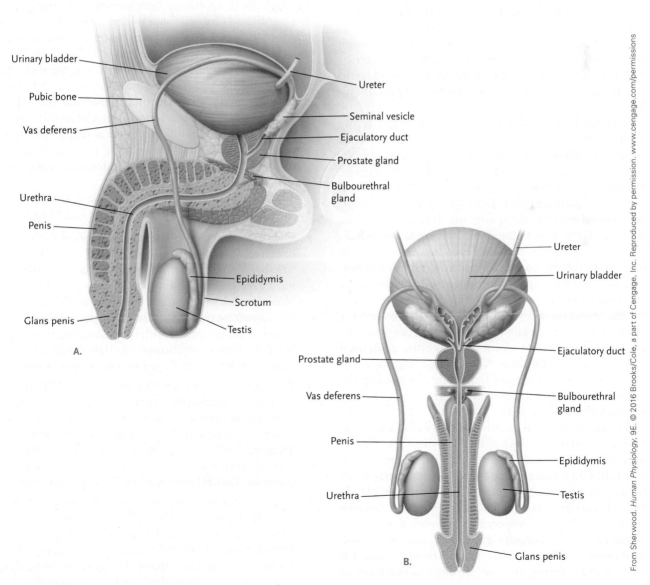

Figure 22.1 Male Reproductive Organs

(A) Lateral view and (B) anterior view

The seminiferous tubules converge to a section of tubule called the **rete testis** that leads to the **efferent ductules**, and then to the coiled **epididymis** where sperm mature. The fully developed sperm travel through the epididymis to two **vas deferens**. The vas deferens are lined with smooth muscle tissue that contracts and moves sperm through the ducts during sexual arousal. Sperm can survive within the vas deferens for several months. The sperm move through the vas deferens, which loops around the anterior and then superior surface of the urinary bladder to the posterior side where the **accessory glands** are located.

The **seminal vesicles** join the vas deferens at the **ejaculatory duct** (Figure 22.1). The seminal vesicles secrete an alkaline viscous fluid that contains fructose, prostaglandins, mucus, proteins, enzymes, and vitamin C; contributes nutrients for the sperm; and makes up approximately 60 to 70% of the volume of the **semen** ejaculated. Sperm use fructose to make ATP, and the prostaglandins contribute to muscle contractions in the female reproductive tract that help to direct the sperm toward the fallopian tubes. The alkalinity is important for neutralizing the acidic environment of the vagina. The ejaculatory duct leads to the **prostate gland**, located below the urinary bladder where the ejaculatory duct merges with the urethra. The prostate gland secretes citric acid, used by the sperm to produce ATP, and prostate-specific antigens (PSA), which are important for helping the sperm enter the cervix. Semen and urine cannot move through the urethra at the same time.

The **bulbourethral glands**, located below the prostate gland and on either side of the urethra, secrete alkaline mucus into the urethra during arousal to neutralize any acid residue from urine that may be in the urethra. Secretions from the bulbourethral glands begin before ejaculation. The urethra travels through the penis, which is composed of cylindrical masses of tissue called the **corpora cavernosa** and the **corpus spongiosum**. These highly vascularized tissues fill with blood, causing erection through stimulation of the autonomic nervous system.

Spermatogenesis

The seminiferous tubules consist of an outer basement membrane, Sertoli cells, germ cells called **spermatogonia**, and cells in various stages of meiotic division (Figure 22.2). Outside the seminiferous tubules are the Leydig cells that produce testosterone. Spermatogonia are the diploid (2n) cells that undergo both mitosis—to continually replenish the spermatogonia population—and meiosis. The spermatogonia that enter meiosis become differentiated: **primary spermatocytes** undergo the first meiotic cell division and become haploid **secondary spermatocytes**. Secondary spermatocytes then undergo a second cell division; sister chromatids separate to become **spermatids**, each containing 23 chromosomes. Spermatids then differentiate into spermatozoa (sperm), but full and complete maturation continues and is completed in the epididymis.

The **Sertoli cells** play an integral role in the development of sperm and have several specific functions.

- They extend from the basement membrane to the lumen of the seminiferous tubules and are held together by tight junctions, forming a closed environment called the **blood-testis barrier,** similar to the blood-brain barrier formed by astrocytes in the brain. It is important that toxins from the bloodstream do not come in contact with the developing sperm. Spermatocytes move through the tight junctions between the Sertoli cells just before meiosis begins.

- They produce **androgen-binding protein (ABP)** that binds to the testosterone secreted by the Leydig cells in the tissue outside the seminiferous tubules. Leydig cells produce testosterone in response to **luteinizing hormone (LH)** that's secreted by the anterior pituitary gland in response to the **gonadotropin-releasing hormone (GnRH)** produced by the hypothalamus.

- They are stimulated by follicle-stimulating hormone (**FSH**) produced by the anterior pituitary gland in response to GnRH produced by the hypothalamus. FSH and testosterone stimulate the Sertoli cells to produce the many signalling molecules that regulate spermatogenesis. Testosterone also

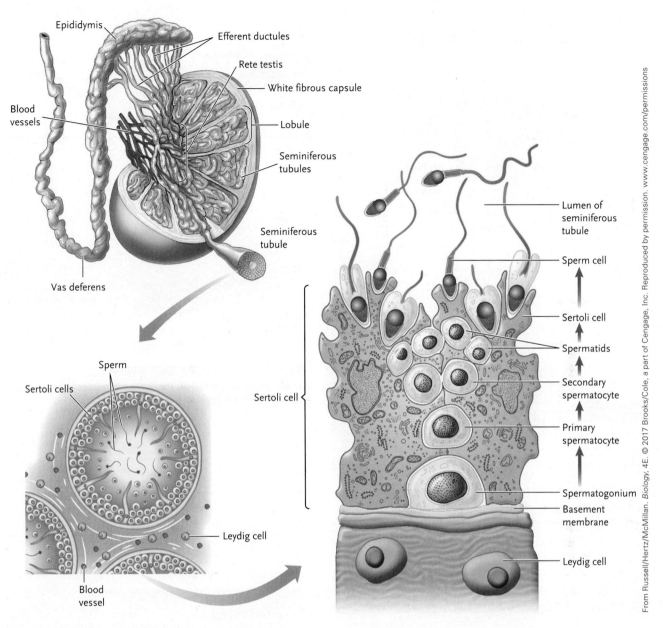

Figure 22.2 Spermatogenesis

Spermatogenesis is the production of sperm through meiosis.

stimulates development of the male secondary sex characteristics.

- They produce the hormone **inhibin**, which signals the anterior pituitary gland to decrease FSH secretion (Figure 22.3).
- They engulf improperly developing sperm through phagocytosis.

Structure of Spermatozoa

Sperm have three sections: the head, mid-piece, and tail (Figure 22.4). The head of the sperm contains the nucleus that has 23 chromosomes and minimal cytoplasm. The head, surrounded by an **acrosome**, contains important enzymes that break down the outer coat of the egg and lead to fertilization. The mid-piece contains mitochondria that produce ATP from the fructose present in the secretions of the accessory glands. The tail is a flagellum composed of microtubules that allow the tail to move and propel the sperm through the female reproductive tract.

22

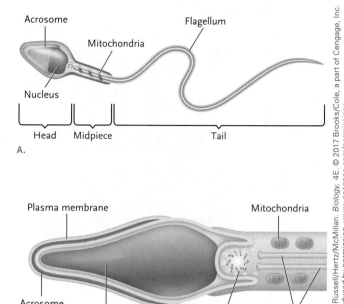

Figure 22.3 Hormonal Regulation of Spermatogenesis

Hormones of the anterior pituitary gland regulate testosterone production in the testes, which regulate spermatogenesis.

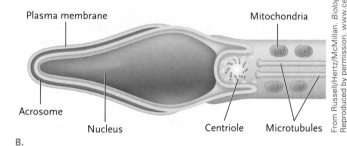

A.

B.

Figure 22.4 Structure of a Spermatozoon

(A) Head, midpiece, and tail. (B) The acrosome contains enzymes that enable entry into the egg during fertilization.

CONCEPT REVIEW 22.2

1. List in order the structures that the sperm travel through from the seminiferous tubules to the urethra.
2. What is the function of the seminal vesicles, the prostate gland, and the bulbourethral glands?
3. Name the cell types that arise during spermatogenesis.
4. Which cell types are diploid and which are haploid?
5. How do Sertoli cells support spermatogenesis?
6. What is the function of the Leydig cells?
7. What is the role of FSH and LH in spermatogenesis?
8. Describe the structure of a spermatozoon.

22.3 FEMALE REPRODUCTIVE SYSTEM

The ovaries are the female gonads. They are located within the lower abdominopelvic region and protected by the sacrum, coccyx, and pelvic bones. The ovaries are located on each side of the uterus (Figure 22.5), where the oocytes develop through oogenesis. An **ovarian ligament** connects each ovary to the outside of the uterus and holds the ovary close to the opening of the fallopian tubes. Oogenesis begins during the embryonic development of the female fetus. As a result, newborn baby girls have already produced all the **oocytes** they will have for their lifetime. The development of the oocytes is halted in mid-meiosis and completed only when fertilization takes place. Ovulation begins during puberty, when approximately every 28 days one egg is released into the **fallopian tube**. The egg is pulled into the **infundibulum** of the fallopian tube by the **fimbriae** where it remains viable for 24 to 48 hours; if no fertilization occurs in that time period,

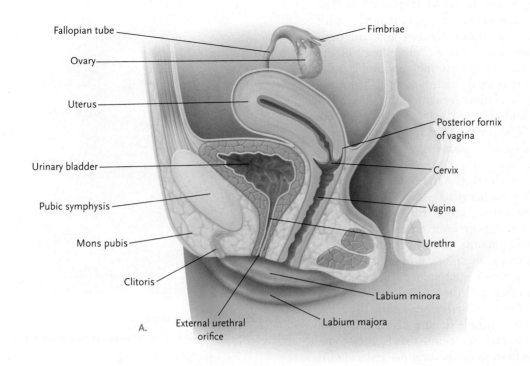

Fallopian tube

Ovary

Uterus

Urinary bladder

Pubic symphysis

Mons pubis

Clitoris

External urethral orifice

Fimbriae

Posterior fornix of vagina

Cervix

Vagina

Urethra

Labium minora

Labium majora

A.

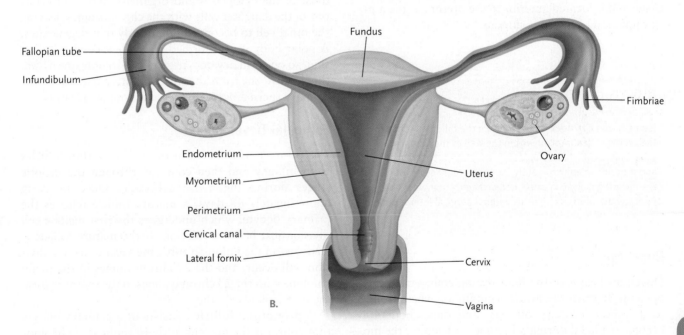

Fallopian tube

Infundibulum

Fundus

Fimbriae

Ovary

Endometrium

Myometrium

Perimetrium

Cervical canal

Lateral fornix

Uterus

Cervix

Vagina

B.

Figure 22.5 Female Reproductive Organs
(A) Lateral view and (B) anterior view

22

the egg will break down. The fallopian tubes extend from the ovaries to the superior edges of the uterus and are lined with ciliated epithelial tissue that helps move the oocyte or zygote through the tubes to the uterus.

If fertilization does occur, the haploid nuclei of the sperm and the egg combine to form the first single diploid cell: the zygote. Successive mitotic divisions occur in the fallopian tube, and after approximately 10 days a hollow ball of cells called a **blastocyst** is formed and can implant in the lining of the **uterus**. The uterus, also called the *womb*, is located between the urinary bladder and the rectum. The larger superior region of the uterus is called the *fundus*, which narrows to the central region called the body, which leads to the inferior opening to the vaginal canal, called the **cervix** (Figure 22.5). The uterus, a thick-walled muscular organ, is made up of three layers: the endometrium, myometrium, and perimetrium. The endometrium is the inner layer of the uterine wall where the blastocyst implants, and it is the mucus membrane that sheds each month, called *menstruation*, when fertilization does not occur. The **myometrium** is the smooth muscle layer that contracts during childbirth. The **perimetrium** is the outer visceral layer composed of simple squamous epithelial and areolar connective tissue.

The area between the thighs and buttocks of males and females is called the **perineum**. The external female genitalia—in combination called the **vulva**—include the **mons pubis**, an adipose layer covering the pubic symphysis; the **labia majora**, which are the outer skin folds; the **labia minora**, the inner skin folds; and the **clitoris**. The labia major and minor surround the opening to the urethra and the vagina; the vagina is posterior to the urethra. The region within the labia minora is called the **vestibule** (Figure 22.6). The labia are analogous to the male scrotum—that is, they come from the same embryonic tissue origin—and contain **sudoriferous glands** (sweat) and sebaceous glands (oil). The clitoris, analogous to the penis, contains erectile tissue and is located anterior to the urethra at the anterior junction of the labia minora.

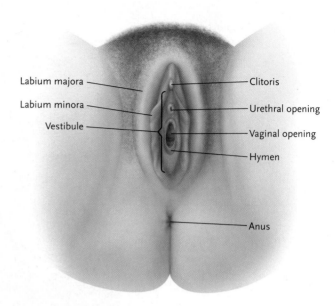

Figure 22.6 Structures of the Vulva
External female genitalia that surround the opening to the vagina

DID YOU KNOW?

If a fertilized egg does not move from the fallopian tubes to the uterus, it can implant within the wall of the fallopian tube. This is called an *ectopic pregnancy* or a *tubal pregnancy*, and it occurs in approximately 5% of all pregnancies. An ectopic pregnancy can cause hemorrhaging that would be fatal to both mother and fetus without surgical intervention.

Oogenesis

Oogenesis begins in the fetus during embryonic development. By birth the ovaries contain anywhere from 2 to 4 million primary oocytes, and only about 400 become ovulated during a lifetime. In females, the diploid germ cells are called *oogonia* and are analogous to the male spermatogonia. Oogonia stop mitotic cell division around the seventh month of gestation and enter the S phase of the cell cycle (Chapter 5) when DNA replicates; at this point they are called **primary oocytes**. Primary oocytes grow during childhood and go through the first meiotic cell division just before ovulation to produce a haploid **secondary oocyte** (Figure 22.7). The second meiotic cell division, to separate the sister chromatids, does not occur unless that secondary oocyte—which develops into a mature follicle right before

ovulation—is fertilized. After the first cell division, most of the cytoplasm and organelles move into only one of the daughter cells with the chromosomes, leaving the other cell to become a **polar body** that degenerates. A polar body is also formed after the second cell division, so one primary oocyte gives rise to only one ovum; this is different from the male counterpart, where one primary spermatocyte gives rise to four spermatids.

Follicular Development

The primary oocytes begin to develop within **follicles** from infancy and then continue through the meiotic stages during the menstrual cycle. Once in every approximately 28 days, a mature follicle releases the primary oocyte, which undergoes the first meiotic cell division just before ovulation. If the mature follicle is fertilized in the fallopian tube, the second meiotic division will occur, and the 23 chromosomes in the ovum combine with the 23 chromosomes in the sperm to form the first diploid cell, the zygote.

A **primordial follicle** consists of a primary oocyte that contains the nucleus with 46 replicated chromosomes (92 total chromosomes). It is surrounded by a layer of glycoproteins, called the **zona pellucida**, and also a single layer of **granulosa cells** (Figure 22.8). As the follicle grows during the follicular phase of the monthly cycle, the oocyte increases in size, the granulosa cells proliferate, and the outer connective tissue cells form the **theca layer**, which generates the preantral follicle. The theca layer continues to increase, and a fluid-filled central cavity begins to form. The cavity is called the *antrum*, and the follicle at this stage is called the **early antral follicle**. The antrum

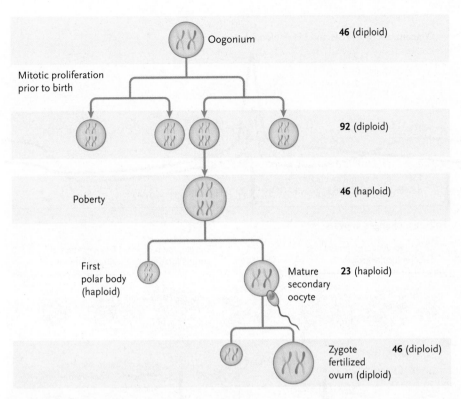

Figure 22.7 Oogenesis

Oogenesis is the production of oocytes through meiosis.

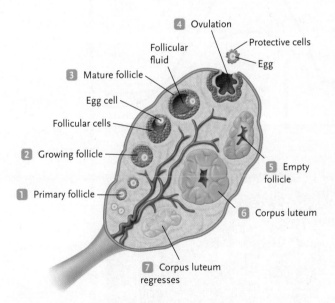

Figure 22.8 Follicular Development

A primary follicle develops into a mature follicle each month, leading to ovulation.

DID YOU KNOW?

There are two types of twins—fraternal and identical. Fraternal twins (also called *dizygotic* because they come from two zygotes) occur when two eggs are ovulated at the same time and are fertilized by different sperm and have different combinations of the parents' chromosomes. Identical twins (also called *monozygotic*) occur when the zygote divides and becomes two separate cells (zygotes), and each then grows into a separate blastocyst that implants and develops. Identical twins have exactly the same DNA.

Hormonal Regulation

The 28-day reproductive cycle (menstrual cycle) can be divided into two phases. The phase just described when the follicle develops is called the *follicular phase*, and the period between ovulation and menses is the luteal phase (Figure 22.9).

The **follicular phase,** which begins on the first day of the cycle, is the first day bleeding begins and the endometrial layer is shed. Early in the cycle, estrogen levels are low; therefore, the hypothalamus stimulates the anterior pituitary gland to release luteinizing hormone (LH) and follicle-stimulating hormone (FSH). LH stimulates the theca cells to produce androgens that diffuse

increases in size, and mitosis of granulosa cells and theca cells continues until a mature follicle is formed. Whichever antral follicle is dominant at the time of ovulation will release the oocyte. Antral follicles that do not release an oocyte regress and undergo apoptosis (programmed cell death).

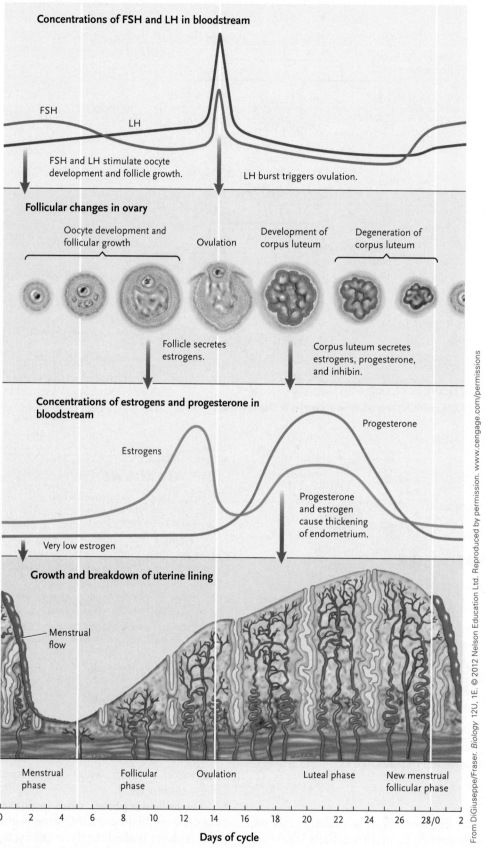

Concentrations of FSH and LH in bloodstream

FSH

LH

FSH and LH stimulate oocyte development and follicle growth.

LH burst triggers ovulation.

Follicular changes in ovary

Oocyte development and follicular growth

Ovulation

Development of corpus luteum

Degeneration of corpus luteum

Follicle secretes estrogens.

Corpus luteum secretes estrogens, progesterone, and inhibin.

Concentrations of estrogens and progesterone in bloodstream

Estrogens

Progesterone

Progesterone and estrogen cause thickening of endometrium.

Very low estrogen

Growth and breakdown of uterine lining

Menstrual flow

Menstrual phase

Follicular phase

Ovulation

Luteal phase

New menstrual follicular phase

Days of cycle

0 2 4 6 8 10 12 14 16 18 20 22 24 26 28/0 2

Figure 22.9 Female Reproductive Cycle

The reproductive cycle has a follicular phase and a luteal phase.

into the granulosa cells, and FSH stimulates the development of the follicle by causing proliferation of the granulosa cells, and they convert androgens into estrogen (Figure 22.8). As follicles develop and granulosa cells proliferate, more estrogen is produced. One follicle becomes the dominant follicle, and then the high level of estrogen has a negative feedback effect on the anterior pituitary gland. This regulation mechanism decreases FSH production, and the non-dominant follicles undergo apoptosis. Estrogen levels reach high levels close to day 14 of the cycle, causing a positive feedback effect on the hypothalamus, which leads to an **LH surge** that causes the ovulation (Figure 22.9).

The **luteal phase** begins when the oocyte is ovulated and the remaining granulosa and theca cells remain within the ovary and transform into a structure called a **corpus luteum**. The corpus luteum functions as a gland, producing the hormones progesterone and some estrogen that support pregnancy if the mature secondary oocyte is fertilized. Progesterone acts on the endometrium to make it suitable for implantation, and it stimulates breast growth. If fertilization occurs, the placenta takes over the role of producing progesterone during the pregnancy. If fertilization does not occur within 24 to 48 hours of ovulation, the ovulated oocyte will break down, and within 10 days the corpus luteum will break down. When fertilization occurs, implantation takes place within 10 days. If the corpus luteum breaks down, the levels of estrogen and progesterone decrease and cause menstruation. The low estrogen levels signal the release of FSH and LH, and a new follicular phase begins.

Granulosa cells have the following important supportive roles in oogenesis:

- They nourish the oocyte; note there is no blood-ovary barrier.

- They produce paracrine-signalling molecules that prevent the oocyte from continuing meiosis. The first meiotic division is stimulated by the LH surge.
- They use androgens produced by theca cells to produce estrogen when stimulated by FSH.
- They produce antral fluid.
- They produce inhibin, which signals the anterior pituitary gland to decrease FSH secretion.

Premenstrual Syndrome (PMS)

The drop in both estrogen and progesterone near the end of the luteal phase can trigger various effects in different women, including irritability, mood changes, increased acne, bloating, fatigue, and food cravings. PMS affects 75% of menstruating women. Symptoms can range from slightly noticeable to very intense, depending on the amount of fluctuation in hormone levels.

DID YOU KNOW?

If fertilization occurs, and the blastocyst implants in the uterine wall (approximately six days after fertilization), the outer blastocyst cells that become the placenta produce a hormone called *human chorionic gonadotropin* (hCG). This hormone stimulates the corpus luteum—which normally degenerates around day 10—to continue to produce progesterone and maintain the endometrium so that implantation and development of the placenta can occur. This hormone is the one measured during a pregnancy test, and it can be detected in the urine as early as the first day a period is missed: approximately 14 days after fertilization occurred. It is also the most likely cause of morning sickness early in pregnancy.

CONCEPT REVIEW 22.3

1. Where does fertilization occur?
2. Describe the three layers of the uterus.
3. What structures make up the vulva?
4. When does meiosis begin in females?
5. When does the first and second meiotic cell division occur?

6. Briefly describe the stages of follicle development.
7. Which hormone causes ovulation?
8. What hormones are produced by the corpus luteum?
9. What causes menstruation?
10. What cell type is stimulated by LH and by FSH?

22.4 DISEASES OF THE REPRODUCTIVE SYSTEM

Endometriosis is involved in up to 45% of cases of female infertility and affects approximately 10% of women of reproductive age. Endometriosis is the growth of endometrial tissue outside the uterus. It can occur on the ovaries, outside of the uterus wall, fallopian tubes, or on other structures in the abdominal cavity. The endometrial tissue breaks down and bleeds during menstruation just like the endometrial tissue inside the uterus, except that it cannot be shed through the vagina. The trapped tissue and blood causes severe pain and inflammation during the menstrual cycle. The cause of endometriosis is not known.

Pelvic inflammatory disease (PID) is an inflammatory condition in females that affects the uterus, fallopian tubes, or uterus. A leading cause of infertility, PID is most often caused by a bacterial infection, such as chlamydia or gonorrhea, but it can also be caused by fungal, viral, or parasitic infection. PID mostly occurs through sexually transmitted infections (STIs), but not all cases. More than 700 000 women are affected by PID in North America each year, and this causes over 100 000 cases of infertility annually. Gonorrhea, the cause of 60% of the cases, is problematic because it can be asymptomatic in women. As a bacterial infection, it can be treated in the acute phase with antibiotics. However, left untreated—since there are no symptoms—gonorrhea can lead to inflammation that requires anti-inflammatory medication to relieve symptoms.

Prostate cancer is the sixth leading cause of death from cancer in males in North America. Prostate cancer can be detected by measuring blood levels of prostate-specific antigen (PSA), which indicates an enlarged prostate that is producing more proteins than normal. The risk of prostate cancer increases with age. However, many men who have prostate cancer do not have symptoms, and they die from something unrelated. Prostate cancer that is left untreated has been shown to spread to other tissues, such as the bones and lymphatic tissues. Prostate cancer can be treated by surgical removal of the prostate, but this frequently leads to erectile dysfunction. Because of the very slow growth of most prostate cancers, surgery may not be required.

Cervical cancer incidence and mortality has decreased in the last 20 years. However, in 2017, it affected approximately 1550 Canadian women, and an estimated 400 women died from this disease. The causes of cervical cancer include infection with human papilloma virus (HPV), giving birth many times, family history, smoking, oral contraceptives, weakened immune system, and history of STIs other than HPV. HPV infection is the most significant risk factor for cervical cancer. Of the more than 100 strains of HPV, approximately 40 are sexually transmitted through skin contact, and only two strains have so far been linked to 70% of cervical cancers. HPV can also cause anal cancer. Cervical cancer can be detected by a Pap test done every one to three years to detect abnormal cells called *dysplasia*. Up to 90% of cervical cancer cells arise from squamous cell dysplasia: called *squamous cell carcinomas*. Treatment of cervical cancer can include cryosurgery, laser surgery, chemotherapy, radiation, or hysterectomy.

CONCEPT REVIEW 22.4

1. What is one of the leading causes of infertility in females?

2. What are the most common causes of pelvic inflammatory disease?

3. What blood test can help to determine a person's risk of prostate cancer?

4. Which type of sexually transmitted infection is highly linked to cervical cancer?

EFFECTS OF AGING

Changes in Males
Males do not undergo as many age-related changes to the reproductive system as females do. The changes that do occur tend to occur gradually, and they include conditions such as a decrease in both testicular mass and testosterone production; the latter can lower sperm production and reduce ability to have an erection. Although men can produce sperm for their whole adult life, the rate and amount of sperm produced can decrease, thereby reducing fertility as they age. It is important to note that the production of testosterone and DHEA (an androgen produced in the adrenal cortex) can be increased with regular, high-intensity exercise; decreased stress, which causes cortisol production that decreases GnRH (Chapter 17); and a healthy diet that includes micronutrients such as zinc and vitamin D.

Changes in Females
Women go through a dramatic reproductive aging process called *menopause*, which occurs when menstrual cycles stop, typically between the ages of 45 and 55. Before menopause occurs, most women experience a phase known as *perimenopause*, where menstrual cycles change, periods can be missed, and periods can have increased or decreased amount of flow. Once menopause occurs the ovaries stop producing estrogen and progesterone and stop producing and releasing eggs, and—after a female has not had a menstrual cycle for one full year—pregnancy can no longer occur. As hormone levels decrease during menopause, several other symptoms can occur, including vaginal dryness, hot flashes, mood changes, trouble sleeping, trouble with short-term memory, decreased breast tissue, decreased libido, and loss of bone density. Symptoms can be treated with hormone replacement therapy, but there are health risks, such as increased risk of stroke, heart attack, and breast cancer. Females also produce the hormone DHEA, which is an androgen produced in the adrenal cortex, and like males, this hormone is produced during exercise. Regular exercise can decrease the severity of menopausal symptoms because DHEA can be converted to estrogen.

CHAPTER REVIEW

Medical Terminology

TERM	DEFINITION	EXAMPLE
amni-	membranous fetal sac	amniocentesis
andr/o	male	androgen
balan/o	glans penis	balanitis
cervic/o	neck, cervix	cervicovaginitis
-cide	killing, agent which kills	spermicide
-clasis, -clast	break, breaking	cranioclasis
colp/o	vagina	colporrhaphy
crypt/o	hidden, concealed	cryptorchism
-cyesis	pregnancy	ovariocyesis
embry/o	embryo	embryopathy
epididym/o	epididymis	epididymectomy
episi/o	vulva	episiostenosis
fet/o	fetus	fetography
galact/o	milk	galactose
genit/o	reproduction	genitourinary
gon/o	genitals	gonorrhea
gonad/o	gonads	gonadogenesis
gravid/o	pregnancy	gravidocardiac
gynec/o	woman/female	gynecology
helc/o	ulcer	helcomenia
hyster/o	uterus	hysterectomy
lact/o	milk	lactorrhea
-lipsis	omit, fail	menolipsis
mamm/o	breast	mammography
mast/o	breast	mastalgia
men/o	menses, menstruation	menorrhagia
nat/o	birth	neonatology
neo-	new	neonatal
nulli-	none	nulliparity
o/o	egg, ovum	oogenesis
obstetr/o	midwife	obstetrician
olig/o	scanty, few, little	oligodendrocyte
oophor/o	ovary	oophorohysterectomy
orch/o, orchi/o, orchid/o	testis	orchidopexy
osche/o	scrotum	oscheoplasty
ov/o, ov/i	egg, ovum	ovicide
ovari/o	ovary	ovariocentesis
-partum	childbirth, labour	postpartum
phall/o	penis	phallodynia

(Continued)

prostat/o	prostate gland	prostatocystotomy
sacchar/o	sugar	monosaccharide
salping/o	fallopian tube	salpingocyesis
semin/i	semen	seminiferous
test/o, testicul/o	testis	testectomy
thel/o	nipple	thelorrhagia
vas/o	vas deferens, vessel	vasovasostomy
venere/o	sexual intercourse	venereologist
vesicul/o	seminal vesicle	vasovesiculitis
viv/i	life, alive	viviparous
vulv/o	vulva	vulvopathy
zo/o	animal	azoospermia

Labelling Activity

Label the following diagrams.

1.   a.

b.

2. a.

b.

22

23 The Immune System

It is our choices that show what we truly are, far more than our abilities.

J. K. Rowling

23.1 OVERVIEW OF THE IMMUNE SYSTEM

The immune system helps the body maintain homeostasis by protecting the body against harmful pathogens. The immune system has mechanisms for recognizing infectious organisms, preventing infection, and killing many different infectious bacteria, viruses, fungi, and parasites. The immune system also develops tolerance for foreign molecules that do not cause harm, such as those in food, and environmental molecules, as in pollen and our intestinal microflora. However, **allergic reactions** occur when the system of tolerance does not develop properly and the immune system reacts to non-harmful molecules. The immune system also builds tolerance for "self" molecules, which means it recognizes the difference between our own cells and those of infectious organisms. During embryonic development and the first few years of life, the immune cells undergo a selection process that destroys cells that would otherwise react against the self-molecules. An autoimmune disease results when the immune system reacts to our own cells. The immune system produces **memory cells** that recognize specific infections. The memory cells remain in the lymphatic tissue, and future exposure to the same microorganism will stimulate a rapid immune response.

Structures of the immune system include the lymphatic vessels, lymph nodes, **thymus**, spleen, tonsils, Peyer's patch of small intestine, and bone marrow (Figure 23.1).

The immune and lymphatic systems perform the following functions in the body:

- Draining excess interstitial fluid from interstitial spaces into lymphatic vessels and then returning it to the bloodstream

- Transporting dietary fats and fat-soluble vitamins from the small intestine to the bloodstream
- Recognizing and killing infectious organisms, including bacteria, viruses, fungi, and parasites
- Recognizing and tolerating our own cells, as well as non-harmful foreign molecules such as food and environmental substances
- Producing immunological memory cells that prevent infection from the same organism in the future

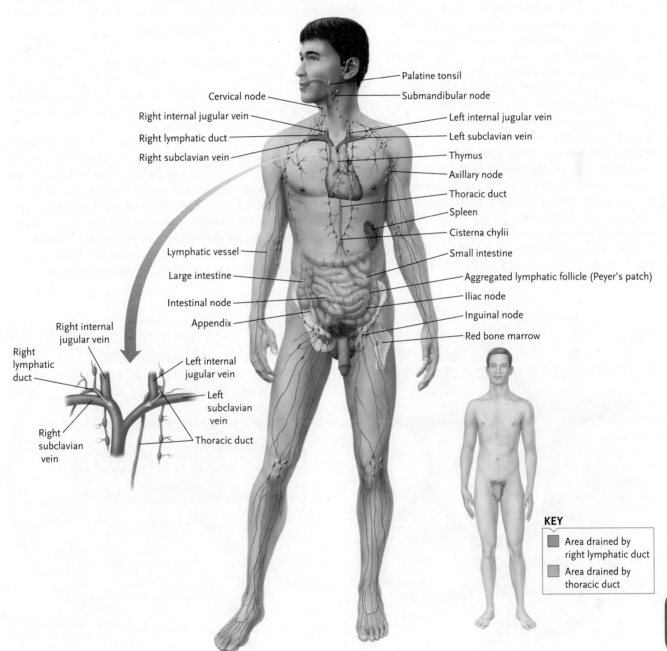

Figure 23.1 Structures of the Immune System

Fluid continually moves from the circulatory system into the lymphatic system and through immune structures such as the lymph nodes and spleen.

1. List the structures of the immune system.

2. What are the functions of the lymphatic and immune system?

3. What are the main types of infectious organisms?

23.2 ANATOMY OF THE LYMPHATIC SYSTEM

Fluid in the bloodstream naturally moves out of capillaries into the interstitial space between cells; this is the interstitial fluid. Interstitial fluid either re-enters the capillary at the venule end or is taken up by lymphatic vessels (Figure 23.2). Lymphatic vessels are similar to capillaries in that they consist of a single layer of endothelial cells. However, lymphatic vessels have closed ends, and fluid is taken up through a one-way opening between the endothelial cells. Fluid does not move out of the vessel because increased fluid volume inside the vessel causes an increase in pressure, which closes the openings between the slightly overlapping endothelial cells (Figure 23.2). Once the interstitial fluid enters the lymphatic vessel, it is called *lymph*. The components of blood plasma, interstitial fluid, and lymph are the same: such as the ions,

vitamins, nutrients, oxygen, CO_2. The only difference in these three fluids is their location. As fluid circulates through blood vessels to the interstitial space and back to the lymphatic system, cells are constantly bathed in water, nutrients, and ions.

Lymph travels through the lymphatic vessels and is returned to the bloodstream through the subclavian veins (Figure 23.1). Lymph moves through lymphatic vessels from the periphery toward the heart. Smaller vessels merge into larger vessels as they progress toward the heart. Lymphatic vessels contain one-way **valves**, as veins do, to prevent the backflow of fluid, and the larger vessels also contain a layer of smooth muscle that contracts and pushes lymph through the vessels, as do the larger blood vessels. In addition, the same mechanisms that help blood flow back to the heart also help lymph flow toward the heart. These mechanisms are the skeletal pump and the respiratory pump. When skeletal muscles contract, they squeeze the lymphatic vessels, thereby altering the pressure

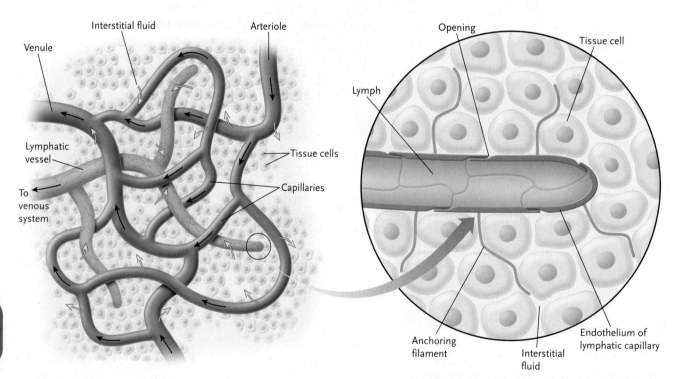

Figure 23.2 Capillary Exchange

Fluid that leaves the capillaries can be taken up by lymphatic vessels.

and causing fluid to move toward the heart. While breathing, inhalation causes the diaphragm to move down, decreasing the pressure within the thoracic cavity, which causes air to move into the lungs, but the lower pressure also causes fluid in lymphatic vessels and blood in the blood vessels to move forward toward the heart. So with each breath, blood and lymph move forward.

The right, upper body lymphatic vessels drain through the right lymphatic duct into the junction of the right internal jugular vein and right subclavian vein to re-enter the circulatory system. Lymph drains from left upper body as well as the right and left lower body through the thoracic duct into the junction of the left internal jugular vein and the left subclavian vein (Figure 23.1).

Fluid accumulation within the interstitial space is called *edema* (Figure 23.2). Many different factors cause edema, including the following:

- Increased **capillary hydrostatic pressure** (hypertension) causes an increase in the amount of fluid, ions, and proteins that leave the vessel and enter the interstitial space, and this increase in proteins or ions causes greater osmotic pressure, which draws more fluid from the capillaries.

- Kidneys do not properly filter or excrete salt and water, as occurs in kidney failure.

- Valve failure, obstruction, and heart failure causes high pressure in the veins and therefore lack of venous return.

- Lack of skeletal muscle pump action due to immobility or paralysis results in more fluid retention in the veins and the lymphatic vessels.

- Medications can affect blood pressure, vascular resistance, the autonomic nervous system, kidney function, and osmotic balance.

- Lack of plasma proteins that help retain fluid in the circulatory system can be caused by wounds, burns, liver disease, or malnutrition.

- Lymphatic vessels can become blocked due to cancer, infection, inflammation, and surgical removal of tissue, such as mastectomy; an inflammatory response also increases capillary permeability.

Structures of the lymphatic system

Bone Marrow Many cell types are derived from stem cells in the bone marrow. The majority of bone marrow consists of **yellow bone marrow**, which primarily stores fats in the medullary cavity. The red bone marrow is located primarily in flat bones, such as the pelvis, scapula, vertebrae, ribs, and cranium, as well as in the epiphyseal ends of long bones. The red bone marrow contains stem cells called **mesenchymal cells** or **stromal cells** that can differentiate into cartilage, osteoblasts, or adipose cells. The **hematopoietic stem cells** of the red bone marrow differentiate into red blood cells, white blood cells, and platelets See Figure 23.3. B cells and T cells are the major cell types involved in the adaptive (also called *acquired*) immune response. B cells become mature cells in the bone marrow and then migrate to the spleen or lymph nodes. Immature T cells migrate to the thymus to differentiate into mature T cells, and then migrate to the spleen or lymph nodes.

Thymus The thymus is located posterior to the sternum and superior to the heart (Figure 23.1). In the thymus, immature T cells—called **thymocytes**—that migrated from the bone marrow go through a maturation process where any cells that recognize self-proteins are destroyed through apoptosis. The thymus also contains macrophages: white blood cells that engulf and remove dead cells through phagocytic activity.

T cells have specific receptors that recognize specific antigens. Every T cell develops a different receptor; theoretically, our entire set of T cells should be able to recognize any possible infectious organism. However, any T cells with a receptor that recognizes one of our own cellular molecules must be eliminated so they do not attack our own tissues. The epithelial cells in the thymus express proteins that are found in other regions of the body, and any thymocytes that bind strongly to these proteins are destroyed; this includes approximately 98% of the T cells that migrate to the thymus. This process of developing a T-cell repertoire that does not attack self-molecules is called *negative selection*. The thymus is most active in newborns and children up to puberty. Then the thymus gradually atrophies, and cells are replaced with fat. Mature T cells leave the thymus and migrate to the spleen and lymph nodes. B cells also have specific receptors and undergo a similar selection process in the bone marrow before they migrate to the spleen and lymph nodes.

Lymph Nodes As lymph travels through the lymphatic vessels, it also passes through many **lymph nodes**. The lymph nodes contain white blood cells that can recognize foreign molecules and initiate an immune response. The greatest number of lymph nodes are located in areas where microorganisms can enter the body, such as the ears, mouth, lungs, digestive tract, urogenital region, and mammary glands (to prevent transfer of infections from mother to infant) (Figure 23.4). Fluid is constantly circulating from the capillaries, to the interstitial space, to the lymphatic system, and back to the circulatory system, ensuring that any pathogen can be detected by the immune system.

Pluripotent stem cell
(Hemocytoblast)

Myeloid stem cell

Lymphoid stem cell

Erythrocyte

Mast cell

Myeloblast

Megakaryocyte

Natural killer cell
(large granular lymphocyte)

Small lymphocyte

Basophil

Neutrophil

Eosinophil

Monocyte

Thrombocytes

T lymphocyte

B lymphocyte

Dendritic cell

Macrophage

Plasma cell

Figure 23.3 Cells of the Immune System

Cells of the immune system originate from pluripotent stem cells in the bone marrow.

Figure 23.4 Lymph Nodes in the Head and Neck Region

Many lymph nodes are located in regions where pathogens more commonly enter the body, such as the mouth, throat, and nose.

Interstitial fluid containing water, ions, proteins, and some types of immune cells enter the lymph vessels and travel to the lymph nodes via the afferent lymphatic vessel. Within the lymph node are many **reticular fibres,** which act as a filtering screen to slow the movement of lymph and to trap particles that should not re-enter the circulatory system (Figure 23.5); these fibres are surrounded by a fibrous capsule. Also within the lymph nodes are the mature B cells and T cells and the phagocytic macrophages and dendritic cells that engulf dead cells and any debris. All T cells and B cells have specific receptors that recognize specific proteins, and as the reticular fibres slow the flow of lymph through the lymph node, a specific immune response—the **adaptive immune response,** also called the *acquired immune response*—may occur. Fluid may flow through multiple lymph nodes and the spleen before it re-enters the circulatory system.

Spleen The spleen is the largest lymphatic organ in the body, located in the superior, lateral abdominal cavity above the stomach and below the diaphragm. The

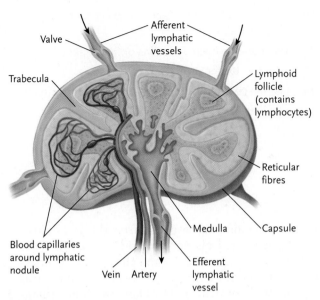

Figure 23.5 Structure of a Lymph Node

Lymph nodes contain immune cells and are located throughout the body.

spleen contains two distinct regions: the red pulp and the white pulp (Figure 23.6). The **red pulp,** which contains macrophages, red blood cells, and platelets, is the location for breakdown of old red blood cells. The liver also breaks down red blood cells. The **white pulp** contains lymphocytes, macrophages, and dendritic cells that are involved in the adaptive immune response. The spleen contains reticular fibres, as do the lymph nodes, and they slow the flow of molecules so that T cells or B cells can initiate an immune response. During embryonic development, the spleen is also involved in hematopoiesis.

Lymphatic Nodules Lymphatic nodules are very similar to lymph nodes, except they are not surrounded by a capsule like the lymph nodes. Lymphatic nodules are found throughout the body and are in higher numbers in regions where infectious organisms can enter the body. Some examples of lymphatic nodules include the **tonsils,** located in the mouth region, and **Peyer's patches,** located around the small intestines. Tonsils, adenoids, appendix, and Peyer's patches are also called the **gut-associated lymphoid tissue (GALT).**

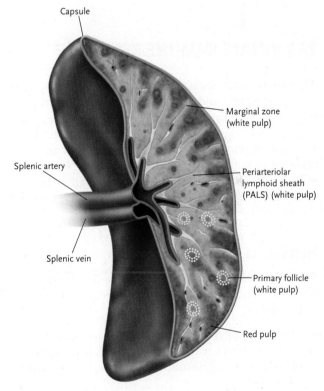

Figure 23.6 Structure of the Spleen

The spleen contains red pulp and white pulp.

CONCEPT REVIEW 23.2

1. Describe the difference between a blood capillary and a lymphatic vessel.

2. How are plasma, interstitial fluid, and lymph the same or different?

3. Where does lymph from the right and left sides of the body re-enter the circulatory system?

4. List reasons why edema can occur.

CONTINUED

5. Where are new red and white blood cells produced? From what type of stem cell?

6. What cell types are produced from mesenchymal cells?

7. What is the function of the thymus?

8. What is the role of reticular fibres in the lymph node?

9. What major cell types are found in lymph nodes?

10. What type of immune response can occur in the lymph node?

11. What are the major functions of the spleen?

12. Give two examples of lymphatic nodules.

13. List the cells of the immune system and their main function.

Application

As you read through the rest of this chapter, consider this scenario. Harold has been exposed to the common cold virus and is starting to show symptoms, including runny nose, sore throat, and coughing. Since the virus has gone past his first line of defence, his innate and acquired immune response must be activated. List the series of events that take place in the immune response after the virus particles have entered his upper respiratory cells. Once his immune system attacks and eliminates the infection, can Harold become sick with the same virus again? Why or why not? What are some things Harold can do to increase his ability to fight off colds in the future?

23.3 INNATE IMMUNE RESPONSE

The **innate immune response** includes mechanisms that prevent infection and non-specific cellular responses to infectious organisms. It does not provide long-term protection like the adaptive immune response. Cells involved in innate immune responses trigger the cells involved in the adaptive immune response. Note that it will be important for you to continually refer to Table 23.1, which lists each type of immune cell and describes their functions.

First Line of Defence—Prevention

The skin, certain epithelial tissues, and secretions are the most important protective barriers to prevent infection:

- The skin, a physical barrier covered with beneficial microorganisms (called *normal flora*), helps fight off infectious organisms, for example, by producing lactic acid.

- Hairs and mucus in the nose block large particles, and movement of the hairs can stimulate the sneezing reflex.

- Cilia and mucus in the upper respiratory tract trap particles and propel them up toward the throat to be swallowed and destroyed by hydrochloric acid in the stomach, or removed by coughing.

- Antimicrobial chemicals such as **lysozyme** are secreted into tears, saliva, and sweat and break down bacterial cell walls.

- Sebaceous glands produce acidic oil that prevents the growth of certain bacteria.

- The stomach produces hydrochloric acid, a very strong acid that kills many organisms taken in with food.

- Resident beneficial bacteria in the digestive tract play an important role in preventing infection.

- The flushing actions of urine, tears, and saliva remove microorganisms from the urogenital region, eyes, and mouth, respectively.

Second Line of Defence—Innate Response

If microorganisms pass through the body's initial physical barriers and their numbers increase in the body, the cells of the innate immune system become activated. Infected cells secrete signalling molecules that alert nearby immune cells by a process called **chemotaxis**. The white blood cells (WBCs) are attracted by the increasing concentration of chemicals in the vicinity of the infection or damaged tissue. **Cytokines** are a large group of chemicals secreted by a variety of cells— including immune cells, epithelial cells, and connective tissue cells—and they have numerous effects. Cytokines mostly function as **autocrine** (act on the same cell that produced it) or **paracrine** (act on nearby cells) signalling molecules. They can act as chemo-attractants, stimulate proliferation or differentiation of other cells types, or cause cells to secrete other cytokines. Often during an immune response, cytokines released by one cell stimulate a cascade of reactions by many other cells, and these result in a positive feedback loop that increases immune cell stimulation until the infection is

TABLE 23.1

Immune System Cell Types and Functions

Cell Type	Function
Neutrophil	Most numerous of all white blood cells, and predominant cell type in pus First-responders that produce chemicals involved in the inflammatory response, such as histamine, vasodilators, reactive oxygen species, and cytokines Phagocytic cells that engulf many types of microbes, mostly bacteria and opsonized pathogens (have antibodies from B-cell response) Migrates from blood vessels to infection area within tissues; attracted to damaged cells by process of chemotaxis (migration of immune cells toward an area of infection)
Basophil	Least common type of white blood cell Releases histamine and several cytokines involved in the inflammatory response Plays a role in allergic reactions and parasitic infections, specifically, ectoparasites such as ticks or lice
Eosinophil	Involved in parasitic infections such as helminths and worms Produces many chemicals, including histamine that is involved in allergies, asthma, and inflammation
Mast cell	Primarily involved in allergic reactions and anaphylaxis by releasing histamine if interaction with IgE antibodies occurs Plays an important protective role in wound healing Prevalent in tissues near openings to the body, such as mouth, nose, skin, lungs, and GI tract
Monocyte	Circulates in the blood and responds to inflammatory signals, migrates into tissues, and differentiates into macrophages or dendritic cells
Dendritic cell	Important antigen-presenting cells (APCs) that stimulate lymphocytes (B cells and T cells) in the spleen, lymph nodes, or lymphoid nodules and initiate the adaptive immune response Phagocytic cells that can engulf pathogens or opsonized pathogens (when antibodies or complement proteins are bound to the pathogen) Present in tissues that are in contact with the environment, such as the skin (Langerhans cells), lungs, and GI tract
Macrophage	Important APCs that stimulate lymphocytes in the lymphatic tissues and initiate the adaptive immune response Phagocytic cells that can engulf pathogens or opsonized pathogens (when antibodies or complement proteins are bound to the pathogen) Secretes toxic chemicals that directly kill invading organisms Secretes cytokines and chemicals involved in the inflammatory response
Natural killer cell	Specifically kills virus-infected cells and cancer cells by lysis with perforin or induction of apoptosis Kills cells opsonized with antibodies Can produce antigen-specific memory cells important for detecting same infection a second time
B cell	Contains a specific receptor for a specific antigen that is a membrane-bound immunoglobulin (antibody) Circulates between blood and lymphatic tissues Phagocytic when antigen matches receptor Antigen-presenting cell Produces cytokines Differentiates into plasma cells when activated by helper T cell and then secretes antibodies Can produce memory cells
Helper T cell	Do not directly kill any invading microorganisms Can only respond to antigen presented by APC (macrophage, dendritic cell, or B cell) Produces cytokines that activate B cells, cytotoxic T cells, macrophages, dendritic cells, and NK cells Can produce memory cells
Cytotoxic T cell	Directly kills virus-infected cells, cancer cells, and transplant tissue Binds to antigen presentation on MHC class I molecules Can produce memory cells
Regulatory T cell	Inhibits other immune cells at the end of an infection and prevents auto-reactive immune cells—the cells that escaped the negative selection process in the bone marrow or thymus—from damaging tissues

23

controlled. The whole system of cytokine function is complex, so this section discusses their function in general and a few cytokines specifically.

The first cells to arrive at an area of infection are most often neutrophils that migrate from the bloodstream into the tissues. These cells are followed closely by other immune cells, such as monocytes, dendritic cells, macrophages, and natural killer cells. Neutrophils secrete cytokines that attract other immune cells, secrete inflammatory molecules that increase vascular permeability, and engulf bacteria by phagocytosis. Monocytes attracted to the area also engulf and kill invading organisms, and they differentiate into phagocytic macrophages and **dendritic cells**. Macrophages and dendritic cells engulf any infectious organisms that are extracellular, including dead cells and cell debris. Once the adaptive immune response has been activated, phagocytic immune cells engulf cells that have antibodies bound to them.

Natural killer cells target virus-infected cells and cancer cells by puncturing the cell membrane with a molecule called **perforin**. Once perforin forms a hole in the membrane, interstitial fluid moves into the cell and causes it to burst (Figure 23.7). Then macrophages and dendritic cells engulf the dead cell.

Some cells infected with a virus produce a signalling molecule called **interferon** (**INF**). Interferon binds to nearby uninfected cells, triggering the production of antiviral proteins that degrade RNA (Figure 23.8). Viruses replicate inside cells by using

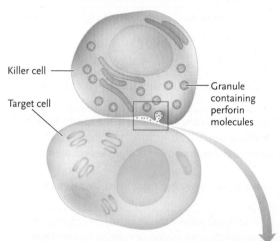

Killer cell

Target cell

Granule containing perforin molecules

1 The killer cell binds to its target.

2 As a result of binding, the killer cell's perforin-containing granules fuse with the plasma membrane.

3 Granules disgorge their perforin by exocytosis into a small pocket of intercellular space between the killer cell and its target.

4 On exposure to Ca^{2+} in this space, individual perforin molecules change from a spherical to a cylindrical shape.

5 Remodelled perforin molecules bind to the target cell membrane and insert into it.

6 Individual perforin molecules group together like staves of a barrel to form pores.

7 The pores admit salt and H_2O, causing the target cell to swell and burst.

Granule

Salt, H_2O

Ca^{2+}

Perforin molecule

Killer cell

Target cell

Perforin molecule

A. Details of the killing process for cytotoxic T cells and NK cells

B. Enlargement of perforin-formed pores in a target cell

Figure 23.7 Perforin

Perforin is a secreted by natural killer cells to kill virus-infected cells or cancer cells.

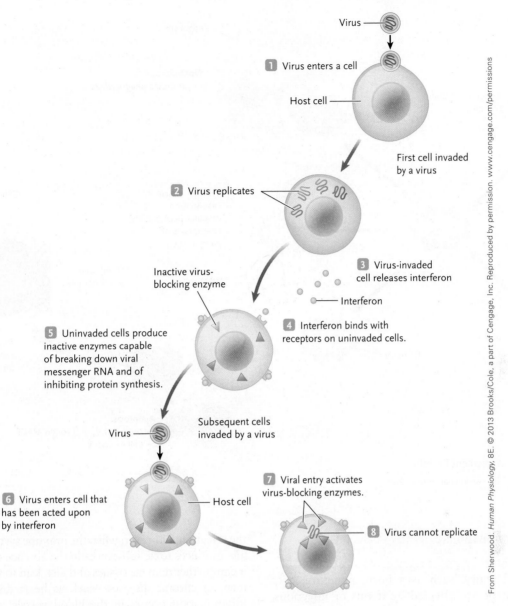

Virus

1 Virus enters a cell

Host cell

First cell invaded
by a virus

2 Virus replicates

Inactive virus-
blocking enzyme

3 Virus-invaded
cell releases interferon

Interferon

5 Uninvaded cells produce
inactive enzymes capable
of breaking down viral
messenger RNA and of
inhibiting protein synthesis.

4 Interferon binds with
receptors on uninvaded cells.

Subsequent cells
invaded by a virus

Virus

7 Viral entry activates
virus-blocking enzymes.

6 Virus enters cell that
has been acted upon
by interferon

Host cell

8 Virus cannot replicate

Figure 23.8   Interferon

Interferon acts as an alarm system to warn nearby cells of a viral infection.

From Sherwood. *Human Physiology*, 8E. © 2013 Brooks/Cole, a part of Cengage, Inc. Reproduced by permission. www.cengage.com/permissions

the host cell's organelles, building blocks, and ATP; however, if all RNA is degraded the virus cannot replicate. Unfortunately, this causes a temporary loss of normal cell transcription and translation. Interferons can also stimulate apoptosis by means of the p53 pathway (see Chapters 5 and 9) in cells that are already infected.

The liver produces complement proteins that circulate in the blood in an inactive form until an infection occurs. The complement cascade can be activated either directly by bacteria or **fungi** (extracellular organisms), or indirectly by any microbe that has antibodies bound to it (Figure 23.9). This activation of complement

proteins causes a cascade of reactions that lead to three main events:

1. Complement proteins bind to microbes and increase phagocytosis.

2. Complement proteins stimulate mast cells to produce histamine, which causes vasodilation, increases vascular permeability, and attracts phagocytic cells (chemotaxis).

3. Complement proteins combine to form a **membrane attack complex (MAC)** that kills bacteria and fungi, similar to how perforin kills virus-infected cells or cancer cells.

23

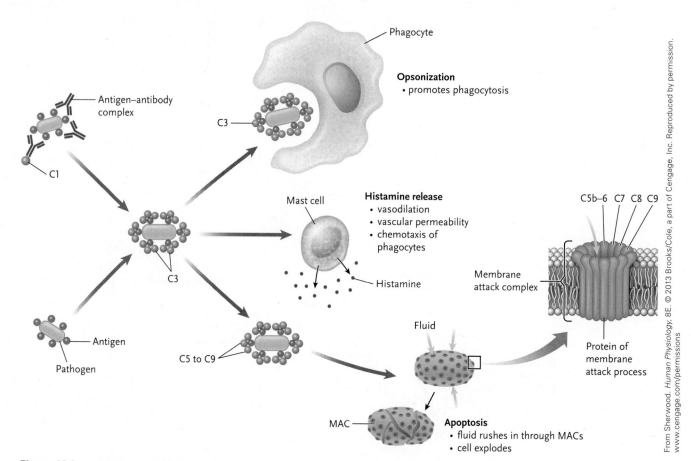

From Sherwood. *Human Physiology*, 8E. © 2013 Brooks/Cole, a part of Cengage, Inc. Reproduced by permission. www.cengage.com/permissions

Figure 23.9 Complement Proteins
Complement proteins can kill bacteria and fungi.

Inflammatory Response

The **inflammatory response** is triggered any time tissues have physical injury, such as a burn, puncture, cut, sprain, break, or are affected by toxins or infectious organisms. Tissue-specific inflammation is often named by the tissue that is undergoing the inflammation: for example, **tendonitis** is inflammation in a tendon, and **appendicitis** is inflammation of the appendix. The function of inflammation is to remove infectious organisms and heal damaged tissue. The classic signs of inflammation are redness, swelling, pain, and heat, which are due to molecules released by cells in the damaged tissue.

Acute inflammation is a healthy response to an injury or infection in the body. Acute inflammation has a rapid onset and includes localized redness, swelling, and pain. The immune system repairs the injury or kills the infection and promotes the regrowth of new healthy cells and stops when the injury or infection is healed.

Chronic inflammation occurs when the immune system continually acts on tissues, producing cytokines that damage tissues over a longer period of time. When fighting infections, the immune system releases toxic substances, such as free radicals, to kill the pathogen. This is a healthy immune response when it is short term and kills the infection. However, when the immune system continually produces toxic substances in the absence of infection, it can further damage tissues and also lead to the development of chronic diseases—such as heart disease, when inflammation occurs in the blood vessels, or arthritis, when inflammation occurs in the joints.

DID YOU KNOW?

Chronic inflammation is implicated in many conditions, including depression, Alzheimer's disease, cancer, diabetes, and an increase in the rate of aging. Many factors contribute to chronic inflammation, but most importantly it is due to consumption of pro-inflammatory foods, such as foods that contain trans fat or sugar and those that are refined or processed.

The primary cells involved in initiating inflammation are cells already in the area that has been damaged; these include phagocytes, basophils, mast cells, and platelets. The main molecules released by these cells are histamine, cytokines, and eicosanoids. Histamine causes vasodilation to increase blood flow to the region, and it increases **vascular permeability** so that more fluid, proteins, and immune cells enter the tissues. Cytokines signal and activate other immune cells. **Eicosanoids** are cytokines produced from the essential fatty acids omega 3 and omega 6, which we need in our diet. Omega 6 fatty acids promote the production of pro-inflammatory eicosanoids, and omega 3 fatty acids promote the production of anti-inflammatory ones; therefore, it is important to ensure a balance of these fats in our diet. One of the major types of eicosanoids is **prostaglandin,** which has the same functions as histamine and also signals pain sensations. Prostaglandin production is inhibited by drugs, such as aspirin and ibuprofen, that decrease the pain and swelling from inflammation.

If an injury has damaged blood vessels, platelets convert the fibrinogen protein that is always circulating in the blood into *fibrin* threads to clot blood, and it also traps pathogens in the tissue to prevent them from spreading and infecting nearby tissue. The combined effects of the chemicals produced during inflammation cause the following classic signs of an inflammatory response (Figure 23.10):

- Redness from increased blood flow
- Swelling from increased vascular permeability and therefore edema, because fluid moves into the interstitial space

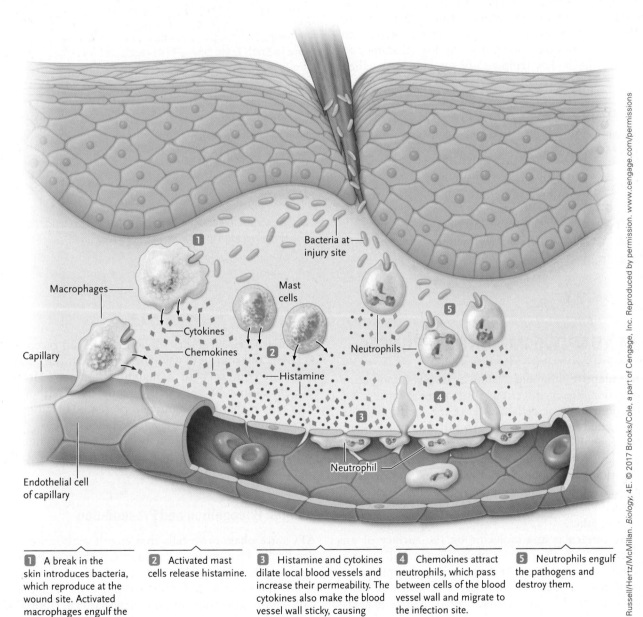

1 A break in the skin introduces bacteria, which reproduce at the wound site. Activated macrophages engulf the pathogens and secrete cytokines and chemokines.

2 Activated mast cells release histamine.

3 Histamine and cytokines dilate local blood vessels and increase their permeability. The cytokines also make the blood vessel wall sticky, causing neutrophils to attach.

4 Chemokines attract neutrophils, which pass between cells of the blood vessel wall and migrate to the infection site.

5 Neutrophils engulf the pathogens and destroy them.

Figure 23.10 Inflammatory Response

The inflammatory response is an important process for killing infections.

From Russell/Hertz/McMillan. *Biology,* 4E. © 2017 Brooks/Cole, a part of Cengage, Inc. Reproduced by permission. www.cengage.com/permissions

- Pain from prostaglandins, due to swelling that puts pressure on pain-sensing neurons, and from direct injury to neurons
- Heat from increased blood flow as well as increased heat production by immune cells producing ATP

Temperature Response

In some cases, an infection causes a fever to develop. A **fever** is a change in the set point of our normal body temperature of 37°C. The body's temperature set point is determined by the hypothalamus. An increase in body temperature by one or two degrees will increase the rate of chemical reactions and inhibit microorganisms that are adapted to reproduce at 37°C, giving our immune system an advantage. However, a fever of 42°C can be fatal because the body's proteins can denature. Molecules that cause the hypothalamus to increase the body's temperature set point are called

pyrogens; these can be cytokines or prostaglandins produced by immune cells, or toxins produced by the infecting organism. When a fever reaches 40°C an **antipyretic**—a medication that reduces fever—should be taken.

Phagocytic cells continue to clean up old dead cells and pathogens, and the damaged tissue heals as new cells are formed through mitosis.

DID YOU KNOW?

People tend to have lots of omega 6 fatty acids in their diet and not enough omega 3. Therefore, many people have chronic inflammatory conditions because they are not getting enough omega 3 fatty acids. Omega 3 fatty acids are high in flaxseeds, fish, and fish oils.

CONCEPT REVIEW 23.3

1. Describe the mechanisms involved in the first line of defence.
2. Which immune cells are phagocytic?
3. Which immune cell will target virus-infected cells?
4. Describe the function of the following chemicals involved in the innate immune response: lysozyme, perforin, interferon, histamine, complement, and prostaglandins.
5. Describe the causes of redness, pain, swelling, and heat involved in an inflammatory response.
6. What are pyrogens?

23.4 ADAPTIVE IMMUNE RESPONSE

The adaptive immune response involves long-term protection from infectious organisms. During an infection, immune cells can leave the cardiovascular system and enter the tissues to kill invading organisms, signal other immune cells, or carry pathogen particles to the lymph nodes, which stimulates the activation of **lymphocytes** involved in the adaptive immune response. The lymphocytes are the B cells and T cells. The innate and adaptive immune responses work together.

An **antigen** is any molecule on the surface of an infectious organism that can be recognized by any immune cell. Antigens can be proteins, glycoproteins, lipoproteins, lipopolysaccharides, or saccharides. Cells of the innate immune system can recognize general molecular features associated with pathogens, whereas T cells and B cells have specific receptors that only recognize specific antigens. There are millions of

possible B-cell receptors (BCRs) and T-cell receptors (TCRs), and some will never be activated if they never come in contact with a matching antigen. Helper T cells cannot be activated directly by a pathogen; they only recognize the antigens presented by **antigen-presenting cells (APCs)**, such as macrophages, dendritic cells, or B cells. B cells can directly recognize antigens; however, further activation by a T cells increases their response.

Antigen Recognition and Presentation

APCs are phagocytic cells that engulf pathogens and dead cells. Phagocytic cells identify pathogens by recognizing that their general surface pattern of molecules is not the same as any found in our cells. These are called **pathogen-associated molecule patterns (PAMPs)**, and they bind to receptors on the phagocyte: the **pattern recognition receptors (PRRs)**. Binding of a PAMP to a PRR stimulates the phagocytic cells to secrete

1 Phagocytosis of microbe

2 Engulfed microbe is digested into fragments

3 MHC II molecules are synthesized and packaged into a vesicle.

4 MHC II-containing vesicles and antigen fragment-containing vesicles fuse.

5 MHC II and antigen fragments bind.

6 Antigen–MHC II complexes insert into plasma membrane.

Microbe   Antigen

MHC II molecule

Dentritic cell

From Russell/Hertz/McMillan/Fenton/Addy/Maxwell/Haffie/Milsom. *Biology*, 2E. © 2013 Nelson Education Ltd. Reproduced by permission. www.cengage.com/permissions

Figure 23.11 Antigen Presentation

Antigens are presented on MHC molecules on the surface of cells.

cytokines that act as autocrine and paracrine signalling molecules, and then the pathogen is engulfed. Once inside the phagocytic cell, lysosomes digest the microbe into many fragments. Those fragments are attached to major histocompatibility (**MHC class II**) molecules and then brought to the surface of the membrane (Figure 23.11). All APCs are phagocytic, engulf microbes in the tissues where the infection originates, present antigen on their cell surface on MHC II molecules, and then circulate to the lymphatic tissues. It is this presented antigen that is recognized by helper T cells in the lymph nodes or spleen.

MHC molecules are encoded by the largest family of genes in humans. The expression of these genes that produce the receptor is both codominant (both maternal and paternal genes are involved) and polygenic (more than one gene are involved). Every person has their own MHC molecules that their immune system recognizes. It is the particular combination of MHC and antigen that tells the T cells to react, or not.

All nucleated cells in the body, including the immune APCs, express **MHC class I** molecules. Since mature red blood cells do not have a nucleus, they do not express MHC. Cells continually present self- or foreign antigens on the surface of the cell. This is a constant process of presenting molecular fragments and then breaking them down and presenting other fragments.

In cells infected with a virus or bacterial toxin, the MHCI molecules present microbial fragments that cytotoxic T cells will recognize and destroy. By contrast, normal cells that express self-proteins will not be harmed by T cells: due to the selection process that occurred in the thymus in early life. When someone has a tissue transplant, the MHC I molecules on the transplanted tissue will be different from their own MHC I molecules and, therefore, cytotoxic T cells will attack the new tissue. This explains why it is important to receive a transplant from a closely related person who has very similar MHC molecules. See Table 23.2 for a summary of MHC molecules.

TABLE 23.2

MHC I and MHC II Expression in and Activation of Specific Cell Types

	Which Cells Express It	Which Cells React/ Respond to It
MHC class I	All body cells, except red blood cells	Cytotoxic T cells, when these cells are infected with a virus
MHC class II	Antigen-presenting cells	Helper T cells that are activated by APCs

23

The adaptive immune response can be divided into two different pathways: the **cell-mediated immune response** that uses T cells, and the **humoral immune response** that uses B cells and antibodies.

Cell-Mediated Immune Response

T cells can be **helper T cells** or **cytotoxic T cells**. Helper T cells do not directly kill invading organisms but do produce many important cytokines that direct the other immune cells. Cytotoxic T cells directly kill cells infected with a virus, cancer cells, and transplant tissue that has foreign MHC I molecules.

Helper T cells recognize only the antigens presented by APCs on MHC II molecules. Helper T cells also have a coreceptor called *CD4* and are, therefore, sometimes called **CD4+ T cells**. HIV (**human immunodeficiency virus**) infects cells that have a CD4 receptor. When an APC engulfs a pathogen in the tissues, which could be a bacterium, virus, fungus, or small parasite, it digests the pathogen and then presents some of the foreign antigen on its cell-surface MHC II molecules. The APC then travels from the tissue to the nearest lymphatic vessel and travels with the lymph through the vessels and lymph nodes until it comes in contact with a helper T cell that has a matching T-cell receptor (TCR) (Figure 23.12). The binding of the APC–antigen–MHC II complex to the TCR–CD4 stimulates the APC to produce the cytokine **interleukin-1 (IL-1)**,

which acts as a paracrine signal on the helper T cell and stimulates the helper T cell to secrete **interleukin-2 (IL-2)**, which acts as an autocrine signal that causes that T cell with that specific TCR to proliferate.

Helper T cells can differentiate into different subtypes based on which cytokines are produced by the APC. Infections that are mostly **intracellular**, such as viruses or bacterial toxins, require cytotoxic T cells to kill infected host cells. **Extracellular** organisms, such as bacteria, fungi, parasites and viruses when they are outside host cells, require antibodies and other extracellular mechanisms such as complement proteins and phagocytosis.

Helper T cells can stimulate one pathway or another by differentiating into subtypes of helper T cells (Figure 23.12):

1. **Th1** cells increase activation of cytotoxic T cells in the case of viral infection.

2. **Th2** cells increase activation of B cells in the case of a bacterial infection. Differentiation depends on the types of cytokines produced by the antigen-presenting cell.

3. Some helper T cells become memory cells that are capable of fighting the same infection in the future.

4. Depending on the cytokines present in the environment, some helper T cells can become **regulatory**

Figure 23.12 Helper T-Cell Activation

T cells (**Th17**) because of the coreceptor they express. Regulatory T cells slow down the immune response at the end of an infection, and in other cases they prevent immune cells from reacting against self-antigens.

5. Helper T cells can also differentiate into non-reactive, *tolerant* cells for certain antigens called **Th0** cells (Figure 23.12). This happens in the case of a high exposure to an antigen, especially through exposure along the GI tract. Molecules that we are constantly exposed to, such as normal flora, food molecules, and environmental molecules, do not stimulate an immune response, unless this immune system regulation does not occur properly and allergies develop. In animal experiments, oral exposure to a pathogenic antigen can cause immune tolerance, and the immune system will not react to future exposure to that pathogen.

Cytotoxic T-cell activation is a little different. Cytotoxic T cells recognize the foreign antigens presented on the MHC I molecules on normal cells that have been infected with a virus or that have produced cancer proteins. Cytotoxic T cells can migrate between the lymphatic system and the tissues; whereas helper T cells tend to stay in the lymphatic tissue until activated. Cytotoxic T cells have the CD8 coreceptor and are therefore also called **CD8+ T cells**.

Cytotoxic T cells each have a specific TCR that recognizes only one antigen. If a cytotoxic T cell recognizes and binds to an infected cell, it becomes fully activated only if a helper cell produces IL-2 to stimulate the cytotoxic T cell to proliferate. Th1 helper cells produce many cytokines that direct the immune response, but IL-2 is the main cytokine that causes the cytotoxic T cells to proliferate—producing many clones expressing the same TCR that can then kill other cells that express the same foreign antigen on MHC I (Figure 23.13). Once activated, some cytotoxic T cells become memory cells.

Cytotoxic T cells kill infected body cells in two different ways: (1) by producing perforin (as do natural killer cells) and other toxic molecules that enter the cell and cause the cell to die; and (2) by producing molecules that trigger apoptosis in the infected cell. Phagocytic cells engulf and clean up any cell debris. As the infection diminishes, regulatory T cells inhibit the immune response and prevent excessive damage to tissues.

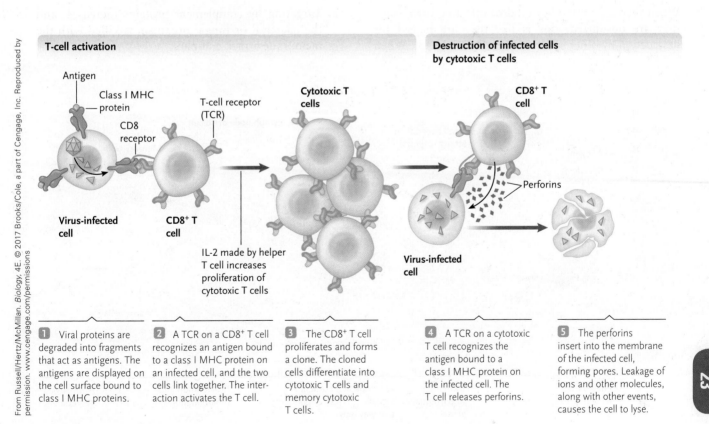

T-cell activation

Antigen

Class I MHC protein

CD8 receptor

T-cell receptor (TCR)

Cytotoxic T cells

Virus-infected cell

CD8+ T cell

IL-2 made by helper T cell increases proliferation of cytotoxic T cells

Destruction of infected cells by cytotoxic T cells

CD8+ T cell

Perforins

Virus-infected cell

1 Viral proteins are degraded into fragments that act as antigens. The antigens are displayed on the cell surface bound to class I MHC proteins.

2 A TCR on a CD8+ T cell recognizes an antigen bound to a class I MHC protein on an infected cell, and the two cells link together. The interaction activates the T cell.

3 The CD8+ T cell proliferates and forms a clone. The cloned cells differentiate into cytotoxic T cells and memory cytotoxic T cells.

4 A TCR on a cytotoxic T cell recognizes the antigen bound to a class I MHC protein on the infected cell. The T cell releases perforins.

5 The perforins insert into the membrane of the infected cell, forming pores. Leakage of ions and other molecules, along with other events, causes the cell to lyse.

Figure 23.13 Cytotoxic T-Cell Activation
Cytotoxic T cells kill virus-infected body cells.

23

Humoral Immune Response

In the same way as T cells, B cells have specific receptors that recognize specific antigens. B-cell receptors are **immunoglobulins** that are attached to the cell membrane. B cells can migrate between the tissues and the lymphatic system and the circulatory system, and when they come in contact with a matching antigen, they can bind directly with that pathogen. B cells are phagocytic cells, and they are also antigen-presenting cells. When an inactivated B cell encounters a matching antigen, it engulfs the organism, digests it, and presents fragments on MHC II molecules. Then the B cell migrates back to the lymphatic system. Upon contact with a helper T cell that recognizes the antigen presented on MHC II molecules, further activation of the B cell occurs (Figure 23.14). Helper T cells that bind with B cells that are presenting an antigen, secrete IL-2 and cause B cells to proliferate, producing many more B cells that express the same receptor. B cells also differentiate into plasma cells that secrete antibodies that match the antigen. Antibodies can circulate throughout the body and bind to the infecting organism; this is called *opsonization*. Some of the B cell population differentiates into memory cells.

Functions of Antibodies

When secreted antibodies bind to infecting organisms, they are **opsonized**; this is like they are tagged for destruction by other immune cells (Figure 23.15). When

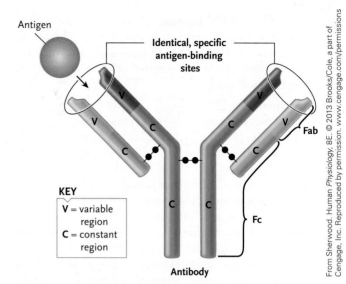

Figure 23.15 Structure of an Antibody

Microbial antigens bind to the variable regions of antibodies

microbes are opsonized with an antibody, the following occurs:

1. Phagocytosis by macrophages and dendritic cells increases.
2. Targeting by complement proteins increases, and the bacterial or fungal organism is killed with the membrane attack complex.

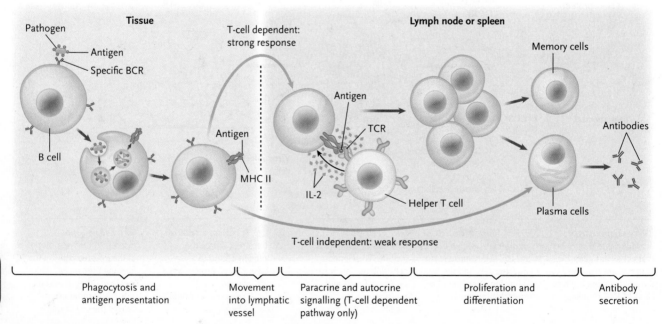

Figure 23.14 B-Cell Activation

B cells can be activated directly by a pathogen and then helper T cells stimulate B cell proliferation

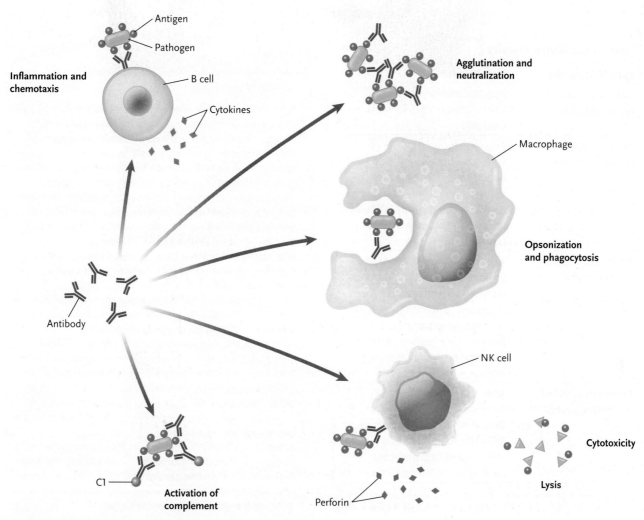

Figure 23.16 Functions of Antibodies

Antibodies are immunoglobulins that identify and neutralize infectious organisms.

3. Microbes become neutralized, so they cannot produce toxins or bind to the body's cells.

4. Bacteria are immobilized, so they cannot spread easily.

5. Agglutination occurs, which means many antibodies and microbes clump together, and so they cannot function (Figure 23.16).

Role of Memory Cells

Memory cells can be formed from the activation of helper T cells, cytotoxic T cells, and B cells. Memory cells are a subpopulation of activated cells with the same receptor that recognizes the antigens from a specific infection. If we were ever infected again with exactly the same or a very similar infection, memory cells would be stimulated to activate, proliferate, and kill the infecting organisms very rapidly, and most often

we wouldn't have any symptoms. This is called a **secondary immune response**. See Table 23.3.

If we cannot be infected by the same organism twice, why do we get sick from the common cold every year? Many viruses, such as those causing the common cold and influenza, change their surface antigens frequently and thereby stimulate our immune system to react to the changes in antigens. Every time the infectious organism changes its surface proteins (microevolution), the cells of the immune system recognize that the antigens are different and respond as if the infection is new. Since some surface proteins stay the same and some change, infectious organisms that have significantly different antigens cause a stronger immune response than organisms that have changed very little. This explains why having a common cold or the flu sometimes causes severe symptoms and other times only minor symptoms. Other factors also affect the severity of symptoms. See the "Did you know box below".

23

TABLE 23.3

Types of Adaptive Immunity

Type of Immunity	How Acquired
Naturally acquired active immunity	Acquired after an infection where memory cells are produced
Naturally acquired passive immunity	Acquired through transfer of antibodies from mother to baby during breast feeding, which temporarily prevents infection in the baby caused by any infection the mother has made antibodies against
Artificially acquired active immunity	Acquired through vaccination—transfer of an antigen into the body from an infectious organism, which stimulates the immune system to produce memory cells related to the infection but without causing disease
Artificially acquired passive immunity	Acquired by transfer of antibodies into the body by injection, which provides temporary immunity to the organism specifically related to the injected antibodies

DID YOU KNOW?

The following factors can alter the effectiveness of an immune response and therefore prolong infection:

- New infections. When organisms that produce infections such as the common cold or influenza change only a few surface antigens from one year to the next, the immune system can fight the infection much more easily than it could if it needed to produce all new memory cells for a completely new infectious organism; this is because memory cells have already been produced for some of the antigens in the previous years.
- Lack of protein, either from malnutrition or lack of available muscle protein that can be broken down into amino acids. The immune system requires a much higher amount of amino acids during illness in order to produce all the cytokines and signalling molecules involved in the innate and adaptive immune responses.
- Lack of sleep, although this connection to immune response is not simple. People suffering from sleep disorders have a significantly increased risk of getting colds and flu.
- Excess iron. Iron supplementation may not be beneficial during an infection since microorganisms require iron for replication and virulence. The liver decreases the level of blood iron during an infection, which decreases the ability of microorganisms to replicate. On the other hand, vitamin A supplementation would be beneficial because viral infections tend to deplete our vitamin A supply.
- Excess stress. Stress causes the release of cortisol, which is normally secreted during infections to act as a negative feedback control system that prevents overstimulation of immune cells. Stress increases cortisol above normal levels and thereby increases the amount of immune inhibition.

CONCEPT REVIEW 23.4

1. What is an antigen?
2. Which cell types are antigen-presenting cells?
3. Why is it important for antigens to be presented on the surface of a cell?
4. Which cell types express MHC I? MHC II?
5. What is the function of IL-1 and IL-2 in activation of helper T cells?
6. What are the various types of cells that helper T cells can differentiate into during activation?
7. What is the main role of each type of helper T cell?
8. Which cells express CD4 and CD8?
9. What is the function of cytotoxic T cells?
10. How are helper T cells activated by B cells?
11. What types of cells can B cells differentiate into?
12. What is opsonization, and how does it help the immune response?
13. What is the difference between the cell-mediated immune response and the humoral immune response?
14. Which cell type is more involved in a viral infection? A bacterial infection?
15. Why are memory cells important?
16. Why do we get colds every year even though we produce memory cells?
17. What is the difference between active and passive acquired immunity?

23.5 VACCINES

As discussed in the previous section, **naturally acquired active immunity** involves the production of memory cells during an infection that can protect the body from future infection from the same organism. **Artificially acquired active immunity** involves the use of **vaccines.** The idea is to try to trigger an immune reaction that causes the production of antibodies so that if the vaccinated person was exposed to that actual infection, they would have immunity similar to a secondary response: meaning, they should be able to fight off the infection without having symptoms.

In 1796, Edward Jenner produced the first vaccine—a vaccine to fight **smallpox** infection. Sometimes one microorganism is similar enough to another one that cross-reactivity can occur, such that the memory cells for one infection protect against another similar infection. This was the case with smallpox (a deadly viral infection) and **cowpox** (a mild viral infection transferred between cows and humans). Edward Jenner determined that people infected with cowpox did not get infected with smallpox. The word *vaccine* that we use today comes from the Latin *vaccinus*, which means "from cows."

Although becoming infected with organisms creates an immune response that produces memory cells that protect us in the future, we might not survive some infections to enjoy the benefits of the memory cells: for example, smallpox, Ebola, tuberculosis, or HIV. It would be ideal to have a vaccine that could stimulate the immune system to produce memory cells without actually causing the disease. It is quite difficult to produce vaccines, and different methods have been used to produce some vaccines, and each has different advantages and disadvantages.

Live Attenuated Vaccines

Viruses can be grown over several generations in a lab until an *attenuated* (less virulent) strain is produced by random DNA mutations. Just as cowpox is less virulent than smallpox, but similar enough to make the memory cells it produces be cross-reactive. A **live attenuated vaccine** stimulates the immune system to mount an adaptive immune response that leads to the production of memory cells, but with very low risk of causing the disease. The vaccine for measles, mumps, and rubella is known as the *MMR vaccine*. The benefits of a live attenuated vaccine is that long-term memory cells are produced even with a very low dose of the vaccine, and immune stimulants called **adjuvant** do not need to be added to the vaccine. The risk with this type of vaccine is possible contamination with other viruses, or the virus may revert back to its virulent form and cause disease. This has not happened with the MMR vaccine;

however, this is not a method of choice for the production of a vaccine for deadly viruses such as HIV or Ebola.

Killed-Virus Vaccines

If the virus used in the vaccine is killed, it cannot cause infection. However, the immune system is difficult to trick because it can recognize the difference in pathogenicity between living and killed organisms. A killed virus still has intact surface antigens that can be recognized by the immune system, and it cannot cause infection. However, an adjuvant needs to be added to the vaccine to increase the immune system's response, and in many cases killed-virus vaccines cannot cause the production of memory cells. They can cause the production of antibodies, but this does not lead to long-term immunity. Killed-virus vaccines have been produced for influenza and cholera, for example.

DID YOU KNOW?

An *adjuvant* is a substance that is added to certain vaccines. It stimulates the immune system to react in order to cause more antibodies to be produced or to activate T cells. Adjuvants can be substances, such as oils (squalene); an irritant, such as toxins (found in tetanus, diphtheria, and pertussis vaccines); aluminum salts; or inflammatory cytokines, such as IL-1 or IL-2. People that have autoimmune diseases or chronic inflammatory conditions may have increased symptoms if vaccinated with vaccines that contain immune-stimulating adjuvants. In some cases, for example, doctors do not recommend the flu vaccine for people who have multiple sclerosis, because the adjuvant can worsen the autoimmune disease symptoms. There have also been concerns raised about the effects of aluminum on brain function.

Toxoids

Some bacteria are pathogenic only when they produce toxins that damage our cells. Vaccines can be made from **inactivated toxins,** so that antigens are present to stimulate the immune system, but bacteria are not present to cause infection. These vaccines are beneficial because they can't cause illness—for approximately 10 years—and they promote the production of memory cells. The toxins themselves act as adjuvant that increases the immune system's reaction to the vaccine. The childhood DPT vaccine for diphtheria, pertussis, and tetanus is an example of this kind of vaccine.

Vaccines Produced Using DNA Technology

New vaccine research is moving into the field of biotechnology to try to produce safer and more-effective

23

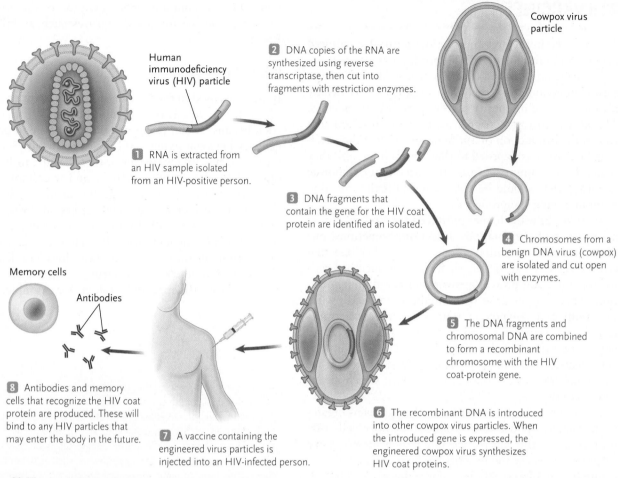

Figure 23.17 Recombinant DNA Vaccines

Recombinant DNA vaccines, produced from genes of the infectious organism, stimulate an immune response without causing an infection.

vaccines against deadly infections such as HIV. Recall from Chapter 10 that genes from one organism can be moved to another, such as moving the insulin gene into bacteria to have bacteria produce human insulin proteins. Likewise, in the lab, virus surface proteins have been expressed in bacteria, yeast, mammalian cells, and low-virulence viruses. These proteins can then be used to make the vaccine without concern that infection from the whole pathogen will result (Figure 23.17). Examples of this type of vaccine include the hepatitis B vaccine and a vaccine against West Nile virus for horses. Many other clinical trials are ongoing.

CONCEPT REVIEW 23.5

1. Describe how Edward Jenner produced the first vaccine.

2. Give three examples of live attenuated vaccines.

3. Give two examples of killed-virus vaccines.

4. How are vaccines produced for diphtheria, pertussis, and tetanus?

5. Explain how DNA technology has been used to produce the hepatitis vaccine.

23.6 AUTOIMMUNE DISEASES AND ALLERGIES

Autoimmune diseases occur when the immune cells attack normal tissues as though they were fighting an infection. During embryonic development and early childhood, our lymphocytes undergo a process called *negative selection* (self-tolerance), where any receptors that recognize self-proteins are eliminated by apoptosis. This process continues into adulthood as we produce new B and T cells from stem cells in the bone marrow, but this occurs at a much lower rate than in early life. Negative selection of B cells occurs in the bone marrow and that for T cells occurs in the thymus. By this process, any lymphocytes that would react with our normal tissues should be eliminated, but this process is not always perfect. Furthermore, there are regulatory T cells that inhibit the immune response and prevent overstimulation of the innate and adaptive immune responses. If the balance of immune regulation is not maintained, overstimulation (chromic inflammation) can occur and damage body tissues.

The exact cause of any autoimmune disease is not known, but there are many theories. One key theory is the "hygiene theory" that states that a lack of exposure to infections as children as well as a lack of exposure to healthy symbiotic bacteria (probiotic foods) increases people's susceptibility to allergies, asthma, and autoimmune diseases. Countries that have higher rates of infections have fewer people with autoimmune diseases, and there is a correlation between a decrease in infection rates over time and an increase in autoimmune diseases (Figure 23.18). The incidence of autoimmune disease has been increasing at an alarming rate in developed countries in the last several decades, and currently one-third of the population of North America lives with an autoimmune disease.

Researchers believe that once a mechanism of action is determined for one of the autoimmune diseases, it will open the door to understanding and effectively treating others. Current treatments for many autoimmune diseases involve **immune-suppression medications** that are not sufficiently effective and do not cure the disease. See Table 23.4 for a list of common autoimmune diseases.

Other Theories of Causes of Autoimmune Diseases

Molecular mimicry. Because the immune system sometimes cross-reacts with very similar antigens (memory cells against cowpox protect against smallpox), it is possible that the immune system will cross-react with a self-tissue after fighting a certain infection.

Improper negative selection during early life. The B and T cells go through a negative selection process in early life to remove cells that react to self-proteins. If this process is not perfect, some self-reactive cells can survive and will move onto the secondary lymphatic tissues that can react with self-proteins.

Genetic factors. Research has shown that some variations of certain MHC genes affect antigen presentation. Many gene mutations have been connected with various autoimmune diseases; for example, some people with Crohn's disease have decreased expression of antimicrobial peptides, such as beta-defensin, and also have decreased expression of the vitamin D receptor.

Lack of vitamin D. People living in equatorial regions with more exposure to UV light have a

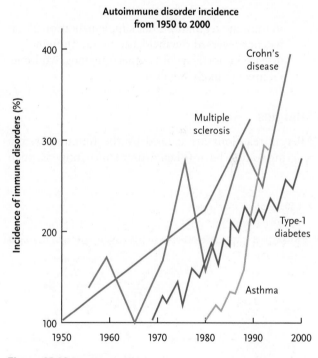

Figure 23.18 Hygiene Theory

Autoimmune diseases are more prevalent in regions that have a low incidence of infections.

Source: Courtesy of Dr. Joel Weinstock

significantly lower incidence of autoimmune diseases than those in northern or southern regions, and this lower incidence also applies to those who lived in equatorial regions in childhood and then moved to northern or southern regions after puberty. Vitamin D does play an important role

TABLE 23.4

Common Autoimmune Diseases

Disease	Tissue Affected
Multiple sclerosis	Myelin on axons in the CNS
Type 1 diabetes	Beta islets cells of pancreas
Celiac disease	Small intestine microvilli in cross-reaction with gluten
Addison's disease	Adrenal cortex
Aplastic anemia	Bone marrow
Crohn's disease	Gastrointestinal tract
Ulcerative colitis	Large intestine
Grave's disease	Thyroid gland (hyperthyroidism)
Hashimoto's disease	Thyroid gland (hypothyroidism)
Lupus	Many tissues
Myasthenia gravis	Skeletal muscles
Rheumatoid arthritis	Joints
Psoriasis	Skin

in immune regulation, but supplementation alone has not reversed the incidence of autoimmune disease in northern or southern regions. Multiple factors are likely involved.

Allergies

Allergic reactions are caused by the immune system reacting to non-harmful environmental antigens, such as pollen, dust, mites, bee venom, or foods. The immune system typically develops tolerance for benign antigens it comes into contact with in very high amounts and on a regular basis, usually early in life. Recall from Section 23.4 that helper T cells can differentiate to become Th1, Th2, Th17, or Th0 helper T cells, depending on the cytokines produced. Th0 cells should recognize but not react to non-harmful substances; when this process fails, allergies and asthma can develop. The exact cause is not known. Allergic reactions are also called *hypersensitivity reactions*. Asthma is considered a form of hypersensitivity reaction due to the release of histamine.

Allergic reactions are primarily humoral immune responses, where B cells are activated to become plasma cells that secrete antibodies against a harmless antigen. The antibody type that is specific to allergic reactions is immunoglobulin E (IgE). B cells circulate throughout the tissues and lymphatic system. The activation of a B cell with a receptor that matches an environmental or food antigen can occur without IL-2 stimulation from a helper T cell (Figure 23.14). The B cell differentiates into a plasma cell that secretes IgE antibodies, which bind to mast cells or basophils (Figure 23.19). During the very first exposure to an antigen, there may not be any symptoms. The second exposure to the same antigen will bind the IgE antibodies already on mast cells or basophils and cause those cells to release large amounts of histamine.

Histamine is the main chemical responsible for allergy symptoms, which include the following:

- Localized vasodilation that causes redness around the eyes and nose

A. Initial exposure to allergen

B. Further exposures to allergen

1 Allergen (antigen) enters the body.

B cell

IgE antibodies

Mast cell

IgE antibodies

Nucleus

Histamine granules

Mitochondrion

Allergen

Histamine release

2 Allergen binds B-cell surface antibodies; the B cell now processes the allergen and, with stimulation by a helper T cell (not shown), proceeds through the steps leading to cell division and antibody production.

3 Activated B-cell clone produces and secretes IgE antibodies active against the allergen.

4 IgE antibodies attach to mast cells in tissues, which have granules containing histamine molecules. Memory B and T cells capable of recognizing the allergen are also produced.

5 After the first exposure, when the allergen enters the body, it binds with IgE antibodies on mast cells; binding stimulates the mast cell to release histamine and other substances.

Figure 23.19 Allergic Reaction

Allergic reactions are primarily humoral immune responses involving IgE, as well as mast cells, basophils, and histamine.

- Increased vascular permeability that causes watery eyes and runny nose
- Itching
- Bronchoconstriction that can cause wheezing

The severity of an allergic reaction depends on how the immune system has been sensitized, the type of allergen, and the mode of entry into body. Very severe reactions can cause increased swelling and broncho-constriction that can be fatal: called **anaphylaxis (anaphylactic shock)**. Generally, the most severe allergic reactions are caused by foods, or venoms such as a bee sting, and less commonly to inhaled antigens. The most common food allergies are to peanuts, eggs, and shellfish.

DID YOU KNOW?

Food allergies have increased by 80% in the last 12 years. The most common food allergies in Canada are shellfish, tree nuts, eggs, milk, fish, soy, and wheat. The most prevalent food allergy is peanut (legume). In the year 2000 North American doctors recommended that children should not have any exposure to allergens such as peanuts until they are three-years-old. The problem is that we produce tolerant T cells very early in life, and to produce tolerant immune cells we need to be exposed to foods containing those allergens. A study in Israel in 2008 showed that feeding small children a peanut puffed snack food prevented peanut allergies, and Israel had the lowest rate of peanut allergies in the world. Now doctors are reassessing their thoughts about waiting to expose children to foods.

CONCEPT REVIEW 23.6

1. Describe how autoimmune diseases cause disease.
2. List the main theories of the causes of autoimmune disease.
3. Where in the world are autoimmune diseases the most prevalent?
4. Describe what tissue is targeted in the following diseases:
   a. type-1 diabetes
   b. multiple sclerosis
   c. myasthenia gravis
   d. Grave's disease
5. What cell types are involved in allergic reactions?
6. Why do no symptoms occur at the time of the very first exposure to a potential environmental antigen?
7. Which type of antigens cause the most severe allergic reactions?
8. What is the main chemical involved in allergic reactions and asthma?

EFFECTS OF AGING

The immune system undergoes a significant decline as we age. People over the age of about 65 to 70 have a significant vulnerability to respiratory infections such as influenza and pneumonia. Other specific things happen to the immune system as we age:

- Hematopoietic stem cells in the bone marrow tend to produce myeloid cells rather than lymphoid cells, thereby causing an overall reduction in the production of mature B cells and T cells.
- The B cells and T cells located in the secondary lymphoid tissues, such as lymph nodes and the spleen, have a reduced ability to proliferate when an infection occurs.
- The thymus becomes increasingly infiltrated with fat cells, and less T-cell maturation occurs.

- When stimulated, T cells tend to differentiate into Th17 (regulatory) rather than the Th1 or Th2 helper cells that are required for effective neutralization of infectious organisms.
- The rate of immune cell apoptosis increases, likely due to increased DNA mutations or exposure to toxins over many years.
- Inflammatory cytokines accumulate more rapidly as we age. Chronic inflammation contributes to tissue damage and is associated with chronic diseases such as cancer, heart disease, and diabetes.
- The production of growth hormone generally decreases with age. Given the evidence that growth hormone stimulates the thymus as well as immune cell proliferation, the age-related reduction in endocrine function contributes to the decline of the immune system. Note that exercise stimulates the production of GH. Some researchers are looking at using exogenous (not made in the body) GH to help stimulate immune function in elderly people.

23

CHAPTER REVIEW

Medical Terminology

Term	Definition	Example
adenoid/o	adenoids	adenoidectomy
anti-	against	antibody
auto-	self	autoimmune
axill/o	armpit	axillary
-coccus	berry-shaped bacterium	staphylococcus
-edema	swelling	lymphedema
-emphraxis	stoppage, obstruction	splenemphraxis
fung/i	fungus, mushroom	fungus
helminth/o	worm	helminthiasis
immun/o	protection, immune	immunogenic
inguin/o	groin	inguinodynia
lien/o	spleen	lienomalacia
lymph/o	lymph	lymphangiophlebitis
myc/o	fungus, mushroom	mycoplasma
nod/o	knot	nodular
nos/o	disease	nosology
-oma	tumour, mass	carcinoma
onc/o	tumour, mass	oncogenesis
-osis	a condition or disease	osteoporosis
path/o	disease	pathogenic
-penia	deficiency	thrombocytopenia
peri-	around	perilymphangitis
-phylaxis	protection	anaphylaxis
ple/o	more	pleocytosis
pyr/o	fire, fever, heat	pyrogen
sarc/o	flesh	lymphosarcoma
-sepsis	putrefaction	antisepsis
splen/o	spleen	splenectasis
tetan/o	tetanus	tetanophilic
thym/o	thymus	thymectomy
tonsill/o	tonsils	tonsillolith
top/o	particular place or area	splenectopy
verm/i	worm	vermiculite
vir/o	virus	virology

23

Labelling Activity

Label the following diagram.

1.

Answer to Application Question

Harold has the common cold virus. Since the virus has evaded his preventative mechanisms, the innate and acquired immune responses are activated.

In the innate response, non-specific immune cells such as natural killer cells can kill virus-infected cells. Macrophages and dendritic cells are phagocytic and are antigen-presenting cells. Viruses also stimulate interferon in the innate response to help prevent virus particles from being able to replicate in nearby cells. Lastly, neutrophils are present in areas of inflammation, which would occur in Harold's upper respiratory tissues as the viruses and immune cells kill epithelial cells.

During the acquired immune response, the antigen-presenting cells circulate to secondary lymphatic tissues such as the spleen and lymph nodes, where they activate helper T cells. APCs secrete IL-1, and helper T cells secrete IL-2 once an antigen–T cell receptor match has been made. This causes the T cells to proliferate and differentiate. Helper T cells then differentiate into Th1 cells, which are the cells that prompt cytotoxic T cells to kill virus-infected cells.

Because all lymphocytes produce memory cells, Harold will not become sick again with the exact same cold virus strain. Instead, a secondary immune response will occur very rapidly and kill the infection before symptoms occur.

Harold can increase the effectiveness of his immune system by attending to diet, exercise, and stress reduction. (See Did You Know? at the end of Section 23.4.)

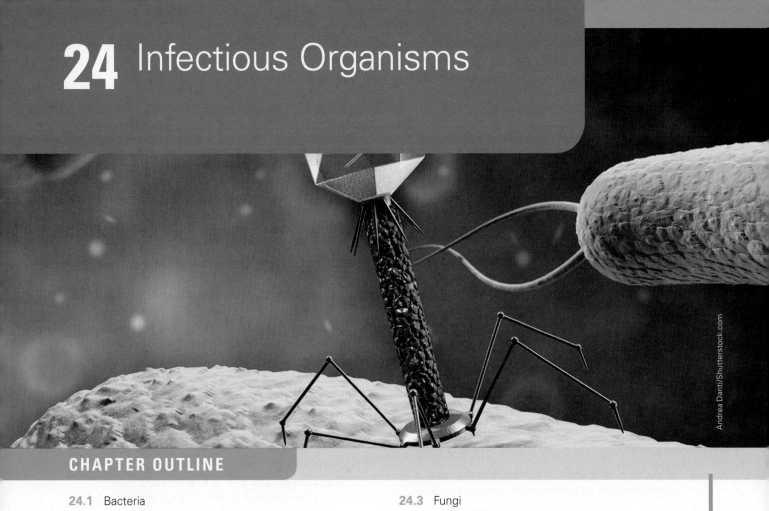

24 Infectious Organisms

Andrea Danti/Shutterstock.com

CHAPTER OUTLINE

24.1 BACTERIA

Bacteria are prokaryotes that make up two of the kingdoms: Archaebacteria and Eubacteria (Chapter 1). Prokaryotes, the first living organisms on earth—beginning approximately 3.5 billion years ago—are still the most abundant life form on the planet. Of all bacterial species, 99% are beneficial and necessary for survival of all living things on earth, and only 1% are disease-causing organisms. Photosynthetic bacteria, called **cyanobacteria**, were responsible for introducing oxygen into Earth's atmosphere and are still a primary source of oxygen, along with plants. Many species of bacteria live in the soil and recycle minerals by breaking down dead, organic, plant material and converting it into inorganic minerals that are used by growing plants, including food crops. Humans have learned to use bacterial organisms to make foods, including fermented vegetables, such as sauerkraut or kimchi, and also sourdough bread, yogurt, and cheese. Eating foods fermented by bacteria is very healthy and contributes to the bacteria population that lives inside our digestive tract and on the surface of our body; these are our normal flora. Without bacteria, no other organisms on earth would be able to survive.

Every great dream begins with a dreamer. Always remember, you have within you the strength, the patience, and the passion to reach for the stars to change the world.

Harriet Tubman

Application

As you read through this chapter and learn about the various bacterial, viral, fungal, and parasitic infections, test your understanding and match the following names (A–Z) with the symptoms or characteristics described below for each infection.

A. MRSA	N. Toxoplasmosis
B. HIV	O. Tapeworm
C. *Aspergillis*	P. *E. coli*
D. Giardia	Q. VRE
E. Malaria	R. Tuberculosis
F. Herpes	S. HPV
G. *Candida albicans*	T. Influenza
H. Rotavirus	U. Pneumocystis pneumonia
I. Chlamydia	V. Hookworm
J. Tetanus	W. Rhinovirus
K. *Salmonella*	X. Epstein–Barr
L. Hepatitis	Y. Strep pyogenes
M. Ringworm	Z. *Campylobacter*

1. Also called *lockjaw*, a bacterial infection that causes muscles to continuously contract _____

2. A bacterial infection that most commonly infects elderly patients on antibiotics and causes diarrhea _____

3. A worm that can infect a person through the skin and can cause iron-deficiency anemia _____

4. A common mould that can grow on bread and fruit and can cause respiratory infection in immunosuppressed people _____

5. A virus that infects the liver and can cause liver cancer or cirrhosis _____

6. A respiratory infection that resembles the common cold, but usually causes fever and muscle aches; resolves in about week in healthy people _____

7. A respiratory disease, prevalent in developing countries, that can kill malnourished people who have the active disease _____

8. A bacterial infection that is multi-resistant; can cause flesh-eating disease, pneumonia, or blood poisoning _____

9. Transmitted by skin-to-skin contact, with outbreaks occurring during periods of stress; infectious organism can "hide" in sensory neurons between outbreaks _____

10. A worm that can be ingested in uncooked meat or fish; commonly causes B12 deficiency _____

11. A yeast infection caused by an overgrowth of yeast, usually due to a lack of healthy normal flora _____

12. A common opportunistic infection in people with HIV _____

13. A viral infection that causes warts _____

14. A bacterial organism that is the most common cause of food poisoning _____

15. A bacterial organism that can cause very severe food poisoning in immunosuppressed people or young children; can be acquired from uncooked chicken or raw eggs _____

16. Food poisoning caused by a bacterial organism that normally resides in the large intestine; can cause illness in people who eat uncooked ground meats or raw foods that have come in contact with this intestinal organism _____

17. A protist infection caused by drinking contaminated water; can cause vomiting and diarrhea for several weeks _____

18. A parasite that can be picked up from animal feces, especially cats, and can cause neurological damage in babies whose mother had been exposed during pregnancy _____

19. A common infection in young children that causes diarrhea and vomiting; usually resolves quickly, but can be fatal in developing countries where children are malnourished or do not have clean drinking water _____

20. A viral infection caused by a retrovirus that infects immune cells and causes immunodeficiency _____

21. A common bacterial infection that is easily sexually transmitted and can cause pelvic inflammatory disease in women (2 answers) _____

22. A bacterial throat infection that causes a painful sore throat, often with fever, and occurs more commonly during periods of stress; may require treatment with antibiotics if not resolved within a week to prevent damage to heart valves _____

23. Virus that infects B cells and causes a range of symptoms from no symptoms to extreme fatigue; 90% of adults infected at some point in their life _____

24. Fungal infection of the skin caused by fungal organism that feeds on keratin protein in the skin; easily transmitted by skin-to-skin contact and resolves with antifungal topical cream. _____

25. A parasitic infection transmitted by mosquitoes that infects red blood cells and liver cells _____

26. A virus that causes common colds _____

24

Normal Flora

In the human body, there are more bacterial cells than human cells, making up approximately 5% of our body weight. Bacteria live on our skin and throughout our digestive tract and play a substantial role in our health. Six square centimetres (or one square inch) of skin contains more than 500 000 microorganisms; these are called *normal flora*. Normal flora contains fungal organisms such as various species of yeast, but bacteria make up the majority of microorganisms. Approximately 1000 species of bacteria exist in our digestive tract, and about 300 of these are the most abundant. The following list describes the most important functions of normal flora (also listed in Chapter 20):

1. Fermenting undigested substrates, mostly soluble fibre
2. Regulating the immune system to tolerate food molecules
3. Preventing growth of harmful bacteria
4. Producing vitamin K and biotin
5. Increasing the growth of intestinal cells and helping maintain a healthy mucosal layer
6. Producing lactic acid during fermentation, which lowers pH and prevents growth of pathogenic species
7. Converting carbohydrates into short-chain fatty acids that can be absorbed and used for energy
8. Releasing important minerals from food, such as magnesium, calcium, zinc, and iron
9. Metabolizing carcinogens, by playing a role in preventing colon, breast, and prostate cancers

Bacterial Cell Structure

Bacterial plasma membranes are similar to the cell membranes of every living organism (Chapter 3); they consist

DID YOU KNOW?

It is very simple to make your own probiotic fermented sauerkraut at home. Thinly slice cabbage, which naturally contains the lactobacillus bacteria that make up a large proportion of our gut bacterial population, and add one tablespoon of sea salt, chopped garlic, one half of a thinly sliced onion, and any other spices you like, such as oregano or chili peppers. Use your hands to mash up the cabbage and spices until the juice of the cabbage is released, then stuff the mixture into a 1 L jar, topping up with tap water. Leave the jar with the lid on loosely on the counter at room temperature for three to four weeks, and allow the bacteria to ferment the cabbage (without oxygen). Then put the jar in the fridge, where the sauerkraut will last for more than a year. The salt is an important ingredient because it prevents the growth of undesirable bacteria. Once the lactobacillus produces lactic acid, no harmful bacteria can grow.

of phospholipids, cholesterol, and integral membrane proteins. Bacterial membranes have additional protection in the form of a cell wall composed of protein and carbohydrates—called *peptidoglycan*. This cell wall is not to be confused with the cellulose structure of plant cell walls. Some bacteria have an extra outer layer on top of the cell wall—called **lipopolysaccharide (LPS)**—that's composed of lipids and sugar molecules. Bacteria that have an LPS layer are called *gram negative* because the outer fatty layer does not absorb crystal violet dye. By contrast, gram positive bacteria have a thick peptidoglycan layer but no LPS layer and, therefore, they do absorb crystal violet dye (Figure 24.1). Although some species of bacteria do not fit into either category, the distinction between gram positive and gram negative helps in determining what is the best

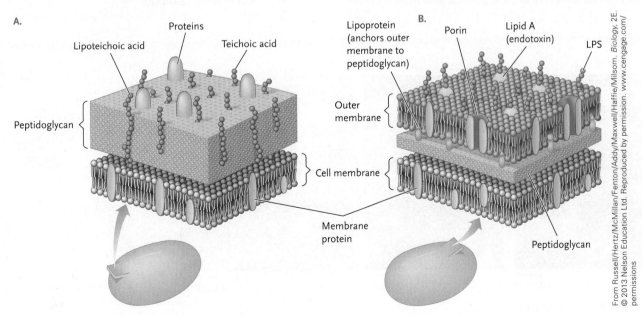

Figure 24.1 (A) Gram Positive and (B) Gram Negative Bacteria
Gram positive and gram negative bacteria have different cell wall structures.

antibiotic to treat a bacterial infection. Gram negative infections are more resistant to treatment with penicillin. Various species of bacteria may have additional structures, such as a protein capsule for protection from environmental stress, flagella for mobility, or pili for transfer of genetic material (Figure 24.2) (see also Chapter 3).

Some bacteria have a mechanism—the formation of **endospores**—that allow them to endure extreme changes in their environment, such as lack of oxygen, extreme heat or freezing temperatures, detergents, radiation, or UV light. An endospore has a core of cytoplasm that contains the DNA and ribosomes; it has a surrounding, protective rigid coat, and it does not undergo metabolism (Figure 24.3). Endospores that are millions of years old have been found. Infectious endospore-forming bacteria include species such as tetanus and anthrax.

Bacteria have one of three general shapes, and they can exist as a single cell or in groups (Figure 24.4).

Spherical bacteria are **cocci**; single spherical bacteria are **monococci**, and pairs are called **diplococci**. When they are grouped together in long chains, they are **streptococci**, and clusters are called **staphylococci**. Strep throat is caused by an overgrowth of streptococci bacteria. The hospital-acquired staph superbug infection is caused by *Staphylococcus aureus*, which is resistant to one of the strongest antibiotics—called *methicillin*—and has become known as **MRSA (methicillin-resistant staphylococcus aureus)**. Many bacteria take on a rod shape, called **bacilli**, and can live as single cells or in long chains called **streptobacilli**. A common intestinal bacterium, *Escherichia coli* (*E. coli*), is rod-shaped. *E. coli* is also present in the intestinal tract of animals and can contaminate food, especially ground beef, which is the last meat pulled off the cow carcass. *E. coli* infection can cause very serious illness and is the reason why all ground beef must be cooked thoroughly to ensure that

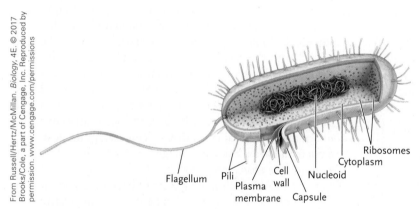

From Russell/Hertz/McMillan. *Biology*, 4E. © 2017 Brooks/Cole, a part of Cengage, Inc. Reproduced by permission. www.cengage.com/permissions

Figure 24.2 Typical Structures Found in Prokaryotic Cells

Prokaryotic cells have a cell membrane, cell wall, and sometimes a capsule.

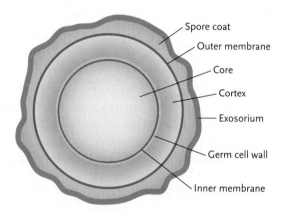

Figure 24.3 Endospores

Endospores allow bacteria to survive extreme environmental conditions.

From Digiuseppe/Fraser. *Biology* 11 U. © 2011 Nelson Education Ltd. Reproduced by permission. www.cengage.com/permissions

Figure 24.4 Shapes of Bacterial Cells

(A) Cocci, (B) bacillus, and (C) spirochete are the most common bacterial cell shapes. (D) Diplococci are a pair of cocci, staphylococci are cocci grouped in clusters, and streptococci are cocci joined in strands.

bacteria are killed. The third common bacterial cell shape is a spiral shape, called **spirochete**: for example, the bacterium that causes syphilis or Lyme disease.

Recall from Chapter 5 that bacteria reproduce asexually by a process called *binary fission*, where the daughter cells produced are identical to the parent cell. However, bacteria have three unique mechanisms for acquiring new DNA that contribute to their ability to have extensive genetic variation and evolutionary success.

1. Conjugation—acquiring DNA directly from another bacterium

2. Transduction—acquiring DNA through viral infection

3. Transformation—acquiring DNA from their surroundings

The quick generation time and genetic variation makes infectious bacteria a constant threat to human health; our immune system must be able to adapt to change equally rapidly (Chapter 23) to prevent the replication of pathogenic organisms inside our body.

Antibiotic Resistance

An example of bacterial evolution that affects human health is the development of antibiotic resistance by some bacteria. Antibiotics are a selection pressure (Chapter 7). They kill many but not all bacteria. The bacteria that have the ability to survive the antibiotic will proliferate, leading to a new population that is resistant to the drug—an example of microevolution. Antibiotics function in two ways: by preventing replication (**bacteriostatic**), or by directly killing the bacteria (**bacteriocidal**).

Since bacteria can acquire new DNA through conjugation, transduction, or transformation, there are countless opportunities for antibiotic resistance genes to be shared between bacteria and **viruses**. Resistance genes most commonly reside on the small extra-chromosomal DNA plasmid within the cell (Figure 24.5). Bacteria have small genomes and keep only the genes that are beneficial to them. The more exposure bacterial populations have to antibiotics, the more important the resistance genes are to them so they can survive exposure to the antibiotics.

Why Antibiotic Resistance Occurs The main cause of antibiotic resistance is the **prophylactic** treatment of food animals. Prophylactic treatment occurs when farmers give a constant low dose of antibiotics to all animals, whether they are sick or not, in attempt to *prevent* infections, which are common when large numbers of animals are housed together in confined areas. This means that all the normal bacteria in the treated animals are constantly exposed to a selection pressure, and they protect themselves by retaining the resistance genes.

In humans, prophylactic treatment, overtreatment, or mistreatment of viral infections through the use of antibiotics exposes the normal flora bacteria to a selection pressure. Antibiotics have no effect on viruses, which

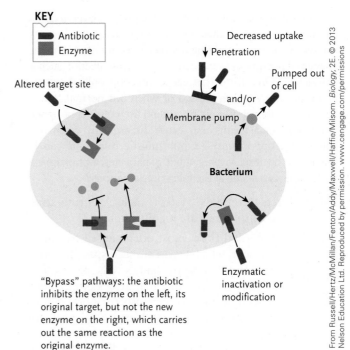

KEY

Antibiotic
Enzyme

Altered target site

Decreased uptake
↓ Penetration

Pumped out
of cell

and/or

Membrane pump

Bacterium

"Bypass" pathways: the antibiotic inhibits the enzyme on the left, its original target, but not the new enzyme on the right, which carries out the same reaction as the original enzyme.

Enzymatic
inactivation or
modification

Figure 24.5 Antibiotic Resistance Genes

Antibiotic resistance genes code for proteins that prevent an antibiotic from killing the bacterium.

lack a cell membrane. So when a person has a cold or flu and takes antibiotics, not only is much of their normal flora bacterial population destroyed—thereby removing the normal healthy, protective, bacterial barrier from their skin and GI tract—but also antibiotic resistance is increased. Resistance genes can be transferred between healthy and **pathogenic bacteria**.

How Resistance Genes Work Resistance genes when transcribed and translated produce proteins that prevent the antibiotic from killing the bacterium (Figure 24.5). There are three main mechanisms:

1. Genes can code for enzymes that *degrade* the antibiotic.

2. Membrane protein pumps can *remove* the antibiotic from the cell.

3. Proteins may *modify* the antibiotic so that it can no longer function.

Bacteria that contain genes that allow them to be resistant to many different antibiotics are called *multiresistant* or *superbugs*, such as MRSA.

Opportunistic Infections

Many bacterial species can live within the human body without causing any illness until the immune system becomes compromised. For example, staphylococcus bacteria are part of our normal flora and are not usually harmful, but an overgrowth can cause serious disease when the immune system does not keep the growth level in check. Sometimes bacteria cause disease only when the immune system is occupied fighting another infection; for example, bacterial pneumonia can occur after a viral infection such as influenza. Organisms that cause infection only when our immune system is suppressed or when our normal flora is damaged (after antibiotic treatment) are considered opportunistic, which means they cause disease only when they have the "opportunity." **Opportunistic infections** can be caused by bacteria, viruses, fungi, or parasites. Opportunistic infections can occur in the following situations:

1. Suppressed immune system, such as during stress when cortisol levels are high (Chapter 17)
2. Altered normal flora, most often caused by antibiotic treatment, which leaves physical space for other organisms to attach to the body's cells, particularly in the digestive tract. Note that after taking antibiotics, it is important to replenish the normal flora with probiotics, either supplements or probiotic food such as fermented vegetables or yogurt.
3. Malnutrition due to lack of nutrients, which exacerbates many infections and is why infection rates are dramatically higher in regions of poverty
4. Cancer patients, who have a greater risk of opportunistic infections, partly because the immune system is occupied with the cancer cells and also because treatment such as chemotherapy or radiation damages healthy cells, including immune cells
5. Immunosuppressant medication, for example, those used when patients have had stem cell therapy (requires irradiating their bone marrow), or corticosteroids used to treat autoimmune disease

6. Immunodeficiency such as occurs in people with HIV
7. Damaged skin barrier, such as punctures, surgery, or burns, which allows organisms to enter the bloodstream where they would normally not be found

How Bacteria Make Us Sick Many bacteria produce **toxins** that either prevent cells from functioning normally, or they kill the cells. Tetanus is caused by a bacterial toxin that prevents the inhibition of muscle contraction, causing severe and painful muscle contractions. **Botulism** is caused by a toxin produced only when the common soil bacterium *Clostridium botulinum* is forced to live in anaerobic conditions (such as in a jar or can); when consumed, that toxin blocks neurotransmitters and causes paralysis. Certain strains of *E. coli* produce a toxin that causes cells to lyse and leads to severe breakdown of tissues. Other bacteria cause disease by invading the cells to reproduce and, ultimately, killing the host cell—a process similar to viruses. Such diseases include chlamydia infection, salmonella poisoning, and tuberculosis (Table 24.1).

DID YOU KNOW?

Botulism is caused by a normal soil bacterium that is toxic only if it lives in an anaerobic environment, for example, in canned vegetables that have not been properly sterilized. You eat a carrot out of the garden, even though botulism bacteria are growing on it. However, if you put a carrot in a jar or can with no oxygen, and without sterilizing, the botulism bacteria make a neurotoxin that causes paralysis. This is the same botulinum toxin that is used to make Botox. It prevents wrinkles by paralyzing the muscles that make you frown.

TABLE 24.1

Pathogenic Bacteria

Bacterial Species	Tissue Infected	Characteristics
Methycillin-resistant *Staphylococcus aureus* (MRSA)	Skin, wounds, blood, respiratory tract, or urinary tract in immunosuppressed people	Staphylococcus, a normal flora bacterium, overgrows in immunosuppressed people and is commonly acquired in hospitals. The condition is often localized to the skin, but can cause necrotizing fasciitis ("flesh-eating" disease), pneumonia, or blood poisoning (Figure 24.6a).
Vancomyocin-resistant *Enterococcus* (VRE)	Small intestine and urinary tract	This is the cause of a diarrhea infection most commonly associated with hospitalized patients taking antibiotics that kill protective normal flora and allow enterococcus species to proliferate.
Chlamydia trachomatis	Genitals and eyes	Chlamydia is one of the most common sexually transmitted infections (STI), known as "silent" because many people do not have symptoms. It is transmitted through oral, vaginal, or anal sexual contact and can be transmitted to babies during childbirth. Chlamydia can cause a serious infection in women, called *pelvic inflammatory disease* (PID), which can cause infertility (Figure 24.6b).
Neisseria gonorrhoeae	Genitals, mouth, throat	Gonorrhea, an STI, can cause symptoms in men, such as a burning sensation during urination and discharge, but many women have no symptoms. Like chlamydia, gonorrhea can cause PID and can be transmitted to a newborn.

Streptococcus pyogenes	Most commonly infects the throat, but can infect skin and cause necrotizing fasciitis; can also cause serious blood infection where bacteria damage red blood cells	Strep throat is caused by an overgrowth of several strains of strep bacteria, usually when the immune system is suppressed, but more likely during high amounts of stress. Symptoms include very sore, red throat with visible white exudate on the tonsils, and fever, and it is usually easily treated with antibiotics. Untreated strep pyogenes infections can cause rheumatic fever that can cause damage to heart valves (Figure 24.6c).
Escherichia coli (*E. coli*)	Gastrointestinal tract	*E. coli* is a normal flora organism that resides in the large intestine of humans and animals. If eaten, *E. coli* in the stomach and small intestine cause food poisoning, thereby causing diarrhea and vomiting. It most commonly results from uncooked ground beef or chicken, or raw fruits and vegetables that have been exposed to farm runoff fertilized with manure.
Species of *Salmonella*	Gastrointestinal tract	*Salmonella* bacteria can cause food poisoning if ingested in large amounts, but usually the acidic stomach environment kills the bacteria, such as the tiny amount that may be in raw eggs. *Salmonella* can cause diarrhea and vomiting due to foods not cooked enough, but it usually does not require any treatment, except in small children or immunosuppressed individuals when severe infection can occur.
Campylobacter jejuni	Gastrointestinal tract	This is the most common cause of food poisoning in developed countries, causing diarrhea and possibly vomiting within hours to one day after eating contaminated food or water, and is often the cause of "travellers' diarrhea." It is found in animal feces and most commonly associated with undercooked poultry, but can also be transmitted from other livestock, contaminated water, unpasteurized milk, and person-to-person.
Mycobacterium tuberculosis	Respiratory system	Tuberculosis (TB) is transmitted by airborne particles from people with an active TB infection, and it kills approximately 50% of people with active infection, approximately 2 million people worldwide per year, mostly in developing countries with high malnutrition and poverty. Of those infected, 90% do not have symptoms. Diagnosis is by a TB skin test. Symptoms include persistent blood-tinged cough, fever, and weight loss. TB can be difficult to treat, requiring multiple antibiotics over a long period of time; antibiotic resistance is a problem (Figure 24.6d).
Clostridium tetani	Central nervous system	Tetanus, also known as lockjaw is caused by a toxin produced by the bacteria that block inhibitory neurotransmitters and therefore causes continuous contraction of muscles. Infection most commonly occurs through deep puncture wounds and is prevented by tetanus vaccination. Severe infections require treatment in an intensive care facility.

Scott Camazine/Alamy Stock Photo

Dr. P. Marazzi/Science Source

Dr. P. Marazzi/Science Photo Library

CNRI/Science Source

Figure 24.6   Bacterial Infections

(A) MRSA, (B) chlamydia, (C), strep throat, and (D) tuberculosis

A.

B.

C.

D.

CONCEPT REVIEW 24.1

1. Name the beneficial functions of bacteria.
2. List the functions of the bacteria in our digestive tract.
3. How is an endospore beneficial for the bacterium?
4. What are the three basic bacterial cell shapes?
5. How do antibiotic resistance genes protect bacteria?
6. What are opportunistic infections?
7. Who is at greater risk of acquiring an opportunistic infection?
8. Which bacteria cause food poisoning?
9. Which bacterial infection is multi-resistant to antibiotics?
10. How does *Clostridium tetani* cause disease?
11. Which bacterial infections are opportunistic?
12. Name the two most common sexually transmitted bacterial infections.

24.2 VIRUSES

Viruses are not considered *living* because they possess only a portion of the properties of living organisms. Viruses do not have a cell membrane or cellular organelles; they are composed of proteins and their hereditary material, which can be DNA or RNA. Viruses cannot replicate outside of host cells, and they can infect only specific cell types that contain receptors that the virus can bind to. For example, hepatitis can infect only liver cells, and the tobacco mosaic virus can infect only tobacco plant leaves. Although viruses are not considered living organisms, they may as well be treated as such, since they evolve and change depending on their environment, infect cells, and act as a selection pressure on all living things. Millions of viruses have infected bacteria, fungi, plants, and animals for as long as there have been living cells on earth.

Viruses are extremely small. A typical human cell is approximately 10 μm, a bacterium is 10 times smaller at 1 μm, and a virus particle is 10 to 100 times smaller than a bacterium (Figure 24.7). Virus particles can take on a variety of structures that contain the genetic material—DNA or RNA—and have a surrounding protective protein capsid (Figure 24.8). Viruses that infect bacteria are called **bacteriophages**. These have a complex structure, including a tail and base, where the genetic material is injected into the bacterium and the protein capsule is discarded. Some viruses have a helical structure with a hollow centre containing DNA or RNA, and some of the viruses with this structure can be very long and filamentous: such as Ebola, one of the deadly hemorrhagic viruses. Many viruses have a spherical structure and a modified cell membrane called an *envelope*, which contains proteins that allow the virus to bind and infect cells: for example, causing influenza and HIV. See Figure 24.9 for examples of common virus structures.

Unlike any other organism, viruses can have RNA as their genetic material. In fact, most viruses are

Red blood cell (10 μm)

Bacterial cell (750 nm)

Smallpox virus (200 nm x 300 nm)

Tobacco mosaic virus (15 nm x 300 nm)

Bacteriophage T4 (50 nm x 225 nm)

Ribosomes (25 nm)

Poliovirus (30 nm)

10 μm   1 μm   100 nm   10 nm

Descending size

Figure 24.7   Cell Sizes

Size comparison between a human cell, a bacterium, and virus particles

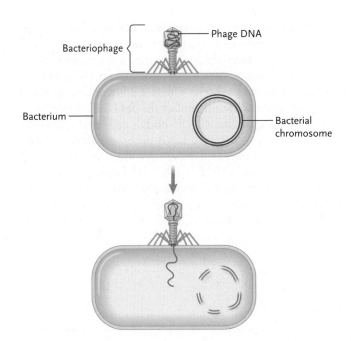

- Phage DNA
- Bacteriophage
- Bacterium
- Bacterial chromosome

Figure 24.8 Bacteriophages

Bacteriophages are viruses that infect bacteria.

RNA viruses. When an RNA virus infects a host cell, it must convert the RNA into DNA before it can replicate and produce new virus particles. RNA is converted into DNA by a viral enzyme called *reverse transcriptase*, which scientists have isolated and used in genetic engineering (Chapter 10). Reverse transcriptase enzymes are notoriously sloppy enzymes and don't copy RNA perfectly into DNA, leaving many point mutations (Chapter 8). The constant small changes in the genetic material of viruses give them the benefit of rapid evolution, which allows them to constantly evade the human immune system. For example,

RNA viruses that cause influenza and the common cold are slightly different each year, which means we can be infected and have symptoms every time we are infected with a new virus. Only the viruses with non-detrimental mutations can continue to infect cells. After our first exposure to a virus that doesn't change every year—such as the one that causes chicken pox—we can develop immunity because our memory cells (Chapter 23) recognize the virus and kill it before we get sick again.

How Viruses Replicate Viruses can replicate only inside host cells because viruses do not have ribosomes or nucleotides or any of the building blocks and ATP that are required to copy their genetic material and produce new virus particles. Viruses that can infect more than one species are more likely to undergo rapid changes than those that can't. For example, type A influenza can infect humans, birds, and pigs, and it is largely responsible for flu pandemics because there are more antigen changes that cause the immune system to mount a stronger response (Chapter 23). Influenza contains two important molecules: **hemagglutinin**, which allows the influenza virus to bind to upper airway cells, and **neuraminidase**, which allows the newly replicated virus particles to be released from the host cell. It is from these two molecules that influenza strains acquire their names, such as H1N1 (Spanish flu in 1918, and swine flu in 2009), or H5N1 (bird flu in 2004).

Viruses have two types of mechanisms for infection and reproduction: a lytic cycle and a lysogenic cycle. The **lytic cycle** occurs when the virus kills the host cell after it reproduces, and the **lysogenic cycle** occurs when the virus integrates its genetic material into the host genome, replicating along with the host cell, but not

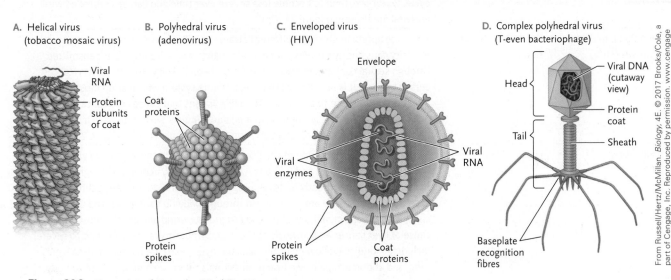

A. Helical virus (tobacco mosaic virus)
- Viral RNA
- Protein subunits of coat

B. Polyhedral virus (adenovirus)
- Coat proteins
- Protein spikes

C. Enveloped virus (HIV)
- Envelope
- Viral enzymes
- Viral RNA
- Protein spikes
- Coat proteins

D. Complex polyhedral virus (T-even bacteriophage)
- Head
- Viral DNA (cutaway view)
- Protein coat
- Tail
- Sheath
- Baseplate recognition fibres

Figure 24.9 Examples of Common Viral Structures

24

killing it (Figure 24.9). Looking at the lytic cycle first, the stages of infection involve the following:

1. **Attachment.** The proteins on the virus capsid bind to a membrane protein on the host cell.
2. **Infiltration.** The host cell takes up the virus, usually by endocytosis (Chapter 3).
3. **Replication.** The virus uses host cell organelles, amino acids, nucleotides, and ATP to replicate genetic material and transcribe and translate virus genes into proteins that make new virus particles.
4. **Assembly.** New virus particles form and are released from the host cell, causing lysis and death of the host cell.

When we have a common cold, such as a rhinovirus or adenovirus, the sore throat symptoms come from viral particles going through the lytic cycle and killing our epithelial cells. Our immune system also contributes to the symptoms when it kills infected cells (Chapter 23). Luckily, epithelial cells regrow quickly and replace the dead cells (Table 24.2).

When a virus infects the body's cells and enters the lysogenic cycle, the viral DNA becomes integrated into the chromosomes. The integrated viral DNA then replicates with each cell division, and because new virus particles are being produced, the virus is hidden from the immune system. For example, herpes viruses will alternate between a lytic cycle and a lysogenic cycle

TABLE 24.2

Viral Infections

Virus	Tissue Infected	Characteristics
Human immunodeficiency virus (HIV)	Immune cells	HIV is an RNA virus that infects CD4+ immune cells and causes immunodeficiency, with death occurring from opportunistic infections.
Rhinoviruses, adenoviruses, and respiratory syncytial virus	Sinuses and upper airways	Over 100 different viruses cause the common cold and are easily killed by the immune system in healthy people. Various cold viruses infect the epithelial cells of the upper airways. Symptoms include stuffed or runny nose, sneezing, sore throat, and coughing, and usually don't last longer than 7 to 10 days.
Influenza	Sinuses and upper airways	There are type A, B, and C influenza viruses, with type A and B being the most common. Symptoms can be similar to a common cold, but typically, pronounced fatigue, fever, and body aches accompany the flu. Influenza viruses are RNA viruses and mutate rapidly, which leads to new flu viruses each year. Often influenza infection can lead to a secondary opportunistic infection, such as viral or bacterial pneumonia, which can be fatal in the elderly and very young children.
Herpes	Lips (cold sores) or genitals, and sensory neurons	DNA viruses that cause cold sores or genital herpes are easily transmitted by skin-to-skin contact. Use of condoms does not necessarily prevent infection if sores in areas not covered by the condom. Herpes infections have a latent period, when the virus enters the sensory neurons, and then it re-enters the skin during the lytic cycle. Outbreaks tend to become less severe over time and can be controlled with antiviral medication (Figure 24.10a).
Human papilloma viruses (HPV)	Keratinocytes of the skin	There are approximately 100 different strains of HPV virus that cause warts, and the majority cause no symptoms. Some strains cause warts: 30–40% are transmitted through sexual contact, and two strains have been linked with 70% of cervical cancers in women. It is now known that the immune system will clear up most HPV infections within one to two years. The HPV viruses that cause cervical cancer do not cause warts or any symptoms, so it is important for women to have a yearly Papanicolaou "pap" test to detect and treat possible cancer cells (Figure 24.10b).
Hepatitis	Liver	Hepatitis is an inflammation of the liver, most commonly caused by five viruses: hepatitis type A, B, C, D, and E. Types A and E are usually acquired through contaminated food or water, mostly in developing countries, and do not cause chronic disease. Types B, C, and D are transmitted through body fluids. Types B and C can cause chronic infections in millions of people around the world and are the most common cause of liver cancer and cirrhosis of the liver. Symptoms range from mild to severe, including jaundice (yellowing of the skin and eyes), extreme fatigue, nausea, vomiting, and abdominal pain. Medical professionals that have high exposure to body fluids can be vaccinated against hepatitis B.

Rotavirus	Gastrointestinal tract	Rotavirus is a common intestinal infection in children, causing vomiting and diarrhea. Almost all children will have at least one infection by age 5, leaving them with antibodies that protect them against future infections. Adults very rarely become infected with rotavirus. Diarrhea can be severe in infants or malnourished children. Rotavirus is usually easily managed with rehydration therapy, but does account for almost 450 000 deaths in developing countries each year.
Epstein–Barr virus (EBV) "Mononucleosis"	B cells and epithelial cells	EBV is a type of herpes virus easily transmitted through saliva. Ninety percent of adults have been infected and have acquired immunity to EBV by age 40. It is one of the most common viral infections. Symptoms can range from asymptomatic (in young children) to severe fatigue that can last for several months (more common in adolescents). Other symptoms in the acute phase include fever, sore throat, swollen lymph nodes, and possibly headaches or abdominal pain, which usually disappear within two weeks. It is thought that EBV infection is associated with a higher risk of autoimmune diseases such as multiple sclerosis, systemic lupus, and rheumatoid arthritis.

(Figure 24.11). When there are herpes sores—cold sores on the lips or genital sores—the virus particles are in the lytic cycle, killing epithelial cells and causing the sores. When the sores go away, the virus is still there; it retreats into a nearby sensory neuron and enters the lysogenic cycle. Herpes infections tend to stay long term within our cells because the lysogenic cycle allows the virus to go unseen by the immune cells.

The same occurs with HIV infection. HIV is an RNA virus that reverse transcribes the RNA into DNA and then integrates into the chromosomes. HIV can also replicate and produce new virus particles that do not kill the cell; this is called **budding** (blue dots in Figure 24.12). HIV binds specifically to immune cells that express the **CD4 membrane receptor** by attachment with the HIV gp120 (glycoprotein with 120 amino acids); therefore, HIV can infect only the immune cells (Figure 24.13). Once HIV infects and kills T cells, the immune system can no longer function sufficiently to prevent infection, and the person becomes severely immune-compromised and has AIDS—acquired immunodeficiency syndrome. With so many new antiviral medications, people can now live for decades with HIV, whereas in the 1980s when the HIV epidemic first began, thousands of people died each year from many opportunistic infections (Section 24.3). HIV is transmitted by sexual intercourse, blood transfusions, and can be transmitted to an infant during childbirth.

DID YOU KNOW?

Different terms are used to define the extent of an infection from a specific organism in a specific area.

- *Endemic* refers to a constant, steady prevalence of an infection in a population. The occurrence of this type of infection does not increase or decrease significantly over time: for example, the common cold is endemic in Canada, and malaria is endemic in Africa.
- *Epidemic* refers to an *increase* in the prevalence of an infection in a specific area, usually in a short period of time: for example, the small pox epidemic that killed North American Indigenous peoples when European explorers arrived.
- *Pandemic* refers to an epidemic that spreads through more than one country, such as the flu pandemic of 1918, also called the *Spanish flu*.

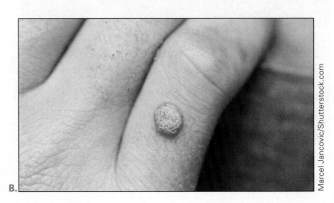

A. Cherries/Shutterstock.com   B. Marcel Jancovic/Shutterstock.com

Figure 24.10 Viral Infections

(A) Herpes and (B) HPV

24

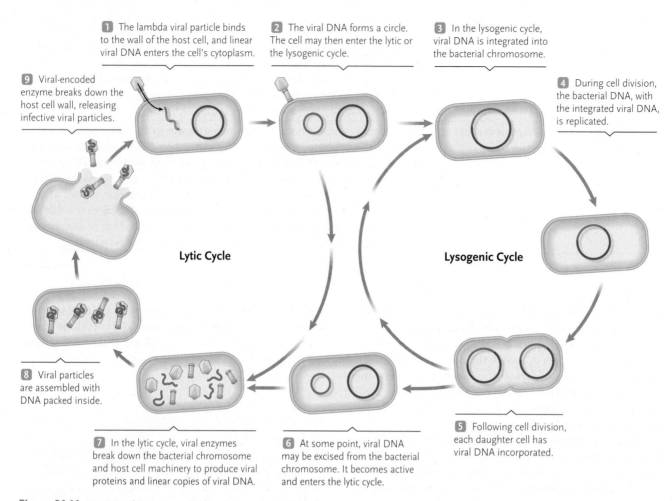

1 The lambda viral particle binds to the wall of the host cell, and linear viral DNA enters the cell's cytoplasm.

2 The viral DNA forms a circle. The cell may then enter the lytic or the lysogenic cycle.

3 In the lysogenic cycle, viral DNA is integrated into the bacterial chromosome.

9 Viral-encoded enzyme breaks down the host cell wall, releasing infective viral particles.

4 During cell division, the bacterial DNA, with the integrated viral DNA, is replicated.

Lytic Cycle

Lysogenic Cycle

8 Viral particles are assembled with DNA packed inside.

5 Following cell division, each daughter cell has viral DNA incorporated.

7 In the lytic cycle, viral enzymes break down the bacterial chromosome and host cell machinery to produce viral proteins and linear copies of viral DNA.

6 At some point, viral DNA may be excised from the bacterial chromosome. It becomes active and enters the lytic cycle.

Figure 24.11 Lytic and Lysogenic Cycles

In the lytic cycle, the virus replicates and kills the host cell. In the lysogenic cycle, the virus integrates its DNA into the host genome.

Cultura RM/Alamy Stock Photo

Figure 24.12 HIV Viruses Budding from a Macrophage

Budding allows HIV to replicate without killing the macrophage.

DID YOU KNOW?

Stress can activate latent viral infections, such as herpes, because cortisol is produced during stress and inhibits the immune system.

DID YOU KNOW?

The stomach flu is not caused by influenza. Influenza is a respiratory illness that can sometimes cause vomiting in small children who have a high fever. The typical 24-hour stomach illness is most commonly caused by campylobacter bacteria or by noroviruses, such as the Norwalk virus. In children, rotaviruses can last anywhere from one to five days.

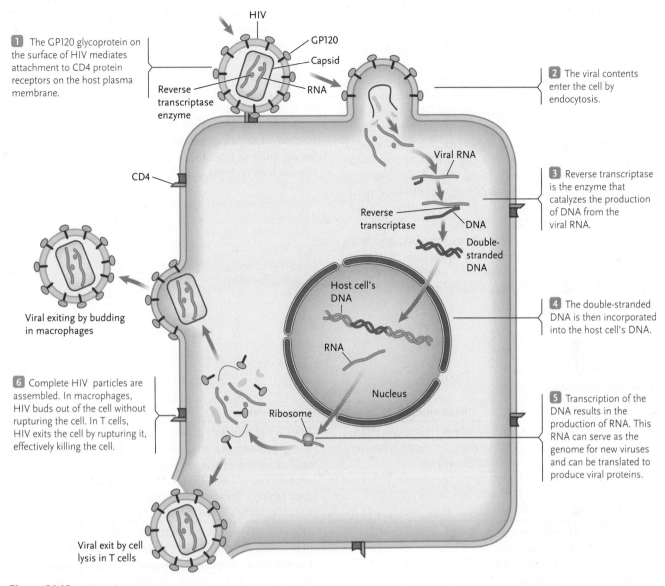

1 The GP120 glycoprotein on the surface of HIV mediates attachment to CD4 protein receptors on the host plasma membrane.

HIV

GP120

Capsid

RNA

Reverse transcriptase enzyme

2 The viral contents enter the cell by endocytosis.

CD4

Viral RNA

Reverse transcriptase

DNA

Double-stranded DNA

3 Reverse transcriptase is the enzyme that catalyzes the production of DNA from the viral RNA.

Host cell's DNA

RNA

Nucleus

4 The double-stranded DNA is then incorporated into the host cell's DNA.

Viral exiting by budding in macrophages

Ribosome

6 Complete HIV particles are assembled. In macrophages, HIV buds out of the cell without rupturing the cell. In T cells, HIV exits the cell by rupturing it, effectively killing the cell.

5 Transcription of the DNA results in the production of RNA. This RNA can serve as the genome for new viruses and can be translated to produce viral proteins.

Viral exit by cell lysis in T cells

Figure 24.13 HIV Infection

HIV infects cells that have a CD1 receptor; it usually enters a lysogenic cycle for a period of years before beginning to lyse helper T cells, causing AIDS.

CONCEPT REVIEW 24.2

1. How are viruses different from living organisms?
2. What is a bacteriophage?
3. How do RNA viruses replicate their genetic material?
4. What is the function of hemagglutinin and neuraminidase in influenza?
5. Explain the difference between the lytic and lysogenic cycles.
6. What are the four stages of a lytic cycle infection?

7. Give an example of a human viral infection that undergoes a lysogenic cycle, making it very difficult for our immune system to get rid of.
8. What cell receptor does HIV bind to? What cell types have this receptor?
9. What viruses can cause liver cancer?
10. Which viral infections can be sexually transmitted?
11. Which virus can cause cervical cancer?
12. Which viral infection is associated with increased risk of autoimmune diseases?

24

24.3 FUNGI

Organisms of the kingdom fungi are eukaryotic and, except for yeast, are multicellular and include organisms such as mushrooms, mould, mildew, and yeasts. Fungi live in soil, on plants, and in the air and water and play an important role in decomposition and nutrient recycling. Many fungi are parasitic and cause extensive damage to food crops. Fungi usually do not cause severe infection in humans unless the immune system is suppressed or the majority of the bacterial normal flora breaks down, such as after antibiotic treatment. Many fungal infections are opportunistic infections (Table 24.3 and Figure 24.14).

Fungi can reproduce asexually or sexually, and both methods commonly involve the production of spores. Single-celled yeast can reproduce by binary fission, like bacteria. Spores expelled from fungi can travel in the air long distances, and infection in humans most often occurs on the skin or in the respiratory tract. In immune-compromised people, fungal infections can be fatal.

Fungi have been used for decades for their antimicrobial properties, for extracting antibiotics that fungi naturally produce to kill bacteria, and for producing vitamins and molecules used to treat cancer and heart disease. Fungal organisms are used to make many foods, such as bread, beer, wine, kombucha, fermented soy such as tempeh, and fermented beans such as miso.

DID YOU KNOW?

Genetic studies have shown that fungi are more similar to animals than to plants.

TABLE 24.3

Fungal Infections

Fungus	Tissue Infected	Characteristics
Candida albicans	Vagina ("yeast infection"), mouth ("thrush"), and also diaper rash	Yeast is a normal flora organism, and infection occurs when it overgrows, such as after antibiotic use that kills protective bacteria, or when a person has suppressed immune function or uncontrolled diabetes (increased blood sugar causes yeast to grow rapidly). Usually this infection can be easily treated with topical antifungal cream or pills.
Pneumocystis jirovecii (previously but incorrectly called *pneumocystis carnii*, which is a species found only in rats)	Lungs	Pneumocystis pneumonia (PCP) is normally found in the lungs of healthy people and does not cause infection, unless a person has a severely compromised immune system. Therefore, it occurs most often in people with HIV, cancer, or those on immunosuppressant medications. PCP symptoms include unproductive cough and severe shortness of breath, which can be fatal if not treated aggressively.
Aspergillis (over 100 species of mould)	Sinuses and upper respiratory tract	This common mould grows on bread and fruits, and its airborne spores are everywhere in the environment. Aflatoxin grows on raw nuts and is carcinogenic, which is one main reason why nuts are roasted. Normally, *Aspergillis* organisms do not cause illness, but in people with weakened immune systems, infection can cause fever, cough, sinusitis, and worsening of asthma or other respiratory problems.
Ringworm	Skin	Ringworm is not caused by a worm; it is a very common fungal infection: dermatophytosis. The fungal organism feeds on the keratin protein in skin, hair, and nails. It is easily transmitted to children by animals, such as pet cats, or through warm, moist environments, such as public showers. When this fungus grows on the feet, it is known as *athlete's foot*; when it grows in the groin area, it's called *jock itch*. Symptoms, which appear as raised red rings on the skin or thickened and yellowish nails, are easily treated with topical antifungal ointment.

Figure 24.14 Examples of Fungal Infections
(A) Ringworm, (B) aflatoxin, and (C) candida

CONCEPT REVIEW 24.3

1. Are fungi prokaryotic or eukaryotic?
2. Name some beneficial characteristics of fungi.
3. How can various fungal species be used in food production?
4. What factors contribute to increased risk of fungal infections?
5. Which fungal infection causes athlete's foot?
6. Which fungal organism is responsible for yeast infection or thrush?
7. Which fungal infections can be fatal for people with HIV?

24.4 PARASITES

Parasites are species of the protist and animal kingdoms, and they cause human infection. There are two general categories of parasites: **protozoans** (protists) and **helminths** (worms). *Parasite* is a general term that indicates an organism acquires its nutrients from a host, which means that some bacteria or fungi can also be considered parasitic. **Ectoparasites** include organisms that live on the blood of other organisms, such as lice, fleas, and ticks. This section focuses on the **endoparasites**: protozoans and helminths.

Protozoa

Protozoa are single-celled, eukaryotic protists that have the animal-like ability to move. These characteristics distinguish them from the spore-producing fungus-like protists—such as slime moulds—or the photosynthetic, plant-like protists—such as algae (see Chapter 1). Protozoa can move around in a number of ways: by extending **pseudopods**, like amoebas; or by using flagella, such as giardia; or by using cilia, such as paramecium. All protozoans digest food in structures called **vacuoles**, similar to a stomach.

Protozoa that infect animal intestines are usually transmitted by the **fecal-oral route** and are common in regions of poverty where sewage treatment is minimal and water is contaminated by human or animal feces. Protozoa that live in the bloodstream or other tissues are most often transmitted by **insect vectors** (carriers), such as sandflies or mosquitoes; they are common in tropical regions.

Some protozoa can form dormant cysts that allow them to survive in harsh environments, such as extreme heat or lack of nutrients or oxygen, for long periods of time outside of a host. Protozoa can reproduce sexually and asexually and often involve a complex life cycle. For example, **malaria** infects over 200 million people each year and causes over 1 million deaths, usually in children; it is most prevalent in Africa. Malaria is caused by *Plasmodium* species—mostly *P. falciparum* and *P. vivax*—that are injected into the bloodstream when a person is bitten by a mosquito carrying the **sporozoite** form of the parasite. From there, they migrate to the liver and reproduce asexually within liver cells. Then thousands of **merozoites** leave the liver cells, killing the liver cells in the process. The merozoites hide from the immune system by wrapping in fragments of liver cell membranes, and then they travel to red blood cells. Once inside a red blood cell, merozoites replicate asexually through cycles of replication, with the result of killing the red blood cell and infecting other red blood cells. Merozoites can develop into gametes that can be taken up by other mosquitoes, where the gametes combine to form new sporozoites. Symptoms of malaria result from damage to the liver and red blood cells, and include severe headache, fever, chills, vomiting, jaundice, and hemoglobin in the urine (Figure 24.15).

24

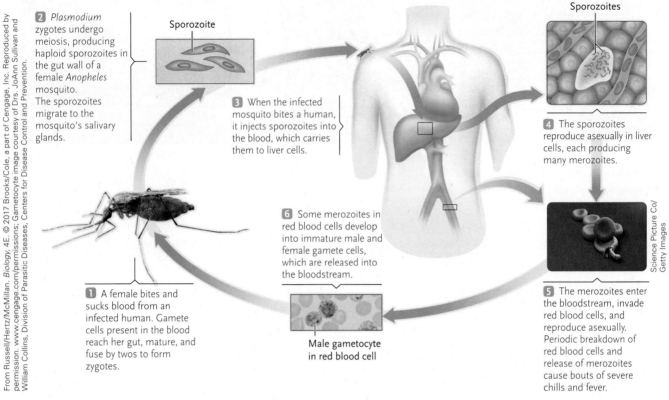

2 *Plasmodium* zygotes undergo meiosis, producing haploid sporozoites in the gut wall of a female *Anopheles* mosquito. The sporozoites migrate to the mosquito's salivary glands.

Sporozoite

3 When the infected mosquito bites a human, it injects sporozoites into the blood, which carries them to liver cells.

Sporozoites

4 The sporozoites reproduce asexually in liver cells, each producing many merozoites.

Science Picture Co/ Getty Images

5 The merozoites enter the bloodstream, invade red blood cells, and reproduce asexually. Periodic breakdown of red blood cells and release of merozoites cause bouts of severe chills and fever.

6 Some merozoites in red blood cells develop into immature male and female gamete cells, which are released into the bloodstream.

Male gametocyte in red blood cell

1 A female bites and sucks blood from an infected human. Gamete cells present in the blood reach her gut, mature, and fuse by twos to form zygotes.

Figure 24.15 Malaria Life Cycle

Malaria has a complicated life cycle that involves infecting the liver and the red blood cells.

Helminths Helminths are multicellular worms that can be free-living or parasitic. Intestinal worms infect hundreds of millions of people worldwide. There are two categories of worms:

1. **Flatworms** (platyhelminths), which include **tapeworms** (cestodes) and **flukes** (trematodes)

2. **Roundworms** (nematodes)

Helminths infect people through contaminated water, food, or soil, and sometimes insect bites. Undercooked pork, fish, and shellfish can carry flukes. Nematodes such as hookworms can penetrate the skin, usually the feet. Helminths infect people throughout the world, but the risk is much higher in developing countries with poor sanitation or undercooked meat, and among children. Infection with worms stimulates a Th2 immune response (Chapter 23) so that the immune system is less active, which allows the parasites to live in the intestine without being killed by the immune system; it also leaves the infected individual at higher risk of acquiring other infections (Figures 24.16 and 24.17).

Due to the inhibited immune response and lack of invasion of any other tissues, early helminth infections

are often asymptomatic. However, people with a significant worm load show signs of malnutrition, weakness, intestinal obstruction, abdominal pain, vitamin

A.

Dr. M.A. Ansary/Science Source

B.

Michael Abbey/Science Source

C.

Juan Gaertner/Shutterstock.com

D.

CDC/Science Source

Figure 24.16 Examples of Parasite Infections

(A) Malaria, (B) giardia, (C) tapeworm, and (D) hookworm

24

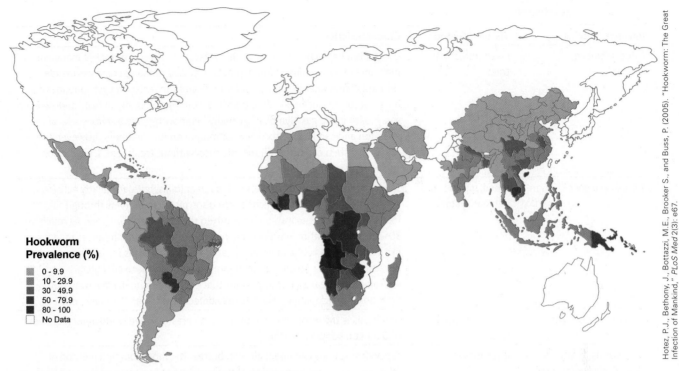

Hotez, P.J., Bethony, J., Bottazzi, M.E., Brooker S., and Buss, P. (2005). "Hookworm: The Great Infection of Mankind." *PLoS Med* 2(3): e67.

Hookworm Prevalence (%)
- 0 - 9.9
- 10 - 29.9
- 30 - 49.9
- 50 - 79.9
- 80 - 100
- No Data

Figure 24.17 Worldwide Prevalence of Hookworm Infection
Hookworm infections are more prevalent in less developed countries.

deficiencies, anemia, cognitive impairment, slowed growth in children, and possibly vomiting or diarrhea that contains worms. Some worms can migrate to other tissues, such as species of *Ascaris* nematodes that migrate to the lungs and can cause hemorrhaging. Helminth infection, as with any other infection, can be particularly detrimental for people with compromised immune systems (Table 24.4).

Public health campaigns and deworming programs began in the early 1900s and have significantly decreased the number of children in North America with helminth infections.

DID YOU KNOW?

Parasitic worms have been used to decrease the inflammation caused by autoimmune diseases because helminths trigger the anti-inflammatory immune response. Regions of the world that have a high rate of intestinal parasites also have the lowest rates of autoimmune disease. And developed countries that now have very low intestinal parasites also have the highest incidence of autoimmune diseases. One study suggests that the high incidence of metabolic syndrome (Chapter 18) in developed countries may be due to a lack of eosinophil stimulation, which normally occurs when immune cells fight parasites (Chapter 23). Eosinophils also play a role in preventing insulin resistance.

TABLE 24.4

Parasites		
Parasite Species	**Tissue Infected**	**Characteristics**
Plasmodium falciparum and *Plasmodium vivax*	Liver and red blood cells	Malaria is a severe protozoan infection (as described in this section) carried by Anopheles mosquitoes. It can be prevented in travellers who use prophylactic medication, but it is an endemic infection in most countries of Africa. Usually, infection by malaria parasites can be effectively treated with antimalarial medication, but in a small percent of cases it causes recurring infection. People who are heterozygous for sickle cell anemia (Chapter 6) are much less likely to die from malaria. This is because their damaged red blood cells are broken down by the spleen much more frequently than normal cells; therefore, the parasite population in the body is significantly decreased.

(Continued)

24

TABLE 24.4 (*continued*)

Parasite Species	Tissue Infected	Characteristics
Giardia lamblia	Gastrointestinal tract	Giardia is a flagellated protozoan that infects the digestive tract of many animals, including humans. Giardia infects over 200 million people worldwide through contaminated food or water and can form resistant cysts outside of a host. Symptoms include violent diarrhea, vomiting, abdominal pain, and excessive bloating and gas. Symptoms usually begin within one to two weeks of infection and can last for six weeks or more without treatment; however, the immune system can often kill the infection without treatment. Dehydration can be fatal in small children.
Toxoplasma gondii	Blood, muscle, and nervous system	Toxoplasmosis infects most warm-blooded animals, but particularly cats (from eating infected rodents), which can be transmitted to humans through the fecal-oral route. It can also be transmitted through uncooked pork. As much as 30% of the population has probably had this infection, which causes mild flu-like symptoms in healthy people. It can be very dangerous for pregnant women because it can be passed onto an unborn infant and cause encephalitis and neurological diseases. It is important that pregnant women stay away from litter boxes, gardening where cats may defecate, and handling raw pork.
Entamoeba histolytica	Gastrointestinal tract	Amebiasis is the most common protozoan parasite in the world, infecting over 600 million people per year.
Tapeworms (many species), *Diphyllobothrium latum* is the longest tapeworm species that infects humans, and can reach 10 metres in length.	Intestines	Tapeworm species can infect all vertebrates. The most common method of infection in humans is ingestion of the larval stage in uncooked fish and beef. Prevalence in the world is highest in populations that eat raw fish. Tapeworms are hermaphrodites, containing both male and female reproductive organs, and can self-fertilize. Infected people are usually asymptomatic, unless a large population of worms develops in the GI tract over a long period of time, which then causes vitamin B12 deficiency and anemia.
Necator americanus and *Ancylostoma duodenale*	Intestines	Hookworm is the most prevalent helminths infection in the world, infecting over 700 million people and causing severe illness in 80 million people, mostly children in developing countries (Figure 24.17). Hookworms are nematodes that enter the body through the skin, usually the feet, and can also be ingested. Hookworm infection can cause iron-deficiency anemia, stunted growth, and impaired cognitive function, but it is rarely fatal. Do not walk barefoot when travelling in regions shown in Figure 24.17.

DID YOU KNOW?

Toxoplasmosis antibodies occur in higher levels in people that have schizophrenia. Minocycline is an antibiotic that happens to be effective in treating toxoplasmosis and has been shown to alleviate schizophrenia symptoms in some people. Other research has shown latent toxoplasmosis infection to be linked with suicidal behaviour, attention deficit disorder, and obsessive-compulsive disorder.

CONCEPT REVIEW 24.4

1. What is the most common method of infection by intestinal protozoans, and by protozoans that live in the bloodstream?

2. What are the two major categories of helminths?

3. How are intestinal worm infections acquired?

4. What populations have the greatest risk of helminths infection?

5. Which helminths infection can be acquired through the skin?

6. Which populations are at highest risk of tapeworm infection?

CHAPTER REVIEW

Labelling Activity

Label the type of bacteria found in each of the three following images.

1.

A. _____ 1.0 μm

_____

2.

B. _____ 3.0 μm

3.

C. _____ 2.0 μm

4.

_____

_____

D. _____

Identify the following infection images.

5.

_____

6.

_____

7.

_____

24

8.

CNRI/Science Source

_____

9.

Cherries/Shutterstock.com

_____

10.

Marcel Jancovic/Shutterstock.com

_____

11.

© iStockphoto.com/Alejandro Rivera

_____

12.

USDA National Peanut
Research Laboratory

_____

13.

Centers for Disease Control and
Prevention/Science Source

_____

14.

Dr. M. A. Ansary/
Science Source

_____

15.

Michael Abbey/Science
Source

_____

16.

Juan Gaertner/
Shutterstock.com

_____

17.

CDC/Science Source

_____

Answer to Application Question

1. Also called *lockjaw*, a bacterial infection that causes muscles to continuously contract. <u>Tetanus</u>

2. A bacterial infection that most commonly infects elderly patients on antibiotics and causes diarrhea. <u>VRE</u>

3. A worm that can infect a person through the skin and can cause iron-deficiency anemia. <u>Hookworms</u>

4. A common mould that can grow on bread and fruit and can cause respiratory infection in immunosuppressed people. <u>*Aspergillis*</u>

5. A virus that infects the liver and can cause liver cancer or cirrhosis. <u>Hepatitis</u>

6. A respiratory infection that resembles the common cold, but usually causes fever and muscle aches; resolves in about week in healthy people. <u>Influenza</u>

7. A respiratory disease, prevalent in developing countries, that can kill malnourished people who have the active disease. <u>Tuberculosis</u>

8. A bacterial infection that is multi-resistant; can cause flesh-eating disease, pneumonia, or blood poisoning. <u>MRSA</u>

9. Transmitted by skin-to-skin contact, with outbreaks occurring during periods of stress; infectious organism able to "hide" in sensory neurons between outbreaks. <u>Herpes</u>

10. A worm that can be ingested in uncooked meat or fish; commonly causes B12 deficiency. <u>Tapeworm</u>

11. A yeast infection caused by an overgrowth of yeast, usually due to a lack of healthy normal flora. <u>Candida</u>

12. A common opportunistic infection in people with HIV. <u>Pneumocystis pneumonia</u>

13. A viral infection that causes warts. <u>HPV</u>

14. A bacterial organism that is the most common cause of food poisoning. <u>*Campylobacter*</u>

15. A bacterial organism that can cause very severe food poisoning in immunosuppressed people or young children; can be acquired from uncooked chicken or raw eggs. <u>*Salmonella*</u>

16. Food poisoning caused by a bacterial organism that normally resides in the large intestine; can cause illness in people who eat uncooked ground meats or raw foods that have come in contact with this intestinal organism. <u>*E coli*</u>

17. A protist infection caused by drinking contaminated water; can cause vomiting and diarrhea for several weeks. <u>Giardia</u>

18. A parasite that can be picked up from animal feces, especially cats, and can cause neurological damage in babies whose mother had been exposed during pregnancy. <u>Toxoplasmosis</u>

19. A common infection in young children that causes diarrhea and vomiting; usually resolves quickly, but can be fatal in developing countries where children are malnourished or do not have clean drinking water. <u>Rotavirus</u>

20. A viral infection caused by a retrovirus that infects immune cells and causes immunodeficiency. <u>HIV</u>

21. A common bacterial infection that is easily sexually transmitted and can cause pelvic inflammatory disease in women (2 answers). <u>Chlamydia or Gonorrhea</u>

22. A bacterial throat infection that causes a painful sore throat, often with fever, and occurs more commonly during periods of stress; may require treatment with antibiotics if not resolved within a week to prevent damage to heart valves. <u>Strep throat</u>

23. Virus that infects B cells and causes a range of symptoms from no symptoms to extreme fatigue; 90% of adults infected at some point in their life. <u>EBV</u>

24. Fungal infection of the skin caused by fungal organism that feeds on keratin protein in the skin; easily transmitted by skin-to-skin contact and resolves with antifungal topical cream. <u>Ringworm</u>

25. A parasitic infection transmitted by mosquitoes that infects red blood cells and liver cells. <u>Malaria</u>

26. A virus that causes common colds. <u>Rhinovirus</u>

Appendix A

CONCEPT REVIEW ANSWERS

Chapter 1 Concept Review Answers

Section 1.1

1. Biology is the study of living things. Anatomy is the study of the structures in an organism. Physiology is the study of how anatomical structures function within the body.

Section 1.2

1. Chart showing characteristics of organisms in each kingdom:

TABLE A.1

Kingdom	Characteristics	Example
Archaebacteria	Unicellular, prokaryotic, live in extreme environments	Thermophiles, methanogens
Eubacteria	Unicellular, prokaryotic, live on human skin and inside digestive tract, can cause infections	*Staphylococcus aureus, Salmonella* bacteria
Protista	Unicellular, eukaryotic, can be multicellular, can have characteristics of other three eukaryotic kingdoms	Kelp, algae
Fungi	Multicellular (except yeast), eukaryotic, not photosynthetic, important for decomposition	Mushrooms, moulds
Plants	Multicellular, eukaryotic, photosynthetic, can live on land or in water	Broccoli, cherry trees
Animals	Multicellular, eukaryotic, have locomotion, can live on land or in water	Fish, humans

2. Examples are included in the chart above.

Section 1.3

1. The human body is organized according to the following levels, from smallest to largest: atoms, molecules, macromolecules, organelles, cells, tissues, organs, organ systems, organism.

Section 1.4

1. Structures in each organ system:
   - Integumentary—skin, hair, nails, glands
   - Skeletal—bones, cartilage, joints
   - Muscular—smooth, cardiac, and skeletal muscles
   - Nervous—brain, spinal cord, receptors
   - Endocrine—hypothalamus, pineal gland, pituitary, thymus, thyroid, adrenals, pancreas, parathyroid, ovaries, testes
   - Cardiovascular—heart, blood vessels, blood
   - Respiratory—mouth, nose, larynx, pharynx, trachea, bronchi, bronchioles, alveoli
   - Digestive—mouth, pharynx, esophagus, stomach, small intestines, large intestine, rectum, anus; also accessory organs—liver, pancreas, gallbladder
   - Urinary—kidneys, ureters, bladder, urethra
   - Reproduction
     a. male—testes, vas deferens, epididymis, prostate, penis
     b. female—mammary glands, ovaries, fallopian tubes, uterus, vagina
   - Immune—white blood cells, bone marrow, lymph, spleen, thymus, lymph nodes, lymph nodules
2. Organs associated with more than one organ system—pancreas, ovaries, testes
3. Main functions of each organ system:
   - Integumentary—protection, temperature regulation, excrete waste, produce vitamin D, detect sensations
   - Skeletal—body movement, store minerals, store fats, produce blood cells
   - Muscular—body movements, produce heat
   - Nervous—detect, interpret, and respond to stimuli
   - Endocrine—important role in homeostasis, different hormones having different functions

- Cardiovascular—transport nutrients, oxygen, hormones, waste
- Respiratory—take up oxygen, get rid of carbon dioxide, regulate blood pH, produce vocal sounds
- Digestive—break down food, absorb nutrients, eliminate waste
- Urinary—filter blood; reabsorb nutrients, water, and ions; eliminate waste
- Reproduction—produce gametes for reproduction, produce hormones; females produce milk and carry embryo
- Immune—fight infections, lymphatic system takes up excess interstitial fluid, absorbing fats in digestive tract

4. Matching:
   a. Respiratory system—D, I
   b. Cardiovascular system—A, G
   c. Endocrine system—C
   d. Muscular system—A, E
   e. Urinary system—D, H
   f. Digestive system—B, C, D
   g. Skeletal system—E, F

Section 1.5

1. Anatomical position is important so that descriptions of locations on the body are consistent between health care professionals.

2. For chart showing common name and anatomical name, see Figure 1.20. The common name is in brackets, and the other name is the anatomical name.

3. Body planes include frontal (cuts body front and back), sagittal (cuts body into left and right), transverse (cuts body into upper and lower)

4. Body cavities are as follows:
   a. Cranial—brain
   b. Vertebral—spinal cord
   c. Thoracic—heart and lungs
   d. Abdominopelvic—digestive organs, spleen, pancreas, liver, kidneys, reproductive organs

5. Pleural membranes surround cavities and visceral organs, pericardial surrounds the heart, and peritoneum surrounds digestive organs.

6. The mediastinum is the area of the thoracic region that contains the heart and trachea

7. Four quadrants are as follows:
   a. Right upper—gallbladder, most of liver, duodenum, some small intestines, upper colon
   b. Right lower—appendix, lower right colon, some small intestine
   c. Left upper—some liver, most of stomach, some small intestines, upper left colon
   d. Left lower—some small intestines, lower left colon

Section 1.6

1. All living things have macromolecules, cells, growth, metabolism, hereditary material, homeostasis, reproduction, evolution.

2. Homeostasis is the regulation of any substance in the body, such as oxygen, glucose, pH, or temperature. Negative feedback is when the initial change is brought back into balance; for example, if blood sugar is increased, a negative feedback response will decrease it. Positive feedback is when the substance continues to increase, such as oxytocin during delivery.

3. Receptors detect stimuli, integrating centre determines if there is a difference between a stimulus and a set point, Effectors exert an effect to cause a change.

4. Different types of receptors are as follows:
   a. Chemoreceptors—chemicals
   b. Osmoreceptor—ions
   c. Tactile receptors—touch, vibration, pressure
   d. Baroreceptors—pressure
   e. Photoreceptors—light
   f. Mechanoreceptors—stretch
   g. Proprioceptors—body position and movement
   h. Nociceptors—pain
   i. Thermoreceptors—temperature

5. Effectors—blood vessels, the heart, bronchiole tubes, muscles, endocrine glands

Chapter 2 Concept Review Answers

Section 2.1

1. Approximately 55%–60% of the human body is composed of water.

2. The difference between a solute and a solvent is that a solvent is a substance that contains dissolved molecules, which are called the *solutes*.

3. The difference between hydrophobic and hydrophilic and polar and nonpolar is that hydrophobic means water-fearing molecules and hydrophilic means water-loving molecules. Polar molecules may have a slight charge or a full charge and dissolve easily in water (hydrophilic) and nonpolar molecules do not have a charge and cannot dissolve easily in water (hydrophobic).

4. Surface tension occurs when hydrogen bonds cause water molecules to form droplets and stay close together.

5. Amphipathic molecules have polar and nonpolar regions and hydrophilic and hydrophobic components.

6. Water is biologically important because it is the universal solvent of the body in which important nutrients, ions, and molecules are dissolved inside and outside of the cell. The specific concentration of different molecules within the cell or surrounding the cell is important to the cell's ability to function properly.

Section 2.2

1. Dehydration synthesis is the process by which macromolecules are made from their building blocks by the loss of a water molecule: for example, the combination of amino acids to build proteins. Hydrolysis is the process of breaking down macromolecules into their building blocks by the addition of a water molecule.

Section 2.3

1. There are 20 different amino acids, and nine are essential amino acids.

2. The three components of an amino acid are a carboxyl group, an amino group, and a functional group. The carboxyl group and amino group are the common components of every amino acid.

3. Peptide bonds, hydrogen bonds, and disulfide bonds. Peptide bonds hold amino acids together. Hydrogen bonds and disulfide bonds are important for the formation of secondary and tertiary structure.

4. The primary structure is the amino acid sequence. The secondary structure is based on the properties of the functional groups of the amino acids, and the way they fold depends on the interactions between the amino acids. Hydrogen bonds form between polar amino acids, and the hydrophobic amino acids are attracted to other hydrophobic amino acids. The tertiary structure is the three-dimensional shape the protein develops due to the primary and secondary structures. This functional structure of the protein is determined by the amino acid sequence. Polar or charged amino acids are attracted to each other and nonpolar amino acids are also attracted to each other. The quaternary structure is sometimes formed when multiple tertiary structures are combined.

5. Alpha helices are a secondary structure of a protein that forms a helix through hydrogen bonding. Beta sheets are a secondary structure of a protein that forms a twisted sheet through hydrogen bonding.

6. Hemoglobin, DNA polymerase, and microtubules

7. Denaturation, an irreversible reaction, occurs when there is enough energy applied to a protein to break the secondary- and tertiary-structure hydrogen bonds; the covalent peptide bonds remain unbroken, and the protein is no longer able to function.

8. Enzymes are important because they facilitate thousands of chemical reactions in our cells. They can form active sites that catalyze chemical reactions and can increase contact between specific regions of molecules to allow chemical reactions to occur.

9. The branched-chain amino acids are valine, leucine, and isoleucine. People take them as supplements to help muscle recovery after exercise.

10. Tryptophan is required for serotonin production.

Section 2.4

1. A chromosome is a single nucleic acid macromolecule called *DNA* that contains hundreds to thousands of genes and is found in every cell except mature red blood cells. Humans have 46 chromosomes.

2. The function of DNA is to store information for protein synthesis. DNA is located in the nucleus.

3. There are three components of a nucleotide: a five-carbon sugar deoxyribose or ribose; a phosphate group; and a nitrogen-containing base that can be double-ring base purines—adenine and guanine—or single-ring base pyrimidines—cytosine, thymine, and uracil.

4. Phosphodiester bonds, glycosidic bonds, and hydrogen bonds

5. TACCGATCAG

6. The function of RNA is to serve as a template to integrate the amino acid sequence of a specific protein.

7. See Table 2.2, DNA compared to RNA.

Section 2.5

1. Carbon, hydrogen, and oxygen

2. A monosaccharide has the formula $C_6H_{12}O_6$ and is a single-ring carbohydrate. A disaccharide consists of two monosaccharides paired together by glycosidic bonds. A polysaccharide is a complex carbohydrate that consists of long chains of glucose.

3. Glucose, galactose, and fructose. Two glucose molecules form maltose, glucose and galactose form lactose, glucose and fructose form sucrose.

4. Starch, glycogen, cellulose. Starch is stored energy found in plant cells; our bodies can digest starch for energy. Glycogen is stored energy in animal cells, and it can be stored in the liver and muscle cells if in excess. Cellulose functions in plant structure; it is not used for energy and cannot be digested by the human body.

5. Maltose—grains; sucrose—table sugar; lactose—milk; starch—rice; cellulose—fruits and vegetables.

6. The glycemic index is a way of measuring how much insulin is produced after eating different foods.

Section 2.6

1. Lipids are important for the functioning of multiple parts of the body. In the brain, fats are essential for brain function, including learning ability, memory retention, and mood. In the cell, fatty acids and cholesterol aid in cell flexibility and build cell membranes. Most of the heart's energy for contraction comes from burning fats. Fats also compose the material that protects and insulates nerves, isolating electrical impulses and speeding up their transmission. Fat in the form of surfactant allows the lungs to function properly. Fats are also used in digestion by slowing down digestion to give the body more time to absorb nutrients and give a constant level of energy. Fat allows fat-soluble vitamins to be absorbed and keeps the body satiated for long periods of time. Fats cushion and protect internal organs and decrease inflammation, as well as helping the body's metabolism and immune system stay healthy.

2. Adipose tissue consists of groups of adipocyte (fat) cells that contain triglycerides.

3. Saturated triglycerides are solid at room temperature and contain only single-bond carbon atoms; unsaturated triglycerides are liquid at room temperature and contain one or more double-bond carbon atoms.

4. Sources of healthy, saturated fats are meats, avocado, coconut oil, and dairy products. Sources of healthy unsaturated fats are plant oils, flaxseed, nuts, seeds, and vegetables. Trans fats should be minimized in your diet and are found in fast foods and deep-fried foods. Trans fats are bad for your health because the enzyme lipase that digests fats is specific only to cis conformation; therefore, trans fats are not broken down, remain circulating in the bloodstream, and can cause plaque in blood-vessel walls, which leads to heart disease.

5. Hydrogenation is the process in which hydrogen atoms are added to the fatty acids to break the double bonds in the carbon atoms to make single bonds.

6. Cholesterol is an important constituent of cell membranes; it is used by the liver to make bile, used to produce LDLs to transport fats through the bloodstream, and used to produce steroid hormones.

7. The primary type of fat that makes up cell membranes is phospholipid.

Section 2.7

1. Water soluble vitamins include vitamin C and all B vitamins.

2. Fat-soluble vitamins include vitamins A, D, E, and K.

3. Matching:
   a. Night blindness—vitamin A
   b. Neural tube defects—folic acid
   c. Scurvy—vitamin C
   d. Osteoporosis—calcium
   e. Fatigue—B vitamins
   f. Muscle cramping—potassium and magnesium
   g. Hypothyroidism—iodine

4. Animal products such as meat, dairy and eggs

5. Fruits and vegetables are high in vitamin C.

6. Vitamin D is made in the skin.

7. We need stomach acid and gut bacteria to absorb minerals.

Chapter 3 Concept Review Answers

Section 3.1

1. Cells have microvilli or dendrites or other cell-surface extensions to increase surface area, which increases the number of molecules that can enter or leave a cell.

2. The cell theory states the following:
   - All living organisms are composed of cells: including bacteria, protists, fungi, plants, and animals, but not including viruses, which are surrounded by protein capsules and are not considered "living."
   - The cell is the basic unit of all living organisms and is the smallest single unit that can survive independently.
   - All cells come from pre-existing cells.
   - All cells contain genetic material—DNA—that is passed on to new cells that arise from cell division and offspring.
   - All cells are either prokaryotic or eukaryotic and are composed of basically the same types of phospholipids, proteins, carbohydrates, and nucleic acid molecules.

Section 3.2

1. Nonpolar molecules are not attracted to water and are also called *hydrophobic*. Polar molecules are attracted to water and are also hydrophilic.

2.

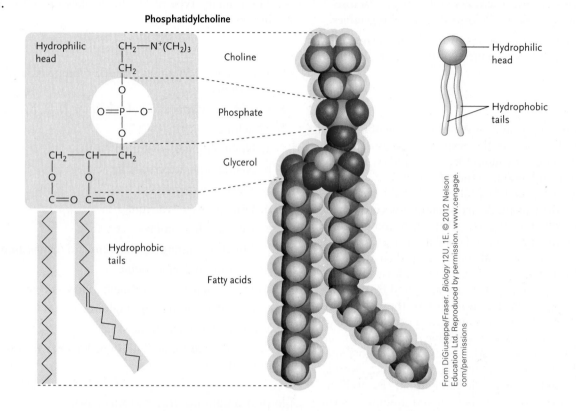

Phosphatidylcholine

Hydrophilic head

$CH_2—N^+(CH_2)_3$ — Choline
|
CH_2
|
O
|
$O{=}P{-}O^-$ — Phosphate
|
O
|
$CH_2—CH—CH_2$ — Glycerol
| |
O   O
| |
$C{=}O$ $C{=}O$

Hydrophobic tails

Fatty acids

Hydrophilic head

Hydrophobic tails

3. The general structure of a cell membrane is termed a fluid mosaic model; it consists of a phospholipid bilayer as well as cholesterol, proteins, glycolipids, and glycoproteins.

4. • Cholesterol—type of fat that increases the strength and decreases the fluidity of the cell membrane
   • Receptors—proteins that bind regulatory molecules such as neurotransmitters, hormones, cell-signalling molecules, or nutrients
   • Antigens—usually glycoproteins that act as cell markers
   • Enzymes—proteins that act as catalysts for assorted chemical reactions, some of which may occur on cell membranes
   • Ion channels—proteins that span the membrane and allow the transport of specific ions into or out of the cell
   • Carrier proteins—proteins that move substances such as nutrients across cell membranes by changing their conformational structure
   • Adhesion molecules—membrane proteins that play an important role in intercellular interactions
   • Gap junction channels—protein membrane channels that play an important role in intercellular communication

Section 3.3

1. Prokaryotic cells contain some or all of the following (depending on the species): nucleoid region of one circular DNA molecule, cell wall, capsule, cytoplasm, ribosomes, pili, flagella.

2. Circular DNA contains the genetic material; the cell wall maintains cell strength and regulates osmotic pressure; the capsule externally protects the cell from environmental changes; cytoplasm contains dissolved ions and nutrients; ribosomes are essential for protein synthesis; pili help the cell adhere to structures and transfer small amounts of DNA to other cells; flagella are used for locomotion.

3. The cell wall and capsule in prokaryotes are important because they provide extra protection for the cell.

4. Gram-positive bacteria stain purple when exposed to crystal violet due to the many layers of peptidoglycan and gram-negative bacteria do not stain due to a smaller layer of peptidoglycan and an extra external layer of lipopolysaccharide.

5. Eukaryotes are approximately 10 times larger than prokaryotes and contain a more complex cell structure and contain membrane-bound organelles.

Section 3.4

1. See Table A.2.

TABLE A.2

Organelle	Function
Centriole	Responsible for organizing spindle fibres during cellular division
Peroxisome	Breaks down lipids for ATP generation in the mitochondria
Lysosome	Breaks down organelles, molecules, and macromolecules as well as bacteria and viruses
Ribosome	Responsible for the production of polypeptides by acting as a site for translation
Mitochondria	Produce ATP through oxidative metabolism
Golgi body	Acts to package and modify proteins
Endoplasmic reticulum	Responsible for the synthesis of proteins on rough ER, as well as carbohydrates and lipids including cholesterol on smooth ER
Nucleus	Houses the genetic information of the cell and is the site of RNA synthesis

2. Mitochondria are thought to have potentially developed as an organelle by means of endosymbiosis due to the fact that they have a double plasma membrane and share many traits in common with bacteria (size, DNA, organelles, and process of replication).

Section 3.5

1. See Table A.3.

TABLE A.3

Cytoskeletal Protein	Function
Microtubule	Maintains cell shape/structure and facilitate intracellular transport
Intermediate filament	Supports cell structure by holding organelles in place
Microfilament	Involved in cell movement

2. See Table A.4.
3. Desmosomes hold the cell membranes of adjacent cells together and prevent damage from shearing. Tight junctions hold the cells together, primarily in epithelial cells where only a limited amount of extracellular fluid can pass through. Gap junctions permit the exchange of small particles between cells.

TABLE A.4

Protein	Function
Integrin	Connects cell membranes to the proteins of the ECF and transmits signals across the membrane
Collagen	Provides strength and resists stretching
Elastin	Allows tissues to stretch and return to original shape
Fibronectin	Connects integrin to ECF proteins

4. Desmosomes are found in stratified epithelial tissue and cardiac muscle. Tight junctions are found in epithelial tissues, such as the blood-brain barrier and gap junctions, are found in cardiac muscle tissue.

Section 3.6

1. Certain molecules are able to pass through while others are not, either due to charge or size of the molecule.
2. Diffusion is the passive movement of particles from areas of high concentration to areas of low concentration. Most small, nonpolar molecules can move easily through the body.
3. Large molecules and polar molecules (except water) typically require proteins to move in and out of the cell.
4. Osmosis is movement of water to an area of low-water concentration from an area of high-water concentration without the input of energy.
5. A solute is a solid substance that is dissolved by the solute. The solute is a liquid substance that when containing a solvent becomes a solution.
6. If the concentration of a solute in a cell is greater than that in the external environment, water will move into the cell by means of osmosis. Conversely, if the concentration of the solute is greater outside the cell, water will move from inside the cell to outside the cell.
7. Hypertonic, hypotonic, and isotonic are relative terms related to different solutions. When a cell is placed in a solution that is hypertonic to the internal fluid of the cell, the water from the cell moves outside the cell, causing the cell to crenate. When the cell is placed in a solution that is hypotonic to the internal fluid of the cell, water moves from outside the cell to inside, causing the cell to swell and perhaps burst. When the cell is placed in an isotonic solution, the concentrations inside and outside the cell are equal, and there will be no net movement of water.
8. Facilitated diffusion is the movement down a concentration gradient without the use of energy, but

aided by a protein channel or carrier. Channels are specific protein openings in the membrane that allow very specific ions to move across; carriers are proteins that bind to a specific molecule and change its shape, which then causes that molecule to move across the membrane.

9. There are no longer any binding sites available for molecules to attach and diffusion will not be aided by the proteins.

10. Active transport is the movement of particles against a concentration gradient by use of proteins. It differs from facilitated diffusion in that the direction of the movement of particles is reversed (against the gradient rather than with the gradient), and this requires energy.

11. The sodium-potassium pump and proton pump are two forms of active transport in the body's cells.

12. Sodium is moved out of cells by means of the sodium-potassium pump. To do this, one ATP is used to alter the shape of the protein that moves three sodium ions out of the cell and two potassium ions into the cell.

13. Sodium-potassium pumps are required for maintaining both concentration and electrical gradients across cell membranes; they are found in all body cells.

14. Proton pumps are found in the mitochondria of cells, and they function to move hydrogen ions into the intermembrane space to create a concentration gradient and electrochemical gradient. The buildup of hydrogen ions causes the protons to move back across the membrane through specific channels, causing ATP synthase to bind a phosphate to ADP creating ATP: a process called chemiosmosis.

15. The natural diffusion of molecules, such as sodium, into the cell creates the energy needed to move larger molecules, such as sugar and amino acids, into the cell.

16. *Endocytosis* (both phagocytosis and pinocytosis) refers to the movement of large molecules into the cell; exocytosis is the movement of large molecules out of the cell.

17. Phagocytosis is the movement of a solid into a cell. Pinocytosis is the movement of a liquid into a cell. Receptor-mediated endocytosis is the movement of a specific substance into a cell that binds to a receptor; therefore, the substance can enter only the cells that have the receptor. For example, HIV binds and enters only white blood cells that have the CD4 receptor.

Chapter 4 Concept Review Answers

Section 4.1

1. Adenosine triphosphate (ATP)

2. ATP yields energy when the last phosphate bond is broken and the phosphate then binds to another molecule, converting ATP to ADP. ATP is used by our cells to perform many functions every second.

3. Glucose, fatty acids, amino acids, and nucleotides

4. Oxidation is a reaction in which a molecule loses electrons. Reduction is a reaction in which a molecule gains electrons.

5. Redox reactions are important for cellular respiration because as glucose becomes oxidized and NAD^+ molecules are reduced, forming NADH.

6. A Calorie is a measurement of food energy, also called a *kilocalorie*. A calorie is the amount of energy required to raise the temperature of 1 mL of water 1°C.

7. The carbon-hydrogen bonds in food molecules are important for the redox reactions because those bonds contain energy, which leads to the production of ATP through cellular respiration.

8. Aerobic respiration is the oxidation of glucose in the presence of oxygen to produce ATP; anaerobic respiration occurs when ATP is produced without oxygen.

Section 4.2

1. Glucose, NAD^+, and two ATP

2. Cytoplasm of all cells

3. Glycolysis is important to produce a small amount of ATP without oxygen.

4. Two pyruvate molecules, two NADH, net two ATP

5. Substrate-level phosphorylation occurs when a specific enzyme in the cytoplasm adds a phosphate to ADP to become ATP during glycolysis.

6. Prokaryotic and eukaryotic organisms can undergo glycolysis.

Section 4.3

1. Pyruvate oxidation and the Krebs cycle occur in the matrix of the mitochondrion.

2. Two acetyl-coenzyme A (acetyl-CoA) molecules

3. Carbon dioxide (CO_2)

4. The production of NADH and $FADH_2$ is the most important result of the Krebs cycle.

5. Two ATP molecules are formed during the Krebs cycle, one for each acetyl-CoA molecule.

6. NADH and $FADH_2$ molecules are important because they provide the hydrogen ions and electrons needed for the electron transport chain.

Section 4.4

1. The purpose of the NADH and $FADH_2$ molecules created during the Krebs cycle is to provide electrons and hydrogen ions required for the electron transport chain to produce ATP.

2. The H^+ ions are transported by a proton pump across the inner mitochondrial membrane (cristae) into the intermembrane space. This forms a high H^+ concentration gradient in the intermembrane space, so the H^+ ions can move down the gradient and back into the matrix, using ATP synthase to produce ATP.

3. NADH and $FADH_2$

4. The electrons combine with hydrogen ions and oxygen molecules (from breathing) to form water molecules at the end of the electron transport chain.

5. ATP synthase is a membrane protein that allows H^+ ions to move back into the matrix of the mitochondrion.

6. Chemiosmosis involves a chemical gradient moving across the membrane and the production of many ATP molecules.

Section 4.5

1. Glucose, fatty acids, glycerol, amino acids, and nucleotides

2. Proteins are broken down into amino acids when digested. Instead of going through glycolysis, amino acids are oxidized by removing a nitrogen group (deamination) to produce pyruvate or acetyl-CoA to enter the Krebs cycle. Nucleic acids are broken down into nucleotides when digested. Like amino acids, nucleotides also contain a nitrogen group and must undergo deamination, not glycolysis, to enter the Krebs cycle. Fats are broken down into fatty acids and glycerol when digested. The carbon-hydrogen bonds in long-chain fatty acids are oxidized into acetyl-CoA, which then enters the Krebs cycle.

3. Deamination is the process of removing the nitrogen group from a molecule in the form of ammonia NH3. Amino acids and nucleotides undergo deamination.

4. CO_2 and water. When amino acids or nucleotides are used to make ATP, ammonia is also a waste product.

5. Lipase is required for lipolysis.

6. Beta-oxidation is the process of breaking down fat for ATP production.

Section 4.6

1. Anaerobic respiration in humans: NADH is converted back to NAD^+ through fermentation, and the H^+ is added to pyruvate to become lactate. Anaerobic respiration in yeasts: pyruvate undergoes decarboxylation, producing acetaldehyde. NADH is converted back to NAD^+ through fermentation, and the H^+ is added to acetaldehyde to form ethanol.

2. Anaerobic respiration occurs in human cells when there is a lack of oxygen, such as during high-intensity exercise.

3. Fermentation is the process of moving the H^+ from NADH to pyruvate to create lactic acid or alcohol.

4. Lactate formed in the muscle cells can be converted back to pyruvate once oxygen is available.

Section 4.7

1. Insulin causes cells to convert nutrients into fat.

2. Hormones that lead to fat breakdown include: cortisol, growth hormone, glucagon, epinephrine, and thyroid hormones

3. Three stimuli that would increase the production of hormones that break down fat include, exercise, deep sleep, and fasting.

Chapter 5 Concept Review Answers

Section 5.1

1. The two main components of bacterial replication are DNA and binary fission, in which the bacterial cell separates into two identical cells.

2. A plasmid is a small extra piece of DNA, in addition to the larger, single, circular DNA molecule found in some bacteria.

3. Step 1. DNA replication begins at the point of origin where enzymes begin to separate the DNA strands.

   Step 2. DNA polymerase adds complementary nucleotides in both directions until two new identical DNA strands are created.

4. Bacteria acquire genetic variation through methods of DNA transfer with other bacteria: conjugation, transduction, and transformation.

5. Conjugation occurs when bacteria transfer DNA through a pilus, as a plasmid or piece of DNA, directly to another bacterium. Transduction occurs when bacteria receive new DNA fragments from viruses. Transformation occurs when bacteria receive DNA fragments or plasmids that are near them (e.g., dead bacteria).

Section 5.2

1. The cell cycle contains two growth phases (G_1 and G_2), DNA replication (S), and cell division (M). Prokaryotes and eukaryotes have a cell cycle.

2. Cells spend the most time in interphase (mostly G_1).

3. In G_1, cells undergo normal cell function and prepare for DNA replication. In S phase, DNA replication occurs. In G_2, cells continue to prepare for mitosis, and DNA is checked for mutations at the

G_2 checkpoint. During mitosis, there is no more cell growth and chromosomes and organelles separate. This results in the formation of two new daughter cells identical to the original cell.

4. Cytokinesis is the process of splitting the cytoplasm.

5. The main purpose of mitosis is to produce two identical daughter cells.

6. Humans have 46 chromosomes in each cell nucleus, 23 chromosomes from each parent.

7. Homologous chromosomes are two similar chromosomes that are not identical. Sister chromatids are the two identical copies of a single chromosome that has been replicated.

8. *Karyotype* refers to chromosomes and their homologous pairs organized from chromosome 1 to chromosome 23.

Section 5.3

1. The main events in the five stages of mitosis are as follows:

   • Prophase. DNA replicated from interphase is condensed, nuclear membrane and nucleolus breaks down, centrioles migrate to opposite ends of the cell, and spindle fibres form.

   • Metaphase. Spindle fibres connect to a kinetochore at the centromere of each set of sister chromatids, which line up at the equatorial plain.

   • Anaphase. Spindle fibres pull the chromosomes toward the opposite poles of the cell.

   • Telophase. Sister chromatids are at opposite poles of the cell, spindle fibres break down, two nuclear membranes form around each set of chromosomes at opposite ends of the cell, chromosomes begin to uncondense, and nucleolus reforms.

   • Cytokinesis. The cytoplasm of the cell divides equally, forming a cleavage furrow in animal cells, and resulting in two somatic, identical, diploid daughter cells.

2. Spindle fibres are microtubules that connect to the sister chromatids at the kinetochore and pull them apart toward opposite ends of the cell.

3. The chromosomes are equally distributed by lining up at the equatorial plane, resulting in the same number of chromosomes moving to opposite ends of the cell when they are pulled apart.

4. Each sister chromatid is identical, so it does not matter which one moves into which new cell.

5. There is a diploid number (2n) of chromosomes in each new cell; this totals 46 in human cells.

6. Examples of somatic cell include liver cells, bone cells, and skin cells.

Section 5.4

1. Apoptosis is programmed cell death.

2. Angiogenesis is the production of new blood vessels.

3. The three checkpoints are as follows:

   • G_1 checkpoint. Cell checks for DNA mutations and, if there are none, proceeds to S phase for DNA replication.

   • G2 checkpoint. Cell checks for mutations that may have occurred during DNA replication, and specific enzymes may correct these mutations.

   • M checkpoint. The final checkpoint occurs during metaphase of mitosis to determine whether the cell should complete cytokinesis.

These three checkpoints are involved in cell regulation to make sure each phase is completed without mistakes before moving on to the next phase. If a mutation is found it will either be fixed by specific enzymes or undergo programmed apoptosis to prevent further mutations.

4. A carcinogen is anything that can cause the development of cancer cells.

5. The factors that can cause cancer include inherited mutations; errors that occur during DNA replication; and environmental factors such as radiation, viruses, and chemicals.

6. Proto-oncogenes regulate the rate of the cell cycle by coding for proteins that increase the cell cycle during normal periods of growth, pregnancy, or repair of damaged tissue. An oncogene, which is a mutated proto-oncogene, abnormally increases the rate of cell division. A tumour-suppressor gene also regulates the rate of the cell cycle by coding for proteins that perform at the three checkpoints to stop the cell cycle if the DNA is damaged. An example of a tumour-suppressor gene is p53.

7. Mutated proto-oncogenes, tumour-suppressor genes, DNA repair enzymes, genes involved in apoptosis, genes that increase angiogenesis, and genes that regulate the immune response can be involved in the development of cancer.

8. The tumour-suppressor gene p53 prevents cancer at development by stopping cell division at the G1 checkpoint if it detects abnormal DNA present before DNA is replicated.

9. Our immune system plays a role in preventing cancer through the presence of natural killer white blood cells that specifically destroy cancer cells.

Section 5.5

1. The purpose of meiosis is to generate haploid sperm or eggs, called gametes, required for an organism to reproduce.

2. Gametes (sperm or eggs) are haploid cells and contain 23 chromosomes. Zygotes and all somatic cells are diploid cells and contain 46 chromosomes.

3. Homologous chromosomes separate during anaphase I, and sister chromatids separate during anaphase II.

4. Gametes are haploid sperm cells formed in male testes and haploid eggs formed in female ovaries.

5. Genetic variation is important because it ensures that every individual is unique and not identical to the parent. A population with high genetic variation is more likely to survive environmental changes.

6. Independent assortment and crossing over.

7. In males there are four sperm cells, each with 23 chromosomes by the end of meiosis II. In females, cell division occurs so that there is more cytoplasm in one cell and the other cell becomes a polar body, which leaves one egg cell. The eggs remain arrested in metaphase II until ovulation.

8. Crossing over is important because it increases genetic variation.

Chapter 6 Concept Review Answers

Section 6.1

1. • Homozygous—alleles the same
   • Heterozygous—alleles different
   • Dominant—allele that is phenotypically prominent (Capital letter)
   • Recessive—allele that is hidden by the dominant allele (Lower-case letter)
   • Allele—one copy of a gene in a diploid cell
   • Phenotype—protein that displays a visible trait or function
   • Genotype—genes that code for the phenotype

2. a. Purple
   b. One gamete containing a purple allele, and one containing a white allele
   c. PP, Pp, and pp
   d. 25%

Section 6.2

1. See Table A.5.

TABLE A.5

	Y	y
y	Yy	yy
y	Yy	yy

a. 50%
b. 50%
c. 50%
d. 0%

2. A testcross can be used to determine if the purple-flowered plant is PP or Pp by crossing the plant with a homozygous recessive white-flowered plant (pp). If the outcome is 100% purple, the genotype is PP. If the outcome is 50% purple and 50% white, the genotype is Pp.

3. Mom is heterozygous dominant, dad is homozygous recessive, and child is homozygous recessive.

4. Yes, if they are both heterozygous dominant for the trait

5. No.

Section 6.3

1. See Table A.6.

TABLE A.6

	RY	Ry	RY	Ry
Ry	RRYy	Rryy	RRYy	Rryy
Ry	RRYy	Rryy	RRYY	Rryy
ry	RrYy	Rryy	RrYy	Rryy
ry	RrYy	Rryy	RrYy	Rryy

a. RRYy, Rryy
b. Each parent can have four possible gametes
c. RRYy, RRYY, RrYy, Rryy
d. Round and Yellow or Round and green
e. ½ Round and Yellow and ½ Round and green

2. There is a 25% chance of having a child with brown hair and freckles.

Section 6.4

1. Polygenic inheritance occurs when a single trait is determined by multiple genes: for example, body height, weight, or skin colour.

2. *Pleiotropic* refers to one single allele causing multiple phenotypes, example, cystic fibrosis.

3. The heterozygous bird's feathers would be purple. If two heterozygotes mated one-fourth of their offspring would be blue.

4. The dad could be either Larry (ii) or Bill, if Bill is $I^B i$. Mom has blood type $I^A i$ because in order to be type O you must receive i from both parents.

5. The possible genotypes are $I^A i$ or $I^B i$, and the possible phenotype is either type A or type B.

6. Epistasis occurs when the effects of one gene are modified by one or more other genes. A Labrador retriever may have the dominant black fur gene (B) but is yellow because it contains the recessive epistasis allele for pigmentation (e) and conceals the gene for black fur.

7. Examples include temperature of incubating reptile eggs, air temperature that affects fur colour, and soil pH of hydrangeas.

8. They will have a 50% chance of having children that are colour blind.

9. Linkage is when genes are inherited together because they are located closely together on the same chromosome.

Section 6.5

1. Chromosome 21 is affected in Down syndrome.

2. XXX (Trisomy X), XXY (Klinefelter syndrome), XO (Turner syndrome).

3. • Sickle cell anemia—autosomal recessive; abnormal red blood cell formation; and reduced capacity to carry oxygen

   • Tay Sachs—autosomal recessive; brain deterioration leading to death before adolescence

   • Cystic fibrosis—autosomal recessive; excessive mucus production impacting the lungs, liver, pancreas, and sweat glands

   • Huntington's disease—autosomal dominant; brain deterioration beginning in the late 30s

   • Hemophilia—recessive X-linked; lack of blood clotting, resulting in the inability to stop bleeding

   • Duchene muscular dystrophy—recessive X-linked; muscle degeneration that affects children in early childhood

   • Familial hypercholesterolemia—autosomal dominant; possible occurrence of cardiovascular disease by age 30

Chapter 7 Concept Review Answers

Section 7.1

1. Evolution is the change in the characteristics of organisms within a population over time.

2. Survival of the fittest means individuals born with traits that are most beneficial for a particular environment will survive, reproduce, and pass on DNA. For example, predators that are better able to catch prey will have more nutrients, better health, and therefore survive and produce offspring with the same traits.

3. The relationship between evolution and ecology is that environmental changes determine which organisms are the "fittest." Changes in the environment, such as the availability of certain foods, affect which organisms survive and pass on their genes.

4. The major factors involved in natural selection are as follows: genetic variation is beneficial, similarities in DNA sequences are more common in closely related organisms, organisms that survive and reproduce pass on their DNA, all organisms over-reproduce so that the fittest survive, and populations change as the environment changes.

5. Examples of environmental factors that can affect evolution are seasons, weather patterns, ice ages, forest fires, availability of food, the presence of predators, or infectious organisms.

6. Examples of human-inherited traits that would be beneficial for survival include intelligence, being fertile, and having the motivation to work and earn money to buy food and shelter. Two hundred thousand years ago, humans needed to be physically fit to hunt for food. People no longer have to be physically fit or be able to hunt to survive because we now live in a technological environment. In different environments, different traits are important for survival.

Section 7.2

1. Bacteria can reproduce every 30 minutes in perfect conditions and therefore can have a very rapid generation time. Bacteria will undergo genetic changes based on changes in the environment, such as lack of a nutrient, change in temperature, or level of moisture, and also recently, exposure to antibiotics. The exposure to antibiotics has created a selection pressure where only those able to withstand exposure to antibiotics are able to survive and therefore pass on the antibiotic resistance gene to future generations of bacteria.

2. As bacterial antibiotic resistance increases, those humans with decreased immune function become more susceptible to infection, and they lack the ability to combat the infections if the antibiotic cannot kill the infection.

3. *Microevolution* refers to the small genotype and phenotype changes within a species, where macroevolution leads to organisms that are so different that they are a new species.

4. Darwin's theory suggests that giraffes were inherently born with longer necks and, therefore, were better adapted to their environment, passing on the "long-neck gene" to future generations. Lamarck would suggest that as adult giraffes were required to stretch their necks, their neck length increased throughout their lifetime. In response, offspring were born with longer necks and, again, they had to stretch, and thus increased the length of their neck throughout their lifespan.

5. Natural selection is the result of a species' reaction to its environment. Artificial selection is the result of a

breeder choosing which plants or animals will reproduce in order to perpetuate specific characteristics.

Section 7.3

1. The finches that Darwin studied are identified as separate species based on the size and shape of their beaks, which are highly adapted to different types of food in South America and the Galapagos. The differences between the birds are enough to make them different species, but there are also a number of similarities between the birds. Darwin hypothesized that microevolution leads to the macroevolution, creating all of the finch species that exist today as a reaction to the environment and food supply. Different finches will be more "fit" than others to eat certain foods, such as large hard seeds, small seeds in small cracks, berries, worms, or insects and, as a result, pass on those traits to their offspring.

2. Gradualism is the slow development of a new species through a series of small changes that lead to a new species over a great deal of time. Punctuated equilibrium in contrast is the theory that little to no changes take place over a great deal of time, with sudden drastic changes appearing all at once.

3. The three primary methods that provide evidence of evolution are homologous anatomy, the fossil record, and homologous molecules

4. Bats are more closely related to humans than to birds because of the similarity of the bones in the forelimb. Bats and humans both have a humerus, ulna, radius, and phalanges. Birds have a completely different structure in their wings.

5. A selection pressure is something that affects the course of evolution of a species because it creates a condition that permits certain individuals in a population to have an increased likelihood of survival compared to others in the same population.

6. Natural selection can be affected by means of the following types of selection (see Table A.7):

TABLE A.7

Type of Selection	Function	Example
Directional selection	Benefits one extreme of the population	Antibiotic resistance in bacteria
Stabilizing selection	Benefits the intermediates of the population	Baby birth size
Disruptive selection	Benefits both extremes of the population	Finch beak size

7. Sickle cell anemia acts through stabilizing selection by benefitting the intermediate phenotype of the population.

8. *Heterozygote advantage* refers to the additive benefits of having both the dominant and recessive alleles, which increases the likelihood of survival for those individuals with both genes.

9. Predation is the primary selection pressure acting on the guppies.

10. Any factors that affect the Hardy-Weinberg equilibrium also affect allele frequencies.

11. The bottleneck effect is random in regard to which individuals it affects—rather than genetics conferring an enhanced survival rate among certain individuals.

12. DNA mutations are the ultimate source of new alleles, though migration may also lead to new alleles in a particular population.

Section 7.4

1. Factors that maintain Hardy-Weinberg equilibrium include random mating, large populations, the lack of mutations, lack of migration, the lack of ecological changes, and the lack of selection pressures.

2. Answer calculation is as follows:
   - $p = 0.5$
   - $p + q = 1$; therefore, $q = 0.5$
   - $p^2 + 2pq + q^2 = 1$
   - $(0.5)^2 + 2(0.5)(0.5) + (0.5)^2 = 1$
   - p^2 is the number of homozygous dominant mice
   - $2pq$ is the number of heterozygous mice
   - q^2 is the number of homozygous recessive mice
   - Short-tail alleles are recessive and therefore will display the phenotype only if homozygous: therefore, $(0.5)^2$ or 25% of the time.
   - Long-tail alleles are dominant and therefore will display the phenotype if either homozygous or heterozygous: therefore, $p^2 + 2pq$ or $(0.5)^2 + 2(0.5)(0.5)$ or 75% of the time.

Chapter 8 Concept Review Answers

Section 8.1

1. Adenine, guanine, thymine, cytosine, and uracil

2. • Purines—adenine, guanine
   • Pyrimidines—thymine, cytosine, uracil

3. A purine always binds to a pyrimidine, adenine binds to thymine, and guanine binds to cytosine in DNA.

4. Hydrogen bonds, phosphodiester bonds, glycosidic bonds

5. A nucleoside consists of a purine or pyrimidine base attached to a sugar; a nucleotide is a purine or pyrimidine attached to a sugar and a phosphate group.

6. We acquire nucleosides from our diet when we eat any plant or animal cells (vegetables, fruit, and meat).

7. DNA forms a double helix.

8. DNA has thymine, RNA has uracil. DNA has deoxyribose, RNA has ribose. DNA is double stranded, RNA is single stranded.

9. The carbons of the sugars are named 1, 2, 3, 4, 5. This is important because carbon 3 and 5 bind to the phosphate group to make up the sugar-phosphate backbone. If there are two hydrogen atoms on carbon 2 the sugar is deoxyribose, and if there is a hydroxyl group on carbon 2 it is ribose.

10. Ribose contains a hydroxyl group on carbon 2, while deoxyribose contains only two hydrogen atoms on carbon 2.

11. ATP is the energy molecule of the cell.

12. Telomeres are a repeated sequence at the ends of each chromosome. They are important because they protect the genes on the ends of the chromosomes from deterioration during DNA replication. Stem cells and germ cells produce telomerase. This is important because it replaces lost telomere sequences; therefore, these cells will always be able to divide.

13. Prokaryotes do not contain telomeres because their DNA molecules are circular and can replace all primers with DNA replication, which allows bacteria to divide indefinitely.

Section 8.2

1. Semi-conservative replication means that each of the original strands of DNA is used to produce the two new strands. Every new DNA molecule contains an old strand and a new strand.

2. • Helicase—breaks the hydrogen bond between the bases of the complementary strands of DNA, starting at the origin and separating the two strands, creating a replication fork

   • Single-stranded binding protein—stabilizes the single-stranded region of the DNA molecule

   • Primase—adds a RNA primer, which is a small sequence of nucleotides with a 3' hydroxyl group

   • Polymerase—adds the complementary nucleotides to each single strand of DNA in the 5' to 3' direction only

   • Ligase—forms phosphodiester bonds to seal the gaps between the nucleotides

   On the leading strand, one RNA primer is needed, and polymerase will follow helicase, adding complementary nucleotides in the 59 to 39 direction on the new strand, and this is continuously replicated. On the lagging strand, replication is done in Okazaki fragments by many new RNA primers in order for polymerase to add nucleotides in the 5' to 3' direction.

3. For polymerase to add complementary nucleotides, it needs a 3' hydroxyl group to proceed in the 5' to 3' direction. Primase adds an RNA primer that has a free 3' hydroxyl group, thus allowing polymerase to start adding the complementary nucleotides.

4. See Glossary.

5. Even though DNA replication enzymes are the same in prokaryotes and eukaryotes, replication differs in prokaryotes because the DNA is circular, only one primer is needed on each strand, and polymerase can replicate continuously on both strands. This is different from eukaryotes, where Okazaki fragments are needed on the lagging strand. Also, since DNA is circular in prokaryotes, supercoiling occurs when the DNA unwinds and therefore the enzyme topoisomerase is needed to release the over-twisting; this is not required in eukaryotes.

Section 8.3

1. Proofreading occurs when the polymerase enzyme recognizes and removes incorrect base pairs and replaces them with the correct nucleotides.

2. Our cells cope with DNA mutations by proofreading. Polymerase can recognize mistakes made during replication and will stop replicating, reverse in the 3' to 5' direction, remove the improper nucleotides by utilizing its exonuclease activity function, and continue replication in the 5' to 3' direction. We also have approximately 130 different genes that code for DNA repair enzymes, and they can locate mistakes missed by polymerase during replication. Tumour-suppressor genes also prevent cell division if mutations occur so DNA repair enzymes can fix the mutation. The p53 gene will induce cell apoptosis if the mutation cannot be corrected, thus preventing cell division with the mutation. If mutation still persists through cell division, our immune system produces natural killer cells to recognize and kill cancerous cells.

3. • Substitution (mismatch) occurs when polymerase adds the wrong nucleotide.

   • Addition (insertion) occurs when polymerase adds an extra nucleotide.

   • Deletion occurs when polymerase skips a nucleotide.

   • Recombination (transposition) occurs when large sequences of DNA are moved to another location within the genome.

4. A frameshift mutation occurs when a nucleotide is added or removed, causing a shift in all of the following nucleotides, which affects every other codon after that mutation.

5. The consequences of DNA mutations occurring in somatic cells can be harmful and possibly lethal to the individual person, but they are not passed on to offspring. Mutations in gametes can be passed on

to offspring; this results in genetic variation among individuals.

6. Mutations contribute to genetic variation by being passed on to offspring and affecting biochemical processes, development, behaviour, morphological traits, or gene expression.

7. Insertional inactivation is the insertion of a sequence of DNA into a location that disrupts a gene and produces either a non-functional protein or no protein.

8. Transposable elements are sequences of DNA that can move themselves (transpose) to another chromosome within the cell. They do not always cause DNA mutations.

9. We acquire DNA mutations by mistakes during DNA replications, transposition, mutagens or carcinogens, viruses, and inherited mutations.

10. The most common types of carcinogens are radiation, environmental toxins, and viruses.

11. See Table A.8.

TABLE A.8

Virus	Cancer
Human papilloma virus (HPV)	Cervical cancer
Human immunodeficiency virus (HIV)	Kaposi sarcoma
Hepatitis B or C	Liver cancer
Epstein–Barr virus (EBV)	Lymphoma

12. See Table A.9.

TABLE A.9

Genetic Disease	DNA Mutation
Cystic fibrosis	Mismatch base in gene that codes for chlorine channels
Huntington's disease	Insertion of multiple CAG repeats in a gene on chromosome 4
Sickle cell anemia	Mismatch in the hemoglobin gene
Cancer	Two or more mutations in genes that code for repair enzymes, cell cycle genes, oncogenes, or tumour-suppressor genes
Phenylketonuria (PKU)	Point mutation in liver enzyme gene that causes brain damage
Nonpolyposis colorectal cancer	Autosomal dominant repeated CA mutation in DNA repair enzyme gene expressed in large intestinal cells

Chapter 9 Concept Review Answers

Section 9.1

1. Adenine and thymine are complementary base pairs and guanine and cytosine are complementary base pairs.

2. The central dogma of gene expression concerns the flow of information from gene to RNA to protein. DNA is transcribed to mRNA (transcription), which then moves out of the nucleus through nuclear pores to a ribosome for protein synthesis to occur (translation).

3. The sense strand is the strand of DNA that codes for the gene and will have the same sequence as the mRNA. The antisense strand is the opposite strand and is the template for the mRNA.

4. The human genome is made up of 25 000 genes.

Section 9.2

1. The differences in transcription between prokaryotes and eukaryotes are as follows:

   - Prokaryotes do not contain a nucleus; therefore, they can undergo transcription and translation simultaneously.

   - Prokaryotes do not contain introns like eukaryotes and therefore have a simpler process.

   - They both have different mechanisms for regulating gene expression.

2. Transcription takes place inside the nucleus in human cells.

3. RNA polymerase binds to a promoter region of the DNA upstream of the gene that initiates transcription.

4. RNA polymerase "knows" where the beginning of the gene is located because the promoter regions are regulatory regions of the DNA that consist of nucleotide sequences that are called the TATA box or CAAT box.

5. AUG is almost always the first codon in an mRNA molecule.

6. The polymerase is using the template (antisense) strand to form the mRNA molecule and can add nucleotides only in the 5′ to 3′ direction on the mRNA molecule.

7. A terminator sequence prevents RNA polymerase from transcribing nucleotides that are not part of the gene sequence. This causes the mRNA to form a hairpin loop, causing RNA polymerase to dissociate.

8. In RNA processing, eukaryotes contain introns that are sequences of DNA that do not code for amino acids and must be removed. The regions of DNA sequences that do code for amino acids are called exons. To remove the introns, transcription first produces a pre-mRNA molecule to be processed by

removing the introns by splicing and adding a 5' cap and 3' tail. Splicing occurs in a spliceosome that consists of pre-mRNA and snRNPs and forms new bonds between the exons. The 5' cap allows mRNA to begin translation at the ribosome, and the 3' tail protects it from rapid enzymatic degradation. The new mRNA molecule that consists only of exons can now move out of the nucleus.

9. Only one strand of DNA is a gene sequence (coding strand). RNA polymerases make RNA, and DNA polymerases replicate DNA. The RNA molecule produced from transcription is single stranded, RNA polymerases do not need a primer, and transcription requires processing.

10. The mRNA sequence is 5' UACGGCUAUG 3'

Section 9.3

1. Ribosomes are composed of a small subunit and a large subunit, which both contain rRNA and proteins.

2. mRNA binds to the small subunit of the ribosome to be translated. rRNA makes it possible for the mRNA to bind to the ribosome. tRNA carries the amino acids to the three binding sites on the large subunit of the ribosome by the anticodon of the tRNA matching the codon on the mRNA.

3. The A site is where the new tRNA enters the ribosome. The P site is where enzymatic activity of the rRNA in the large subunit catalyzes the formation of peptide bonds between amino acids. It is also where the first tRNA molecule binds. The E site is where the empty tRNA molecule moves after the P site and then leaves.

4. Ribosomes are also called ribozymes because the RNA in ribosomes have enzymatic activity.

5. The role of activating enzymes is to recycle the tRNA once it leaves the ribosome in the cytoplasm and attaches the correct amino acid to it.

6. Translation works by synthesizing a protein from an mRNA sequence that binds to a ribosome. tRNA that contains the anticodon then brings the amino acid to the ribosome to bind to the correct codon forming peptide bonds in the P site of the ribosome between the amino acids until the protein is complete.

7. When a stop codon enters the A site, no tRNA molecule can bind, and the ribosome will dissociate, thus completing translation.

8. Protein tertiary structure is complete when translation is complete.

9. If translation occurs on a ribosome of the rough ER, proteins are sent to the Golgi bodies to be processed, such as proteins combined with lipids to produce lipoproteins, or sugars combined with proteins to produce glycoproteins.

10. Leu-Asp-Thr-Glu

11. A silent mutation occurs when a mismatch mutation happens outside a coding region or does not have an effect on the amino acid sequence; therefore, the DNA mutation has no impact on the function of the protein.

Section 9.4

1. Regulation of gene expression is related to homeostasis because molecules involved in regulating body systems—such as blood pressure or utilization of glucose—have to be transcribed and translated in the correct amounts to properly regulate the system and maintain homeostasis.

2. Two types of regulatory proteins that affect transcription in prokaryotes are activators and repressors.

3. The lac operon is found in bacterial species where three genes transcribed together in the presence of lactose.

4. The repressor binds to the operator sequence. The CAP activator binds to the CAP-binding sequence. The RNA polymerase binds to the promoter.

5. The repressor is removed from the DNA when lactose is present and binds to it.

6. Transcription only occurs if lactose is present and glucose is not present.

7. If glucose and lactose are present transcription does not occur. Since glucose is present, the CAP activator does not bind to the CAP-binding sequence; therefore, RNA polymerase will not bind to the promoter site, even though lactose is present and the repressor is removed.

8. See Table A.10.

TABLE A.10

Mechanism of Gene Regulation	Description
Methylation	Methyl groups are added to prevent gene transcription, and they are maintained during DNA replication.
RNA processing	Introns are removed to form a mature mRNA molecule, but not necessarily in the same manner every time the gene is transcribed; this is called alternate splicing.
mRNA degradation	Many mRNA transcripts have specific lifespans, and they may be broken down before translation occurs.
Transcription factors	This is the most common mechanism. Many factors can bind to regulatory regions of the DNA and promote transcription of certain genes.

9. Gene silencing occurs when a cell prevents the transcription of a gene.

10. Alternate splicing occurs when only certain exons become part of the mature mRNA molecule after the introns are removed during RNA processing, and other exons at other times. This is important because protein function depends on which exons are used, and this allows over 100 000 different proteins from 25 000 genes.

Chapter 10 Concept Review Answers

Section 10.1

1. Recombinant DNA technology is the process of using DNA from more than one species to create organisms that have useful traits.

2. Recombinant DNA contains DNA sequences from more than one species.

3. Genetically modified organisms are organisms such as plants and animals that have specific DNA sequences inserted into them. For example, the papaya tree is genetically modified to be resistant to the ringspot virus.

4. A gene, restriction endonucleases, a vector, ligase, and a host cell.

Section 10.2

1. Palindromes are short, four-to-six nucleotide, sequences of DNA that are the same on the forward strand as on the reverse strand. Palindromes are sequences of DNA recognized by restriction enzymes.

2. Restriction enzymes come from many species of bacteria.

3. Bacteria protect themselves from cutting their own DNA by using methylase enzymes that block recognition sites on their own chromosome with methyl groups.

4. Restriction enzymes break DNA phosphodiester bonds.

5. Sticky ends are important to allow DNA that was cut from a restriction enzyme to anneal with other DNA that was cut from the same enzyme.

6. Restriction enzymes are useful to genetic researchers because they can be used to cut DNA from two different organisms, such as bacterial DNA and human DNA, to create recombinant DNA.

Section 10.3

1. Vectors have the following important features: able to replicate within the host cell, contain restriction enzyme cut sites, have a selectable marker, be easy to recover from the host cell, and the inserted gene must be expressed as a protein.

2. Plasmids contain an origin of replication; a selectable marker, commonly an antibiotic resistant gene; and multiple restriction enzyme cut sites, which are usually placed into a multiple cloning site sometimes found within another selectable marker.

3. Antibiotic resistance gene and lac Z gene are used as selectable markers because it allows the correct colonies to be located.

4. The lac Z gene is used as a selectable marker because it is responsible for the breakdown of lactose (or the very similar molecule x-galactosidase, x-gal). When a bacterium is grown on x-gal, it turns blue, thus making it a selectable marker.

5. Only white bacterial colonies contain recombinant DNA because if a gene is inserted into the MCS, then the bacteria will be white.

6. • Bacteriophage. Genes involved in lysogenic phase are removed and replaced with foreign DNA, and the recombinant DNA can then infect other bacterial cells: called transduction.

   • Cosmid. This vector is used to carry larger sequences of DNA than plasmids, and is inserted into the bacterial cell by transformation.

   • YAC. These vectors can be grown in yeast cells, which are eukaryotic.

Section 10.4

1. Gene cloning is a method for making millions of copies of a gene using bacteria.

2. Transformation occurs when bacteria take up DNA from their surroundings, such as a recombinant plasmid.

3. A restriction map is useful for determining the appropriate restriction enzymes to cut your DNA with so that you have a whole gene.

4. You can determine which bacterial colonies have the gene of interest by growing the bacteria on plates containing growth media and x-gal. Any bacteria that did not take up the plasmid will not grow on the growth media because the plasmid contains the growth media antibiotic resistant gene. Any bacteria that did not take up the recombinant DNA will turn blue due to the lac Z gene in the presence of x-gal. Therefore, the bacteria with the gene of interest will be white.

5. a. Isolate plasmid DNA and human DNA.

   b. Look at restriction map to determine which restriction enzymes to use that will not ruin your insulin gene.

   c. Cut both plasmid DNA and human DNA with the same restriction enzyme.

   d. Mix cut plasmid and human DNA (that contains insulin gene) together with ligase to form recombinant DNA.

e. Mix recombinant DNA with *E. coli* bacterial cells, and heat shock to transform bacteria.

f. Grow transformed bacteria on plate overnight.

g. Use DNA hybridization to screen colonies.

h. Grow specific bacterial colony containing the gene of interest and isolate synthesized protein.

6. • Insulin—type-1 diabetes
   • Erythropoietin—anemia
   • Growth hormone—prevents dwarfism
   • Anticoagulants—prevents blood clotting
   • Clotting factors—hemophilia
   • Interferons—certain autoimmune diseases

Section 10.5

1. Gel electrophoresis is a method used to separate strands of DNA based on the length of the DNA sequence by using an electrical current. Its main use is to determine if the correct fragment has been inserted into a plasmid to ensure the gene of interest has been cloned.

2. PCR is useful because it generates multiple copies of a specific DNA sequence and amplifies a gene much faster and more reliably than using bacteria.

3. The main ingredients for PCR are template DNA, primers, dNTPs, and taq polymerase. The steps in PCR are denaturation, primer annealing, and primer extension.

4. VNTRs are varying lengths of repeated DNA sequences in the introns and are different in each person. Since VNTRs are unique to each individual person, then we can use them to identify specific people.

5. Gene therapy is the process of inserting a gene into an individual's cells to replace a mutated gene.

6. It is not frequently successful because it is difficult to target specific cell types and have the new gene inserted into the right location in the genome.

Section 10.6

1. Blue roses, blue strawberries, *flavr savr* tomatoes

2. Expression is dependent on the promoter sequence the gene is attached to.

3. They are eukaryotic and can process RNA as well as modify proteins on the Golgi body.

4. a. Isolate the spider silk gene.

   b. Ligate into a plasmid containing a promoter sequence that is involved in the regulation of milk production in goats.

   c. Microinject the recombinant plasmid into stem cells that can be injected into the blastocyst of a goat, produced by in vitro fertilization.

   d. Implant the blastocyst into a pseudopregnant goat. A pseudopregnant goat is a female that has been given hormone injections to cause the uterus to be prepared for implantation.

   e. The goat that is born will have the spider silk gene connected to the promoter involved in lactation. When lactation occurs in that goat, the milk will contain spider silk that can be isolated and used to build many very strong substances. This process is already being used by a company called BioSteel, and they produce bullet-proof vests.

5. Knockout mice are valuable to researchers because genes can be deleted from mouse genomes so that researchers can find out the functions of those genes and determine what may happen to humans with mutations of those genes.

6. To resist herbicides, to produce pesticides, to resist cold temperatures, or to add nutrients

7. bt toxin is produced by a natural soil bacterium; it kills insects that eat it by making pores in their digestive tract, which leads to cell death. Crop plants are genetically engineered to contain bt toxin to kill harmful insects.

Chapter 11 Concept Review Answers

Section 11.1

1. Diffusion, secretion, filtration, and absorption
2. By cell shape and the number of layers
3. See Tables A.11 through A.18.

TABLE A.11

Simple Squamous Functions	Simple Squamous Location
This type is a single layer of flattened cells that allows for diffusion, osmosis, filtration (glomeruli of kidneys), and secretion (surfactant in the alveoli).	Simple squamous epithelial cells line the heart, blood vessels, lymphatic vessels, alveoli in lungs, and the glomeruli in the kidney. They form part of any serous membrane the covers internal organs such as the heart (pericardium), lungs (pleura), and abdominal organs (peritoneum).

TABLE A.12

Simple Cuboidal Functions	Simple Cuboidal Location
Single layer of cuboidal cells that allows for secretion and absorption.	This type of cells is found primarily in the ducts of exocrine glands, the lumen of the nephron in the kidney, the surface of the ovaries, and parts of the thyroid gland.

TABLE A.13

Simple Columnar Functions	Simple Columnar Location
Ciliated simple columnar epithelial cells function to *move* substances, such as foreign particles trapped by mucus in the upper respiratory tract, or to move the egg along the fallopian tubes. Secretion of mucus.	Ciliated simple columnar epithelial cells are found in the upper respiratory tract, sinuses, fallopian tubes, and uterus.
Non-ciliated simple columnar epithelial cells usually contain specialized extensions called microvilli that allow for absorption of nutrients. This tissue contains goblet cells (considered unicellular glands) that secrete mucus.	Non-ciliated simple columnar epithelial cells are found primarily in digestive tract (stomach to anus), also found in the gallbladder; mucus protects lining of the digestive tract.

TABLE A.14

Pseudostratified Columnar Functions	Pseudostratified Columnar Location
Ciliated pseudostratified columnar epithelial functions to secrete mucus and to trap foreign particles and move substances along the surface of the tissue. Notice how the cells *appear* to be stratified but they are just one layer with some cells not extending to the surface (*pseudo* means "false").	Ciliated pseudostratified epithelial cells are found in similar locations as ciliated simple columnar, primarily the respiratory tract.
Non-ciliated type functions in absorption and secretion.	Non-ciliated pseudostratified epithelial cells can be found in the epididymis and many glands.

TABLE A.15

Stratified Squamous Functions	Stratified Squamous Location
Stratified squamous is the most abundant of the stratified tissue types. The multiple layers function to protect the underlying tissues from abrasion and from microorganisms. Note that the apical cells are squamous and the deeper basal cells are cuboidal.	Keratinized stratified epithelial cells are found in the skin; the outermost layers contain keratin in dead cells which provides a tough barrier (Chapter 12). Non-keratinized stratified epithelial cells are located in the mouth, tongue, esophagus, pharynx, and vagina.

TABLE A.16

Stratified Cuboidal Functions	Stratified Cuboidal Location
Stratified cuboidal epithelial cells can be two or more layers, and they function to protect underlying tissues and have some secretion and absorption properties.	Found mainly in esophageal ducts, sweat glands, mammary glands, and salivary glands

TABLE A.17

Stratified Columnar Functions	Stratified Columnar Location
A less common cell type, it functions as protection and secretion.	Found in the ducts of some exocrine glands, such as esophageal glands, and some parts of the urethra

TABLE A.18

Transitional Functions	Transitional Location
Having characteristics of both squamous and cuboidal, transitional epithelial cells allow for a hollow organ to distend and stretch. Cell shape is squamous when stretched and cuboidal when relaxed.	Found in bladder, ureters, and urethra

Section 11.2

1. Connective tissue contains various cell types, protein fibres, and ground substance.

2. The main functions of each type of connective tissue are as follows:

   - Loose—is found in almost every body structure, connects epithelial tissue to muscle layers (skeletal or smooth), gives tissues strength and elasticity, stores fat (loose adipose), acts as a filter in spleen and lymph nodes (loose reticular).

   - Dense—forms strong attachments between structures such as ligaments, connecting bones within a joint, and allows organs to stretch and recoil, such as within arteries or the lungs.

   - Cartilage—a very strong tissue that acts as a smooth surface for joint movement and gives structures shape, such as the ear.

- Bone—forms our skeleton, stores calcium and other minerals, produces red and white blood cells, stores some fat, and protects internal organs.
- Blood—transports oxygen, nutrients, waste products, and hormones throughout the body.

3. The main functions of each cell type found in connective tissue are as follows:

- Fibroblasts produce collagen, elastin, reticular fibres and ground substance, and are important in wound healing.
- Adipocytes store fat as triglycerides when extra calories are not utilized within the body.
- Mast cells are immune cells that produce histamine when tissue damage is present or during an allergic reaction.
- Reticular cells produce reticular fibres to filter pathogens and old red blood cells in the lymph, spleen, and liver.
- Chondrocytes make up the cartilage tissue.
- Osteocytes make up bone tissue.
- Blood cells include red blood cells that carry oxygen and carbon dioxide, and white blood cells that are involved in the immune response.

4. The types and importance of the protein fibres in connective tissue are the following:

- Collagen is the most abundant protein in the body, and makes tissue very strong.
- Elastin gives tissues the ability to stretch and recoil to their original shape.
- Reticular fibres are thin collagen fibres surrounded by glycoproteins; they form a framework for many tissues and act as a filter in lymph nodes and spleen.

5. Ground substance is important because it is the fluid part of the connective tissue that supports the functions of the cells and the fibres. It is what oxygen, nutrients, and hormones move through between blood vessels and cells of the body.

6. See Tables A.19 through A.29 29 for the main function and locations of each type of connective tissue.

TABLE A.19

Loose Areolar Functions	Loose Areolar Location
This is the most widely distributed connective tissue; it functions to strengthen, support, and provide elasticity to every body structure.	This type is found around all epithelial tissues, such as the subcutaneous layer and dermis layer of the skin, mucous membranes, around blood vessels, nerves and all organs.

TABLE A.20

Loose Adipose Functions	Loose Adipose Location
Adipose tissue stores fat as triglycerides that can be used to produce energy (ATP), insulate the body, and protect and cushion organs.	This type is found wherever there is areolar connective tissue: in the subcutaneous layer of the skin and around blood vessels, nerves and organs.

TABLE A.21

Loose Reticular Functions	Loose Reticular Location
This type provides structural support and forms part of the basement membranes that surround organs, blood vessels, and muscles. It acts as a filter to remove old red blood cells in the spleen and pathogens in the spleen and lymph nodes.	This type is found in liver, spleen, lymph nodes, bone marrow, blood vessels, all organs, and smooth muscle.

TABLE A.22

Dense Regular Functions	Dense Regular Location
Dense regular connective tissue provides very strong attachments for connecting muscle to bone (tendons), and connecting bone to bone (ligaments). It contains a high number of parallel strands of collagen, making it very strong. This tissue does not contain blood vessels, so injuries heal very slowly.	This type is found in tendons, ligaments, and aponeuroses (the sheet-like tendons that connect muscle to muscle) (Chapter 14).

TABLE A.23

Dense Irregular Functions	Dense Irregular Location
This type consists primarily of collagen fibres, like dense regular tissue, except that the fibres are not parallel; they are arranged randomly and form a strong sheet-like connective tissue that provides strength when pulled in many directions.	This type makes up the fascia that surrounds muscles, organs such as the heart (pericardium), digestive tract, bones (periosteum), and forms part of the lower dermis region of the skin.

TABLE A.24

Dense Elastic Functions	Dense Elastic Location
This type allows a great amount of stretch and recoil.	It is found in structures that undergo stretching forces, such as arteries, lungs, trachea, bronchial tubes, and vocal cords.

TABLE A.25

Hyaline Cartilage Functions	Hyaline Cartilage Location
This type provides a strong, smooth surface for joint movement. Cartilage does not contain any blood vessels. Hyaline cartilage is the weakest of the three types and can fracture.	It is found at the ends of long bones, ribs, nose, trachea, larynx, bronchi, and is also the majority of the fetal skeleton.

TABLE A.26

Elastic Cartilage Functions	Elastic Cartilage Location
Provides strength as well as elasticity and shape of certain structures.	It is found in the epiglottis, external ear, and the Eustachian tubes (tubes that connect the inner ear to the throat).

TABLE A.27

Fibrocartilage Functions	Fibrocartilage Location
This is the strongest type of cartilage, and it contains thick collagen fibres that connect certain structures together and act as shock absorbers.	It is found in the intervertebral discs, pubic symphysis, and knee joint.

TABLE A.28

Bone Functions	Bone Location
Bones move when muscles contract and produce movement. Bones are important in mineral homeostasis and provide a reservoir for calcium and phosphorus, and they protect internal organs. Red marrow produces red and white blood cells; yellow marrow stores some fat.	The skeleton includes long bones, skull, ribs, vertebrae, and pelvis. See Chapter 13 for details.

TABLE A.29

Blood Functions	Blood Location
Blood transports oxygen, nutrients, hormones, and waste products throughout the body. Red blood cells transport oxygen and some carbon dioxide (Chapters 18 and 19). White blood cells include a variety of immune cells (Chapter 23). Platelets are important for blood clotting.	Blood is located within blood vessels, although red and white blood cells, as well as platelets are produced in red bone marrow.

Section 11.3

1. Movement, production of heat, posture protection, regulation of movement
2. Skeletal, smooth, and cardiac (see Table A.30)

TABLE A.30

Skeletal Muscle	• Connected to bones and allows movement; voluntary contraction or unconscious reflexes • Multinucleated, striated muscle fibres that are fast twitch or slow twitch, or intermediate
Smooth Muscle	• Found in walls of hollow structures; involuntary contractions • Single nucleus, non-striated muscle fibres, single unit or multiunit
Cardiac Muscle	• Found only in the walls of the heart and controls blood circulation; involuntary contraction and autorhythmic • One to three nuclei per cell, branched striated muscle fibres connected by intercalated discs that contain gap junctions for signalling

3. Walls of hollow structures, such as blood vessels, lymphatic vessels, stomach, intestines, airways, fallopian tubes, ureters, bladder, uterus, vas deferens, erector pili muscles in the skin, and iris of the eye
4. Single-unit smooth muscle is found in sheets that surround visceral organs where contraction happens simultaneously.
5. Multiunit smooth muscle is found in blood vessels, the eye, erector pili, airways, and is made up of multiple cells that contract individually.
6. The special feature of cardiac muscle is that it contains striated, branched muscle fibres that are

connected by intercalated discs that contain gap junctions.

7. Intercalated discs that contain gap junctions are important because they allow signalling molecules to pass quickly between the cells, which is necessary to maintain control and simultaneous contraction of the heart chambers for blood to be circulated throughout the body.

Section 11.4

1. Detect or transmit sensations, interpret environmental stimuli, respond to stimuli

2. Muscle cells, glands, or other neurons

3. Neurons and neuroglia

4. Three types of neurons and their location are as follows:
   - Multipolar—motor neurons
   - Unipolar—sensory neurons
   - Bipolar—located in the olfactory sensory system and the retina

5. The types of neuroglia and their main functions are as follows:
   - Oligodendrocytes (CNS) and Schwann cells (PNS) produce the myelin covering of neurons.
   - Astrocytes (CNS) and satellite cells (PNS) maintain the ion concentrations surrounding neurons, and astrocytes form the blood-brain barrier.
   - Microglia are immune cells in the brain and spinal cord.
   - Ependymal cells produce cerebral spinal fluid.

Chapter 12 Concept Review Answers

Section 12.1

1. The components of the integumentary system include the skin, nails, hair, and glands.

2. The integumentary system is responsible for protecting the body from the external environment, regulating body temperature, synthesizing Vitamin D, and enabling the excretion of certain waste products.

3. Vitamin D is used by the body to help with certain processes such as bone mineral regulation, hormone synthesis, and immune system regulation.

4. By constricting blood vessels in the dermal layer of the skin, the body restricts the amount of blood that can transfer heat energy to the external environment. Conversely, the body can increase blood flow by dilating the blood vessels in the dermis to cool down the body; the increased blood flow allows for greater heat loss to the external environment. The skin also produces sweat, which, during the process of evaporation allows the body to dissipate heat.

Section 12.2

1. The layers of the epidermis in order from deepest to most superficial are the stratum basale, the stratum spinosum, the stratum granulosum, the stratum lucidum, and the stratum corneum.

2. The primary types of cells found in the various layers of the epidermis include keratinocytes, Langerhans cells, melanocytes, and Merkel cells.

3. See Table A.31.

TABLE A.31

Cell Type	Function
Merkel cells	Mechanoreceptors that allow for touch sensation
Langerhans cells	A type of immune cell that helps prevent infections by the phagocytosis of various pathogens
Keratinocytes	The most numerous cells in the epidermis; produce keratin—a strong, structural, fibrous protein that gives strength to the tissues and helps protect the body from damage
Melanocytes	In response to ultraviolet light, produces melanin, which acts as a photo-protectant, preventing damage to other skin cells caused by UV light

4. See Table A.32.

TABLE A.32

Product	Function
Keratin	Structural protein the gives strength to the tissue and helps prevent damage
Melanin	Photo-protectant pigmentation that prevents damage caused by a cell's exposure to ultraviolet light
Lamellar granules	Organelles that release lipids and proteins into the extracellular matrix that act as a waterproof barrier and help break down bacterial cell walls

5. In order from deepest to most superficial, the layers of the dermis include the reticular layer and the papillary layer. The dermis includes blood vessels,

hair follicles, sebaceous glands, sweat glands, neurons, sensory receptors and the arrector pili muscles.

6. See Table A.33.

TABLE A.33

Receptor Name	Function
Merkel cells	Mechanoreceptors that detect touch
Pacinian corpuscles	Detect pressure
Nerve endings	Detect pain
Thermoreceptors	Detect temperature
Meissner corpuscles	Detect touch and vibration

7. The hypodermis acts primarily to connect the skin to the underlying tissues and to store energy in the form of fat.

8. See Table A.34.

TABLE A.34

Molecule	Effect
Melanin	Black or brown pigment secreted by melanocytes
Hemoglobin	Protein present on red blood cells that becomes bright red when bound to oxygen
Carotene	Orange or yellow pigment in many foods (e.g., carrots and egg yolk)

Section 12.3

1. The two types of sudoriferous glands are eccrine and apocrine. Eccrine are found throughout the body, though particularly on the forehead, the palms of the hands, and soles of the feet. Eccrine glands produce sweat and are particularly important for body temperature regulation. Apocrine glands are found primarily in the axilla, the areola of the nipples, and in the genital regions. The sweat produced in apocrine glands is thicker because it contains fatty acids along with water and ions.

2. Sebaceous glands cover the body surrounding hair follicles. Sebaceous glands release sebum which coats the hair and skin, providing lubrication and helping prevent infection.

3. Acne is caused by increased sebum production, which gets trapped in the gland. When bacteria ingest the sebum, the result is a minor inflammatory response.

4. Ceruminous glands are found in the ear canal and produce earwax (cerumen). Mammary glands are found in the breast skin tissue and produce milk during lactation.

Section 12.4

1. Burns are divided into three classifications based on severity. Superficial partial-thickness (first-degree) burns involve only the epidermis and are characterized by redness and slight inflammation. Deep partial-thickness (second-degree) burns involve damage to the epidermis and the papillary region of the dermis. Full-thickness (third- and fourth-degree) burns result in damage through the epidermal and dermal layers and often include damage to the subcutaneous layer and possibly underlying muscle.

2. See Table A.35.

TABLE A.35

Type of Cancer	Cells Affected
Basal cell carcinoma	Basal cells
Squamous cell carcinoma	Keratinocytes
Melanoma	Melanocytes

3. See Table A.36.

TABLE A.36

Type of Cancer	Risks
Basal cell carcinoma	Rarely spreads, often appears as a raised smooth bump that can sometimes ulcerate
Squamous cell carcinoma	Can spread quickly, may bleed and ulcerate
Melanoma	Most dangerous form of skin cancer, caused by uncontrolled growth of melanocytes

a. Risk factors include the following:
   - Exposure to ultraviolet light
   - Chronic inflammation
   - Continual wound healing
   - Use of immunosuppressant medication
   - Certain cosmetics, lotions, or sunscreens

b. Melanoma can be identified using the mnemonic ABCDE:

- A = Asymmetrical
- B = Borders irregular
- C = Colour dark and inconsistent
- D = Diameter greater than 6 mm
- E = Evolving over time

Chapter 13 Concept Review Answers

Section 13.1

1. Total is 206: 80 bones in the axial skeleton, 126 in the appendicular skeleton.
2. The axial skeleton is made up of the skull, vertebrae sternum, and ribs. The pelvis, shoulders, and limbs make up the appendicular skeleton.
3. The skeletal system is responsible for providing structure; allowing movement; protecting the brain, heart, lungs, and spinal cord; producing blood cells; storing mineral and fat; and hearing.

Section 13.2

1. Bones are made up of connective tissue, bone marrow, blood vessels, nerves, and epithelial tissue. Bone tissue consists of cells, extracellular matrix collagen, and various minerals.
2. Osteoblasts are responsible for bone tissue production. Osteocytes maintain bone tissue. Osteoclasts break down the mineral portion of the extracellular matrix to release calcium, magnesium, and phosphate.
3. Minerals provide the bone with strength, and collagen provides flexibility.
4. Five categories of bone types and examples are as follows:
   - Sesamoid bones—patella
   - Long bones—humerus, femur, etc.
   - Short bones—wrists and ankles
   - Irregular bones—vertebrae and facial bones
   - Flat bones—ribs, sternum, scapulae, etc.

Section 13.3

1. There are four main sutures: (1) coronal, between the frontal bone and the anterior portion of the parietal bones; (2) sagittal, between the medial edges of the parietal bones; (3) lambdoid, between the occipital bone and the posterior edges of the parietal bones; and (4) two squamous sutures, between the temporal bone and the lateral parietal bone, as well as the edges of the occipital and sphenoid bones, on each side of the skull.

2. The condylar process of the mandible sits in a grove in the temporal bone called the *mandibular fossa*, forming the temporomandibular joint (TMJ).
3. Fontanels are "soft spots" in an infant skull, where the bones have not yet fused together.
4. Scoliosis is an S-shaped lateral curve of the spinal column; kyphosis is an over-curvature of the thoracic vertebrae, also called *hunchback*; and lordosis is an over-curvature of the lumbar region, also called *sway back*.
5. Bones that articulate with the frontal bones are the parietal bones, sphenoid, ethmoid, lacrimal, zygomatic, nasal, maxilla.
6. Bones that have sinuses are the frontal, sphenoid, ethmoid, and maxilla.
7. The foramen magnum is in the occipital bone.
8. The optic foramen is in the ethmoid bone.
9. The pituitary gland sits in the sella turcica of the sphenoid bone.
10. Vertebrae—cervical (7), thoracic (12), lumbar (5), sacral (5 fused)
11. The largest and strongest vertebrae are the lumbar vertebrae.
12. The vertebral processes are spinous, transverse, superior articular, and inferior articular.
13. The floating ribs are 11 and 12.
14. The three parts of the sternum are the manubrium, xiphoid process, and the body.

Section 13.4

1. The scapula, clavicle, humerus, radium, ulna, carpals (8), metacarpals (5), and phalanges (14) are all found in the upper limbs.
2. The pelvis, femur, tibia, fibula, tarsals (7), metatarsals (5), and phalanges (14) are all found in the lower limbs.
3. The ischia (singular ischium) are the pelvic bones we sit on.
4. The area bound by the sacrum and pubis bones and extending to the pubic symphysis is called the *true pelvis*.
5. The patella protects the knee joint from insult.
6. The radius articulates at the lateral epicondyle, and the ulna articulates at the medial condyle.
7. The medial wrist bone is styloid process of the ulna, and the lateral wrist bone is the styloid process of the radius.
8. The olecranon is the elbow.
9. The wrist bones are the carpals.
10. The glenoid cavity is found in the scapula.
11. The acetabulum is formed by the connection between the ilium, ischium, and pubis.

12. Medial and lateral condyles are the protrusions of the distal femur and form the knee.

Section 13.5

1. Hormones that increase bone growth include calcitonin, growth hormone, IGF, thyroid hormones, testosterone, estrogen, and insulin. The only hormone that decreases bone growth and causes bone minerals to be released is parathyroid hormone.

2. Osteoblasts are the osteocytes that promote bone growth and production.

3. Osteoclasts are targeted by parathyroid hormone to increase the breakdown of bone.

4. The key minerals in the bone matrix are calcium, magnesium, and phosphorus.

5. Vitamin D improves the absorption of calcium in the digestive tract. Vitamin A is used to regulate osteoclasts, and Vitamin C is used to help move collagen from the cells to the ECM.

Section 13.6

1. The main classifications of fractures are as follows:

   - Transverse fracture—break across the transverse section of a bone

   - Spiral fracture—a twisted fraction

   - Commuted fracture—crushed bone

   - Stress fracture—typically longitudinal cracks, but not full breaks, in a bone

   - Greenstick—a fracture where the bone has bent significantly prior to breaking on one side; most commonly found in children

2. During healing from a fracture, phagocytic cells consume the dead cells, chondrocytes form cartilage to temporarily replace the broken section, osteoblasts create new bone, and osteoclasts assist with remodelling the new bone.

3. Diet and exercise are the primary factors involved in preventing osteoporosis.

4. Osteoporosis occurs in menopausal women due to the decreased production of hormones.

Section 13.7

1. Structural classifications include fibrous, cartilaginous, and synovial. Functional classifications refer to *syndesmosis* or no movement; *amphiarthrosis* or slight movement; and *diarthrosis*, meaning highly movable.

2. • Fibrous—for example, skull sutures, teeth and sockets, connections between radius and ulna or tibia and fibula

   - Cartilaginous—for example, joints between ribs and sternum, pubic symphysis, joint between the manubrium and the sternum, joints between the vertebrae and the intervertebral discs

   - Synovial—for example, shoulders, knees, elbows

3. Syndesmosis joints are immovable and include joints such as the suture joints. Amphiarthrosis joints have limited movement and include joints such as the connections between the radius and the ulna. Diarthrosis joints are fully movable and include most joints in the limbs and between phalanges.

4. Synovial fluid provides nutrients, prevents friction between the ends of the connecting bones, and absorbs shock to the joint.

5. Bursa sacs encapsulate synovial fluid, containing it within an area and creating a cushion between bones or tendons and muscle.

6. See Figure 13.32

7. • Plane joint—includes the carpals of the wrist or tarsals of the ankles

   - Hinge joint—include the knees, ankles, and elbows

   - Pivot joint—includes the radioulnar joints and the atlanto-axial join between C1 and C2 that allow for rotational movement of the head

   - Condyloid—includes the wrists and elbows

   - Saddle joint—joint between the proximal phalange thumb and the metacarpal

   - Ball-and-socket joint—includes the hips and the shoulders

8. Ball-and-socket joints are able to produce angular movements.

9. Pivot joints allow for rotational movement.

10. Pronation and supination are opposing movements that allow a greater degree of dexterity and flexibility.

Chapter 14 Concept Review Answers

Section 14.1

1. The three types of muscle are skeletal, smooth, and cardiac.

2. Muscle functions include the following:
   - Producing body movements
   - Stabilizing body positions and posture
   - Movement of internal organs, such as the constriction of the blood vessels, digestive tract, fallopian tubes, bronchial tubes, and ureters
   - Regulating capillary blood flow and peristalsis
   - Producing heat

Section 14.2

1. Tendons connect muscle to bone.

2. The fascia is dense connective tissue that envelops the muscles and connects them to the subcutaneous layer of the skin.

3. Muscles are held by the endomysium (around each fibre), the perimysium (around each fascicle), and the epimysium.

4. Thick and thin filaments—myofibrils—sarcolemma—endomysium—fascicle—perimysium—epimysium

Section 14.3

1. See Table A.37.

2. Lifting the knee up uses the hip flexors, extension of the leg uses the quadriceps, pushing the leg back uses the gluteus maximus, pushing the ground with the toes uses the calves, bringing the leg up behind with the knee bent uses the hamstrings.

Section 14.4

1. Actin and myosin are responsible for muscle contraction.

2. Troponin and tropomyosin are the two regulatory proteins. Troponin binds with calcium, causing tropomyosin to roll off the myosin-binding sites on the actin filament.

3. Thin filaments and thick filaments slide across each other during contraction. Neither the thick or thin filament shortens, but as the filaments slide across each other, the whole sarcomere shortens. The myosin heads bind to actin, then pivot and pull the actin toward the M line. Each myosin head then pivots toward the M line, pulling the actin of the thin filament toward the M line so that the thick and thin filaments slide across one another.

4. During sarcomere shortening, the H zone and I bands contract while the A bands remain unchanged.

5. Calcium is stored in the sarcoplasmic reticulum.

6. The action potential from the movement of sodium into the muscle causes the release of calcium from the sarcoplasmic reticulum.

7. The following describes the four steps of the cross-bridge cycle:

   a. Myosin is unbound from the actin and contains an ATP molecule in the ATP-binding site on the myosin head. The ATP hydrolyzes, the ADP and phosphate stay bound, and the action of the bond breaking in the ATP energizes the myosin.

   b. The phosphate molecule leaves the myosin, and the energized myosin binds to the myosin-binding site on the actin proteins, as long as calcium is present to bind troponin and move the tropomyosin off the binding sites. The binding of myosin to actin forms the cross-bridge.

   c. The ADP molecule leaves the myosin, and the myosin head moves toward the M line; this is called the *power stroke*. After the power stroke, the myosin remains bound to the actin.

TABLE A.37

Muscle	Origin	Insertion	Action
Occipitofrontalis frontal belly	Epicranial aponeurosis	Skin superior to orbit	Raises eyebrows
Occipitofrontalis occipital belly	Occipital and temporal bone	Epicranial aponeurosis	Draws scalp backward
Orbicularis oculi	Wall of orbit	Eyelid	Closes eyes
Orbicularis oris	Muscles around mouth	Skin around mouth	Closes and protrudes lips as in kissing
Zygomaticus major	Zygomatic bone	Skin at corners of mouth	Draws edges of mouth up as in smiling
Platysma	Fascia over deltoid	Mandible, skin of lower face	Depresses mandible as in pouting
Masseter	Zygomatic arch	Mandible	Closes the mouth as in chewing
Temporalis	Temporal bone	Mandible	Elevates mandible as in chewing
Sternocleidomastoid	Manubrium and clavicle	Occipital bone and mastoid process	Turns head side to side
Trapezius	Occipital bone and thoracic vertebrae	Clavicle and scapula (acromion anteriorly, and spine posteriorly)	Raises shoulders as in shrugging

d. A new ATP molecule binds to the myosin head, which allows the myosin to be released from the actin. Then the cycle repeats, with ATP hydrolysis in step 1.

8. Muscles use ATP to bind myosin and cause power stroke; they also require ATP for the active transport of calcium back into the sarcoplasmic reticulum.

9. The load required determines the force of contraction and, therefore, how many fibres are activated.

10. Muscle tone prevents muscles from overstretching and helps maintain both balance and posture.

11. Sodium causes action potentials in both nerves and muscles.

12. Calcium causes acetylcholine to be released into the synaptic cleft.

13. Acetylcholine (Ach) causes the sodium channels to open in the muscle membrane, stimulating an action potential.

14. After stimulating an action potential, acetylcholine is broken down and recycled by an enzyme called *acetylcholinesterase.*

15. The following order of events, beginning with stimulation by the motor neuron, occurs during muscle contraction:

a. Action potential travels down motor neuron to axon terminal.

b. Action potential triggers movement of calcium into the axon terminal.

c. Calcium triggers release of Ach neurotransmitters into synaptic cleft.

d. Ach binds to specific receptors on the sarcolemma that trigger the opening of Na^+ channels.

e. Action potential occurs in muscle cell membrane as Na^+ ions cross the membrane from the T-tubules and enter the sarcoplasm.

f. Action potential in the sarcolemma causes the sarcoplasmic reticulum to release Ca^{2+} ions into the sarcoplasm.

g. Ca^{2+} ions bind to troponin, which moves tropomyosin off the myosin-binding sites of the actin proteins.

h. ATP is bound to myosin in the relaxed state. Once an action potential occurs and the myosin-binding sites on the actin are available, ATP hydrolyzes, releasing the phosphate group, and myosin binds to actin—cross-bridge formation.

i. ADP leaves the myosin, causes the power stroke, and sarcomeres shorten.

j. A new ATP molecule binds to myosin to release myosin from actin; the cycle will continue as long as the motor neuron stimulates the muscle cell, then the muscle relaxes.

Section 14.5

1. See Table A.38.

TABLE A.38

Characteristic	Slow Oxidative Fibre	Fast Oxidative Fibre	Fast Glycolytic Fibre
Speed of contraction	Slow	Fast	Fast
Resistance to fatigue	High	Intermediate	Low
Oxidative phosphorylation capacity	High	High	Low
Enzymes for anaerobic glycolysis	Low	Intermediate	High
Mitochondria	Many	Many	Few
Capillaries	Many	Many	Few
Colour of fibre	Red	Red	White
Glycogen content	Low	Intermediate	High
Diameter of fibre	Smallest	Intermediate	Largest
Strength of contraction	Weakest	Intermediate	Strongest

2. The following exercises use the following types of muscle fibre:

a. squatting 50 lb (fast)

b. jogging 1 hr (slow)

c. running 1500 m (intermediate)

d. sitting in a chair with good posture (slow)

e. typing (slow)

f. eating (slow)

g. downhill skiing (fast)

h. walking up a flight of stairs (intermediate)

i. running long jump (fast)

Section 14.6

1. The primary difference between aerobic and anaerobic respiration is the presence of oxygen.

2. Creatine phosphate is stored as creatine kinase and used as a rapid source of energy.

3. Two ATP are used as the net product of anaerobic respiration (glycolysis). In total, four are created in

the process, but two are required, resulting in a gain of only two ATP.

4. Approximately 36 ATP result from aerobic respiration starting with one glucose molecule.

5. Aerobic respiration

6. The two primary factors in maintaining a healthy body weight are diet and exercise.

7. Exercise also helps with increasing bone density, preventing osteoporosis, increasing balance and coordination, decreasing the risk of fracture, improving mood (affect), decreasing inflammation in the body, and preventing type-2 diabetes and cardiovascular disease.

Chapter 15 Concept Review Answers

Section 15.1

1. The nervous system is divided into (1) the central nervous system (CNS), composed of the brain and spinal cord; and (2) the peripheral nervous system (PNS), composed of the cranial nerves, spinal nerves, and the neurons outside the CNS.

2. The visceral organs, blood vessels, endocrine glands, and exocrine glands are innervated by the autonomic nervous system.

3. The submucosal plexus stimulates the release of acids, digestive enzymes, and other substances required by the digestive system; the myenteric plexus innervates the smooth muscle needed for peristalsis.

4. The somatic nervous system is primarily involved in voluntary muscle control.

5. See Table A.39.

TABLE A.39

Receptor	Stimulus
Chemoreceptor	Chemical concentrations, such as hormones or neurotransmitters
Osmoreceptor	Osmolarity, concentration of water and ions
Baroreceptor	Blood pressure
Photoreceptor	Light
Mechanoreceptor	Stretch and physical movement
Proprioceptor	Body position
Nociceptor	Pain
Thermoreceptor	Temperature
Tactile receptors in the skin	Touch, pressure, vibration

Section 15.2

1. The two categories of cells that make up nervous tissue are neurons and glial cells.

2. Dendrites extend from the cell body and detect various stimuli by means of membrane receptors. (See Table A.38.) Cell bodies contain the organelles required for typical cell functions such as ATP production and protein synthesis. Axons are the main extensions from the cell body, and serve primarily as a conduit for action potentials ending at the axon terminal. The axon terminal is where neurotransmitters are released and trigger the receptors for muscles, glands, or other neurons.

3. Neurons can be either multipolar, bipolar, or unipolar.

4. See Table A.40.

TABLE A.40

Type of Glial Cell	Function
Astrocyte	• Forms the blood-brain barrier • Regulates ion concentrations • Stores glycogen and releases glucose as needed by the neurons • Removes waste products • Produces scar tissue following insult to the CNS as part of the healing process
Oligodendrocyte	Forms the myelin sheathe around the axons of the neurons
Microglia	Immune cells specifically for the nervous system
Ependymal cell	Secretes cerebrospinal fluid used to provide nutrients to the brain
Satellite cell	Functions similar to astrocytes, but found only in the PNS
Schwann cell	Forms the myelin sheathe around the axons of afferent and efferent neurons in the PNS

Section 15.3

1. In a polarized resting cell, the sodium and chloride are found in higher concentration outside the cell, and potassium is in higher concentration inside the cell.

2. Ion channels that open or close in response to a chemical are considered chemically gated, and those that open or close in response to a change in the charge of a membrane are referred to as *voltage-gated*.

3. The movement of sodium into the cell causes depolarization as the membrane potential is lost. Repolarization takes place once the sodium channels have closed and potassium leaves the cell.

4. It is important because it results in a refractory period and ensures that the action potential remains unidirectional.

5. The movement of calcium into the cell triggers the exocytosis of neurotransmitters into the synapse.

6. Saltatory—from Latin *saltare*, "to jump"—conduction occurs when the action potential bypasses segments of the axon, resulting in depolarization at the nodes of Ranvier and speeding transmission along the axon.

7. Myelination of the axon and an increased diameter of the axon result in faster transmission along the axon.

8. a. Dopamine plays a large role in the reward pathway of the brain and is largely responsible for addictive behaviours.

   b. Serotonin elevates mood/affect.

   c. Acetylcholine stimulates muscle contraction.

Section 15.4

1. Pia mater, arachnoid mater, and dura mater

2. The subarachnoid space contains the cerebrospinal fluid (CSF) and blood vessels; the epidural space contains blood vessels, lymphatic vessels, nerve roots, and fats.

3. White matter is composed of myelinated axons. Grey matter is composed of unmyelinated interneurons, axon terminals, cell bodies, and dendrites.

4. Sensory (afferent) information enters the spinal cord from the dorsal roots and travels up the ascending tracts of the dorsal horn. Motor (efferent) information travels through the ventral horn through the descending tracts and leaves the spinal cord via the ventral roots.

5. The outermost layer of the nerve is the epineurium and is composed of connective tissue. Inside the nerve are bundles of axons referred to as *fascicles*. Each fascicle is composed of groups of axons wrapped in the endoneurium.

6. There is a total of 31 pairs of spinal nerves. Eight pairs are found in the cervical region, 12 in the thoracic, five in the lumbar, five in the sacral, and one pair of coccygeal nerves.

7. See Table A.41.

TABLE A.41

Nerve	Vertebral Region	Tissue Innervated
Phrenic	Cervical	Diaphragm
Axillary	Brachial	Deltoid, rotator cuff
Musculocutaneous	Brachial	Biceps, brachialis
Median	Brachial	Forearm, thumb, and first three fingers
Radial	Brachial	Thumb and first finger
Ulnar	Brachial	Last two fingers
Obturator	Lumbar	Hip and abductors
Pudendal	Sacral	External genitalia and sphincters of both bladder and rectum
Femoral	Lumbar	Quadriceps and hip flexors
Sciatic	Sacral	Hamstrings
Tibial	Sacral	Calves and foot
Fibular	Sacral	Tibialis anterior
Superior and inferior gluteal	Sacral	Gluteal muscles

Section 15.5

1. A monosynaptic reflex has a single synapse in the spinal cord linking a sensory neuron to a motor neuron. Polysynaptic reflexes have multiple synapses from sensation to response.

2. Reflexes reduce time and cognitive energy required for certain tasks, whether they help prevent damage to the body or facilitate certain tasks.

3. Tapping on the quadriceps muscle causes the muscle to stretch. In response to the stretch, the patellar reflex causes contraction of the quadriceps muscles and the leg extends.

4. Sensory neurons detect a stimulus (heat, pain, etc.) and send the message to interneurons in the spinal cord. A motor neuron then sends a response to the appropriate muscle cell, causing it to contract.

Section 15.6

1. The brain is protected by the blood-brain barrier, skull bones, meninges, and CSF.

2. Ependymal cells produce cerebrospinal fluid in the choroid plexus located in each of the ventricles of the brain.

3. At the arachnoid villi, CSF is reabsorbed into the bloodstream into the superior sagittal sinus.

4. The four regions are the diencephalon, the brainstem, the cerebellum, and the cerebrum. The diencephalon consists of the thalamus, hypothalamus, and pineal gland. The brainstem consists of the midbrain, pons, and medulla oblongata. The cerebellum is located at the base of the brain, posterior to the brainstem. The cerebrum consists of the cerebral cortex (the outermost portion) and the subcortical region located beneath the cortex.

5. The thalamus operates as a relay station through which signals pass to various parts of the brain to be properly interpreted. The pineal gland is responsible for maintaining circadian rhythms and is responsible for the production of the hormone melatonin.

6. The reticular activating system is located in the central core of the brainstem and is responsible for consciousness and the ability to filter extraneous stimuli.

7. Substantia nigra in the midbrain are responsible for regulating motor movements.

8. The pons is responsible for bridging signals from the motor cortex of the cerebrum to the cerebellum and, in turn, sending signals from the cerebellum to the muscles in a smooth manner. The pons also plays a role in voluntary breathing control.

9. The medulla oblongata is primarily responsible for control of the respiratory and circulatory systems.

10. The cerebellum is responsible for coordinating, learning, and remembering complex motor functions.

11. Basal nuclei regulate skeletal muscle, posture, and complex behaviours. The amygdala is involved in emotions, the memories linked to emotions, and complex social and sexual interactions.

12. The limbic system plays a major role in emotions, learning, memory, and social interactions.

13. Lobe and region are as follows:

    a. Temporal lobe, auditory cortex

    b. Parietal lobe, somatosensory cortex

    c. Frontal lobe, Broca's area

    d. Frontal lobe, pre-frontal cortex (also called the *pre-motor cortex*)

    e. Parietal lobe, left side

    f. Occipital lobe, visual cortex

    g. Temporal lobe, hippocampus

    h. Frontal lobe

    i. Temporal lobe, Wernicke's area

    j. Insula

14. See Table A.42.

Section 15.7

1. The somatic nervous system innervates skeletal muscle; the autonomic nervous system innervates glands, smooth muscle, and cardiac muscle.

TABLE A.42

Nerve	Function	Origin
I Olfactory	Sense of smell	Cerebrum
II Optic	Vision	Cerebrum
III Oculomotor	Movement of eyelid, eyeball, lens, pupil constriction	Midbrain
IV Trochlear	Movement of eyeball	Midbrain
V Trigeminal (ophthalmic, maxillary, and mandibular branch)	Sensations in head, face, and jaw; motor control of chewing	Pons
VI Abducens	Movement of eyeball	Pons
VII Facial	Taste, facial expression, tears, salivation, sensations	Pons
VIII Vestibulocochlear	Equilibrium and hearing	Lateral to facial nerve
IX Glossopharyngeal	Taste, swallowing, speech, salivation	Medulla oblongata
X Vagus	Taste, sensations from pharynx, swallowing, coughing, voice, GI tract smooth muscle contraction, heart rate reduction, digestive secretion	Medulla oblongata
XI Accessory	Swallowing and movement of head and shoulders	Medulla oblongata
XII Hypoglossal	Speech and swallowing (tongue muscles)	Medulla oblongata

2. Acetylcholine is released by the preganglionic nerve at the synapse.

3. Norepinephrine and epinephrine are released by the sympathetic postganglionic neuron, and acetylcholine is released by the parasympathetic postganglionic neuron.

4. Five tissues are (1) smooth muscle in bronchi, (2) cardiac muscle, (3) smooth muscle in peripheral blood vessels, (4) sweat glands, and (5) kidneys.

5. The primary role is stimulation of processes used for digestion and relaxation.

6. The sympathetic nervous system triggers the breakdown of glycogen and fat; glycogen production and fat storage are promoted by the parasympathetic system.

7. Preganglionic nerves leave the CNS either through a cranial or spinal nerve.

8. The adrenal gland produces hormones that cause the same effects as the autonomic nervous system, mimicking a sympathetic response but having more long-lasting effects.

Chapter 16 Concept Review Answers

Section 16.1

1. Receptors found in skin are as follows:
    a. Meissner's corpuscle—detects touch and pressure
    b. Merkle's corpuscle—detects touch and pressure
    c. Free nerve endings—some are nociceptors that detect pain, others detect temperature
    d. Pacinian corpuscles—detect vibration and deep pressure
    e. Ruffini corpuscles—detect skin stretching
    f. Hair follicles—detect touch and vibration

2. Proprioceptors detect body position and movement.

3. When a stimulus is increased, the frequency of action potentials increases.

4. The smaller receptive fields enable us to more precisely detect the location of a stimulus. The body parts with the smallest receptive fields have the most sensitivity, and these include lips, tongue, and fingers.

5. The anterolateral tract carries pain and temperature information. The dorsal-medial tract carries information about touch, pressure, limb position, and body movements.

6. The dorsal-medial tract crosses over in the medulla oblongata.

7. The somatosensory cortex is in the parietal lobes.

8. Nociceptors detect pain.

9. Two neurotransmitters involved in transmission of pain are glutamate and substance P.

10. An analgesic is a drug that reduces the transmission of pain.

11. An example of an NSAID is Ibuprofen.

12. Gallbladder pain can be felt in the right shoulder.

Section 16.2

1. We can see colours in the 400–700 nm range.

2. The choroid layer is highly vascular and contains melanin.

3. The lens changes shape.

4. Rods detect black, white, and grey scale in low light. Cones detect colour in brighter light.

5. The anterior compartment contains aqueous humor and the posterior compartment contains vitreous humor.

6. Rod cells contain rhodopsin.

7. The ganglion cells play a role in the production of melatonin in low light, which promotes sleep.

8. The lens becomes long and thin when focusing on objects far away and becomes thick and round when focusing on objects close up.

9. Nearsightedness can be caused by being born with an elongated eye, or if the lens is too concave shaped.

Section 16.3

1. Humans can hear sounds in the 20–20 000 Hz range.

2. Sounds louder than 85 dB can damage the ear.

3. The eardrum is the tympanic membrane.

4. The three bones of the inner ear—called *ossicles*—that amplify sound are the malleus, incus, and stapes.

5. The Eustachian tube connects the middle ear to the throat and allows air to move in or out of the middle ear; this regulates pressure inside the ear when atmospheric pressure changes.

6. The vestibular apparatus detects body position and movement.

7. In the cochlea, the base detects high-pitch sounds and the apex detects lower-pitch sounds.

8. The organ of Corti hair cells are mechanoreceptors.

9. Glutamate is the neurotransmitter that sends information through the auditory nerve.

10. The primary auditory cortex is in the temporal lobe.

11. The utricle detects information about head position and orientation; the saccule detects acceleration.

Section 16.4

1. Taste receptors are located on taste buds, and there are hundreds of taste buds on every tongue papilla.

2. Humans can detect sweet, salty, sour, bitter, and savory.

3. Basal cells continually divide and replace gustatory cells.

4. The trigeminal nerve sends information about texture, spiciness, and temperature.

5. The brain perceives capsaicin in hot peppers to be hot.

Section 16.5

1. Odours are detected by chemoreceptors.

2. Olfactory cilia are located in the olfactory epithelium of the superior nasal concha.

3. Olfactory information is the only sensory information that does not travel to the thalamus.

4. Three regions of the brain that receive olfactory information include the limbic system, reticular formation, and the hypothalamus.

Chapter 17 Concept Review Answers

Section 17.1

1. The nervous system and the endocrine system are similar because they communicate with various tissues in the body to maintain homeostasis. They are different because the nervous system uses neurotransmitters as the signalling molecule and the endocrine system uses hormones.

2. Endocrine glands secrete hormones into the body. Exocrine glands are epithelial cells that excrete hormones either to the surface of the body or into a hollow cavity or duct in the body.

3. See Table A.43.

TABLE A.43

Gland	Location
Pituitary gland	Base of the hypothalamus
Pineal gland	Centre of the brain
Thyroid	Anterior to the trachea
Parathyroid	On the surface of the thyroid gland
Thymus	Between the heart and the sternum
Pancreas	Inferior to the stomach
Adrenal glands	Superior to the kidneys
Testes	In the scrotum
Ovaries	In the abdomen

4. General functions of the endocrine system are as follows:
   - Homeostatic balance
   - Regulation of growth, metabolism, and energy production
   - Reproductive functions, such as secondary sex characteristics, lactation, childbirth, development of sperm or eggs
   - Stress response
   - Regulation of digestion
   - Regulation of circadian rhythms such as sleep-wake cycles

Section 17.2

1. Hormones target specific receptors either on the surface of the cell or within the target cell.

2. Water-soluble hormones cannot diffuse through the cell membrane and, therefore, target receptors on the surface of the cell. These membrane receptors then make use of an internal secondary messenger system to alter the cell activity. Fat-soluble hormones can diffuse through the cell and, as a result, can easily bind directly to the target receptor located inside the cell.

3. Secondary messengers relay signals from the membrane receptor to a target molecule inside the cell.

4. Hormones affect transcription and function as transcription factors.

5. Amino acids, small peptides, and proteins make up water-soluble hormones.

6. Cholesterol makes up fat-soluble hormones called *steroids*: except does not make up the thyroid hormones.

7. See Table A.44.

TABLE A.44

Water Soluble	Fat Soluble
Catecholamines	Thyroid hormones
Pancreatic hormones	Cortisol
Pituitary hormones	Aldosterone
Hypothalamic hormones	Androgens
	Calcitriol

8. Thyroid hormones are fat-soluble and are made from tyrosine.

Section 17.3

1. Oxytocin and antidiuretic hormone (ADH) are made in the hypothalamus and released by the posterior pituitary gland.

2. The hypothalamus produces releasing hormones or inhibiting hormones which, in turn, signal the anterior pituitary to produce specific hormones.

3. Oxytocin causes contractions during labour and promotes the release of milk from the mammary glands. ADH acts on the distal tubule and collecting ducts of the kidneys to increase water reabsorption.

4. Growth hormone causes bone growth, protein synthesis (particularly in skeletal muscle), increased metabolism, and stimulation of the immune system. During adulthood, growth hormone maintains healthy cell division to replace old and damaged cells.

5. Excess growth hormone can cause gigantism, and a deficit of growth hormone can result in dwarfism.

6. The thyroid, adrenal glands, and gonads are stimulated by TSH, ACTH, and FSH and LH, respectively.

7. The primary function of prolactin is to stimulate milk production in the mammary glands.

Section 17.4

1. Follicular cells produce T3 and T4 (triiodothyronine and thyronine) and the parafollicular cells produce calcitonin.

2. T3 contains three iodine atoms, while T4 contains four. Most T4 is converted to T3 in the body as T3 is the more active hormone.

3. Tyrosine is used with iodine and other micronutrients to create both T3 and T4.

4. Iodine and selenium are both required for thyroid hormone production. Iodine is found in iodized table salt, and selenium is found in foods such as nuts, fish, wheat bran, pork, shellfish, eggs, and mushrooms.

5. Thyroid hormones are not water soluble and must be transported through the blood attached to plasma proteins.

6. Thyroid hormones have the following functions:
   - Regulating oxygen use by all cells during cellular respiration
   - Increasing basal metabolic rate (BMR) by increasing the production of ATP from carbohydrates and fats
   - Increasing the effects of epinephrine and norepinephrine
   - Stimulating the production of growth hormone
   - Increasing heart rate, breathing rate, and cardiac output

7. Hypothyroidism leads to decreased energy levels and body temperature, and to weight gain, fatigue, and depression. Hyperthyroidism causes insomnia, nervousness, irritability, increased heart rate, weight loss, and protruding eyes.

8. The most common causes of hypothyroidism are iodine or selenium deficiencies (sometimes secondary to anorexia). The most common cause of hyperthyroidism is an autoimmune disease called *Graves' disease.*

9. Calcitonin inhibits osteoclast activity—prevents breakdown of bone matrix—thereby preventing the increase of calcium levels in the blood.

10. Parathyroid hormone stimulates osteoclasts to release calcium, and kidney cells to increase reabsorption of $Ca^{2+}Mg^{2+}$. It also increases the excretion of HPO_4^{2-} by the kidney and increases the production of active vitamin D (calcitriol) by the kidneys, which then increases the absorption of calcium in the small intestines.

11. Low calcium levels cause increased parathyroid hormone, which stimulates osteoclasts to break down bone to increase blood calcium levels.

12. Dairy products, leafy green vegetables, nuts, beans, and fish all contain calcium.

Section 17.5

1. The adrenal cortex releases cortisol, androgens, and aldosterone. The adrenal medulla releases epinephrine and norepinephrine.

2. The zona glomerulosa secretes mineralocorticoids, mainly aldosterone; the zona fasciculate secretes glucocorticoids, mainly cortisol; and the zona reticularis secretes androgens.

3. The most common mineralocorticoid is aldosterone, the most common glucocorticoid is cortisol; and the most common androgen is dehydroepiandrosterone (DHEA).

4. The ANS stimulates the adrenal medulla to secrete epinephrine and norepinephrine.

5. Aldosterone causes more sodium to be reabsorbed back into the bloodstream, which increases blood pressure and causes nephrons to secrete potassium ions.

6. Cortisol stimulates glycogen and fat breakdown in the liver and fat cells and, therefore, increases blood sugar. It also inhibits the immune response. Cortisol is secreted as part of the stress response.

7. Cushing's syndrome results from too much cortisol and causes the following symptoms: hyperglycemia, muscle weakness, osteoporosis, type-2 diabetes, immune suppression, poor wound healing, thin skin, bruising, depression, moodiness, fat redistribution from limbs to trunk and face, "moon face," "buffalo hump," hair loss, decreased fertility, fatigue, insomnia, menstrual irregularities, water retention, and high blood pressure.

8. Androgens contribute to the development of secondary sex characteristics in males along with sperm production. In females, androgens contribute to libido and are converted into estrogen.

9. Epinephrine and norepinephrine are released as part of the stress response.

10. Epinephrine causes the following symptoms:
   - Bronchiole dilation
   - Pupil dilation
   - Increased blood pressure
   - Vasodilation and increased blood flow in liver, skeletal muscles, cardiac muscle, and adipose tissue
   - Vasoconstriction in digestive tract
   - Increased blood sugar

Section 17.6

1. The exocrine functions include the release of sodium bicarbonate and enzymes into the small intestine.

2. The alpha cells of the pancreas secrete glucagon; the beta cells release insulin.

3. Insulin acts on all cells of the body to increase glucose uptake.

4. The *glycemic index* refers to the variable amount of insulin triggered by the consumption of different foods.

5. Glucagon is secreted between meals and acts to increase blood sugar by causing the body to break down glycogen and triglycerides in the liver.

6. Glucose is stored as glycogen in both the liver and muscle cells. Fats, glucose, and proteins are stored in both the liver and adipose tissue.

7. Type-1 (or insulin-dependent diabetes mellitus) is an autoimmune disease that leads to a lack of insulin production. Type-2 (or non-insulin-dependent diabetes mellitus) is the result of insulin resistance caused by constant high blood insulin levels.

8. Non-insulin dependent diabetes mellitus (type-2) can be prevented through regulation of diet and exercise.

Section 17.7

1. Testosterone, estrogen, and progesterone are all stimulated by luteinizing hormone.

2. Aromatase

3. Testosterone causes increased muscle mass; increased bone density; bone maturation; development of sex organs and secondary sex characteristics, such as facial and axillary hair growth, and deepening voice. Estrogen causes the LH surge leading to ovulation, growth of uterine lining, increased bone formation, increased blood clotting, and many other functions. Progesterone supports pregnancy and plays many roles in fetal development.

4. FSH promotes spermatogenesis, oogenesis, and egg development.

5. Ovulation is caused by an increase in luteinizing hormone, which is called the *LH surge*.

Chapter 18 Concept Review Answers

Section 18.1

1. The cardiovascular system functions to transport gases to and from the lungs, distribute nutrients to the tissues, and carry waste to the kidneys. It also delivers defensive cells and signalling molecules, regulates blood pressure, and assists with body temperature regulation.

2. The main components of the cardiovascular system include the heart, blood vessels, and the blood.

3. The pulmonary system carries deoxygenated blood in the arteries to the lungs, whereas the systemic circuit carries oxygenated blood from the heart to the body tissues.

Section 18.2

1. Cardiac muscles tissue is composed of cardiac myocytes, and it is held together by intercalated discs containing desmosomes and gap junctions.

2. Desmosomes provide a physical connection between cells, and gap junctions provide the chemical connection that permits the action potential to move from one cardiac cell to another. This ability to transmit the action potential allows for the coordinated contraction of the heart.

3. The heart contains specialized cells called the *SA node* that spontaneously depolarize, acting as a "pacemaker" for the heart.

Section 18.3

1. The external features of the heart include the coronary sinus, which receives blood from the inferior and superior vena cavae. The blood is pumped from the heart through either the pulmonary arteries or into the aortic arch. The heart is surrounded by a membrane structure called the *pericardium*, which functions to reduce friction during contraction.

2. The coronary arteries are located in the coronary sulcus: a groove between the right and left chambers of the heart.

3. The upper chambers that receive blood from the two circuits are the atria. The right atrium receives blood from the inferior and superior vena cavae of the

systemic circuit, and the left atrium receives blood from the pulmonary vein. The lower chambers called the *ventricles* pump blood through the circuits; the right ventricle is connected to the pulmonary arteries, and the left ventricle supplies the aorta.

4. The pericardium attaches to the diaphragm and the great vessels, which secures the heart in the mediastinum.

5. The myocardium and the endocardium are the middle and innermost layers of the heart, respectively.

6. The four valves of the heart are (1) the mitral (bicuspid) valve, also called the *left atrioventricular valve*; (2) the tricuspid valve, also called the *right atrioventricular valve*; (3) the aortic semilunar valves; (4) the pulmonary semilunar valves.

7. The atrioventricular valves (AVs) are connected to the papillary muscles by chordae tendinae. The chordae tendineae prevent the valves from opening in the wrong direction and thereby prevent backflow during contraction.

8. The first sound—lub—is caused by the sudden closing of the AV valves. This is followed by the second heart sound, dub, which results from the closing of the semilunar valves.

9. The valves are forced open or closed by a pressure gradient created by the force of contraction of the heart and the resisting pressure in the vessels.

Section 18.4

1. Right atrium—tricuspid valve—right ventricle—pulmonary arteries—pulmonary arterioles—pulmonary capillaries—pulmonary venules—pulmonary veins—left atrium—mitral (bicuspid) valve—left ventricle—aorta—arteries—arterioles—capillaries—venules—veins—vena cava—right atrium

2. The "atrial kick" provided by atrial systole accounts for an increase of 10–20 mL of blood per contraction.

3. Increased pressure in the atria during atrial systole causes the AV valves to open, allowing blood to move into the ventricles.

4. Increased pressure in the ventricles during ventricular systole causes the SL valves to open, allowing blood to move into the aorta and pulmonary arteries.

5. Systole is the contraction of the cardiac muscle. Diastole is the relaxation of the same muscle.

Section 18.5

1. The sinoatrial node, atrioventricular node, bundle of His, bundle branches and Purkinje fibres make up the conduction system of the heart.

2. Autorhythmicity is the heart's ability to regulate its own depolarization spontaneously.

3. Sodium, calcium and potassium

4. The pacemaker potential does not have a stable resting membrane potential. Therefore, it slowly depolarizes after each action potential, unlike the action potential found in other cells, such as neurons, that depolarize only if stimulated.

5. During the P wave of the ECG, the atria are contracting. During the QRS, the ventricles are contracting and the atria are relaxing; note that the atrial relaxation is hidden by the more prominent electrical impulse in the ventricles. During the T wave, the ventricles are relaxing.

6. The first heart sound is caused by the closing of the AV valves. The second heart sound is caused by the closing of the semilunar valves.

7. Arrhythmias are caused by malfunctions of the conduction system of the heart.

8. During the first isovolumetric phase, the blood volume of the ventricles is not changing as the AV valves are closed and the semilunar valves have not yet opened. During the second isovolumetric phase, the ventricular volume isn't changing because the semilunar valves are closed and the AV valves have not yet opened.

9. Most of the blood enters the ventricles in the time between the relaxation of the ventricles and the contraction of the atria.

10. End diastolic volume is the volume of blood in the ventricles immediately prior to ventricular contraction. The end systolic volume is the volume of blood remaining in the ventricles immediately following ventricular contraction.

Section 18.6

1. The potential increase in cardiac output in response to stimuli is called the *cardiac reserve*, and it is usually two to three times higher than the resting cardiac output.

2. 130 mL – 50 mL = 80 mL (stroke volume); 80 mL/beat × 75 beats/min = 6000 mL/min OR 6 L/min

3. The sympathetic nervous system increases the rate of cardiac contraction by releasing norepinephrine. Acetylcholine released by the parasympathetic nervous system decreases the heart rate.

4. Epinephrine and thyroid hormones both function to increase heart rate, thereby increasing total cardiac output.

5. Increased vascular resistance leads to increased afterload and a decrease in stroke volume.

6. Preload is the blood volume returning to the atria from the circulation. Afterload is the resistance created by the vasculature.

7. The Frank-Starling law dictates that the stretch of the myocardial fibres is directly proportional to the force of contraction. In essence, increased stretching means a stronger force of contraction.

8. Venous return is affected by the skeletal pump of the body in conjunction with the total fluid volume of the body.

9. The skeletal and respiratory pumps function to move blood toward the heart during muscular contraction.

10. Factors that increase cardiac output include the following:
   • Increased sympathetic stimulation (epinephrine and norepinephrine)
   • Decreased parasympathetic stimulation
   • Exercise
   • Increased sodium, potassium, and calcium levels
   • Increased body temperature
   • Increased preload
   • Increased blood volume
   • Increased ventricular contractility and distensibility
   • Decreased afterload

Section 18.7

1. The three layers of the arteries from innermost to outermost are as follows:
   • Tunica intima, composed of simple squamous epithelial tissue (also called the *endothelium*)
   • Tunica media, composed of smooth muscle tissue and elastic connective tissue
   • Tunica externa, composed of loose connective tissue

2. Arteries are under greater pressure, have a smaller lumen, and are more elastic. Veins contain valves, have a larger lumen, and are less elastic.

3. Arteries withstand more pressure than veins.

4. Only veins contain valves.

5. Gas exchange occurs in the capillaries.

6. Precapillary sphincters control blood flow and assist with thermo-regulation.

7. The difference is the location. Plasma is located in the vasculature, interstitial fluid is in the tissues, and lymph is found in the lymph ducts and nodes.

8. Edema is the buildup of fluid in the interstitial space.

9. Lymph is moved through the vessels by skeletal pumps, akin to the skeletal pumps in the veins. Larger lymphatic vessels contain smooth muscle that contracts and helps push lymph through the vessels. Lymphatic vessels also have valves, like those in veins, which prevent backflow.

Section 18.8

1. The circle of Willis both regulates the intercranial pressure and provides redundancy should blood vessels in the brain become damaged.

2. The CSF drains through the superior sagittal sinus and the transverse sinus—the two largest veins in the brain—down into the jugular veins.

3. The hepatic portal system allows the blood passing from the digestive system to be filtered by the liver. The liver acts to store and process various nutrients.

Section 18.9

1. Flow $\propto$ change in pressure/ resistance

2. Vessel diameter, constriction and dilation, blood viscosity, and turbulent blood flow all increase resistance.

3. Factors that affect blood flow include the following:
   • Metabolic requirements of tissues
   • Body temperature
   • Total blood volume
   • Blood viscosity
   • Cardiovascular disease
   • Medications

4. When the body temperature decreases, peripheral blood vessels constrict to maintain temperature. When the body temperature increases, blood vessels dilate to dissipate heat.

5. An increased hematocrit increases blood viscosity and therefore increases resistance.

6. Laminar flow is even flow through vessels with a smooth lumen. Turbulent flow results from an uneven surface in the lumen, which causes random blood flow patterns, typically caused by plaque buildup in the vessels called *atherosclerosis*. Laminar blood flow can cause blood clots to form.

7. Diuretics decrease blood pressure by decreasing the total blood volume. Vasodilators reduce resistance in the vessels and also cause decreased blood pressure.

Section 18.10

1. A normal, healthy blood pressure is 120/80.

2. A negative feedback loop consists of sensory receptors, sensory neurons, an integrating centre, motor neurons, and effectors.

3. Baroreceptors detect pressure; chemoreceptors detect O_2, CO_2, and H^+; proprioceptors detect body movement.

4. Information from the carotid arteries is sent through the afferent nerves that travel through the

glossopharyngeal nerve to the brain. Information from the aortic arch is transmitted via the vagus nerve.

5. The cardiovascular centre is located in the medulla oblongata.

6. Neuronal stimulation increases the heart rate and pressure, as well as stimulating the hormonal response that causes epinephrine to be released from adrenal medulla.

7. See Table A.45.

TABLE A.45

Hormone	Where Produced	Effect on Blood Pressure
Antidiuretic hormone (ADH)	Hypothalamus	Increases water reabsorption in the kidneys
Aldosterone	Adrenal cortex	Increases sodium reabsorption in the kidneys, thereby increasing water reabsorption
Epinephrine	Adrenal medulla	Causes peripheral vasoconstriction, meaning increased vascular resistance
Angiotensin II	Converted from angiotensin I by ACE	Causes aldosterone release and also stimulates vasoconstriction
Atrial natriuretic peptide (ANP)	Heart	Lowers blood pressure by causing decreased sodium and therefore water reabsorption in the kidneys

Section 18.11

1. Atherosclerosis is the buildup of plaque, fats, and minerals in the walls of the arteries.

2. Plaque is primarily made up of LDLs ("bad cholesterol"): cholesterol-protein molecules produced by the liver to transport triglycerides from the liver to the tissues. HDLs ("good cholesterol") remove fat from blood vessel walls to be processed in the liver for energy. HDLs can be increased and LDLs decreased with exercise.

3. See Table A.46.

4. Atherosclerosis in the coronary vessels can lead to a decreased blow flow to the myocardium. Initially, this can cause ischemia and eventually may lead to myocardial infarction: heart attack. These can also lead to transient ischemic attacks and strokes (also called *CVAs* (cerebrovascular accident). Chronic hypertension can also lead to kidney failure.

TABLE A.46

Risk Factor	Effect
High blood LDLs and triglycerides	Forms plaque in the vessels, leading to atherosclerosis
Low blood HDL levels	
Obesity	Associated with higher LDL numbers (see above)
Diet	Trans fats increase LDL numbers more than any other type of food (see above)
Lack of physical activity	Contributes to obesity and lower HDL levels
High levels of C-reactive protein	Associated with chronic inflammation
High blood homocysteine levels	High levels linked to CV disease
Smoking	Causes hypertension
Stress	Not clinically proven; may be due to association with other behavioural risk factors
Menopause	Believed due to decreased estrogen levels
Alcohol abuse	Excessive alcohol intake is associated with many diseases, among them CV disease.
Genetic predisposition	Various DNA mutations could affect fat metabolism, transport of fats in the bloodstream, immune response, or any aspect of CV disease.
Ethnicity	Those of African or South Asian descent and also First Nations populations are most likely to suffer from CV disease and diabetes.

5. Metabolic syndrome is identified as central obesity coupled with high blood triglycerides, low HDLs, hypertension, or hyperglycemia.

6. Angina and a heart attack (MI) are very similar in symptoms, but angina pain subsides when the oxygen demand of the heart decreases.

7. TIAs, also called *mini-strokes*, have similar symptoms to strokes but tend to resolve in a relatively short period of time.

8. Atherosclerosis can be significantly decreased by eliminating trans fats, daily exercise, controlling blood sugar, not smoking, losing weight, and managing stress.

9. See Table A.47.

TABLE A.47

Medication Type	Effect
Diuretic	Decreases blood pressure by causing increased water excretion
Beta blocker	Inhibits the effects of the sympathetic system, including epinephrine, which decreases the heart rate and force of contraction
Alpha blocker	Decreases the effects of norepinephrine, which decreases vascular tone (leads to vasodilation) and preload
Calcium channel blockers	Decreases force of contraction
Vasodilators	Decreases vascular resistance and preload
ACE inhibitors	Decreases conversion of angiotensin I into angiotensin II (Chapter 21), which causes the production of aldosterone

TABLE A.48

Type of Leukocyte	Function
Neutrophil	Secretes chemicals that kill bacteria in the nearby region
Basophil	Releases histamine during an allergic reaction
Eosinophil	Fights off multicellular parasites and assists with the allergic response
Natural killer cell	Directly kills virus-infected cells and cancer cells
Monocyte	Differentiates into macrophages and dendritic cells, which act as antigen-presenting cells
B Cell	Involved in adaptive immune response and differentiates into plasma cells that release antibodies
T Cell	Involved in adaptive immune response and differentiates into cytotoxic T cells and helper T cells

Section 18.12

1. The blood functions to transport nutrients and oxygen to the tissues, carry waste products to the kidneys, regulate body temperature, and assist with the immune response.

2. Electrolytes are ions used by the cells required to maintain all cell functions.

3. Plasma proteins such as albumins regulate the osmolarity of the blood. Antibodies and complements provide an immune response. Fibrinogen aids in clotting. Hormones are transported in the blood to signal cells throughout the body.

4. Platelets form clots when the vessels become damaged.

5. In a healthy person, approximately 45% of the total blood volume is composed of red blood cells. If the red blood cell count drops below this value, the oxygen-carrying capacity of the blood decreases, leading to decreased oxygen delivery. The decreased oxygen can lead to fatigue.

6. Hematopoiesis is the formation of red blood cells; it is stimulated by the hormone erythropoietin.

7. Red blood cells contain the protein hemoglobin and use the mineral iron to transport oxygen.

8. When red blood cells expire, the majority of the cells' parts are recycled to be reused, but the heme group is broken down by the liver, released along with bile, and excreted into the small intestine.

9. See Table 18.3

10. See Table A.48.

Chapter 19 Concept Review Answers

Section 19.1

1. Respiratory system functions include the following:
   - Taking in oxygen and removing carbon dioxide
   - Regulating blood pH
   - Warming and moistening air
   - Filtering particulate matter from the inhaled air
   - Providing a sense of smell
   - Producing sound through the vocal cords

Section 19.2

1. The conducting zone transports air, and the respiratory zone is responsible for gas exchange.

2. Nasal conchae help warm the air, possess scent receptors to pick up smells, and cause the air to become turbulent, increasing the amount of filtration that takes place.

3. The pharynx is divided into the nasopharynx, the oropharynx, and the laryngopharynx.

4. Eustachian tubes help equalize pressure within the inner ear.

5. As air passes by the vocal cords it causes them to vibrate, producing sound.

6. Hair and cilia act as physical barriers, filtering various particles from the air prior to the air entering the lungs.

7. The visceral pleura surrounds the lungs; the parietal pleura lines the inside of the thoracic cavity.

8. The fluid in the pleura provides lubrication that reduces friction during the expansion and contraction of the lungs.

9. Surfactant is produced by type II alveolar cells.

10. Surfactant reduces surface tension and prevents alveolar collapse.

Section 19.3

1. Relaxed inspiration is caused by the diaphragm and the internal intercostal muscles. Relaxed exhalation requires no muscular contraction.

2. Accessory muscles may assist the inspiratory effort. These muscles include the sternocleidomastoid muscles, the scalenes, and the pectoralis muscles.

3. The external intercostals, external obliques, rectus abdominis, transverse abdominis, and internal obliques contract during forced exhalation, which increases abdominal and thoracic pressure.

4. Boyle's law states that, in a closed container with a constant number of molecules at a constant temperature, the pressure is inversely proportional to the volume.

5. Volume and pressure are inversely proportional. As the volume increases, the pressure decreases and vice versa.

6. At sea level, the atmospheric pressure is 760 mmHg. If there is no change in volume, the pressure in the lungs—or alveolar pressure—would be 760 mmHg, and the intrapleural pressure would be 756 mmHg.

7. If the intrapleural pressure wasn't lower than the alveolar pressure the lungs would collapse.

8. Compliance is the ability for the lungs to expand and take in air. Someone with decreased compliance has less ability to take in air. Recoil is the elasticity of lungs that facilitates movement of air out of the lungs.

9. The lung volumes are as follows:

    a. Tidal volume

    b. Inspiratory reserve volume

    c. Expiratory reserve volume

    d. Residual volume

10. Lung volume is affected by age, sex, altitude, physical fitness, and disease.

11. *Anatomical dead space* refers to the air that does not contribute to gas exchange.

12. Alveolar ventilation = [tidal volume (mL/breath) − dead space (mL/breath)] × respiration rate (breaths/min)

    a. alveolar ventilation = [600 mL/breath − 180 mL/breath] × 14 breaths/min

    b. alveolar ventilation = [420 mL/breath] × 14 breaths/min

    c. alveolar ventilation = 5880 mL/min OR 5.88 L/min

Section 19.4

1. Dalton states that the sum of the pressures exerted by each gas individually comprises the total pressure. Boyle's law states that the pressure of a gas is inversely proportional to the volume.

2. See Table A.49.

TABLE A.49

Molecule	mmHg
Nitrogen	593.408
Oxygen	158.84
Argon	6.84
Carbon dioxide	0.228
Ozone	0.076
Other trace elements	3.8

3. As the percentage of oxygen from the atmosphere becomes higher than that in the lungs, the oxygen moves toward the alveoli. Once it reaches the alveoli, the pressure in the alveoli is greater than that of the dissolved gas in the capillaries of the lungs, and the oxygen moves into the bloodstream. Once in the bloodstream, the pressure of the dissolved gas is greater than that of the tissues, and the oxygen dissolves into the cells.

4. Atmospheric pressure, ventilation, and metabolism

5. Cells produce carbon dioxide as a waste product, and it moves into the blood to the right side of the heart. Blood circulates to the lungs where the carbon dioxide moves down the concentration gradient to the alveoli and diffuses into the dead space, and is eventually expelled into the atmosphere. The partial pressure of carbon dioxide is higher in venous blood and lower in arterial blood.

6. The following factors cause an increase in the amount of oxygen that dissociates from hemoglobin:

    a. Decrease in pH

    b. Increase in partial pressure gradient

    c. Increase in body temperature

7. Unlike oxygen, carbon dioxide can be transported in three ways (Figure 19.13):

    a. Dissolved in plasma, 7–10%

    b. Bound to hemoglobin, 25–30%

    c. In the form of bicarbonate ions, 60–65%

8. Chloride shift is important to help stabilize the blood pH.

Section 19.5

1. The dorsal respiratory group controls inspiration, and the ventral respiratory group acts as the pacemaker, managing the breathing rhythm. The ventral respiratory group also signals the muscles used for forceful exhalation.

2. The upper pons smooths the transition from inhalation to exhalation. The lower pons helps regulate the respiratory rate by stimulating the medulla oblongata.

3. Stretch receptors in the lungs signal the medulla oblongata to stop the dorsal respiratory group from further triggering the contraction of inspiratory muscles.

4. The cerebral cortex permits the voluntary control of breathing, and the limbic system controls breathing in response to emotional stimuli. The autonomic nervous system increases the rate and depth of breathing when stimulated by the sympathetic nervous system; conversely, the parasympathetic system decreases the rate and depth of breathing. Increased body temperature can increase respiratory rate. Other factors that can alter breathing include medications, irritants, and a change in the concentrations of oxygen, carbon dioxide, and hydrogen ions in the blood.

5. Peripheral chemoreceptors are located in the aortic arch and carotid arteries. Central chemoreceptors are located in the medulla oblongata.

6. Chemoreceptors detect changes in O_2, CO_2 and H^+ in the blood.

7. • Stimulus—changes in blood oxygen and carbon dioxide levels
   • Receptor—peripheral chemoreceptors detecting changes in blood pH, and central chemoreceptors detecting increased carbon dioxide in the cerebrospinal fluid
   • Integrating centre—signals being received by medulla oblongata
   • Effectors—motor neurons causing the contraction of inspiratory muscles, alternating with the stimulation of the expiratory muscles

Section 19.6

1. Emphysema and chronic bronchitis are the two primary diseases under the classification of COPD.

2. Emphysema is the breakdown of the alveoli into larger sacs with decreased elasticity. Chronic bronchitis results from chronic inflammation and leads to increased mucus production and a chronic cough.

3. The most common cause of COPD is the inhalation of irritants and pollutants such as cigarette smoke.

4. Respiratory difficulty and the use of additional muscles to aid the respiratory effort (referred to as *accessory muscles*) are common symptoms of asthma.

5. Bronchodilators and anti-inflammatory medications can be used to treat asthma.

6. Hypoxia can be caused by low hematocrit or hemoglobin, poor circulation, poisoning such as by cyanide, and decreased ventilation caused by disease.

Chapter 20 Concept Review Answers

Section 20.1

1. The organs of the gastrointestinal tract are the esophagus, the stomach, the small intestine, the large intestine, and the anus.

2. The accessory organs are the gallbladder, liver, pancreas, and salivary glands.

3. Ingestion, mechanical digestion, chemical digestion, secretion, absorption, and defecation.

4. The six main processes of the digestive system are as follows:
   • Physically and chemically breaking down ingested food
   • Absorbing digested nutrients into the bloodstream
   • Removing undigested waste
   • Producing intrinsic factor, which is required for the absorption of vitamin B12
   • Housing symbiotic bacteria, the normal flora, mostly in the large intestine

Section 20.2

1. The layers of the GI tract in order are the mucosa, submucosa, muscularis, and serosa.

2. The three components of the mucosa are the epithelial cells that form the mucous membrane, the connective tissue of the lamina propria, and the thin layer of smooth muscle called the *muscularis mucosa*.

3. The submucosal plexus nerves are responsible for detecting stimuli such as stretching; the myenteric plexus is responsible for coordinating smooth muscle contraction.

4. The circular layer contracts to mix food within the GI tract, and the longitudinal layer acts to move food through the GI tract.

5. The peritoneum is made up of areolar connective tissue covered by simple squamous epithelial tissue.

6. The visceral peritoneum is the outermost layer of the digestive tract, whereas the parietal peritoneum lines the interior of the abdominal cavity.

7. The greater omentum is a collection of fat that provides protection for the viscera (internal organs), and the mesentery holds the organs in place.

Section 20.3

1. The maxilla, the mandible, and the palatine bones make up the mouth.

2. The *gingival* are commonly known as the *gums*.

3. Adults have a total of 36 teeth: eight incisors, four canines, eight premolars, 12 molars, and four wisdom teeth.

4. The tongue aids in chewing and swallowing by rolling food into a bolus. The tongue also contains taste receptors that detect food flavours such as sweet, salty, bitter, sour, and savoury (or umami).

5. *Peristalsis* refers to wavelike contractions that move food through the gastrointestinal tract.

6. The uvula and epiglottis are physical barriers that prevent food from entering the nasopharynx and trachea, respectively.

7. From inner to outer, the muscle layers of the stomach are the oblique, circular, and longitudinal layers.

8. The pyloric sphincter closes to prevent the passage of contents from the stomach to the small intestine.

9. Goblet cells produce mucus, chief cells produce pepsinogen, and parietal cells secrete hydrochloric acid.

10. The liver performs the following functions:
    - Synthesizing bile
    - Storing nutrients
    - Breaking down nutrients
    - Breaking down old red blood cells
    - Breaking down many organic and inorganic toxins
    - Excreting excess trace metals
    - Synthesizing the plasma proteins
    - Synthesizing fibrinogen
    - Synthesizing angiotensinogen
    - Activating vitamin D
    - Producing insulin-like growth factors (IGFs)
    - Secreting complement proteins

11. Bile is made of bile salts, water, mucus, and other inorganic salts.

12. The pancreatic endocrine functions include the production and release of insulin and glucagon. The exocrine functions include the production and excretion of pancreatic juice, water, salts, and enzymes.

13. The majority of digestion and absorption of nutrients occurs in the duodenum.

14. Folds, villi, and microvilli line the walls of the small intestine.

15. In order, the large intestine moves from the ileocecal sphincter through the cecum to the ascending colon, transverse colon, descending colon, sigmoid colon, rectum, anal canal, and anus.

16. The large intestine absorbs ions, vitamins, and water, as well as storing waste products until defecation.

17. Normal flora or gut bacteria serve the following functions:
    - Fermenting undigested substrates
    - Regulating the immune system
    - Preventing growth of harmful bacteria
    - Producing vitamin K and biotin
    - Increasing growth of intestinal cells
    - Producing lactic acid
    - Converting carbohydrates into short-chain fatty acids
    - Releasing important minerals from food
    - Metabolizing carcinogens

Section 20.4

1. Proteins consist of amino acids, carbohydrates are formed from monosaccharides, triglycerides are broken into fatty acids, and nucleic acids are broken into nucleotides.

2. The enzymes that break down proteins and peptides are pepsin, trypsin, chymotrypsin, and peptidases.

3. Enzymes that break down disaccharides are produced in the small intestine.

4. Amylase is produced by the salivary glands and the pancreas.

5. Lipase breaks down fat.

6. Bicarbonate is released to neutralize the acidic content before it can cause damage to the small intestine.

Section 20.5

1. Water and ions move either into the cells by osmosis, or ion channels, or between epithelial cells.

2. Vitamin D is required for calcium absorption in the small intestine.

3. Intrinsic factor is required for the absorption of vitamin B12.

4. Amino acids, glucose, and galactose all enter the cell by means of cotransport with sodium.

5. Sodium-potassium pumps are required to move the sodium back out of the cell after being used for co-transport of glucose and galactose.

6. Fats are broken down into very small droplets called *micelles*. Once broken down, fats can be absorbed by simple diffusion.

7. Water-soluble vitamins are cotransported across the membrane with sodium, with the exception of vitamin B12, which requires intrinsic factor to be absorbed. Fat-soluble vitamins are absorbed by simple diffusion across the membrane and are packaged into chylomicrons with other fats and enter the lacteal.

8. Injury or inflammation can cause damage to the mucosa and alter the path of absorption, causing triglycerides to be absorbed directly into the blood-stream rather than through the lymph vessels.

9. The *thermogenic effect* refers to the amount of energy (calories) used by the for digestion.

Section 20.6

1. The cephalic phase is the initiation of the secretion of gastric juices and muscle contractions in anticipation of ingesting food. The gastric phase begins when food enters the stomach and causes gastric secretions and muscle contractions. The intestinal phase triggers the release of pancreatic secretions, bile, intestinal enzyme secretion, and motility.

2. See Table A.50.

TABLE A.50

Hormone	Where Produced	Function
Gastrin	Pyloric region of stomach	Stimulates the production of hydrochloric acid, intrinsic factor, pepsinogen, mucus; also promotes muscular contractions
Secretin	Intestinal mucosa	Stimulates the pancreas to release bicarbonate ions as well as pancreatic enzymes
Cholecystokinin	Intestine	Causes the stomach to slow gastric emptying, causes the gall bladder to release bile, and signals the hypothalamus to create the feeling of fullness

3. Cholecystokinin triggers in the hypothalamus the feeling of being full.

Section 20.7

1. Crohn's disease and ulcerative colitis are chronic inflammatory diseases.

2. A high-fibre diet can help prevent diverticulitis and hemorrhoids.

3. Gastritis and ulcers

Chapter 21 Concept Review Answers

Section 21.1

1. The urinary system functions are as follows:
   - Regulating blood, water, and ion composition
   - Excreting waste products of metabolism
   - Regulating blood pressure
   - Regulating blood pH by excreting excess H^+ ions
   - Producing calcitriol and erythropoietin

2. The urinary system is made up of the kidneys, the ureters, the bladder, and the urethra.

Section 21.2

1. The cortex is the outer portion of the kidney; the medulla is the inner portion.

2. Urine flows from the collecting ducts through the minor calyx to the major calyx and collects in the renal pelvis. Each renal pelvis drains through the ureters and into the bladder. From there the bladder empties via the urethra.

3. Renal artery—segmental arteries—interlobar arteries—cortical radiate arteries—afferent arteriole—glomerulus—efferent arteriole—peritubular capillaries—cortical radiate veins—arcuate veins—interlobar veins—renal vein

Section 21.3

1. Renal corpuscle—proximal convoluted tubule—descending loop of Henle—ascending loop of Henle—distal tubule—collecting duct

2. The majority of reabsorption takes place in the proximal convoluted tubule.

3. The descending loop of Henle is impermeable to ions, but permits the passage of water. The ascending loop of Henle is impermeable to water, but permits the passage of ions through the epithelial layer.

4. In a healthy individual, 100% of glucose and amino acids are reabsorbed by the kidneys.

5. PTH causes the kidneys to reabsorb calcium. ADH causes the kidneys to reabsorb water. Aldosterone causes the kidneys to reabsorb sodium.

6. Ammonia, urea and uric acid are formed when proteins or nucleic acids are broken down.

7. Secretion occurs in the proximal convoluted tubule, the distal convoluted tubule, and the collecting ducts.

8. Nitrogenous wastes, creatinine, hydrogen ions, potassium ions, and some medications are secreted in the urine.

Section 21.4

1. • Filtration. Molecules move from glomerulus into tubule at Bowman's capsule.

   • Reabsorption. Molecules move out of tubule and into capillaries.

   • Secretion. Molecules move from peritubular capillaries into tubule.

2. Water can be reabsorbed by osmosis, through aquaporins, or via paracellular transport.

3. In the apical membrane, glucose and amino acids cotransport with sodium; in the basolateral membrane, glucose and amino acids move by facilitated transport with membrane carrier proteins.

4. The sodium-potassium pump is located on the basolateral membrane of the tubule cells.

Section 21.5

1. See Table A.51.

2. Renin is used to cleave angiotensinogen, turning it into angiotensin I.

3. Angiotensin-converting enzyme (ACE) converts angiotensin I into angiotensin II.

4. Angiotensin II acts in the following ways:

   • Stimulating the adrenal cortex to produce aldosterone

   • Causing vasoconstriction to increase blood pressure

   • Stimulating the posterior pituitary to release ADH

   • Stimulating the sympathetic nervous system to increase blood pressure

   • Stimulating the reabsorption of sodium, chloride and, therefore, water in the kidneys

   • Stimulating thirst and cravings for salt

Section 21.6

1. Hypertension and diabetes are the most common causes of chronic kidney disease (CKD).

2. Factors that help prevent CKD or mitigate its complications involve maintaining a healthy blood pressure (through diet, exercise, and medications) along with maintaining proper blood sugar levels (through diet, exercise, and medications).

3. Increased salt intake leads to a greater accumulation of ions in the kidneys. The increased ion concentration leads to increased water reabsorption, resulting in higher blood volumes. The increased blood volume will exacerbate the pre-existing hypertensive state.

4. The symptoms of advanced chronic kidney disease include, but are not limited to, the following:

   • Nausea

   • Weakness

   • Fatigue

TABLE A.51

Hormone	Stimulus for Production	Production Site	Target Tissue	Functions
ADH	Low blood pressure	Hypothalamus, released from posterior pituitary gland	Distal tubules and collecting duct	Producing aquaporin membrane proteins to increase water reabsorption
Aldosterone	Increased angiotensin II or high blood potassium	Adrenal cortex (zona glomerulosa)	Distal tubule and collecting duct	Increasing sodium reabsorption, therefore water reabsorption, and increasing potassium secretion
PTH	Low blood calcium or magnesium	Parathyroid gland	Distal tubule	Increasing calcium and magnesium reabsorption
ANP	High blood pressure	Atria	Distal tubule and collecting duct	Decreasing sodium reabsorption; therefore, more water excreted to decrease blood pressure

- Inability to concentrate
- Decreased appetite
- Muscle cramping
- Increased blood pressure
- Itching of the skin
- Edema in the soft tissues, usually including the feet and around the eyes

5. Dialysis and kidney transplant are the only treatments currently available for advanced CKD.

6. Calcium-oxalate and uric acid are the two most common types of kidney stones.

7. Apart from kidney stones, gout can result from a buildup of uric acid

Chapter 22 Concept Review Answers

Section 22.1

1. The male reproductive system is responsible for the following:
   - Spermatogenesis
   - Producing steroid hormones, including testosterone
   - Storing and transporting sperm
   - Producing additional secretions that form semen

2. The female reproductive system is responsible for the following:
   - Oogenesis
   - Producing steroid hormones, primarily estrogen and progesterone
   - Storing and transporting ova
   - Providing a location for the fertilization of eggs (fallopian tubes)
   - Providing a site for the implantation and growth of the fetus
   - Producing milk in the mammary glands for feeding the newborn

Section 22.2

1. Seminiferous tubules—rete testis—efferent ductules—epididymis—vas deferens— seminal vesicles—ejaculatory duct—prostate gland—urethra

2. The seminal vesicles secrete an alkaline fluid that helps neutralize the acidic environment of the male urethra. The prostate secretes citric acid, used by the sperm to produce ATP and create prostate-specific antigens that help the sperm enter the cervix. Bulbourethral glands secrete alkaline mucus that neutralizes and lubricates the urethra.

3. Spermatogonia, primary spermatocytes, secondary spermatocytes, spermatids, and finally spermatozoa

4. Spermatozoa, secondary spermatocytes, and spermatids are haploid. Primary spermatocytes and spermatogonia are diploid.

5. Sertoli cells support spermatogenesis in the following ways:
   - Forming the blood-testis barrier
   - Providing nutrients and secreting fluid that surrounds the developing sperm
   - Stimulating and regulating spermatogenesis
   - Producing the hormone inhibin, which prevents the release of FSH
   - Engulfing the improperly developing sperm

6. Leydig cells produce testosterone in response to luteinizing hormone.

7. Luteinizing hormone causes the release of testosterone. Testosterone and FSH work together to stimulate Sertoli cells to release signalling molecules to regulate spermatogenesis.

8. The spermatozoon consists of three sections:
   - Head—contains 23 chromosomes and minimal cytoplasm
   - Mid-piece—contains mitochondria that metabolize fructose for energy
   - Tail—a flagellum composed of microtubules that propel the sperm

Section 22.3

1. Fertilization occurs in the infundibulum of the fallopian tube.

2. From innermost to outermost layer, the endometrium is where the blastocyst implants and is shed each month if no fertilization takes place. The middle layer, or myometrium, is the smooth muscle layer that contracts during childbirth. Finally, the perimetrium is the outermost layer, composed of simple squamous epithelial and areolar connective tissue.

3. The vulva is made up of an adipose layer covering the pubic symphysis called the *mons pubis*; the external labia majora; the internal labia minora; and the clitoris.

4. In females, meiosis begins in utero during fetal development.

5. The first meiotic division occurs just before ovulation, and the second takes place only if fertilization occurs.

6. Primary oocytes form in the follicles during fetal development, and undergo their first meiotic division before ovulation. If the oocyte is fertilized, the cell undergoes a secondary division, and the 23 chromosomes combine with the 23 chromosomes from the sperm to form a diploid zygote.

7. A luteinizing hormone surge causes the oocyte to undergo the first meiotic division and start ovulation.

8. Progesterone and estrogen are produced by the corpus luteum.

9. Decreased levels of estrogen and progesterone are caused by a breakdown of the corpus luteum and cause menstruation.

10. LH stimulates the theca cells, and FSH stimulates granulosa cells.

Section 22.4

1. Endometriosis

2. Chlamydia and gonorrhea infections

3. A blood test to look at prostate-specific antigen (PSA) levels

4. Human papilloma virus (HPV) is highly linked to cervical cancer.

Chapter 23 Concept Review Answers

Section 23.1

1. Lymphatic vessels, lymph nodes, thymus, spleen, tonsils, the Peyer's patch of small intestine, and bone marrow make up structures of the immune system.

2. The functions of the immune system include the following:

   • Returning fluid from the interstitial space to the bloodstream

   • Transporting dietary fats and fat-soluble vitamins from the small intestine to bloodstream

   • Recognizing and killing infectious organisms

   • Recognizing cells and non-harmful foreign molecules, such as food and environmental substances

   • Producing immunological memory cells to prevent infection from the same organism in the future

3. The main infectious organisms are bacteria, viruses, fungi, and parasites.

Section 23.2

1. Lymphatic vessels have closed ends, and fluid is taken up through one-way opening in between the endothelial cells.

2. They are all the same except for the location in which each is found.

3. The right upper body drains into the right internal jugular vein and right subclavian vein to re-enter the circulatory system. Lymph drains from left upper body into the junction of the left internal jugular vein and the left subclavian vein.

4. Edema can be caused by the following:

   • Increased capillary pressure or interstitial osmotic pressure

   • Improper kidney filtration

   • High venous pressure

   • Lack of skeletal muscle use

   • Medications that affect blood pressure, vascular resistance, the autonomic nervous system, kidney function, or osmotic balance

   • Lack of plasma proteins

   • Blocked lymphatic vessels

5. Red and white blood cells are produced in the bone marrow by hematopoietic stem cells.

6. Mesenchymal cells differentiate into adipose cells, osteoblasts, and cartilage.

7. In the thymus, thymocytes (immature T cells) that recognize self-proteins are destroyed by apoptosis.

8. Reticular fibres act as a filter to prevent particles from entering the bloodstream.

9. Phagocytic macrophages, dendritic cells, and lymphocytes are all found in lymph nodes.

10. The adaptive or acquired immune response occurs in the lymph node.

11. The spleen acts to break down red blood cells in the red pulp, and the white pulp of the spleen assists with the immune response. The spleen can also assist with haematopoiesis in embryos.

12. The tonsils, adenoids, appendix, and Peyer's patches (of the small intestine) are all examples of lymphatic nodules.

13. See Table A.52.

Section 23.3

1. The first line of defence is prevention and includes the chemical and physical barriers that prevent infection:

   • The skin provides a physical barrier that is covered with beneficial normal flora organisms that help fight off infectious organisms.

   • Hairs and mucus in the nose functions to block large particles, and movement of the hairs can stimulate the sneezing reflex.

   • Cilia and mucus in the upper respiratory tract trap particles and propel them up toward the throat to be swallowed, or removed by coughing.

   • Antimicrobial chemicals such as lysozyme are secreted into tears, saliva, and sweat, and break down bacterial cell walls.

TABLE A.52

Cell Type	Function
Neutrophil	• Most numerous of all white blood cells, and predominant cell type in pus • First-responders that produce chemicals involved in the inflammatory response, such as histamine, vasodilators, reactive oxygen species, and cytokines • Phagocytic cells that engulf many types of microbes, mostly bacteria and opsonized pathogens (have antibodies from B cell response) • Migrates from blood vessels to infection area within tissues; attracted to damaged cells by process of chemotaxis (migration of immune cells toward an area of infection)
Basophil	• Least common type of white blood cell • Releases histamine and several cytokines involved in the inflammatory response • Plays a role in allergic reactions and parasitic infections, specifically, ectoparasites such as ticks or lice
Eosinophil	• Involved in parasitic infections such as helminths and worms • Produces many chemicals, including histamine that is involved in allergies, asthma, and inflammation
Mast cell	• Primarily involved in allergic reactions and anaphylaxis by releasing histamine if interaction with IgE antibodies occurs • Plays an important protective role in wound healing • Prevalent in tissues near openings to the body, such as mouth, nose, skin, lungs, and GI tract
Monocyte	• Circulates in the blood and responds to inflammatory signals • Migrates into tissues and differentiates into macrophages or dendritic cells
Dendritic cell	• Important antigen-presenting cells (APCs) that stimulate lymphocytes (B cells and T cells) in the spleen, lymph nodes, or lymphoid nodules and initiate the adaptive immune response • Phagocytic cells that can engulf pathogens or opsonized pathogens (when antibodies or complement proteins are bound to the pathogen) • Present in tissues that are in contact with the environment, such as the skin (Langerhans cells) lungs, and GI tract
Macrophage	Important APCs that stimulate lymphocytes in the lymphatic tissues and initiate the adaptive immune response Phagocytic cells that can engulf pathogens or opsonized pathogens (when antibodies or complement proteins are bound to the pathogen) Secretes toxic chemicals that directly kill invading organisms Secretes cytokines and chemicals involved in the inflammatory response
Natural killer cell	Specifically kills virus-infected cells and cancer cells by lysis with perforin or induction of apoptosis Kills cells opsonized with antibodies Can produce antigen-specific memory cells important for detecting same infection a second time
B cell	Contains a specific receptor for a specific antigen that is a membrane-bound immunoglobulin (antibody) Circulates between blood and lymphatic tissues Phagocytic when antigen matches receptor Antigen-presenting cell Produces cytokines Differentiates into plasma cells when activated by helper T cell and then secretes antibodies Can produce memory cells
Helper T cell	Do not directly kill any invading microorganisms Can only respond to antigen presented by APC (macrophage, dendritic cell, or B cell) Produces cytokines that activate B cells, cytotoxic T cells, macrophages, dendritic cells, and NK cells Can produce memory cells
Cytotoxic T cell	Directly kills virus-infected cells, cancer cells, and transplant tissue Binds to antigen presentation on MHC class I molecules Can produce memory cells
Regulatory T cell	Inhibits other immune cells at the end of an infection and prevents auto-reactive immune cells that escaped the selection process in the bone marrow or thymus from damaging tissues

- Sebaceous glands produce acidic oil that prevents the growth of certain bacteria.
- The stomach produces hydrochloric acid, a very strong acid that kills many organisms taken in with food.
- Resident beneficial bacteria in the digestive tract play an important role in preventing infection.

2. B Cells, macrophages, dendritic cells and neutrophils are all phagocytic.

3. Cytotoxic T-cells specifically target virus-infected cells.

4. See Table A.53.

TABLE A.53

Chemical	Function
Lysozyme	Breaks down bacterial cell walls
Perforin	Causes lysis in cells by puncturing membranes
Interferon	Released when cells are infected by viruses and triggers nearby cells to produce antiviral proteins
Histamine	Causes vasodilation, increased capillary permeability, and attracts phagocytic cells
Complement	Stimulates mast cells to produce histamine, binds to microbes to increase phagocytosis, and forms a membrane attack complex that kills bacteria and fungus
Prostaglandins	Same effects as histamine and also cause pain

- Redness is caused by increased blood flow.
- Swelling is caused by increased vascular permeability leading to edema.
- Pain is caused by the release of prostaglandins and the effect of swelling putting pressure on the nerves.
- Heat is caused by increased blood flow.

5. Redness, pain, swelling, and heat all result from chemicals released by damaged cells.

6. Pyrogens are molecules that increase body temperature by increasing the set point determined by the hypothalamus.

Section 23.4

1. An antigen is any molecule on the surface of an infectious organism that can be recognized by any immune cell.

2. Macrophages, B cells, and dendritic cells are all types of antigen-presenting cells.

3. Antigens presented on MHC I allow immune cells to recognize self-proteins and for immune cells to target specific microbial antigens.

4. All nucleated cells express MHC I. MHC II is found only on antigen-presenting cells.

5. Interleukin-I causes T cells to secrete interleukin-2, which causes proliferation of the T cell with that specific T cell receptor.

6. See Table A.54.

7. See Table A.54.

TABLE A.54

Helper T Cell Subtype	Main Role of Each Subtype
Th0	Non-reactive, tolerant cells for certain antigens
Th1	Increases activation of cytotoxic T cells
Th2	Increases the activation of B cells
Memory cell	Helps fight future infections of the same type
Th17	Slows down the immune response at the end of the infection

8. $CD8^+$ cytotoxic T cells express CD8. Helper T cells are called $CD4^+$ if they express the CD4 coreceptor.

9. Cytotoxic T cells each have a specific TCR that recognizes only one antigen that, when recognized, binds to an infected body cell, but it will only become fully activated if a helper T cell produces IL-2 and stimulates proliferation.

10. Helper T cells bind to B cells that present the proper matching antigen on MHCII and cause them to proliferate. Th2 helper cells bind to B cells and secrete IL-2

11. B cells may differentiate into plasma cells that secrete antibodies matching the antigen.

12. When an infectious organism is marked by antibodies, it's called *opsonized*. An opsonized microbe results in the following:

- Increased phagocytosis by macrophages and dendritic cells
- Increased targeting by complement proteins
- Neutralized microbes that cannot produce toxins or bind to body cells
- Immobilized bacteria that cannot spread as easily
- Agglutination—clumping of many antibodies and microbes together so they cannot function

13. Cell-mediated immune responses use T cells; humoral responses use B cells and antibodies.

14. Intracellular infections such as viral infections are fought by cytotoxic T cells, and bacterial or extra-cellular infections are fought by antibodies and B cells.

15. Memory cells improve the body's ability to rapidly fight off the same, or sometimes similar, infection and may even prevent symptoms from appearing.

16. Viruses, such as those causing the common cold, change surface antigens frequently and this is always seen by the immune system as a new infection.

17. Active immunity occurs when antibodies are produced by the individual. Passive immunity results from antibodies being transferred to the individual either by injection or ingestion.

Section 23.5

1. Edward Jenner recognized that people with cowpox didn't become infected by smallpox.

2. Live attenuated viruses can be produced by producing many generations until a non-virulent strain appears. Examples include measles, mumps, and rubella—known as the *MMR vaccine.*

3. A killed virus can also be injected directly into a host without causing infection. Examples include the influenza and cholera vaccines.

4. Vaccines for diphtheria, pertussis, and tetanus are created by using inactivated toxins that contain the antigens, but lack the bacteria to cause an infection.

5. DNA technology has been shown to be effective in making vaccines by causing non-pathogenic bacteria, fungi, mammalian cells, or low-virulence viruses to exhibit surface proteins that match those of the actual pathogen.

Section 23.6

1. In any autoimmune disease, the immune cells target a specific type of tissue and attack it as though it were a pathogen, leaving that tissue no longer capable of functioning normally.

2. The main theories of autoimmune disease are as follows:

   • Molecular mimicry. The immune system loses the ability to differentiate between foreign and self-antigens.

   • Hygiene theory. Countries with lower rates of infection exhibit a higher rate of autoimmune disease.

   • Lack of beneficial bacteria exposure in child-hood. Pasteurization and sterilization of foods

prevents children from fully developing the normal flora required to fight off infection.

   • Genetic factors. Research has shown some variations of certain MHC genes affect antigen presentation.

   • Vitamin D. It is known that vitamin D does play an important role in immune regulation, but supplementation alone has not reversed autoimmune disease.

3. Autoimmune diseases are most prevalent in developed countries.

4. The tissues targeted are as follows:

   a. Beta islets of the pancreas

   b. Myelin on the axons of the ANS

   c. Skeletal muscle

   d. Thyroid gland

5. Mast cells and basophils are responsible for the release of histamine, while B cells that are activated become plasma cells and then release antibodies.

6. Upon exposure for the first time, no IgE antibodies have been created that are specific to the particular allergen, and as such, no symptoms result.

7. The most severe allergic reactions tend to be caused by venoms and foods.

8. Histamine is the main chemical involved in both allergic reactions and asthma.

Chapter 24 Concept Review Answers

Section 24.1

1. Bacteria provide the benefits of producing oxygen, recycling minerals, fermenting foods, and acting as normal flora

2. The functions of bacteria in the gastrointestinal tract are as follows:

   • Fermenting undigested substrates, mostly soluble fibre

   • Regulating the immune system to tolerate food molecules

   • Preventing growth of harmful bacteria

   • Producing vitamin K and biotin

   • Increasing growth of intestinal cells and helping maintain a healthy mucosal layer

   • Producing lactic acid during fermentation, lowering pH, and preventing growth of pathogenic species

   • Converting carbohydrates into short-chain fatty acids

   • Releasing important minerals from food, including magnesium, calcium, zinc, and iron

   • Metabolizing carcinogens

3. Endospores form a protective physical barrier that protects bacteria from changes to the environment.

4. The three basic bacteria shapes are spherical (cocci), rod (bacilli), and spiral (spirochete).

5. Antibiotic resistance genes work by
   - degrading the antibiotic,
   - modifying the antibiotic into a non-functional form, and
   - forming membrane pumps that remove the antibiotic from the cell.

6. Infections are considered opportunistic when they live in the human body without causing harm or illness until the host's immune system becomes weakened.

7. Children, the elderly, and the immune-suppressed are most at risk for opportunistic infections.

8. *E.coli*, *Salmonella*, and *Campylobacter* are the bacteria most to blame for food poisoning.

9. *Methicillin-resistant Staphylococcus aureus* (MRSA) is a multi-resistant strain of a normal flora bacterium.

10. *Clostridium tetani* releases a toxin that blocks inhibitory neurotransmitters, leading to continuous nervous stimulation of muscle.

11. A number of normal flora bacteria can become pathogenic when the host becomes immune-compromised; therefore, they are considered opportunistic. *Staphylococcus* is one such example.

12. Chlamydia and gonorrhea are the two most common bacterial sexually transmitted infections (STIs).

Section 24.2

1. Viruses are not true cells as they do not have a cellular membrane or cellular organelles.

2. A bacteriophage is a virus that infects bacteria.

3. RNA viruses make use of an enzyme called *reverse transcriptase* to convert their RNA into DNA so it can be integrated into their host.

4. Hemagglutinin allows the influenza virus to bind to upper airway cells, and neuraminidase allows the newly replicated virus particles to be released from the host cell.

5. During the lytic cycle the virus kills the host during replication, but in the lysogenic phase the viral DNA is integrated and replicated along with that of the host cell, without killing it.

6. Attachment—infiltration—replication—assembly

7. The herpes virus undergoes a lysogenic cycle, allowing it to go undetected by the immune system and thereby making it extremely difficult to combat.

8. HIV binds to the CD4 membrane receptors on human immune cells.

9. Hepatitis B and C are the most likely strains to cause liver cancer.

10. Herpes, HPV, hepatitis, and EBV are all examples of sexually transmitted viruses.

11. HPV causes genital warts and two strains have been identified as potential causes of cervical cancer.

12. The Epstein–Barr virus (EBV) has been linked to increased rates of multiple sclerosis.

Section 24.3

1. Fungi are eukaryotic organisms.

2. Fungi have antimicrobial properties and are used in the treatment of heart disease and cancer.

3. Fungi can be consumed whole (mushrooms) or used for the fermentation process of foods and beverages, such as bread, fermented rice, wine, and beer.

4. Immunosuppression, protracted use of antibiotics, or a breakdown of the normal flora can all lead to fungal infections.

5. Dermatophytosis is responsible for conditions such as athlete's foot.

6. *Candida albicans*

7. Fungal infection caused by *Pneumocystis jirovecii*

Section 24.4

1. The most likely method of protozoan transmission for intestinal protozoans is the fecal-oral route; protozoans that live in the bloodstream are most often transmitted by insect vectors.

2. The two major categories of helminths are flatworms and roundworms.

3. While some worms can penetrate the skin, most helminth infections result from consuming undercooked and contaminated soil, water, or food.

4. As with any infection, those with a decreased immune system are most susceptible to helminth infections. Those in developing countries, particularly children, are most at risk for infection from these organisms.

5. Helminths called *hookworms* are able to penetrate the skin to infect the host.

6. The highest prevalence for tapeworms is in populations that eat raw fish.

PERIODIC TABLE

Group	Element

Uranium
92 ----- Atomic number
U ----- Symbol

Legend:
- MAIN GROUP METALS
- TRANSITION METALS
- METALLOIDS
- NON-METALS
- PROPERTIES UNKNOWN

1																	18
Hydrogen 1 H																	Helium 2 He

	1	2											13	14	15	16	17	
2	Lithium 3 Li	Beryllium 4 Be											Boron 5 B	Carbon 6 C	Nitrogen 7 N	Oxygen 8 O	Fluorine 9 F	Neon 10 Ne
3	Sodium 11 Na	Magnesium 12 Mg	3	4	5	6	7	8	9	10	11	12	Aluminum 13 Al	Silicon 14 Si	Phosphorus 15 P	Sulfur 16 S	Chlorine 17 Cl	Argon 18 Ar
4	Potassium 19 K	Calcium 20 Ca	Scandium 21 Sc	Titanium 22 Ti	Vanadium 23 V	Chromium 24 Cr	Manganese 25 Mn	Iron 26 Fe	Cobalt 27 Co	Nickel 28 Ni	Copper 29 Cu	Zinc 30 Zn	Gallium 31 Ga	Germanium 32 Ge	Arsenic 33 As	Selenium 34 Se	Bromine 35 Br	Krypton 36 Kr
5	Rubidium 37 Rb	Strontium 38 Sr	Yttrium 39 Y	Zirconium 40 Zr	Niobium 41 Nb	Molybdenum 42 Mo	Technetium 43 Tc	Ruthenium 44 Ru	Rhodium 45 Rh	Palladium 46 Pd	Silver 47 Ag	Cadmium 48 Cd	Indium 49 In	Tin 50 Sn	Antimony 51 Sb	Tellurium 52 Te	Iodine 53 I	Xenon 54 Xe
6	Cesium 55 Cs	Barium 56 Ba	Lanthanum 57 La	Hafnium 72 Hf	Tantalum 73 Ta	Tungsten 74 W	Rhenium 75 Re	Osmium 76 Os	Iridium 77 Ir	Platinum 78 Pt	Gold 79 Au	Mercury 80 Hg	Thallium 81 Tl	Lead 82 Pb	Bismuth 83 Bi	Polonium 84 Po	Astatine 85 At	Radon 86 Rn
7	Francium 87 Fr	Radium 88 Ra	Actinium 89 Ac	Rutherfordium 104 Rf	Dubnium 105 Db	Seaborgium 106 Sg	Bohrium 107 Bh	Hassium 108 Hs	Meitnerium 109 Mt	Darmstadtium 110 Ds	Roentgenium 111 Rg	Copernicium 112 Cn	Ununtrium 113 Uut	Flerovium 114 Fl	Ununpentium 115 Uup	Livermorium 116 Lv	Ununseptium 117 Uus	Ununoctium 118 Uuo

Lanthanides

Cerium 58 Ce	Praseodymium 59 Pr	Neodymium 60 Nd	Promethium 61 Pm	Samarium 62 Sm	Europium 63 Eu	Gadolinium 64 Gd	Terbium 65 Tb	Dysprosium 66 Dy	Holmium 67 Ho	Erbium 68 Er	Thulium 69 Tm	Ytterbium 70 Yb	Lutetium 71 Lu

Actinides

Thorium 90 Th	Protactinium 91 Pa	Uranium 92 U	Neptunium 93 Np	Plutonium 94 Pu	Americium 95 Am	Curium 96 Cm	Berkelium 97 Bk	Californium 98 Cf	Einsteinium 99 Es	Fermium 100 Fm	Mendelevium 101 Md	Nobelium 102 No	Lawrencium 103 Lr

At the date of publication, elements 113, 115, 117, and 118 had not been named and have been given temporary names.

From Mahaffy/Tacker/Bucat/Kotz/McMurry. *Chemistry*, 2E. © 2015 Nelson Education Ltd. Reproduced by permission. http://www.cengage.com/permissions.

Appendix C

SYSTÈME INTERNATIONALE/METRIC SYSTEM

Metric Measures and Imperial Equivalents

Unit	Measure	Symbol	Imperial Equivalent
LINEAR MEASURE			
1 kilometre	= 1000 metres	10^3 m km	0.62137 mile
1 metre		10^0 m m	39.37 inches
1 decimetre	= 1/10 metre	10^{-1} m dm	3.937 inches
1 centimetre	= 1/100 metre	10^{-2} m cm	0.3937 inch
1 millimetre	= 1/1000 metre	10^{-3} m mm	Not used
1 micrometre (or micron)	= 1/1 000 000 metre	10^{-6} m μm (or μ)	Not used
1 nanometre	= 1/1 000 000 000 metre	10^{-9} m nm	Not used
MEASURES OF CAPACITY (FOR FLUIDS AND GASES)			
1 litre		L	1.0567 U.S. liquid quarts
1 millilitre	= 1/1000 litre	mL	
	= volume of 1 g of water at STP*		
MEASURES OF VOLUME			
1 cubic metre		m^3	
1 cubic	= 1/1000 cubic metre	dm^3 = L	
1 decimetre	= 1 litre	L	
1 cubic centimetre	= 1/1 000 000 cubic metre	cm^3 = mL	
	= 1 millilitre (mL)		
1 cubic millimetre	= 1/100 000 000 cubic metre	mm^3	
MEASURES OF MASS			
1 kilogram	= 1000 grams	kg	2.2046 pounds
1 gram		g	15.432 grains
1 milligram	= 1/1000 gram	mg	0.01 grain (about)
1 microgram	= 1/1 000 000 gram	μg (or mcg)	

*STP = standard temperature and pressure

Imperial to Metric Conversions

Length

IMPERIAL		METRIC
1 inch	=	2.54 centimetres
1 foot	=	0.30 metre
1 yard	=	0.91 metre
1 mile (5280 feet)	=	1.61 kilometre

TO CONVERT	MULTIPLY BY	TO OBTAIN
inches	2.54	centimetres
feet	30.0	centimetres
centimetres	0.39	inches
millimetres	0.039	inches

Mass/Weight

IMPERIAL		METRIC
1 grain	=	64.80 milligrams
1 ounce	=	28.35 grams
1 pound	=	453.60 grams
1 ton (short) (2000 pounds)	=	0.91 metric ton

TO CONVERT	MULTIPLY BY	TO OBTAIN
ounces	28.3	grams
pounds	453.6	grams
pounds	0.45	kilograms
grams	0.035	ounces
kilograms	2.2	grams

Volume and Capacity

Imperial		Metric
1 cubic inch	=	16.39 cubic centimetres
1 cubic foot	=	0.03 cubic metre
1 cubic yard	=	0.765 cubic metres
1 ounce	=	0.03 litre
1 pint	=	0.47 litre
1 quart	=	0.95 litre
1 gallon	=	3.79 litres

TO CONVERT	MULTIPLY BY	TO OBTAIN
fluid ounces	30.0	millilitres
quarts	0.95	litres
millilitres	0.03	fluid ounces
litres	1.06	quarts

Appendix D

MOST COMMON AMINO ACIDS

Amino Acid	Three-Letter Abbrev.	One-Letter Symbol
Alanine	Ala	A
Arginine	Arg	R
Aspartic acid	Asp	D
Asparagine	Asn	N
Cysteine	Cys	C
Glutamic acid	Glu	E
Glutamine	Gln	Q
Glycine	Gly	G
Histidine	His	H
Isoleucine	Ile	I
Leucine	Leu	L
Lysine	Lys	K
Methionine	Met	M
Phenylalanine	Phe	F
Proline	Pro	P
Serine	Ser	S
Threonine	Thr	T
Tryptophan	Trp	W
Tyrosine	Tyr	Y
Valine	Val	V

Note: The nine essential amino acids are highlighted.

Appendix E

FURTHER READING

Chapter 1

http://www.unep.org/newscentre/default.aspx?DocumentID=2649&ArticleID=8838

http://news.nationalgeographic.com/news/2011/08/110824-earths-species-8-7-million-biology-planet-animals-science/

http://www.pnas.org/content/87/12/4576.long

http://nutritiondata.self.com/facts/vegetables-and-vegetable-products/2482/2

http://www.fda.gov/Food/FoodborneIllnessContaminants/CausesOfIllnessBadBugBook/ucm071092.htm

Chapter 2

http://ajcn.nutrition.org/content/33/1/27.full.pdf

http://www.bio.davidson.edu/genomics/2011/Piper/Background.html

http://www.journals.elsevierhealth.com/periodicals/yjada/article/S0002-8223(03)00294-3/fulltext

http://biology.stackexchange.com/questions/546/whats-the-maximum-and-minimum-temperature-a-human-can-survive

http://www.medbio.info/Horn/Time%201-2/carbohydrate_metabolism.htm

http://www.mayoclinic.com/health/fiber/NU00033

http://culinaryarts.about.com/od/culinaryreference/a/smokepoints.htm

http://www.glycemicindex.com/

https://www.diabetes.ca/diabetes-and-you/healthy-living-resources/diet-nutrition/the-glycemic-index

https://medlineplus.gov/vitamins.html

https://medlineplus.gov/minerals.html

https://www.canolacouncil.org/oil-and-meal/what-is-canola/how-canola-is-processed/steps-in-oil-and-meal-processing

Chapter 3

http://www.niaid.nih.gov/topics/antimicrobialResistance/Examples/gramNegative/Pages/default.aspx

http://faculty.ccbcmd.edu/courses/bio141/lecguide/unit1/prostruct/flag.html

http://www.medicalnewstoday.com/articles/216798.php

http://users.rcn.com/jkimball.ma.ultranet/BiologyPages/J/Junctions.html

https://www.nature.com/scitable/topicpage/aging-and-cell-division-14230076

Chapter 4

http://www.sciencedirect.com/science/article/pii/S0009279706000998

Chapter 5

http://www.ncbi.nlm.nih.gov/books/NBK10008/

http://www.bscb.org/?url=softcell/centrioles

http://users.rcn.com/jkimball.ma.ultranet/BiologyPages/C/CellCycle.html

http://www.lef.org/magazine/mag2009/nov2009_Horseradish-Protection-Against-Cancer-And-More_01.htm

http://www.sciencedirect.com/science/article/pii/S1044579X98900980

http://www.webmd.com/cancer/features/seven-easy-to-find-foods-that-may-help-fight-cancer

Chapter 6

http://www.rch.org.au/bloodtrans/about_blood_products/Blood_Groups_and_Compatibilities/

http://www.nature.com/scitable/topicpage/rare-genetic-disorders-learning-about-genetic-disease-979

http://www.orangutanssp.org/wp-content/uploads/2013/07/ga_blood_typing.pdf

http://mbe.oxfordjournals.org/content/14/4/399.full.pdf

http://www.redcrossblood.org/learn-about-blood/blood-types

Chapter 7

http://enaliaphysis.org.cy/educational-initiatives/public-lectures/natural-selection/

http://www.sciencedaily.com/releases/2005/10/051026090636.htm

http://mbe.oxfordjournals.org/content/14/4/399.full.pdf

https://www.sciencenews.org/article/human-blood-types-have-deep-evolutionary-roots

https://www-ncbi-nlm-nih-gov.proxy1.lib.uwo.ca/pubmed/19539396

Chapter 8

http://www.annualreviews.org/doi/abs/10.1146/annurev.biochem.70.1.369

http://www.uniprot.org/uniprot/P11387

http://www.ucsf.edu/news/2013/09/108886/lifestyle-changes-may-lengthen-telomeres-measure-cell-aging

http://www.sciencemag.org/content/338/6108/758

http://genetics.thetech.org/about-genetics/mutations-and-disease

Chapter 9

http://www.ndsu.edu/pubweb/~mcclean/plsc431/prokaryo/prokaryo2.htm

http://www.nature.com/scitable/topicpage/regulation-of-transcription-and-gene-expression-in-1086

Chapter 5

http://www.mayoclinic.com/health/calories/WT00011

http://www.ncbi.nlm.nih.gov/pubmed/3593498

http://www.nature.com/scitable/topicpage/the-role-of
-methylation-in-gene-expression-1070

http://www.ncbi.nlm.nih.gov/pubmed/14650573

http://www.ncbi.nlm.nih.gov/pubmed/22155943

Chapter 10

http://www.biotechniques.com/multimedia/archive/00036/
BTN_A_04361RR02_O_36116a.pdf

http://www.news-medical.net/health/What-is-Gene-Therapy.aspx

http://www.forbes.com/sites/stevensalzberg/2012/09/24/
does-genetically-modified-corn-cause-cancer-a-flawed-study/

http://www.ncbi.nlm.nih.gov/pubmed/12369203

http://www.sciencedaily.com/releases/2011/08/110811201523.htm

http://www.dailymail.co.uk/sciencetech/article-1080042/
Meet-Mr-Green-Genes--worlds-glow-dark-cat.html

http://sciencereview.berkeley.edu/bridges-made-of-spider-silk
-you-can-thank-the-goats-for-that/

Chapter 11

http://www.mhhe.com/biosci/ap/histology_mh/simpleep.html

https://www.hormone.org/hormones-and-health/hormones/leptin

Chapter 12

http://www.medindia.net/education/familymedicine/Burns
-Classification.htm#

http://www.lpch.org/DiseaseHealthInfo/HealthLibrary/burns/
classify.html

http://www.mayoclinic.com/health/melanoma/DS00439/
DSECTION=causes

http://www.ncbi.nlm.nih.gov/pubmed/19519827

http://www.nlm.nih.gov/medlineplus/ency/article/004014.htm

http://lpi.oregonstate.edu/infocenter/skin/vitaminE/

http://www.nlm.nih.gov/medlineplus/ency/article/004014.htm

http://www.webmd.com/beauty/aging/myth-vs-reality-on
-anti-aging-vitamins

https://www.medicalnewstoday.com/articles/286496.php

Chapter 13

http://www.diabetes.org/news-research/research/access-diabetes
-research/physical-activity-remains-key.html

http://www.joanvernikos.com/pages/sitting-kills-moving-heals.php

http://www.mayoclinic.com/health/tmj-disorders/DS00355

http://www.webmd.com/menopause/osteoporosis-menopause

http://www.niams.nih.gov/Health_Info/Bone/Bone_Health/
Nutrition/vitamin_a.asp

http://www.canorth.org/en/patienteducation/Default.aspx?
pagename=Long%20Bone%20Fractures

http://www.osteoporosis.ca/

http://www.osteoporosis.ca/osteoporosis-and-you/what-is-osteoporosis/

https://medlineplus.gov/ency/article/004015.htm

Chapter 14

http://www.ncbi.nlm.nih.gov/pubmed/7286246

http://lpi.oregonstate.edu/infocenter/othernuts/choline/

http://link.springer.com/article/10.1007%2Fs00109-002-0384-9

http://www.ncbi.nlm.nih.gov/pmc/articles/PMC1133170/pdf/
biochemj00125-0011.pdf

http://www.livestrong.com/article/423722-workouts-that
-increase-speed-train-fast-twitch-muscles/

https://medlineplus.gov/ency/article/004015.htm

Chapter 15

http://pubs.niaaa.nih.gov/publications/aa63/aa63.htm

http://www.health.harvard.edu/press_releases/benefits-of
-exercisereduces-stress-anxiety-and-helps-fight-depression

http://pubs.niaaa.nih.gov/publications/arh21-2/144.pdf

http://www.mayoclinic.org/diseases-conditions/glioma/basics/
definition/con-20035538

http://www.sciencedirect.com/science/article/pii/000689937490208X

Chapter 16

https://nei.nih.gov/healthyeyes/myopia

https://vestibular.org/news/09-24-2013/rethinking-motion
-sickness

https://imagine.gsfc.nasa.gov/science/toolbox/emspectrum1.html

http://www.sciencemag.org/news/2014/09/sounds-you-cant
-hear-can-still-hurt-your-ears

https://www.ncbi.nlm.nih.gov/pubmedhealth/PMHT0028128/

https://www.sciencedirect.com/topics/neuroscience/olfactory
-nerve

Chapter 17

http://www.ncbi.nlm.nih.gov/pmc/articles/PMC3079864/

http://www.ncbi.nlm.nih.gov/pmc/articles/PMC329619/

http://www.mayoclinic.org/healthy-living/stress-management/
in-depth/stress/art-20046037?footprints=mine

http://www.nlm.nih.gov/medlineplus/ency/article/000353.htm

http://onlinelibrary.wiley.com/doi/10.1111/j.1469-445X.2000
.tb00014.x/abstract

http://diabetes.diabetesjournals.org/content/43/2/212.short

https://www.physiology.org/doi/abs/10.1152/physrev.00030.2013

Chapter 18

http://www.nlm.nih.gov/medlineplus/ency/article/001179.htm

http://www.mayoclinic.com/health/cholesterol-levels/CL00001

http://circ.ahajournals.org/content/111/5/e89.full

http://www.mayoclinic.com/health/trans-fat/CL00032

http://www.heart.org/HEARTORG/Conditions/More/
MyHeartandStrokeNews/Menopause-and-Heart-Disease
_UCM_448432_Article.jsp

http://www.heartandstroke.com/site/c.ikIQLcMWJtE/
b.3484033/k.2811/Heart_disease__Excessive_alcohol
_consumption.htm

http://diabetes.niddk.nih.gov/dm/pubs/stroke/#connection

http://www.nhlbi.nih.gov/health/health-topics/topics/hbp/

file:///C:/Users/wroscoe/Downloads/clinical-lab-tests-reference
-values-e.pdf

Chapter 19

http://www.nhlbi.nih.gov/health/health-topics/topics/copd/

http://www.nhlbi.nih.gov/health/health-topics/topics/asthma/causes.html

https://www.lung.ca/diseases-maladies/copd-mpoc/what-quoi/index_e.php

http://www.asthma.ca/

http://www.ncbi.nlm.nih.gov/pubmed/19399004

https://www.ncbi.nlm.nih.gov/pmc/articles/PMC2695176/

Chapter 20

http://www.webmd.com/digestive-disorders/understanding-ulcers-basic-information

http://www.ncbi.nlm.nih.gov/pubmed/8822613

http://www.ncbi.nlm.nih.gov/pubmed/23384445

http://www.ncbi.nlm.nih.gov/pubmed/23041079

http://www.ncbi.nlm.nih.gov/books/NBK9847/

http://www.ncbi.nlm.nih.gov/books/NBK54116/#s4.2

http://www.ncbi.nlm.nih.gov/pubmed/12362464

http://www.ccfc.ca/site/c.ajIRK4NLLhJ0E/b.6349429/

http://www.ncbi.nlm.nih.gov/pubmed/19607778

https://www.merckmanuals.com/en-ca/home/digestive-disorders/biology-of-the-digestive-system/effects-of-aging-on-the-digestive-system

Chapter 21

http://www.kidney.org/kidneydisease/aboutckd.cfm

http://www.mayoclinic.org/diseases-conditions/kidney-stones/basics/definition/con-20024829

http://www.med.umich.edu/intmed/nephrology/docs/stones.pdf

http://kidney.niddk.nih.gov/kudiseases/pubs/highblood/

https://medlineplus.gov/ency/article/004010.htm

Chapter 22

http://www.ncbi.nlm.nih.gov/pmc/articles/PMC2941592/

http://www.medicinenet.com/endometriosis/article.htm

http://www.cdc.gov/std/pid/stdfact-pid.htm

http://www.prostatecancer.ca/

https://medlineplus.gov/ency/article/004017.htm

Chapter 23

http://www.ncbi.nlm.nih.gov/pubmed/15877894

http://www.ncbi.nlm.nih.gov/pubmed/20066002

http://www.ncbi.nlm.nih.gov/pubmed/9553779

https://www.ncbi.nlm.nih.gov/pmc/articles/PMC3582124/

https://medlineplus.gov/ency/article/004008.htm

Chapter 24

http://www.ncbi.nlm.nih.gov/pubmed/22291721

http://www.cdc.gov/mmwr/preview/mmwrhtml/mm5021a2.htm

http://www.phac-aspc.gc.ca/lab-bio/res/psds-ftss/necator-americanus-eng.php

http://sabre.ucsf.edu/docs/Science-2011-Wu-243-7.pdf

http://www.wildfermentation.com/making-sauerkraut-2/

Glossary

3′ tail several hundred adenine nucleotides attached to the mRNA molecule, also known as a *poly A tail*. p. 152

5′ cap a methylated GTP that enables mRNA to efficiently begin translation at the ribosome. p. 152

A band the overlapping region of the sarcomere. p. 271

abdominal aorta (descending aorta) segment of the aorta located in the abdomen. p. 405

abdominal muscles muscles at the anterior aspect of the abdomen. p. 263

abdominopelvic cavity contains organs of the digestive system, the spleen, bladder, parts of the large intestine, and the internal reproductive organ. p. 16

absorption movement of molecules from the lumen of the digestive tract, or through the skin, into the bloodstream. p. 179

absorptive phase stage in which digestive end products are absorbed from the gut and sent to storage; period after eating when insulin levels are highest. p. 376

accessory glands glands that add to the secretions of the digestive tract or aid in digestion but are not part of the alimentary canal. p. 500

accessory organs organs that secrete substances required for digestion into the alimentary canal. p. 458

acetaldehyde molecule produced when yeast undergoes fermentation. p. 83

acetylcholine (ACh) a neurotransmitter. p. 274

acetylcholinesterase enzyme that breaks down acetylcholine. p. 275

acini cluster of cells that resembles a berry; found in the pancreas and secretes digestive enzymes. p. 467

acromion a process at the lateral end of the scapular spine. p. 224

acrosome a caplike structure that covers the head of a sperm cell. p. 501

actin a globular protein that forms microfilaments and is found in all eukaryotic cells. p. 61

action potential a short-lasting event wherein the voltage of a cell membrane is altered in sequence. p. 275

activating enzyme enzyme that recycles the empty tRNA molecule in the cytoplasm and adds the appropriate amino acid. p. 156

activator a protein that initiates transcription in bacteria. p. 159

active sites grooves on an enzyme where a reaction is catalyzed. p. 32

active transport movement of molecules against a concentration gradient through a membrane protein and with input of energy. p. 79

acute inflammation a short-term response to tissue injury that is characterized by redness, pain, swelling, heat, and reduced mobility. p. 522

adaptive immune response also referred to as *acquired immune system*. p. 516

adductor longus a muscle that helps move the knee toward the midline. p. 267

adhesion molecules membrane proteins that play an important role in intercellular interactions. p. 55

adipocytes cells that store fat as triglycerides. p. 43

adipose tissue composed of adipocytes; also called *fatty tissue*. p. 43

adjuvant a pharmacological or immunological substance that increases the efficacy of a vaccine or other agent. p. 531

adrenal medulla innermost part of the kidney that produces epinephrine and norepinephrine. p. 322

aerobic respiration production of ATP with oxygen. p. 75

afterload pressure of the arteries that the heart must work against to move blood into the vessels. p. 396

agonist the primary muscle contributing to a particular motion. p. 257

agranular leukocytes lymphocytes, monocytes, and plasma cells that don't form membrane granules. p. 423

albumin a plasma protein used to help regulate blood pressure. p. 401

aldosterone a hormone produced by the adrenal cortex that causes the kidneys to reabsorb sodium, which causes more water to be absorbed and therefore increases blood pressure. p. 417

allele frequency proportion of each allele of a gene in a population. p. 135

allele one of multiple forms of a gene; different alleles of a gene lead to different phenotypes, such as blond hair or brown hair. p. 135

allergic reactions (hypersensitivity reactions) immune reactions to an otherwise harmless substance. p. 512

alpha blockers medications that block the effects of the sympathetic nervous system and result in decreased blood pressure. p. 420

alpha helices helices formed by hydrogen bonds as part of the secondary structure of a polypeptide. p. 32

alternate splicing some exons forming part of the mature mRNA, other exons doing so at other times, to form alternate amino acid sequences and therefore different proteins. p. 160

alveolar ducts ducts that connect the alveoli to the respiratory bronchioles and allow for gas exchange. p. 436

alveolar pressure pressure held in the alveoli at any point during breathing. p. 441

alveolar processes and sockets contain both the upper and lower teeth. p. 218

alveolar ventilation movement of gas between the alveoli and the external environment during breathing. p. 444

alveoli air sacs that permit gas exchange. p. 436

amino acid organic compounds that contain an amine group, carboxyl group, and a functional group; the building blocks of proteins. p. 27

amino group a nitrogenous functional group consisting of a nitrogen atom bound to two hydrogen atoms. p. 28

ammonia (NH_3) molecule that consists of one nitrogen atom and three hydrogen atoms, formed in the body when proteins are broken down; is extremely alkaline and must be converted into urea in the body to be excreted. p. 81

amniocentesis test that takes a sample of amniotic fluid surrounding an embryo to check for abnormalities in the chromosomes. p. 99

amphiarthrosis slightly movable joint connected by fibrocartilage, such as the articulations between the vertebrae. p. 234

amphipathic molecules that contain hydrophobic and hydrophilic regions. pp. 27, 418

ampullae regions within each of the semicircular canals that contain endolymph and the cupula, which contain the hair cells that bend during movement; also called *crista ampullaris*. p. 353

amygdala groups of nuclei in the temporal lobes that control emotion and memory. p. 315

amylase an enzyme that breaks down starch. pp. 465, 470

anaerobic respiration production of ATP in the absence of oxygen. p. 75

anal canal terminal part of the large intestine. p. 467

analgesic a substance that blocks the transmission of pain information. p. 344

anaphylaxis (anaphylactic shock) a severe allergic reaction characterized by effects to multiple body systems. p. 535

anatomical dead space portion of air in the respiratory tract that in the conducting zones doesn't permit gas exchange. p. 443

anatomical neck narrow groove of the humerus bone where the head meets the body. p. 224

anatomical position human body that is standing up, facing forward, with toes and palms forward. p. 14

anatomy the study of structures within the human body. p. 1

androgen-binding protein (ABP) a protein produced in the seminiferous tubules that binds to testosterone. p. 500

androgens testosterone-like hormones produced in the adrenal glands that stimulate growth. p. 205

aneuploidy one gamete having more or fewer chromosomes than it is supposed to have. p. 116

angina chest pain brought on by ischemia or hypoxia in the heart. p. 419

angiotensin II a hormone that helps increase blood pressure by stimulating the production of aldosterone. p. 417

angiotensinogen a precursor hormone formed in the liver. p. 466

angular motion of a synovial joint in a single plane around a fixed point. p. 237

animals eukaryotic, multicellular organisms capable of locomotion. p. 4

anneal formation of complementary base pairs, such as when a primer binds to a single strand of DNA. p. 166

antagonist (muscular) the opposing muscle to a particular motion. p. 257

anterior fontanel the space between the skull bones located at the joining of the frontal bone and both parietal bones; does not close until nearly age two. p. 218

anterior median fissure longitudinal fissure the length of the spinal cord, approximately 3 mm in depth. p. 305

anterior tibial artery artery at the ventral aspect of the tibias. p. 408

anterolateral fontanel the space between the skull bones located at the joining of the temporal and sphenoid bones. p. 218

antibiotic resistance a form of drug resistance where bacteria are not killed by antibiotics. p. 124

antibiotics a drug that kills bacteria. p. 124

antibodies proteins that act as markers for the immune system during a humoral immune response. p. 421

anticipatory response physiological changes initiated by thinking about a change that may occur, such as breathing rate increase prior to exercise that prepares the body for increased oxygen and nutrient demand when exercise occurs. p. 416

anticodon a complementary sequence on the tRNA molecule that binds to the codon on the mRNA. p. 152

antidiuretic hormone a hormone produced by the hypothalamus when blood pressure decreases; causes the kidneys to reabsorb water and therefore increases blood pressure; also called *vasopressin*. p. 417

antigen substance that triggers an immune response. p. 524

antigen-presenting cell (APC) cell that displays foreign antigens on its surface as markers. p. 524

antigens cell-recognition markers. p. 55

antiparallel two strands of DNA running in opposite directions. p. 138

antiport when two substance move across the membrane in opposite directions. p. 67

antipyretic a pharmacological agent that reduces fever. p. 524

antisense strand *see* template stand. p. 149

anus the ring at the distal end of the gastrointestinal tract. p. 467

aorta primary artery of the body, located where the blood leaves the left ventricle of the heart. p. 387

aortic arch most superior part of the aorta. p. 387

aortic semilunar (SL) valve valve between the aorta and the left ventricle. p. 390

apical side of a cell that is facing the surface or cavity. pp. 179, 489

apocrine glands type of sweat gland found predominantly in the axilla (armpits), areola, and genital regions. p. 201

apoptosis programmed cell death. p. 93

appendicitis inflammation of the appendix. p. 522

appendix accessory structure of the large intestine proposed to be a vestigial structure. p. 467

aquaporin a membrane protein that allows the transport of water across a cell membrane. p. 489

aqueous humor a transparent, watery fluid similar to plasma that fills the space between the lens and the cornea. p. 346

arachnoid mater middle layer of the meninges. p. 305

arachnoid villi protrusions of the arachnoid mater through the dura mater. p. 312

archaebacteria prokaryotic unicellular organisms that survive in extreme environments. p. 2

arcuate veins blood vessels of the renal circulation. p. 485

aromatase an enzyme that converts testosterone into estrogen in females. p. 368

arteries blood vessels that carry blood away from the heart. p. 387

arterioles small arteries that bridge the flow of blood from arteries to capillaries. p. 390

artificial selection occurs when humans choose the breeding of traits considered the most desirable in plants or animals. p. 125

artificially acquired passive immunity short-term immunization granted by the transfer of antibodies. p. 530

ascending colon the part of the large intestine located in both the right-upper and right-lower quadrants of the abdomen, and leading to the transverse colon. p. 467

ascending tracts nerve tracts that carry signals from the body toward the brain. p. 306

asexual reproduction from a single parent to produce an identical offspring. p. 87

asthma disease condition that causes an increase in mucus secretion and bronchoconstriction. p. 452

astrocytes glial cells that maintain the ion concentration surrounding neurons in the CNS and that form the blood-brain barrier. p. 192

atherosclerosis cardiovascular disease that results in the hardening of the blood vessels. p. 418

atlas first cervical vertebra, designated C1. p. 219

atmospheric pressure the pressure created by the gases of the atmosphere. p. 441

ATP adenosine triphosphate; energy. pp. 66, 138

ATP synthase a specific membrane protein that allows protons to move back across the mitochondrial membrane into the matrix of the mitochondrion. p. 67

atrial diastole atrial relaxation. p. 391

atrial natriuretic peptide (ANP) a hormone released by the atria of the heart. p. 417

atrial systole atrial contraction. p. 391

atrioventricular bundle *see* bundle of His. p. 392

atrioventricular node (AV node) a collection of cells that transmit an action potential from the atria to the bundle of His. p. 392

auditory canal *see* external auditory meatus. p. 217

auditory nerve the nerve that is connected to the cochlea and that sends information to the brain about sound. p. 350

AUG first codon of mRNA molecule; codes for the amino acid, methionine. pp. 152, 156

auricle *see* **pinna**. p. 349

autocrine a substance secreted by a cell that acts on the same cell. p. 518

autonomic ganglion a collection of cell bodies in the autonomic nervous system. p. 321

autonomic involuntary. p. 291

autonomic nervous system (ANS) portion of the nervous system that controls involuntary functions. p. 190

autorhythmic automatic contractions in the heart. p. 190

autorhythmicity ability of the electrically active cells to establish their own action potential. p. 392

autosomal refers to a gene found on chromosomes 1 to 22. p. 117

avascular does not contain blood vessels. p. 179

axillary area the region beneath the arms. p. 408

axillary artery artery found in the axillary region. p. 408

axillary nerve nerve located in the axillary region. p. 308

axis second cervical vertebra, designated C2. p. 219

axon terminal the end of a nerve cell branch that contains neurotransmitters. p. 274

B cells immune cells. p. 423

bacilli a class of bacteria that is rod-shaped. p. 542

bacteriocidal a substance that kills bacteria. p. 543

bacteriophage a virus that infects bacteria. pp. 165, 546

bacteriostatic a chemical, biological, or physical agent that prevents the replication of bacteria. p. 543

ball-and-socket joint synovial joint that allows the greatest freedom of motion of any joint. p. 235

baroreceptors detect pressure changes. p. 20

basal cell carcinoma cancerous growth composed of cells from the stratum basale. p. 202

basal cells cells in papillae that replace old gustatory cells. p. 355

basal metabolic rate (BMR) the calorie requirement to maintain basic cellular function. p. 281

basal nuclei collections of nerve cell bodies at the base of the forebrain. p. 314

basal side of the membrane opposite the apical side. p. 179

base pair a coupled purine and pyrimidine. p. 149

basement membrane connective tissue layer underlying epithelial cells that consists of laminin and integrin proteins and that connects the epithelial cells to the loose connective tissue. p. 179

basilica vein superficial vein of the upper arm. p. 409

basophils a type of white blood cell. p. 423

belly (muscular) middle of the muscle. p. 257

beneficial bacteria *see* **normal flora**. p. 517

benign tumour a growth of cells that is not cancerous. p. 94

beta blocker medication that blocks the effects of epinephrine and causes a decrease in blood pressure. p. 420

beta oxidation the breakdown of fatty acids into two-carbon acetyl groups that can enter the Krebs cycle and produce ATP. p. 82

beta sheet a three-dimensional structure formed by the hydrogen bonds between amino acids as part of the secondary structure of a polypeptide. p. 32

bicep muscle that works to flex the forearm at the elbow. p. 265

bifidobacteria a type of normal flora bacteria. p. 517

bilayer a double layer of phsopholipids. p. 54

bile substance produced in the liver that helps break down fats in the small intestine. p. 421

bilirubin a yellow pigment formed by the breakdown of the hemoglobin in red blood cells. p. 421

binary fission method of reproduction in bacteria. p. 86

biology the study of all living things. p. 1

biotechnology the use of living organisms to produce biological products. p. 164

bipolar neurons neurons that produce two extensions branching from the cell body. p. 192

bitter the sensation of taste produced from substances such as coffee, cocoa, and alcohol. p. 354

blastocyst an early embryonic structure that contains a fluid-filled space that implants into the uterine wall. p. 503

blind spot region of the eye where there are no rods or cones; another term for the optic disc. p. 346

blood cells include red blood cells that carry oxygen and some carbon dioxide and white blood cells that are involved in the immune response. p. 184

blood pH the measure of alkalinity or acidity in the blood; normal blood pH is 7.35 to 7.45. p. 353

blood pressure the force of blood pushing against the walls of the blood vessels. p. 318

blood viscosity thickness of the blood. p. 415

blood volume total amount of blood. p. 396

blood-testis barrier a physical barrier between the blood and the seminiferous tubules in animals. p. 500

body cavity a space in the body that contains specific organs or tissues. p. 16

bolus a mass of fluid or semi-fluid. p. 463

bottleneck effect random change in allele frequency after a large portion of a population dies, such as in a natural disaster. p. 134

botulism a disease resulting from exposure to the botulinum toxin that causes paralysis. p. 544

Bowman's space the space between the glomerulus and the Bowman's capsule. p. 487

Boyle's law a physical law that states the volume of a gas in a container is inversely proportional to its pressure. p. 441

brachial arteries the arteries located in the upper arms. p. 408

brachial plexus network of nerve fibres running from the spine. p. 308

brachial vein a vein located in the upper arm. p. 408

brachialis muscle of the upper arm that works to flex the forearm. p. 265

brachiocephalic trunk artery of the mediastinum that supplies the right arm and head. p. 405

brainstem posterior part of the brain that connects to the spinal cord. p. 313

Broca's area region of the brain's frontal lobe that allows the physical ability to speak. p. 317

bronchi large airway that branches from the trachea. p. 435

bronchiole small airway that branches from the bronchi. p. 435

bronchoconstriction a decrease in the lumen size of the bronchi. p. 452

bronchodilator medication that increases the lumen size of the bronchi. p. 453

brush border the surface of a cuboidal cell covered in microvilli. p. 467

budding production of viral cells that does not kill the infected cell. p. 549

bulbourethral glands glands that produce pre-ejaculate in most mammals. p. 500

bundle branches two branches of the bundle of His. p. 392

bundle of His cells that are part of the fast-conduction system in the septum of the heart and that transmit the action potential to the apex of the heart. p. 392

bursa sacs small sac lined with a synovial membrane and filled with fluid found in synovial joints. p. 235

butyric acid metabolic by-product of fatty acids released in sweat produced by apocrine glands. p. 201

calcitonin hormone released by the parafollicular glands of the thyroid to decrease blood-calcium levels. p. 209

calcitriol increases calcium levels by increasing the amount of calcium absorbed in the small intestine. pp. 231, 482

calcium channel blockers medications that work to decrease blood pressure by preventing the movement of calcium ions into the heart muscle cells. p. 420

callus a protective mechanism resulting from increased keratin production. p. 198

calorie amount of energy required to raise the temperature of 1 mL of water 1°C, which is 1/1000th of a Calorie. p. 74

Calorie kilocalorie measurement of food. p. 74

canaliculi channels that connect lacunae, allowing osteocytes to communicate with one another. p. 210

cancer disease that occurs when mutated cells continue to divide and are not regulated by cellular checkpoints. p. 94

CAP-binding protein catabolite activator protein, an activator that binds to the CAP-binding sequence and initiates transcription in prokaryotes. p. 159

capillaries small blood vessels that provide oxygen and nutrients to the body's cells and pick up waste. p. 390

capillary pressure *see* **hydrostatic pressure**. p. 515

capitulum lateral aspect of the distal end of the humerus that articulates with the radius. p. 224

capsule a protein or carbohydrate external layer that protects bacteria from external environmental changes. p. 56

carboxyl group a functional group consisting of a carbon atom with a double bond to oxygen and a single bond to a hydroxyl group. p. 28

carboxypeptidase a pancreatic enzyme that breaks peptide bonds. p. 470

carcinogen anything that causes cancer. p. 95

cardia pertaining to the heart. p. 464

cardiac cycle repeated sequence of both systole and diastole for the atria and ventricles of the heart. p. 390

cardiac muscle muscles of the heart. p. 20

cardiac myocytes heart muscle cells. p. 386

cardiac notch a lateral deflection of the anterior wall of the left lung. p. 437

cardiac output the amount of blood pumped by the heart in a given amount of time. p. 394

cardiac reserve residual blood in the ventricles following ventricular systole. p. 395

carotene pigment that the body converts to vitamin A. p. 200

carotenemia condition caused by having a high level of beta-carotene in the blood. p. 200

carotid foramen opening in the styloid process through which the carotid arteries pass. p. 217

carpal tunnel small concave region at the front of the wrist. p. 224

carpals wrist bones. p. 224

carrier proteins membrane proteins that regulate the movement of certain molecules into or out of the cell down a concentration gradient and without input of energy. p. 55

cartilage very strong connective tissue that acts as a smooth surface for joint movement and gives structures their shape. p. 183

cartilaginous joints joints composed of, or containing, cartilage at the articulation. p. 234

catalyze increase the speed of chemical reactions. p. 32

cauda equina distal end of the spinal cord. p. 307

caudate nucleus one of the structures of the basal ganglia. p. 315

CD4 membrane receptor a coreceptor that assists in communication with an antigen-presenting cell. p. 549

CD4$^+$ T cells mature T cells that express the CD4 membrane receptor. p. 526

CD8$^+$ T cells mature T cells that express the CD8 membrane receptor. p. 527

cecum proximal end of the large intestine. p. 467

celiac trunk first major branch of the abdominal aorta. p. 405

cell membrane a membrane that consists of a phospholipid bilayer, contains proteins and cholesterol, and surrounds the cell. p. 46

cell smallest unit of life. p. 5

cell types different varieties of cells. p. 183

cell-mediated immune response immune response that involves cytotoxic T cells. p. 526

cellular metabolism process by which the cell generates energy from nutrients. p. 59

cellular respiration process of oxidizing food molecules and reducing NAD$^+$ molecules to produce ATP in various steps. p. 75

cellulose structural plant carbohydrate not used to produce energy. p. 38

central canal middle of the osteon, which contains blood and lymphatic vessels; also called the *Haversian canal*. p. 210

central chemoreceptors chemical receptors in the medulla oblongata that detect oxygen and carbon dioxide. p. 450

central dogma of gene expression the flow of information from gene to RNA to protein. p. 150

central nervous system (CNS) brain and spinal cord. p. 291

central sulcus border between the anterior and posterior portions of the brain. p. 315

centrioles organelle found in most eukaryotic cells, except plants and fungi, and involved in the organization of spindle fibres. p. 60

cephalic phase first stage of digestion, which involves thinking about eating and causes salivation and the beginning of smooth-muscle contractions along the GI tract. p. 475

cephalic vein a vein that runs up the lateral side of the arm. p. 409

cerebellar disease damage in the cerebellum resulting in the inability to coordinate muscle movements. p. 315

cerebellum the part of the brain located inferior to the occipital lobe, and that facilitates fine motor control, coordination of multi-muscle movements, and balance. p. 315

cerebral cortex outermost layer of the cerebrum divided into lobes with distinct functions. p. 315

cerebrum largest and most superior part of the brain. p. 315

cerumen waxy substance, found in the auditory canal, that traps foreign particles and inhibits bacterial growth. p. 201

ceruminous glands the glands that produce earwax. p. 200

cervical cancer a form of cancer that occurs in the cells of the cervix, which is located at the lower part of the uterus and connects to the vaginal canal. p. 508

cervical enlargement region of the spinal cord with increased diameter due to the addition of the nerves for the upper limbs. p. 307

cervical plexus (C1–C5) branched network of nerves in the neck. p. 308

cervix most inferior portion of the uterus. p. 504

Charles Darwin credited with the modern theory of natural selection that's based on his book *On the Origin of Species by Means of Natural Selection*. p. 120

checkpoints points at which the cell cycle is controlled to ensure that a previous phase is fully completed before advancing to the next stage. p. 93

chemical digestion chemical breakdown of food and nutrients into smaller components. p. 459

chemical energy the energy that results from breaking bonds of ATP in cellular reactions. p. 74

chemical synapse junctions that permit the neurons to pass on a signal to other cells through gap junctions. p. 386

chemically gated channel membrane proteins that open and allow passage in response to a chemical stimulus. p. 298

chemicals individual elements that combine to form molecules. p. 5

chemiosmosis process of using a chemical gradient across a membrane to produce many ATP molecules. p. 67

chemoreceptor receptor that detects chemical concentrations, such as neurotransmitters, drugs, or hormones. p. 20

chemotaxis movement in response to chemical stimuli. p. 518

chloride shift exchange of bicarbonate and chloride across the membranes of red blood cells. p. 448

chlorophyll pigment found in chloroplasts in plants. p. 4

chloroplast membrane-bound organelle in plant cells that contains the pigment chlorophyll that undergoes photosynthesis. p. 4

cholecystokinin (CCK) hormone that causes the gallbladder to release stored bile into the GI tract. p. 476

chondrocytes cells that make up cartilage connective tissue. p. 184

chondroitin sulfate sulfated chains of polysaccharides found in ground substance of connective tissue, usually bound to proteoglycans. p. 184

choroid plexus location in the ventricles of the brain where cerebrospinal fluid is produced. p. 312

choroid the pigmented vascular layer of the eyeball between the sclera and the retina. p. 345

chromatin DNA in loose strands during regular growth phase. p. 57

chromosome a single condensed nucleic acid macromolecule called *DNA*. p. 33

chronic bronchitis one of the disease conditions that make up chronic obstructive pulmonary disease (COPD). p. 451

chronic inflammatory disease a persistent condition that involves a chronic immune response. pp. 373, 418

chronic kidney disease (CKD) a progressive loss of renal function. p. 493

chronic obstructive pulmonary disease (COPD) a respiratory disease that decreases the body's ability to manage gas exchange. p. 451

chylomicron a type of lipoprotein that delivers fats from the digestive system to the bloodstream. p. 474

chyme partially digested food expelled from the stomach into the small intestine. p. 464

chymotrypsin digestive enzyme that is part of the pancreatic juice that digests proteins. p. 470

cilia cells that contain hair-like extensions. p. 179

ciliary body part of the eye, including the ciliary muscle, that controls the shape of the lens. p. 345

circle of Willis a cerebral circular flow of blood vessels that supply the brain. p. 407

circulatory system organ system consisting of the heart, blood vessels, and the blood. p. 384

cis Golgi the side of the Golgi that is closest to the endoplasmic reticulum. p. 58

clavicles bones that comprise what is called the *collarbone*. p. 224

cleavage furrow occurs when animal cell membranes indent due to contraction of actin microfilaments in order for the cell to divide. p. 92

clitoris a female sex organ and most sensitive erogenous zone. p. 504

cocci bacteria with a spherical shape. p. 542

coccyx vertebrae four fused vertebrae inferior to the pelvis, commonly referred to as the *tailbone*. p. 218

cochlea snail-shaped organ in the inner ear that contains stereocilia, which convert sound waves into action potentials. p. 350

coding strand DNA strand that codes for a gene. p. 149

codominance a form of dominance in which both alleles are expressed in a heterozygote. p. 111

codon a three-nucleotide segment that determines the amino acid. p. 152

collagen extracellular protein that surrounds connective tissue, cartilage, skin, tendons, and ligaments, and that provides strength to many tissues and resists stretching. p. 62

collecting duct a series of tubules and ducts that connect the nephrons to the ureters. p. 483

columnar epithelial cells that are rectangular or shaped in columns. p. 179

comminuted fracture a fracture that results in two or more separate bone components. p. 232

common bile duct a tubelike structure formed by a combination of the cystic and common hepatic ducts. p. 465

common fibular nerve a nerve in the lower leg. p. 308

common hepatic artery an artery that provides blood to the liver. p. 405

common iliac arteries large arteries that originate from the aorta. p. 407

compact bone tightly arranged bone composed of osteons. p. 210

complement proteins proteins that exist in the blood and form part of the innate immune system response. p. 467

compliance tendency of an organ to resist returning to its original shape and size. p. 441

concentration gradient the difference in concentration of molecules across a cell membrane. p. 298

concentric contraction (muscular) shortening of a muscle during contraction. p. 258

conducting system path by which the electrical signal of the heart is transmitted. p. 391

conducting zone organs of the respiratory system that filter, warm, and moisturize the air and conduct it toward the lungs. p. 434

condylar joint synovial joint that permits motion in two planes. p. 235

condylar process rounded extension of a bone that articulates with another bone. p. 217

cones photoreceptors that detect colours in bright light. p. 345

conjugation transfer of DNA directly from one bacterium to another through a pilus. p. 56

connexin proteins that make up gap junction channels. p. 63

consciousness awareness of self or surroundings. p. 313

constipation infrequent or difficult bowel movements where stool is dry and hard. p. 40

continuous variation occurs when a phenotype shows multiple variations in a population through multiple genes. p. 110

contractile proteins proteins that allow for contraction of a muscle cell (e.g., actin and myosin). p. 269

conus medullaris tapered end of the spinal cord. p. 306

coracoid process protrusion on the scapula. p. 224

core muscles abdominal and back muscles that function to stabilize the body during movement. p. 258

coronal suture suture that connects the frontal bone to the anterior portion of the parietal bones. p. 213

coronary arteries arteries that provide oxygenated blood to the heart. p. 387

coronary sinus collection of veins that return blood from the heart to the right atrium. p. 387

coronary sulcus (atrioventricular groove) groove separating the atria from the ventricles. p. 387

coronoid fossa indentation in the humerus that articulates with the ulna and increases the range of motion when the arm is bent. p. 224

corpora cavernosa erectile tissue of the penis. p. 500

corpus luteum structure made up of granulosa cells after ovulation; secretes progesterone to maintain a potential pregnancy. p. 507

corpus spongiosum *see* **corpora cavernosa.** p. 500

cortical radiate arteries renal blood vessels. p. 485

cortical radiate veins renal blood vessels. p. 485

cortisol glucocorticoid hormone produced in the adrenal gland during stress. p. 84

cosmid synthetic plasmid containing the *cos* site from the lambda phage. p. 167

costal (hyaline) cartilage cartilage that gives the chest wall elasticity and is found at the anterior end of the ribs. p. 223

coupled transport the use of an electrochemical gradient to move nutrients along with an ion across the plasma membrane into the cell; also called *cotransport*. p. 66

covalent bond bond in which electrons are shared between two atoms. p. 26

cowpox virus used in humans to build immunity to similar viruses that have more harmful effects. p. 531

cranial cavity area of the head that contains the brain. p. 16

cranial nerves nerves that originate in the brain or brainstem. p. 291

craniofacial bones bones that make up the skull and face. p. 213

c-reactive protein (CRP) protein produced in the liver during an inflammatory response. p. 419

creatine kinase enzyme used to create creatine phosphate in muscle cells. p. 279

creatinine a breakdown of creatine phosphate. p. 488

cribriform plate bone that forms the base of the nasal cavity. p. 217

crista ampullaris *see* **ampullae.** p. 353

cristae inner membrane of the mitochondria that is folded and is the location of the electron transport chain. p. 59

cross-bridge cycle that uses the binding of actin and myosin to generate force during muscle contraction. p. 272

crossing over process where homologous chromosomes pair up and fragments of non-sister chromatids break and are swapped in the same place, resulting in hybrid chromosomes. p. 99

cross-sectional plane *see* **transverse plane.** p. 16

cuboidal epithelial cells in a shape closely resembling a cube. p. 179

cupula structure within the ampullae of the semicircular canals that contains hair cells. p. 353

cyanobacteria bacteria that undergo photosynthesis and produce oxygen. p. 539

cytokines proteins that function in cell-signalling during an immune response. p. 518

cytokinesis process of splitting the cytoplasm. p. 88

cytoplasm water-filled fluid inside a cell. p. 46

cytoskeleton intracellular proteins that hold organelles in place, give cells shape, and help transport other molecules within the cell. p. 61

cytotoxic T cells immune cells that target cancer cells and virus-infected cells. p. 526

Dalton's law principle that states that the total pressure of a mixed gas equals the sum of the partial pressures. p. 444

deamination process in which an amino group is removed from an amino acid. p. 81

decarboxylation process of removing a carbon from a molecule to release CO_2. p. 77

decomposition breakdown of organic substances into smaller forms. p. 4

defecation elimination of solid or semi-solid waste from the digestive tract. p. 459

dehydration synthesis removal of a water molecule to form a covalent bond. p. 27

deletion a nucleotide skipped during replication. p. 145

deltoid triangular-shaped muscle superior and lateral to the shoulder. p. 263

deltoid tuberosity attachment site for the insertion of the deltoid muscle. p. 224

denaturation irreversible unfolding of a protein in which hydrogen bonds are broken. p. 32

dendrites branches from the cell body of a neuron that detect stimuli. p. 295

dendritic cells antigen-presenting cells of the immune system. pp. 423, 520

dens protuberance on the axis that permits horizontal rotation of the skull. p. 219

dense connective tissue the tissue that forms strong attachments between structures, and allows organs to stretch and recoil. p. 183

deoxyribonuclease an enzyme that breaks down DNA. p. 470

deoxyribose five-carbon sugar in a DNA nucleotide. p. 138

depolarization movement of sodium or calcium into a cell that causes the inside of the cell to become positively charged. p. 298

dermal papillae projections of the dermis that extend into the epidermis to provide increased surface area for diffusion of oxygen and nutrients. p. 199

dermal ridges contiguous papillae that form ridges into the epidermis and form fingerprints. p. 199

dermis middle layer of the skin that contains blood vessels, nerves, and other cell types. p. 197

descending aorta segment of the aorta that's distal to the aortic arch. p. 388

descending colon the part of the large intestine located in the left-upper and left-lower quadrants of the abdomen, leading to the sigmoid colon. p. 467

desmosomes proteins that adhere cells together. p. 63

detrusor muscle smooth muscle in the wall of the bladder. p. 483

DHEA (dehydroepiandrosterone) a testosterone-like hormone produced by the adrenal cortex in males and females. p. 84

dialysis a medical process for removing waste products from the blood when the kidneys no longer function properly. p. 493

diaphragm muscle at the base of the thoracic cavity that facilitates respiration. p. 258

diaphysis middle region, or shaft, of the long bone. p. 213

diarthrosis *see* **synovial joints.** p. 235

diastole relaxation of the heart muscles. p. 416

diencephalon located between the brainstem and the cerebrum and includes the thalamus, hypothalamus, and the pineal gland. p. 312

dihybrid cross a cross between two individuals that carry two different traits that determines the probability of the traits being passed onto the offspring. p. 109

diplococci paired, spherical bacteria. p. 542

diploid (2n) twice the amount of haploid (n) chromosomes found in somatic and germline cells. p. 90

directional selection favouring one extreme trait over another in a population so that the opposite extreme is eliminated. p. 133

directional terms terms used to describe regions of the body in relationship to each other. p. 14

disaccharide two monosaccharides joined by glycosidic bonds. p. 37

disruptive selection a form of selection that favours each extreme in a population so that the intermediate traits are decreased or eliminated. p. 133

distensibility ability to be stretched under pressure. p. 396

diuretics medications that increase urination by decreasing water reabsorption in the nephron. p. 420

diversity variety. p. 2

DNA polymerase enzyme that adds complementary nucleotides to DNA in the 5′ to 3′ direction during replication. p. 86

DNA repair enzymes enzymes that correct DNA mutations. p. 146

DNA replication a process in which DNA is copied before cell division. p. 86

DNA transfer method by which genetic variation is acquired among bacteria. p. 87

dominant the allele that is phenotypically expressed. p. 104

dopamine a neurotransmitter involved in task-based concentration and the reward pathway. p. 314

dorsal horn dorsal grey matter of the spinal cord. p. 305

dorsal root afferent sensory root of a dorsal nerve that is outside the spinal cord. p. 305

double helix two strands of DNA bound together by nitrogenous bases that forms a helical (spiral) structure. p. 35

Down syndrome delayed development and mental impairment in an individual with trisomy 21. p. 117

duodenum the initial section of the small intestine. p. 467

dura mater outermost layer of the meninges. p. 304

dystrophin muscle protein that holds myofibrils together to prevent muscle degeneration. p. 118

early antral follicle follicle formed by granulosa cells that surround the fluid-filled cavity referred to as the *antrum*. p. 504

eccentric contraction lengthening of the muscle during contraction. p. 258

eccrine glands type of sweat gland, found predominantly on the forehead, palms, and soles of the feet. p. 200

ecology the study of the relationship between organisms and their environment. p. 122

ectoparasites parasites that live on the surface of the host. p. 553

edema an accumulation of fluid in the interstitial space. p. 402

effectors any part of the body that responds to stimulation from the integrating centre that will bring an altered factor back to homeostasis. p. 20

efferent ductules connect the rete testis to the epididymis. p. 500

eicosanoids signalling molecules involved in inflammation. p. 523

ejaculatory duct a duct formed from the union of the vas deferens and the seminal vesicle. p. 500

elastin extracellular matrix protein that allows tissues to be flexible and stretchy, and to reform to their original shape after being stretched. p. 62

electrical gradient difference in electric potential across a cell membrane. p. 67

electrocardiogram (ECG) a recording of the electrical activity of the heart. p. 392

electrolytes ions in the fluids of the body that conduct an electrical charge. p. 421

electromagnetic spectrum the range of all electromagnetic waves arranged according to wavelength and frequency. p. 345

electron transport chain electrons transferred through membrane proteins, producing many ATP molecules. p. 79

emphysema one of the disease conditions that make up chronic obstructive pulmonary disease (COPD). p. 451

encapsulated refers to benign tumours surrounded by a layer of cells. p. 94

end diastolic volume (EDV) maximum volume of blood in the ventricles after diastole. p. 393

end stage renal disease progressive loss of kidney function to the point where the kidneys no longer function; caused primarily by chronic hypertension. p. 420

endocardium innermost layer of the heart. p. 388

endocrine gland an organ that secretes hormones. p. 20

endocytosis movement of large molecules or fluid into the cell. p. 68

endometriosis a gynecological condition where endometrial cells grow outside of the uterus. p. 507

endometrium innermost layer of the uterus. p. 307

endomysium connective tissue that surrounds individual muscle cells. p. 256

endoparasites parasites that live inside their host. p. 553

endoplasmic reticulum membrane network within the cytoplasm of every cell type that can be smooth or rough. p. 58

endospore a highly resistant state that some bacteria enter where they are dormant and can survive harsh environments. p. 542

endosteum dense, irregular, connective tissue that lines the surface of bones and contains bone precursor cells. p. 213

endosymbiosis a process whereby primitive bacteria were engulfed by another cell, giving rise to eukaryotic organisms with organelles. p. 59

endothelium layer of cells that lines the interior surface of vessels. p. 388

enteric nervous system nervous system that controls the digestive system. p. 292

enzymes globular proteins that facilitate thousands of chemical reactions within cells. pp. 32, 55

eosinophils white blood cells that fight parasitic infections. p. 423

ependymal cells glial cells that produce cerebral spinal fluid. p. 192

epicardium *see* **visceral pericardium**. p. 388

epidermal ridges extensions of dermal ridges into the epidermis, which form fingerprints and footprints. p. 199

epidermis most superficial layer of the skin. p. 197

epididymis a part of the male reproductive system connecting the efferent ducts to the vas deferens. p. 500

epidural space anatomical space outside the dura mater. p. 304

epigastric region central-upper region of the abdomen. p. 467

epiglottis a small cartilaginous structure that prevents the passage of solids or fluids into the trachea during swallowing. p. 435

epinephrine a stimulatory hormone produced in the adrenal medulla in response to sympathetic nervous system stimulation. p. 281

epineurium outermost layer of connective tissue surrounding neurons. p. 307

epistasis effects of one gene modified by effects of one or more other genes. p. 113

epithelial tissue type of tissue that lines and covers organs and body cavities, and also makes up outermost layer of the skin. p. 178

equatorial plane place where chromosomes line up in the centre of a cell during cell division. p. 92

erector spinae deep muscles that extend the length of the back. p. 264

erythrocytes red blood cells. p. 421

erythropoietin hormone that stimulates the production of red blood cells. pp. 421, 482

esophageal arteries a group of arteries that supply the esophagus. p. 405

essential amino acids nine amino acids that cannot be synthesized by the body and must be acquired in the diet. p. 30

essential fatty acids fats that cannot be synthesized by the body and must be acquired in the diet. p. 45

estrogen a sex hormone produced in the ovaries. p. 205

ethanol two-carbon alcohol molecule. p. 83

ethmoid bone bone that forms the anterior portion of the base of the skull and forms the nasal septum. p. 217

eubacteria prokaryotic unicellular organisms that survive in many different environments. pp. 2, 56

eukaryotic complex cellular structure that contains membrane-bound organelles. p. 2

Eustachian tube tube that connects the middle ear to the throat and is important for equalizing air pressure between the middle ear and the atmosphere. p. 435

evolution the change in the inherited traits in a population of organisms over time. p. 18

excitatory postsynaptic potentials (EPSPs) graded membrane potentials that increase the likelihood of an action potential in the postsynaptic neuron. p. 302

excretion process of eliminating waste from an organism. p. 486

exhalation (or expiration) movement of air from the lungs into the external environment. p. 433

exocrine substance such as sweat or enzymes that is secreted through a duct into a hollow organ or to the outside of the body. p. 467

exocytosis the movement of large molecules out of vesicles and across the cell membrane. p. 68

exon shuffling *see* **alternate splicing**. p. 160

exons sequences of DNA that translate into amino acid sequences for protein synthesis. p. 152

exonuclease activity an enzyme function that allows polymerase to reverse direction and remove incorrect nucleotides. p. 146

expiratory reserve volume amount of air in the lungs following normal exhalation. p. 442

extensors muscles in the posterior of the forearm that allow extension of the hand at the wrist. p. 265

external auditory meatus opening in the skull where sound waves enter the inner ear. p. 217

external environment everything outside the body that is not regulated, such as temperature, nutrients in food, content of air inhaled; also includes the inside of the GI tract, urethra, vagina, ear canal, nasal and oral cavities, and respiratory tract. p. 19

external intercostal muscles the superficial muscles between the ribs that contract to increase thoracic volume during inspiration. p. 440

external jugular veins large veins that receive blood from the head. p. 408

external obliques largest and outermost muscle of the lateral and anterior abdomen; allows the body to twist and move forward during contraction. p. 441

external urethral sphincter a ring of muscle that can be voluntarily controlled to contract and prevent urination. p. 485

extracellular fluid the fluid outside a cell; includes interstitial fluid and blood plasma. p. 68

extracellular matrix the proteins and connective tissue located outside a cell. p. 183

extracellular outside the cell. p. 526

extracellular proteins proteins produced by each cell, secreted or integrated into the cell membrane, and involved with intercellular interactions such as adhesion, signalling, or cellular communication. p. 62

facilitated diffusion movement of molecules across a membrane, through a membrane protein, down a concentration gradient, and without input of energy. p. 65

fallopian tube tubes that lead eggs cells from the ovaries to the uterus. p. 502

farsightedness occurs when a person has the ability to focus on objects that are far away but not close up; also known as *hyperopia*. p. 348

fascia layer of fibrous tissue that surrounds muscles and groups of muscles. p. 256

fascicle a bundle or cluster of neurons or muscle cells. p. 256

fasting phase stage of digestion between meals. p. 376

fat-soluble hormones nonpolar hormones that can move into the target cell and bind to a receptor that affects gene expression. p. 364

fatty acid chains of carbon and hydrogen; three fatty acid chains plus a glycerol molecule form a triglyceride. p. 43

fecal-oral route infections acquired through consuming contaminated food or water. p. 553

femoral nerve a nerve located along the femur. p. 308

femur longest bone in the human body; located in the thigh. p. 227

fermentation process of anaerobic respiration in animals, bacteria, and yeast that allows small amounts of ATP to be produced in the absence of oxygen. p. 83

fever an elevation in body temperature, often in response to infection. p. 524

fibre *see* **cellulose**. p. 38

fibrinogen a protein that helps form clots. p. 421

fibroblasts most abundant connective tissue cell type, which produces collagen, elastin, reticular fibres, and ground substance. p. 183

fibrocartilage type of cartilage composed of fibrous connective tissue. p. 232

fibronectin protein that connects integrins to other extracellular matrix proteins. p. 63

fibrous joints structural joints that are connected by fibrous connective tissue and are immovable, such as the joint between the radius and the ulna. p. 234

fibula the lateral bone in the lower leg. p. 227

filtrate blood plasma that has passed through the glomerulus into the Bowman's capsule. p. 487

filtration movement of substances from the glomerular capillaries into the Bowman's capsule in the kidney. p. 179

fimbriae fringed tissue surrounding the opening of the fallopian tube. p. 502

fixators muscles that stabilize the primary muscle of a motion. p. 258

flagella external structures composed of microtubules; used for locomotion and feeding (singular, *flagellum*). p. 56

flat bones bones such as the sternum, ribs, and scapulae. p. 211

flatworms (platyhelminths) a type of animal parasite, such as tapeworms. p. 554

flexors muscles in the anterior of the forearm that allow flexion of the hand at the wrist. p. 265

flow movement of a fluid. p. 413

fluid mosaic model term used to describe a typical cell membrane composed of unsaturated phospholipids that are liquid rather than solid, cholesterol, and proteins. p. 55

flukes (trematodes) a type of animal parasite. p. 554

follicle-stimulating hormone (FSH) a sex hormone that stimulates the production of testosterone and estrogen in males and females. p. 377

follicle an egg surrounded by granulosa cells. p. 504

follicular phase phase in the menstrual cycle in which the follicles of the ovaries mature. p. 505

food energy energy from digested food, such as glucose that can be used to make ATP. p. 74

foramen any opening in a bone that allows blood vessels and nerves to pass through. p. 219

foramen magnum opening through which the spinal cord exits the skull. p. 217

foramen ovale (1) opening in the base of the skull that allows the mandibular nerve to travel through the skull; (2) a fetal cardiac opening that allows blood to enter the left atrium from the right atrium. pp. 217, 388

force (muscular) a power that's generated by a muscle to overcome the load. p. 272

fossa ovalis a depression in the right atrium of the heart, the remnant of a thin fibrous sheet that covered the foramen ovale during fetal development. p. 388

fovea *see* **macula lutea**. p. 346

frameshift mutation a change in multiple codons that results from an insertion or deletion mutation. p. 145

Frank-Starling law the principle that increased stretch in the muscle cells will increase the force of contraction. p. 396

free nerve endings a sensory neuron that detects pain. p. 337

frontal bones bones that form the forehead. p. 213

frontal lobe anterior lobe of the brain. p. 316

frontal plane an anatomical division of the body into anterior and posterior sections. p. 15

fructose monosaccharide found in fruit. p. 37

full-thickness burn third- and fourth-degree burn that extends either into or through the subcutaneous layer of the skin to the underlying muscle. p. 202

fundus the part of a hollow organ opposite from its opening, such as the left portion of the body of the stomach. p. 464

fungi eukaryotic multicellular organisms (except yeast) that are not photosynthetic and that absorb nutrients from other organisms. p. 521

G_1 phase first phase of interphase; the main gap or growth phase when the cell undergoes normal cell functions and prepares for DNA replication. p. 88

G_2 checkpoint point at which DNA is checked for DNA mutations after replication. p. 88

G_2 phase third phase of interphase; the second gap or growth phase where the cell continues to prepare for mitosis. p. 88

galactose monosaccharide found in milk. p. 37

gallbladder digestive organ that stores bile. p. 465

gametes sperm or egg cells that contain a haploid number of chromosomes. p. 90

gametogenesis formation of the gametes. p. 498

ganglia a collection of neuron cell bodies. p. 292

ganglion cells cells that transmit action potentials from the bipolar cells in the retina to the brain, and play an important role in circadian rhythms. p. 347

gap junction channels protein membrane channels that play an important role in intercellular communication by allowing the movement of molecules directly between adjacent cells. p. 63

gap phase *see* **G_1 phase**. p. 88

gas exchange movement of oxygen from the lungs to the bloodstream and carbon dioxide from the bloodstream to the lungs. p. 400

gastric phase a phase of digestion that involves digestion in the stomach. p. 476

gastrin a hormone released by the stomach that stimulates the production of hydrochloric acid. p. 476

gastrocnemius calf muscle in the lower leg that allows for flexion of the foot at the ankle. p. 268

gastrointestinal tract organs that form the alimentary canal. p. 458

gel electrophoresis method that uses an electrical current to separate strands of DNA based on the length of the DNA sequence. p. 172

gene cloning replication of a gene sequence in bacteria. p. 168

gene DNA sequence of hundreds of nucleotides that code for a specific protein. p. 33

gene expression transcription and translation of a gene that results in the synthesis of a protein. p. 150

gene flow occurs when phenotypes in a population become similar. p. 133

gene pool all the genes and allele variations in a population. p. 135

gene silencing addition of methyl groups to prevent gene transcription. p. 160

gene splicing *see* **recombinant DNA technology**. p. 164

genetic code universal chart used to determine the amino acid based on the mRNA codon. p. 156

genetic engineering *see* **recombinant DNA technology**. p. 164

genetic modification any kind of alteration in the genome, including deleterious mutations caused by radiation or chemicals. p. 164

genetic variation a means for individuals to acquire new alleles through sexual reproduction. p. 121

genetically modified organisms (GMOs) organisms such as plants or animals that contain recombinant DNA. p. 164

genome all genes in an organism. p. 164

genotype genes that code for a phenotype. p. 135

germ cells cells in the testes or ovaries that contain a diploid number of chromosomes and produce gametes during meiosis. p. 90

gingiva mucosal tissue that lines the inside of the mouth. p. 463

gingivitis inflammation of the gingiva. p. 463

glenoid cavity socket in which the humerus sits. p. 224

glial cells cells that support the neuron functions. p. 294

gliding movement of a joint in a single plane running parallel to the articulating structure. p. 237

glomerular filtration rate (GFR) the amount of fluid moved from the glomerulus to the Bowman's capsule in a set amount of time. p. 487

glomerulus a network of capillaries where fluid leaves and enters the Bowman's capsule during filtration. p. 485

glossopharyngeal nerve the ninth cranial nerve that carries motor and sensory information to the mouth and tongue. p. 417

glucagon a hormone produced in the pancreas that stimulates the liver to break down glycogen and increase blood sugar. pp. 281, 467

gluconeogenesis formation of glucose from the breakdown of fat or amino acids. p. 466

glucosamine sulfate a sugar–amino acid found in ground substance in connective tissues. p. 184

glucose monosaccharide that is the building block of starch, also found in fruit and table sugar. p. 37

gluteal tuberosity attachment site for the gluteal muscles. p. 227

gluteus maximus largest and strongest muscle in the body; located on the posterior pelvis. p. 268

glycated cell-surface proteins that become coated in sugars. p. 69

glycerol a sugar alcohol that connects three fatty acids in a triglyceride. p. 43

glycogen polysaccharide that is stored energy in liver and muscle cells. p. 38

glycogenesis the production of glycogen from excess glucose in the liver and muscle cells. p. 466

glycoprotein a molecule composed of protein and sugar. pp. 55, 156

glycosidic bond bond between a sugar and another molecule, such as the base in a nucleotide. p. 35

goblet cells cells in the respiratory and digestive tract that produce mucus. p. 435

Golgi bodies a separate membrane network found in the cell near the endoplasmic reticulum, where newly synthesized proteins are further processed and packaged into vesicles. p. 156

gomphosis joint the joint that houses the teeth. p. 234

gonadal arteries arteries that provide oxygenated blood to the gonads. p. 407

gonadotropin-releasing hormone (GnRH) a hypothalamic hormone that stimulates the anterior pituitary gland to produce FSH and LH in males and females. p. 500

gradualism rate of evolution that is a slow and gradual process where very small changes occur in a population over millions of years. p. 126

gram-negative bacteria bacteria that have a small layer of peptidoglycan and an external layer of lipopolysaccharide that does not stain when exposed to crystal violet. p. 56

gram-positive bacteria bacteria that have many layers of peptidoglycan but do not have a lipopolysaccharide layer, and stain purple when exposed to crystal violet. p. 56

granular leukocytes a type of white blood cells that contain granules and are vital to the immune system. p. 423

granulosa cells cells that surround the egg, forming the follicle, and produce sex hormones and growth factors. p. 504

Graves' disease an autoimmune disease caused by antibodies binding to the TSH receptor, causing increased production of thyroid hormones. p. 370

great saphenous vein a large vein in the leg. p. 409

greater omentum a large, protective fold of the visceral peritoneum. p. 462

greater sciatic notch indentation in the ilium. p. 227

greater trochanter lateral projection of the femur that acts as an attachment site for muscles. p. 227

greenstick fractures bone fractures where the bone bends, and partially breaks; more common in children than adults. p. 232

grey matter tissue of the brain that appears grey and contains cell bodies, dendrites, and axon terminals of neurons. p. 305

ground substance fluid part of connective tissue that supports the functions of cells and fibres. p. 183

growth hormone (GH) hormone produced by the anterior pituitary gland that stimulates growth of bone and muscle and the regeneration of cells throughout the body. p. 231

growth phase *see* G$_1$ **phase.** p. 88

gustatory cell taste bud cell that contains receptors that bind to various molecules and can depolarize and send action potentials to the olfactory bulb. p. 355

gut-associated lymphoid tissue (GALT) lymphatic nodules and lymph nodes located along the digestive tract; important for preventing and fighting infections that enter via the GI tract. p. 515

gyrus a ridge of the cerebral cortex. p. 315

H zone central region of the sarcomere; contains the central portion of the thick filaments, but no thin filaments. p. 271

hair follicles can detect touch and vibration. p. 337

hamstrings muscles that flex the leg at the knee. p. 268

haploid (n) the number of chromosomes found in gametes, which is half of the diploid number. p. 90

Hardy-Weinberg equilibrium condition when a population is stable and not evolving and allele frequencies stay the same over successive generations. p. 135

heart murmur an extra heart sound during the cardiac cycle made by blood flowing backward through a valve. p. 390

helicase an enzyme that breaks hydrogen bonds between the bases on complementary strands of DNA, beginning at the origin of replication. p. 141

helminths (worms) a type of animal parasite that feeds on a living host for nourishment. pp. 553, 554

helper T cells part of the adaptive immune response that stimulates other immune cells to fight bacterial and viral infections, but do not themselves directly kill infections. p. 526

hemagglutinin a substance that causes red blood cells to agglutinate, and is also the membrane protein that influenza viruses bind to during infection. p. 547

hematocrit level percentage of the blood that is red blood cells; normal level, 45%. p. 421

hematoma an accumulation of blood, commonly called a *bruise*. p. 232

hematopoiesis process of creating new blood cells. p. 421

hematopoietic stem cells stem cells that differentiate into blood cells and are located in the bone marrow. p. 515

heme chemical group that contains Fe^{2+} that binds oxygen. p. 421

hemoglobin blood protein that contains heme. p. 117

hemorrhagic stroke a stroke caused by a ruptured blood vessel in the brain. p. 419

hepatic portal vein blood vessel that takes blood from the GI tract to the liver. p. 407

hepatic sinusoids low-pressure channels that take blood from the hepatic artery and portal vein and deliver it to the central vein and then to the hepatic vein; includes Kupffer cells inside the sinusoids that take up and destroy foreign material. p. 465

hepatopancreatic duct duct that carries secretions from both the common bile duct and the pancreatic duct to the small intestine during digestion. p. 465

heredity the passing of genes from parents to offspring. p. 102

heterozygote advantage the benefits of having a heterozygote genotype compared to having homozygous dominant or recessive genes; for example, the benefit of being heterozygous for sickle cell anemia when living in an environment with a high rate of malaria infection. p. 133

heterozygous alleles that are different. p. 135

hinge joint joint where the articular surfaces allow for motion only in a single plane, such as the elbow. p. 235

hip flexors (iliopsoas) a group of muscles that work to flex the thigh at the hip, raising the knee up toward the body. p. 265

histology the study of tissues. p. 178

HIV (human immunodeficiency virus) a virus that targets the cells of the immune system that express the CD4 receptor. p. 526

homeostasis the regulation of multiple factors within the internal environment of the body, such as temperature, blood sugar, ions, water, and pH. p. 8

homocysteine a non-protein amino acid that is associated with an increased risk for cardiovascular disease. p. 422

homologous chromosomes two similar, but not identical, chromosomes where one comes from the mother and one comes from the father; also called *homologues*. p. 89

homologous structures term used by evolutionists to compare anatomy among organisms to show that similarities indicate a common ancestor. p. 128

homologues *see* **homologous chromosomes.** p. 89

homozygous alleles that are the same. p. 135

homunculus a type of map of the brain that shows regions that detect sensory stimuli from various regions of the body. p. 341

horizontal fissure separates the superior and middle lobe of the lung. p. 437

horizontal plane *see* **transverse plane.** p. 16

host cell a cell that's used to take up the recombinant DNA. p. 165

humerus bone of the upper arm. p. 224

humoral immune response a response involving the secretion of antibodies by plasma cells. p. 526

hyaluronic acid component of extracellular matrix of connective, epithelial, and nerve tissue that plays a key role in signalling cells to proliferate; often a component in anti-aging creams to stimulate the growth of new skin cells. p. 184

hyaluronidase enzyme that breaks down hyaluronic acid. p. 184

hydrochloric acid (HCl) an acid formed by the elements hydrogen and chlorine dissolved in water, produced by parietal cells in the stomach. p. 465

hydrogen bond a slightly positive charge on the hydrogen of one molecule attracting a slightly negative charge on another molecule, most often oxygen. p. 26

hydrogen peroxide H$_2$O$_2$, used to break down substances such as alcohol and formaldehyde and can be effective in destroying infectious organisms. p. 60

hydrogenation process in which hydrogen is added to fatty acid molecules that contain double-bonded carbon atoms and converts them to single-bonded carbon atoms, such as the production of margarine from vegetable oil. p. 44

hydrolysis the process of breaking covalent bonds with the addition of a water molecule. p. 38

hydrophilic water loving. p. 26

hydrophobic water fearing. p. 27

hydrostatic pressure mechanical pressure generated by a fluid on the wall of its container. pp. 401, 487

hyoid bone attachment site for the tongue and neck muscles; does not articulate with any other bones. p. 218

hypercapnia increased amount of carbon dioxide in the blood, which can result from hypoventilation or lung disease. p. 446

hyperpolarization increased charge of a cell beyond the normal resting membrane potential. p. 299

hypertension condition of increased blood pressure. p. 418

hypertonic concentration of the solute higher in the interstitial fluid than inside the cell. p. 65

hypodermis innermost layer of the dermis. p. 197

hypothalamus a portion of the brain located superior to the brainstem, part of the diencephalon; is responsible for regulating the autonomic nervous system and controls the endocrine functions of the pituitary gland. p. 20

hypothyroidism underproduction of thyroid hormones. p. 370

hypotonic concentration of the solute lower in the interstitial fluid than inside the cell. p. 65

hypoxia decreased oxygen level. pp. 419, 421

I band region between the A bands of adjacent sarcomeres. p. 271

ileocecal sphincter a ring of muscle that controls the opening to the large intestine from the small intestine. p. 467

ileum the final section of the small intestine. p. 467

iliac crest top of the hip bones. p. 225

iliacus one of the hip flexor muscles. p. 265

iliocostalis muscles muscles of the back that attach the ribs. p. 265

ilium largest and most superior bones in the pelvic girdle. p. 225

immune suppression medications medications that decrease the ability of the immune system to function and are used to treat chronic inflammatory conditions, such as rheumatoid arthritis. p. 533

immunoglobulins proteins made by the immune system that attach to foreign substances in the bloodstream and neutralize them. p. 528

inactivated toxin a toxin that has been altered either through heat or chemical means to no longer be pathogenic. p. 531

incus anvil-shaped bone that connects the incus to the stapes. p. 349

independent assortment random separation of homologues into gametes during meiosis. p. 99

inferior vena cava inferior portion of the largest vein in the body; brings blood to the right atrium. p. 387

inflammatory response part of an initial reaction to tissue damage that fights infection and involves vasodilation, swelling, redness, and pain. p. 522

infundibulum funnel-shaped cavity or organ, such as the stalk that connects the hypothalamus with the posterior pituitary gland. p. 502

ingestion consumption of food. p. 458

inhalation (or inspiration) movement of air from the external environment into the lungs. p. 433

inhibin a hormone that inhibits FSH secretion in males and females. p. 501

inhibitory neurotransmitters neurotransmitters that reduce the likelihood of an action potential occurring in the postsynaptic neuron by causing IPSPs. p. 301

inhibitory postsynaptic potentials (IPSPs) graded membrane potentials that decrease the likelihood of an action potential in the postsynaptic neuron. p. 302

innate immune response a non-specific immune response that acts as the second line of defence against infection and involves cells such as neutrophils, macrophages, dendritic cells, and natural killer cells. p. 518

insect vectors insects that carry or transmit a pathogen, such as mosquitoes that transmit malaria; also known as *carriers*. p. 553

insertion (DNA replication), an extra nucleotide added during replication; also known as *addition*. p. 145

insertion (muscular) connection point of a muscle to a moving bone. p. 238

insertional inactivation occurs when a transposable element inserts into a DNA sequence in the middle of a gene, producing a non-functional protein. p. 140

insoluble fibre fibre that does not attract water and adds bulk to the stool. p. 41

inspiratory capacity total amount of air that can be drawn into the lungs; includes the tidal volume. p. 442

inspiratory reserve volume amount of air that can be drawn into the lungs after normal inspiration, not including the tidal volume. p. 442

insula the part of the brain believed to be involved in consciousness. p. 288

insulin hormone that is secreted by the pancreas and lowers blood sugar after a meal. p. 364

insulin resistance when cells no longer respond to insulin to take up blood sugar. p. 376

insulin-like growth factors (IGFs) a hormone, stimulated by GH and secreted by the liver, that stimulates growth of many cell types, such as bone and muscle. p. 231

integrating centre part of the body that interprets signals coming from the receptors and determines if there is any deviation from the set point. p. 20

integrins membrane-spanning proteins involved in attaching the cell membrane to proteins in the extracellular matrix and in transferring signals from outside to inside the cell, and vice versa. p. 62

integumentary system the skin, which is the organ system that prevents damage to the body and provides a physical barrier to outside the body. p. 196

interatrial septum the barrier between the right and left atria of the heart. p. 388

intercalated discs feature of cardiac muscle that connects the branched, striated muscle fibres and contains desmosomes and gap junctions. p. 191

intercondylar fossa indentation at the distal end of the posterior femur. p. 227

intercostal arteries arteries that provide oxygen to the area between the ribs. p. 405

interferon (INF) a cytokine released by virus-infected cells that cause nearby uninfected cells to block viral replication. p. 520

interleukin-1 (IL-1) a cytokine produced by antigen-presenting cells that signal helper T cells to proliferate. p. 526

interleukin-2 (IL-2) a cytokine-signalling molecule that signals immune cells such as cytotoxic T cells and B cells to proliferate. p. 526

interlobar arteries blood vessels that carry oxygenated blood in the kidneys. p. 485

interlobar veins blood vessels that carry deoxygenated blood in the kidneys. p. 485

intermediate filaments intracellular cytoskeletal proteins composed of intertwined proteins that support cell structure and help organelles remain in place. p. 61

intermembrane space the space between the inner membrane and the outer membrane of the mitochondria. p. 79

internal environment everything that is in close contact with the body's cells and regulated by homeostasis, such as interstitial fluid and blood. p. 19

internal iliac artery main artery of the pelvis. p. 407

internal intercostal muscles deep muscles between the ribs. p. 263

internal jugular veins veins that collect blood from the brain, superficial face, and neck. p. 408

internal obliques innermost anterolateral abdominal oblique muscle, which acts as an antagonist to the diaphragm in forced exhalation. p. 441

internal urethral sphincter a ring of muscle that controls the flow of urine. p. 485

interneurons short neurons used to transmit signals within the central nervous system. p. 295

interosseous membrane the membrane between bones composed of connective tissue, such as between the radius and ulna, and the tibia and fibula. p. 224

interphase growth phase of cell, consisting of the G1, S, and G2 phases that occur before mitosis. p. 88

intertubercular groove the groove in the humerus that houses the biceps tendon; also called the *bicipital groove*. p. 221

interventricular septum barrier between the right and left ventricles of the heart. p. 388

intervertebral disc cylindroid structure made of cartilage that separates and cushions the vertebrae. p. 218

intestinal phase a phase of digestion that occurs in the intestines. p. 476

intracellular inside the cell. p. 526

intrapleural pressure pressure in the pleural cavity (usually negative compared to the alveolar and environmental pressures) that prevents the lungs from collapsing. p. 441

intrinsic factor a protein secreted by the stomach that is required for the absorption of vitamin B12 in the ileum. p. 465

introns sequences of DNA in genes of eukaryotes that do not translate into part of an amino acid sequence. p. 152

involuntary movement movement that is unconsciously controlled, such as the knee jerk reflex. p. 190

ion channels proteins that span the membrane and allow the transport of specific ions into or out of the cell. p. 55

iris the coloured part of the eye. p. 345

irregular bones bones that do not belong to the classification of long, short, flat, or sesamoid; include the vertebrae and the bones of the face and skull. p. 211

ischemia decreased blood supply. p. 419

ischemic stroke a stroke caused by an occlusion of a blood vessel in the brain. p. 419

ischial spines protrusions that form the narrowest portion of the pelvic opening. p. 227

ischial tuberosities protrusions at the inferior and posterior aspects of the pelvis; the bones that we sit on. p. 225

islet cells endocrine cells of the pancreas that produce either glucagon (alpha islet cells) or insulin (beta islet cells). p. 467

isokinetic contraction change in tension of a contraction without changing the velocity. p. 258

isometric contraction muscle contraction that doesn't result in a change of muscle length. p. 258

isotonic concentration of the solute the same both in the interstitial fluid and inside the cell. p. 258

isotonic contraction contraction that causes muscle to shorten with even tension throughout the motion. p. 258

jaundice yellow pigmentation that builds up in the skin—caused by an increased level of bilirubin (produced from the breakdown of red blood cells)—that occurs if the liver is not functioning properly. p. 421

Jean Baptiste de Lamarck French scientist who believed traits were acquired by individuals during their lifetime, depending on their behaviours. p. 123

jejunum the middle component of the small intestine. p. 467

joint an articulation between bones. p. 234

juxtaglomerular cells cells in the kidney that synthesize, store, and secrete renin when blood pressure decreases. p. 487

karyotype organization of chromosomes and their homologous pairs from chromosome 1 to 23, from longest to shortest. p. 90

keratin intermediate-filament protein found in hair and nails. p. 61

keratinized cells that contain the protein keratin, found in the outermost layer of the skin, hair, and nails. p. 181

keratinocytes skin cells identified by their ability to produce the protein keratin. p. 198

ketone an organic compound produced when fat is broken down to produce ATP. p. 466

kidney stones solid, crystalline accumulations that form in the kidneys and may prevent the passage of urine. p. 493

kinetochore protein located at the centromere region that holds sister chromatids together. p. 92

kingdom categorization of all living organisms; six kingdoms—Archaebacteria, Eubacteria, Protista, Fungi, Plants, and Animals. p. 2

knockout mice mice in which a gene has been blocked or deleted so a specific functional protein is not produced. p. 176

Kupffer cells specialized macrophages in the liver that engulf foreign material and cell debris. p. 465

kyphosis convex curvature of the thoracic vertebrae, commonly referred to as *hunchback*. p. 222

labia majora longitudinal folds of skin that extend between the mons pubis and the perineum. p. 504

labia minora extensions of skin on either side of the vaginal opening that are situated between the labia majora. p. 504

lac operon three genes involved in lactose metabolism in bacteria. p. 159

lacrimal bones smallest facial bone, which forms the medial portion of the eye sockets. p. 213

lactase enzyme that breaks down lactose. p. 37

lactate three-carbon molecule produced from pyruvate during anaerobic respiration. p. 83

lacteal lymphatic vessels that absorb fats from the small intestine. p. 467

lactic acid a by-product of anaerobic metabolism in humans derived from pyruvate. p. 83

lactobacillus bacteria a type of gram-positive, rod-shaped bacteria that is part of our normal flora. p. 517

lactose a disaccharide formed from glucose and galactose. p. 37

lacunae small spaces inside the bone that house the osteocytes and are connected by canaliculi. p. 210

lacZ gene a gene found in the *E. coli* operon that's used as a selectable marker in genetic engineering. p. 166

lambdoid sutures suture that connects the occipital bone to the posterior edges of the parietal bones. p. 213

lamella ring of bone matrix. p. 210

lamellar granules organelles that release compounds into the extracellular fluid to waterproof the skin and prevent infection. p. 199

lamina propria a layer of dense connective tissue under the epithelial layer of mucus membranes. p. 461

Langerhans cells immune cells found in the epidermis. p. 198

lanugo fur that humans have in certain stages of embryonic development. p. 129

large intestine segment of the intestine consisting of the cecum, and the ascending, transverse, descending, and sigmoid colon. p. 458

laryngopharynx cavity at the posterior aspect of the upper respiratory tract. p. 435

larynx commonly referred to as the *voice box*. p. 435

lateral condyles lateral attachment site that articulates with the fibula. p. 227

latissimus dorsi largest muscle in the back. p. 264

law of independent assortment Mendel's law that homologous pairs separate into gametes completely randomly, so any possible combination can occur. p. 109

law of segregation Mendel's law that the two alleles of a trait separate during the formation of gametes, so that half the gametes carry one copy and half carry the other copy. p. 109

LDLR (low-density lipoprotein receptor) the receptor on fat cells that recognizes and takes in LDLs; *see also* **low-density lipoprotein (LDL)**. p. 119

leading strand DNA strand in which only one RNA primer is required for polymerase to follow helicase in the 5′ to 3′ direction. p. 142

left carotid artery artery in the left side of the neck that feeds the brain. p. 405

left gastric artery artery that supplies the superior portion of the stomach. p. 407

left subclavian artery artery beneath the left clavicle. p. 405

left subclavian vein vein beneath the left clavicle. p. 403

lens transparent structure that changes shape and helps refract light so it becomes focused on the retina. p. 345

leukocytes white blood cells. p. 423

LH surge an increase in luteinizing hormone that triggers ovulation in females. p. 507

ligament connective tissue that connects bone to bone. p. 235

ligand-gated *see* **chemically gated channel**. p. 298

ligase an enzyme that forms phosphodiester bonds after the RNA primers have been removed during DNA replication. p. 142

limbic system a set of brain structures located under the cerebrum. p. 315

linked genes located close together on the same chromosome. p. 115

lipase an enzyme that breaks down lipids. p. 82

lipogenesis formation of lipids from acetyl-CoA. p. 466

lipolysis the breakdown of lipids into fatty acids to be used for energy production. p. 82

lipopolysaccharide (LPS) a polysaccharide that contains a fat, such as the external cell wall layer of gram-negative bacteria. p. 541

lipoprotein lipase (LPL) a group of enzymes that break down fat into components that can be used for energy. p. 281

lipoproteins proteins that have been processed and bound to lipids. pp. 156, 421

live attenuated vaccine a virus that has had the virulence decreased, but still prompts an immune response. p. 531

liver organ in the upper right quadrant of the abdomen. p. 465

load (muscular) the resistance to a muscle contraction. p. 272

lobes anatomical or functional division of an organ. p. 436

lobule a division of a tissue that is only visible histologically, such as the lobules of the liver. p. 437

locomotion movement from one location to another. p. 4

long bones type of bone found in the arms, legs, fingers, and toes. p. 211

longissimus muscles that connect the transverse processes of various vertebrae to the sacrum. p. 264

longitudinal fissure a deep groove that separates the two hemispheres of the brain. p. 315

loose connective tissue tissue that connects to muscle layers, gives strength and elasticity, stores fat, or acts as a filter in the spleen and lymph nodes. p. 183

lordosis concave curvature of the spine; also known as *sway back*. p. 222

low-density lipoprotein (LDL) a type of cholesterol molecule made by the liver to transport fats to the adipose tissue; also known as "bad" blood cholesterol. p. 418

lumbar enlargement an enlargement of the spinal cord at the separation to the lower limbs. p. 307

lumbar plexus branched connections of nerves in the lower back. p. 308

luteinizing hormone (LH) a sex hormone that stimulates the production of sex hormones in the gonads of males and females; also causes ovulation in women. pp. 377, 500

lymph fluid within the lymphatic system. p. 401

lymph node an oval-shaped organ of the lymphatic system that contains lymphocytes, the T cells and B cells. p. 515

lymphocytes the B cells and T cells. p. 524

lysogenic cycle the viral cycle in which the host is not killed, but the viral DNA is integrated into the host DNA. p. 547

lysosome small membrane-bound organelle that contains digestive enzymes that break down molecules, macromolecules, and old organelles inside the cell. p. 59

lysozyme enzyme that damages bacterial cell walls; found in human tears, sweat, and saliva. p. 518

lytic cycle the viral cycle in which the host is destroyed and new viral particles are produced. p. 547

M line centre of the H zone of the sarcomere. p. 271

M phase mitosis occurs, with no more protein synthesis and cell growth. p. 88

macroevolution large-scale changes that may give rise to a new species and that occurs over a very long period of time. p. 120

macromolecule a combination of small molecules; the four macromolecules that make up cells are proteins, carbohydrates, fats, and nucleic acids. pp. 5, 27

macrophages white blood cells that engulf bacteria, viruses, and any dead cells by phagocytosis to break down and destroy invading microorganisms. pp. 68, 425

macula densa an area of cells that line the distal convoluted tubule of the nephron in the kidney; detects sodium levels and regulates blood vessel diameter to regulate blood pressure. p. 487

macula lutea the part of the eye that contains a concentrated number of cone cells that are responsible for detailed central vision; also called the *fovea*. p. 346

major calyx the point at which multiple minor calyces converge. p. 483

malaria a disease resulting from a parasite commonly found in Africa that infects the liver and red blood cells and can be fatal. p. 553

malleus hammer-shaped bone that connects the tympanic membrane to the incus. p. 349

maltase an enzyme that breaks down maltose. p. 470

maltose a disaccharide formed from two glucose molecules and found in grains. p. 37

mammary glands glands that produce milk in females. p. 200

mandible forms the lower jaw bone, the largest and strongest of the facial bones. p. 213

mandibular fossa groove in the temporal bone that houses the condylar process of the mandible. p. 217

manubrium the broad superior portion of the sternum. p. 222

maxilla bones the bones that form the upper jawbone and house the upper teeth. p. 213

mechanical digestion the physical breakdown of food into smaller components, such as chewing. p. 458

mechanoreceptors receptors that detect stretching. p. 20

medial and lateral epicondyles medial and lateral processes of the elbow. p. 224

median antebrachial vein vein that drains the venous plexus of the hand. p. 411

median cubital vein superficial vein of the upper limb. p. 411

median nerve one of the main veins originating at the brachial plexus. p. 308

mediastinum anatomical central region of the thoracic cavity. p. 16

medulla oblongata the part of the brain responsible for maintaining basic life functions such as breathing and heart rate. pp. 20, 306

meiosis the process of cell division that produces haploid gametes from diploid germ cells. p. 121

Meissner corpuscles sensory receptors that detect light touch and vibration in the skin. p. 200

melanin skin pigment produced by melanocytes in response to sunlight; protects the skin cells from DNA damage caused by ultraviolet light. p. 198

melanocytes skin cells identified by their ability to produce melanin. p. 198

melanoma a tumour originating from mutated melanocytes. p. 202

melatonin a hormone produced by the pineal gland in low light; stimulates sleep. p. 312

membrane attack complex (MAC) formed from complement proteins circulating in the blood that can directly kill bacteria and fungi. p. 521

memory cells B cells and T cells that recognize antigens from previous infections. p. 512

meninges protective covering of the brain and spinal cord. p. 304

Merkel cells touch receptors in the skin. p. 198

merozoites one of the life-cycle stages of malaria. p. 553

mesenchymal cells cells that can differentiate into a variety of cell types. p. 515

mesentery a double layer of the peritoneum. p. 462

messenger RNA (mRNA) the product of transcription of a gene that will be the template for protein synthesis on a ribosome. p. 152

metabolic syndrome a number of conditions that combine to increase the risk of heart disease, obesity, and type-2 diabetes. p. 421

metastasis the spread of cancerous cells to other regions of the body. p. 61

methionine (Met) first amino acid in the protein molecule that is coded by AUG mRNA codon. p. 156

methylase enzyme enzyme that adds methyl groups to DNA that affect the regulation of transcription. p. 165

MHC class I a class of molecules found on all cells of the body that are the major reason why the immune cells can recognize our own cells versus foreign cells, such as pathogens or transplant tissue; *MHC* refers to major histocompatibility. p. 525

MHC class II a class of molecules found on antigen-presenting cells (APCs) that activate helper T cells in the lymphatic tissue; *MHC* refers to major histocompatibility. p. 525

micelles a collection of lipids, often spherical, that have outer hydrophilic heads aggregating around a hydrophobic core. p. 472

microevolution the change in inherited traits in a population of organisms within a single species over time. p. 120

microfilaments the smallest intracellular cytoskeletal proteins, composed of actin proteins; have a role in cellular movement, such as the "crawling" movement of immune cells through the tissues. p. 61

microglia glial cells that function as immune cells in the CNS. p. 192

microtubules the largest intracellular cytoskeletal proteins, composed of large, hollow tubes; used to transport molecules throughout the cell. p. 61

microvilli microscopic extensions of the cellular membrane that act to increase surface area. p. 179

micturition urination. p. 485

midbrain upper portion of the brainstem primarily associated with motor control, but also involved in consciousness (contains the reticular formation), vision, hearing, and temperature regulation. p. 314

midsagittal plane an anatomical division of the body into equal left and right sections. p. 15

minor calyx the part of the kidney that collects urine from the collecting ducts. p. 483

mitochondria membrane-bound organelle that produces the cell's energy in the form of ATP in the presence of oxygen. p. 59

mitosis process of cell division in somatic cells to produce two new identical diploid daughter cells. p. 88

mitral valve valve between the left atrium and the left ventricle. p. 389

modifier genes genes that modify the effects of another gene. p. 113

monococci a form of cocci that exist as individual cells rather than as clusters or chains. p. 542

monocytes immune cells that differentiate into macrophages during an infection. p. 423

monomers single molecules. p. 27

monosaccharide simple carbohydrate monomer occurring as three types: glucose, fructose, and galactose. p. 37

monosynaptic reflex reflex consisting of one sensory neuron and one synapse to the motor neuron. p. 309

monounsaturated fat fats that contain one double-bonded carbon in the fatty acid chain. p. 44

mons pubis a mass of fatty tissue over the pubic symphysis, more prominent in females. p. 504

mouth the opening through which most animals take in food. p. 465

MRSA (methicillin-resistant staphylococcus aureus) an antibiotic-resistant form of the bacterium *Staphylococcus aureus*. p. 542

mucosa mucus membrane covered in epithelial cells that are often involved in secretion or absorption. p. 461

mucus a sticky, fat-containing substance that is secreted in the digestive system and respiratory system to protect epithelial cells and trap foreign particles. p. 435

multinucleated cells having several nuclei, such as skeletal muscle. p. 190

multiple cloning site (MCS) usual location in a plasmid where the restriction enzyme cut sites are placed, usually within another selectable marker. p. 166

multipolar neurons a type of neuron that has multiple dendrites attached to the cell body and receives stimuli. p. 192

multiunit smooth muscle smooth muscle that consists of many cells contracting individually, which allows for precise control. p. 190

muscle fibre single muscle cell. p. 256

muscle tone residual tension in muscles during relaxation. p. 273

muscular system organ system composed of smooth, cardiac, and skeletal muscle. p. 7

muscularis refers to the muscular area of a tissue or organ. p. 461

muscularis mucosa the muscular layer of a mucus membrane. p. 461

musculocutaneous nerve nerve that originates in the brachial plexus and innervates the lateral arm. p. 308

mutagen a substance that mutates DNA. p. 95

myenteric plexus part of the enteric nervous system that controls smooth muscle contractions during digestion. p. 292

myocardium middle layer of the heart consisting of the cardiac muscle tissue. p. 388

myofacial pain muscular facial pain that is a common symptom of the improper alignment of the condylar process in the mandibular fossa. p. 217

myofibrils bundles of contractile and regulatory proteins in muscle cells. p. 256

myoglobin a muscle protein that binds to oxygen, similar to hemoglobin. p. 277

myometrium muscular layer of the uterus. p. 504

myosin a protein contained in the thick filaments of the myofibrils that binds to actin during contraction. p. 257

Na$^+$/K$^+$ pump an active transport membrane protein, found in every cell of the body, that maintains the electrical and ion concentration gradient across the cell membrane by moving three sodium ions out of the cell for every two potassium ions that move in. p. 298

nasal conchae long, narrow, curved bones in the nasal passageway that provide a pathway for inhaled air; warms and moistens air before it enters the lungs. p. 434

nasal septum bone that separates the right and left sides of the nasal cavity. p. 217

nasopharynx most superior portion of the pharynx. p. 435

natural killer cells white blood cells that recognize cancer cells and virus-infected cells and destroy them. pp. 95, 147, 425

natural selection means by which a population evolves based on environmental influences; makes the organisms most "fit" in a particular environment more likely to survive and reproduce. p. 121

naturally acquired active immunity immunological memory formed after exposure to an infectious pathogen. p. 530

naturally acquired passive immunity a type of immunity passed from a mother to a fetus or baby as antibodies are transferred through the placenta or milk. p. 531

nearsightedness occurs when a person has the ability to focus on objects that are close up but not far away; also known as *myopia*. p. 348

negative feedback a process by which the body reacts to a deviation from a normal set point to return the body to within normal limits. p. 19

nephron the functional unit of the kidney. p. 483

nerve a collection of neurons surrounded by connective tissue. pp. 291, 307

neuraminidase a type of enzyme expressed by the influenza virus that allows the virus to reproduce inside respiratory cells. p. 547

neurofilament intermediate filament found in the axons of neurons. p. 61

neuroglia non-neuronal cells located within the central and peripheral nervous systems; include oligodendrocytes, microglia, astrocytes, ependymal cells, Shwann cells, and satellite cells. p. 191

neurons cells of the nervous system that conduct action potentials. p. 191

neurotransmitter a chemical-signalling molecule that is released from axon terminals and allows communication between neurons and other cells. p. 274

neutrophils white blood cells that are the most abundant type of white blood cell and are usually the first responders during an infection, killing pathogens by secreting neutralizing chemicals. p. 423

niacin vitamin B_3; produces NAD^+. p. 77

nitrogen group amino group that contains nitrogen and hydrogen. p. 81

nitrogenous wastes nitrogen-containing substances that are excreted by the body once they are no longer usable, including ammonia, urea, and uric acid. p. 488

nociceptors detect pain; also known as *nerve endings*. p. 20

non-ciliated epithelial cells that do not contain cilia. p. 180

nondisjunction failure of chromosomes to separate correctly during meiosis I or meiosis II. p. 116

nonpolar molecules that do not have a charge and cannot dissolve easily in water. p. 27

non-striated muscle fibres that are not striated, such as smooth muscle. p. 190

norepinephrine (NE) a neurotransmitter released from ANS neurons that is a hormone produced in the adrenal medulla in response to sympathetic nervous stimulation; also called *noradrenaline*. p. 322

nuclear envelope the double-layer membrane that surrounds the nucleus and contains pores that allow large molecules, such as mRNA, to move out of the nucleus. p. 58

nuclear membrane *see* **nuclear envelope.** p. 57

nuclei plural of *nucleus*. p. 292

nucleolus a small non-membrane-bound organelle found in the nucleus; contains DNA that transcribes ribosomal RNA that is used to make ribosomes. p. 58

nucleoplasm fluid within the nucleus. p. 57

nucleoside a molecule that contains a sugar and a base. p. 138

nucleoside triphosphate nucleoside with three phosphates added by DNA polymerase to the 3′ end of a nucleotide on the new strand; includes the nucleoside triphosphate ATP. p. 142

nucleotide an organic monomer of DNA or RNA; made up of a five-carbon ring sugar, a phosphate group, and a nitrogen-containing base. pp. 34, 137

nucleus accumbens part of the basal ganglia; plays an important role in laughter and reward. p. 315

nucleus membrane-bound organelle that contains genetic material, DNA. p. 57

oblique fissure a depression or groove that separates the inferior lobe of the lungs from the rest of the lungs. p. 437

oblique plane an anatomical division of the body on an angle. p. 16

obturator foramina openings in the ischium that allow the passage of nerves and blood vessels. p. 227

obturator nerve nerve that receives sensory information from the medial thigh and is responsible for motor innervation of the lower extremity. p. 308

occipital bone posterior portion of the skull that protects the occipital lobe of the brain. p. 217

occipital condyles two processes at base of the skull that sit on the first cervical vertebra (atlas) and allow movement of the head. p. 217

occipital lobe most posterior lobe of the brain that contains the primary visual cortex. p. 318

occludens proteins that make up tight junctions. p. 63

olecranon (elbow) a process at the proximal end of the ulna that articulates with the humerus. p. 224

olecranon fossa a large indentation in the posterior of the humerus that allows the arm to straighten. p. 224

olfactory bulb a neural structure located in the base of the frontal lobe that connects with other brain structures, such as the amygdala and the hippocampus, and is involved in perception of odours. p. 357

olfactory cilia hair-like structures that protrude from the olfactory epithelium. p. 356

olfactory foramina openings for the olfactory nerves. p. 217

olfactory system sensory system that allows the brain to perceive odours. p. 356

oligodendrocytes glial cell that produce the protective myelin that covers neurons in the CNS. p. 192

oligosaccharide small chain of monosaccharides. p. 59

omega 3 fatty acid essential fatty acid; also called *alpha-linoleic acid.* p. 45

omega 6 fatty acid essential fatty acid; also called *linoleic acid.* p. 45

omega 9 fatty acid essential fatty acid; also called *oleic acid.* p. 45

oncology the study of cancer. p. 95

oocyte a female gamete. p. 502

opioids endogenous or exogenous substances that block the receptors that bind to

the neurotransmitters that send pain information. p. 344

opportunistic infections a type of infection that remains dormant unless the affected individual is immunosuppressed. p. 544

opsonize the process by which a pathogen is marked for destruction, such as when an antibody binds to an infectious organism and increases the ability of other cells to target and kill the infection. p. 528

optic disc the region of the retina where the neurons carrying information from the photoreceptors join the optic nerve. p. 346

optic foramen opening for the optic nerve. p. 217

organ combination of tissues that function as a unit, such as the stomach or the heart. p. 5

organ of Corti structure located on the basilar membrane within the cochlea that contains the hair cells that convert sound waves into action potentials and transmits information to the brain about loudness and pitch of sounds. p. 350

organ system organs that function together with other organs and tissues, such as the digestive system or the cardiovascular system. p. 5

organic molecules molecules of a living organism, such as a plant or animal. p. 27

organism a living entity composed of one or more cells. p. 5

origin the connection point of a muscle to a non-moving bone. p. 257

oropharynx part of the pharynx bordered by the uvula and the hyoid bone. p. 435

osmoreceptor receptor that detects changes in osmolarity. p. 20

osmotic composition components of the blood that determine its osmolarity. p. 482

osmotic concentration the concentration of a solute in the interstitial fluid surrounding a cell relative to the concentration inside the cell. p. 65

osmotic gradient difference in concentration between two solutions across a membrane. p. 483

osmotic pressure the pressure created by the movement of water due to an osmotic gradient. pp. 401, 487

ossicles small bones of the middle ear that include the malleolus, incus, and stapes. p. 349

osteoblasts immature osteocytes that build bone. p. 209

osteocytes mature bone cells. p. 184

osteon smallest unit of bone tissue, composed of concentric rings of bone matrix that surround a central canal and contain osteocytes. p. 210

osteoporosis a disease characterized by the loss of bone density that can lead to increased risk of fractures. p. 371

otolithic organs the saccule and utricle. p. 353

ova (eggs) egg cells. p. 498

oval window membrane that connects the middle ear to the inner ear. p. 349

ovarian ligament ligament that connects the ovaries to the uterus. p. 502

oxalate an organic compound that binds to minerals; found in many plants. p. 49

oxidation reactions in which electrons are donated from one molecule to another. p. 74

P wave a positive inflection on an ECG that indicates atrial depolarization. p. 392

p53 tumour-suppressor gene that prevents cell division when DNA mutations are recognized; recruits DNA repair enzymes, and can initiate apoptosis during the G_2 phase. p. 88

pacemaker cells specialized cells of the heart that control the rate of contraction. p. 191

pacemaker potential a slow change in voltage across the cell's membrane caused by sodium and calcium ions. p. 392

Pacinian corpuscles a type of mechanoreceptor that detects pressure in the skin. p. 200

palatine bones bones that form the posterior portion of the hard palate, which is the roof of the mouth. p. 218

palindrome a specific symmetrical sequence of nucleotides that are the same in opposite directions. p. 165

pancreatic duct a vessel through which pancreatic enzymes are secreted into the gastrointestinal tract. p. 465

pancreatic juice a liquid secreted by the pancreas containing a combination of enzymes and bicarbonate. p. 467

pantothenic acid vitamin B5; produces coenzyme A. p. 77

papillae rounded protuberances on the tongue that contain taste buds. p. 354

paracellular transport the movement of molecules *between* cells as opposed to *through* cells, which is transcellular transport. p. 489

paracrine a form of cellular communication where cells produce a signalling molecule that affects nearby cells without circulating through the bloodstream. p. 518

parafollicular cells cells of the thyroid gland that produce calcitonin. p. 373

paranasal sinuses openings in the skull that are lined with mucosal membranes that warm and moisturize the incoming air before it enters the lungs. p. 217

parasagittal plane an anatomical division of the body into unequal left and right sections. p. 15

parasite an organism that benefits at the expense of its host. p. 553

parasympathetic nervous system a division of the autonomic nervous system that is involved in resting and digesting. p. 321

parathyroid hormone (PTH) hormone that acts as an antagonist to calcitonin by raising blood calcium levels. p. 231

parietal bones bones that form the superior and lateral portions of the skull. p. 214

parietal lobe the superior-lateral lobe of the brain. p. 318

parietal pericardium outer layer of the pericardium. p. 388

parietal peritoneum segment of the peritoneum that's bound to the wall of the abdominal cavity. p. 462

parietal pleura segment of the pleura that is bound to the chest wall and surrounds the lungs. p. 437

Parkinson's disease a degenerative disorder of the central nervous system that involves the loss of dopaminergic neurons that regulate muscle movements. p. 314

parotid salivary gland largest and most posterior of the salivary glands. p. 465

partial pressure the pressure created by one particular gas as a part of the total air composition. p. 444

partial-thickness burn deep burn that extends into the dermis, characterized by the formation of blisters; also known as *second-degree burn*. p. 202

patella sesamoid bone located between and anterior to the distal femur and the proximal tibia; also called the *knee cap*. p. 227

patellar ligament a continuation of the quadriceps tendon that inserts on the tibia. p. 227

patellar reflex a monosynaptic reflex arc; also called the *knee-jerk reflex*. p. 309

pathogen-associated molecule patterns (PAMPs) molecules associated with pathogens that are recognized by the immune system. p. 524

pathogenic bacteria bacteria that cause harm or suffering to the host. p. 543

pattern recognition receptors (PRRs) proteins expressed by immune cells that identify pathogen-associated molecular patterns. p. 524

pectoralis the primary chest muscles. p. 258

pelvic brim opening formed by the pubis bones and the sacrum. p. 225

pelvic inflammatory disease (PID) refers to the condition of inflammation of the uterus, fallopian tubes, and ovaries and can cause infertility. p. 508

pepsin an enzyme that breaks proteins down into component peptides. p. 465

pepsinogen a compound activated by hydrochloric acid to become pepsin. p. 465

peptidases enzymes that break down polypeptides. p. 470

peptide bond covalent bond that links amino acids together. p. 30

peptidoglycan eubacteria cell wall composed of sugars and amino acids. pp. 56, 541

perforating canal opening in the bone matrix that allows blood and lymph vessels to connect with vessels outside the bone. p. 210

perforin a protein that opens a pore in the membrane of a target cell, which allows water to rush into the cell, which kills the cell. p. 520

pericardial cavity space between the pericardial membranes that surround the heart. pp. 16, 388

pericardium a double-walled connective tissue membrane that surrounds the heart. p. 388

perimetrium outermost layer of the uterus. p. 504

perimysium sheath of connective tissue that surrounds a bundle of muscle fibres. p. 256

perineum area between the anus and the scrotum (in males) or the anus and the vagina (in females). p. 504

perineurium protective sheath around a bundle of nerves. p. 307

periosteum dense, irregular connective tissue that covers the bones. p. 211

peripheral chemoreceptors chemical receptors located in the aortic arch and carotid bodies that detect changes in the chemical composition of the blood. p. 450

peripheral nervous system (PNS) all nerves outside the central nervous system. p. 291

peristalsis rhythmic contraction of smooth muscle that causes movement through a tube. p. 464

peritubular capillaries small blood vessels that surround the nephron. p. 485

peroxisomes organelles that break down oxygen radicals. p. 60

Peyer's patch a bundle of lymphatic tissue surrounding the digestive system. p. 517

phagocytosis endocytosis that involves the movement of particles into the cell. p. 68

pharyngeal pouches grooves in the lateral neck region in all vertebrate embryos that disappear in mammals later in embryonic development but differentiate into gills in fish. p. 129

phenotype an organism's visible traits. p. 102

pheromones molecules that act to attract a mate in most species. p. 201

phosphodiester bond type of bond between a sugar and a phosphate in nucleic acids. p. 138

phospholipid primary fats found in cell membranes and containing two nonpolar fatty acid tails and a polar head. p. 46

phosphorylation enzymes enzymes in the cell that add phosphate groups to some molecules. p. 59

photopsins the photopigment in cone cells. p. 347

photoreceptors receptors that detect light and are found only in the retina. p. 20

photosynthesis process in which plants combine carbon dioxide from the air with hydrogen from water and energy from the sun to produce glucose. p. 4

phrenic nerve nerve that innervates the diaphragm. p. 308

physiology the study of how anatomical parts function. p. 1

phytate an organic compound that binds to minerals; found in beans, legumes, nuts, and seeds. p. 49

pia mater innermost layer of the meninges. p. 305

pili hair-like structures that help bacteria adhere to structures and transfer small pieces of DNA (singular, *pilus*). p. 56

piloerection contraction of the arrector pili muscles that cause the phenomenon commonly called *goosebumps*. p. 200

pineal gland endocrine gland found in the brain that produces melatonin. p. 312

pinna the visible outer part of the ear; also called the *auricle*. p. 349

pinocytosis endocytosis that involves the movement of fluids into a cell. p. 68

pituitary gland endocrine gland located at the base of the brain that controls many important functions, such as the stress response, growth, metabolism, and reproduction. p. 312

pivot joint joint that permits only rotation. p. 235

planar joint synovial joint that permits only a gliding motion. p. 235

plants eukaryotic, multicellular organisms that are photosynthetic. p. 4

plasma membrane *see* **cell membrane**. p. 54

plasma the fluid component of the blood. p. 401

platelets cell fragments that assist the process of blood clotting. p. 421

pleiotropic a single allele causing multiple phenotypes in a population. p. 110

pleura fluid-containing membranes that surround the lungs. p. 437

pleural cavity space between the pleural membranes that surround the lungs. p. 16

podocytes cells in the Bowman's capsule that wrap around the capillaries in the glomerulus. p. 487

point mutation one or a few nucleotides that are incorrectly placed in the DNA sequence during replication. p. 145

polar body a small cell produced during meiosis in females and that undergoes apoptosis. p. 504

polar molecules that contain a charge and dissolve in water. p. 26

polyadenylation addition of many adenine nucleotides to mRNA to prevent mRNA from degradation. p. 152

polygenic multiple genes contributing to one trait, such as height or weight. p. 110

polymer combination of monomers, such as amino acids combining to produce a protein. p. 27

polymerase chain reaction (PCR) technique used to generate multiple copies of a specific DNA sequence—faster and more reliable than using bacteria for this process. p. 172

polypeptide protein with many amino acids covalently bonded together by peptide bonds. p. 30

polysaccharide complex carbohydrates composed of long chains of glucose. p. 37

polysynaptic reflex a reflex that involves more than two neurons. p. 309

polyunsaturated fat fats that contain two or more double-bonded carbons. p. 44

pons the part of the brain that connects the thalamus to the medulla oblongata and that regulates breathing. p. 314

popliteal artery an artery located in the knee. p. 408

population a group of the same species of organisms that live in the same geographical area. p. 120

positive feedback a process that continues to increase the response caused by an initial stimulation, such as during labour. p. 19

post-central gyrus the portion of the parietal lobe that is the sensory cortex. p. 316

posterior fontanel space between the skull bones located at the joining of the occipital bone and both parietal bones. p. 218

posterior median sulcus a narrow, longitudinal groove that runs along the posterior aspect of the spinal cord. p. 305

posterior tibial artery artery located at the dorsal aspect of the lower leg. p. 408

posterolateral fontanel space between the skull bones located where the temporal and occipital bones meet. p. 218

postganglionic nerve fibres that carry signals from the ganglion to the effector in the autonomic nervous system. p. 321

postsynaptic neuron neuron that receives neurotransmitters at the synapse. p. 301

post-translational modification any change made to a protein after translation, such as methylation or addition of sugars to form a glycoprotein. p. 156

power stroke movement of the myosin head toward the M line in a sarcomere; causes muscle contraction. p. 272

prebiotic fibre from foods that feeds intestinal bacteria. p. 40

pre-central gyrus the portion of the frontal lobe that is the motor cortex. p. 316

pre-frontal cortex anterior part of the frontal lobe of the brain involved in planning muscle movements. p. 317

preganglionic nerve fibres that carry signals from the central nervous system to the ganglion in the autonomic nervous system. p. 321

preload pressure generated to return blood to the heart. p. 396

pressure force generated against an area. p. 413

primary oocytes an egg cell that has not yet undergone the second meiotic division and has only a single layer of granulosa cells. p. 504

primary spermatocytes sperm cells that have not yet undergone the second meiotic division. p. 500

primase an enzyme that adds a short sequence of nucleotides that initiates DNA replication. p. 142

primer annealing the binding of primers to single-stranded DNA. p. 174

primer extension occurs when DNA polymerase replicates the DNA using the primers as a starting point. p. 174

primordial follicle first stage of folliculogenesis in females. p. 504

probability the likelihood that some event will occur. p. 105

promoter region a region of DNA upstream from the gene that is not transcribed and that RNA polymerase binds to. p. 152

proofreading a function in which polymerase is able to recognize some mistakes that occur during replication. p. 146

prophylactic a preventative measure or means. p. 543

proprioceptors receptors that help determine body position. p. 20

prostaglandins a subclass of eicosanoids involved in inflammation. p. 523

prostate cancer cancer of the prostate gland. p. 508

prostate gland a ring-shaped exocrine gland found in males, which produces an alkaline fluid that makes up 50–75% of seminal fluid. p. 500

proteoglycan a glycosylated protein that makes up a large part of the extracellular matrix of most tissues and is present in high amounts in connective tissue. p. 184

proteome all the proteins produced in an organism. p. 164

proton pump an active transport membrane protein that moves protons against its concentration gradient with input of energy. p. 67

proto-oncogenes genes that normally increase the rate of cell division. p. 95

protozoan single-celled, eukaryotic protists that have locomotion. p. 553

pseudopods extensions of the cytoplasm that resemble fake feet. p. 553

pseudostratified cells that are a single layer but appear to be stratified, and the nuclei may appear layered. p. 179

psoas major one of the hip flexor muscles. p. 265

pubic symphysis connection between the right and left pubis bones that form the anterior portion of the pelvic girdle. p. 225

pulmonary arteries arteries that take blood from the right ventricle to the lungs. p. 387

pulmonary circulation the circuit by which blood is pumped to the lungs for gas exchange before returning to the systemic circulation. p. 384

pulmonary semilunar valve valve between the pulmonary artery and the right ventricle. p. 390

punctuated equilibrium an evolutionary process where periods of no change are followed by periods of rapid change. p. 127

Punnett square method used to predict the phenotype ratios for any combination of genotypes. p. 104

pupil opening in the centre of the iris where light enters the eye. p. 345

purines nitrogenous bases—adenine and guanine—that have a double-ring structure. p. 35

Purkinje fibres extensions of highly excitable cells that pass through the ventricles from the bundle of His. p. 392

pyloric sphincter sphincter between the distal end of the stomach and the beginning of the small intestine. p. 464

pylorus region of the stomach that connects to the duodenum. p. 464

pyrimidines nitrogenous bases—cytosine, thymine, and uracil—that have a single-ring structure. p. 35

pyrogens molecules that promote a fever. p. 524

pyruvate dehydrogenase an enzyme that removes the carbon atom from pyruvate. p. 77

pyruvate three-carbon molecule formed through glycolysis. p. 77

QRS section of the ECG that indicates the depolarization of the ventricles and contains the (hidden) repolarization of the atria. p. 392

radial artery artery that runs along the medial forearm. p. 408

radial fossa indentation in the humerus that articulates with the radius and increases the range of motion when the arm is bent. p. 224

radial nerve nerve that runs along the radius in the forearm. p. 308

radial notch point at which the head of the radius articulates with the ulna. p. 224

radial vein vein located in the lateral aspect of the forearm. p. 408

radioulnar joint joint between the head of the radius and the ulna. p. 224

radius bone of the forearm that's located laterally in anatomical position. p. 224

reabsorption the movement of molecules back into the bloodstream. p. 487

receptors extensions of a sensory neuron or a specialized cell that detects stimuli, for example, thermoreceptors that detect changes in temperature. pp. 19, 55

recessive the allele that is hidden by the dominant trait. p. 104

reciprocal innervation occurs when one muscle contracts and the antagonistic muscle on the opposite side of the body relaxes. p. 309

recoil the return of a tissue to its original state after being stretched. p. 441

recombinant bovine somatotropin (rBST) bovine growth hormone, grown in bacteria and given to cows to increase their milk production or growth rate. p. 170

recombinant DNA also known as a *hybrid molecule*, contains nucleotide sequences from two different species. p. 164

recombinant DNA technology process of moving specific sequences of DNA from one organism to another, also known as *genetic engineering* or *gene splicing*. p. 164

recombination mutation large sequences of DNA moved to another location within the genome. p. 145

rectum most distal part of the large intestine. p. 467

rectus abdominis paired muscles running vertically at the anterior aspect of the abdomen. p. 441

rectus femoris the largest of the quadriceps muscles in the thigh. p. 265

red blood cells biconcave-shaped cells that contain the protein hemoglobin and carry oxygen through the bloodstream. p. 421

red bone marrow bone marrow responsible for the production of blood cells. p. 210

red pulp a segment of the spleen that breaks down old red blood cells. p. 517

redox reaction oxidation-reduction reaction. p. 75

reduction reactions in which a molecule accepts electrons from another molecule. p. 74

referred pain when pain is felt in a region of the body where the stimulus is not originating, such as left-arm pain during a heart attack. pp. 16, 344

reflex an involuntary and nearly instant reaction to a stimulus. pp. 190, 309

reflex arc pathway taken by a nerve impulse from sensation to effect. p. 309

regulatory proteins proteins that regulate the function of other molecules (e.g., regulation of muscle contraction by troponin and tropomyosin). p. 256

regulatory region region of DNA not transcribed, but that helps regulate transcription, such as a promoter region. p. 152

regulatory T cells (Th17) a population of T cells that decrease the immune response after an infection. p. 526

renal arteries arteries that supply blood to the kidneys. pp. 407, 483

renal capsule outermost connective tissue layer of the kidney. p. 486

renal columns a medullary extension of the renal cortex in between the renal pyramids. p. 483

renal failure a disease state in which the kidneys fail to filter waste from the blood. p. 493

renal hilum a recessed central fissure in the kidney. p. 483

renal lobe a segment of the kidney. p. 483

renal pelvis the funnel-like, dilated, proximal part of the ureter. p. 483

renal pyramid cone-shaped tissue of the kidneys. p. 483

renal vein vein that drains the kidney. p. 485

renin enzyme secreted by juxtaglomerular cells of the kidney that cleaves angiotensin to produce angiotensin I and is involved in the production of aldosterone, which increases blood pressure. p. 492

replicate to produce identical copies. p. 166

replication fork open portion of the DNA molecule during replication. p. 141

repolarization movement of potassium ions out of the cell to bring the cell membrane back to its resting membrane potential. p. 299

repressor a protein found in bacteria that prevents transcription of a gene. p. 159

residual capacity amount of air in the lungs following normal exhalation. p. 443

residual volume amount of air in the lungs following forced exhalation. p. 443

resistance force generated in opposition to pressure. p. 124

respiration the transportation of oxygen and carbon dioxide between cells and the external environment. p. 433

respiratory alkalosis an increase in blood pH due to a decrease in CO_2 and hydrogen ions. p. 448

respiratory bronchioles bronchioles that are involved in gas exchange. p. 436

respiratory zone area of the lungs where gas is exchanged. p. 434

resting membrane potential electrical gradient of a cell, usually –70mV. p. 67

restriction endonuclease enzymes that cut DNA at specific palindrome sequences. p. 165

restriction enzyme cut sites *see* **restriction endonuclease.** p. 166

restriction map a map of all restriction enzyme sequences in a DNA sequence. p. 168

rete testis a network of tubules in the hilum of the testes. p. 500

reticular cells cells that produce reticular fibres in loose, reticular connective tissue and that filter pathogens and old red blood cells in the spleen and lymph nodes. p. 184

reticular fibres thin collagen fibres produced by reticular cells and surrounded by glycoproteins that form a framework for many tissues; function as a filter in the spleen and lymph nodes. pp. 184, 516

reticular formation region of the brainstem that is involved in consciousness. p. 313

reticular layer deep layer of the dermis, which contains many elastin and fibrin layers that give the skin strength and elasticity. p. 197

retina thin layer of tissue at the back of the inside of the eyeball, where light is focused onto photoreceptors that trigger action potentials that transmit visual information to the brain. p. 345

rhodopsin the photopigment in rod cells. p. 347

riboflavin vitamin B2; produces FAD^+. p. 77

ribonuclease an enzyme that breaks down RNA. p. 470

ribose five-carbon sugar contained in RNA, differentiated from deoxyribose by a hydroxyl group bound to carbon 2. p. 138

ribosomal RNA (rRNA) makes up part of the ribosome and ensures that mRNA can bind to the ribosome. p. 152

ribosome an organelle composed of a large and a small subunit of protein and ribosomal RNA and is the site of translation. p. 56

ribozymes alternative name for ribosomes because of the enzymatic activity of rRNA. p. 156

rickets a bone-deforming disease caused by a vitamin D deficiency in childhood. p. 231

right carotid artery artery located in the right side of the neck. p. 405

right lymphatic duct lymphatic duct that drains fluid from the superior and right portions of the body. p. 403

right subclavian artery artery running along the inferior portion of the right clavicle. p. 405

right subclavian vein vein running along the inferior portion of the right clavicle. p. 403

RNA polymerase an enzyme that transcribes DNA into mRNA. p. 152

RNA ribonucleic acid. p. 35

rods photoreceptors that detect black, grey, and white images in low light. p. 345

rotation movement around a central axis. p. 237

rotator cuff muscles that allow for the full range of motion of the shoulder joint. p. 263

rough ER endoplasmic reticulum that contains ribosomes where protein synthesis occurs. p. 58

roundworms (nematodes) an animal parasite. p. 554

Ruffini corpuscles a type of mechanoreceptor that detects the stretching of skin. p. 337

rugae a series of ridges in the wall of an organ, such as in the stomach. p. 464

S phase second phase of interphase when DNA replication occurs. p. 88

saccule otolithic organ of the inner ear that is sensitive to vertical movement. p. 353

sacral plexus collection of branched nerves near the sacrum. p. 308

sacrum vertebrae the five fused vertebrae that form the posterior aspect of the pelvic girdle. p. 218

saddle joint synovial joint that permits motion in two planes and results from the articular surfaces being reciprocally concave and convex. p. 235

sagittal plane an anatomical division of the body into left and right sections. p. 15

sagittal suture the suture that connects the medial edges of the two parietal bones. p. 213

salivation process of secreting saliva. p. 465

saltatory conduction transmission of an action potential along a myelinated axon. p. 296

salty the sensation produced from food molecules that contain ions such as sodium, chloride, or potassium. p. 354

sarcolemma cell membrane of a muscle cell. pp. 256, 269

sarcomere smallest contractile unit of the myocyte. p. 257

sarcoplasm fluid, similar in nature to the cytoplasm of non-muscle cells. p. 256

sarcoplasmic reticulum smooth endoplasmic reticulum of muscle cells that contains calcium. p. 256

sartorius muscle extends the knee and rotates the lower leg. p. 267

satellite cells glial cells that maintain the ion concentration surrounding neurons in the PNS. p. 192

saturated fat fats that do not contain double-bonded carbons and are solid at room temperature. p. 43

saturation when membrane proteins are full and cannot transport more molecules into or out of the cell. p. 66

savoury (umami) the sensation produced from molecules such as the amino acids found in meat and cheese. p. 355

scalene muscle of the neck that contracts during forced inhalation. p. 440

scapula bones located posterior to the ribs; also called the *shoulder blades.* p. 224

Schwann cells glial cells that produce the protective myelin that covers neurons in the PNS. p. 192

sciatic nerve longest nerve in the human body, extending from the spine along the posterior thigh. p. 227

sclera the white outer layer of the eyeball. p. 345

scrotum a dual-chambered mammalian body part that contains the testes. p. 416

sebaceous glands glands that produce the oil sebum. p. 200

secondary immune response an immune response to a previously encountered infectious agent that is much more rapid than the primary immune response. p. 529

secondary oocyte a female gamete immediately following the second meiotic division. p. 504

secondary spermatocyte male gamete immediately following the second meiotic division. p. 500

secretin a hormone that regulates secretions in the stomach and pancreas. p. 476

secretion movement of substances out of the blood or the body. p. 179

segmental arteries branches of the renal arteries. p. 485

selectable marker method used to identify the host cells that contain the recombinant DNA. p. 166

selection pressure an environmental factor that kills a portion of a population, such as predators. p. 124

sella turcica bone surrounding the pituitary gland. p. 217

semen a fluid that contains sperm and secretions from the prostate and bulbourethral glands. p. 500

semi-conservative replication method of DNA replication where the original strands of DNA separate and act as a template for two new strands. p. 141

seminal vesicles tubes that secrete the fluid that becomes semen. p. 500

seminiferous tubules tubules in the testes where meiosis occurs. p. 499

senescence when a cell can no longer replicate due to aging the process. p. 69

senescent a cell unable to divide. p. 139

sense strand *see* **coding strand.** p. 149

sensory neurons neurons that transmit sensory information to the brain. p. 192

serosa smooth membrane composed of a thin layer of cells that secrete serous fluid. p. 462

serous fluid a lubricating fluid contained within a membrane. p. 388

Sertoli cells cells that act to nourish developing sperm cells. p. 500

serum the fluid portion, plasma, of the blood that does not contain fibrinogen. p. 421

sesamoid bones bones that develop within a tendon, such as the patella. p. 211

sex chromosomes X and Y chromosomes that determine gender. p. 117

sex hormones hormones produced by the gonads. p. 231

short bones bones of the ankles and wrists. p. 211

sigmoid colon distal, S-shaped portion of the large intestine. p. 467

silent mutation mutation that has no effect on phenotype. p. 158

single-unit smooth muscle smooth muscle, found in sheets, that surrounds visceral organs, where many cells contract simultaneously. p. 190

single-strand binding protein protein that stabilizes the single-stranded region of the DNA molecule during replication. p. 141

sinoatrial node (SA node) region of highly excitable cells found in the right atrium of the heart; regulates heart rate. p. 386

sister chromatid identical replicated chromosomes. p. 89

skeletal muscle striated muscle connected to bone and usually under voluntary control. p. 20

sliding filament theory a theory to describe the movement of actin sliding across myosin during muscle contraction. p. 271

small intestine organ consisting of the ileum, jejunum, and duodenum; where absorption of nutrients occurs. p. 458

small saphenous vein superficial vein of the upper leg. p. 409

smallpox a viral infection caused by a strain of Variola virus that infects the blood vessels and skin. p. 531

smooth ER endoplasmic reticulum that contains no ribosomes and is where lipids and carbohydrates are produced. p. 58

smooth muscle involuntary, non-striated muscle. p. 20

sodium-potassium pump a membrane protein that uses one molecule of ATP to move three sodium ions out of the cell and two potassium ions into the cell. p. 66

soleus muscle in the lower leg that allows for plantar flexion of the foot, such as standing on your toes. p. 268

soluble fibre fibre that attracts water and turns to a gel during digestion. p. 40

solute molecules that are dissolved in a solvent. p. 26

solvent fluid that contains dissolved molecules. p. 26

somatic cells cells of the body not destined to become sperm or eggs. p. 90

somatic nerves under voluntary control. p. 291

somatosensory cortex the main sensory receptive area of the parietal lobe for touch sensation. p. 318

sour the sensation produced from molecules such as citric acid in fruit, lactic acid in yogurt, or acetic acid in vinegar. p. 354

species a group of organisms that can breed only with each other and produce viable offspring. p. 123

spermatid gamete formed by the division of the secondary spermatocytes. p. 500

spermatogonia an undifferentiated male germ cell. p. 500

spermatozoa (sperm) male gametes. p. 498

sphenoid bone the bone that forms the middle base of the skull and is behind the eye sockets. p. 217

sphincter of Oddi a muscular ring that controls the flow of pancreatic juices and bile. p. 466

sphincter specialized muscle tissue that regulates the movement of substances within the body. p. 190

spinalis muscles that originate and insert into the spinous processes of the vertebrae, connecting them. p. 264

spindle fibres microtubules that transport chromosomes during cell division. p. 92

spinous processes posterior extensions of the vertebrae. p. 219

spiral fracture a fracture that's caused by torsional forces along the length of the bone. p. 232

spirochete helically shaped bacteria. p. 543

spleen an organ that functions to filter the blood and break down old red blood cells; contains immune cells that fight infections. p. 407

spliceosome six different snRNPs and their associated proteins that remove introns and form new bonds between exons. p. 152

spongy bone made up of trabeculae, but not arranged into osteons in the same way as compact bone; also called trabecular bone. p. 210

sporozoite a stage in the life cycle of Plasmodium species. p. 553

squalene a 30-carbon oil that is used in vaccines to stimulate the immune system. p. 201

squamous cell carcinoma cancerous growth originating in the keratinocytes of the stratum spinosum. p. 202

squamous flattened epithelial cells. p. 179

squamous sutures sutures that connect the temporal bones to the lateral portion of the parietal bones. p. 213

stabilizing selection favouring the average of the common traits in a population. p. 131

stapes stirrup-shaped bone that connects the stapes to the oval window. p. 349

staphylococci round clusters of bacteria. p. 542

stereocilia hair cells located in the auditory and vestibular system that convert energy into action potentials so the brain can perceive sound and movement. p. 353

sternal notch visible depression at the base of the neck. p. 223

sternocleidomastoid muscles that allow for flexion of the neck. p. 440

sternum a flat, long bone in the centre of the chest that's joined with the ribs to form the anterior portion of the rib cage; also called the breastbone. p. 222

sticky ends the overhanging ends of non-paired nucleotides. p. 166

stratified refers to multilayered epithelial cells. p. 179

stratum basale deepest layer of the epidermis, composed of a single layer of rapidly dividing keratinocytes. p. 197

stratum corneum most superficial layer of the epidermis that's constantly being replaced; the cytoplasm is replaced by keratin. p. 199

stratum granulosum middle layer of the epidermis. p. 199

stratum lucidum extra layer of the epidermis, found in areas of the body that endure stress; composed of dead keratinocytes and large amounts of keratin. p. 199

stratum spinosum greatest proportion of the epidermis, primarily composed of keratinocytes. p. 198

streptobacilli rod-shaped bacteria that grow in chains. p. 542

streptococci small circular bacteria that live in chains and can cause throat infections. p. 542

stress fractures small cracks in the bone resulting from overuse. p. 232

stretch receptors receptors that detect mechanical pull on the cells. p. 449

stretch reflex muscular contraction in response to stretching. p. 309

striated sequence of muscle fibres that have a light and dark banding pattern. p. 190

striations dark and light stripes in skeletal and cardiac muscle cells, as seen with a microscope. p. 269

stromal cells connective tissue cells in an organ. p. 515

styloid process extension of the temporal bone that contains a small opening to permit the passage of the carotid artery. p. 217

subarachnoid space anatomical space between the arachnoid mater and the pia mater. p. 305

subcortical region structures located inferior to the cortex. p. 315

subcutaneous fat adipose tissue found in the subcutaneous layer under the dermis of the skin. p. 43

sublingual salivary gland salivary gland found under the tongue that produces saliva. p. 465

submandibular salivary gland major salivary gland in the base of the mouth. p. 465

submucosa connective tissue that supports the mucosa. p. 461

submucosal plexus part of the enteric nervous system. p. 292

substance P a neurotransmitter that is released by neurons that send pain information. p. 343

substantia nigra structure in the brain that plays an important role in regulating body movements and is involved in Parkinson's disease. p. 314

substitution an incorrect nucleotide added during replication; also known as a *mismatch*. p. 145

substrate-level phosphorylation the process in which a kinase enzyme in the cytoplasm adds a phosphate to ADP, which occurs during glycolysis. p. 77

sucrase an enzyme that breaks down sucrose. p. 470

sucrose formed by dehydration synthesis reaction between glucose and fructose. p. 37

sudoriferous glands *see* **sweat glands**. p. 200

sulcus a fissure in the surface of the brain. p. 315

superficial burn burn that affects only the epidermal layer of the skin; also known as *first-degree burn*. p. 202

superior and inferior articular facets articular processes on vertebrae where they connect with adjacent vertebrae and have hyaline cartilage to prevent friction. p. 219

superior and inferior gluteal nerves nerves that signal the gluteal muscles. p. 308

superior and inferior mesenteric arteries arteries that supply the small and large intestines as well as the pancreas. p. 405

superior phrenic arteries arteries that supply the superior portion of the diaphragm. p. 405

superior sagittal sinus region that collects blood and CSF from the brain. p. 312

superior vena cava the superior portion of the largest vein in the body that brings blood to the right atrium. p. 387

surface tension the adhering force created by hydrogen bonds formed between water molecules. p. 26

surfactant an amphipathic molecule the reduces surface tension in the lungs. pp. 42, 437

surgical neck part of the humerus most commonly fractured. p. 224

survival of the fittest individuals born with traits that are most beneficial for them to survive, reproduce, and pass on DNA within a particular environment. p. 121

sutures immovable joints between bones; categorized as either fibrous or synarthrosis. p. 213

sweet the sensation produced from molecules such as glucose, sucrose, aldehydes, ketones, and artificial sweeteners. p. 354

sympathetic nervous system a division of the autonomic nervous system that controls the fight-or-flight response. p. 321

symport when two substances move across the membrane in the same direction. p. 67

synapse the connection between the axon terminal and the target cell. p. 274

synaptic cleft gap between the axon terminal and the target cell. p. 300

synaptic end bulb part of the axon terminal that stores the vesicles containing neurotransmitters. p. 274

synarthrosis a type of joint that produces little or no movement. p. 213

syndesmosis an immovable articulation where bones are connected by ligaments. p. 234

synergistic (muscular) muscles working together to produce a specific motion. p. 257

synovial cavity space between bones containing synovial fluid. p. 235

synovial fluid thick, viscous fluid that lubricates joints and helps absorb shock. p. 235

synovial joints highly movable joints, also called *diarthrosis*, that contain synovial fluid; include the shoulder, knee, and elbow. p. 235

systemic circulation the circuit of blood from the heart to the tissues and back to the heart. p. 384

systole refers to the contraction portion of the cardiac cycle. p. 416

T cells lymphocytes—immune cells—that are involved in the acquired immune response and react to specific antigens. p. 423

T wave portion of an ECG that represents ventricular repolarization. p. 392

tactile receptors receptors that detect touch, pressure, and vibration in the skin. p. 20

tapeworms (cestodes) an animal parasite. p. 554

taste type of perception that allows the brain to interpret chemicals that bind to receptors in the mouth; also called *gustatory perception*. p. 354

telomerase an enzyme found in stem cells and germ cells that prevents the loss of telomeres during replication. p. 148

telomere a repeated DNA sequence at the end of each eukaryotic chromosome that shortens after each cell division and protects genes from deterioration. p. 69

template strand the strand of DNA that is copied or transcribed into mRNA. p. 149

temporal bones bones that form the lateral portions of the skull inferior to the parietal bones and protect the temporal lobes of the brain. p. 217

temporal lobe inferior-lateral lobe of the brain that contains the auditory cortex and the hippocampus, which is involved with long-term memory. p. 318

temporomandibular joint (TMJ) the articulation between the mandible and the temporal bones. p. 217

tendon connective tissue between muscles and bone. p. 235

tendonitis inflammation of the tendons. p. 522

terminator sequence a sequence of DNA at the end of the gene that causes the mRNA molecule to form a hairpin loop, causing the polymerase to dissociate from DNA. p. 152

testcross method devised by Mendel to determine the genotype of individuals that express the dominant phenotype. p. 107

testes male gonads in animals. p. 498

thalamus part of the brain responsible for sense perception and motor function; considered the "sensory relay station." p. 312

theca layer the layer of ovarian tertiary follicle cells that indirectly plays a role in the production of estrogen. p. 504

thermogenic effect term that describes the energy required for digestion and absorption of nutrients in the digestive tract. p. 474

thermoreceptors receptors that detect changes in temperature. p. 20

thick filaments part of a myofibril that contains myosin proteins. p. 257

thin filaments part of a myofibril containing actin, troponin, and tropomyosin. p. 257

thoracic pertaining to the chest. p. 16

thoracic vertebrae the 12 vertebrae that are connected to ribs and form the posterior aspect of the rib cage. p. 220

thymocytes cells that differentiate into mature T lymphocytes. p. 515

thymus a specialized organ of the immune system where T cells mature; located posterior to the sternum. p. 513

thyroid hormones (T3 and T4) hormones produced by the thyroid that regulate metabolic rate. p. 231

tibia the large anterior bone in the lower leg. p. 227

tibial nerve nerve that runs along the tibia in the lower leg. p. 308

tibial tuberosity attachment site on the tibia for the patellar ligament. p. 227

tibiofibular joint point of articulation between the tibia and the fibula at the distal end of the leg. p. 227

tight junctions connections between endothelial cells that prevent the movement of molecules between endothelial cells, such as in the digestive tract or the blood-brain barrier. p. 63

tissue a group of cells that is a combination of cells and proteins that function together in a specialized role. p. 5

tongue a muscular organ in the mouth used for talking and moving food around during chewing. p. 463

tonsils collection of lymphoid tissue in the pharynx. p. 517

topoisomerase an enzyme in prokaryotes that relieves the supercoils (over-twisting) that occurs during the unwinding of the circular DNA during replication. p. 141

total lung capacity maximum volume of the lungs. p. 443

toxin a poisonous substance. p. 544

trabeculae osteocytes inside lacunae and surrounded by matrix; found in spongy bone. p. 210

trachea largest part of the lower respiratory tract and part of the conducting zone. p. 435

tracts bundles of neurons in the central nervous system. p. 291

trans fats incompletely hydrogenated fats in the *trans* conformation. p. 44

trans Golgi the side of the Golgi that is farthest from the endoplasmic reticulum. p. 58

transcription factors factors that promote or inhibit the transcription of RNA in a eukaryotic cell. pp. 161, 364

transcription process in which DNA produces a single-strand nucleic acid called *RNA*. p. 149

transduction process by which bacteria can acquire new DNA from viruses. p. 167

transfer RNA (tRNA) molecules that carry or transfer amino acids to the ribosome during protein synthesis. p. 152

transformation process by which bacteria can acquire DNA that is near them. p. 166

transient ischemic attack (TIA) a temporary condition that shows signs and symptoms similar to a stroke; also called *mini stroke*. p. 420

transitional epithelial cells epithelial cells that can change shape, such as the cells of the bladder that change as the bladder fills with urine. pp. 179, 483

translation protein synthesis. p. 59

transposable elements sequences of DNA that can move themselves to another chromosome within a cell; also called *transposons* or *jumping genes*. p. 145

transverse abdominis a muscle of the abdomen that runs horizontally. p. 441

transverse colon the part of the large intestine that runs parallel to the transverse plane from the right to left and leads to the descending colon. p. 467

transverse fracture a fracture that's perpendicular to the length of the bone. p. 232

transverse plane an anatomical division of the body into superior and inferior sections; also called a *cross-sectional* or *horizontal plane*. p. 15

transverse process lateral protuberance of the vertebrae. p. 219

transverse sinus areas beneath the brain that allow blood to drain from the back of the head. p. 408

transverse tubules (T-tubules) tunnel-like extensions of the cell membrane that extend into muscle cells and allow an action potential to penetrate into the cell. p. 256

tricep muscle that works to extend the forearm at the elbow. p. 265

tricuspid the valve that separates the right atrium from the right ventricle. p. 389

trisomy 21 a nondisjunction mutation where the cells have an extra chromosome 21; also called *Down syndrome*. p. 117

trochlea medial aspect of the distal end of the humerus that articulates with the ulna. p. 224

tropomyosin a regulatory protein that binds to the myosin-binding sites on the actin and prevents contraction unless calcium is present. p. 257

troponin a regulatory protein of the thin filaments that binds calcium and moves tropomyosin. p. 257

true pelvis bones that protect the internal, lower abdominal organs. p. 227

trypsin a pancreatic enzyme that breaks down proteins. p. 470

tubulin protein that forms microtubules. p. 61

tumour a growth of cells that is either benign or malignant. p. 94

tunica externa outermost layer of the blood vessels. p. 398

tunica intima innermost layer of the blood vessels. p. 398

tunica media middle layer of the blood vessels. p. 398

turbinates *see* nasal conchae. p. 434

tympanic membrane the part of the ear that transmits sound waves from the outer ear to the bones in the middle ear; also called the *eardrum*. p. 349

type I alveolar cells simple epithelial cells that form the alveolar wall. p. 437

type II alveolar cells specialized epithelial cells that secrete surfactant. p. 437

ulna bone of the forearm that's located medially in anatomical position. p. 224

ulnar artery artery running along the ulna in the forearm. p. 408

ulnar vein vein running along the ulna in the forearm. p. 408

unipolar neurons *see* sensory neurons. p. 192

unsaturated fat fats containing one or more double-bonded carbons and are liquid at room temperature. p. 43

urea waste molecule produced from ammonia and carbon dioxide, and excreted by the kidneys. pp. 81, 488

ureters vessels that carry urine from the kidneys to the urinary bladder. p. 483

urethra vessel that carries urine from the urinary bladder to the external environment. p. 485

urine a nitrogenous, liquid-waste product in animals. p. 482

uterus a female sex organ. p. 503

utricle otolithic organ that is sensitive to horizontal movement. p. 353

vaccine an agent that stimulates immunity to a particular infectious disease. p. 531

vacuole a membrane-bound organelle. p. 553

valve a physical barrier that controls or directs the movement of a fluid. p. 514

variable amplitude differences in membrane potential. p. 302

variable number tandem repeats (VNTRs) varying lengths of repeated DNA sequences in the introns; unique to every individual and used in DNA fingerprinting. p. 174

vas deferens ducts that connect the right and left epididymis to the ejaculatory ducts. p. 500

vascular permeability ability of a blood vessel wall to permit the flow of small molecules in and out of the vessel. p. 522

vasoconstriction a decrease in the lumen size in the blood vessels. p. 400

vasodilation an increase in the lumen size in the blood vessels. p. 400

vasodilators medications that work to decrease blood pressure by dilating the blood vessels. p. 420

vastus intermedius one of the quadriceps muscles; located underneath the rectus femoris. p. 265

vastus lateralis one of the quadriceps muscles located in the lateral thigh. p. 267

vastus medialis one of the quadriceps muscles in the medial thigh near the knee. p. 265

vector also known as a *vehicle*, such as a plasmid or virus that carries the inserted gene. p. 165

veins blood vessels that carry blood toward the heart. p. 387

venous return the return of blood to the heart through the venous system. p. 396

ventilation movement of air between the lungs and the external environment. p. 433

ventricular diastole relaxation of the ventricles. p. 391

ventricular systole contraction of the ventricles. p. 391

venules small veins that bridge the flow of blood from the capillaries to the veins. p. 390

vertebrae 33 bones that make up the spine, sacrum, and coccyx. p. 219

vertebral foramen an opening in each vertebra through which the spinal cord is able to extend. p. 219

vesicles sacs within a cell that are surrounded by a membrane and contain cellular products such as newly formed proteins, hormones, or neurotransmitters. pp. 58, 156

vestibular apparatus the part of the inner ear that consists of the utricle, saccule, and semicircular canals, and that detects sensory information about motion, equilibrium, and spatial orientation. p. 350

vestibular nerve the nerve that is connected to the vestibular system and that sends information to the brain about motion and spatial orientation. p. 350

vestibule an entrance or passageway. p. 504

villi extensions from the lining of the intestinal wall. p. 467

viruses small infectious agents that reproduce inside living cells. p. 543

visceral pericardium segment of the pericardium attached to the heart. p. 388

visceral peritoneal membrane layer of the peritoneum that covers the internal organs; also called the *peritoneum*. p. 462

visceral pleura segment of the pleura connected to the outer surface of the lungs. p. 437

vital capacity maximum volume of air that can be expelled after a full inhalation. p. 442

vitamin A derivative of beta-carotene, used by the body to maintain healthy skin and mucus membranes and improve vision; also called *retinol*. p. 200

vitamin D a compound that can be synthesized by the body when exposed to sunlight and be absorbed from the diet; required for calcium absorption in the small intestine and important for immune regulation. p. 231

vitreous humor a clear gel that fills the space between the lens and the retina. p. 346

vocal cords membranes that vibrate to generate sound. p. 435

voluntary muscles controlled consciously. p. 190

vomer a small part at the base of the skull that is a component of the nasal septum. p. 213

vulva external female genital organs. p. 504

water-soluble hormones polar hormones that act by binding to a receptor on the cell membrane, rather than inside the cell as fat-soluble hormones do. p. 364

white matter segments of the nervous system composed of myelinated axons. p. 305

white pulp segments of adenoid tissue in the spleen that contain white blood cells. p. 517

withdrawal reflex a spinal reflex designed to prevent damage to the body. p. 309

xiphoid process inferior portion of the sternum. p. 223

X-linked genes on the X chromosome unrelated to sex characteristics. p. 115

yeast artificial chromosome (YAC) chromosome that's easily grown in eukaryotic yeast cells and is used as a vector in genetic engineering because it can carry large sequences of DNA and yeast cells can process RNA. p. 167

yellow bone marrow bone marrow composed primarily of fat cells. p. 515

Z discs ends of a sarcomere. p. 269

zona pellucida membrane that surrounds the plasma membrane of an oocyte. p. 504

zygomatic arch part of the temporal bone that connects to the zygomatic bones to form the cheek bones. p. 217

zygomatic bones bones that articulate with the temporal bones to form the cheek bones. p. 213

zygote the initial cell formed when two gametes combine. p. 97

Index

Note: Page numbers with "f" indicate figures; those with "t" indicate tables.